COMPARATIVE PHYSIOLOGY OF ANIMALS

COMPARATIVE PHYSIOLOGY OF ANIMALS

AN ENVIRONMENTAL APPROACH

RICHARD W. HILL
Michigan State University

Harper & Row, Publishers
New York, Hagerstown, San Francisco, London

Sponsoring Editor: Joe Ingram
Project Editor: Holly Detgen
Designer: Michel Craig
Production Supervisor: Will C. Jomarrón
Compositor: V & M Typographical, Inc.
Printer and Binder: Halliday Lithograph Corporation
Art Studio: J & R Technical Services Inc.

COMPARATIVE PHYSIOLOGY OF ANIMALS: An Environmental Approach
Copyright © 1976 by Richard W. Hill

Library of Congress Cataloging in Publication Data
Hill, Richard W
    Comparative physiology of animals: an environmental approach.

    Includes bibliographies and index.
    1. Physiology, Comparative. 2. Adaptation
(Physiology) I. Title.
QP31.2.H54   591.1   75-20174
ISBN 0-06-042827-9

To my mother and father

CONTENTS

PREFACE

I have endeavored in this text to provide a basic introduction to those aspects of comparative animal physiology that bear particularly on the physiological interrelationships between animals and their environments. The physiological attributes of animals are discussed with a view to their significance in the ecological theater, and I have emphasized an evolutionary perspective inasmuch as it is the historical interaction between animal and environment that has conditioned the form and physiology of modern species. Because comparative physiology draws its great strength as a discipline from broad study of the differences and similarities among animals, I have attempted to treat the entire animal kingdom, with the exception of the protozoans and metazoan parasites. It is my firm belief that this broadly comparative approach provides a far deeper and more interesting understanding than would be obtained by study of only the vertebrates or some other limited set of taxa. Background in vertebrate and invertebrate zoology will be useful in reading the text, but I have tried to familiarize the reader with the animal groups as they are discussed. In particular, form and function are so intimately interrelated that I have often found it appropriate to outline anatomical features in some detail prior to discussing physiological attributes.

My emphasis throughout is deliberately conceptual, for I feel that a proper conceptual frame of reference should be the primary goal of an introductory study. Physiological concepts are developed from fundamental principles of biology, chemistry, and physics; and this *conceptual background is then applied to study of the animal groups. I have tried to portray the diversity of animal physiology without at the same time obscuring the fundamentals or attempting to duplicate more advanced and detailed treatments that are already available. It is my hope that the reader, with the foundation developed here, will find that he or she can readily use the more advanced physiological literature to pursue topics in which he or she becomes especially interested. Frequently I have developed particular examples in considerable depth. I only wish that space and the reader's time would allow more examples to be treated similarly, for it is in exploring all the ramifications of an animal's responses that environmental physiology provides its greatest insight and finds much of its excitement as a discipline.*

The preface is often the last part of a text to be written. This is well, for the writer, having struggled with many topics, should have become fully imbued with a sense of humility before the vast diversity not only of the animal kingdom but also of our human attempts to understand it. In this vein, I can only hope that my errors have been small ones.

I wish to express my gratitude to William R. Dawson, who provided inspiration and guidance in my first ventures into comparative physiology and reviewed parts of this manuscript in their preliminary stages. Albert F. Bennett reviewed virtually the entire manuscript, and to him I am grateful not only for his helpful suggestions on this project but also for innumerable hours of insightful professional discussion. Other reviewers have also provided valuable advice. Naturally I take full responsibility for whatever deficiencies remain and hope that my readers will advise me of areas in which the text could be improved. I thank Margaret Beaver for her skillful help in preparing the manuscript and the editors at Harper & Row, particularly Joe Ingram and Holly Detgen, for assistance and moral support in all stages of bringing this project to completion. I also express my appreciation to my students for helping me to clarify many issues. Most of all, I owe my gratitude to my wife Susan and my son David for their support, encouragement, and patience. They have been a source of strength over the three years of long hours the writing has demanded, and their contribution to this work, though largely indirect, has been substantial.

R. W. H.

COMPARATIVE PHYSIOLOGY OF ANIMALS

1 ORGANISM AND ENVIRONMENT

The physiological ecologist attempts to understand those aspects of physiology that are important to the relationship between an organism and its environment. He should have a thorough appreciation of what is meant by *organism* and *environment*. As a preliminary definition, we shall say that the environment includes all chemical, physical, and biotic parameters of the organism's surroundings.

The great physiologist Claude Bernard pointed out about a century ago that organism and environment are defined in terms of each other. It is only by saying what is "organism" that we can say what is "environment." This is illustrated nicely by consideration of a parasite inside a human being. From our usual perspective, the human being is "organism," but from the perspective of the parasite, the human being is "environment." Thus we have a tantalizing, though somewhat semantic, suggestion that organism and environment may not be as clearly distinct as we may have thought them to be.

Because we are accustomed to seeing the world macroscopically, to some extent we may see it incompletely. If for a moment we reduce our scale of vision to the atomic-molecular level, we shall see that the distinction between organism and environment is blurry in more than a semantic way. Suppose we look at a woodland. We see atoms and molecules everywhere. Here and there we see self-sustaining physicochemical systems of high organization. Organization is reflected both in the presence of large, complex molecules and in the patterned orientation of these molecules relative to each other. The organization persists through time. These self-sustaining organized systems are what at a macroscopic level we call organisms.

We see atoms and molecules moving into and out of each organized system. Some atoms and molecules have a rapid passage; they enter at one particular point, follow a particular path through the organized system, and exit at a particular point. These atoms and molecules are being ingested and egested. More importantly, we see other atoms and molecules entering and exiting the *structure* of the organized system. They are incorporated into or removed from the organization itself. Thus in a human being iron atoms are incorporated into red blood cells, and later some of the atoms are excreted when the blood cells break down. Calcium atoms enter the skeleton and later are withdrawn. Tissues break down for a variety of causes and are rebuilt, in part, with new atoms from the outside world.

From such a view of the world we gain a number of important insights. (1) Organisms are in a state of dynamic exchange with their environment. In this way they differ from objects such as telephones, which may be highly organized at an atomic-molecular level but which do not exchange material with the outside world except through surface wear. (2) Atoms and molecules can change between being part of organism and being part of environment. Consider a carbon atom in an amino acid in a piece of meat. When the meat is sitting on the table, we have little trouble in saying that the carbon atom is part of the environment. Most of us are still inclined to assign it to environment when it is in our small intestine. But how do we assign it when it is being transported in the bloodstream? And how should it be assigned when it has been absorbed by some cell? When finally the carbon atom is built into our own muscle protein, we can firmly say that it is part of us. But we should realize from this exercise that it is not easy to say where environment stops and organism begins. The material "boundaries" between the one and the other are not sharp. (3) Most importantly, an organism is not an entity. It is not an object. Suppose you were to mark every atom in an adult animal's body at one point in time and then you examined the animal two years later. Some of the marked atoms would be gone. They would have passed into the outside world and been replaced with new, unmarked atoms taken in from the outside world. Unlike the case of an object, the precise material construction of an organism does not persist through time. Then what does persist? The *organization*. Fundamentally, an organism is a self-sustaining organization, and it is its self-sustaining nature that is referred to by the concept of *homeostasis*.

Scientists have found it useful both semantically and conceptually to separate organism and environment despite the indistinctness of the boundaries between them. Such a separation may be perfectly suitable at all except philosophical levels, but once the separation is made we are still faced with interactions complex enough to weld organism and environment into a remarkably intertwined whole. Thus all organisms are part of the environment of other organisms. Sometimes, as with the host of a parasite, an organism may constitute the entire environment of another organism. Organisms may be altered anatomically and physiologically by agents of the environment. Thus temperature, light, infection, and pollutants may influence the nature of an organism. Finally, organisms may modify the parameters of their environment. Consider a mouse that builds a nest in a small cavity of a tree. From one viewpoint, the mouse has, by its own actions, added a component to its environment, namely, the nest. We may reasonably ask, however, whether the nest is part of the environment. It is built to provide insulation, just as hair provides insulation; and it would not be there if the mouse had not put it there. Perhaps it is almost as much a part of the mouse as hair is. Such musing aside, we may note that the mouse in winter will warm its tree cavity to temperatures above those in the outside world. The mouse will then respond physiologically to the temperatures in the cavity. Clearly, the animal has been able to modify the thermal parameter of its own environment. In like manner, wood lice may lose water by evaporation to whatever minute cavities they occupy. Their rates of evaporative water loss are then governed, in part, by the elevated humidities produced. Fish may deplete the water of oxygen and then must cope with low "environmental" oxygen levels.

The environment has traditionally been divided into factors such as temperature, humidity, light, wind, pH, salt concentration, oxygen concentration, food supply, competition, and predation. Such a factorial approach is taken, at least in part, because our human minds must partition complex things in order to study and understand them. We must be wary, however, of at least two dangers.

The factors we recognize vary greatly in the extent to which they may be quantified. Temperature, for example, may be measured accurately with a very inexpensive instrument; light requires a more expensive instrument; and food supply can be measured only roughly and then only after much effort. There is a temptation to study that which we can measure readily. The danger is that we will overemphasize the importance to the animal of factors we can quantify easily and underemphasize the significance of less tractable factors.

The organism, of course, is exposed to all factors together, and the influence of any one factor may depend on the simultaneous influences of other factors. Most studies in physiological ecology have dealt with one factor in isolation; a good many have dealt with interactions of two factors; but relatively few have dealt quantitatively with interactions of three or more factors. Ultimately, we must appreciate at least the more important interactions if we are to understand the organism in its encounter with the total environment. Two examples will serve to illustrate some of the gamut from simplicity to complexity that is realized in studies of factor interactions. Carp at 1°C die when the water oxygen level falls below 0.8 mg O_2/liter. At 30°C they die at levels below 1.3 mg O_2/liter. This interaction between temperature and oxygen level is readily understandable because fish have higher metabolic rates at higher temperatures and thus need to take in more oxygen per unit time. Studies by Wellington on the spruce budworm have elucidated a great complexity of interactions among temperature, life stage, nutritional status, light, and type of light. Spruce budworms pass through a series of instars separated by molting of the exoskeleton. All instars move toward light at low temperatures and away from it at high temperatures, but the temperature at which their reaction changes depends on the particular instar. At 25°C at least three of the instars are photopositive to all light sources when fed, but when starved they are photopositive to diffuse light sources and photonegative to discrete light sources. The physiological basis for these factor interactions is not understood. Unfortunately, a similar statement must be made for a great many other cases that have been studied.

The physiological ecologist seeks to understand the organism in relation to *its* environment. Defining the environment of an animal is frequently more difficult than it may sound and involves studies of (1) perception, (2) habitat and distribution, and (3) microclimate.

Different animals perceive their surroundings in different ways, and to some extent, physiological and behavioral responses are dependent on perception. There is always a danger that we will superimpose our own human perceptions on other species and in so doing fail to appreciate the interaction between organism and environment. We, for example, have moderately good senses of smell. Birds are less sensitive to odors than we are. Many mammals are much more sensitive and live in a world where odors are exceedingly important and pervasive environmental cues. Kangaroo rats and some other rodents exhibit high sensitivity to low-frequency sounds. They can detect vibrations from the wings of attacking owls of which we are not aware. Honeybees and certain other insects can perceive light in the ultraviolet range, and recent pictures taken with an ultraviolet-sensitive television camera have shown that these animals probably see the world very differently from the way we do. Species of flowers that to us appear almost identical are quite distinct in the ultraviolet, and the insects can presumably distinguish them readily in their foraging for pollen and nectar.

Understanding of the total environment of a species is also gained by study of habitat and distribution. The habitat is, loosely, where the animal spends its time. Star-nosed moles do not simply live under the ground. They usually live in the dark, moist soils of boggy regions, and they occasionally swim in ponds and lakes in search of insect larvae and other prey. One can hardly begin to study the physiological ecology of a species until one appreciates the conditions to which the species is actually exposed, and such appreciation must originate with painstaking natural history work. The distribution or range of a species is the total area in which the species may be found. Certain lizards, for example, are limited to arid regions of our Southwest, and chickadees may be found overwintering as far north as Alaska. Knowledge of distribution gives some idea of the extremes of macroclimatic conditions that are encountered successfully.

Man is a relatively large animal. As he walks about, most of his sense organs are located several feet above the ground. His conception of climate in an area may thus have little relation to the actual climate experienced by other animals. George Bartholomew has expressed this important point especially well:

> Most vertebrates are much less than a hundredth of the size
> of man and his domestic animals, and the universe of these
> small creatures is one of cracks and crevices, holes in logs, dense
> underbrush, tunnels and nests—a world where distances are
> measured in yards rather than miles and where the difference
> between sunshine and shadow may be the difference between
> life and death. Climate in the usual sense of the word is, therefore,
> little more than a crude index to the physical conditions in which
> most terrestrial animals live.*

*From Bartholomew, G. A. 1964. Symp. Soc. Exp. Biol., No. 18: 7–29. Academic, New York.

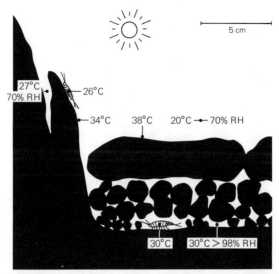

Figure 1–1. Diagrammatic section of the base of a red sandstone cliff and shingle inhabited by *Ligia oceanica*, showing microclimatic temperatures and relative humidities (RH) and body temperatures of the animals. (From Edney, E. B. 1953. J. Exp. Biol. 30: 331–349.)

The actual climate of a species has been termed its *ecoclimate* or *microclimate.* Three examples will serve to indicate the kind of essential information that is gained in studies of microclimatology. Figure 1–1 depicts temperature and humidity conditions in a habitat occupied by *Ligia oceanica*, a semiterrestrial marine isopod crustacean. Note that the entire area in the figure is less than 8 in. across. Temperature above the ground is 20°C, and relative humidity is 70%. Rocks exposed to the sun in *Ligia*'s microhabitat have surface temperatures that far exceed the temperature of the air, whereas the small crevice on the left is heated to a lesser degree. Small chambers among the pebbles near the ground are at 30°C and have relative humidities elevated substantially by evaporation of water from the ground. In these small chambers, *Ligia*'s body temperature equals air temperature, for there is little direct radiative input of heat from the sun, and evaporative cooling of the animal is minimal, because of the nearly saturated ambient humidity. *Ligia* emerging onto exposed rock surfaces experience a much different microclimate: they are warmed by the sun's rays and are cooled by the 20°C air and by evaporative loss of water across their body surfaces. In balance, the heat gains and losses result in a body temperature 4°C lower than in the concealed chambers. Here we see a striking example of how climate may change markedly over distances of only a few centimeters.

Figure 1–2 shows the annual maximal and minimal temperatures at various depths below the ground surface in portions of the Arizona desert. The annual temperature excursion in the air at 1 m above the ground is about 40°C. The excursion on the desert surface is twice as great. The surface heats up more than the air during the daytime because of absorption of solar radiation, and it cools below air temperature at night because of radiative loss of heat to the cold sky (see Chapter 3). In the soil, total annual temperature excursion becomes rapidly smaller with increasing depth

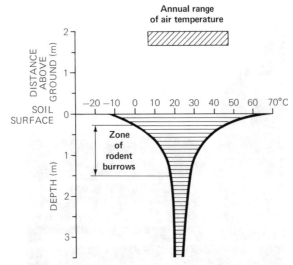

Figure 1–2. Annual range of temperatures in the soil of the Arizona desert near Tucson. Curve to the left depicts minimal temperatures recorded over the year; curve to the right depicts maximal temperatures recorded. The annual range of temperatures in the air is also shown. (From Misonne, X. 1959. Mem. Inst. Roy. Sci. Natur. Belg., 2me ser., vol. 59.)

down to 2 m. At 1 m the excursion is only 16 °C. Small rodents in their burrows find temperatures far below those on the surface during summer and much above those on the surface during winter.

Finally, in Figure 1–3, we see the ameliorating influence of a deep snow cover on temperatures in a cold northern region. At the time the data were collected, temperatures near the ground surface were around 20°C higher than those on the surface of the snow or in the air immediately above the snow. Clearly, animals that burrow under the snow realize a considerable thermal advantage.

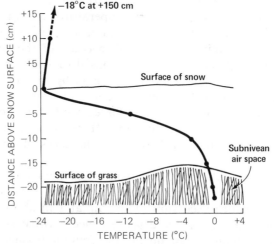

Figure 1–3. Temperature at points above, in, and under a 20-cm-thick snow cover in Sweden on March 5, 1962, at 4 A.M. Various factors lead to development of a subnivean air space among the grass, between the ground and lower edge of the snow cover. It is postulated that small mammals can move about relatively freely in this air space. (From Coulianos, C.-C. and A. G. Johnels. 1963. Ark. Zool. 15: 363–370.)

2 ENERGY METABOLISM

The natural tendency in the universe is for order to change to disorder; nonrandom states tend to be reduced to random states. This basic principle is evident all around us. A cup of hot coffee, for example, tends to cool until it is at the same temperature as the surrounding air; initially the coffee is at thermal disequilibrium with its surroundings, but cooling will continue until a thermal equilibrium has been established. Similarly, an inflated tire tends to come to pressure equilibrium with the surrounding atmosphere, and the materials in a new home—so nonrandomly organized at first—tend to assume a random organization over the years as bricks crumble, plaster cracks, and paint peels and powders. Living organisms are no less subject to these processes than inanimate objects. In the absence of contravening forces, a man or other mammal will cool to the same temperature as his surroundings; and the blood of a freshwater fish, normally having a far higher concentration of salts than the surrounding water, will tend to become as dilute as fresh water. Complex structural and physiological molecules in organisms tend to break down spontaneously, and blood coursing through the circulation tends to come to a halt. In both the animate and inanimate worlds, the maintenance of order and organization in the face of the natural tendencies toward disorder and disorganization requires energy. A cup of coffee can be kept hot by using electrical energy on the stove, and a handyman can keep his home in order through use of electrical energy and the mechanical energy generated by his own arm muscles. Organisms must use energy to support and maintain the complex organization that is such a central feature of being alive. Energy is required for a mammal to maintain its body temperature or for a fish to maintain the normal salt concentration of its blood. Similarly, energy is required to restructure molecules that spontaneously break down, keep the blood in motion, and perform many other functions. Animals obtain the energy they need from the chemical bonds of ingested foodstuffs, and the requirement for this energy is vividly illustrated by the consequences of starvation. The use of energy by animals is termed *energy metabolism* and is the theme of this chapter.

Basic principles of energy utilization by animals

Because energy is neither created nor destroyed in the ordinary world, one might wonder why an animal could not take in a certain amount of energy early in life and use that energy to maintain its life processes thereafter, thus obviating the need to eat. Similarly, though we know that a continuing input of solar radiant energy is necessary to the survival of most of the biosphere, we might question why this is so. In the case of both the individual animal and the biosphere, the crux of the matter is that only certain *forms* of energy can be utilized to sustain the living organization. In using these forms of energy, organisms convert them to other forms that cannot be so utilized; as a consequence, a continuing input of the useful forms is necessary. Green plants require energy in the form of light, and animals require it in the form of chemical-bond energy. To appreciate these principles more fully, we must briefly review the forms of energy and the types of energy conversions that take place within organisms.

Energy is the capacity to do work, in the sense of causing a force to act through a distance. Among the forms of energy that are significant in biological systems are light, chemical energy, electrical energy, mechanical energy, and heat. Chemical energy is energy liberated or required when atoms are rearranged into different assemblages. By mechanical energy we refer here to energy of organized motion, such as that possessed by a moving arm or the circulating blood. Heat is molecular kinetic energy, or the energy of random molecular motion. The molecules of any object above absolute zero move randomly and constantly, colliding into each other and moving on to new collisions. The temperature of an object is a function of its molecular kinetic energy.

Light, chemical energy, electrical energy, and mechanical energy are said to possess a high degree of orderliness, in contrast to heat, which is energy of completely random movement. Light waves oscillate in an orderly manner; chemical compounds are highly organized; electrical energy results from an ordered separation of electrical charge; and the mechanical energy of motion has a particular direction. The ordered forms of energy have a greater capacity actually to perform work than does energy in the form of heat. Heat can be caused to do work, as in the steam turbine, but, comparatively speaking, little work is accomplished relative to the molecular kinetic energy available.

It is impossible to convert heat to work in a system of uniform temperature. Because thermal differences within cells are small and transient, we would expect that cells can make little or no use of heat to perform work, and a great abundance of evidence indicates that cells, in fact, cannot use heat to do work. Nor, as a corollary, can they convert heat into chemical energy or some other form of energy which they can use to do work. Thus although heat may assume a vital role in warming the body, it cannot be utilized as a source of energy for muscular motion, nervous electrical activity, chemical synthesis, or other such essential life processes. We all recognize that a starving animal cannot be kept alive by providing it with energy in the form of heat.

To summarize, forms of energy can be placed into two groups insofar as living organisms are concerned. On the one hand, there are the "high-grade" forms, which are ordered and capable of doing physiological work, namely, light, chemical energy, electrical energy, and mechanical energy. On the other hand, there is a disordered form incapable of performing work, namely, heat. It is common to say that energy is "degraded" upon transformation from a high-grade form to heat.

When organisms transform high-grade energy from one form to another, the conversion is characteristically incomplete, and some energy is degraded to heat. The efficiency of such a conversion is defined as the ratio:

$$\frac{\text{output of high-grade energy or output of work}}{\text{input of high-grade energy}}$$

The efficiency is typically substantially less than 1. In part, the loss of high-grade energy during conversion is an unavoidable attribute of the physico-chemical universe. The second law of thermodynamics tells us that any closed system will tend, over time, toward a greater degree of disorder; or, put another way, high-grade energy will always tend to become degraded to heat. Inefficiencies in biological systems are typically greater than those predicated by the second law because of frictional factors and other aspects of the mechanisms that couple energy release and energy uptake. Some examples will serve to emphasize the pervasive importance of inefficiencies in living things. When a plant chloroplast converts light energy to chemical energy, only a small fraction of the light energy absorbed is actually incorporated into chemical bonds. The remainder appears as heat. When a cell converts the chemical energy of glucose into the chemical energy of adenosine triphosphate (ATP), only somewhat less than half the energy released from the glucose is incorporated into the bonds of ATP (see Chapter 11). When the chemical energy of ATP is converted to the mechanical energy of muscular motion, the transformation is likewise incomplete.

Animals ingest high-grade chemical energy in the form of plant or animal protoplasm. They transform such energy in myriad ways within their bodies, using it to accomplish the tasks of being alive. Bond energy of an ingested sugar molecule, for example, may first be transferred to bond energy of ATP. Then the bond energy of ATP may be used to structure an enzyme (some of the energy appearing, then, as bond energy of the protein); or it may be used to establish an electrical potential across the membrane of a nerve cell (some of the energy appearing as electrical energy); or it may be used to drive the contraction of a muscle fiber (some of the energy appearing as mechanical energy). With each transformation some energy lost from the high-grade, work-capable forms to heat. This heat is dissipated to the outside world.

Another factor to be recognized in the energetics of animals is that the conversion of chemical energy to any other high-grade form, such as electrical or mechanical energy, is a one-way process. So long as energy is retained in the form of chemical energy, it can be directed along many, diverse paths of utilization; but once converted to another high-grade form, it is irreversibly committed to a certain path of utilization and cannot be returned to the pool of chemical energy that is the body's root source of all energy. Thus energy used to contract a muscle fiber accomplishes work in setting the fiber in motion but is destined to be degraded to heat in overcoming frictional forces. Energy used to establish an electrical potential across the membrane of a nerve cell is likewise irreversibly committed and will ultimately be lost as heat when the cell fires. Although it would be theoretically possible for organisms to convert electrical or mechanical energy to chemical energy, mechanisms for doing so have not evolved.

We see at this point that energy is not cycled within the individual organism or, by extension, within the biosphere as a whole. The animal takes in high-grade chemical energy and, in using it to accomplish physiological work, degrades it to heat, which cannot be used to do physiological work. Accordingly, animals are dependent on a continuing input of chemical energy. For the vast majority of living things the ultimate source of high-grade energy is the sun. Green plants fix this energy as chemical energy. Some of the chemical energy is incorporated into plant protoplasm, and much of it is degraded to heat in the metabolic processes of the plants. Herbivorous animals consume that portion of the chemical energy which is incorporated into plant protoplasm and, in turn, incorporate some of it into their own protoplasm while degrading much of it to heat in their metabolic processes. As the chemical energy is passed further up the food chain, each trophic level extracts its toll in the form of heat until finally the original pool of chemical energy is entirely dissipated. Heat produced in the biosphere is radiated to the cold reaches of outer space, this being the final disposition of the photic energy that arrived on earth from the sun.

With this overview, we may now examine in somewhat more detail the fate of chemical energy ingested by the individual animal. As depicted in Figure 2–1, this energy is destined to leave the animal in one of three forms: (1) chemical energy, (2) heat, or (3) external work.

A portion of the ingested chemical energy is simply egested as chemical energy in the feces. Such energy never really entered the animal and remains available to drive the fires of life of those many organisms that consume fecal matter. The chemical energy taken up through the gut wall is termed the *absorbed* or *assimilated energy*. Some of this energy is actually exported from the animal as biologically useful chemical energy. Urine contains compounds that other organisms can metabolize to obtain energy. Similarly, gametes, milk, mucoid secretions, shed skin, shed exoskeletons, antlers, and other such materials composed of organic compounds contain chemical energy that may be utilized by other living things. The chemical energy ingested by the animal must, of course, be passed through a number of conversions before it is incorporated into the bonds of compounds in skin, mucus, or the other substances under discussion. Inefficiencies are involved all along the way, and some of the original energy is degraded to heat. The growing animal builds up stores of chemical energy in its own protoplasm. Thus, again, some of the ingested chemical energy is preserved as chemical energy, though inefficiencies in the catabolic and anabolic reactions involved

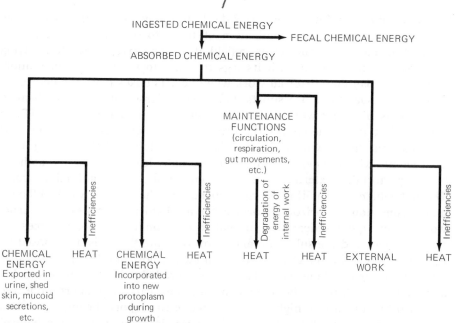

Figure 2–1. Diagrammatic representation of the utilization of energy in the animal. Lower line shows forms in which energy "leaves" the animal. Energy enters as chemical energy and leaves as heat, chemical energy, or external work. See text for full discussion.

in growth imply a considerable loss of energy as heat in the process. The chemical energy incorporated in the protoplasm of an animal is available, within limits, to provide the energetic needs of the animal in times of food shortage and will ultimately supply the needs of predators or saprotrophs.

As indicated in Figure 2–1, much assimilated energy is used for maintenance functions such as circulation, respiration, nervous coordination, activity of the gut, and tissue repair. Energy used in these maintenance functions is nearly entirely degraded to heat within the animal. A brief consideration of circulation will illustrate the progressive degradation of energy in the performance of an important maintenance function. Chemical energy from food substances must first be converted to chemical energy of ATP, with concomitant inefficiency. When ATP is used for contraction of the heart muscle, some energy is again lost as heat owing to inefficiencies in the energy coupling between ATP and the contractile process and also in overcoming viscous and frictional forces within the contracting muscle. Finally, a fraction of the chemical energy originally obtained from food molecules appears as mechanical energy of motion in the blood ejected from the heart. Even this energy is progressively degraded to heat within the body, however, because of viscous resistances within the moving blood. This is clear from the fact that without a fresh contraction of the heart, the blood will quickly slow and stop. Thus all the chemical energy taken from food substances in one way or another gets converted to heat. Work that takes place entirely within the body, such as the acceleration of the blood by the heart, is termed *internal work,* and it must be strongly emphasized that the energy of internal work is always degraded to heat within the body and thus appears as heat at the body surface. Frictional and viscous resistances are as ubiquitous within the body as without.

A final manner in which energy is used by the animal is in the performance of *external work* (Figure 2–1). External work is performed by the muscles and involves the application of force to objects in the external world. When a mouse runs across a field, it is doing external work, applying force against the ground with its legs in order to set its body in motion. Similarly, a man does external work when he lifts boxes from the floor onto a shelf. As usual there are substantial internal losses of energy as heat in the process of doing work; there are inefficiencies in generating ATP, in using ATP to drive the contractile process, and in overcoming viscous and frictional resistances within the muscles and the joints of the moving limb. That energy which is degraded to heat internally leaves the animal as heat, but, in performing external work, some energy leaves the animal as work. The immediate fate of this energy depends on the type of work. When a mouse runs along a horizontal surface, the mechanical energy of its external work is very quickly degraded to heat in overcoming frictional forces between its feet and the substrate and in overcoming the resistance posed by the air to motion. Similarly, when a fish swims horizontally or a bird flies horizontally, external mechanical energy is degraded to heat about as quickly as it is generated. If, on the other hand, a mouse carries an acorn up a tree or a bird flies upward to high altitude, some of the exported energy is stored as potential energy of position. Such energy is converted to mechanical energy, and thence to heat, upon descent. Regardless of the type of external work, energy exported as work cannot be reconverted to chemical energy for reuse by the organism. It is destined ultimately to be degraded to heat and thus lost to the biosphere.

The definition, significance, and direct measurement of metabolic rate

We have seen that the animal assimilates chemical energy and in turn puts out chemical energy, external work, and heat. That energy which is converted to heat or external work is said to be *consumed*. This term is somewhat unfortunate because it may imply that the energy has been destroyed, which is impossible. The term alludes to the fact, however, that such energy has been removed from the chemical state that is the currency for physiologically useful energy transfer within the biosphere. In comparative and environmental physiology, *metabolic rate* is taken to mean the rate of energy consumption, that is, the rate at which the organism converts chemical energy to heat and external work. It is obviously a very important parameter, for the heat and external work exported from an animal reflect closely and quantitatively the overall activity of the animal's physiological machinery. They indicate the amount of energy utilized to support the state of disequilibrium between organism and environment that we call life. In an ecological context the metabolic rate of an animal, being the rate of consumption of chemical energy, represents the drain placed by that animal on the physiologically useful energy supplies of the biosphere.

Metabolic rate can be determined directly in a device termed a *calorimeter*, which measures the rate of heat dissipation from the organism. The device is named after the calorie, a unit of measurement for heat, and the method is termed *direct calorimetry*. One type of calorimeter for small mammals consists of an animal chamber surrounded by a water jacket; the water jacket in turn is surrounded by heavy insulation. Water is pumped through the water jacket. Heat generated by the animal in the inner chamber is picked up by the circulating water, and the amount of heat is determined by measuring the rise in water temperature. The layer of insulation surrounding the water jacket prevents the water from exchanging heat with the surrounding room. The animal in the inner chamber must, of course, be supplied with oxygen, and air is accordingly pumped through the chamber. Heat generated by the animal and picked up by the air stream must also be measured and added to the heat absorbed by the water jacket to obtain total heat production. The air stream can pick up heat in two ways. First, the air can literally be warmed in its passage through the chamber. Second, the air will pick up water vapor dissipated by the animal. Because the conversion of water to water vapor requires heat (see Chapter 3), the evaporative loss of water vapor represents a heat loss and must also be taken into account. The basic rationale in direct calorimetry, regardless of the particular devices used, is to measure all heat dissipated from the animal. The rate of heat dissipation is equal to the rate of heat production in steady state. If the rate of dissipation should not equal the rate of production, this would be reflected in a rise or fall of the body temperature of the animal.

Thus far we have not mentioned the measurement of external work, the second component of metabolic rate. If the animal is at rest, no external work will be performed, and measurement of heat production alone will reflect the entire metabolic rate. In many cases when external work is being performed, the energy of external work is rapidly degraded to heat, as in the case, discussed earlier, of an animal walking about on a horizontal floor. When this is true, a measure of heat production will encompass the energy of external work and thus, again, will be sufficient in itself for determination of metabolic rate. If some energy of external work should be stored as potential energy of position, it then becomes necessary to measure this energy directly and add it to heat production to obtain metabolic rate. Thus if a man were to be stacking boxes on a shelf in the metabolic chamber, the potential energy imparted to the boxes by virtue of their new position would have to be determined independently of the measure of heat production and added in.

It cannot be emphasized too much that production of heat is an inevitable attribute of life and that all organisms produce heat (or, more rigorously, convert chemical energy to heat). The average person, sitting relatively quietly and reading this book, will be producing and dissipating heat at a rate of about 1400 calories per minute—approximately equivalent to the rate of heat production of a 100-watt light bulb. This is the rate at which the reader's body is degrading the chemical energy obtained from ingested food. Should you go out and walk at a brisk pace, your rate of heat production might rise to about 4000 cal/min, reflecting the increased metabolic effort of exercise. Although it is relatively obvious that birds and mammals produce heat and keep their bodies warm by virtue of this heat production, there is a common misconception that animals such as frogs, clams, and fish do not produce heat. Their bodies are cool to the touch and often at nearly the same temperature as the air or water. The temperature of the body is not, however, an indication of the rate of heat production. It is true, of course, that the addition of heat to an animate or inanimate object will raise its temperature. However, organisms are not only producing heat but losing it to the environment, and their temperature is a dynamic function of the rate of heat production and various factors that affect heat loss. In a frog, clam, or fish, heat is produced far less rapidly than in a bird or mammal of equivalent size and is lost so readily owing to poor insulation that metabolic heat production usually does not result in more than a slight elevation of body temperature over the prevailing ambient temperature (see Chapter 3). The fact that their bodies are not warmed cannot be allowed to obscure the fact that they do produce heat. A 90-g goldfish resting at an ambient temperature of 15°C might produce 18 cal/hr, and a 60-g toad resting at 15°C might produce 16 cal/hr.

Indirect methods for measuring metabolic rate (indirect calorimetry)

Metabolic rate is defined to be the rate of heat production plus external work. Direct calorimetry has the advantage of measuring metabolic rate immediately in terms of the definition and constitutes the standard by which all indirect methods must be judged. A difficulty with direct calorimetry is that, for high accuracy, it requires a complex apparatus and considerable technical expertise. Often metabolic rate is measured by indirect methods that have proved to be more convenient in many types of applications.

Metabolism is most commonly determined by measurements of respiratory gas exchange. To understand why this is feasible, it will be helpful first to review certain stoichiometric and thermodynamic relationships in the oxidation of foodstuffs. If a mole of glucose is oxidized completely to carbon dioxide and water in the laboratory, simple stoichiometry dictates that 6 moles of molecular oxygen will be used and 6 moles of carbon dioxide will be produced: $C_6H_{12}O_6 + 6O_2 \rightarrow 6CO_2 + 6H_2O$. We also know that a considerable amount of energy will be released as heat. This *heat of combustion* is typically measured in a device known as a bomb calorimeter. When a mole of glucose is burned in a bomb calorimeter, the heat of combustion is found to be 673,000 cal. Because the oxidation of the glucose also uses a known amount of oxygen, we clearly can establish a relationship between oxygen utilization and heat production, namely, 673,000 cal of heat per 6 moles of O_2. Similarly, we can establish a relationship between CO_2 production and heat production: 673,000 cal of heat per 6 moles of CO_2 evolved. From principles of stoichiometry and thermodynamics, we are led to predict that these same relationships will hold whenever a mole of glucose is oxidized completely to CO_2 and water, regardless of the particular pathway by which the oxidation is accomplished. Thus we predict that an animal that biochemically oxidizes a mole of glucose and uses the energy released in physiological work will also require 6 moles of oxygen and produce 6 moles of CO_2 and 673,000 cal of heat. In the bomb calorimeter all the energy released in the oxidation appears immediately in the form of heat. An animal will temporarily harness some of the energy as chemical energy of ATP. However, as we have seen earlier, this energy will ultimately be degraded to heat as it is used in physiological work. Experiments have verified that oxidation of carbohydrates by animals does involve the same quantitative relationships between oxygen consumption, CO_2 production, and heat production as obtain in the bomb calorimeter. Thus by measuring oxygen consumption or CO_2 production (respiratory gas exchange), we can predict the amount of

Table 2–1. Heat production per unit weight, per unit oxygen consumption, and per unit CO_2 production in the aerobic catabolism of carbohydrates, lipids, and proteins. Values given are for representative mixtures of each of the three foodstuffs. In the case of proteins, values depend on the metabolic disposition of nitrogen (see Chapter 6), and those given apply to mammalian catabolism. As will be true throughout this text, gas volumes are expressed under standard conditions of temperature and pressure (0°C, 760 mm Hg of pressure).

	Heat Production (cal) per Milligram Oxidized	Heat Production (cal) per cc O_2 Consumed	Heat Production (cal) per cc CO_2 Produced
Carbohydrates	4.1 cal/mg	5.05 cal/cc	5.05 cal/cc
Lipids	9.3	4.74	6.67
Proteins	4.2	4.46	5.57

heat production. This is the fundamental principle in the use of respiratory gas exchange to measure metabolic rate.

Table 2–1 summarizes relationships between oxygen consumption, CO_2 production, and heat production for the three major classes of foodstuffs: carbohydrate, lipid, and protein. It is immediately apparent that the relationships depend on the nature of the foodstuff being oxidized. The ratio of heat production to oxygen utilization varies less with the type of foodstuff than does the ratio of heat production to CO_2 production. These observations have important implications, which we will outline below.

Suppose that an investigator has measured the rate of oxygen consumption of an animal. What, then, does he know about the metabolic rate? First we must recognize that oxygen consumption is an indirect measure of metabolic rate. To obtain the true metabolic rate, the rate of oxygen consumption should be translated into a rate of heat production, but this cannot be done unless the investigator knows what foodstuffs were being oxidized by the cells. From Table 2–1, we see that at an oxygen consumption of 10 cc/min, the metabolic rate would be $10 \times 5.05 = 50.5$ cal/min if carbohydrates were being oxidized but would be only $10 \times 4.74 = 47.4$ cal/min if lipids were being oxidized. Generally a mixture of foodstuffs is being used, and generally the investigator does not know the proportions of each. Accordingly, there will be uncertainty about the correct conversion factor to use in calculating metabolic rate from oxygen consumption. Commonly a "representative" conversion factor of about 4.8 cal/cc O_2 is used; this is the approximate factor to be expected in an animal that is metabolizing a representative mixture of carbohydrate, lipid, and protein. One of the strong points of using oxygen consumption to measure metabolism is that because the true conversion factor varies relatively little with foodstuff (Table 2–1), the potential errors produced by using an approximate factor are not large. To illustrate, suppose that an investigator uses the conversion factor of 4.8 cal/cc O_2 but that the animal was oxidizing only carbohydrate and thus that the true conversion factor was 5.05 cal/cc O_2. The investigator will then have underestimated the metabolic rate, but only by 5%. If, on the other hand, the animal was oxidizing only protein, meaning that the true conversion factor was 4.46 cal/cc O_2, the investigator will have overestimated the metabolic rate, but again to only a relatively small extent, 7.5%. It is on these grounds that oxygen consumption is often used as a measure of metabolic rate even though the nature of the foodstuffs is unknown. Nonetheless, though errors cannot be very large, it must not be forgotten that oxygen consumption is an indirect measure, and small but sometimes significant errors are possible. Frequently, measures of oxygen consumption are reported as such in the scientific literature rather than being converted to equivalent rates of heat production. Thus metabolic rate, though technically expressed in calories per unit time, is often expressed as oxygen consumption per unit time.

Carbon dioxide production was commonly used to measure metabolic rate several decades ago but is seldom used any longer because of two factors that render it distinctly inferior to oxygen consumption. The first of these is evident from Table 2–1, namely, that the conversion factor between heat production and CO_2 production depends very strongly on foodstuff. Accordingly, if the foodstuffs being oxidized are unknown and a "representative" conversion factor is used, metabolic rate can be misestimated by as much as 15–20% if the actual foodstuffs differ strongly from those assumed by the "representative" factor.

The second limitation of using CO_2 production to measure metabolic rate arises from a basic presumption of all methods based on respiratory gas exchange. Oxidation of foodstuffs occurs within the cells of the animal, and it is there that oxygen utilization and CO_2 production bear strict relationships to heat production. We, of course, do not measure oxygen consumption and CO_2 production at the level of the cells. Rather we must measure both the rate of oxygen consumption and the rate of CO_2 elimination at the respiratory organs. Accordingly, our measures of gas exchange relate strictly to metabolic rate only when a steady state exists, that is, only when gas exchange at the respiratory organs proceeds at the same rate as at the level of the cells. This consideration presents problems in certain situations even when oxygen consumption is being used. If a mammal, for example, suddenly starts to exercise, the muscle cells at first meet some of their increased oxygen demand by drawing oxygen from oxygen stores pre-existing in the body. Respiratory oxygen uptake lags behind cellular oxygen demand for a minute or more, and it is only after this period of adjustment that a steady state is again established such that the rate of respiratory oxygen consumption corresponds to the rate of cellular utilization. During the adjustment period, the respiratory rate of oxygen consumption does not provide an accurate measure of metabolic rate. Carbon dioxide production is far more liable to these types of errors than oxygen consumption. Carbon dioxide dissolves in body fluids more readily than oxygen, and, unlike oxygen, carbon dioxide can react to form other compounds such as bicarbonate (see Chapters 8 and 9). Although respiratory CO_2 elimination must in the long run correspond to cellular CO_2 production, fairly sustained periods can occur over which the necessary steady state does not exist.

It is apparent that measures of oxygen consumption or CO_2 production could be converted entirely accurately to measures of metabolic rate if the foodstuffs being utilized by the cells were known (Table 2–1). There are methods for determining the nature of the foodstuffs, which we may now briefly examine. First it must be clearly recognized that the relationships between gas exchange and metabolic rate depend on the foodstuffs being catabolized *in the cells*. Because animals interconvert and store food materials, the recent diet of the animal does not reveal the nature of the food-stuffs being oxidized at any given time. A respiratory parameter that varies with foodstuff is the *respiratory quotient*: the steady-state ratio of CO_2 production to oxygen consumption. The respiratory quotient over a period of time is calculated by dividing the volume of CO_2 produced by the volume of oxygen consumed, both volumes being corrected to standard conditions of temperature and pressure (STP). As is evident from the analysis of glucose oxidation presented earlier, oxidation of carbohydrates requires and produces equivalent molar quantities of oxygen and CO_2, respectively. Because equivalent molar quantities of gases occupy virtually equivalent volumes at STP, we see that oxidation of carbohydrate produces a volume of CO_2 equivalent to the volume of oxygen used. Accordingly, the respiratory quotient (RQ) for carbohydrate is 1.0. Oxidation of lipid produces a smaller volume of CO_2 than the volume of oxygen consumed, and the RQ for lipids is 0.71. Protein has an intermediate RQ of 0.80. We see that if *both* oxygen consumption and CO_2 production are measured so that the RQ can be calculated, the types of foodstuffs being oxidized can sometimes be estimated. An RQ near 1.0, for example, would indicate a preponderance of carbohydrate, and an RQ near 0.71 would indicate a preponderance of lipid. Knowing such RQ's, the investigator would have good insight as to which conversion factor from Table 2–1 he should use to calculate metabolic rate from oxygen consumption. Unfortunately RQ's that fail to approximate one of these extreme values are not so useful. An RQ of 0.8, for example, could indicate catabolism of only protein, or catabolism of a particular mixture of lipid and carbohydrate, or catabolism of a mixture of all three foodstuffs. The problem is simple. There are three unknowns (the quantities of the three foodstuffs being oxidized), but RQ encompasses only two knowns (CO_2 production and oxygen consumption). To solve for three unknowns, three knowns are needed. For certain applications the problem can be solved by measuring the elimination of waste nitrogen (nonprotein nitrogen) in the urine. Because nitrogen is derived only from protein, urinary nitrogen loss can be used to estimate the amount of protein catabolism. This provides the third known and, used in conjunction with the RQ, allows calculation of the amounts of all three foodstuffs being oxidized. Then metabolism can be calculated precisely from oxygen consumption. In many situations investigators feel no need for the additional complications of measuring RQ and urinary nitrogen. They measure oxygen consumption alone and "live with" the relatively small potential errors that are involved in estimating metabolic rate from oxygen utilization without knowledge of the foodstuffs being used.

A final point that deserves emphasis is that measures of oxygen consumption (or CO_2 production) relate to metabolic rate only when foodstuffs are being oxidized aerobically. As will be discussed in detail in Chapter 11, animals sometimes resort to anaerobic mechanisms of ATP production which do not require oxygen (as when glucose is catabolized to lactic acid by anaerobic glycolysis). When the energy resources of foodstuffs are being tapped by such mechanisms, oxygen consumption is clearly an unsuitable measure of metabolic rate, and other procedures (such as direct calorimetry) must be used.

Because oxygen consumption is by far the most common method of measuring metabolic rate, some brief comments on techniques are appropriate. One approach is to seal the animal in a closed container for a period and measure the oxygen extraction from the air or water in the container. Another approach is to pump air or water through the container at a known rate and measure the drop in oxygen content between the inflowing and outflowing streams. The former approach (involving a closed container) is termed a *closed-system technique*, whereas the latter is termed an *open-circuit technique*. Because oxygen consumption is proportional to the ultimate yield of heat from catabolism of foodstuffs, external work that results in storage of potential energy need not be measured independently; the heat equivalent of the potential energy is included in the metabolic rate computed from oxygen utilization.

A final indirect technique for measuring metabolic rate is the so-called *material balance method*. The approach is to measure the amount of food eaten and the amounts of urine and feces eliminated over a period of time. Using a bomb calorimeter, the available energy content of the ingested food is determined; the available energy content of the excreted material is also measured. The latter is then subtracted from the former to estimate metabolic rate. The logic of this method is straightforward: the animal must consume that chemical energy which it ingests but does not in turn void. Additional considerations are involved if the animal is increasing or decreasing its biomass. If, for example, the animal is growing and thus increasing the chemical energy content of its body, then some of the chemical energy ingested but not voided is nonetheless not being consumed, and an estimate of this quantity must enter the calculation of metabolic rate. A shortcoming of the material balance method for many types of applications is that measurements must extend over an appreciable period of time (typically a day or more), and the metabolic rate obtained is the average metabolic rate over the entire test period. The long time interval is mandated by the need to obtain *steady-state* measures of the rates of ingestion and egestion. An animal might, for example, eat but not defecate over a given short period, such as an hour, and it would obviously be inaccurate to assume an egestion rate of zero. A long test period is needed to assure that the measured rates of ingestion and egestion are representative of the true average rates. One of the great advantages of using oxygen consumption is that it provides, more or less, a minute-by-minute indication of metabolism.

Factors that affect metabolic rate

Metabolic rate represents the amalgamation of innumerable and diverse energy-consuming processes within the body. It is not surprising then to find that the metabolic rate of an animal can be influenced by an immense variety of parameters, including age, sex, reproductive condition, hormonal balance, psychological stress, nutritional condition, time of day, race, and oxygen availability. Two factors that exert particularly potent influence are environmental temperature and activity level. These will receive detailed attention in Chapters 3 and 14. A quick impression of the strong effects exerted by activity can be gained by looking at some representative values for man. Asleep, a man might expend 1,000 cal/min; sitting quietly, he might expend 1,400 cal/min; values during walking, bicycling, and full-speed running might be 4,000, 7,000, and 17,000 cal/min, respectively. If a measurement of the metabolic rate of an animal is to be meaningful, care must be taken to define prevailing circumstances as closely as possible.

Besides all the factors that can affect the metabolic rate of a given species, there are also strong differences among species. At rest, for example, fish and amphibians typically have metabolic rates at least an order of magnitude below those of comparably sized birds and mammals. Accordingly, they have far lower food requirements and, as individuals, place smaller energetic demands on the ecosystem. Interspecific differences in metabolism will receive more quantitative attention later in this chapter and in Chapter 3.

Specific dynamic action

When a previously fasting animal consumes food, metabolic rate increases for a period of time, though all other conditions are kept constant. This effect is known as the calorigenic effect or specific dynamic action (SDA) of the food. The *magnitude* of the SDA is defined to be the total *excess* metabolic heat production induced by the meal, integrated from the time metabolism first rises to the time that it falls back to the "background" level. To illustrate, suppose that a resting, fasting man has a metabolic rate of 1,400 cal/min. If he consumes some protein, his metabolic rate will rise within the first hour after the meal and will then remain above 1,400 cal/min for several hours. Suppose that it takes 5 hours for his metabolic rate to return to 1,400 cal/min, and suppose that his total metabolic heat production over the 5 hours is 500,000 cal. If he had not eaten the meal, his total heat production over 5 hours would have been only $1,400 \times 60 \times 5 = 420,000$ cal. Accordingly, the total *excess* heat production induced by the meal, the SDA, is $500,000 - 420,000 = 80,000$ cal. SDA is commonly expressed as a percentage of the total caloric value of the food ingested. To illustrate, if the protein consumed by our hypothetical man would yield a total of 320,000 cal upon complete physiological oxidation, the measured SDA would represent 25% of the caloric value of the protein. The implication is that the equivalent of 25% of the available energy from the protein is obligatorily dissipated as heat in the process of assimilating the protein. Only the remaining 75% is then available for other physiological uses.

Proteins typically exhibit SDA's of around 25–30%. Lipids and carbohydrates display far lower SDA's, about 5–10%. The explanation for SDA remains uncertain. Experiments have demonstrated that the activity of the digestive tract makes only a very minor contribution to the increase in metabolic rate. The consensus is that the SDA largely arises from the biochemical processing of absorbed nutrient materials within the cells of the body. The especially great SDA of protein may be associated with deamination reactions.

In physiological studies an animal is said to be *fasting* or *postabsorptive* when it has not eaten for a time sufficient to eliminate any SDA.

Standardized measures of metabolic rate

In consideration of the many influences on metabolic rate, physiologists have long recognized the need for some standardized measure of this parameter. Such a measure is essential for meaningful comparison of metabolic levels in different individuals, species, or higher taxonomic groups. Some animals (the poikilotherms) allow their body temperature to fluctuate over a wide range according to environmental circumstances. When such animals are postabsorptive and at rest, their metabolic rate is said to be the *standard metabolic rate* for the prevailing body temperature. Other animals (the homeotherms) regulate their body temperature. For these animals there is a certain range of environmental temperatures, termed the *thermoneutral zone*, in which metabolic rate turns out to be minimal (see Chapter 3), and the rate of a resting, postabsorptive individual in the thermoneutral zone is termed the *basal metabolic rate*. It is not surprising that both of the standardized measures of metabolic rate here discussed call for a resting condition. Activity level, as mentioned earlier, exerts a strong effect on metabolism, and, of all the levels of activity, the only one that is easily defined in reasonably quantitative terms for all species is rest. Unfortunately, however easy the experimenter may find it to rest, it is often no small challenge to get an experimental animal to rest completely. *Rest* thus has somewhat different meanings in different studies, and increasingly there has been an attempt at formal recognition of different gradations of the "resting condition." Some workers apply the term *routine metabolic rate* to reasonably quiet animals exhibiting only relatively small, spontaneous movements and reserve *standard metabolic rate* for animals that have been coaxed to a truly minimal activity level. According to these definitions, many "standard" metabolic rates in the biological literature are, in fact, routine metabolic rates. When activity is truly minimal under standard or basal conditions, metabolic rate approximates the rate necessary for simple physiological maintenance of life.

Relationships between metabolic rate and body weight

Figure 2–2A shows the general relationship between basal metabolic rate (BMR) and body weight in placental mammals and illustrates a most significant characteristic: Although metabolic rate increases steadily with weight, it does not increase in proportion to weight. Large animals have higher BMR's than small mammals, but not proportionally higher. An average 10-g mammal, for example, will exhibit a BMR near 20 cc O_2/hr. The dashed line in the figure shows how metabolic rate would increase with weight if larger mammals retained the same proportional relationship between BMR and weight as the 10-g mammal. Under such circumstances a 100-g mammal would have a BMR near 200 cc O_2/hr, but in actuality the average BMR of 100-g mammals is nearer 110 cc O_2/hr. This same trend persists throughout the entire range of mammalian weights. Thus given a BMR of 110 cc O_2/hr for a 100-g animal, we would expect a 400-g animal to have a BMR of 440 cc O_2/hr on the basis of proportionality, but an average 400-g mammal will, in fact, have a BMR of only about 300 cc O_2/hr.

Another way to see this relationship is to examine the metabolic rate per unit of body weight, termed the *weight-specific metabolic rate*. The curve in Figure 2–2A is replotted in weight-specific terms in Figure 2–2B, and it can be seen that weight-specific metabolic rate decreases steadily as weight increases. Under basal conditions a 3700-kg elephant produces only about 40% as much metabolic heat per gram as a 60-kg human and only about 5% as much as a 20-g mouse. Stated differently, it costs far less for an elephant to sustain a gram of elephant than for a mouse to sustain a gram of mouse.

Similar relationships between metabolic rate and body weight are

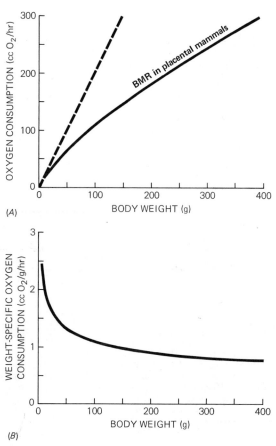

Figure 2–2. The general relationship between basal metabolic rate and body weight in placental mammals. (*A*) Solid curve shows actual relationship between rate of oxygen consumption and weight. Dashed line shows how oxygen consumption would vary with weight if all mammals exhibited the same proportional relationship between metabolism and weight as is seen in 10-g mammals. See text for discussion. (*B*) Weight-specific rate of oxygen consumption as a function of weight. (Curves based on equation given in the legend of Figure 2–3.)

found in virtually all animal groups. Typically, basal or standard metabolic rate is related to weight according to the following equation:

$$M = aW^b$$

where M is the total metabolic rate, W is weight, and a and b are constants. If b were equal to 1.0, this equation would reduce to $M = aW$, indicating a proportional relationship between metabolism and weight. However, b is usually less than 1.0 (commonly 0.6–0.9), meaning loosely that as W increases, M does not increase "as fast." Dividing both sides of the preceding equation by W, we get

$$\frac{M}{W} = aW^{(b-1)}$$

The expression on the left, M/W, is the weight-specific metabolic rate. Because b is usually between 0.6 and 0.9, $(b-1)$ is usually negative and be-

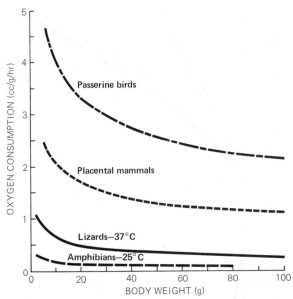

Figure 2-3. Weight-specific basal or standard metabolic rate as a function of body weight in four groups of vertebrates according to the following equations, where M/W is weight-specific metabolic rate expressed in cc O_2/g/hr and W is body weight expressed in grams: (a) Basal metabolic rate in placental mammals: $M/W = 3.8W^{-0.27}$. (b) Basal metabolic rate in passerine birds: $M/W = 7.54W^{-0.276}$. (c) Standard metabolic rate in lizards at a body temperature of 37°C: $M/W = 1.33W^{-0.35}$. (d) Standard metabolic rate in temperate-zone amphibians (anurans and salamanders) at a body temperature of 25°C: $M/W = 0.355W^{-0.33}$. (Sources for equations—mammals: Brody, S. 1945. *Bioenergetics and Growth*. Reinhold, New York; birds: Lasiewski, R. C. and W. R. Dawson. 1967. Condor 69: 13–23; lizards: Templeton, J. R. 1970. Reptiles. *In:* G. C. Whittow (ed.), *Comparative Physiology of Thermoregulation*. Vol. I. Academic, New York; amphibians: Whitford, W. G. 1973. Amer. Zool. 13: 505–512.)

tween −0.4 and −0.1. The negative value of $(b - 1)$ signifies what we have already said, namely, that weight-specific metabolic rate decreases with increasing body weight.

The coefficients a and b depend on the animal group under consideration. In those organisms (the poikilotherms) that allow their body temperature to vary with ambient conditions, the standard metabolic rate is a function of prevailing body temperature; and a, and sometimes b, also depend on the body temperature under consideration. Figure 2–3 provides four examples of metabolism-weight equations. Note that the equations are plotted and written (in the legend) in terms of weight-specific metabolic rate; accordingly, the exponents in the equations are, in each case, equal to $(b - 1)$. The values of b for the four groups of vertebrates shown are rather similar, ranging only from 0.65 to 0.73. The values of a, however, are quite disparate. In essence, a reflects the absolute level of metabolism (being, in fact, equal to the theoretical metabolic rate of a 1-g animal). Of the groups shown, the passerine birds have the highest metabolic rates for given weight and also have the highest value for a; contrariwise, the amphibians at 25°C have the lowest metabolic rates and the lowest value for a. The figure illustrates dramatically a point made earlier, that birds and mammals have far higher resting metabolic rates than reptiles or amphibians of equivalent size.

The approach to determining a and b for a particular animal group is to obtain metabolic data for animals of a wide range of sizes and then determine, by statistical procedures, the values of a and b that best fit the equation $M = aW^b$ to the animal data. The resulting equation, when plotted on a graph, will never pass directly through all the data points; it is simply the best average statistical "fit" to the data. Accordingly, although the equation for an animal group can be used to approximate the metabolic rate expected for an animal of given size, it is not a substitute for actual data on the animal. Also, because a and b are determined statistically by using available data, they are always subject to revision as new data are obtained.

If we take the logarithm of both sides of the equation $M = aW^b$, we get

$$\log M = \log a + b \log W$$

This indicates that a plot of log M against log W will be linear, for if we set $Y = \log M$ and $X = \log W$, we get $Y = \log a + bX$, a linear equation (b and log a are constants). Similarly, a plot of log (M/W) against log W will be linear. Data relating metabolic rate to weight are nearly always graphed on a log-log plot, where these linear relationships pertain. Figure 2–4 provides two examples.

The fact that weight-specific metabolic rate usually decreases with weight within a given taxonomic group has numerous practical implications. Small animals, for example, require more food per unit of their body weight than larger, related animals. Some small shrews, to illustrate, require an amount of food (wet weight) equivalent to their body weight each day, but humans certainly do not demand 100–200 lb of food per day. The uninitiated might expect that the basal energy requirements of 3,500 mice, each weighing 20 g (total weight: 70,000 g), would place no greater demands on a woodland ecosystem than a single 70,000-g deer. Because the weight-specific basal metabolic rate of a 20-g mouse will be about eight times greater than that of a deer, however, the total basal metabolism of only about 440 mice would be equivalent to that of a deer. In these days of widespread pollution, the relationship between metabolism and weight also has potential implications for the effects of environmental toxins. Because small animals must eat more food per unit of body weight, they also are exposed to greater weight-specific doses of toxins in the food. The same holds for airborne toxins because small animals also must respire more air per unit of body weight to meet their greater weight-specific oxygen demands. The relationship of metabolism to weight is frequently a significant factor in the design of experiments. Suppose that an investigator, for instance, were studying the effect of temperature on the metabolic rate of a species. If he inadvertently studied predominantly small individuals at one temperature and predominantly large individuals at another, weight-specific metabolic rate would be affected not only by temperature, but also by the differences in size between the two test groups. To see the effects of temperature alone, the effect of weight should be controlled by studying individuals of similar size at both temperatures.

Ever since the relationship between metabolic rate and weight was discovered, much interest has centered on explaining it. Early students of this problem worked on mammals and enunciated the so-called surface law as an explanation. Their reasoning may be outlined as follows. Mammals maintain a fairly constant and high body temperature (near 37°C) through

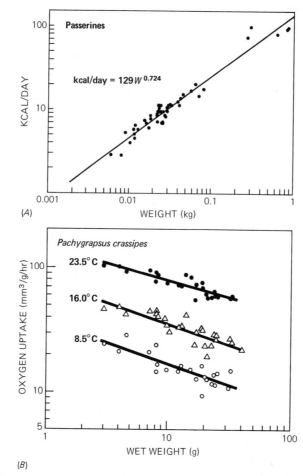

Figure 2-4. Relationships between metabolic rate and body weight in two groups of animals plotted on double logarithmic coordinates. (*A*) Total basal metabolic rate as a function of body weight in passerine birds. The line drawn through the data points conforms to the equation shown. Data represent over 30 different species. (*B*) Weight-specific metabolic rate as a function of body weight at three different body temperatures in the crab *Pachygrapsus crassipes*. (*A* from Lasiewski, R. C. and W. R. Dawson. 1967. Condor 69: 13–23; *B* from Roberts, J. L. 1957. Physiol. Zool. 30: 232–242, used by permission of The University of Chicago Press. *B* © 1957 by the University of Chicago.)

production of metabolic heat. Heat is dissipated to the environment in proportion to the surface area of the body. As with all geometrically similar objects, the ratio of surface area to body volume or body mass decreases with increasing size. Thus, other things being equal, larger animals lose less heat to the environment per unit weight than smaller animals, and it follows that larger animals need to produce less heat per unit weight. The argument is quite plausible. Early estimates of *b* placed it at about 0.67 for mammals, and surface also is expected to increase as the two-thirds power of weight for objects of similar geometry.* Metabolism, therefore, was thought to increase in direct proportion to surface area, just as the surface law would predict. The surface law ran into trouble as more comprehensive data on

---

*To illustrate, the weight (or volume) of a sphere is proportional to r^3, where r is the radius. The surface area of a sphere is proportional to r^2. Since $(r^3)^{2/3} = r^2$, we see that surface area increases as the two-thirds power of weight.

mammals and other groups accumulated, and there may be few areas of biology that can match the enormous verbiage that has been devoted to discussion of its merits. A most serious challenge to the surface law was raised by the discovery that animals that allow their body temperature to vary with ambient conditions (poikilotherms) exhibit the same sort of relationship between metabolism and weight as mammals. The arguments behind the surface law are clearly not applicable to such animals, for they do not keep their bodies warm through metabolic heat production. Also, revision of the estimate of b for mammals now places it at about 0.73 or 0.75; it does not equal the 0.67 predicted by the surface law. An evaluation of the surface law is at present difficult, especially because the relationship between metabolism and weight has yet to be explained in a definitive manner in any group of animals. Whereas smaller birds and mammals unquestionably do need to produce more heat per unit weight than larger birds and mammals in order to maintain a high body temperature, the ubiquity of the metabolism-weight relationship in the animal kingdom might nonetheless suggest an explanation in birds and mammals more fundamental than the one postulated by the surface "law." There is the possibility, on the other hand, that the metabolism-weight relationship may find different explanations in different animal groups. Hypotheses that could have broader application than the surface law are numerous. One, for example, points out that if the tissues of large animals demanded as much oxygen per unit weight as those of small animals, an inordinate development of the respiratory and circulatory structures required to supply oxygen might be required. We can illustrate this argument in an elementary way with reference to the lungs. If one animal were to be four times the weight of another and if metabolic rate per unit weight were to be the same for both, the larger animal would require four times the oxygen of the smaller animal. Other things being equal, this might require four times as much surface area for respiratory exchange in the lungs. Because surface area increases less rapidly than volume in geometrically similar structures, the lungs of the larger animal might then have to be more than four times larger in volume than those of the smaller animal and might consequently have to pre-empt a larger portion of the total body volume. Such a circumstance might be un- _- defendable_ tenable in terms of overall demands on available space in the larger animal. Unfortunately, there are many complexities in assessing the applicability of this type of reasoning to the real world, and despite an immense amount of effort devoted to understanding metabolism-weight relationships, they as yet remain largely enigmatic. _- baffling, puzzling_

Postscript

The reader who has persevered to this point will be interested to know that concerted mental effort is not very expensive in energetic terms. The physiologist F. Benedict calculated that the mental activity of one hour's hard study demands only the energy contained in an oyster cracker or that in one half of a salted peanut.

SELECTED READINGS

Animal Nutrition. (A Symposium.) 1968. Amer. Zool. 8: 70–174.
Bertalanffy, L. Von. 1957. Quantitative laws in metabolism and growth. Quart. Rev. Biol. 32: 217–231.

Brody, S. 1945. *Bioenergetics and Growth*. Reinhold, New York.

Brown, A. C. and G. Brengelmann. 1965. Energy metabolism. *In:* T. C. Ruch and H. D. Patton (eds.), *Physiology and Biophysics*. Saunders, Philadelphia.

Kleiber, M. 1961. *The Fire of Life*. Wiley, New York.

Lasiewski, R. and W. R. Dawson. 1967. A re-examination of the relation between standard metabolic rate and body weight in birds. Condor 69: 13–23.

Zeuthen, E. 1953. Oxygen uptake as related to body size in organisms. Quart. Rev. Biol. 28: 1–12.

See also references in Appendix.

3 THERMAL RELATIONSHIPS

The temperature of an animal's body generally has profound effects on function. The cells, tissues, and organs of all organisms have upper and lower lethal temperatures, and within the thermal range compatible with life, rates of function are typically highly dependent on temperature, just as rates of simpler physical and chemical processes are thermally sensitive. Animals display several different types of relationships with their thermal environment. These relationships and their physiological and ecological implications are the subject of this chapter. Two particularly prevalent types of thermal relationships are *homeothermy* and *poikilothermy*. Homeothermy is characterized by the physiological maintenance of a relatively constant internal temperature regardless of external temperature. The tissues of the homeothermic animal are permitted to function in a more or less stable thermal milieu, but the organism must expend energy to maintain a state *surroundings,* of thermal disequilibrium between itself and its environment. In poikilo- *environment* thermy, body temperature is allowed to vary with ambient conditions; thus the body temperature will be low in a cold environment and high in a warm environment. The poikilothermic animal realizes a considerable energy saving compared to the homeothermic animal in not maintaining a thermal disequilibrium between its body and the environment, but its cells, tissues, and organs must cope with a changing internal temperature. Organisms that predominantly or exclusively display one or the other of these thermal relationships are termed *homeotherms* and *poikilotherms*, respectively. Although these terms have limitations, which we shall discuss later in this chapter, we may recognize here that the birds and mammals are classed as homeotherms, and many of the lower vertebrates and invertebrates are classed as poikilotherms.

The effects of temperature on individual organisms have much ecological significance. We mentioned in Chapter 2 that metabolic rate in both homeotherms and poikilotherms is often strongly dependent on ambient temperature. Thus the energetic demands placed on the environment by an organism increase and decrease with thermal circumstances. Temperature is also significant in that the animal must be able to survive the various thermal challenges imposed upon it throughout the year. The distribution and habitat of species may thus be influenced by thermal effects. Temperature per se may exceed viable limits at certain times or in certain places, or temperature may interact with other factors to produce a physiologically or ecologically stressful or untenable situation. Mammals and birds in deserts, for example, may face critical dehydration problems if required to expend large amounts of water in evaporative cooling to keep their body temperature from rising to threatening extremes during exposure to the intense environmental heat. Fish in summer may have high metabolic rates because their body temperatures are elevated in the warm water. At the same time, they are faced with relatively low oxygen availability because warm water holds less dissolved oxygen than cold water. The interaction of these factors may prove critical. The profound ecological impact of temperature may nowhere be more strikingly illustrated than in our temperate woodlands. In the warm seasons of the year we witness sustained activity of birds, mammals, reptiles, insects, amphibians, and other organisms. Plants are actively photosynthesizing, producing chemically fixed energy that then passes through food webs, fueling the metabolic fires of other living things. In the winter, plants and poikilothermic animals such as reptiles, amphibians, and insects go into a state of quiescence. Their body temperatures are lowered, and their functional rates are correspondingly depressed. Activity in the woodland is largely limited to those birds and mammals that, by virtue of their relatively constant internal temperatures, can continue to move about and forage for food despite lowered ambient temperature. We see the profound influence that temperature may exert on the nature of the whole ecological community, an influence that is also displayed by the changes in community types evident as one travels from the equator to the poles.

MECHANISMS OF HEAT TRANSFER

Some of the principles of heat transfer relevant to animal thermobiology may be outlined with reference to the simplified model depicted in Figure 3–1. The body core is considered to be at a uniform body temperature T_B, and the environment is at uniform ambient temperature T_A. Separating the core from the environment are the outer layers of the organism, wherein temperature gradually changes from T_B on the inside to T_A on the outside. Heat moves among parts of the organism and between organism and environment via one or more of four possible routes: conduction, convection, radiation, and evaporation of water. The core gains heat continuously from metabolism and may lose heat to or gain heat from the environment. If total gains and losses are equal over a period of time, T_B will be constant. If gains and losses are unequal, T_B will rise or fall. To appraise an organism's ther-

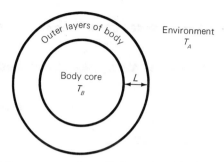

Figure 3–1. A schematic representation of the animal body. The outer layers of the body have thickness L. The body core is at temperature T_B and the surrounding environment is at temperature T_A.

mal situation therefore, one must examine metabolic rate and the several possible modes of heat exchange with the outside world.

Conduction is diffusional transfer of heat without movement of the medium in which transfer is occurring. *Convection*, on the other hand, is transfer that involves movement of the medium. We are all familiar with these modes of heat transfer from our everyday experience. If a man sits on warm sand, he will receive heat from the sand through the seat of his pants largely by conduction. If he stands in a hot desert wind, he will receive heat from the air largely by convection.

The rate of heat movement via conduction is proportional to the prevailing *thermal gradient*. The thermal gradient for any given object through which heat is moving conductively is defined as the difference in temperature between the two sides of the object divided by the thickness of the object. Thus for our simplified animal in Figure 3–1, the thermal gradient across the outer layer of the body is

$$\frac{T_B - T_A}{L}$$

Heat will move conductively from the warmer to the cooler side and at a rate proportional to the preceding ratio. It can be seen that heat will move faster, the greater the temperature difference. Thus a camel lying on sand will receive more heat per unit time when the sand is at 50°C than when it is at 45°C. Also, for a given temperature difference, the thinner the conductive layer, the faster heat will move. One function of hair is to trap a layer of relatively static air around the organism. Heat must move conductively through such a layer, and the thicker the layer of air trapped by a mammal's pelage, the slower the animal will lose heat in the cold of winter. The rate of movement of heat by conduction depends on the properties of the conductive medium as well as on the thermal gradient. Some media, such as air, are characterized by low conductivity or high insulative value. Others, such as water, have higher conductivity or lower insulative value.

Heat transfer by convection is much more rapid than that by conduction for a given temperature difference. A horizontal surface that is 10°C warmer than the surrounding air may lose heat 70 times faster if the air is moving at 10 mph than if the air is perfectly still. We are made quickly aware of this basic phenomenon if we dip our fingers in cold water. The sen-

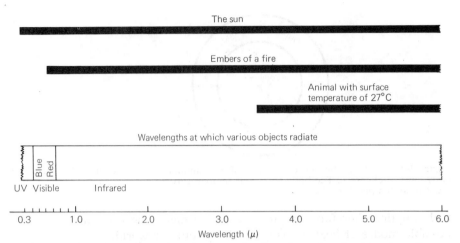

Figure 3–2. The electromagnetic spectrum, showing the approximate range of wavelengths at which various objects radiate by virtue of their surface temperatures. All three objects radiate out to long wavelengths in the infrared (not shown here).

sation of cold is less severe if our fingers and the water are kept still than if we swirl our fingers around. It is common to think only of external media in discussions of convective transfer. In fact, one of the important functions of the circulatory system in many organisms is to move heat convectively from one part of the body to another. Man in the desert is strikingly dependent on such heat transfer. Death often results from an explosive rise in body temperature following a slowdown in circulation. Convective transfer of heat from the body core to the skin and environment is reduced by the decreased rate of blood flow; conductive transfer, being much slower, cannot take up the slack, and heat accumulates in the core, driving T_B up.

When water changes from its liquid to its gaseous state, a certain quantity of heat is absorbed at the surface at which the change in state occurs. This _latent heat of vaporization_ amounts to 570–595 cal/g of water at physiological temperatures. When organisms sweat or pant they take advantage of this property of water. The change of state of water in a sweating camel, for example, occurs at the skin surface, and heat is taken away from the surface. Such heat may have been brought to the skin from the body core by the blood. Organisms such as frogs and wood lice that passively lose water fairly freely through their integument may be slightly cooler than the surrounding air because of evaporative cooling.

The mode of heat transfer that has often been least appreciated is _radiation_. We perceive electromagnetic radiation of wavelengths between 0.4 and 0.72 μ with our eyes as light (Figure 3–2). Radiation of longer wavelengths is less apparent to us but, if strong enough, may be sensed as heat. Such radiation is known as _infrared_ or _thermal_ radiation. All physical objects above a temperature of absolute zero emit electromagnetic radiation. The higher the surface temperature of an object, the shorter the wavelengths at which it radiates (Figure 3–2). Organisms and most of their surroundings, being fairly cool, radiate only at infrared wavelengths, which are too long to be perceptible to the eye. The embers of a fire are hot enough to radiate at the longer wavelengths of the visible spectrum and thus appear red-orange. The sun, being still hotter, radiates at the shorter as well as the

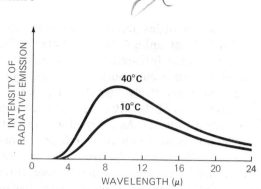

Figure 3-3. Intensity of radiative emission as a function of wavelength for an object at two different surface temperatures, 10°C and 40°C.

longer wavelengths of the visible spectrum and therefore produces a more nearly white light. It is to be emphasized that the radiative emissions from organisms are of the same basic nature as those from the sun or a fire but are at such long wavelengths that our eyes do not perceive them and are of sufficiently low intensity that our thermal sensors generally do not signal their immediate presence.

When electromagnetic radiation at either visible or infrared wavelengths impinges on an object, some of the radiant energy is absorbed, some is reflected, and if the object is transparent, some is transmitted. The organs of vision of man and other organisms perceive visible radiant energy that has been reflected from other organisms or objects. When we perceive a tree, we see solar radiation of appropriate wavelengths that has been reflected from the tree; we do not perceive the radiation actually being emitted by the tree as a function of its surface temperature. Radiant energy that impinges on an object and is absorbed is converted, upon absorption, into heat at the surface of the absorbing object; we are familiar with the way in which solar radiation or radiation from a fire warms our skin. Radiation thus provides a route of heat transfer among organisms and objects in the biosphere. Heat energy radiated by one organism or object may be absorbed by other organisms or objects in the surrounding environment. Any animal, in fact, experiences a veritable multitude of radiative exchanges with objects in its environment. It receives radiation from the sun, sky, trees, rocks, soil, grass, and so on. It also emits radiant energy to these various objects. The complex interplay of all these inputs and outputs of energy may result in either a net gain or a net loss of heat.

Warm objects not only emit down to shorter wavelengths than cooler objects (Figure 3-2) but also tend to emit more radiation per unit time. This is expressed in the Stefan-Boltzmann equation, which says that the intensity of radiation emitted by a given object over all wavelengths combined increases in proportion to the fourth power of the absolute surface temperature of the object. To illustrate, Figure 3-3 provides emission curves for an object at two different surface temperatures. The area under each curve reflects the total, integrated intensity of emission at all wavelengths and can be seen to increase with temperature. Surface temperature is not the sole determinant of intensity of emission, for the intensity also depends on certain qualities of the surface. Without going into this in detail, we may state that the surface temperature determines the maximum possible intensity of emission, and other surface qualities determine to what extent this maximal intensity will be realized. Most animals and objects in the biosphere are relatively "good" emitters in the infrared, meaning that they emit at an intensity that approaches the maximum possible for their prevailing surface temperature. They are also relatively "good" absorbers in the infrared, meaning that a high proportion of the infrared radiation impinging on them is absorbed (and converted to heat). As a broad generalization, we can say that if an animal and an object in the environment are exchanging radiation, the net transfer of heat will typically be from the warmer to the cooler. Both will emit well for their respective surface temperatures, and both will absorb most of the radiation impinging upon them. Because the warmer will emit more radiation per unit time than the cooler, the transfer of heat from warmer to cooler will exceed that from cooler to warmer. A cool lizard standing beside a sun-heated rock will gain heat from the rock in net fashion by radiation, and a bird in a woodland in winter will lose heat in net fashion by radiation to the cold trunks of surrounding trees. In nature, of course, the animal is in radiant exchange with many different objects and may be gaining heat from some and losing it to others. The overall net flow of heat by radiation, whether to or from the animal, is a complex function of all prevailing exchanges with individual objects.

Whereas animal body surfaces tend uniformly to show good absorption in the infrared, they vary considerably in their absorption of visible wavelengths. This is evident from variations in color. Black animals, for instance, tend to absorb radiant energy across the entire visible spectrum, whereas white animals tend to reflect across the entire visible spectrum. These considerations become significant in analyzing the effects of solar radiation, for although about half of the solar energy reaching the earth is in the infrared, about half is in the visible. All animals tend to absorb the infrared portion well, but absorption of the visible portion varies considerably.

Radiant exchanges of heat integrate with conductive, convective, and evaporative exchanges to determine the overall thermal flux between animal and environment. We quickly appreciate how a lizard emerging in the cool desert morning may warm to well above air temperature by basking in the sun. Despite a tendency for the warming lizard to cool through convective, conductive, and evaporative loss of heat to the air, the radiant input of heat is sufficient to assure a net heat gain and thus raise the body temperature. We are perhaps less familiar with situations where radiant heat exchange tilts the balance toward an increased loss of heat. To cite an example from common experience, in our houses we usually feel quite comfortable if the air temperature is near 22°C (72°F); yet if you have ever spent an evening in a cabin or other poorly insulated building in winter, you have probably experienced a sense of chill despite a similar air temperature in the cabin. The difference between the cabin and a well-insulated house is the temperature of the interior walls. Because the interior wall surfaces are colder in a poorly insulated building, they act as a more powerful radiative "heat sink"; that is, net radiative loss of heat to the walls is increased. Thus despite the fact that conductive, convective, and evaporative relationships may be similar in the cabin and a house, your overall tendency to lose heat is increased in the cabin. Out of doors the clear sky typically has a far lower effective radiant temperature than objects on the surface of the earth, and the radiant temperature is characteristically increased by clouds. These properties are manifest in our everyday experience; in winter, clear nights feel colder than cloudy nights even though the air temperature at the level of the ground is the same, and in late fall frosts are more likely on clear than cloudy nights because the clear sky acts as a strong heat sink, radiatively lowering the surface temperature of the ground and plants. It is commonly recognized that small mammals living in burrows under the snow in winter are protected from rapid conductive and convective loss of heat to the cold air above. It is less well known that they also avoid radiative loss of heat to the exceedingly cold nighttime sky by interposing the snow as a barrier to radiant exchange between themselves and the atmosphere.

Schmidt-Nielsen has offered the interesting hypothesis that diurnal desert mammals such as the jackrabbit might be able to dissipate heat to the clear sky during the heat of the day when the air temperature is so high that, conductively and convectively, the animal suffers a heat load. Measurements showed that the effective radiant temperature of the blue sky was only 16°–21°C during the afternoon in June in the Arizona desert. Now if the jackrabbit were to sit out in the open, it could lose heat by radiation to the sky, but the radiant input of heat from the sun would more than make up the difference. However, if the jackrabbit were to shade itself from the sun by sitting under a tree, it might then be able to take advantage of the heat sink provided by the sky without incurring a large solar heat load. Under such circumstances, Schmidt-Nielsen suggested that the jackrabbit might engorge its immense ears with blood and use them as radiators to help dissipate excess heat. The radiant situation of the jackrabbit in the shade involves more than just the sky, however. The ground can be extremely hot during the day (60°–70°C) and then imposes a high radiant heat load; further, much solar radiation can reach the animal in the shade by reflection from the ground. Measurements made by Schmidt-Nielsen and his colleagues revealed that during much of the hot part of the day, the overall radiant situation was such as to impose a net radiant heat load even in the shade. By midafternoon, however, the radiant and reflected heat load from the ground had ameliorated sufficiently that, with the heat sink provided by the sky, a net radiant heat loss could be realized, and this was adequate for the animal to lose heat in net fashion to the environment even though the air temperature near the ground was equal to or greater than the body temperature of the jackrabbit. The extent to which jackrabbits actually exploit these radiative possibilities for cooling is not well known, but this work nicely exemplifies the need to consider radiant heat exchanges as well as conductive, convective, and evaporative exchanges in analyzing the thermal situation of an animal.

In the case of a terrestrial animal, we can predict the direction of conductive and convective heat exchange with the air by measuring the body temperature of the animal and the air temperature. The comparison of air temperature and body surface temperature may not, however, provide any indication at all of the direction of radiant heat exchange, for that depends on the surface temperatures of objects in the environment, and they may differ substantially from air temperature. Although the radiant environment may be very complex in nature, it is usually simplified in the laboratory. In a laboratory animal chamber, for example, the walls of the container and other objects with which the animal is in radiant exchange often will be nearly at thermal equilibrium with the air and have radiant temperatures near air temperature. In this special case, radiant exchanges are typically in the same direction as conductive and convective exchanges with the air; an animal, for example, that is warmer than the air will be losing heat not only conductively and convectively, but also radiatively because environmental surfaces, in equilibrium with the air, will be uniformly cooler than the animal's surfaces. This is an important point because most laboratory studies are performed under such conditions. An environment in which the radiant temperatures of all environmental surfaces approximate air temperature may be termed a *uniform* thermal environment.

Thermal exchanges are simplified in aquatic environments for two reasons. The first is that, except for those aquatic forms that breathe air, animals in water do not lose heat evaporatively. The second reason arises from the different properties of air and water in transmitting thermal radiation. Air is quite transparent to infrared wavelengths; thus terrestrial animals can exchange heat directly by radiation with distant objects. Water, on the other hand, is very opaque to infrared wavelengths. Accordingly, radiant energy leaving an aquatic animal is absorbed by the water and converted to heat within a short distance of the animal's surface, and the animal receives thermal radiation only from the water very close to its surface. Whereas the sun can warm a terrestrial animal directly and thus elevate the animal's surface temperature well above air temperature, solar radiation is largely intercepted by the water in aquatic environments and warms the water, not the animal, directly. Thus, to a considerable extent, solar radiation can warm an aquatic animal only indirectly; the water is warmed, and the animal then receives heat primarily by conduction and convection from the water. The result of these effects is that we can often accurately predict the direction of overall heat exchange with the environment in aquatic animals merely by knowing the body temperature and the temperature of the immediately surrounding water.

POIKILOTHERMY

In this and the following sections we shall review some of the properties of poikilothermy, homeothermy, and other thermal relationships with the environment. It is to be strongly emphasized from the outset that these are best viewed as *types of thermal relationships* displayed by certain animals under certain conditions; they are not necessarily deterministic properties of particular animal groups. We can readily ascertain, for example, if a species under certain conditions is responding poikilothermically, but it is more onerous to take the next step and declare, in the strict sense of the word, that the species is a poikilotherm. Saying that a species is a poikilotherm implies that it always, or at least nearly always, responds poikilothermically, and this can only be ascertained by exhaustive study under a wide variety of conditions. Traditionally all animals except the birds and mammals have been viewed as poikilotherms. Whereas most groups in this vast assemblage do appear to respond largely poikilothermically, we recognize more and more that many do not on certain occasions, and some never, or hardly ever, do so. These cases will be discussed mainly toward the end of this chapter.

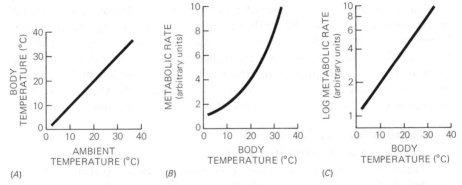

Figure 3-4. Common relationships between body temperature, resting metabolic rate, and ambient temperature in poikilotherms. (*A*) Body temperature as a function of ambient temperature. (*B*) Resting metabolic rate as a function of body temperature. (*C*) The logarithm of resting metabolic rate as a function of body temperature. Metabolic rate is expressed in the same arbitrary units in parts *B* and *C*.

In the poikilothermic animal, body temperature is determined by equilibration with the thermal conditions of the environment and varies as environmental conditions vary. In aquatic poikilotherms such as clams, starfish, crayfish, and fish, body temperature is largely determined by conductive and convective equilibration with the surrounding water, and body temperature therefore typically approximates water temperature, as illustrated in Figure 3-4*A*. The animal, of course, produces internal heat metabolically, and this may result in some elevation of body temperature over water temperature. Water, however, absorbs heat so effectively and these animals are so poorly insulated that the difference between body and water temperature is usually slight. In terrestrial poikilotherms, such as frogs, snails, and many insects, body temperature does not necessarily approximate the temperature of the surrounding air, for it is determined by the totality of environmental thermal conditions. Radiant input of heat from the sun or other sources, for example, may elevate body temperature well above air (ambient) temperature, and evaporation of water from the integument and respiratory organs may depress body temperature a few degrees below ambient temperature. In a uniform thermal environment, where the radiant temperatures of environmental surfaces are close to air temperature, body temperature does approximate air temperature. Thus in this special case (which frequently applies in the laboratory) the relations of Figure 3-4*A* again apply as approximations. Even here, though, metabolic heat production may tend to elevate body temperature slightly, and evaporation may tend to depress it slightly below air temperature.

Poikilothermic animals have often been called "cold-blooded," in reference to their coolness to the touch under certain conditions. In point of fact, this is quite an inappropriate term because body temperature may rise to high levels in many species, provided only that there is adequate input of heat from the environment. Lizards in the desert, for example, may have body temperatures higher than those maintained in man. The term *poikilotherm* (*poikilos* = "manifold" or "variegated") refers to the lack of regulated constancy of body temperature and is more descriptive of the actual physiological status of these organisms. Some authors object to the term *poikilotherm*, in part because it has sometimes been taken to imply that there is no control whatsoever over body temperature, when in fact many species do exert control behaviorally by selecting their thermal environment (see below). Recently the animals traditionally called poikilo-therms have increasingly been termed *ectotherms*, in reference to the fact that their body temperature is determined primarily by external thermal conditions. This term emphasizes the mechanism by which body tempera-ture is determined, whereas *poikilotherm* emphasizes the variation of body temperature with external conditions. Both terms have value.

Resting metabolic rate in the poikilothermic animal

Figure 3–4B depicts the usual relationship between resting metabolic rate and *body* temperature in poikilotherms. It is important to recognize that body temperature is the appropriate independent variable. A similar relationship is found between metabolic rate and *ambient* temperature only in those situations where body temperature approximates the air or water temperature (see the preceding discussion). As depicted in Figure 3–4B, resting metabolic rate typically increases approximately exponentially with body temperature. An exponential relationship signifies that metabolic rate increases by a given *multiplicative factor* for given *additive increment* in temperature. For example, metabolic rate might double (increase by a *factor* of 2) for each *increment* of 10°C in temperature. Then if metabolic rate were 0.1 cc O_2/hr at 0°C, it would be 0.2 cc O_2/hr at 10°C, 0.4 cc O_2/hr at 20°C, and 0.8 cc O_2/hr at 30°C. The relation between metabolic rate and body temperature is usually, in fact, only approximately exponential. That is, the factor by which metabolism increases for a given increment in temperature is usually not precisely constant from one temperature range to the next but might, for example, be 2.5 between 0°C and 10°C but only 1.8 between 20°C and 30°C.

The approximately exponential increase in metabolism with body temperature in animals is in good measure a manifestation of the same type of thermal sensitivity displayed by simple chemical reactions. It has long been recognized that simple chemical reactions "in a test tube" tend to increase exponentially in velocity as temperature is elevated, a behavior described by the well-known Arrhenius equation and reflecting the fact that the number of molecules with sufficient kinetic energy to react in-creases with temperature. Metabolism represents the sum total of a com-plexity of chemical reactions, and it is therefore not surprising to find that it also increases in rate as the temperature of the cells is elevated.

If metabolic rate M is truly exponentially related to temperature T, the relationship can be described by an equation of the form

$$M = a\,10^{bT}$$

where a and b are constants. Taking the common logarithm of both sides of this equation, we get

$$\log M = \log a + bT$$

This latter equation demonstrates that if M is an exponential function of T, then $\log M$ is a linear function of T (for $\log a$ and b are constants). This

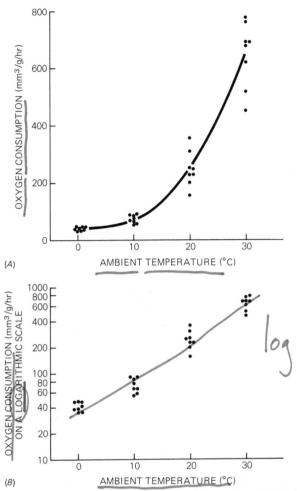

Figure 3–5. Oxygen consumption of tiger moth caterpillars (Arctiidae) as a function of temperature. The same data are plotted on rectangular coordinates (*A*) and on semilogarithmic coordinates (*B*). Body temperature approximates ambient temperature. (Data from Scholander, P. F., W. Flagg, V. Walters, and L. Irving. 1953. Physiol. Zool. 26: 67–92. Used by permission of The University of Chicago Press.)

is the rationale behind the common practice of plotting metabolism-temperature data on semilogarithmic coordinates. The logarithm of metabolic rate is plotted on the ordinate, and temperature itself is plotted on the abscissa; and a linear relationship is expected if metabolic rate is a truly exponential function of temperature. The curve of Figure 3–4*B* is replotted on semilogarithmic coordinates in Figure 3–4*C*, illustrating this "linearizing" effect. A similar comparison is provided in Figure 3–5, using data on tiger moth caterpillars. As emphasized earlier, metabolic rate commonly proves to be related to temperature in only an approximately exponential fashion. When this is the case, the semilogarithmic plot will not be precisely linear, an effect that is evident in Figure 3–5*B*.

A parameter commonly used in describing the thermal sensitivity of metabolic rate is the *temperature coefficient, Q_{10}.* It is defined as the ratio of the rate at one temperature to the rate at a temperature 10°C lower:

$$Q_{10} = \frac{R_T}{R_{(T-10)}}$$

where R_T is the rate at body temperature T and $R_{(T-10)}$ is the rate at body temperature T minus 10°C. To illustrate, if the metabolic rate of an animal were 0.22 cc O_2/hr at a body temperature of 15°C and 0.10 cc O_2/hr at 5°C, Q_{10} would be 2.2. Clearly, the magnitude of Q_{10} reflects thermal sensitivity. A Q_{10} of 3, indicating that metabolic rate triples for a 10°C increase in temperature, would reflect far greater thermal sensitivity than a Q_{10} of only 2, indicating just a doubling of metabolic rate for the same increment in temperature. Q_{10}'s for metabolic rate in poikilotherms tend usually to be in the neighborhood of 2, though substantially lower or higher values are sometimes found. If metabolic rate is a truly exponential function of temperature, then Q_{10} is a constant over all temperature ranges. When, as is usual, metabolic rate is not a strictly exponential function of temperature, then Q_{10} varies with the particular range of temperature considered. The Q_{10} between 0°C and 10°C, for example, might be 2.3, but the Q_{10} for the same animal between 20°C and 30°C might be only 1.9. One value of the semilogarithmic plot of metabolic rate against temperature is that its *slope* reflects changes in Q_{10}. (The slope is not, however, equal to Q_{10}.) A constant slope indicates a constant Q_{10}. If the slope is increasing or decreasing as temperature is raised, Q_{10} (and thermal sensitivity) is also increasing or decreasing, respectively. The slope in Figure 3–5B, for example, is greater between 10°C and 20°C than between 0°C and 10°C, indicating a higher Q_{10} in the former thermal range than in the latter.

When Q_{10} changes with temperature, it typically does so continuously. To illustrate, if an animal exhibits decreasing thermal sensitivity with increasing temperature between 0°C and 20°C, the Q_{10} will not assume a fixed, high value between 0°C and 10°C and then drop suddenly to a lower value between 10°C and 20°C; rather there will usually be a steady drop in Q_{10} over the whole range. The Q_{10} between 2°C and 4°C might be a little lower than that between 0°C and 2°C, and that between 4°C and 6°C might be a little lower yet. We often need, then, to compute Q_{10} for ranges of temperature of less than 10°C in order to describe thermal sensitivity adequately. The defining formula for Q_{10} given earlier does not suffice for these purposes, for it demands that metabolic rate be measured at two temperatures that are 10°C apart. The formula used is the van't Hoff equation,

$$Q_{10} = \left(\frac{R_2}{R_1}\right)^{10/(T_2 - T_1)}$$

where R_2 is the rate at any temperature T_2 (measured in °C) and R_1 is the rate at any lower temperature T_1 (again measured in °C). T_2 and T_1 need not differ by exactly 10°C. If they should differ by 10°C, the van't Hoff equation reduces to $Q_{10} = R_2/R_1$, which is identical to the defining equation given earlier; but this is simply a special case. To illustrate use of the van't Hoff equation, suppose that the metabolic rate of an animal is 0.22 cc O_2/hr at 15.4°C and 0.17 cc O_2/hr at 13.1°C. We then get

$$Q_{10} = \left(\frac{0.22}{0.17}\right)^{10/(15.4 - 13.1)} = 1.294^{.35} = 3.1$$

It is important to understand the meaning of this result. It obviously does not mean that metabolic rate increases by a factor of 3.1 between 13.1°C and 15.4°C. What it does mean (and always means) is that if thermal sensitivity were to remain constant over a 10°C interval, metabolic rate would increase by a factor of 3.1 over the whole 10°C interval. This is emphatically not to imply that a Q_{10} calculated for a narrow range of temperature should be used to extrapolate to a 10°C interval; if thermal sensitivity is changing with temperature, such extrapolation would lead to false conclusions. The issue is simply that if measures of thermal sensitivity are to be compared meaningfully, they must be calculated in some standardized way, and the standardization that has come to be accepted is to compute the thermal effect on metabolic rate over a hypothetical interval of 10°C.

Just as metabolic rate increases with body temperature in poikilotherms, such other rate functions as heart rate and breathing rate also generally show a strong thermal dependence. Although not so readily quantified as these parameters, behavior is also affected; it is a matter of common observation that poikilotherms, such as frogs, insects, and fish, are often strikingly more lethargic at low body temperatures than at high temperatures. The temperature coefficient Q_{10} can be used to describe the thermal sensitivity of any quantifiable physiological rate function (e.g., heart rate or breathing rate as well as metabolic rate).

Reductions in thermal sensitivity: introduction and cases of low sensitivity to acute changes in temperature

The strong dependence of functional rates on body temperature led Barcroft to the apt remark that poikilotherms are subject to a certain "tyranny of the Arrhenius equation." Organisms, as chemicophysical systems, must simply cope, to some extent, with the thermal sensitivity inherent in such systems. Life is, however, an adaptively evolving kind of chemicophysical system, and we have come to expect it not to show the static inflexibility of a test-tube reaction. We can presume that there would sometimes be selective advantage to escaping the tyranny of the Arrhenius equation, and at least two dramatic kinds of "escape" are known among poikilotherms.

To appreciate these phenomena fully it is first necessary to draw a clear distinction between the effects of *acute* and *chronic* exposure to temperature changes. If an animal that has been living at 20°C is suddenly, within a few hours' time, exposed sequentially to 15°C, 10°C, and 5°C, we say that its exposure to 20°C has been chronic, whereas its exposure to 15°C, 10°C, and 5°C has been acute. The difference lies in the amount of time allowed for the animal's response. When exposure to temperature changes is acute, the animal cannot bring into play any forms of physiological adjustment that require a substantial amount of time for their manifestation (such as synthesis of new enzymes that might aid its life at the new temperatures). On the other hand, chronic exposure to a temperature allows the manifestation of long-term responses. Returning to our earlier example, if the animal were allowed to live at 5°C for a number of weeks, we might well find that its responses were different from those initially displayed upon acute exposure to 5°C. Further, if the animal, having chronically adjusted to 5°C, were then acutely exposed to 10°C and 15°C, we might well find that

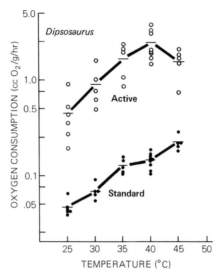

Figure 3–6. Standard (resting) and active rates of oxygen consumption as functions of temperature in the desert iguana (*Dipsosaurus dorsalis*). Standard rates were determined after at least eight hours of undisturbed adjustment to the test chamber. Active rates were determined during two minutes of maximal activity induced by electrical stimulation. Horizontal bars depict means. It is important to note that anaerobic processes accounted for 60–80% of ATP production during activity; thus the rate of oxygen consumption during activity, indicating only the rate of aerobic ATP production, reflects less than half of energy expenditure during activity (see Chapter 11). Total energy expenditure during activity varied with temperature more or less in parallel with the curve for oxygen consumption that is shown. During rest, metabolism is aerobic and thus accurately reflected by oxygen consumption alone. (From Bennett, A. F. and W. R. Dawson. 1972. J. Comp. Physiol. 81: 289–299.)

its responses at 10°C and 15°C were different from those displayed earlier when the animal had been living at 20°C. In short, the response to acute exposure can depend on the previous chronic exposure, for animals that have chronically been living at different temperatures can differ physiologically because of their different long-term adjustments. In our earlier discussion of metabolic rate as a function of temperature (see Figures 3–4 and 3–5), we have been looking at responses to acute temperature changes.

Some poikilotherms display very low sensitivity to temperature within certain ranges of temperature during acute exposure. Figure 3–6 provides an example for a lizard, the desert iguana (*Dipsosaurus dorsalis*). The Q_{10} for standard metabolic rate over each of the temperature ranges, 25°–30°C, 30°–35°C, and 40°–45°C, is between 2.1 and 3.5. However, the Q_{10} between 35°C and 40°C is not significantly different from 1.0, indicating virtual independence of temperature. Similar narrow ranges of low thermal dependence have been reported in a number of other lizards. The significance of these ranges is not fully clear. When active during the day, *Dipsosaurus* typically regulates its body temperature behaviorally (discussed later) to be in the high 30s or low 40s; even at night in its burrow in the summer, body temperature may be in the high 30s. The low thermal dependence between 35°C and 40°C may, then, represent the evolution of homeostatic mechanisms operative within part of the usual range of body temperatures.

Perhaps the most dramatic examples of low thermal dependence during acute exposure are reported among a variety of intertidal invertebrates, animals that live between the low- and high-tide marks on the seashore and

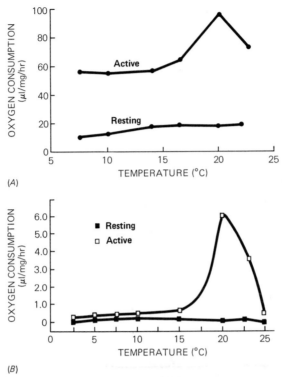

Figure 3–7. Oxygen consumption as a function of ambient temperature in resting and active intertidal invertebrates: (*A*) Barnacles (*Balanus balanoides*), weighing 1.5 mg (dry weight). (*B*) Periwinkles (*Littorina littorea*), weighing 100 mg (dry weight). Active barnacles were undergoing normal cirral beating; resting barnacles maintained a small opening between their opercular valves but showed no cirral beating. Active periwinkles were crawling; resting periwinkles were quiescent. (*A* from Newell, R. C. and H. R. Northcroft. 1965. J. Mar. Biol. Ass. U.K. 45: 387–403; *B* from Newell, R. C. 1969. Amer. Zool. 9: 293–307.)

accordingly are alternately immersed in the water and exposed to the air as the tides ebb and flow. Two examples are provided in Figure 3–7, where it can be seen that standard metabolic rate in a barnacle, *Balanus balanoides*, is virtually independent of temperature between 14°C and 20°C, and that in a snail, *Littorina littorea*, is little affected between at least 10°C and 23°C. Ranges of thermal independence in standard metabolic rate covering 10°C or more have similarly been reported in certain intertidal anemones, annelids, clams, and crustaceans. As with *Dipsosaurus*, active metabolic rate in these intertidal animals is typically more broadly thermally sensitive than standard metabolic rate.

In their natural habitat, intertidal animals can be exposed to wide thermal extremes within relatively short periods of time. In England during the summer, for example, barnacles were found to have body temperatures as high as 30°C when exposed to the warm air and sun at low tide; but the ocean temperature was only around 14°C, indicating that their body temperature would fall rapidly by as much as 16°C upon immersion at high tide. Species of intertidal animals may have evolved mechanisms for metabolic homeostasis to assure reasonable functional stability during such rapid thermal fluctuations. Usually their range of thermal independence corres-

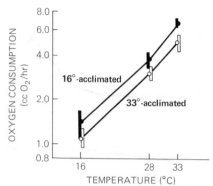

Figure 3–8. Metabolic responses to acute changes in temperature in two groups of resting, fasting fence lizards (*Sceloporus occidentalis*). One group (upper curve and solid symbols) was acclimated for five weeks to 16°C prior to testing; the other group (lower curve and open symbols) was acclimated to 33°C for five weeks prior to testing. Circles indicate means; vertical bars indicate ±2 standard deviations of the mean. Note that oxygen consumption is plotted on a logarithmic scale. (From Dawson, W. R. and G. A. Bartholomew. 1956. Physiol. Zool. 29: 40–51. Used by permission of The University of Chicago Press. Original figure © 1956 by the University of Chicago.)

ponds fairly closely to the actual range of temperatures experienced in their habitat. Unfortunately, we have little knowledge of the mechanisms by which they hold their standard metabolic rates rather constant over broad ranges of temperature.

Acclimation and acclimatization

Thermal independence during acute changes in temperature represents one form of escape from the "tyranny of the Arrhenius equation." We may now turn to responses to chronic changes in temperature, where forms of "escape" are even more widespread. Certain basic concepts may be introduced by using the data for fence lizards, *Sceloporus occidentalis*, presented in Figure 3–8. Two groups of lizards were maintained for five weeks at 16°C and 33°C, respectively. Their resting metabolic rates were then determined during short-term, or acute, exposure to 16°C, 28°C, and 33°C, and it can be seen that the responses of the two groups were different. Those that had been living at the cooler temperature, 16°C, had higher metabolic rates at a given test temperature than those that had been living at the warmer temperature, 33°C. The only known difference between the two groups was the temperature at which they had been living. When differences in physiological state appear after exposure to environments differing in only one or two well-defined parameters (e.g., temperature), we say that *acclimation* has occurred.

It is important to appreciate certain implications of this type of acclimation. Figure 3–9 presents a hypothetical example resembling the data for the lizards in basic outline. The upper solid line shows the *acute* response of cold-acclimated individuals, whereas the lower solid line shows the *acute* response of warm-acclimated animals. The dashed line connects the point for warm-acclimated animals at warm temperature with the point for cold-acclimated animals at cold temperature. The dashed line thus shows the responses of the animals to *chronic* temperature changes; it is the metabolism-temperature curve for animals that are allowed to live at each temperature for a long period of time before being tested. On a semi-

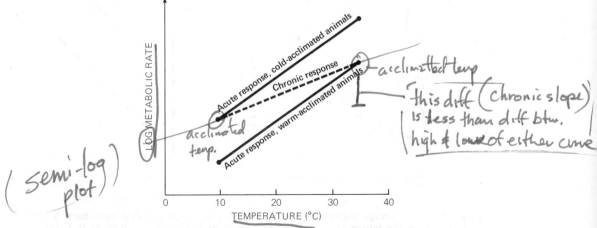

(semi-log plot)

Handwritten annotations: *acclimated temp*; *acclimated temp*; *this diff (chronic slope) is less than diff btw. high & low of either curve*

Figure 3–9. A hypothetical example of acclimation to temperature. See text for explanation.

logarithmic plot, as in Figure 3–9, the slope of the metabolism-temperature curve reflects Q_{10} or thermal sensitivity. Thus we see the very important point that the chronic response (dashed line) is less thermally sensitive than the acute response of either acclimation group. Put another way, the difference in metabolism between warm and cold test temperatures is less if the animals are permitted to acclimate to each test temperature than if they are changed suddenly from one temperature to another. This same principle is evident in the data for lizards in Figure 3–8. The difference in metabolism between 16°C-acclimated lizards at 16°C and 33°C-acclimated lizards at 33°C is less than the difference displayed by either acclimation group alone upon sudden transfer from 16°C to 33°C. In essence, then, the type of acclimation shown by the lizards represents a mechanism for relative stabilization of metabolic rate—a type of escape from the tyranny of the Arrhenius equation.

[margin note: acclimation: Chronic escape from tyr. of A.E.]

An alternative way to look at these results is provided in Figure 3–10A. Suppose we start with 33°C-acclimated lizards at 33°C and within a few hours lower their temperature to 16°C. Metabolism will fall, as shown, according to the acute response curve for 33°C-acclimated animals and will be much reduced. If the lizards are then left at 16°C for several weeks so that acclimation can occur, metabolism will *rise* back toward its original level. This rise represents metabolic *compensation,* an adjustment tending to reduce the effect of temperature. We say that the compensation is only *partial* because metabolism at 16°C does not return to the original level seen at 33°C. *Complete* compensation would take the form shown in Figure 3–10B. Usually compensation is only partial.

[margin note: Compensation]

Earlier we defined acclimation to represent changes in physiological state resulting from long-term adjustment to environments differing in only one or two well-defined parameters, such as temperature. As such, acclimation is a laboratory phenomenon. In nature, environments probably never differ in only one or two well-defined parameters. Winter in temperate woodlands, for example, implies not only lower temperatures than those in summer, but also shorter days, lowered atmospheric humidities, altered food sources, and a good many other changes. Differences in physiological

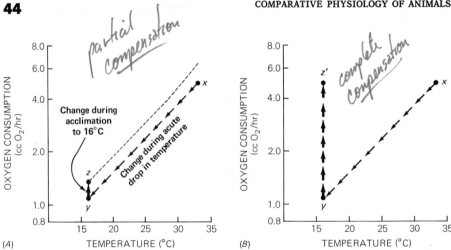

Figure 3–10. (*A*) Average changes in resting metabolism of lizards that have been living at 33°C when the temperature is lowered acutely to 16°C and then held at 16°C for five weeks, based on Figure 3–8. Point *x* is the metabolic rate of 33°C-acclimated lizards at 33°C. During the acute drop in temperature, metabolism falls to point *y*, the metabolic rate for 33°C-acclimated lizards at 16°C. Then metabolism rises over the period of acclimation to point *z*, the rate for 16°C-acclimated lizards at 16°C. Thin arrows show change in metabolism during acute temperature change and follow the curve for 33°C-acclimated animals in Figure 3–8. Thick arrow shows change in metabolism during acclimation to 16°C. Dashed curve shows the acute response of 16°C-acclimated lizards (see Figure 3–8). (*B*) Average changes in metabolism that would occur if animals were treated as in part *A* but displayed complete metabolic compensation during acclimation to 16°C. Acclimation (thick arrows) would then lead to a metabolic rate (*z'*) identical to that initially shown at 33°C (*x*).

state that appear after exposure to different natural environments are said to represent *acclimatization*. A classic example of acclimatization is illustrated in Figure 3–11. Mussels from colder, higher-latitude waters show greater water-pumping rates at any temperature of acute exposure than mussels from warmer, lower-latitude waters. The pumping rate is significant because mussels extract both food and oxygen from water pumped across their gills. As in the metabolic acclimation of fence lizards, the acclimatization of these mussels is of a compensatory nature. The pumping rates of mussels from northern waters at cold temperatures are higher than those of mussels from more southerly waters and thus more closely approximate the pumping rates seen at warm temperatures.

It is useful to distinguish acclimatization from acclimation because we cannot immediately pinpoint the environmental factors responsible for differences observed between populations in different natural environments, whereas in laboratory environments we can identify and control the factors of interest. In the case of the fence lizards, we are reasonably confident that temperature alone was responsible for the differences observed. In the case of the mussels, differences in temperature between northern and southern waters may have been largely responsible for the differences seen in the physiology of the mussels, but without controlled experimentation we cannot be sure. Perhaps temperature was inconsequential, and differences in daylength or nutrient supplies were critical. Similarly, if we found differences between summer and winter populations of a species, we would not know without experimentation whether temperature or other factors were most significant. With these provisos, we may recognize that poikilotherms that show thermal acclimation in the laboratory usually show

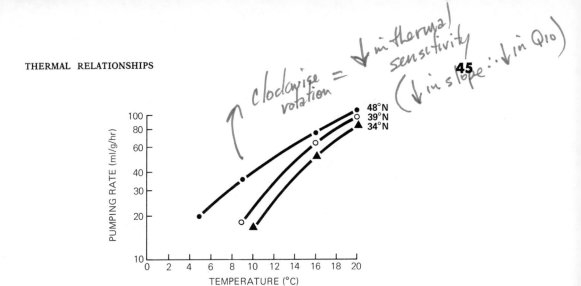

Figure 3-11. Pumping rate in milliliters of water per gram of body weight per hour as a function of temperature for 50-g mussels (*Mytilus californianus*) from three latitudes along the west coast of the United States. [From Bullock, T. H. 1955. Biol. Rev. (Cambridge) 30: 311–342.]

qualitatively similar responses in natural habitats that differ in temperature. If, for example, metabolic rate at given temperature is elevated by acclimation to low temperatures in the laboratory, it will usually also be elevated during cool seasons in nature. This observation indicates that, whatever other factors may be influential in nature, their effects usually do not qualitatively mask the effects of temperature. Because quantitative differences are likely and even qualitative differences can occur, however, it is necessary to exercise caution in extrapolating from thermal acclimation experiments to the natural situation.

One system of classifying patterns of acclimation and acclimatization rests on a description of results plotted on semilogarithmic coordinates (log rate plotted on the ordinate, temperature plotted on the abscissa). When, as in Figure 3–8, the curve for cold-acclimated animals is shifted upward but remains parallel to the curve for warm-acclimated animals, acclimation is said to represent an *upward translation* of the rate-temperature curve. Simple upward translation of the curve is also seen in comparing mussels from 39°N with those from 34°N (Figure 3–11). Comparing mussels from 48°N with those from 39°N, however, the curve for the animals from colder waters is not only translated upward, but also *rotated* relative to the curve for the animals from warmer waters. The rotation is evident from the fact that the curves are not parallel, and in this case is said to be a clockwise rotation because the curve for animals from 48°N could be obtained by moving the curve for 39°N upward and then rotating it clockwise. The presence or absence of rotation has explicit physiological significance. With no rotation, as in the fence lizard example, cold acclimation may result in an increase in rate at all temperatures, but thermal sensitivity as reflected by Q_{10} is not changed, for the slope on the semilogarithmic plot is unchanged. Clockwise rotation, on the other hand, indicates a decrease in thermal sensitivity and Q_{10}, whereas counterclockwise rotation would indicate an increase in Q_{10}. It is theoretically possible in acclimation studies to obtain all possible combinations of translation and rotation. Specifically, there can be upward, downward, or no translation and, in addition, clockwise, counterclockwise, or no rotation. The pattern depends on the species and experi-

mental conditions. Some species fail altogether to show acclimation or acclimatization. The most common type of response to cold in species that
show a response is an upward translation, often accompanied by clockwise
rotation. These common responses are compensatory in the sense discussed
earlier and are very widespread.

When a physiological rate or some other capacity to perform a function
is altered by long-term exposure to certain conditions, the process is termed
capacity acclimation or *acclimatization*, or sometimes capacity adaptation.
In contradistinction, when the upper or lower temperatures that are lethal
or incapacitating are altered, we speak of *resistance acclimation* or *acclimatization*, or resistance adaptation. Resistance adaptation, like capacity
adaptation, is widespread. To illustrate, if young sockeye salmon are acclimated to 5°C, the upper and lower temperatures that will kill 50% of the
test animals over periods of prolonged exposure are 21°C and 0°C, respectively. By contrast, if the fish are acclimated to 15°C, the upper and lower
lethal temperatures are 22°C and 4°C, respectively. Insects are often im-
mobilized by cold before they are actually killed, and often the temperature
that induces this "chill coma" is affected by past thermal experience. Thus,
for example, cockroaches (*Blatta*) acclimated to 30°C enter chill coma
at an average temperature of 7.5°C, but animals acclimated to 14°–17°C
enter chill coma at a lower mean temperature, 2.0°C. As suggested by
these illustrative examples, the range of temperatures compatible with adequate function is typically shifted downward by cold acclimation and upward by warm acclimation.

Acclimation and acclimatization, whether of the capacity or resistance
type, require appreciable amounts of time to become manifest, but it is
difficult to generalize about the actual amount of time required inasmuch
as it varies considerably with the species and function under study. Sometimes at least a partial expression of acclimation is evident within the first
day after animals have been placed under new environmental conditions,
but full expression more usually requires days, weeks, or even months. As
we shall see repeatedly in this text, temperature is by no means the only
parameter to which acclimation occurs. Animals may show long-term
physiological adjustments to such diverse parameters as humidity, salinity,
oxygen supply, photoperiod, and food supply, to name just some. In some
cases the response to temperature depends in part on other prevailing environmental conditions. A most important point for the experimental scientist is that the environmental past of the animals studied will often influence the results obtained. Thus unless acclimation itself is under investigation, it is usually prudent in many types of physiological work to acclimate all test animals to the same conditions prior to study so that differences due to environmental history will be minimized.

Genotypic versus phenotypic differences between populations

Up to this point we have discussed two time courses of response:
responses to acute changes in temperature and the longer-term responses
known as acclimation and acclimatization. Both of these types of responses,
by definition, occur within the lifespan of the individual and, although
conditioned by the genotype, are phenotypic. A third significant time frame

is that of evolutionary time, in which natural selection operating on alternative genic alleles can result in genotypic physiological differences between populations.

If differences should be found between populations in different environments, it is important to determine whether the differences are due to basic genetic differences or whether they reflect acclimatization in genetically similar populations. Genetic differences can be presumed only if the physiological differences persist after steps have been taken to eliminate differences in acclimatization. A classic example of this approach is provided by the work of Segal on limpets, *Acmaea limatula*, occupying the intertidal zone. Limpets are molluscs with a single dome-shaped shell which attach firmly to the rocks in their habitat and move around to only a small extent. Segal examined the heart rates of two groups of limpets gathered, respectively, from the lowest reaches of the intertidal zone and approximately the midpoint of the intertidal zone. In their natural habitat those from the low intertidal zone were immersed in the sea at least 90% of the time, whereas those from higher in the zone were immersed only about 50% of the time. It was postulated that the "high" animals, being more often exposed to the air and sun, experienced a higher mean body temperature than the "low" animals. When tested at temperatures from 9°C to 29°C, heart rates of "low" animals proved to be consistently higher than those of "high" animals at a given test temperature. The question, then, was: Were the differences in heart rate due to acclimatization, or were the "low" and "high" animals distinct microgeographic physiological races? To answer the question, reciprocal transplants were performed; "high" animals were transferred to the low habitat, and vice versa. Within two to four weeks the transplanted animals assumed heart rates similar to those of the natives in their new habitat. Thus this proved to be a case of acclimatization, a phenotypic adjustment to prevailing environmental conditions. It seems likely that temperature was a significant parameter in the acclimatization.

Populations of oysters (*Crassostrea virginica*) from waters off Virginia spawn at a higher temperature than oysters from Long Island Sound. When adult Virginia oysters, which spawn at about 25°C, were transplanted to Long Island Sound, they failed to spawn over a period of two years when water temperature in the Sound failed to reach 25°C. Here, then, we would seem to have an example of genotypic, racial divergence; but it is conceivable that the differences in spawning temperature are the result of acclimatization during preadult stages of development. To be certain that the differences are truly genotypic, larval oysters should be transplanted to see if they also retain the spawning temperature of their home habitat. In a recent study of resistance adaptation, lower lethal temperatures were determined for related species of marine fish from either side of the Panamanian isthmus. Average water temperature on the Pacific side is lower than that on the Atlantic side, and lethal temperatures were lower for the Pacific fish. In the case of two related gobies (genus *Bathygobius*), differences in lethal temperature were abolished after fish had been acclimated to the same temperature in the laboratory. Similar treatment failed to eliminate differences between two species of cardinal fish (genus *Apogon*). Thus divergence of lethal temperatures in the former genus is based on acclimatization, whereas that in the latter appears to find a genetic basis.

Adaptation to different climates

There have been many studies of potential physiological differences among related species along latitudinal gradients. Experiments designed to determine if identified differences are genotypic or phenotypic have not, unfortunately, always been performed. Latitudinal differences in lethal temperatures among related forms are widely reported and are often unequivocally genetic. In experiments on 15 species of toads acclimated at 23°C prior to testing, for example, the mean low temperature at which 50% of test individuals failed to survive after 24 hours was near 11°C for equatorial species but near 0°C for species from far northern latitudes (50°–60°N). By contrast, species of ranid frogs collected from 10°N to 60°N latitude all had similar lower lethal temperatures, near 0°C. Although tropical and temperate-zone fiddler crabs (*Uca*) do not show consistent differences in upper lethal temperature, the temperate-zone species can survive considerably lower temperatures than the tropical species. Fish of high latitudes tend to have lower upper lethal temperatures and lower lower lethal temperatures than tropical fish. An extreme example is offered by some fish of the antarctic seas that occupy perpetually cold waters and have upper lethal temperatures of only 5°–10°C.

Looking at rate functions, it is a common field observation that related poikilotherms in cold and warm climates seem to function behaviorally at much the same level despite large differences in body temperature. Thus, for example, fish, crabs, and starfish in the frigid waters of northern Maine are not strikingly more lethargic than their relatives in the warm waters of Bermuda. Such observations, although obviously subjective, suggest the presence of physiological compensations to prevailing thermal conditions. Studies of standard or routine metabolism have frequently indicated that aquatic poikilotherms of far northern waters have considerably higher metabolic rates at low temperatures than related species from temperate or tropical waters. Data of this nature are available, for example, for a variety of crustaceans and molluscs. Similar results on fish have indicated that arctic and antarctic species have resting metabolic rates three to five times higher than those of temperate species at low temperatures. However, new data gathered recently on a number of species of arctic fish have challenged this concept; with special effort being taken to assure minimal activity levels, resting metabolic rates far closer to those expected of temperate fish were obtained. Accordingly, at least among fish, the question of differences in resting metabolism between polar and temperate species requires further study before a definitive conclusion can be reached.

Biochemical aspects of thermal relations

Changes in such parameters as metabolic rate and heart rate, which we observe in studies of whole animals, must ultimately reflect changes at the biochemical and biophysical level within cells of the organism. Thus we are led to such questions as: What biochemical and biophysical alterations underlie the processes of acclimation and acclimatization? What attributes of biochemical organization permit many of the intertidal invertebrates to have relatively stable resting metabolic rates over such broad ranges of temperature? What are the biochemical differences between an antarctic fish that is killed by heat at 5°C and a tropical fish that is killed by cold at 10°C?

Research into such questions has been proceeding at an ever-increasing tempo over the last two decades, and although answers can still be provided in only a general and tentative way, many basic outlines are becoming clear.

The basic "problem" of decreasing body temperature is that the average kinetic energy possessed by molecules in the cells decreases. For a particular molecule to undergo a particular enzymatically catalyzed reaction, the molecule must possess energy equal to or greater than the necessary energy of activation, as determined by the enzyme. Only a fraction of all molecules will at any one time possess the needed energy to react, and it was Arrhenius who pointed out that this fraction decreases disproportionately as the average kinetic energy of the molecules falls with temperature. Thus the average falls by only about 3% when temperature is reduced by 10°C, but the fraction of molecules possessing the needed activation energy may fall by 200 or 300%. The effect is to slow the rate of the reaction. Much of the effort of biochemists has been directed to the question of what the cell can do to ameliorate these effects. The most obvious possibilities lie in modifying the amount of enzyme or the catalytic properties of the enzyme responsible for catalyzing the reaction.

One of the experimental approaches used extensively has been the measurement of enzymatic activity. To illustrate, suppose we wish to know the activity of succinic dehydrogenase in muscle tissue of a fish. This enzyme is an important member of the Krebs citric acid cycle, catalyzing the oxidation of succinate to fumarate. One approach would be to take a sample of muscle tissue, homogenize it, and then add a known concentration of the substrate, succinate. We could then determine the rate at which fumarate is formed, this being a measure of the activity of the enzyme. Using carefully controlled technique, we could compare the activities of succinic dehydrogenase in muscles of fish that have been acclimated to two different temperatures. This comparison has actually been performed on goldfish, and the activity at given test temperature proved to be higher in cold-acclimated than in warm-acclimated fish, indicating changes at the enzymatic level in the cold-acclimated animals that would tend to compensate for the reduced kinetic energies prevailing at cold temperatures. Similar comparisons have been performed on a considerable variety of enzymes, mostly in fish but also in some invertebrates. In general, enzymes associated with glycolysis, the Krebs cycle, and the electron transport chain—that is, enzymes involved in the generation of ATP—have proved to show increased activities in cold-acclimated individuals. This is entirely in line with the whole-animal data showing that metabolic rate is frequently elevated at given test temperature in cold-acclimated animals; an elevated metabolic rate implies increased rates of ATP generation and utilization. Data on metabolic rates of isolated tissues also complement these findings. Excised brain of goldfish acclimated to 20°C, for example, exhibits a greater rate of oxygen consumption at given test temperature than brain of fish acclimated to 27°C, though a similar difference is not seen in excised muscle. Muscle from 5°C-acclimated frogs has a higher metabolic rate than muscle from 25°C-acclimated animals, and isolated gills of cold-acclimated bivalve molluscs (*Venus* and *Mytilus*) consume oxygen at a greater rate than gills from warm-acclimated individuals. Interestingly, enzymes involved in the degradation of metabolic products, such as those associated with the lysosomes and peroxisomes, frequently show no change or a decrease in activity in cold-acclimated individuals. It has been postulated that because metabolic rate is typically lowered in the cold (despite partial compensation through acclimation), the cells produce metabolic breakdown products at a lower rate and thus are not in need of as much degradative activity as at warm temperatures.

A shortcoming of information on enzyme activity is that it does not tell us what properties of the enzyme have changed. An increase in activity during cold acclimation, for example, could reflect an increase in enzyme concentration, a change in the cellular milieu effecting increased enzymatic efficiency, or a change in the chemical structure of the enzyme (synthesis of a new isozyme). Recently attention has increasingly been directed to specifying which properties of enzyme systems change during acclimation. By far the most progress has been made in analysis of enzymatic affinity, and we shall now turn to examining that important parameter. Although properties other than affinity, such as enzyme concentration, could be of much significance in acclimation, we simply know little about them at this time.

The substrate of an enzyme is the initial reactant; succinate, for example, is the substrate of succinic dehydrogenase. If substrate concentration is varied, we find that, within limits, the velocity of the reaction, that is, the rate at which substrate is converted to product, also varies. Commonly, reaction velocity varies with substrate concentration as in Figure 3–12A. At relatively low substrate concentrations velocity increases as substrate concentration increases, but a point is reached where velocity is no longer affected by substrate concentration. This behavior follows from the fact that a substrate molecule must enter into a complex with an enzyme molecule in order to react. At low substrate concentrations the amount of substrate available is the limiting factor in determining velocity; all available enzyme molecules are not "occupied" at any one time, and greater substrate concentration increases velocity by allowing a fuller utilization of the available enzyme. At some point, however, substrate concentration becomes adequate to utilize fully or *saturate* the available enzyme. Then the amount of enzyme present becomes limiting, and further increases in substrate cannot enhance velocity.

Suppose now that we have a fixed amount of enzyme, meaning that the maximal or saturation velocity is also fixed. As substrate concentration is raised, velocity could approach the maximal velocity along many possible trajectories, as shown in Figure 3–12B. The actual trajectory reflects the *affinity* of enzyme for substrate. If the enzyme, upon contacting a substrate molecule, very readily complexes with the substrate molecule, we say it has a high affinity for substrate. If the enzyme is less effective in forming a complex, we say it has a lower affinity. Affinity is a very important attribute of enzyme function, for the enzyme and substrate must complex before the enzyme can catalyze the conversion of substrate to product. Line *a* in Figure 3–12B would be characteristic of an enzyme with high affinity, whereas line *c* would reflect low affinity. Note that at given subsaturating substrate concentration, the reaction velocity is much closer to the maximal velocity if the enzyme has high affinity than if it has low affinity. Affinity is a factor in determining reaction rate only at subsaturating substrate concentrations; if substrate concentration is high enough, even the low-affinity enzyme can

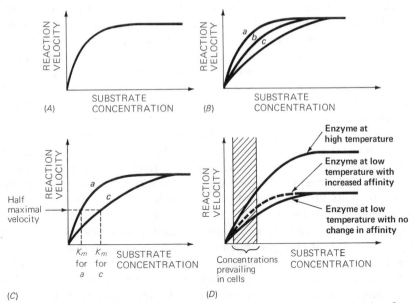

Figure 3–12. Diagrammatic illustration of several concepts of enzyme kinetics. See text for discussion. Shaded vertical bar in (*D*) shows a range of substrate concentrations that might prevail in cells.

become saturated and operate at peak rate. Because the evidence is that substrate concentrations in cells are typically subsaturating, affinity is a most significant parameter in determining reaction velocities in living animals. A convenient numerical expression of affinity is the apparent Michaelis constant, K_m, defined to be the substrate concentration required to attain one half of the maximal velocity. K_m is derived in Figure 3–12*C* for lines *a* and *c* from Figure 3–12*B*. Note that the low-affinity enzyme (line *c*) has the greater K_m. *Thus K_m and affinity are related inversely. A high K_m means low affinity, and a low K_m means high affinity.*

Figure 3–12*D* illustrates how an increase in enzyme affinity can aid the maintenance of high reaction rates at low temperatures. Suppose that an enzyme behaves according to the upper solid line at high temperatures. Suppose, then, that we lower the temperature and that the affinity of the enzyme remains unchanged. The enzyme would then behave according to the lower solid line, and reaction velocity at prevailing substrate concentrations would be reduced. Suppose, now, that instead of having the same affinity at low temperature as at high temperature, the affinity at low temperature is increased. The enzyme would then behave according to the dashed line, and we see the important point that an increase in affinity occurring along with a decrease in temperature can help to limit the decrease in reaction rate caused by temperature. That is, it can help to *compensate* for the effect of temperature, for reaction rate will be greater with the increase in affinity than if the affinity remained unchanged. Animals have evolved enzymes that very often increase in affinity as temperature is dropped over a broad range. When this is the case, each drop in temperature is accompanied by a partially compensatory increase in affinity, the result being that the effect of temperature on reaction rate is far less than it would be if affinity were invariant. The degree of compensation depends in part on prevailing substrate concentrations in the cells, for the compensatory effect of a given change in affinity is greater at low substrate concentrations than at concentrations approaching saturation (see Figure 3–12D). There are known examples of enzymes showing sufficient changes in affinity that at low substrate concentrations, the reaction velocity remains nearly *constant* ($Q_{10} \cong 1$) over broad ranges of temperature. The issue has not been studied, but it is conceivable that it is this type of process that allows some intertidal invertebrates to have resting metabolic rates so little affected by temperature.

Figure 3–13 presents data on changes in affinity with temperature for acetylcholinesterase in two species of fish. Remember in interpreting these results that K_m varies *inversely* with affinity. Looking at the curve for rainbow trout acclimated to 17°C, we see that affinity increases as temperature falls from about 30°C to about 17°C. Over this range, changes in affinity act to compensate for the effects of temperature as outlined in the previous paragraph. Below about 17°C, however, affinity decreases with decrease in temperature so that changes in affinity act to reduce reaction rate and thus aggravate the effects of temperature. At low temperatures, a fish with this enzyme would suffer very low, perhaps life-threatening, reaction rates. When rainbow trout are acclimated to 2°C, they develop a new form of acetylcholinesterase that has affinity properties much more suitable to life at low temperatures. As can be seen, this enzyme has its highest affinity near 2°C (but would have quite low affinity at high temperatures). We see, then, that rainbow trout have the capacity to make two types of acetylcholinesterase, each well suited to a particular range of temperatures. Different chemical forms of an enzyme are termed *isozymes*. It has been shown that the two forms of acetylcholinesterase in rainbow trout are, in fact, different proteins and thus represent true isozymes. Trout acclimated to 17°C have one isozyme; those acclimated to 2°C have the other; and animals acclimated to an intermediate temperature, 12°C, have both. Because isozymes are different proteins, it takes time to convert from one to another; a trout transferred from 17°C to 2°C, for example, must synthesize the new isozyme de novo. Accordingly, adaptive changes in isozymes do not provide for adjustment to acute changes in temperature but rather are part of longer-term responses (acclimation). Of course, animals can maintain mixtures of isozymes (as trout at 12°C do) and then can immediately benefit from the special properties of each particular isozyme.

Changes in isozymes that provide for adaptive alteration of affinity-temperature relationships have been demonstrated for several enzymes of rainbow trout acclimated to 17°C and 2°C: pyruvate kinase, phosphofructokinase, and citrate synthase as well as acetylcholinesterase. On the other hand, some enzymes such as malate dehydrogenase fail to show isozyme changes; these tend to have affinities which are relatively insensitive to temperature. A most interesting phenomenon has been found in the case of muscle pyruvate kinase (a glycolytic enzyme) in the Alaskan king crab, *Paralithodes camtschatica*. This enzyme shows "cold" and "warm" variants, much as seen earlier for trout acetylcholinesterase. However, they are not true isozymes but rather are interconverted, one to another, instantaneously, apparently by a thermal effect on the conformation of one and the same protein. Thus if a king crab is warmed rapidly from cold temperatures to 15°C, its pyruvate kinase immediately assumes a new form ("instant iso-

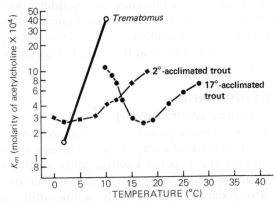

Figure 3–13. The apparent Michaelis constant K_m as a function of temperature for acetylcholinesterase from the brains of three groups of fish: (1) rainbow trout (*Salmo gairdneri*) acclimated to 17°C—closed circles; (2) rainbow trout acclimated to 2°C—closed squares; and (3) an antarctic fish, *Trematomus borchgrevinki*, collected from waters at −2°C—open circles. Acetylcholinesterase is the enzyme responsible for breaking down acetylcholine, the latter being an important chemical transmitter substance of neural synapses. When an impulse arrives at a synapse on one nerve fiber (the presynaptic), the fiber releases acetylcholine into the synapse, and the acetylcholine activates the other (postsynaptic) nerve fiber. It is important that acetylcholine be destroyed by acetylcholinesterase soon after each release, for otherwise it would continue to stimulate the postsynaptic fiber rather than simply relaying a discrete impulse across the synapse. (From Somero, G. N. and P. W. Hochachka. 1971. Amer. Zool. 11: 159–167.)

zyme") having quite different affinity-temperature relationships from those seen at cold temperatures. These data on the trout and the king crab provide great insight into adaptive possibilities at the biochemical level. We can expect to learn much in the next decade about the extent to which these possibilities are put to use across the whole spectrum of the animal kingdom.

Returning to Figure 3–13, we may note briefly the properties of acetylcholinesterase in another fish, *Trematomus borchgrevinki*. This is an antarctic species that lives in waters perpetually colder than 0°C and that dies at an upper lethal temperature of only 6°C. The affinity-temperature relationships of its acetylcholinesterase are dramatically different from those of either of the trout isozymes. Indeed, affinity declines so rapidly at elevated temperatures that it is conceivable that the reaction rate might be sufficiently retarded to be a factor in causing the demise of the animal. However that may be, we see here a good example of the evolution of quite different enzyme forms in species occupying different thermal environments.

In concluding this section, we may turn briefly to a consideration of changes in lipids that commonly accompany acclimation in poikilotherms. The temperatures at which lipids solidify are in good measure a function of their degree of unsaturation. Highly saturated lipids solidify at higher temperatures than less-saturated lipids with equivalent fatty-acid chain lengths. It has been shown in a number of groups of animals, as well as in some bacteria and plants, that individuals living at low temperatures deposit lipids of lower saturation than individuals living at high temperatures. This is believed to be an adaptation to preventing deleterious "hardening" of lipids at low temperatures. Particular importance is attributed to the lipids of the cellular and intracellular membranes. Stiffening of these membranes could adversely alter their important biophysical properties. Further, many enzymes are bound to membranes, and solidification of membrane lipids could adversely affect the function of such enzymes (as by deforming the enzyme proteins). There is evidence, in fact, that the activity of some membrane-bound enzymes is significantly curtailed at temperatures below the lipid solidification point.

Behavioral interactions with the thermal environment

Whereas the physiological properties of poikilotherms demand much attention, one cannot forget that many of these animals are mobile and can therefore select among the thermal environments available. The poikilotherm thus often can and does exert some control over its body temperature despite a strong dependence of body temperature on immediately prevailing environmental conditions. Thermally stressful situations are typically avoided. Many species—including some insects, arachnids, fish, amphibians, and reptiles—exhibit fairly well-defined thermal preferenda, choosing certain temperatures if they are available. Body temperature may be regulated also by moving back and forth between cooling and warming environments. Fiddler crabs, for example, periodically return to their cool burrows when temperatures at the surface of the ground are high. During the day some lizards control their body temperatures at high levels, and to within a range of 5°C or less, by moving back and forth among sun, shade, and burrows and by orienting so as to expose greater or lesser areas of their bodies to direct solar illumination. This type of process is termed *behavioral thermoregulation* (sensu stricto) and will be discussed at more length at the end of this chapter.

Behavioral and physiological responses to freezing conditions

When temperatures drop below freezing, poikilotherms are faced with a potentially lethal freezing of their body fluids. Some escape such danger by migration. Others may enter microclimates where freezing is not a problem. Frogs, turtles, fish, crayfish, freshwater mussels, and many other aquatic species, for example, spend the winter at the bottoms of lakes and ponds, where the water temperature often stays at 4°C. Snakes, salamanders, toads, earthworms, and many insects seek refuge below the frost line in soil. Some enter animal burrows, rock crevices, or caves. Many poikilotherms, on the other hand, are exposed to freezing temperatures and must cope physiologically.

Before discussing some of the physiological responses, we should briefly examine the potential dangers of freezing. The aspects of freezing that cause death are not well understood, though certain potential effects are readily enumerated. Low temperatures can simply depress metabolic functions sufficiently that essential maintenance functions cannot be performed satisfactorily, leading to death if the conditions are prolonged. Also, they may deleteriously affect the catalytic properties of certain enzymes (as by lowering affinity) or, in a more general sense, cause disorganization of metabolic systems by unequally affecting the function of component enzymes. Freezing temperatures can cause denaturation of some enzyme proteins.

The formation of ice crystals can rupture or distort cell structures. Although some animals can tolerate ice formation in extracellular water, there are no known examples of animals that can survive intracellular freezing under natural conditions; intracellular ice formation probably damages the cellular ultrastructure. Commonly, ice forms first in the extracellular fluid; in an unknown way, cell membranes seem to offer some protection against the invasion of ice to the intracellular spaces. Even if freezing is entirely limited to the extracellular water, however, it may still pose a critical danger to cellular function by causing dehydration of the cells. This, in fact, is thought to be a common mode of death caused by freezing, and it is important to understand the mechanism. The body fluids are solutions of inorganic and organic compounds, and, as will be discussed at more length in Chapters 4 and 5, dissolved solutes act to lower the freezing point of the body fluids in direct relation to their total concentration (strictly, osmotic pressure). Thus the body fluids of animals do not freeze at 0°C, but somewhat below; the fluids of a marine clam, for example, might freeze at −1.8°C, whereas those of a fish, having lower total solute concentration than those of the clam, might freeze at −0.8°C. Now when extracellular fluids freeze under natural conditions, they usually do not do so all at once, but freeze gradually, in stages. Further, water tends to freeze out in *relatively* pure form. That is, when ice crystals form, they tend to contain proportionately more water and proportionately less solute than the original body fluids. The obvious corollary is that the body fluids that remain in liquid form have elevated solute concentrations, containing not only their original solutes but also the solutes left behind by the frozen fluids. This process, in fact, acts to lower the freezing point of the remaining fluids and thus tends to impair further freezing. In the unfrozen animal the (osmotically) effective solute concentrations of the intracellular and extracellular fluids are about equal; accordingly, there is little or no tendency for water to enter or leave the cells by osmosis. However, when some of the extracellular water has frozen, we have seen that the concentration of the remaining extracellular fluids tends to be elevated. This disrupts the balance between intracellular and extracellular fluids, and water is lost by osmosis from the less-concentrated intracellular fluids to the more-concentrated extracellular fluids. The process, as we shall see shortly, can proceed to a profound state of cellular dehydration as temperatures fall and more and more of the extracellular water freezes.

Certain animals are able to survive substantial freezing of their extracellular fluids. Mussels (*Mytilus*), barnacles (*Balanus*), and periwinkles (*Littorina*) in the intertidal zone, for example, are frequently exposed to freezing conditions on northern shores and can survive freezing at −10°C to −20°C. Under such conditions, as much as two thirds of their body water may be in the form of ice, and solute concentrations in the unfrozen fluids may be elevated by a factor of three or four. Some insects are also known to survive freezing. A recently reported example is that of an Alaskan carabid beetle (*Pterostichus*) found frozen in large numbers in tree stumps in winter. Experiments revealed that freezing occurred at about −10°C, but the animals could survive nonetheless down to −35°C. The mechanisms of freezing tolerance in these animals are not known. It is interesting that the intertidal invertebrates and *Pterostichus* are not so tolerant to freezing in summer as in winter, indicating that physiological acclimations are involved.

Some animals prevent freezing by producing physiological antifreezes. These are solutes manufactured and added to the body fluids during cold seasons, increasing the total solute concentration and thus depressing the freezing point. Many insects, for example, produce high concentrations of glycerol, sorbitol, or mannitol in winter. To cite an especially striking example, overwintering larvae of *Bracon cephi* may exhibit freezing points as low as −17°C, as compared to a freezing point of only about −1°C in summer individuals; glycerol concentrations in their body fluids may be as high as 4–5 molal. Overwintering eggs of the tent caterpillar (*Malacosoma*) have glycerol equivalent to 35% of their dry weight.

Protective antifreezes are also found in certain marine teleost (bony) fish of polar regions. In comparison to most other aquatic animals, the marine teleosts can face particular problems of freezing under cold conditions, for reasons that we should briefly examine before looking at the properties of the polar fish. Marine invertebrates generally have freezing points close to those of seawater and accordingly are not threatened with freezing unless the water in which they are living freezes. Freshwater animals have freezing points below those of fresh water and thus also do not freeze unless the water freezes. The sessile marine invertebrates and the sessile freshwater forms typically are found in bottom waters that are least likely to freeze, and the mobile ones can seek out the bottom waters even though the surface waters freeze. By contrast, marine teleost fish generally have freezing points in summer (−0.6°C to −1.1°C) that are higher than those of seawater (about −1.8°C). Thus unlike most aquatic animals, they can potentially freeze even though in unfrozen water.

The production of protective antifreezes in marine fish of polar regions was first reported in certain shallow-water species of Labrador that had summer freezing points of about −0.8°C but winter freezing points of as low as −1.5°C. The antifreezes produced by these fish are organic compounds but have not as yet been identified. Recently studies have been performed on a number of species that occupy the shallow waters of the antarctic seas, such as *Trematomus borchgrevinki* and *T. bernacchii*. These have been found to have freezing points near −2°C. The shallow waters in the antarctic seas remain near the freezing point year round (unlike those of the Arctic), and these fish have perpetually low freezing points. The presence of special antifreezes is indicated, nonetheless, by the fact they have quite ordinary blood concentrations of NaCl and the other usual solutes of fish blood. The antifreezes have been identified as a number of glycoproteins that show remarkable effects on freezing point. Common antifreezes (such as glycerol in insects or ethylene glycol in the radiator of a car) lower the freezing point by a concentration effect; as will be discussed in Chapter 4, the degree of freezing-point depression caused by such solutes is a function of the concentration of solute molecules (a colligative property) and can be closely predicted from basic principles. The glycoproteins of *Trematomus*, however, lower the freezing point to a much greater degree than would be predicted on the basis of their concentration; in fact, the effect can be as much as 500 times that predicted at low concentrations. As yet there is no satisfactory explanation of this phenomenon.

A final common means by which animals survive potentially freezing conditions is through supercooling; that is, they cool to temperatures below the freezing point of their body fluids without freezing. The supercooled state is to some extent inherently unstable. A fraction of supercooled individuals will freeze spontaneously over time, the probability of freezing increasing with the degree of supercooling and the length of time supercooling is maintained. Also, a supercooled animal will freeze if its body fluids are exposed to ice, just as a test tube of supercooled water will immediately freeze solid if seeded with only a small ice crystal. Because most of the animals that supercool die if frozen, these principles assume considerable significance.

A number of deep-water marine fish of both the Arctic and Antarctic have been shown to have freezing points of $-0.9°C$ to $-1.0°C$, yet swim about unfrozen in waters that are at $-1.7°C$ to $-1.9°C$. Because the deep waters are near the freezing point of seawater year round, these fish, in fact, spend their entire lives in a supercooled condition. The degree of supercooling, about $1.0°C$, is apparently low enough that there is little probability of spontaneous freezing. The fact that it is deep-water fish that supercool is significant, for it is in the deeper waters that there is little chance of encountering ice crystals that would precipitate freezing. Shallow-water species, which commonly encounter ice, have antifreezes that depress their freezing point, as discussed earlier. Though some arctic species that occupy shallow waters supercool to a slight extent, the degree of supercooling is apparently not sufficient for ice seeding to be a serious problem. That it would be a problem for some of the deep-water species is demonstrated by the fact that they often freeze solid when exposed to floating ice.

Among other vertebrates, many species of lizards, turtles, and snakes survive supercooling to $3.5°$–$7.2°C$ below zero for at least brief periods; the freezing point of their body fluids is $-0.25°C$ to $-0.75°C$, depending on species. A similar phenomenon is reported in certain torpid bats of the Soviet Union.

The greatest capacities to supercool are reported in insects. Most insects of temperate and cold climates enter a particular "cold hardy" life stage as winter approaches. This is a resting stage, typically characterized by cessation of feeding, growth, mobility, and reproduction. Capacities to withstand chilling and to supercool are often enhanced. The physiological basis for these latter changes is but vaguely understood. Some species enter diapause, a genetically determined and obligatory resting state characterized by cessation of development and protein synthesis and by especially great supression of metabolic rate. The winter diapausing insect may actually require exposure to cold before it can emerge from its quiescent condition. For cold hardy or diapausing insects, supercooling to $20°$–$25°C$ below the freezing point of the body fluids is about average, and $30°$–$35°C$ is not uncommon. Those that produce antifreezes combine the advantages of a depressed freezing point and the capacity to supercool below their freezing point. Thus overwintering larvae of *Bracon cephi*, which we noted earlier have freezing points as low as $-17°C$ because of high concentrations of glycerol, are able to supercool as much as $30°C$ and thus can live, unfrozen, at a body temperature of as low as $-47°C$.

Supercooling in animals, in and of itself, is not particularly surprising in view of the fact that simple aqueous solutions can be supercooled even to the great extent seen in insects. The real challenge to the physiologist is to account for such phenomena as changes in the capacity to supercool over the life cycle and for differences among animal groups in their capacity to supercool and survive the supercooled state. Thus far, we have not achieved a definitive understanding of these intriguing problems.

Hibernation and estivation in poikilotherms

Many poikilotherms that rest through the winter are said to be in hibernation. Use of this term is somewhat unfortunate because hibernation in these animals bears only superficial resemblance to that in birds and mammals. Nonetheless, it is appropriate to have a special term because many poikilotherms do enter a special physiological condition in winter; they are not simply summer animals that happen to be cold. We have referred already to cold hardiness and diapause as specialized resting stages in insect life cycles. Insects in these states are often said to be in hibernation. Hibernating reptiles and amphibians have lowered metabolic rates, are inactive, and do not feed. Fat development is at a maximum prior to hibernation, and such reserve nutrients supply energetic needs through the winter. During hibernation, frogs and toads may show such manifold physiological changes as decreased blood sugar, increased liver glycogen, altered concentration of blood hemoglobin, altered oxygen and carbon dioxide content in the blood, darkened skin, altered muscle tonus, and so on. The precise adaptive significance of these changes is largely unexplored. This is an area for research that deserves much more attention than it has received.

Many poikilotherms enter a resting state, often called estivation, in response to heat or drought. This is seen, for example, among desert snails, where it may last for years until a heavy rain occurs. The animals seek out cracks in the soil or rocks and seal off the opening of their shell. Many species of earthworm dig deep into the soil and line a small chamber with mucus, wherein they spend summer months. Some species of insects from arid regions enter diapause in the summer and may thus realize a greatly enhanced resistance to the effects of heat and dryness. The African lungfish (*Protopterus aethiopicus*) is well known for its ability to live in a mud cocoon for over a year in times of drought. We shall explore this and other examples of estivation more extensively in the discussion of water relations in Chapter 5.

At least some advantages to entering a quiescent stage such as hibernation or estivation during times of environmental stress are obvious. The animal minimizes its energetic needs and lives off of stored nutrients. The reduced respiration that accompanies reduced metabolism results in considerable curtailment of respiratory water loss in many terrestrial species, an important benefit in times of drought. The animal frequently enjoys enhanced ability to withstand the stresses of high or low body temperature or dehydration as a result of its special physiological state, and it can remain continuously in a relatively nonstressful microhabitat, never exposing itself to the full harshness of the outside world.

HOMEOTHERMY IN BIRDS AND MAMMALS

Homeothermy is regulation of body temperature by physiological means. Although now known to occur in several animal groups, it has been studied most exhaustively in the birds and mammals, and an examination of these forms will serve to introduce many basic principles before we examine the phenomenon as it occurs elsewhere in the animal kingdom. All birds and mammals are not homeothermic all the time. The alternative thermal relationships sometimes displayed by these forms will be discussed toward the end of the chapter.

Placental mammals typically maintain body temperatures near 37°C. Temperatures in birds are usually 3°–4° higher. Birds and mammals possess physiological capacities both to maintain such body temperatures in cool environments and to limit or prevent any rise in temperature in hot environments. Some poikilotherms at times have body temperatures as high as those in birds and mammals but do so only when warmed by external sources of heat. In contrast, birds and mammals maintain their high temperatures in cool environments by virtue of metabolic heat production. Accordingly, they have sometimes been termed *endotherms* rather than homeotherms. Endothermy is the maintenance of an appreciable difference between body temperature and ambient temperature through internal production of heat. Indisputably, then, the birds and mammals are endothermic, but there are two good reasons for calling them homeotherms rather than endotherms. First, in common usage endothermy implies only the elevation of body temperature through metabolic heat production; it does not necessarily imply maintenance of a *stable* body temperature by this means. Thus birds and mammals, which maintain stable temperatures in cool environments through adjustment of metabolic heat production, display a particular *type* of endothermy. Second, birds and mammals often have mechanisms for keeping their body temperature from rising in hot environments. These mechanisms do not in any immediate sense rest on internal production of heat for their effectiveness, and accordingly the mechanisms of homeothermy in these forms include more than just endothermy. In short, avian and mammalian homeothermy, although including endothermy, goes well beyond the austere implications of endothermy alone.

It is important to recognize from the outset that the deep-body temperatures of birds and mammals are not held absolutely constant but are allowed to vary within certain limits characteristic of the species. Under thermally nonstressful circumstances, there is commonly a daily cycle in temperature; man, for example, exhibits daily variations in rectal temperature with an amplitude of about 1.5°C, the low point occurring during sleep. Exercise is often accompanied by some elevation of deep temperature; again citing man as an example, an individual with a resting temperature near 37°C might show an increase to 38° or 39°C during moderate to heavy exercise. Exposure to hot or cold environments is not uncommonly accompanied, respectively, by some elevation or depression of deep temperature. Thus although reasonably well-hydrated humans allow only slight increases in temperature (1.0°C or less) under hot conditions, many birds and mammals of desert regions readily allow their temperature to rise by 4°C or more when confronted with high heat loads. In short, each species permits its body temperature to vary within a restricted range, the actual temperature at any given time being a function of physiological, behavioral, and environmental circumstances. This is not to say that body temperature is uncontrolled; on the contrary, temperature typically is not permitted to cross well-defined upper and lower limits, and there is every reason to believe that most, if not all, variations within the typical range for the species are subject to control.

As illustrated in Figure 3–14, the relationship between resting metabolic rate and ambient temperature in avian and mammalian homeotherms is very different from that observed in poikilotherms (compare Figure 3–4). There is a range of temperatures, the *thermoneutral zone*, over which metabolic rate does not vary with temperature. The upper and lower limits of this range are termed the *upper* and *lower critical temperatures*, respectively. Metabolism rises as ambient temperature falls below the lower critical temperature or rises above the upper critical temperature. This is because the animal must perform physiological work to maintain its stable internal temperature as ambient temperatures become relatively cold or warm. The metabolic rate of a resting, postabsorptive animal in the thermoneutral range is termed its *basal metabolic rate*.

Basic properties of homeothermy

As we have seen earlier, the body temperature of an animal is a function of its rate of metabolic heat production and the factors affecting its rate of heat exchange with the environment. In this section we shall examine the basic interactions of these parameters in vertebrate homeotherms as ambient temperature is varied.

The bird or mammal in a uniform thermal environment cooler than its body temperature tends to lose heat to the environment passively by conduction, convection, radiation, and evaporation. As a first approximation, its total rate of heat loss will increase in proportion to the difference between its body temperature and ambient temperature $(T_B - T_A)$, for conductive, convective, and radiative losses (and sometimes evaporative losses as well) will increase in rate as the environment becomes cooler relative to the animal. We can express this mathematically with the equation

$$\text{rate of heat loss} = C(T_B - T_A)$$

where C is a proportionality factor. The factor C has readily apparent biological meaning. Suppose, to illustrate, that we have two animals with the same difference between body temperature and ambient temperature $(T_B - T_A)$ but that one is losing heat more rapidly than the other. Clearly, C would be greater for the animal with the greater rate of heat loss. We would also say that heat could escape more readily from this animal. Putting these two ideas together, we see that C is a measure of the facility with which heat leaves the organism; accordingly, it is termed the thermal *conductance* of the animal. Animals with high conductance tend to lose heat rapidly for a given thermal difference between themselves and the environment, whereas animals with low conductance tend to lose heat relatively slowly. Conductance provides a measure of all the factors that affect the

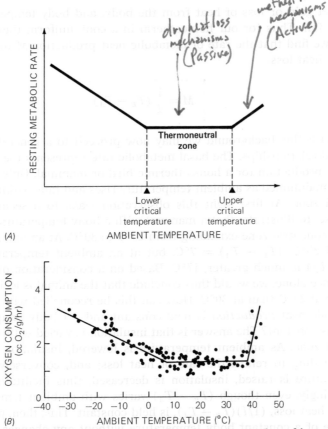

dry heat loss mechanisms (Passive)

wet heat loss mechanisms (Active)

Figure 3-14. Metabolic responses of birds and mammals to changes in ambient temperature. (A) General type of relation between resting metabolic rate and ambient temperature, indicating terminology used in description of response curve. (B) Resting metabolic rate of fasting white-tailed ptarmigan (*Lagopus leucurus*) as a function of ambient temperature. (From Johnson, R. E. 1968. Comp. Biochem. Physiol. 24: 1003–1014. Reprinted by permission of Pergamon Press.)

rate of heat loss, including such things as posture and the insulatory effectiveness of the pelage or plummage.

Insulation varies inversely with conductance. We would say that an animal with high conductance has low insulation, and one with low conductance has high insulation. Thus we can define the insulation of the animal I to be equal to $1/C$ and get

$$\text{rate of heat loss} = \frac{1}{I}(T_B - T_A)$$

This equation simply says that for given $(T_B - T_A)$, the rate of heat loss will be greater if I is low than if I is high. Again, this measure of insulation, like the measure of conductance, reflects all factors affecting the rate of heat loss. It is not, for example, a measure of the insulation of the pelage or plumage alone but also includes such other factors as posture. (A thorough discussion of factors affecting I and C is provided later.)

If an animal is to have a constant body temperature, its rate of metabolic heat production must equal its rate of heat loss. Otherwise there would be a net gain or loss of heat from the body, and body temperature would rise or fall. Thus for our homeotherm in a cool, uniform thermal environment, we find that the rate of metabolic heat production M must equal the rate of heat loss:

$$M = \frac{1}{I}(T_B - T_A)$$ @ steady state

With this background we may now proceed to a general overview of thermal relationships. The basal metabolic rate represents the minimal rate of heat production for a homeothermic bird or mammal. Only this minimal rate is maintained as ambient temperature rises and falls within the thermoneutral zone. At first sight this observation leads to a seeming paradox. Suppose, to illustrate, that a mammal with a body temperature of 37°C has a thermoneutral zone extending from 20°C to 30°C. At an ambient temperature of 30°C, $(T_B - T_A) = 7$°C, but at an ambient temperature of 20°C, $(T_B - T_A)$ is much greater, 17°C. Based on a consideration of the thermal difference alone, we would thus conclude that the animal is losing heat more rapidly at 20°C than at 30°C. How can this be reconciled with the fact that metabolic heat *production* is held constant and yet body temperature also remains constant? The answer is that insulation is varied within the thermoneutral zone. As ambient temperature is lowered, insulation is increased, thus tending to retard the rate of heat loss; and, conversely, as ambient temperature is raised, insulation is decreased, thus facilitating heat loss. Accordingly, even though $(T_B - T_A)$ varies with ambient temperature, the rate of heat loss, $(1/I)(T_B - T_A)$, is held constant. This, then, permits maintenance of a constant body temperature without any changes in metabolic rate over the thermoneutral zone.

All animals have limits to the degree to which they can increase their insulation. Thus as ambient temperature is lowered, a point is reached where insulation is approximately maximized, meaning that further decreases in ambient temperature cannot be countered by insulatory responses. The lower critical temperature (see Figure 3–14) is the ambient temperature at which insulatory adjustments become inadequate to compensate fully for further increases in the thermal difference between animal and environment. Thus as ambient temperature falls below the lower critical temperature, the rate of heat loss from the animal increases and must be countered with an increase in the rate of heat production if body temperature is to be held constant. This is the reason that metabolic rate rises. Putting these concepts in terms of the equations developed earlier, we have seen that as $(T_B - T_A)$ is increased within the thermoneutral zone, the rate of heat loss, $(1/I)(T_B - T_A)$, is held constant by increasing I. But once I is maximized at the lower critical temperature, $(1/I)$ assumes a fixed value, and further increases in $(T_B - T_A)$ imply increases in the rate of heat loss that demand increases in the rate of heat production. The increase in metabolic rate with decreasing temperature below thermoneutrality is one of the most striking differences between homeotherms and poikilotherms. For the poikilotherm, low ambient temperatures often imply low body temperatures and correspondingly depressed metabolic rates. For the homeotherm, however, low ambient temperatures imply increased metabolic effort to maintain a high, stable body temperature.

Looking at the other end of the thermoneutral zone, we must first remember that the animal in the thermoneutral range is producing metabolic heat at the basal rate. As ambient temperature rises in a uniform thermal environment, the thermal difference between animal and environment $(T_B - T_A)$ falls, and accordingly the tendency to lose metabolic heat to the environment falls. This is compensated within the thermoneutral zone by a decrease in insulation, but once insulation has been minimized, further increases in ambient temperature result in problems of heat dissipation. That is, the difference between body temperature and ambient temperature becomes so small that even with minimal insulation, the passive mechanisms of heat loss (conduction, convection, radiation, passive evaporation) do not carry heat away at the rate that it is being produced metabolically. Many birds and mammals respond by sweating, panting, or otherwise *actively* augmenting their rate of evaporative water (and heat) loss. These mechanisms demand metabolic effort, and it is this which frequently is the basis for the increase in metabolism that commences at the upper critical temperature. It may seem paradoxical that an animal under heat stress would increase its internal heat production. This is the price that must be paid, however, to accomplish the effort of active evaporative cooling. Evaporative cooling is such an effective means of dissipating heat that the increase in rate of heat loss exceeds the increase in rate of heat production, thus accomplishing a net increase in the rate of heat loss. It is important to recognize that the upper critical temperature in a uniform thermal environment is always *below* the body temperature. That is, at upper critical temperature, $(T_B - T_A)$ is still positive, and there is still a passive tendency to *lose* heat from the body. The problem is simply that the passive tendency to lose heat becomes inadequate to dissipate heat at the rate at which it is produced metabolically.

Another factor often involved in the elevation of metabolic rate at high ambient temperatures is that many birds and mammals allow their body temperature to rise to some extent in hot surroundings, a process we shall examine in more detail later. Just as in poikilotherms, the rise in body temperature is typically accompanied by a rise in metabolic rate.

With this brief introduction to the properties of homeothermy, it should be instructive now to examine certain basic differences between the vertebrate homeotherms and poikilotherms. We have seen that ambient temperatures within the zone of thermoneutrality are below body temperature (note, for example, Figure 3–14*B*). Thus homeotherms can maintain their body temperature appreciably above ambient temperature with only their *minimal* rate of heat production. If this is the case, why do not reptiles and amphibians likewise keep warm by virtue of their inherent metabolic heat production? The answer is basically twofold. (1) The resting metabolic rates of reptiles and amphibians are low by comparison to those of birds and mammals. This is rigorously demonstrated by comparing the basal (minimal) metabolic rates of birds and mammals with the standard metabolic rates of lizards that have been warmed to the same body temperatures as birds and mammals. As indicated in Figure 2–3, lizards at 37°C have standard metabolic rates only one third to one tenth as high as the basal rates of mammals of equal size. Indeed, we must say that an increase in the level of metabolism was one of the most dramatic elements in the evolution of vertebrate homeothermy. (2) The reptiles and amphibians have much less well-insulated bodies than the birds and mammals. Taking this factor in combination with their relatively low metabolic rates, we see that they do not produce heat rapidly enough or retain it well enough to warm their bodies much above ambient temperature. Conversely, birds and mammals produce a great deal of heat and retard its dissipation to the environment through good insulation. It may be noted that the low insulation of reptiles and amphibians is actually often a benefit in their ectothermic way of life; they warm themselves using external sources of heat (such as the sun), and high insulation would interfere.

The metabolism-temperature curve below thermoneutrality: determinants and implications

As an extension of the discussion in the preceding section, we may now examine in more detail the factors that affect the shape of the metabolism-temperature curve below thermoneutrality. Metabolic rate commonly increases approximately linearly as ambient temperature falls below the lower critical temperature, as illustrated in Figure 3–14. The slope of this line has important biological significance, which can be elucidated according to a model of thermal exchange outlined by Scholander and his colleagues.

Consider a sphere consisting of a concentric layer of insulation surrounding a heating element (resembling Figure 3–1). Suppose that we can control the rate of heat production by the heating element and that we wish to maintain a constant temperature of 40°C inside the sphere as we vary the ambient temperature in a uniform thermal environment. This, of course, is a simple model of a homeothermic animal, and we can call 40°C the "body temperature" of the sphere and call the rate of heat production the "metabolic rate" of the sphere. We shall assume that for any given sphere, the insulation separating the core of the sphere from the environment is constant. As a first approximation, the rate of heat loss from the sphere will be proportional to the difference between core temperature and ambient temperature ($T_B - T_A$). Thus we get, as in the preceding section,

$$\text{rate of heat loss} = C(T_B - T_A) = \frac{1}{I}\,(T_B - T_A)$$

An important distinction from our previous discussion of this equation is that for any given sphere, C and I are *constants*. This equation for heat loss is derived from physical laws of heat transfer, but unfortunately there has been confusion in the literature over the historical origins of the laws that apply. The equation is probably best termed the *linear heat transfer equation* but has commonly been referred to as Newton's law of cooling or, more recently, Fourier's law of heat flow.

If we are to maintain a constant temperature inside the sphere, we must vary the rate of heat production M so that it equals the rate of heat loss; thus we must assure that

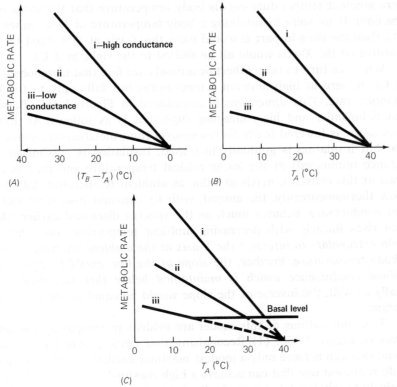

Figure 3–15. Behavior of a metabolic model of the homeothermic animal. See text for discussion.

$$M = C(T_B - T_A) = \frac{1}{I}\,(T_B - T_A)$$

Figure 3–15A is a graph of this latter equation for three different values of conductance C. It shows, very simply, that for a given value of C, the rate of heat production must increase in proportion to the thermal difference between the sphere and the environment $(T_B - T_A)$, for the rate of heat loss also increases in proportion to this difference. The slope of each line in Figure 3–15A is equal to the prevailing conductance C. Thus a steep slope (curve i) reflects a higher conductance than a shallow slope (curve iii). Because insulation I is the inverse of conductance $(I = 1/C)$, a steep slope (curve i) reflects a lower degree of insulation than a shallow slope (curve iii). These are important points to remember.

The curves of Figure 3–15A are reproduced in Figure 3–15B, with the X-axis relabeled so that we now have metabolism-temperature curves plotted in the usual way, with the X-axis representing ambient temperature rather than $(T_B - T_A)$. Inspection will reveal that the graphs in parts A and B have identically the same meaning. The important point demonstrated in part B is that the lines relating metabolism to temperature intersect the X-axis at the *ambient temperature that equals body temperature* (40°C). This obviously must be the case. We have seen in part A that the lines meet the X-axis at $(T_B - T_A) = 0$; that is the point where there is no thermal difference and thus neither heat loss nor a need for heat production. In part B it is where ambient temperature equals body temperature that the same conditions hold. If we were maintaining a body temperature of 37°C rather than 40°C, then the lines in part *B* would meet the *X*-axis at 37°C (and the temperatures on the *X*-axis would all be shifted to the right by 3°C).

When we turn to homeothermic animals, we find that metabolism does not fall to zero at high ambient temperatures but falls only to the basal metabolic rate. This dimension is introduced in Figure 3–15*C*. Basically, what Scholander and his colleagues suggested is as follows. As ambient temperature is lowered *in the thermoneutral zone,* metabolism remains constant but insulation is gradually increased. Insulation is maximized (conductance minimized) at the lower critical temperature and then is maintained at this constant, maximal value as ambient temperature falls lower. Below thermoneutrality, the animal, with its constant insulation and constant conductance, behaves much as the spheres discussed earlier. Metabolism rises linearly with decreasing ambient temperature *on a line that would extrapolate to intersect the X-axis at the ambient temperature equal to body temperature.* Further, the *slope of the line would be equal to the minimal conductance which is maintained below thermoneutrality,* and, equally as well, the inverse of the slope would be equal to the *maximal insulation.*

Two implications of this model are evident in comparing the several curves of Figure 3–15*C*. (1) Recognizing that curve i would represent an animal that can achieve only a modest maximal insulation and that curve iii would represent one that can achieve a high maximal insulation, we see that relatively poorly insulated animals display a steeper increase in metabolism below thermoneutrality than relatively well-insulated animals. The energetic advantages of high insulation at low temperatures are obvious. (2) We also see that the relatively well-insulated animal (curve iii) displays a lower lower critical temperature than the relatively poorly insulated animal, meaning that it does not have to increase its metabolic rate at all until the ambient temperature has fallen to a lower level than in the case of the more poorly insulated animal—again an energetic advantage. It is important to recognize that the effect on lower critical temperature evident in Figure 3–15*C* is in part conditioned by two assumptions inherent in the figure: first, that basal metabolic rate (BMR) is the same in all three animals depicted and, second, that body temperature is also the same. In a more general framework, lower critical temperature is a function of both BMR and body temperature as well as minimal conductance. Thus in comparing animals with different BMR's and body temperatures, relatively low conductance does not necessarily imply a relatively low lower critical temperature. It does have this implication when BMR and body temperature are constant, meaning that lower critical temperature tends to vary in tandem with minimal conductance in comparisons of animals with similar BMR's and body temperatures.

With the implications of the model in mind, it is possible to assess certain attributes of homeotherms by simple inspection of their metabolism-temperature curves. Looking at Figure 3–22, for example, it is readily apparent that the winter fox was able to achieve a higher insulatory value than the summer fox, for it showed a lower metabolic slope below thermoneutrality and also a lower lower critical temperature. (In this comparison BMR and body temperature were about the same in both winter and summer.)

As applied to the thermal relations of the individual animal, the model of Scholander and his colleagues implies that metabolism will rise linearly from the basal level at low ambient temperatures and that this line will extrapolate to the ambient temperature equal to body temperature. These conclusions are predicated by the assumptions of the model; and although the model serves as a most useful paradigm, it must be recognized that its assumptions are not always fulfilled in practice, meaning that the model does not always describe the actual thermal relations of animals. Three of the assumptions deserve discussion here. (1) As we have stressed earlier, the linear heat transfer equation, which says that the rate of heat loss is proportional to $(T_B - T_A)$, is in fact only a first approximation. For one thing, it makes simplifying assumptions about the physical laws of heat transfer. Each of the mechanisms of heat loss (conduction, convection, radiation, and evaporation) adheres to its own physical laws, and the rate of each is dependent not only on properties of the animal, but also on prevailing ambient conditions. The simplifications of physical laws inherent in the linear heat transfer equation may be reasonable approximations of reality if, as ambient temperature is varied, factors such as wind speed are held constant and the radiant temperatures of environmental surfaces remain close to air temperature. These conditions are usually met, at least approximately, within the confines of a laboratory chamber, where metabolism-temperature data are generally gathered. Such controls are not maintained in nature, however; and in nature, as well as sometimes in the laboratory, the simplifying assumptions of the equation may be inappropriate even as approximations. The operation of the physical laws of heat transfer may introduce significant nonlinearities in the relationship between rate of heat loss and ambient temperature, and then metabolism will also be a nonlinear function below thermoneutrality. (2) A second assumption of the model that is not always met in practice is that physiological and behavioral mechanisms of increasing insulation are maximized at the lower critical temperature and remain maximized below thermoneutrality. Sometimes, to illustrate just one type of departure from this assumption, insulatory responses continue to increase to some extent after metabolism has started to rise below the lower critical temperature. If insulatory responses are not constant below thermoneutrality, then heat loss and metabolic rate will again vary nonlinearly with ambient temperature. (3) A final assumption of the model that is sometimes violated is that body temperature is constant. Some animals allow their body temperature to fall to a certain extent at cool ambient temperatures. A white-footed mouse, for example, may show a body temperature of 37°C at an ambient temperature of 30°C but allow its body temperature to decrease to 34°C at low ambient temperatures. Changes in body temperature can, again, cause departures from linearity.

The slope of the metabolism-temperature curve below thermoneutrality is a rigorous measure of minimal conductance only if the assumptions of the model are reasonably met; that is, metabolism should rise linearly, and an extrapolation of the line should intersect the X-axis at the ambient temperature that equals body temperature. If the assumptions are violated, the slope commonly provides a *qualitative* indication of minimal conductance, but for *quantitative* purposes other methods of calculating conductance should be used. To determine minimal conductance as the slope of the metabolism-temperature curve, we calculate the increase in metabolic rate for a given decrease in ambient temperature and divide the former by the latter (both as absolute values):

$$C = \frac{\Delta \text{ metabolic rate}}{\Delta \text{ temperature}}$$

Metabolic rate is usually expressed in weight-specific terms. If expressed in cc O_2/g/hr, then conductance assumes dimensions of cc O_2/g/hr/°C, and insulation, the inverse of conductance, would have dimensions of °C/cc O_2/g/hr. As stressed earlier, these measures of conductance and insulation reflect *all* the parameters affecting the rate of heat loss from the animal, not any particular parameters (such as pelage) alone.

Because the rate of heat loss from the animal is a function of all parameters affecting heat loss, it is affected by environmental conditions as well as the properties of the animal. Accordingly, the lower critical temperature and the slope of the metabolism-temperature curve below thermoneutrality are, in part, functions of prevailing conditions. To illustrate, suppose that we study one and the same animal in the presence of high and low wind speeds. Because high wind velocity increases the rate of convective heat loss at given ambient temperature, we will typically find that the animal exposed to high wind velocity will have to produce heat at a higher rate for given ambient temperature than the animal exposed to low wind velocity and will show an increased metabolic slope below thermoneutrality and often a higher lower critical temperature as well. Another example of the effect of the environment is provided by studies using nonuniform radiant conditions. Roadrunners (*Geococcyx californianus*), when studied in dark, uniform thermal environments, showed a lower critical temperature of about 27°C; but when provided with an artificial sun, they were observed to "sun" themselves and showed a lower critical temperature as low as 10°C. Thus with the added radiant heat provided by the artificial sun, they did not have to increase metabolic heat production until air temperature had fallen far lower than in the dark and displayed a very different metabolism-temperature curve.

These examples illustrate the great importance of environmental conditions to thermal relationships and make it clear that any measure of minimal conductance is a function not only of the animal's responses, but also of prevailing conditions. Because this is true, measures of conductance will be of value in comparing the thermal responses of different animals only if the animals are studied under comparable conditions. Usually metabolism-temperature data are gathered in a "metabolism chamber" where wind speeds are low and the radiant temperatures of all environmental surfaces are close to air temperature. Provided such standardized conditions pertain, differences in conductance between animals can be taken to reflect differences in the animals themselves and not in the environment. For comparative purposes, therefore, conductance measured under the standardized conditions can be treated as a property of the organism, and it is thus that we often speak of the "conductance (or insulation) of the animal" without reference to the environment.

Factors that affect insulation

We may now proceed to examine the actual physiological mechanisms of homeothermy, looking sequentially at insulatory responses, responses at temperatures below thermoneutrality, and responses at temperatures above thermoneutrality.

Birds and mammals employ several mechanisms to modify their insulation. As we have seen, these are the predominant mechanisms of thermoregulation within the thermoneutral zone. One means of varying insulation is elevation or compression of the hairs or feathers; these responses are termed *pilomotor responses* in the case of mammals and *ptilomotor responses* in the case of birds. The hairs or feathers are raised or fluffed out as temperature falls, thus trapping a thicker layer of stagnant air around the animal. Another mechanism is alteration of the peripheral or superficial blood flow (*vasomotor responses*). Constriction of the peripheral vessels at cooler temperatures results in retarded convective movement of heat to the body surface via the blood. Vasodilation at warmer temperatures results in enhanced heat loss. Insulation may also be modified by changes in *posture* that alter the amount of body area directly exposed to ambient conditions. Many birds, for example, hold their wings away from their body when temperatures are high. At low temperatures, mammals may curl up, and some birds tuck their heads under their wings or squat so as to enclose their legs in the ventral plumage.

In addition to these variable insulatory parameters, there are parameters that affect insulation but are more or less fixed for any given animal. Outstanding among these is body size. Because small birds and mammals have higher surface-to-weight ratios than large ones, they tend to lose heat at a greater rate per unit of weight for a given difference between body and ambient temperature. Metabolic rate expressed in weight-specific terms tends to increase more steeply below thermoneutrality in small forms than in larger forms, placing the small bird or mammal in a more energetically demanding situation at low ambient temperatures. In both birds and mammals minimal weight-specific conductance (realized roughly at lower critical temperature and below) tends to decrease systematically as a power function of weight: $C = aW^{-0.5}$, where a is a constant depending on units of measure and the group under consideration. The implications of this equation may be exemplified as follows. On the average, minimal conductance in 10-, 100-, and 1000-g rodents is, respectively, about 0.3, 0.1, and 0.03 cc O_2/g/hr/°C. According to the equation $M = C(T_B - T_A)$, this means that for all three rodents to maintain the same body temperature at a given ambient temperature below thermoneutrality, the 10-g animal will have to maintain a resting weight-specific metabolic rate about three times that required in the 100-g animal and about ten times that required in the 1000-g animal.

Besides surface-to-weight ratio, another factor contributing to the increase in maximum potential insulation with size is the fact that larger animals can carry a thicker coat of hair or feathers than smaller animals. Hair thicknesses of 5–6 cm, for example, are common among larger animals such as the white fox and caribou but clearly would be untenable for smaller mammals. A mouse with 5 cm of hair would be ensconced in a ball of fur, unable even to move effectively.

Although such "fixed" insulatory factors as surface-to-weight ratio and hair or feather thickness give rise to clear trends for insulation to increase with size, the relationship between insulation and size is hardly deterministic. There are significant interspecific differences in the insulatory capabilities of animals of similar size, as we shall see later. Mechanisms of varying insulation such as pilomotor, vasomotor, and postural responses operate against the insulatory "background" of the more fixed parameters.

Modes of increasing heat production below thermoneutrality

Below the lower critical temperature, heat production must be elevated as ambient temperature falls. Although all metabolic processes result indirectly in production of heat, birds and mammals have evolved processes that have the specific function of generating heat for thermoregulation. These *thermogenic* processes accomplish little or no meaningful physiological work in the strict sense of work but instead emphasize the conversion of chemical energy to heat.

The mechanism of thermogenesis with which we are most familiar is shivering, and it appears that all adult mammals and birds utilize this mechanism. Shivering is a high-frequency contraction of skeletal muscle mediated via the nervous system. All muscular contraction liberates heat, and here the conversion of chemical energy to thermal energy becomes the primary function of the contraction.

Curare is a drug that prevents transmission of nerve impulses across neuromuscular junctions. Animals injected with curare are unable to move or, more importantly in the present context, shiver. If cold-acclimated laboratory rats are injected with curare, they continue to exhibit an increase in metabolic rate in response to cold, demonstrating the existence of mechanisms for thermogenesis other than shivering. Such *nonshivering thermogenesis* is known to be widespread among mammals, but its existence in adult birds remains controversial.

There appear to be several sites and mechanisms of nonshivering heat production, but, unfortunately, most are not well understood. The site of nonshivering thermogenesis that is best understood at present is *brown fat*. This is a type of lipid tissue, found widely in mammals, that differs greatly from the "white" fat with which we are more familiar. It is distinguished by its great numbers of relatively large mitochondria and by other cytological characteristics. It receives a rich supply of blood vessels and is well innervated by the sympathetic nervous system. The rich blood supply and the yellowish cytochrome pigments of its dense supply of mitochondria are in large part responsible for its characteristic brownish-red color. The function of brown adipose tissue was obscure to early anatomists and physiologists, and it was often considered to be a gland (the "hibernation gland"). Only since 1961 have we appreciated its function as a site of thermogenesis. Release of norepinephrine into the tissue by the sympathetic nervous system results in a great increase in oxidation of lipid, with consequent liberation of heat. Recent evidence indicates that uncoupling of oxidative phosphorylation occurs; thus to a greater extent than usual in lipid oxidation, energy released from the lipid appears directly as heat rather than being bound into ATP. This is further indication that the tissue evolved to produce heat rather than to supply energy for physiological work.

Brown fat is particularly prominent in three groups of mammals: cold-acclimated adults, hibernators, and newborn. It has been reported in such diverse species as hedgehogs, bats, laboratory rats, laboratory mice, hamsters, guinea pigs, ground squirrels, marmots, rabbits, monkeys, and man (in neonates). It tends to be located in rather discrete masses in the axillae, cervical area, interscapular area, abdomen, and other parts of the body. Some newborn are heavily dependent on brown fat for thermogenesis in cool environments. The neonatal guinea pig, for example, can shiver but ordinarily does not, relying totally on nonshivering thermogenesis, including that of brown fat; shivering thermogenesis replaces nonshivering thermogenesis as the animal grows older, and brown fat deposits diminish in size. Neonates of some species may not be able to shiver at all. The thermoregulatory significance of brown fat in newborn humans is not clear; most of it disappears by puberty. Brown fat plays an important role in emergence from hibernation, during which the hibernator must warm itself up from relatively low body temperatures. Brown fat may constitute 1–3% of total body weight in these animals, and vascular arrangements direct heat from the fat primarily to the regions of the heart and brain during the onset of emergence from the hypothermic condition. Cold acclimation is often accompanied by enlargement of brown fat deposits. Rats placed at 6°C shiver violently at first, but after several weeks, shivering is supplanted completely by nonshivering thermogenesis. Development of auxiliary sources of heat production during cold acclimation has advantages that will be discussed at more length shortly.

There is no question that other tissues besides brown fat are considerably involved in nonshivering thermogenesis, but the exact sites of thermogenesis and the mechanisms are poorly known in comparison to the wealth of information now available on brown fat. Nonshivering thermogenesis in muscle tissue seems to be clearly established; in liver, kidney, and some other tissues it has been claimed and denied. Activation of nonshivering thermogenesis in such tissues is thought to be under control of hormones or the autonomic nervous system, with both norepinephrine and thyroid hormone being implicated.

To this point we have discussed mechanisms of augmenting heat production (shivering and nonshivering thermogenesis) that serve specifically in thermoregulation; when activated at low ambient temperatures, production of heat is their main function. All metabolic processes produce heat at least as a by-product, and we must consider whether heat generated by processes that are not specifically thermoregulatory serves to keep the animal warm in cold surroundings. The process that immediately comes to mind is exercise, for, as we have seen earlier, it can greatly increase metabolic heat production. At first sight, one might presume with little ado that the exercising homeotherm could use this heat to keep warm in a cold environment and thus reduce dependence on shivering and the nonshivering mechanisms of thermogenesis discussed above. The issue is not really so clear, however, for activity may reduce insulation in a variety of ways and thus increase the rate of heat dissipation. One must consider both the heat produced by activity and the increase in rate of heat loss caused by activity if one is to appraise the overall influence of exercise on thermoregulatory status. Insulation is decreased through a number of effects. The cover of pelage or plumage may be disrupted by movement. The animal must typically extend its appendages during exercise and thus fully expose these structures, which have high surface-to-volume ratios and are therefore sites of especial heat loss. Movement may increase the relative role of convective, as opposed to conductive, heat transfer between organism and environment, and convective transfer is more rapid. Evaporative loss of heat is enhanced by the increased ventilation that accompanies exercise. Peripheral blood vessels commonly dilate.

Large animals are sufficiently well insulated that they typically gain a net thermoregulatory advantage through exercise; thus shivering and nonshivering thermogenesis may be reduced or even eliminated. We are all familiar with the fact that we can keep warm in cold environments without shivering when we are active. Some animals retain the heat of exercise so well that its dissipation actually becomes a problem. Fur seals, for example, are exceedingly well insulated, possessing both a coat of hair and thick subcutaneous layers of fat. They exercise without overheating in the water because water has a high specific heat and is not a good insulator. On land, however, heat dissipation is reduced because of the insulating properties of air, and the animals may die of overheating if herded too rapidly, even though ambient temperature remains below 10°C. Small animals face quite a different situation than large animals because of their greater surface-to-volume ratios and typically thinner coverings of pelage or plumage. Small rodents realize little, if any, thermoregulatory advantage through exercise. Their decrease in insulation resulting from activity is sufficient to offset, substantially or entirely, the increased rate of heat production resulting from activity, and heat production by shivering or nonshivering thermogenesis must continue more or less unabated for maintenance of a stable body temperature. In fact, at relatively cold ambient temperatures, body temperature may actually fall somewhat when the animal becomes active despite an increase in metabolic rate resulting from the combined thermal inputs of shivering, nonshivering thermogenesis, and exercise.

One advantage of nonshivering thermogenesis is that it is not inhibited by activity. Shivering heat production, on the other hand, may be reduced or even eliminated by exercise, for shivering ceases in muscles directly involved in the coordinated movements of walking, running, flying, or climbing. The advantage of nonshivering thermogenesis during exercise is illustrated by experiments on laboratory rats. When acclimated to 30°C, the rats possess little capacity for nonshivering thermogenesis and must rely primarily on shivering to maintain body temperature when at rest in the cold. Shivering is suppressed during running, and exercising warm-acclimated rats could maintain their body temperature only down to an ambient temperature of 10°C. Rats develop a considerable ability for nonshivering thermogenesis when acclimated to cold. When 6°C-acclimated rats were exercised, they could maintain their body temperature down to −20°C, for they could simultaneously generate heat through exercise and nonshivering thermogenesis.

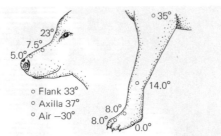

Figure 3–16. Skin temperatures (°C) on the head and foreleg of an arctic sled dog at an air temperature of −30°C. (From Irving, L. and J. Krog. 1955. J. Appl. Physiol. 7: 355–364.)

Regional heterothermy

Structures such as legs, tails, and ears have relatively high surface-to-volume ratios and are potentially major sites of heat loss. Keeping the appendages warm in cold surroundings presents much the same problem we have repeatedly recognized in comparisons of large and small species, namely, that the surface area of small structures is so great relative to their size that their rate of heat loss per unit of weight is high. Thus in a bird or mammal that keeps its appendages at the same temperature as the body core, the appendages contribute disproportionately, for their weight, to the overall weight-specific metabolic demands of homeothermy. Another factor leading to relatively high heat loss from the appendages is that they are often more thinly covered with pelage or plumage than the rest of the body.

Commonly, the appendages of birds and mammals in cold surroundings are allowed to cool below the temperature maintained in the head, thorax, and abdomen. To illustrate, reindeer in the Arctic were found to have a deep abdominal temperature of around 38°C, but the temperature deep in their lower legs was only about 8°C. Because heat is lost in direct relation to the thermal difference between animal and environment, allowing the deep temperature of the appendages to fall closer to ambient temperature serves to reduce the rate of heat loss per unit of surface area and thus compensates for the relatively high surface area and relatively thin insulatory covering. The phenomenon of maintaining different temperatures in different parts of the body is termed *regional heterothermy*. Effectively, maintenance of lowered temperatures in the appendages in cold surroundings reduces the overall weight-specific metabolic cost of thermoregulation. It therefore decreases the animal's overall thermal conductance and increases its insulation.

Two examples of regional heterothermy are provided in Figures 3–16 and 3–17. In one case, subcutaneous temperatures along the foreleg and snout of a sled dog are shown. Note that the temperature under the thick fur of the upper leg was 35°C (near deep abdominal temperature), whereas temperatures recorded from the lower leg and foot were very much lower. The bare footpads were near 0°C. Figure 3–17 is an infrared radiograph of the head of an opossum. You will recall that the intensity of radiative emission from a surface depends on the temperature of the surface; thus by measuring infrared emission we can obtain a measure of surface temperature. The radiograph is not a simple photograph of the animal but is es-

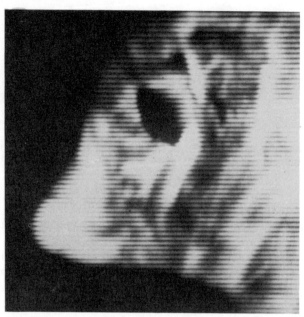

Figure 3–17. An infrared radiograph of the head and neck of a Virginia opossum (*Didelphis marsupialis*) at an ambient temperature of 10°C. See text for description of methodology. The animal is facing to the left, and its left ear appears as a black ellipse near the center of the radiograph. Note that the surfaces of the snout are much warmer than air temperature, whereas the surface temperature of the ear is virtually the same as ambient. Over the thick fur of the posterior head and neck, surface temperatures are variegated but generally relatively low. Regions of high surface temperature within these thickly furred parts (as, for example, below the ear) represent places where the fur has parted, thus allowing relatively free flow of heat to the environment.

sentially a thermal map generated by measuring the intensity of radiation from each point of the body. Surfaces that were close to ambient temperature are colored black, whereas those that were very much warmer than ambient are colored white; intermediate temperatures are represented by appropriate gray tones. The surface temperature of the ear of the opossum is seen to be very close to ambient temperature, indicating that the ear is not kept warm. Being little more than a thin flap of naked skin, it would be a significant site of heat loss if maintained at the same temperature as the body core.

Appendages, especially their distal parts, often are sparsely muscularized and do not sustain a sufficient endogenous heat production to maintain themselves at high internal temperatures in cold environments. They are then dependent on heat delivered from the thorax and abdomen by the circulation to keep warm, and, as we shall see in the next section, special vascular arrangements that curtail circulatory influx of heat are commonly involved in establishing a heterothermic condition in the appendages. Frequently, circulatory influx of heat is adaptively modified according to the thermal circumstances of the animal, being limited when there is a premium on heat conservation but augmented when heat is to be dissipated. To illustrate, the tails of such diverse animals as the muskrat, laboratory rat, and opossum are allowed to cool when the animals are at rest in a cool environment, but heat delivery and tail temperature are increased during exercise, when metabolic heat production is elevated. Apparently, at the ambient temperatures at which the animals were studied (e.g., 10°C in the opossum, 18°–29°C in the muskrat), exercise produces a surfeit of heat and increased delivery of heat to the tail, with its high surface area, serves to balance heat dissipation and production.

In species that have evolved in frigid climates, it appears that temperatures within the tissues of heterothermic appendages are not permitted to go below freezing even if the environment gets substantially colder. Thus here again the thermal status of the appendages is under adaptive control. A dramatic example has recently been provided in studies of two arctic species, the arctic fox and gray wolf. In cold surroundings, the deep foot temperature in these species remains relatively high (perhaps 30°C), but the surfaces of the footpads, which actually contact the substrate during walking and running, are allowed to cool to near 0°C. Even when the foot was immersed in a liquid bath at −35°C, the surface temperature of the footpads was maintained at an average of 1.5°–4.0°C. The evidence is that footpad temperature is controlled by circulatory adjustments in vessels that bring heat to the pads from the warm, inner parts of the foot and leg. Species that have evolved in warm climates do not always show the adaptive responses needed to prevent freezing when exposed to frigid conditions. The opossum, for example, is a species that has only recently invaded northern climates, and when exposed to frigid conditions, it frequently suffers frostbite of its ears and tail.

The surface of the animal that actually contacts the environment is the site where conductive, convective, and radiative heat losses ultimately occur. The smaller the thermal differential between this surface and the environment, the lower the rate of heat loss. In terrestrial birds and mammals covered with pelage or plumage, the relevant surface is that of the hair or feathers. Over the trunk of the body, their *skin* temperature, even in very cold air, is often within a few degrees of deep body temperature in northern species; that is, even the superficial tissues of the thorax and abdomen are maintained at high temperatures. There is a steep drop in temperature through the layer of relatively stagnant air trapped by the pelage or plumage so that the surface of the pelage or plumage—the actual exchange surface with the environment—is much cooler than the skin or deep tissues. In those mammals lacking appreciable hair, the skin itself is the exchange surface with the environment, and the only way in which the temperature of this surface can be lowered is to allow living tissues to cool. In contrast with the well-haired mammals, many of these forms exhibit steep thermal gradients through the outer tissue layers of their thorax and abdomen when in cold environments, a form of regional heterothermy. To illustrate, harbor and harp seals in water at 0°C were found to have skin temperatures near 0°C. Within the layer of blubber under their skin, there was a steep thermal gradient extending inward for about 5–6 cm; that is, the temperature at 5–6 cm below the skin was near core temperature (over 30°C), but in the more superficial layers of blubber, temperature fell steadily with increasing proximity to the skin surface. A similar phenomenon is reported in pigs. The appendages of marine mammals in cold waters also exhibit regional heterothermy, similar to that seen in the appendages of terrestrial species.

We might expect tissues that sometimes function at low temperatures to show special adaptation for this condition. One of the earliest studies along these lines was performed on the conducting properties of nerve axons from the legs of herring gulls. The legs of gulls are subject to regional heterothermy in cold surroundings; although skin temperatures under the thick feathers of the upper leg may be over 35°C, those on the naked parts of the lower leg and webbing of the foot can be below 10°C. Axons from the legs of cold-acclimated gulls were tested for the temperature at which they would cease to respond to electrical stimulation. Axon segments from the naked lower leg failed at about 4°C, but segments from the upper leg failed at a higher temperature, about 12°C. This result is made particularly interesting by the fact that the segments from the upper and lower leg were parts of the same nerve fibers. Thus one and the same cell can show different responses in different parts of the leg, indicating that differences in membrane properties are likely involved. Other studies of nerve function have been performed on mammals, and it has similarly been found that fibers from heterothermic parts of the body are more effective at low temperatures than fibers from parts of the body that are always kept warm. Spinal nerves from the tail of the beaver, for example, were found to be excitable down to −5°C, at which temperature they were supercooled; in contrast, spinal nerves from the thorax failed at about +4.5°C. Along a different line, arteries from the flippers of winter-acclimatized harbor seals exhibited their normal vasoconstrictive response to adrenaline down to 1°C, but arteries from the kidneys, deep within the trunk of the body, failed at 15°C. Unfortunately, we know little as yet of the biochemical basis for the regional differences in responses of arteries or nerves. In the legs of arctic caribou, wolves, and fox, fats from the distal extremities have been shown to have low solidification temperatures by comparison to fats from the proximal portions of the leg or trunk. Visceral fats and marrow fats from the femur of caribou, for example, are solid at room temperature and become brittle hard at freezing temperatures, but marrow fats from the phalanges remain soft at near-freezing temperatures. Fats with low solidification temperatures are probably important to adequate function in tissues that are allowed to become cold, but they do not appear to have evolved as a specific response to life in northern climates inasmuch as marrow fats from the leg of a tropical deer proved to show much the same properties as those from caribou.

Countercurrent heat exchange

Figure 3–18A depicts one possible arrangement of the arteries and veins in the limb of a mammal or bird. The arteries are located relatively deep within the limb, but the veins are superficial. With this arrangement, as blood flows through the limb there is a steady loss of heat to the environment. The blood returns to the body core at a much-reduced temperature, indicating a substantial loss of heat that must be made up through metabolic heat production.

Figure 3–18B depicts another vascular arrangement, in which the veins are closely juxtaposed to the arteries. In this case much of the heat lost from the outgoing arterial blood is picked up by the returning venous blood rather than being lost to the environment. As a result, the venous blood is steadily warmed on its return and re-enters the body core only slightly cooler than outgoing arterial blood, reflecting substantial conservation of the heat carried out into the limb by the arterial blood. The system shown in Figure

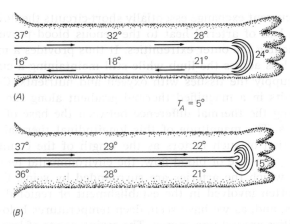

Figure 3–18. A diagrammatic representation of circulation in a limb of a mammal showing hypothetical temperature changes of the blood in the absence (*A*) and presence (*B*) of countercurrent heat exchange. Arrows indicate direction of blood flow.

3–18*B* is known as a *countercurrent heat exchange system* because it depends on heat exchange between two closely juxtaposed fluid streams flowing in opposite directions.

The thermal dynamics of countercurrent heat exchangers are perhaps most readily understood by artificially subdividing the system to consider, first, the uptake of heat by the venous blood and, second, the loss of heat from the arterial blood. In fact, these two processes are intimately interrelated. The venous blood, as illustrated in Figure 3–18*B*, is quite cool as it starts its return from the outer extremities. As it moves proximally, it soon encounters arterial blood somewhat warmer than itself and accordingly picks up heat. If the arterial blood all along the limb were at the same temperature, the venous blood would soon come to thermal equilibrium with the arterial blood, and heat transfer to the venous blood would stop for want of the necessary thermal gradient. Instead, as the venous blood moves along the limb it steadily encounters warmer and warmer arterial blood. Thus even though it is itself being warmed, it steadily comes to be adjacent to still warmer arterial blood, and thermal gradients favorable to heat transfer from the arterial to the venous blood are maintained all along the limb (as illustrated). This permits the venous blood to take up arterial heat throughout its return and to re-enter the body at a high temperature. Looking now at the other side of the coin, we see that the arterial blood flowing out into the limb steadily encounters cooler and cooler venous blood. Thus though it is itself becoming cooler, favorable thermal gradients for loss of heat to the venous blood are maintained all along the limb. The cool venous blood in such close proximity to the arterial blood typically, in fact, acts to cool the arterial blood considerably more than if the arterial blood were simply losing heat to the environment (through the insulating tissue layers of the limb). Thus, as shown in Figure 3–18, temperatures at the outer extremities are characteristically lower in the presence of countercurrent exchange than in its absence. (This is a simplified explanation of the phenomenon, but a more rigorous analysis must be done mathematically. Note that the returning venous blood is simply arterial blood that has been "turned around"; thus the initial temperature of the venous blood is itself partly a function of the degree of cooling of the arterial blood.)

Countercurrent exchange essentially short-circuits the flow of heat into the limb; the loss of arterial heat to the venous blood prevents that heat from ever reaching the outer extremities. It thus provides a mechanism for limiting heat loss across the limb while still maintaining a circulatory rate adequate to supply the tissues with oxygen and nutrients. Countercurrent exchange results in a magnified thermal gradient along the length of the limb, increasing the thermal difference between the base of the limb and the extremities. The magnitude of this longitudinal thermal gradient is a dynamic function of such factors as the length of the countercurrent exchange system, the rate of blood flow, the environmental temperature, and the extent to which the arteries are surrounded by veins. Countercurrent systems are often involved in the establishment of regional heterothermy in appendages, and, as we have seen, deep temperatures below 10°C at the outer extremities are not uncommon. The maintenance of low temperatures at the outer extremities in itself aids heat conservation. These parts of the body tend to be especially poorly insulated, and the reduction of the thermal difference between tissue and environment retards the rate of heat dissipation.

Vascular arrangements meeting the prerequisites for countercurrent exchange (close juxtaposition of arteries and veins) are widely reported in appendages that display regional heterothermy. They are known, for example, in the arms of man; in the legs of many mammals and birds, including some tropical species such as sloths and armadillos; in the flippers and flukes (tail fins) of porpoises; in the tails of a considerable number of rodents (beaver, rat, squirrel); and in the ears of rabbits. Anatomically they vary from the relatively simply to the complex. There may simply be a close intermingling of rather ordinary veins and arteries, as is found in the human arm. In the flippers of porpoises we find arteries that are entirely surrounded by venous channels; with this arrangement heat leaving the arteries cannot help but pass to venous blood. In some animals (such as armadillos and sloths) the main arteries and veins of the limb split up to form many fine vessels that intermingle in a complex network—this being termed a *rete mirabile* ("wonderful net"). Commonly (as, for example, in the human arm and the flipper of the porpoise), there are two sets of veins, one superficial and not in close proximity to the major arteries, the other deep and part of a countercurrent exchange system. By modulating the return of blood along these two venous systems, the animal can adaptively emphasize heat dissipation or conservation in the limb according to its thermal status (see Figure 3–18). As we are aware from common experience, return of blood via the superficial veins of our arms is augmented in warm surroundings and reduced in the cold.

Mechanisms of thermoregulation
in hot environments: evaporative cooling

As we have often noted, birds and mammals produce heat metabolically at a considerable rate even in the thermoneutral zone and exchange heat with the environment by conduction, convection, radiation, and evaporation. As air temperatures rise toward the body temperature, conductive and convective rates of heat loss become lower; and if air temperatures exceed the body temperature, conduction and convection contribute a heat gain to the animal. In a uniform thermal environment, where the radiant temperatures of environmental surfaces approximate air temperature, radiant heat transfers likewise become less favorable to heat loss as ambient temperature rises and begin to contribute a heat gain when the ambient temperature exceeds body temperature. The radiant environment in nature is seldom so simple as this. In particular, the sun often acts as a powerful radiant source of heat, not only by impinging directly on the animal but also by warming the ground and other objects, thereby raising their radiant temperatures. In warm environments the problem faced by the animal is to assure an adequate rate of heat loss to balance metabolic and environmental heat gains so that the body temperature does not rise to deleteriously high levels. The solution can involve manipulation of any or all of the four basic mechanisms of heat transfer. We shall consider manipulations of conductive, convective, and radiative circumstances in subsequent sections but here emphasize evaporative cooling. Evaporation from the integument or respiratory tract always acts to carry heat away from the animal, and augmentation of evaporation is frequently utilized in balancing heat loss against gains in warm circumstances. If conduction, convection, and radiation should all be contributing heat gains, evaporation becomes the only mechanism available for maintenance of homeostasis.

When water changes state to become water vapor, nearly 600 cal of heat is absorbed per gram. This latent heat of vaporization is absorbed from the body surface at which the change of state occurs. The water vapor produced must be dissipated rapidly to the environment, or the air next to the evaporative surfaces will become saturated with vapor, and evaporation will cease (see Chapter 4). Work must be performed by the animal in bringing water and heat to the evaporative surfaces and, frequently, in dissipating water vapor; this is reflected in a rise in metabolic rate when mechanisms of augmenting evaporative cooling are brought into play. Clearly, the efficiency of these mechanisms must be such that more heat is lost in evaporation than is produced metabolically in the course of achieving evaporation. Otherwise there would be augmented evaporation but no net loss of heat.

The mechanism of augmenting evaporative cooling that is most familiar to us is sweating. A saline fluid is secreted onto the body surfaces by the sweat glands, and water vapor is carried away by diffusion and convection into the surrounding air. Birds do not sweat. All placental mammals except rodents and lagomorphs (rabbits and hares) have integumentary sweat glands, but in some—such as the dog and pig—secretion rates are low, and sweating plays little or no demonstrable role in thermoregulation. Other mammals, such as man, horses, cattle, camels, and donkeys, sweat a great deal in response to heat stress. Men doing strenuous work in the desert, for example, have been reported to dissipate as much as 2 liters of water per hour, representing a heat loss of over a million calories per hour.

Sweating is not the only mechanism by which water is lost across the integument. Because the skin of birds and mammals is somewhat permeable to water, water is constantly lost by diffusion through the integument, and its vaporization near the skin surface absorbs the usual amount of heat per gram. This form of water loss is termed *transpirational* or *insensible water loss* and represents a path of passive heat dissipation at all ambient temperatures. Sweating differs in that it involves an active and controlled secretion of water onto the skin surface, and in mammals that sweat, the rate of water vaporization during sweating may exceed the transpirational, background rate by 50 or more times. There is reason to believe that some birds and mammals can actively enhance transpirational losses when under thermal stress, thus implicating transpiration as a mechanism of thermoregulation.

Other modes of evaporative cooling are centered on the respiratory tract. The surfaces of the respiratory tract are very moist, and heat is lost via evaporation during normal breathing. Respired air becomes fully saturated with water vapor during its passage through the respiratory tract, and it is warmed at most ambient temperatures. Warming of the air enhances the loss of water vapor inasmuch as warm air carries more vapor at saturation than cooler air (see Chapter 4).

Many species have evolved modes of augmenting respiratory loss of heat. The mechanism with which we are probably most familiar is panting, an increase in rate of breathing in response to heat stress that is found in many birds and mammals. In some species, breathing rate increases progressively as the extent of heat stress increases. In others there is an abrupt change in respiratory frequency at the onset of panting, and the rate of panting is independent of the degree of thermal stress. Dogs in cool air, for example, breathe around 10 to 40 times per minute, and the breathing rate jumps abruptly to 200 or more breaths per minute when panting begins. Various analyses have indicated that animals with such a stepwise change in respiratory rate often pant at the resonant frequency of their thoracic respiratory structures. This is of advantage because breathing requires considerable muscular work and therefore entails substantial heat production. Panting at the resonant frequency reduces the muscular effort required and thus reduces the amount of heat that must be produced in the process of augmenting heat dissipation.

A mechanism of increasing respiratory loss of heat that is found in a great many birds is gular fluttering. The gular area, or floor of the mouth, is vibrated rapidly while the mouth is held open. Rates of 70 to 1000 vibrations per minute have been reported. Here the respiratory tract proper is not ventilated to an increased extent. Rather, air is simply driven across the thin, moist, and highly vascular oral membranes. Less muscular effort is required than in panting, and there can be a significantly more equitable relationship between the amount of heat dissipated and the heat production necessary to achieve dissipation. Pelicans possess especially large gular areas, but effective gular fluttering has been reported in many birds of several orders, including cormorants, herons, ducks, quail, domestic fowl, goatsuckers, doves, and owls. In many species fluttering apparently occurs at the resonant frequency of the structures involved, whereas in others the rate of fluttering varies with heat load. Panting and gular fluttering are often employed synchronously by birds.

Respiratory cooling, whether through panting or gular fluttering, has at least two potential advantages over sweating. First, air saturated with water vapor is driven away from the evaporative surfaces through the activities of the animal itself, rather than by less predictable or controllable external forces. Second, there need be no loss of salts to the external world because evaporation occurs within the body and only water vapor is dissipated. Sweat always contains some salt, and salt loss in sweating man may be so severe as to be threatening to life. In some other species, the salt content of sweat is sufficiently low that this problem is less menacing.

Respiratory cooling also has at least two potential disadvantages as compared to sweating. In the first place, movements of the respiratory structures at a rapid rate may involve considerable metabolic heat production. As discussed earlier, gular fluttering poses less of a problem in this regard than panting, and resonant frequencies are often used to advantage. In the second place, increased ventilation may cause increased loss of carbon dioxide and lead to elevation of blood pH, or alkalosis. Ordinarily, ventilation of the respiratory exchange membranes deep in the lungs—the alveolar membranes of mammals or the air capillary membranes of birds—is closely regulated so that the rate of dissipation of CO_2 is equal to the rate of metabolic production of CO_2 (see Chapter 8). This ventilatory regulation acts to maintain stable concentrations of CO_2 and bicarbonate in the body fluids. When the respiratory tract becomes involved in evaporative cooling, the potential exists for the rate of CO_2 loss to exceed the rate of CO_2 production, for the rate of ventilation is no longer dictated simply by the need for exchange of oxygen and CO_2. In short, although ventilation rate is ordinarily closely coupled to needs for oxygen and CO_2 exchange, the appropriation of the respiratory tract to evaporative cooling can uncouple these functions. Excessive dissipation of CO_2, by lowering CO_2 concentrations in the body fluids, shifts the following reaction sequence in the body fluids to the left:

$$CO_2 + H_2O \rightleftharpoons H_2CO_3 \rightleftharpoons H^+ + HCO_3^-$$

The result is a lowering of the H^+ concentration and increase in pH (see Chapter 9). This can have major deleterious effects, for many cellular processes are acutely sensitive to pH; we have probably all observed how a person can render himself dizzy and nauseous by deliberately breathing rapidly. For alkalosis to develop, the respiratory exchange membranes themselves must be hyperventilated. Accordingly, gular fluttering, which involves increased ventilation of only the oral membranes, does not present problems of alkalosis. Panting, on the other hand, often causes at least a mild elevation of pH. A factor limiting this alkalosis, at least in mammals, is that the breaths in panting are relatively shallow, and much of the enhanced ventilatory volume is not drawn to the level of the alveoli but rather passes in and out of the nonrespiratory tracheal and bronchial airways. Thus the increase in overall ventilatory rate exceeds the increase in alveolar ventilatory rate. It has long been hypothesized that birds could avoid alkalosis by passing the increased ventilatory volume during panting primarily or exclusively to the air sacs, thus avoiding hyperventilation of the lungs. There is now good evidence that this occurs in the ostrich, but the extent to which it takes place in most species remains uncertain; many birds have been shown to become alkalotic during panting (see Chapter 8 for further discussion). Resistance to alkalosis may be enhanced in species that pant. Man, for example, experiences cramps and incipient unconsciousness if blood CO_2 concentration drops by 30% from ordinary levels. Dogs, however, tolerate a drop of 75% quite well.

A mechanism of increasing evaporative cooling reported in many rodents and marsupials is moistening of the body, limbs, or tail with saliva. Spreading of saliva on furred portions of the body is not believed to be an efficient mechanism of cooling because the underlying pelage acts to insulate the evaporative surface, where heat is absorbed, from the body. Further, heat coming in from the environment impinges directly on the evaporative surface; thus much water is lost in carrying away heat that never really entered the animal. For rodents, which neither sweat or pant, spreading of saliva—however inefficient—is thought to provide an emergency cooling mechanism that can offer help for a relatively short time while a more favorable environment is sought. There is good evidence that saliva spreading significantly aids thermoregulation in hot environments in some species. In marsupials, panting appears to offer the primary defense against heat stress; the contribution of saliva spreading, which is often much in evidence, has not been quantitatively assessed.

All methods of evaporative cooling have the disadvantage of drawing on bodily water resources. This factor becomes particularly important when we recognize that many of the more thermally stressful environments (such as deserts) are also environments where water can be in short supply. This dimension of thermoregulation by evaporative cooling is given special attention in Chapter 5. Here we may simply note that there is often an advantage to minimizing reliance on evaporative cooling by behavioral or physiological means. Some of these are discussed in the following sections.

Insulation in hot environments

Conduction, convection, and radiation are known collectively as mechanisms of dry heat transfer; for unlike evaporation, they do not involve the participation of water. The rate of dry heat transfer between the animal and environment is affected by such insulatory parameters as pilomotor, ptilomotor, vasomotor, and postural responses. In an environment imposing heat stress if the net effect of conduction, convection, and radiation is to cause loss of heat from the animal, there is typically an advantage to minimizing insulation against dry heat transfer, thereby enhancing the rate of heat dissipation by mechanisms of heat exchange that place no demands on bodily water resources. Lowering of insulation is thus the usual response of birds and mammals in uniform thermal environments as ambient temperature is raised toward the body temperature; so long as body temperature exceeds ambient, heat loss is favored by the thermal gradient. But if the net effect of conduction, convection, and radiation is to cause heat uptake from the environment, there is typically, within limits, an advantage to increasing insulation against dry heat transfer; for just as high insulation acts to retard heat loss in a cold environment, it can retard heat uptake in an environment where the thermal gradient between the body core and surface of the pelage or plumage is favorable to heat gain. We thus are not surprised to find that some animals increase their insulation in hot environments by ptilomotor, pilomotor, vasomotor, or postural means, and some species that have evolved in hot environments have remarkably well-developed coverings of pelage or plumage.

A few examples will illustrate. Jackrabbits (*Lepus alleni*) were exposed to a wide range of temperatures in a uniform thermal environment, and their overall conductance for conductive, convective, and radiative heat exchange with the environment was calculated. At ambient temperatures *below* thermoneutrality, conductance was minimized. Conductance increased as ambient temperature was raised toward body temperature above the lower critical temperature and reached 2–3 times the minimal value when ambient temperature was only slightly below body temperature; that is, insulation decreased substantially as the environment became warm, so long as there was still a thermal gradient favorable to heat loss. The most interesting response for our present discussion was that as soon as ambient temperature exceeded body temperature, thus establishing a gradient favorable to heat gain, conductance fell to only slightly above the minimal value that had been seen at temperatures below thermoneutrality. The mechanisms by which jackrabbits increase insulation at high ambient temperatures are not well known. Although many birds appear to exhibit little or no increase in insulation as ambient conditions become favorable to heat gain, the ostrich has been observed to erect its plumage under such circumstances. Some of the larger desert animals, which have difficulty finding shade owing to their size, have strikingly thick insulatory coverings. The dorsal pelage of the dromedary camel in summer, for example, can be at least 5–6 cm thick, and the plumage of the ostrich can be 10 cm thick when erected. The importance of this heavy insulation is emphasized by observations of pelage temperatures in animals exposed to the sun in warm environments. The surfaces of the dorsal hair have been reported to reach 70°–80°C in camels under such conditions, and temperatures as high as 85°C have been reported in Merino sheep. The hair insulates the body from these enormous heat loads; without hair, the skin itself would be exposed to such temperatures.

When conduction, convection, and radiation all impose a heat load, the animal's only recourse to maintain a stable body temperature is evaporation. Thus all heat reaching the body from the environment must be dissipated at a cost in body water. Good insulation against dry heat transfer under such circumstances reduces demands on water resources by decreasing the rate of influx of heat. In addition to this general advantage of insulation, which applies whether sweating, panting, or gular fluttering is employed for evaporative cooling, we should also look at some of the particular implications of pelage insulation in sweating mammals such as the camel.

Because the evaporative surface in sweating is the general integument, the potential exists for a rapid influx of environmental heat directly to the surface—a circumstance that would result in rapid evaporation. Pelage serves to insulate the evaporative surface from the influx of environmental heat; in an animal such as the camel, which has thick pelage, incident heat directly warms the outer surfaces of the hair and can only pass to the evaporative surfaces relatively slowly through the insulating layers. In fact, when solar heat warms the surfaces of the pelage to high temperatures, as it often does, heat can be lost from the pelage to the environment by conduction, convection, and radiation, thus never reaching the skin, where evaporation of body water must come into play. The benefit of the thick pelage is perhaps best illustrated by an experiment performed on camels in the summer desert, where evaporative water loss was found to increase by 50% when animals were shorn to a hair thickness of 0.5–1.0 cm. It is important in a sweating mammal that the pelage not be so thick that it interferes with adequate dissipation of vaporized water from the skin to the surrounding air.

It is also important that the hair not become wet with sweat and matted, for then the evaporative surface would be transferred to the outer surface of the hair, with the disadvantages cited earlier in the discussion of saliva spreading. Men living in the desert have long benefited by wearing loose clothing, which, like the hair of the camel, allows proper dissipation of water vapor from the skin but insulates the skin from environmental heat input.

Body temperature in hot environments

Camels provide a good example of another mechanism that reduces reliance on use of body water for evaporative cooling. When dehydrated and exposed in the summer desert, dromedaries have been found to allow their body temperature to fall to 34°–35°C overnight and then to rise to over 40°C during the day. By not rigidly maintaining a particular body temperature, they thus simply absorb a considerable amount of incident heat during the day rather than dissipating it evaporatively. Knowing that it takes about 0.8 cal to warm a gram of camel by 1°C, we can readily calculate that a 400-kg camel will absorb about 1.9 million cal by allowing its body temperature to rise by 6°C; dissipation of this heat by evaporation would require over 3 liters of water. When night falls, conditions become favorable for cooling by conduction, convection, and radiation, and these mechanisms (plus passive respiratory and integumentary water loss) provide for the nocturnal fall in body temperature. Thus, in the end, much of the heat load of the day is dissipated by nonevaporative means at night. Another advantage of allowing the body temperature to rise during the day is that the thermal difference between the animal and the hot environment is reduced, thus reducing the rate of heat flow into the body. This reduction in the thermal difference assumes particular significance when we recognize that camels do not allow their temperature to go above certain limits, 40°–41°C. On hot days they reach their peak temperature—and thus cease storing heat—before the heat of the day is over and then must thermoregulate by evaporative cooling. The heat load during this period is less with a body temperature of 40°–41°C than it would be if the body temperature were lower, and the demands on water resources are accordingly diminished.

Many small animals also sustain elevated body temperatures during exposure to heat. Increase in body temperature is a most significant and virtually universal response in birds, elevations of up to about 4°C being readily tolerated. Because birds typically have body temperatures near 40°–41°C in the absence of thermal stress, such a degree of warming will place their temperature at quite a high level, 44°–45°C. The actual storage of heat during warming is not generally of great significance inasmuch as a small body absorbs relatively little heat per unit rise in temperature. The elevated body temperature itself is highly significant, for it establishes more favorable conductive, convective, and radiative relationships with the environment and consequently reduces, or even eliminates, the need for active evaporative cooling. Most small mammals of hot regions, unlike birds, are nocturnal and are thus not exposed to the heat of the day. An interesting example of a phenomenon much like that shown by the camel is found in the antelope ground squirrel, a diurnal resident of our southwestern deserts. This species, like other rodents, can neither sweat nor pant. It alternately ventures out onto the hot desert floor to forage and then returns to its cool burrow. Body temperature may rise as high as 43°C during foraging but then is reduced to 38°–39°C by conductive, convective, and radiative heat loss in the burrow. Again, as in the camel, heat storage serves to obviate the need for augmented evaporation of water. The small body of the squirrel warms much more rapidly than that of the camel, however, and the cycle of heating and cooling is repeated at short intervals rather than extending over the length of the day.

A number of other examples of animals that exploit elevation of body temperature to ameliorate their situation in hot environments will be discussed in Chapter 5.

The control of homeothermy

As we have seen, the regulation of body temperature in birds and mammals involves a great diversity of effector mechanisms, such as pilomotor and vasomotor responses, shivering, nonshivering thermogenesis, panting, and sweating. These mechanisms must be controlled in an integrated fashion so as to maintain appropriate body temperatures. Thus in addition to the effector mechanisms, which actually modify rates of heat production and dissipation, the homeotherm must also possess thermal sensors, which provide information on its current thermal status, and control centers, which process inputs from the sensors and activate the effectors as needed.

The subject of thermoregulatory control has been receiving increasing attention in recent years, and though this is not the place for an extended treatment, some of the major findings deserve note. Many experiments indicate that the major centers responsible for integration of sensory information and activation of the thermoregulatory effectors are located in the hypothalamus and its associated preoptic tissues. Lesions in these parts of the brain disrupt or abolish normal regulation of body temperature. Information used to assess the current thermal status of the animal and thereby govern appropriate thermoregulatory responses comes from many parts of the body. Significantly, the preoptic-hypothalamic region is itself thermally sensitive, thus providing for sensation of brain temperature. Additionally, thermal information is known to be derived from thermal receptors or thermally sensitive neurons in the skin, deep abdomen, and spinal cord of various species. In certain ungulates, information from receptors in the scrotum or udder exerts a powerful effect on thermoregulatory responses, and receptors have been implicated in the lining of the upper respiratory tract of dogs. Although our knowledge of thermal receptors is incomplete, enough is thus known to show that the hypothalamus gains information on the animal's thermal status not only by being responsive to its own temperature, but also by receiving inputs from receptors located in diverse parts of the body. A major topic of current interest is how all these sensory inputs are integrated in determining the thermoregulatory responses to be elicited at any given time.

One working hypothesis is that of modulated hypothalamic set-point temperatures. This may be outlined by a review of experiments performed on dogs equipped with an apparatus for artificial control of their hypothalamic temperature. In an animal resting at a constant air temperature of 23°C, hypothalamic temperature was gradually lowered, starting from a level in excess of 38°C. At first no response was observed, but when hypothalamic temperature reached about 36.8°C, shivering was elicited, and metabolic heat production showed an increase. As hypothalamic temperature was dropped below 36.8°C, the rate of heat production increased steadily as the difference between hypothalamic temperature and 36.8°C became larger. This experiment demonstrates several important points. (1) The hypothalamus is sensitive to its own temperature in regulating shivering. (2) It behaves as if there is a threshold temperature or set point for eliciting shivering; shivering is stimulated only when the hypothalamic temperature is 36.8°C or below. (3) The response is not a simple "on-off" response; that is, heat production by shivering does not jump from zero to a high level as soon as the threshold temperature is crossed. Instead, the response is graded, being small at first and increasing as hypothalamic temperature falls further below the threshold.

Recall now that the preceding experiments were performed at an air temperature of 23°C. When the experiments were repeated at an air temperature of 13.5°C, the results were qualitatively similar, but there was one most significant difference: the threshold or set-point temperature was 38.8°C, not 36.8°C. In other words, the dog started to shiver at a higher hypothalamic temperature when in a cooler environment. This type of result gives rise to the idea that receptors in the skin, responding to air temperature, provide information that is used to modify, or modulate, the hypothalamic set point. The direction of the modulation is that which would be expected intuitively. Low air temperatures, sensed by receptors in the skin, signify a greater potential for cooling than higher air temperatures. It makes intuitive sense that the animal in the cooler environment should not allow its deep (hypothalamic) temperature to fall as low before eliciting a homeostatic response (shivering) as the animal in a warmer environment. This, in fact, is precisely the effect of the modulation of the hypothalamic set point.

According to the modulated set-point hypothesis, each type of thermoregulatory response has its own hypothalamic set point, which is modified by inputs from peripheral thermoreceptors. Returning to the dog discussed above, it was found, for example, that panting was elicited by warming of the hypothalamus above 38.8°C when the dog was in air at 23°C. The shivering threshold at this same air temperature, you will recall, was different (36.8°C). After transferring the dog to air at 13.5°C, the panting threshold was raised to about 41°C. That is, the dog in the cooler air allowed its hypothalamic temperature to rise to a higher level before panting than the dog in the warmer air—again a result that would follow intuitive expectations.

The concept of modulated hypothalamic set points provides a model for understanding many of the results that have been obtained on interactions between hypothalamic and peripheral temperatures in the control of thermoregulation. Studies on man, for example, have indicated much the same type of interaction between deep head temperature and skin temperature as in the dog; both shivering and sweating are elicited at lower hypothalamic temperatures when the skin is warm than when it is cool. There is also evidence that information other than that from thermal receptors can affect hypothalamic set points. This is illustrated by results on exercising dogs. Upon the onset of exercise in a warm environment, dogs immediately increase their evaporative water loss through panting even though hypothalamic temperature has not as yet been affected by the increased rate of metabolic heat production. It is postulated that exercise, perhaps acting through proprioceptors in the joints, acts to lower the hypothalamic panting set point. Thus the prevailing hypothalamic temperature, though unaltered, comes to be further above the set point than during rest, thereby eliciting more panting.

Experiments on resting cats and dogs have revealed that under normal circumstances, hypothalamic temperature does not rise or fall appreciably when ambient temperature is changed. Dogs, for example, showed no significant change in hypothalamic temperature when transferred from a hot (35°C) to a cold (10°C) environment. These results indicate that panting and shivering are not elicited under normal circumstances by a rise or fall of the deep, hypothalamic temperature. How, then, are these thermoregulatory responses triggered? The modulated set-point hypothesis postulates much the same type of interactions between hypothalamic temperature and set points as indicated earlier in discussion of exercise. In a hot environment, information from the skin and perhaps other peripheral regions causes a lowering of the set point for panting, as we have seen. Thus it is postulated that the set point comes to be below the stable hypothalamic temperature, and panting begins. Conversely, peripheral inputs cause both the panting and shivering set points to rise in a cold environment. Thus it is postulated that upon transfer to a cold environment, panting ceases as the panting set point comes to be above hypothalamic temperature, and shivering begins as the shivering set point moves above the hypothalamic temperature. In short, deep hypothalamic temperature can remain constant while the set points for thermoregulatory responses are shifted around it through the effects of peripheral inputs. Indeed, it is this process which allows deep temperatures to remain constant; inputs from the skin rapidly signal changes in the environment and elicit thermoregulatory responses appropriate to maintenance of internal homeostasis.

Although the modulated set-point hypothesis is compatible with many experimental observations on the control of thermoregulation, alternative hypotheses have been put forward. Further, information that allows rigorous analysis of thermoregulatory control is available for only a relatively few species, and even in those species only certain aspects of thermoregulation have as yet been studied. Thus there is much to be learned, and we can expect significant new developments over the decades to come.

It is worth noting before leaving this topic that experiments on blue-tongued lizards (*Tiliqua scincoides*) have revealed that they have thermal sensors in both peripheral areas of the body and the preoptic region of the brain, and information from these sensors is integrated to control behavioral responses to temperature. Given a choice between a cold (15°C) and hot (45°C) environment, the lizards would move back and forth, typically remaining in the hot environment until rectal and preoptic temperatures had risen to 37°C, then retiring to the cold environment until these temperatures had fallen to 30°C. When the preoptic temperature was artificially raised to 41°C, the lizards left the hot environment at a rectal temperature somewhat lower than 37°C; and when the preoptic temperature was lowered to 25°C, they left the hot environment at a rectal temperature somewhat higher than 37°C. These results indicate that both preoptic and peripheral (rectal or skin) temperatures are sensed and interact in the control of behavioral thermoregulation. This suggests that the controlling and sensory mechanisms seen in homeotherms had evolved in vertebrates in at least their rudimentary forms before homeothermy itself appeared on the scene. If so, these mechanisms could simply have been appropriated, with refinement, to the control of physiological processes of thermoregulation.

Behavioral responses to temperature in birds and mammals

Behavior is no less important to thermoregulation in homeotherms than it is in poikilotherms. Many mammals and birds in cold surroundings seek refuge in tree cavities or other holes and crevices. Burrowing under the ground or snow is common among small mammals, and a few birds, such as the willow ptarmigan and ruffed grouse, tunnel in snow. Recent studies have shown that some birds may reduce the metabolic effort required to keep warm in cold environments by exposing themselves to the sun; presumably, many homeotherms capitalize on this possibility. Nest building is common, and huddling is reported in some species. Numerous species of birds and some mammals migrate to warmer regions during the cold of winter.

Many behavioral responses to heat stress are also known. Small mammals in the desert are generally nocturnal, spending the hot daytime hours in burrows. An interesting consequence of this behavior is that the greatest thermal challenge to these animals may well be the cold of winter nights rather than the summer heat that we usually associate with deserts. Larger mammals cannot burrow, and their encounter with hot environments is thus fundamentally different. Most birds, both small and large, are diurnal and must also face, to some extent, the maximal thermal stresses of their surroundings. Shade-seeking behavior is common. Many small birds spend the hottest parts of the day in cavities or crevices. Soaring birds may find less stressful conditions at high altitudes. Camels often sit with their legs tucked beneath their bodies and orient themselves so as to face the sun, turning as the sun moves across the sky. In this way they expose as small an area of their body as possible to direct solar radiation. Many birds keep their backs oriented toward the sun. Here again they avoid the broadside orientation that would maximize radiative heat input, and they also keep the exposed, vascular membranes of their gular area and feet in the shadow created by their head, neck, and body. Many birds and mammals restrict activity during the hottest parts of the day. This response not only minimizes metabolic heat load, but allows them to remain continuously in favorable microenvironments or postures.

Effects of body size on thermal relationships

We have often alluded to the fact that small and large homeotherms find themselves in different situations both physiologically and behaviorally when confronting warm and cold environments. It is the purpose of this section to summarize these important relationships and contribute several additional concepts.

In a strictly physiological sense, large size often has clear advantages in coping with the exigencies of heat or cold. To a substantial extent this

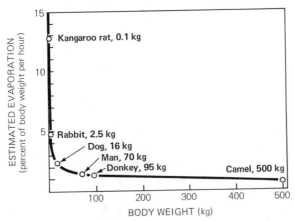

Figure 3–19. Relation between body size and evaporative water loss estimated to be necessary for mammals to maintain a stable body temperature near 37°C during exposure to the hot conditions of the day in the Nevada desert. Note that the unit of measure for evaporation (percent of body weight per hour) is a weight-specific expression of evaporative water loss (g H_2O/100 g body weight/hr). (From Schmidt-Nielsen, K. 1964. *Desert Animals*. Clarendon, Oxford.)

follows from the frequently mentioned relationship between size and surface-to-weight ratio. Because the rate of thermal exchange with the environment by conduction, convection, and radiation is, to a considerable extent, dependent on surface area and because larger animals have less surface area per unit of weight, it follows that—other things being equal—the larger animal loses or gains heat at a lower rate per unit of body mass than the smaller animal. In cold surroundings the larger animal can offset heat losses with a lower weight-specific metabolic rate, whereas in hot surroundings it can offset heat gains with a lower weight-specific loss of evaporated water.

We have earlier discussed in some detail the effects of body size on metabolism and conductance in moderate to cold environments. Looking at hot environments, we find that smaller animals are at a less advantageous physiological position than larger animals not only because of their higher surface-to-weight ratios, but also because of their higher weight-specific metabolic rates. That is, the weight-specific heat load of the small animal under hot conditions is enhanced relative to that of the large animal both by a greater weight-specific input of heat from the outside world and by a greater weight-specific internal production of heat. By quantifying these factors, Schmidt-Nielsen developed the theoretical relationship shown in Figure 3–19 between body size and the weight-specific loss of evaporated water needed to dissipate total heat load under exposed conditions during the summer day in the Nevada desert. In interpreting this relationship it is important to note that many mammals die under hot conditions when they have lost water equivalent to about 15% of their body weight, though camels, donkeys, Merino sheep, and many other inhabitants of hot climates can tolerate dehydration of over 25%. We see that a small mammal, such as the 100-g kangaroo rat, would have to dissipate water at an inordinate rate in order to maintain a stable body temperature through evaporative cooling under the hot conditions of the desert day; even if it could tolerate dehydration of 25% of its body weight, it would succumb in a matter of only two hours. Clearly, small desert inhabitants cannot afford to expose themselves to the heat of the day for any length of time and keep cool evaporatively. This perhaps helps to explain why rodents, even in the desert, have failed to evolve mechanisms for great augmentation of evaporative cooling through sweating or panting. Most are nocturnal and remain in their burrows during the heat of the day, thus avoiding rather than confronting the thermal extremes of their environment. Larger mammals are in a far more advantageous position, and species such as the dog, donkey, and man are known to maintain stable body temperatures evaporatively in the desert at a cost of body water equivalent to 1–3% of their body weight per hour. Camels do considerably better than indicated in Figure 3–19, for they not only have the advantages of large size, but simply store a great deal of heat during the day by allowing their body temperature to rise. In birds, as in mammals, the weight-specific water cost of maintaining a given difference between body temperature and ambient temperature in a hot environment increases with decreasing size. Small birds, which are mostly diurnal, routinely allow their body temperature to rise to 44°–45°C in hot surroundings, however, and this reduces the heat load and may even establish a thermal gradient favorable to conductive, convective, and radiative heat loss. Hyperthermia is fundamental to their ability to survive in hot environments without incurring fatal demands on their water resources for evaporative cooling.

Although larger birds and mammals frequently meet the stresses of heat and cold from a more advantageous physiological position than smaller forms, the smaller species have the outstanding behavioral advantage of being more readily able to avoid thermal extremes by moving into burrows, cavities, shade, and other more equitable microhabitats. The larger animal is often required to meet the full severity of the thermal environment. Further, the large animal must find and consume more food and water than the small animal, and, unlike its smaller relatives, it cannot readily establish food stores adequate to meet its needs over stressful periods of summer or winter. These considerations are important in view of the relative difficulty of finding food or water in many thermally stressful circumstances, such as the desert or arctic. An ameliorating factor is the greater mobility of large species as compared to small species: Larger animals can travel more extensively in search of the food and water they need. Large desert mammals undoubtedly benefit from their ability to seek out water holes which may be scattered widely; this is an important factor in their capacity to rely on evaporative cooling as a protection against overheating. The potential advantages of small size, on the other hand, are illustrated by some animals living in areas of severe cold. Deep snow cover may be life threatening to large herbivores, which cannot simply burrow down to the plants that they depend on for food. In the Rocky Mountains, elk, bighorn sheep, and other large species routinely leave the highest altitudes during winter, moving to areas below timberline where the trees protect them from high winds and food can be found above the surface of the snow. Small pikas and some small rodents, on the other hand, remain active all winter above timberline, burrowing under the snow to avoid the intense cold above and consuming food stored from the previous summer. In total, the interplay of physiological and behavioral factors is such that no clear net advantage rests with large or small size, though different sizes have different implications for the interaction between the organism and its thermal environment.

It should be recognized that the size of an animal, although important, is not deterministic in its effects on thermal relationships. Animals of similar size can differ in many parameters that affect success in coping with thermal extremes, including insulatory capabilities, capacities to increase metabolic heat production in cold surroundings, capacities to augment evaporative water loss, capacities to exploit variations in body temperature, and behavioral responses. Some of these differences will receive attention in the next sections.

Climatic adaptation

Species of birds and mammals have become genetically adapted in their evolution to the thermal conditions of their habitats, a phenomenon known as climatic adaptation. The types of adaptation that have occurred are best elucidated by comparing related forms of similar size that live under different thermal regimes.

One parameter that has received attention is the insulation provided by the pelage or plumage. Mammals of the Arctic typically have thicker pelage than similarly sized mammals of the warm tropics, and it has been demonstrated that the insulation provided by the pelage is correspondingly greater in the arctic forms. The 5-kg arctic white fox (*Alopex*) in winter, for example, was found to have pelage about 5 cm thick, whereas a small Panamanian deer (*Mazama*) of higher weight had pelage less than 0.5 cm thick; the insulatory value of the pelt proved to be five to six times as high in the fox as in the deer. Although medium-sized and large arctic mammals have winter pelts that range from 3 to 7 cm thick, small arctic ground squirrels, lemmings, and weasels may have pelts only 0.5–2 cm thick; in general, the differences between arctic and tropical forms are greater in species of medium to large size than in small species. These differences between large and small forms probably reflect the fact that small species are limited in the maximal pelage thickness they can support without encumbrance. Most birds are small and thus suffer similar constraints on plumage development in cold climates; further, the conformation of the plumage is of aerodynamic importance during flight, a factor that can place constraints on the development of the plumage as a purely insulatory covering. Although the insulation of the plumage has defied rigorous quantitative analysis, various workers have expressed the opinion that arctic and tropical birds do not appear to differ as much in plumage development as mammals do in pelage development. There is evidence that birds that spend the winter in far northern climates have thicker plumage than species that migrate southward. Some larger birds of the north, such as ptarmigan and brant, have strikingly thick plumage, and feathered legs and feet are more common among northern resident birds (such as ptarmigan and various owls) than in species that migrate to the south in winter.

Figure 3–20 shows the results of a classic comparison of metabolic relationships to temperature in arctic and tropical mammals. It is apparent that the tropical forms have higher lower critical temperatures than the arctic forms and must increase their metabolism proportionately more above the basal level for any given drop in temperature below lower critical temperature. Because the metabolic rates in Figure 3–20 are expressed relative to basal rates rather than in absolute units, differences among species

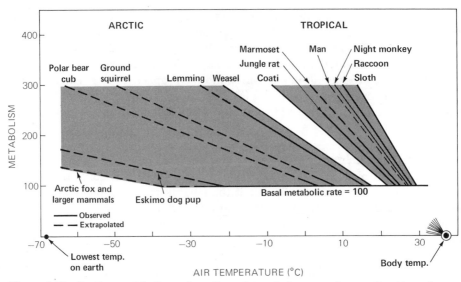

Figure 3–20. Resting metabolism of arctic and tropical mammals as a function of ambient temperature. All curves are adjusted to a common standard, such that basal metabolism equals an arbitrary value of 100. [From Scholander, P. F., R. Hock, V. Walters, F. Johnson, and L. Irving. 1950. Biol. Bull. (Woods Hole) 99: 237–258.]

in the slope of the metabolism-temperature curve below thermoneutrality do not quantitatively reflect differences in maximal overall insulation. It is clear, though, that some of the arctic forms attain very high levels of insulation, and the energetic advantages accruing to them as a result are obvious. In the extreme cases of the eskimo dog and arctic fox, metabolism need not be elevated above the basal level until ambient temperature has fallen to −20°C or below, and resting metabolic rate would apparently need to exceed basal by less than a factor of two even at the lowest temperature ever recorded in the arctic, around −70°C. Birds do not match these mammalian extremes; even such well-insulated forms as black brant and ptarmigan (see Figure 3–14) have lower critical temperatures somewhat above 0°C. Lest the misconception be gained that all northern species avoid the need for high metabolic rates in winter, it should be noted that many smaller species have rather high lower critical temperatures. That of the lemmings in Figure 3–20, for example, is about 15°C. The evening grosbeak has a lower critical temperature near 16°C, and redpolls and black-capped chickadees have critical temperatures near 24°C; all three of these small birds overwinter at high latitudes. Clearly, such animals must spend much of the winter at substantially elevated metabolic rates, with the concomitant need to find adequate food.

It might be expected that body temperature and basal metabolic rate would be adapted to climatic conditions. Animals in cold surroundings might benefit by sustaining relatively low body temperatures and by having relatively high basal metabolic rates. Converse arguments could be made for species in warm surroundings. Body temperatures do vary somewhat among species and larger taxonomic groups. Monotremes have lower body temperatures than placental mammals; and penguins, grebes, and petrels have lower temperatures than passerines and galliforms. There is, however, no clear correlation between body temperature under thermally moderate conditions and climate. It seems that temperature became fixed rather early in the evolution of the various taxa and was not adaptively modified during subsequent radiation into thermally different habitats. It must be recognized, nonetheless, that some mammals of hot environments have evolved an ability to tolerate higher degrees of hyperthermia than species of more temperate climates. In birds, on the other hand, the ability to tolerate hyperthermia appears to be about the same in species from both hot and temperate climates.

There is no evidence for elevation of basal metabolic rate (BMR) in either avian or mammalian residents of cold climates. In fact, there is little or no reason to expect such an elevation, for the normal basal rate is adequate for thermoregulation in the thermoneutral zone, and an elevation of BMR, although it might extend the thermoneutral zone to some extent, would needlessly increase energy demands at temperatures where it is not required. By contrast, there are cogent theoretical reasons for believing that a lowered BMR could be of advantage in hot climates; a species that could allow its metabolic heat production to fall to unusually low basal levels would reduce its overall heat load under thermally stressful conditions. The approach used to determine if a species has an unusual BMR is to compare its BMR with that expected for its body size according to the standard relationship between BMR and weight for its taxonomic group (see Chapter 2). By this criterion, many avian and mammalian residents of hot climates have ordinary BMR's, but others do, in fact, have significantly lower BMR's than expected for their size. Relative to the standard relationship between metabolism and weight for placental mammals, for example, BMR's of desert and semidesert rodents average about 10% below expected values, whereas those of mesic rodents average about 10% above expected values; thus there is a significant tendency for the desert and semidesert forms to have reduced BMR's. Perhaps the most extreme case is found in the naked mole rat of Africa (*Heterocephalus*), which has a BMR only 20–40% as high as expected for its size. Among birds, the poorwill (*Phalaenoptilus*), a caprimulgid of our western deserts, has a BMR only about half as high as would be predicted. This low rate of metabolic heat production, coupled with an efficient gular flutter mechanism, allows it to dissipate heat evaporatively at a rate that exceeds metabolic heat production by a factor of over three in dry, hot air. Accordingly, a substantial environmental heat load can be dissipated in addition to the endogenous heat load. Other caprimulgids of both hot and temperate climates also tend to have low BMR's, indicating that low basal rates may not be so much a specific adaptation to desert life as a general property of the taxonomic group.

Acclimation and acclimatization

Acclimation and acclimatization to thermally different environments often result in adaptive alteration of thermoregulatory parameters in birds and mammals, though it cannot be emphasized too much that the effects of laboratory acclimation are commonly different from those of seasonal acclimatization.

Figure 3–21 provides a basic frame of reference for interpreting results on acclimatization to cold seasons in birds and mammals. One possible response, shown in part *A*, is for the capacity to produce heat to increase, with

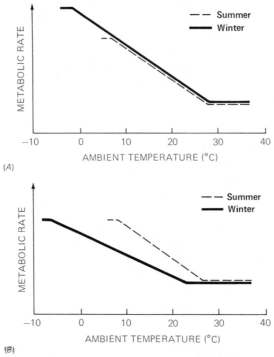

Figure 3–21. Simple metabolic acclimatization (*A*) and simple insulatory acclimatization (*B*). The plateau of each curve at the left indicates where metabolic rate has been maximized.

little or no change in insulatory characteristics. In this case the resting metabolic rate required for thermoregulation at any given temperature is unaltered, but the cold-acclimatized animal can maintain its body temperature down to lower ambient temperatures than the warm-acclimatized animal by virtue of an increased ability to augment heat production. This type of response is termed simple *metabolic acclimatization*. The enhanced ability to increase metabolic rate might, for example, result from an increase in capacity for nonshivering thermogenesis elicited by cold acclimatization. Another possible response, shown in part *B,* is for maximal insulation to increase, with little or no change in the peak capacity to produce metabolic heat. Here the resting metabolic rate required for thermoregulation at any given temperature below thermoneutrality is reduced, and even though metabolic capacity remains unchanged, the cold-acclimatized animal can again maintain body temperature down to lower ambient temperatures than the warm-acclimatized animal through its increased ability to retard heat loss to the environment. This is termed simple *insulatory acclimatization*. The responses illustrated in Figure 3–21 represent only certain simplified possibilities. Thus, for example, a species could exhibit *both* insulatory and metabolic acclimatization and thereby extend its viable ambient temperature range to lower limits by virtue of the advantages conveyed by both increased insulation and increased metabolic capacity.

    Many species of mammals and some species of birds exhibit clear insulatory acclimatization during winter, either with or without metabolic acclimatization. An example is provided in Figure 3–22. Changes in vaso-

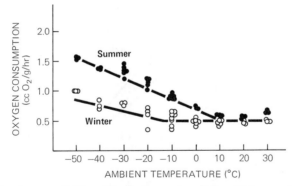

Figure 3–22. Resting metabolism of a single arctic red fox (*Vulpes vulpes*) during summer and winter. (From Hart, J. S. 1957. Rev. Can. Biol. 16: 133–174; based on data of Irving, L., H. Krog, and M. Monson. 1955. Physiol. Zool. 28: 173–185.)

motor control of peripheral heat flow have sometimes been implicated in insulatory acclimatization. A factor more widely known is that many species molt to a thicker, more insulating pelage or plumage in winter. As a generalization, the increase in insulatory value of the fur is greater in large mammals than in small; again the constraints of small body size appear to come into play. In a series of mammals of high latitudes, black bears and wolves exhibited an increase of 40–50% in the insulatory value of their fur in winter, whereas smaller forms such as deer mice and hares exhibited increases of only 15–20%. The molt of the pelage or plumage as winter approaches is, to a large extent, under photoperiodic control.

As illustrated in Figure 3–23, some birds and mammals display only metabolic acclimatization in winter. Among mammals, metabolic acclimatization without insulative changes occurs predominantly in small species. In birds, however, the picture is mixed; the chickadee and cardinal show insulatory acclimatization in winter, but a much larger species, the black brant, as well as some small species exhibit no significant changes in insulation as judged from the metabolism-temperature curve.

As winter approaches, animals in nature are exposed to gradually de-

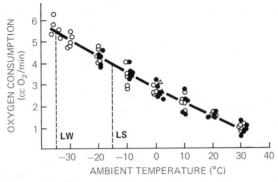

Figure 3–23. Resting metabolism of deer mice (*Peromyscus maniculatus*) during summer (closed symbols) and winter (open symbols). Vertical dashed lines indicate temperatures at which survival is limited to 200 minutes in summer (LS) and winter (LW); note that winter animals survive for this time at a temperature 20°C lower than that for summer animals. (From Hart, J. S. 1957. Rev. Can. Biol. 16: 133–174.)

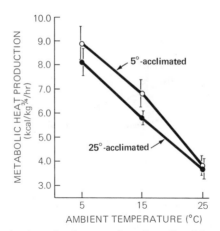

Figure 3–24. Metabolic heat production as a function of ambient temperature in laboratory rats (*Rattus norvegicus*) acclimated for 60 days to 5°C (upper line) or 25°C (lower line). Points represent mean values; vertical bars depict ±1 standard deviation around the mean. (From Cottle, W. and L. D. Carlson. 1954. Amer. J. Physiol. 178: 305–308.)

clining temperatures, to shortening daylength, and often to changes in nutritional regime and other factors. Any or all of these changes can exert effects on acclimatization. By contrast, cold acclimation in the laboratory generally involves just change in temperature, and the responses of cold-acclimated animals are commonly different from those of winter-acclimatized animals of the same species. Frequently, as in acclimatization, there is an increase in metabolic capacity during cold acclimation that acts to extend the viable range of ambient temperatures. However, overall insulation below thermoneutrality is generally either unchanged or *decreased*. This latter phenomenon, which is quite common, is illustrated in Figure 3–24; note that the cold-acclimated individual expends *more* energy at a given temperature below thermoneutrality than the warm-acclimated individual. Among small mammals there is not uncommonly an increase in resting metabolic rate at thermoneutral temperatures as well during cold acclimation. Several factors can be involved in explaining these phenomena. Commonly, there is little or no change in the thickness of the pelage or plumage over several weeks of cold acclimation (though there may be an increase over longer periods); among other considerations, appropriate photoperiodic stimuli for molting are absent. Although the lack of increase in pelage or plumage thickness helps to explain why insulation does not increase during cold acclimation of several weeks, it does not account for the decrease in insulation that is commonly observed. In rats and rabbits, increased blood flow has been observed in such peripheral areas as the ears, appendages, and tail during cold acclimation. This maintains these parts at somewhat higher temperatures than would otherwise be the case but also increases loss of heat to the environment and decreases insulation. Recently evidence has accumulated to indicate that the elevation of metabolism at given temperature in cold-acclimated rats is due to increased thyroid activity and can be abolished experimentally without apparent detriment to the animal. The latter result raises the question of whether the increased rate of heat production and loss provides any adaptive advantage or is, in itself, a nonadaptive by-product of hormonal or other responses to cold. At present, no definitive conclusion can be reached.

To summarize, both acclimatization and acclimation enhance abilities to live under cold conditions. The chief factor in cold acclimation is an increase in metabolic capacity, the metabolic expenditure at given temperature being either unaltered or increased. Winter acclimatization also is commonly accompanied by an increased metabolic capacity but, in addition, frequently involves insulatory changes promoting metabolic economy.

Acclimation and acclimatization to heat have been less well studied than the responses to cold, most effort having been directed to man and his domesticated animals. A few examples will serve to illustrate the types of insight obtained thus far. Humans exposed to heat develop a greater capacity to work under hot conditions and experience less discomfort than individuals acclimated to cooler temperatures. Capacity to sweat may double; the salt content of sweat decreases; and sweating commences at lower skin temperatures. Cattle, sheep, and chickens become more resistant to heat during summer months. Reduced pelage insulation is important in cattle and sheep, and there is evidence that the summer pelage of some breeds of cattle absorbs a lesser fraction of incident radiation than winter pelage. Among chickens, White Leghorn hens were found to pant at a greater frequency under hot conditions during summer than during winter, presumably reflecting an increased capacity for evaporative cooling; the summer birds became hyperthermic less rapidly and survived longer than the winter birds. In rodents there is evidence for thinning of the hair and a reduction of basal metabolic rate during heat acclimation in some species. There is also evidence for a reduction of basal metabolic rate during heat acclimation in some birds, but whether this occurs during summer acclimatization remains uncertain.

The ontogeny of homeothermy

Apparently, no birds or mammals are born with the full thermoregulatory capabilities of the adults of their species. Thus thermal and energetic relationships to the environment change during postnatal development. Some species, such as the domestic pig, domestic sheep, and caribou, exhibit substantial capabilities of thermoregulation at birth. Three newborn caribou, for example, remained homeothermic during direct exposure to a 12 mph wind at 1°–3°C for at least 12 hours; one animal, in fact, was thoroughly wet with amniotic fluid. Nonetheless, such animals exhibited a rise in metabolic rate when ambient temperature dropped from 20°C to 0°C in dry, still air, whereas a 9-month-old calf exhibited no rise down to at least −55°C. Thus there is a considerable improvement in overall insulation and metabolic economy in this species over the early months of life. At the other extreme, young marsupials function essentially as poikilotherms for a month or two after birth; their body temperature equals ambient temperature, and their metabolic rate falls as temperature falls. Young opossums (*Didelphis*), for example, cannot maintain high body temperatures even at the mild ambient temperature of 27°C until about 80 days of age. Thereafter, homeothermic capabilities increase rapidly, and at 92 days of age they can thermoregulate at 5°C for two hours.

Most young birds and mammals display at least some thermoregulatory competence within the first weeks after birth, but the capacities of the newborn are meager in diverse groups, including small rodents, passerine birds, herons, rhesus monkeys, and domestic rabbits. The type of gradual transition in thermoregulatory characteristics found in such forms is illustrated by the data on vesper sparrows in Figure 3–25. Note that nestlings aged zero to two days exhibit essentially poikilothermic responses in both body temperature and resting metabolic rate. Gradually, over the period from three to seven days of age, they become able to thermoregulate at lower and lower ambient temperatures. A week after hatching they can maintain high body temperatures at 10°C and show a typical homeothermic relationship between metabolism and ambient temperature down to 10°C. To provide another example, two-day-old white-footed mice (*Peromyscus leucopus*) can maintain high body temperatures for 2.5–3.0 hours only at ambient temperatures of 30°C and above; their body temperature falls when they are tested at 25°C and closely approximates ambient temperature at 20°C and below. The capacity to thermoregulate is extended only to 25°C by eight days of age but thereafter improves rapidly, and thermoregulation at 0°C is possible at 18 days of age. Their development of homeothermy over the nestling period rests largely on considerable increases in both insulation and metabolic capacity. As might be expected, young animals that have only meager abilities to thermoregulate generally display a high tolerance to reduced body temperatures. Again to use white-footed mice as an example, young animals readily survive cooling to body temperatures of 1°–2°C for two hours even though they cease breathing. Such tolerance to lowered body temperatures often diminishes with age; adult white-footed mice, for example, are killed by body temperatures below 13°–15°C.

As is true in so many areas of biology, data gathered on the newborn under laboratory conditions can be appraised fully only by having a thoroughgoing understanding of circumstances in nature. At present, our understanding of these circumstances is instructive but incomplete. Young with meager thermoregulatory capabilities are typically incubated by their parents and thus may spend much of their lives at adult body temperatures despite their own inadequacies. Sometimes parental care is immediately essential to life. Certain birds with altricial young, such as some herons and pelicans, for example, nest in exposed sites in warm climates. The young quickly die of overheating if not shaded from solar radiation by their parents. Recent experiments on white-footed mice indicate that typical laboratory studies may often not provide a proper understanding of thermoregulation in young independent of their parents. Four-day-old mice studied alone without a nest were unable to maintain adult body temperatures at ambient temperatures of 25°C or lower. The same young studied in groups of four with a nest, however, were able to thermoregulate for at least 2.5 hours at ambient temperatures down to 5°C. Thus litters of white-footed mice developing within a nest under cool conditions in nature have considerably greater capabilities for maintaining elevated body temperatures than studies of individual young would suggest. When it is considered that they are incubated by their mother for much of the time, it becomes clear that they are exposed to the possibility of hypothermia for, at most, only a few hours of each day despite the very limited abilities of young mice to

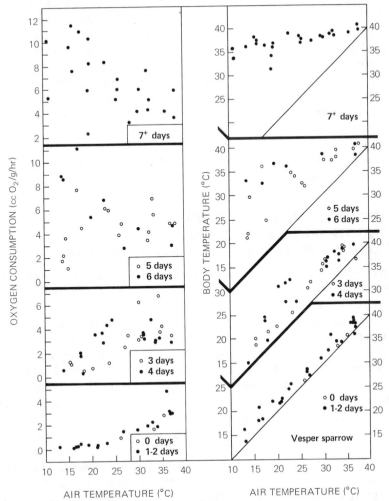

Figure 3–25. Rate of oxygen consumption (left) and body temperature (right) as functions of ambient temperature in nestling vesper sparrows (*Pooecetes gramineus*) at various ages after hatching. Birds were studied individually, and all measures were taken after about two hours of exposure to experimental ambient temperatures. The diagonal lines on the righthand side of the figure are lines of equality between body temperature and ambient temperature. Vesper sparrows fledge at an average of about ten days of age. (From Dawson, W. R. and F. C. Evans. 1960. Condor 62: 329–340.)

maintain high body temperatures when tested alone.

There has occasionally been a tendency to view the immaturity of thermoregulation in many young homeotherms as a "primitive" characteristic. This has been especially evident in some discussions of birds in which species, such as passerines and herons, with meager capabilities for thermoregulation at hatching have been considered, in some sense, inferior to species, such as certain gulls and ducks, having better-developed capabilities. This is one of those areas of biology in which it seems shortsighted (and often anthropocentric) to attempt to make a value judgment. Without ascribing any superiority to the less mature thermoregulatory condition, it is worthwhile to note some of the potential advantages of incomplete homeothermy during early development. As we have repeatedly emphasized, small animals must typically, in weight-specific terms, expend more energy to thermoregulate in the cold and more evaporated water to thermoregulate under hot conditions than large animals. The parent must, of course, provide food and water resources for the young. If the young carry out their own thermoregulation, they do so at considerable expense to their resource supplies. Nutrient energy that could be diverted to growth and development must be channeled to temperature regulation, and demands for resources from the parent may be increased. But if the parent assumes responsibility for maintaining proper temperatures in the young, it often does so with but a trifling added demand on its own resources because it would have to thermoregulate anyway. And the resources it delivers to the young may be directed more exclusively into growth and development.

THERMOREGULATORY AND ENDOTHERMIC PHENOMENA IN LOWER VERTEBRATES AND INVERTEBRATES

As stressed at the beginning of this chapter, homeothermy and poikilothermy are *types of thermal relationships* between the animal and its environment. Some animals so predominantly show one type of relationship or the other that we call them poikilotherms or homeotherms. It has become increasingly clear, however, that many animals exhibit both types of relationship at different times or exhibit thermal relationships that are not adequately described as either poikilothermy or homeothermy. In this and the next section we examine some of these phenomena. Before proceeding, it should be helpful to distinguish clearly between the two concepts mentioned in the title of this section: thermoregulation and endothermy. Thermoregulation is the maintenance of a more or less stable body temperature, whereas endothermy is the maintenance of a difference between body temperature and ambient temperature through metabolic heat production. In birds and mammals these two go hand in hand; thermoregulation at low ambient temperatures is dependent on controlled endothermy. However, the two are not necessarily coupled. Some animals thermoregulate without depending on internal heat production, and some maintain their body temperature above ambient by virtue of metabolic heat but do not control the rates of heat production and loss in such a way as to maintain a stable body temperature.

We have noted that many poikilotherms, though lacking physiological means of thermoregulation, can exhibit some control over their body temperature by behaviorally selecting certain thermal environments. Frequently these behavioral efforts result in some stabilization of body temperature despite variations in ambient conditions, and the process is then termed *behavioral thermoregulation*. It is a common phenomenon, being well known, for example, in such diverse groups as insects, fish, amphibians, and reptiles. Here we may pay special note to the lizards, which exhibit some of the more sophisticated capabilities now known. Placed in a uniform thermal environment in the laboratory, lizards typically have body temperatures close to ambient—the characteristic response of poikilothermy. In nature, however, they typically exploit possibilities for heating and cooling during the day to maintain relatively stable body temperatures that may be considerably higher than air temperature. Desert lizards, for example, emerge in the morning and bask in the sun until their temperature has risen to within the usual range maintained during daily activity. Thereafter, they maintain their temperature within this range by a variety of mechanisms. Frequently they are observed to shuttle back and forth between sun and shade. They also can modify the amount of their body surface exposed to the direct rays of the sun by changing their posture and orientation to the sun. They may flatten themselves against the substrate to lose or gain heat, depending on substrate temperature, and when the substrate has become very hot during midday, they may minimize body contact by elevating all but their feet off the ground or even climbing on bushes. If circumstances above ground become so uniformly hot that body temperature cannot be held below threatening levels, the lizard retreats underground. By exploiting the numerous opportunities for heating and cooling, many species can maintain body temperatures that vary by only a few degrees centigrade over long periods during the day. The mean temperature depends on species. Desert iguanas (*Dipsosaurus*), for example, maintain very high body temperatures, averaging about 42°C, whereas desert spiny lizards (*Sceloporus magister*) maintain temperatures near 35°C. Poikilotherms that behaviorally thermoregulate clearly possess one of the characteristics that has perhaps been more commonly associated with homeothermy: a *thermal control system* which allows them to monitor their own thermal status and call into action mechanisms that serve to maintain temperature between certain upper and lower limits.

Behavioral thermoregulators are at the mercy of the thermal opportunities available in the outside world. The lizard in the desert, for example, may be unable to attain temperatures within its usual activity range on a cool, cloudy day. Physiological mechanisms of thermoregulation provide increased independence from the vagaries of the environment, and it has been most interesting to discover that such mechanisms are found among several groups of lower vertebrates and invertebrates.

The lizards are of some particular interest because they may provide particular insight into the thermoregulatory potentialities of the reptiles that gave rise to birds and mammals millions of years ago. Panting or vigorous gular pumping movements in response to heat stress have been reported in many species and can provide a sufficient rate of evaporative water loss to dissipate heat somewhat more rapidly than it is produced in metabolism. Another type of physiological response reported not only among lizards but also in turtles is a capacity to warm more quickly than they cool for a given thermal gradient between the body and environment. In the marine iguana (*Amblyrhynchus*) of the Galapagos Islands, to cite a particularly marked example, cool animals in a warm environment increase in body temperature about twice as fast as warm animals in a cool environment decrease in body temperature. In general, such responses allow lizards to remain at high body temperatures for more time each day than would otherwise be possible; they heat up relatively rapidly in the morning but cool down relatively slowly in the evening. Marine iguanas spend much of their time under hot conditions on land but forage for food in the sea, where water temperatures may be only 10°–15°C. Their relatively slow rate of cooling helps them to maintain high body temperatures during foraging, and their relatively high rate of warming aids re-establishment of high body temperatures on return to land. The current evidence indicates that differential heating and cooling rates are produced by vasomotor and cardiac adjustments. It appears that during heating the heart circulates blood to the periphery more rapidly and the cutaneous vascular beds are more dilated than during cooling. A final type of physiological response widely seen in lizards is alteration of skin color; commonly the skin is darkened when it is of advantage to increase absorption of solar radiant energy and lightened when absorption is to be limited. Variation of color is also often important in permitting the animal to match the color of its background, and the demands of thermoregulation and concealment are balanced in different ways by different species.

We have seen in the discussions of behavioral thermoregulation and hypothalamic control of thermal relationships that modern lizards possess a thermal control system that, like that of birds and mammals, integrates information from both central and peripheral thermal sensors. We have now also seen that they have certain capacities for augmenting evaporative water loss and controlling thermal exchange by cardiovascular means. These observations suggest that, in the evolution of the vertebrates, certain of the physiological mechanisms involved in avian and mammalian homeothermy had made their appearance, at least in preliminary form, at the reptilian level.

Recent studies on the Indian python (*Python molurus*) have elucidated an intriguing example of physiological temperature regulation. Female pythons that are not brooding eggs are typical poikilotherms. When a female is brooding, however, she coils around her eggs and keeps them warm through a process of spasmodic contractions of the body musculature acting to increase production of metabolic heat. As ambient temperature falls from 32°C to 25°C, the metabolic rate of the brooding snake *increases*, as does the frequency of muscular contractions. Body temperature is maintained at 30°–32°C down to the ambient temperature of 25°C, with body-ambient temperature differences of up to 7°C having been reported. Below an ambient temperature of 25°C, body temperature falls with ambient temperature while still maintaining a differential of around 5°C. It might be tempting to conclude from these observations that a muscular thermogenic process analogous to shivering had evolved in the reptilian ancestors of birds and mammals. However, such a process has not been reported in other reptiles and would appear to be a specialized development in certain pythons.

Although the rate of metabolic heat production by lower vertebrates is usually low by avian and mammalian standards and heat is typically lost so rapidly that there is little metabolic elevation of body temperature, it is possible for body temperature to be elevated if circumstances favor unusually good heat retention. One factor that could promote heat retention is an exceptionally large body size, acting to reduce surface-to-volume ratio. In this light, it was recently found that a leatherback turtle (*Dermochelys*) kept in water at 7.5°C for a day had a body temperature of around 25°C. The leatherback is the largest living species of turtle, and the individual studied weighed nearly half a ton. The retention of metabolic heat, apparently resulting at least in part from immense size, may be of importance in allowing these animals to remain normally active in the cold waters of

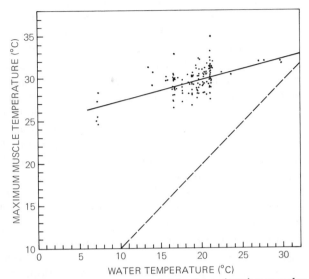

Figure 3–26. Maximum temperatures recorded in the swimming muscles of bluefin tuna (*Thunnus thynnus*) captured in waters of various temperatures. Maximum muscle temperatures were characteristically found within the dark muscles (see text). The dashed line is a line of equality between body temperature and ambient temperature. Note that muscle temperatures were maintained increasingly above ambient as the waters became colder. (From Carey, F. G. and J. M. Teal. 1969. Comp. Biochem. Physiol. 28: 205–213. Reprinted by permission of Pergamon Press.)

the northern Atlantic seaboard. Along somewhat similar lines, rattlesnakes often aggregate in groups of 100 or more in protected microhabitats during winter months; and a recent analysis indicates that the collective metabolism of such groups, coupled with the substantial diminution of total surface-to-volume ratio that results from close aggregation, may serve to elevate body temperatures 10°C or more above ambient temperatures. An empirical test of this hypothesis has not been carried out.

It is now known that in several species of tunas and lamnid sharks, temperatures within the deep swimming muscles may be substantially above the temperature of the surrounding water. These relatively large fish swim very rapidly (up to 70 km/hr) and thus are able to prey on such fast-swimming animals as squid and herring. It is believed that the maintenance of elevated muscle temperatures allows the muscles to contract and relax more rapidly, thus delivering more power and promoting the high swimming speeds that are such an important feature in the ecological relations of the species. In skipjack and yellowfin tuna, body temperature is not controlled at a particular level but rather is elevated by a relatively fixed amount over water temperature regardless of the prevailing water temperature. Thermal differences of 5°–10°C between the water and the deeper muscles are reported, for example, in the skipjack. In these species we see examples of endothermy without thermoregulation. The bluefin tuna, which occurs over a broader range of water temperatures than the skipjack or yellowfin, differs in that it maintains fairly constant deep muscle temperatures over a wide range of ambient temperatures, as illustrated in Figure 3–26.

Both the tunas and the sharks possess unusual vascular arrangements, including elaborate countercurrent heat exchangers, which promote retention of metabolic heat and are responsible for their endothermic condition. In most fish the major longitudinal arteries and veins carrying blood to and from the muscles run through the center of the body, just under and parallelling the spinal cord. In bluefin and some other tuna as well as in the lamnid sharks, the major longitudinal arteries and veins run, instead, just under the skin on either side of the body. Small arteries branch off from the longitudinal arteries and penetrate inward through the muscles; in turn, blood is brought outward from the muscles in small veins which discharge their contents into the longitudinal veins for return to the heart. This is illustrated schematically in Figure 3–27. What is most important is that the small arteries and veins carrying blood inward and outward from the muscles are closely juxtaposed, thus forming countercurrent exchange networks. These networks are especially elaborate in the vessels supplying the dark muscles; there, huge numbers of minute arteries and veins, each measuring only about 0.1 mm in diameter, are closely intermingled in thick layers of vascular tissue—a true *rete mirabile*. With these countercurrent exchange arrangements, heat picked up by the venous blood in the muscles is transferred to the ingoing arterial blood rather than being carried to the periphery of the animal, where it would be readily lost to the environment (especially across the thin membranes of the gills when the venous blood reached the gills). Essentially, then, the metabolic heat produced by the exercising muscles tends to be retained in the muscles. The dark muscles, which play a predominant role at steady cruising speeds, are, as noted, provided with the most elaborate vascular exchange networks, and it is within them that the highest muscle temperatures are found. Bluefin tuna, you will recall, are able to exercise some degree of thermal control, maintaining larger temperature differentials between their muscles and the water at low than at high ambient temperatures. This is apparently accomplished not by varying the rate of heat production but by altering the rate of heat dissipation, presumably through modifications in the heat-retaining efficiency of the countercurrent networks.

Many examples of endothermy and physiological thermoregulation are now known among insects. Colonial bees, wasps, termites, and ants often elevate or control temperature in their colonies. Mound-dwelling termites, for example, may raise the temperatures of their mounds as much as 18°C above ambient temperature through metabolic heat production. Honeybees have long been known to regulate hive temperature. When ambient temperature is low, they cluster together within the hive and become active, thus augmenting heat production. When ambient temperature is high, they disperse within the hive, collect water and spread it on the combs, and fan the moistened surfaces to cause evaporative cooling. Air that has been cooled by evaporation may be directed into the brood combs. By utilizing these various mechanisms, honeybees achieve a remarkable degree of thermoregulation in their hives. In one study, temperature in the brood area of a hive varied by only 0.7°C over a 24-hour period in which the outside air temperature varied by 17°C. In another study, hive temperature was found to be at 38°C despite an ambient temperature of 50°C. Thermoregulation assumes especial significance when it is realized that the brood requires a range of 32°–36°C for survival.

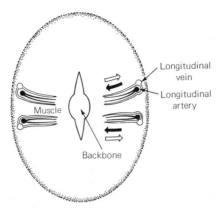

Figure 3-27. A diagrammatic cross section of a bluefin tuna showing the nature of the blood supply to the swimming muscles in highly schematic form. The longitudinal arteries, carrying blood along the length of the body, give off small arteries that penetrate inward through the muscles toward the backbone. Small veins running in close juxtaposition to the small arteries return blood peripherally to the longitudinal veins, which lead back to the heart. Dark vessels and dark arrows refer to arterial flow; light vessels and light arrows refer to venous flow.

The solitary insect at rest metabolizes at a sufficiently low rate that, given its small body size and low insulative capacity, body temperature is not appreciably elevated over ambient temperature. Insects in flight, however, may exhibit very high metabolic rates; strong flyers may actually release more heat per gram than active birds or mammals. Thus it is not surprising to find that many insects have body temperatures several to many degrees higher than air temperature when in flight. In some of these species, such as migratory locusts, body temperature is not regulated. Instead, over a wide range of ambient temperatures it is simply kept a rather fixed number of degrees above ambient—about 6°C, for example, in the locust. As in the skipjack and yellowfin tuna, this represents endothermy without thermoregulation.

It has become increasingly apparent in recent years that a good many larger insects thermoregulate during flight and some other active states. This phenomenon has been most thoroughly studied in certain sphinx moths, which are among the largest of insects, weighing as much as several grams and therefore reaching equivalence in weight to some of the smallest birds and mammals. In these forms the temperature of the thorax, but not the abdomen, is regulated; the thorax, of course, is the location of the flight muscles and thus is the site where most heat is liberated during flight. To illustrate, sphinx moths, *Celerio lineata*, were studied during extended periods of flight at ambient temperatures of 16°–27°C. They alternately flew and alighted, maintaining thoracic temperature continuously in the range of 34.5°–38.0°C. Other sphinx moths, *Manduca sexta*, were examined during two-minute flights. At air temperatures from 17°C to 32°C, thoracic temperature was maintained at 40°–42°C. When air temperature fell to 12.5°C, thoracic temperature also declined but was still kept around 38°C. Thermoregulation is not limited to just these very large forms. Worker bumblebees (*Bombus vagans*), averaging 0.12 g in weight, for example, were collected while foraging on flowers in the field and displayed stable thoracic temperatures over a broad range of ambient temperatures. The thoracic temperature of workers collected in shade at air temperatures from 9°C to 24°C was regulated at about 32°–33°C. At higher air temperatures, 26°–31°C, thoracic temperature rose approximately in parallel with air temperature, being consistently 6°–8°C higher. With these introductory examples, we may now look at some of the implications and mechanisms of insect thermoregulation.

The flight muscles of insects must be able to achieve a certain minimal power output, depending on species, before flight is possible. The temperature of the muscles is important in determining whether the power output required for flight can be realized. Small insects such as fruit flies and midges have such high surface-to-volume ratios that the activity of their flight muscles cannot warm them more than slightly above the temperature dictated by ambient conditions. It is interesting, then, that some of them are able to fly at low thoracic temperatures, about 6°C in certain midges, for example. By contrast, many larger insects, including those known to thermoregulate, require high thoracic temperatures for flight. Sphinx moths (*M. sexta*) from the Mojave Desert, for instance, require temperatures near 35°–38°C, and worker bumblebees (*B. vagans*) require about 29°C. There are also maximum thoracic temperatures compatible with flight; these vary with species but are usually between 40°C and 45°C. We see, then, that many species of insects must, either behaviorally or physiologically, maintain thoracic temperatures within relatively restricted limits if they are to fly. Maintenance of such temperatures is one of the important functions of physiological thermoregulation in those species in which it occurs.

Insects frequently find themselves at air temperatures below the body temperatures required for flight. Basking in the sun can help to maintain adequate body temperatures, but diurnal species may alight in shade or other cool microhabitats, and nocturnal species, such as the sphinx moths, cannot in any case use the solar heat. As noted earlier, insects at rest cool to the temperature dictated by ambient conditions. Frequently, then, physiological heat production must be invoked to warm the thorax sufficiently to take to the air. Such preflight warm-up, a form of endothermy, has been reported in insects of at least four orders (Lepidoptera, Hymenoptera, Coleoptera, and Orthoptera). In some cases thermoregulation during flight has also been demonstrated, but in others it has not been investigated; thus our concrete knowledge of preflight endothermy is more widespread than that of actual thermoregulation. Both bumblebees and sphinx moths show the phenomenon.

The mechanisms of preflight warm-up in insects involve high-frequency activation of the flight muscles and are often termed shivering. In some forms, such as moths and butterflies, the muscles responsible for the upstroke and downstroke of the wings are contracted more or less synchronously during shivering (rather than alternately as in flight), thus working against each other. Heat is evolved, and the wings are seen to go through high-frequency, low-amplitude vibrations. In other forms, such as bees, the flight muscles are mechanically uncoupled from the wings during warm-up, and, though the muscles may be activated in more or less their normal manner, no wing movements are seen. As would be expected, it takes longer to warm to flight temperature, the lower the initial body temperature. Sphinx moths, *M. sexta*, for instance, may require only a minute to warm to flight temperature from 30°C but may take 15 minutes to warm from 15°C. As they warm from low temperature, the rate of heat production by shivering gradually increases as the muscles are warmed and contract more rapidly. By the time thoracic temperature has reached flight temperature, the metabolic rate of the shivering moth is nearly as high as that of the moth in flight. Then suddenly the pattern of muscular contraction changes, the wings are driven through the flapping motions of flight, and the animal takes to the air.

The mechanisms of actual thermoregulation during flight and other activities are not fully understood. One possible mechanism is to increase the rate of metabolic heat production as ambient temperature is lowered. It is well accepted that the rate of heat production by *shivering* can be modulated according to thermoregulatory needs, but it is thought that the flight muscles can be used for shivering only when the insect has alighted, not when it is flying. Honeybees and bumblebees often maintain high and stable body temperatures for long periods when working in the hive. Under these circumstances the flight muscles are available for shivering, and it is well established that their rate of heat production is modulated according to demands for thermoregulation. Individual honeybees under these conditions, for example, show a rise in metabolic rate as ambient temperature is lowered from 35°C to 15°C—a response entirely like that of birds and mammals. Queen bumblebees (*Bombus vosnesenskii*) incubate their first brood cluster day and night and exhibit an inverse relationship between metabolic rate and ambient air temperature. They place their abdomen next to the brood cluster and keep the abdomen warm by shunting heat into it from the thorax; at air temperatures of 3°–33°C the thorax is kept at 34.5°–37.5°C, and the abdomen is maintained only a few degrees cooler. The brood is kept warm by this interplay of maternal physiology and behavior.

Modulation of shivering can also be used to thermoregulate during intermittent flight; that is, each time the animal alights it can shiver (or not shiver) according to its current thermal needs. This phenomenon has been reported, for example, in bumblebees foraging for food. They may visit many flowers per minute, alternately flying and alighting; and it is important that thoracic temperature always be maintained at the flight level. At air temperatures (in the shade) of about 24°C, the heat generated just from the intervals of flight appears adequate to keep body temperature at about 32°–33°C. At lower air temperatures, additional heat is required and is produced by shivering while the animal is collecting nectar on flowers. Accordingly, the overall metabolic rate averaged over periods of flight and nectar collection would be higher at low than at high temperatures.

Sphinx moths, unlike foraging bumblebees, commonly hover while feeding, thus remaining continuously in flight for long periods. Although there are mechanisms by which the metabolic rate of the flight muscles can be altered during flight, there is serious question about whether such modulation occurs in response to thermoregulatory needs. Some authors have argued that the metabolic rate during flight is more or less fixed by the demands for flight itself and that an increase in metabolism with decrease in ambient temperature would not therefore be possible. The issue has not as yet been clearly resolved experimentally. Thus although metabolism is modulated for thermoregulation in insects that are in a position to shiver, the same may not hold true for insects in steady flight.

One aspect of thermoregulation during flight in sphinx moths has been elucidated. To appreciate its significance, it is first necessary to look at certain basic properties of these animals as thermal-exchange systems. The thorax in sphinx moths (as well as in bees) is covered densely with scales ("fur"), a factor that is demonstrably important in increasing insulation and thus aiding retention of metabolic heat at cool ambient temperatures. The thorax in these large forms also has a relatively low surface-to-volume ratio (for insects), and the rate of heat production by the thoracic muscles during flight is enormous. Treating the thorax as an isolated system, we can calculate that its temperature during flight may be elevated by 20°C or more above ambient as a purely *passive* consequence of its insulatory properties and the rate of heat production within. To a large extent, then, endothermy is an obligatory outcome of the evolved properties of the animal, and the maintenance of large body-ambient temperature differentials requires no elaborate explanation. Instead, we must wonder how sphinx moths in continuous flight keep their thoracic temperature from going *too high* at high ambient temperatures; at air temperatures above 25°C, an elevation of thoracic temperature by 20°C would be incompatible with flight. In *M. sexta* this problem is prevented by increasing the rate of heat dissipation from the thorax at high ambient temperatures through circulatory adjustments. The abdomen in these animals is relatively thinly covered with scales and does not have a high intrinsic rate of heat production. At low ambient temperatures, the heart beats weakly, and blood circulates only slowly from the thorax to the abdomen; thus heat tends to be retained in the thorax, and the abdomen is near ambient temperature. At high ambient temperatures, however, the heart beats more vigorously, thus increasing the transport of heat from the thorax to the abdomen. This increases the total surface area across which the heat produced in the thorax can be lost to the environment. The result of these responses is thermoregulation through modulation of heat dissipation. At air temperatures of 12.5°–20°C, the thorax is held 22°–25°C above ambient; but as air temperature is raised above 20°C, the difference between thoracic and ambient temperature falls steeply, being just 8°C at an air temperature of 35°C.

CONTROLLED HYPOTHERMIA
IN BIRDS AND MAMMALS

Although birds and mammals frequently function in the homeothermic mode and maintain high, relatively stable body temperatures, many mammals and some birds have the ability to relax their homeothermic responses and allow their body temperature to fall, even to close approximation with low ambient temperatures. There are four major forms of such *controlled hypothermia:* hibernation, estivation, daily torpor, and "winter sleep."

Winter sleep, seasonal lethargy, carnivorous lethargy, and partial hibernation are among the many terms that have been applied to the condition seen in bears, raccoons, skunks, and some other mammals during winter. These animals sleep for long periods in protected microhabitats such as caves or tree cavities and allow their body temperature to fall a few degrees while still maintaining it far above ambient levels. Body temperatures around 30°C, for example, have been reported in the black bear. The reduction in body temperature, by diminishing the thermal gradient between animal and environment, reduces the rate of heat loss and thus allows the animal to survive longer on its fat reserves. Winter sleep is apparently a different phenomenon from hibernation, and even calling it partial hibernation would appear, at present, to be unwarranted from a physiological point of view.

Hibernation, estivation, and daily torpor seem to be different manifestations of a basically similar physiological process. In these states the animal relaxes its homeothermic processes more or less completely within a certain range of ambient temperatures and, like the poikilotherm, allows its body temperature to approximate ambient temperature. As will be discussed more below, the range of temperatures over which body temperature is allowed to follow ambient temperature varies with species. Many hibernators will allow their temperature to go below 10°C or even 5°C. As in poikilotherms, metabolic rate decreases with body temperature, and the demand for nutrient resources decreases in tandem. At the low body temperatures reached by many hibernators, metabolic rate may be equivalent to only a few percent of the basal level. Heart rate and breathing rate also fall with body temperature; at cold temperatures they become very low and often irregular. Although the animal may retain some ability to move and respond behaviorally to its environment at body temperatures well below normal, there is increasing lethargy as body temperature falls, and at low temperatures the lethargy becomes extreme.

The chief benefit conferred by these hypothermic states is a reduction in energy demands. As we have seen earlier, homeothermy is energetically costly; by abandoning homeothermy and allowing its body temperature to follow ambient temperature, the animal may expend far less energy than it would have, had it continued to maintain a high, stable body temperature. More will be said of this later.

When body temperature is freed to vary with ambient temperature for periods of several days or longer during winter, the process is termed *hibernation*. When this occurs during summer, it is called *estivation*. If body temperature is freed to follow ambient temperature for only part of each day, generally on many consecutive days, the process is termed *daily torpor*, regardless of season. These three forms of controlled hypothermia are thus distinguished largely, if not completely, on the basis of their duration and time of occurrence.

The types of changes in body temperature and metabolic rate that occur during these forms of hypothermia may be illustrated with data on small rodents undergoing daily torpor. Figure 3–28 shows a record of body temperature in a pigmy mouse (*Baiomys*) during an episode of daily torpor. Note that the animal's temperature fell steeply on entry into torpor, when the homeothermic response was relaxed, and declined to within a few degrees of ambient temperature; then, during arousal from torpor, it returned rapidly to a high level as the animal renewed metabolic heat production at a rate commensurate with homeothermy. Figure 3–29 shows metabolic rate over three successive days for a white-footed mouse (*Pero-*

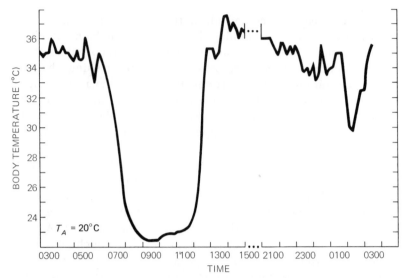

Figure 3-28. Body temperature over 24 hours in a pigmy mouse (*Baiomys taylori*) exposed without food to an air temperature of 20°C. Note the profound drop in body temperature during the episode of torpor (0700–1230). (Reprinted from Hudson, J. W. 1965. Physiol. Zool. 38: 243–254, by permission of The University of Chicago Press. © 1965 by the University of Chicago.)

myscus) undergoing daily torpor. At the ambient temperatures prevailing in this study, 13°–15°C, this animal required a resting metabolic rate of about 3.0 cc O_2/g/hr for maintenance of high body temperatures. During torpor body temperature fell to around 17°C, and the metabolic rate remained near 0.5 cc O_2/g/hr for many hours. Note the strong overshoot of metabolic rate during each arousal; it rose well above the minimal resting rate required for homeothermy and then subsequently declined. Although such an overshoot is not always observed, it serves to emphasize a general and important point: rewarming from the hypothermic state, be it torpor or hibernation, is metabolically costly. Because arousal is a necessary sequel of hypothermia, the metabolic cost of arousal must always be considered in computing the net metabolic savings realized by an episode of torpor or hibernation. Taking this into account, the animal in Figure 3–29 expended only 40–50% as much energy over each period of torpor and arousal as it would have expended had it remained homeothermic and at rest over the same period—a substantial energy savings.

Though Figures 3–28 and 3–29 pertain specifically to daily torpor, they also reflect the types of changes that occur in body temperature and metabolic rate during hibernation and estivation. Although the latter extend over more than a day, daily torpor occupies just a part of each day, and accordingly the animal undergoing daily torpor is also homeothermic and fully active for a period of each day. White-footed mice (Figure 3–29), for example, are a nocturnal species, and their episodes of torpor extend over the hours of early morning and daylight; in the evening they are homeothermic and thus able to venture out for their nocturnal activities.

Two different types of hypothermia are sometimes found in the same species. Some bats and the birch mouse (*Sicista betulina*), for example, exhibit daily torpor during the summer and hibernation during the winter. Occasionally, one form of hypothermia blends into another. Periods of

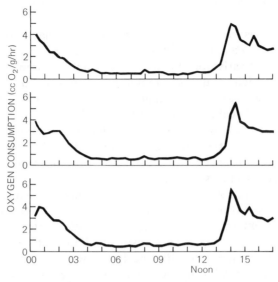

Figure 3–29. Oxygen consumption over three consecutive days (from top to bottom) in a white-footed mouse (*Peromyscus leucopus*) studied at an air temperature of 13°–15°C with an abundance of food. The animal required a resting metabolic rate of about 3.0 cc O_2/g/hr for maintenance of high body temperatures. It underwent a prolonged episode of torpor on each day, as indicated by the drop in metabolic rate. Torpor began with a gradual decline in metabolic rate after midnight (hour 00) and terminated on each day with a sharp rise in metabolic rate between 1:00 and 2:00 in the afternoon (hours 13–14). Body temperature determined during an episode of torpor similar to those shown was 16.8°C. (From Hill, R. W. 1975. Comp. Biochem. Physiol. 51A: 413–423. Reprinted by permission of Pergamon Press.)

hypothermia, for instance, may last from part of a day to several days, depending on food supply, in several species of pocket mice of the genus *Perognathus*. Such examples support the concept that hibernation, estivation, and daily torpor are variations on a common theme—different manifestations of a continuum of physiologically similar hypothermic responses. Recognizing this, it must also be noted that there are innumerable differences in physiological detail from one species or higher taxonomic group to another. Stimuli for entry into and exit from the hypothermic state vary widely; there are differences in the interplay of physiological mechanisms controlling heat gain and loss during entry and exit; and tissue responses to the hypothermic condition exhibit considerable variation. By and large, such differences do not appear to indicate basic distinctions between hibernation, estivation, and daily torpor. Rather, they at least partly reflect the fact that all these forms of hypothermia probably have polyphyletic origins, having arisen repeatedly in several to many independent evolutionary lines of homeothermic vertebrates.

Hibernation is known in a great many mammals, including hamsters, many ground squirrels, jumping mice (*Zapus*), dormice, woodchucks, marmots, hedgehogs (which are in the order Insectivora), some bats, some monotremes, and some marsupials. Such animals typically store considerable quantities of body fat during the months preceding entry into hibernation and then utilize this stored nutrient material over the winter. Hibernators arouse periodically (often frequently), and at such times they may void urine and feces and consume food they have stored in their burrow or den.

Estivation has received much less attention than hibernation, partly because it is not as easy to detect. Body temperature approximates ambient temperature in the estivator, but in summer ambient temperatures are generally relatively high, and body temperature may be only slightly below values in the nonestivating individual. As a consequence, behavioral lethargy may be hardly noticeable, quite unlike the situation found among hibernators. The pigmy possum (*Cercaertus nanus*), for example, can eat and move about in an essentially normal manner if disturbed while in estivation at body temperatures near 30°C. Estivation has been reported mostly in species of desert ground squirrels.

Daily torpor is found in a great many mammals and birds in both warm and cold situations. Many bats enter a torpid condition during daylight hours and emerge to forage at night. Pocket mice, pigmy mice, and kangaroo mice are among desert residents known to show daily torpor under certain conditions. Certain species of white-footed mice from temperate climates enter daily torpor if exposed to moderate starvation, and at least some do so in the presence of adequate food and water. Among birds, controlled hypothermia appears generally to occur on a daily basis. Many hummingbirds become torpid each night when food supply is not fully adequate. Daily torpor has also been reported in various caprimulgids (nightjars, poorwills, nighthawks) and swifts. Certain caprimulgids are known to exhibit hibernation or estivation.

The physiological state of birds or mammals in estivation, hibernation, or daily torpor is strongly reminiscent of poikilothermy. Nevertheless, the hypothermic condition in these animals differs from poikilothermy in at least two significant ways: (1) the animal can arouse from its torpid condition by physiological means, and (2) the animal frequently monitors its body temperature and asserts physiological control if temperature should fall too low. Both of these attributes deserve some discussion.

Capacity to arouse from hypothermia is essential because foraging for food, reproduction, and other vital functions occur only during homeothermic periods. There is quite a bit of variation from species to species in the minimal body temperature from which successful arousal is possible. Many hibernators can arouse from body temperatures of around 5°C. The California pocket mouse (*Perognathus californicus*) and some white-footed mice (*Peromyscus*) cannot arouse from temperatures lower than approximately 15°C, and the pigmy mouse (*Baiomys*) requires around 22°C. Arousal is accomplished by a rather sudden and intense metabolic effort, involving shivering and, among mammals, nonshivering thermogenesis. The importance of brown fat to many hibernators was discussed earlier.

It is possible to demonstrate that many animals in torpor monitor their body temperature and exert control if needed. The hibernating hamster, for example, allows its body temperature to vary freely with ambient temperature provided body temperature does not fall below 4°C. If body temperature should start to go below that level, many individuals quickly arouse. Others do not respond, and if temperature remains much below 4°C, they die. Still others respond by regulating their body temperature at 4°C even though ambient temperature may descend to substantially lower levels. These latter animals are particularly interesting because they are homeothermic, but their thermal control center has shifted its set point from the usual 37°C. The fact that hamsters and many other species do often arouse or thermoregulate when temperature falls to a certain low level indicates clearly that they retain a degree of control over body temperature while hypothermic despite their poikilothermic behavior over a range of ambient conditions. In general, the temperature at which such responses are elicited is close to the minimal temperature from which the species is capable of arousal.

Controlled hypothermia generally appears to have evolved as a mechanism of reducing energy demands during times of actual or potential food shortage. The energy savings that may be realized are great. A California pocket mouse, for example, that undergoes a 10-hour bout of torpor, including both entry and arousal, at an ambient temperature of 15°C consumes only 19% as much energy as a mouse that remains homeothermic at high body temperatures over the same period of time. Whereas the homeothermic mouse must maintain its metabolism continuously well above the basal level in order to keep warm at 15°C, the torpid individual enjoys a metabolic rate well below basal except during the early stages of entry into torpor and during arousal. Clearly, energetic savings are potentially very high in hibernators that, except for occasional arousals, remain at body temperatures of 10°C or less for weeks on end.

Some small species, such as certain white-footed mice and insectivorous bats, are known sometimes to undergo daily torpor even when food is in abundance. Thus they utilize torpor as a routine mode of reducing their demands for energy. Other forms, at least in laboratory settings, undergo torpor only when faced with an inadequacy of food. In some, such as the California pocket mouse, bouts of torpor are known to increase in duration as the food supply is rendered more and more inadequate. The seasonal forms of hypothermia, hibernation and estivation, are often stimulated by changes in daylength or temperature. In these cases animals have evolved to respond to cues which, in essence, provide a signal that a stressful time of year is approaching. One advantage of utilizing such advance cues rather than responding to immediate inadequacy of food is that physiological preparation may be made. Hibernators typically develop stores of fat as winter approaches.

A benefit of hypothermia that may assume importance for animals under conditions of drought is the reduction in water loss that accompanies a reduced metabolic rate and lowered body temperature. Because less oxygen is needed during hypothermia, ventilatory rate is decreased, and because body temperature is reduced, exhaled air is cooler and carries less water vapor. Total respiratory water loss may be substantially diminished as a consequence.

Those birds and mammals that are capable of controlled hypothermia are often termed *heterotherms* in recognition of the fact that they sometimes regulate their body temperature and sometimes do not. The heterotherm can enjoy, in a sense, the best of both the homeothermic and poikilothermic worlds. When thermoregulating at high body temperatures, the animal is able to move about with the independence of external thermal conditions that we have previously emphasized as a prime advantage of homeothermy. When in torpor, on the other hand, it benefits from the decreased demand on food and water resources that is characteristic of poikilothermy.

CONCLUDING COMMENTS

In concluding this chapter, it is perhaps well to offer a rejoinder to the common belief that homeothermy is in some sense superior to poikilothermy. Although we may be led to this conclusion by the fact that we ourselves are homeotherms, this is hardly a valid basis for objective judgment. In nature's benign indifference (Camus' phrase), the only criterion of success is perpetuation of the species, involving the subsidiary elements of survival and reproduction at the individual level. The living world provides the incontrovertible evidence that, by this criterion, poikilothermy can be as successful a mode of life as homeothermy; many poikilothermic taxa have persisted for hundreds of millions of years. In fact, poikilothermy and homeothermy are simply extremes in a continuum of thermal responses that have been exploited in the evolution of different niches within which success is possible, and our scientific explorations into the phenomena are directed to understanding the place of thermal relationships in the ecological life of each species. Although we may speak of advantages and disadvantages of various thermal relationships, these are usually highly relative judgments made in hindsight. Homeothermy, for example, is not only an "advantage" but an essential for terrestrial animals that carry out their life functions unabated in the dead of winter. Extending the argument a step further, however, we recognize that only some animals remain fully active in winter, and the others persist perfectly well. Thus homeothermy is simply one element of a complex of attributes involved in a particular kind of niche specialization; it is not, in itself, of absolute benefit.

Animals were once viewed as falling into two rather rigid classes: the birds and mammals—homeotherms—on the one hand, and all other groups —poikilotherms—on the other. One of the greatest insights that has emerged with full clarity in the last few decades is that this is a grossly inaccurate view of the world. There are many types of thermal relationships with the environment, ranging from simple poikilothermy to behavioral thermoregulation to endothermy without thermoregulation to full-fledged physiological homeothermy. Simple poikilothermy was undoubtedly the earliest type of relationship, but other types of relationship have evolved in many groups, not just in birds and mammals; and species, including birds and mammals, often exploit different relationships at different times.

SELECTED READINGS

Bartholomew, G. A. and J. W. Hudson. 1961. Desert ground squirrels. Sci. Amer. 205: 107–116.

Bligh, J. 1973. *Temperature Regulation in Mammals and Other Vertebrates.* North-Holland Publishing Company, Amsterdam.

Brown, G. W., Jr. (ed.). 1968. *Desert Biology.* Academic, New York.

Bullock, T. H. 1955. Compensation for temperature in the metabolism and activity of poikilotherms. Biol. Rev. 30: 311–342.

Carey, F. G., J. M. Teal, J. W. Kanwisher, K. V. Lawson, and J. S. Beckett. 1971. Warm-bodied fish. Amer. Zool. 11: 137–145.

Dill, D. B. (ed.). 1964. *Handbook of Physiology. Section 4: Adaptation to the Environment.* American Physiological Society, Washington, D.C.

Fisher, K. C., A. R. Dawe, C. P. Lyman, E. Schönbaum, and F. E. South, Jr. (eds.). 1967. *Mammalian Hibernation III.* Elsevier, New York.

Gates, D. M. 1962. *Energy Exchange in the Biosphere.* Harper & Row, New York.

Hart, J. S. 1957. Climatic and temperature induced changes in the energetics of homeotherms. Rev. Can. Biol. 16: 133–174.

Heinrich, B. 1974. Thermoregulation in endothermic insects. Science 185: 747–756.

Hochachka, P. W. and G. N. Somero. 1973. *Strategies of Biochemical Adaptation.* Saunders, Philadelphia.

Irving, L. 1972. *Arctic Life of Birds and Mammals.* Springer-Verlag, New York.

Maloiy, G. M. O. (ed.). 1972. *Comparative Physiology of Desert Animals.* Symposia of the Zoological Society of London, No. 31. Academic, New York.

Milstead, W. W. (ed.). 1965. *Lizard Ecology. A Symposium.* University of Missouri Press, Columbia, Mo.

Newell, R. C. 1973. Factors affecting the respiration of intertidal invertebrates. Amer. Zool. 13: 513–528.

Prosser, C. L. (ed.). 1967. *Molecular Mechanisms of Temperature Adaptation.* American Association for the Advancement of Science, Washington, D.C.

Schmidt-Nielsen, K. 1964. *Desert Animals.* Oxford University Press, London.

Schmidt-Nielsen, K. 1972. *How Animals Work.* Cambridge University Press, London.

Scholander, P. F., R. Hock, V. Walters, F. Johnson, and L. Irving. 1950. Heat regulation in some arctic and tropical mammals and birds. Biol. Bull. (Woods Hole) 99: 237–258.

Somero, G. N. and P. W. Hochachka. 1971. Biochemical adaptation to the environment. Amer. Zool. 11: 159–167.

Whittow, G. C. (ed.). 1970–73. *Comparative Physiology of Thermoregulation.* Vols. I–III. Academic, New York.

See also references in Appendix.

4 EXCHANGES OF SALTS AND WATER: MECHANISMS

Animals are composed in large part of solutions. The universal biological solvent is water, and among the important solutes are a great variety of ions such as sodium, potassium, calcium, chloride, sulfate, and phosphate. This and the following chapter are concerned with the exchanges of these vital materials between the organism and its outside world.

Animals occupy all the major environments of the earth—land, oceans, fresh waters, and the interfaces among them. The body fluids of all animals that have been studied differ in salt and water composition from the environmental medium. The differences may be relatively small, as in many marine invertebrates. Or they may be large. The concentrations of the major physiological ions in freshwater animals are much higher than those in fresh water. The concentrations in marine bony fish are much lower than those in seawater. Terrestrial animals are not even surrounded with an aqueous solution. We see that in this dimension, as in so many others, animals are at disequilibrium with their surroundings. From experience and theory, we know that there are always forces tending to bring disequilibria to equilibrium. In particular, there are forces tending to bring animal body fluids to the same composition as the surrounding environmental medium. The body fluids of freshwater animals tend to be diluted. Those of marine bony fish tend to be concentrated. The water resources of terrestrial animals tend to be vaporized. The animal must combat these tendencies, must maintain the disequilibria, by active, energy-demanding intervention. Some animals are very sensitive to changes in the composition of their body fluids, whereas others are relatively insensitive; inevitably there are limits beyond which life itself is threatened.

The present chapter is basically a review of principles regarding the mechanisms by which water and salts are exchanged. The presentation is predicated on the belief that the student and prospective investigator must understand mechanisms before he can fully appreciate the problems faced by animals and the evolutionary adaptations that permit homeostasis despite those problems.

The first two mechanisms to be discussed are relevant only to animals living in air: evaporation and uptake of water vapor.

EVAPORATION

The meaning of fractional concentration and partial pressure in gas mixtures

On land, the animal is surrounded by a mixture of gases, one of which is water vapor. The representation of any given component gas in a mixture can be expressed as its mole fractional concentration, or that fraction of the total moles of gas present that is represented by the component in question. The mole fractional concentration of oxygen in air at sea level, for example, is about 0.21. According to Avogadro's principle, the mole fractional concentration of a given component is equal to its volume fractional concentration, or that fraction of the total volume that is occupied by the component in question. Thus if we remove the oxygen from a known volume of air at a given temperature and pressure and subsequently restore the residual gas mixture to the same temperature and pressure, we will find that the original volume has been reduced by approximately 21%.

The atmosphere exerts a total pressure that varies with altitude and meteorological conditions and is typically expressed as the height of the mercury column that is supported by the pressure in a mercury barometer. The "standard" atmospheric pressure at sea level is 760 mm of mercury (mm Hg), though actual pressure at sea level varies above and below this level. The atmospheric pressure is exerted not only on solids and liquids at the earth's surface, but also on volumes of air and other gases near the surface. According to Dalton's law of partial pressures, each gas in a volume of mixed gases behaves as if it alone occupied the entire volume, and it exerts a pressure, its *partial pressure*, that is independent of the other gases present. Thus the total pressure exerted by a gas mixture can be viewed as the sum of the partial pressures exerted by its individual components. The partial pressure of any given gas is equal to the total pressure of the mixture times the volume or mole fractional concentration of the gas in question. In other words, each gas in the mixture contributes to the total pressure in proportion to its representation. The partial pressure of oxygen in air at standard sea level pressure is approximately 21% of 760 mm Hg, or 159 mm Hg.

Water vapor pressure, saturation pressure, and measures of humidity

The partial pressure of water vapor is generally termed the *water vapor pressure*. If air of a given temperature is exposed to pure water of the same temperature in a closed system, water will evaporate into the air until a certain maximum, equilibrium water vapor pressure is realized. This vapor pressure is called the *saturation vapor pressure* of the air and represents the maximum amount of water vapor that air at that temperature will hold. The air is said to be saturated. The saturation vapor pressure is strongly and, in the range of physiological conditions, exclusively dependent on temperature. (It does not depend, for example, on the composition or total pressure of the gas mixture.) The saturation vapor pressure at 0°C is 4.6 mm Hg, that at 20°C is 17.5 mm Hg, and that at the temperature of the human respiratory tract, 37°C, is 47.1 mm Hg. It is convenient to remember that the saturation vapor pressure approximately doubles for every 10°–12°C rise in temperature in the physiological range of temperatures.

Some microhabitats approach saturation or are saturated (e.g., air spaces within soil or leaf litter in many regions of the world and the spaces among the rocks in Figure 1–1). Most microhabitats are characterized by water vapor pressures lower than the saturation pressure.

Humidity can be expressed as water vapor pressure or as the weight of water per unit volume of air, the latter being termed the *absolute humidity*. Absolute humidity is approximately related to water vapor pressure by the following equation:

$$\text{grams of water per cubic meter} = \frac{289}{T} \times \text{water vapor pressure (mm Hg)}$$

where T is the absolute temperature of the air. This equation or conversion tables that eliminate the small errors inherent in the equation are important, for example, when we wish to compute the actual respiratory water loss of an animal from information on the volume and water vapor pressure of exhaled air. Humidity of air at a given temperature is often expressed relative to the saturation vapor pressure of air at the same temperature. The common modes of expressing humidity in this way are *relative humidity* and *saturation deficit*. Relative humidity gives the prevailing vapor pressure as a percentage of the saturation vapor pressure. The saturation deficit is the difference between prevailing vapor pressure and saturation vapor pressure. For example, if the vapor pressure of some air at 20°C (saturation pressure: 17.5 mm Hg) is 12.0 mm Hg, the relative humidity is 68.6% and the saturation deficit is 5.5 mm Hg. Both relative humidity and saturation deficit reflect the drying power of the air. At a given temperature, drying power increases as relative humidity decreases and as saturation deficit increases. Given air with a certain concentration of water vapor, the extent of departure from saturation varies with temperature because the saturation pressure varies with temperature. If the air is warmed, the relative humidity decreases and the saturation deficit increases. If the air is cooled, the converse is true. If cooling continues until the relative humidity is 100% (or the saturation deficit is zero), further cooling will cause condensation of water into the liquid form, or dew formation.

The concept of vapor pressure difference between animal and environment

In the physiological literature the rate of evaporation has generally been related to the saturation deficit. The observed facts will probably be better understood, however, in terms of the *vapor pressure difference* between the animal and air, a concept that has been developed fully by physical chemists but that has received little usage in the hands of physiological ecologists. The problem with saturation deficit is that it describes only one member of the evaporative system, the air, when in fact both the air and the animal require attention.

A body of water at given temperature will tend to evaporate to establish an equilibrium vapor pressure in the air above that *is identical to the saturation vapor pressure of air at the same temperature, whether or not the air surrounding the water is in fact of the same temperature.* To illustrate, evaporation from water at 20°C will tend to establish an equilibrium vapor pressure of 17.5 mm Hg whether or not the air is also at 20°C; in a closed system the equilibrium vapor pressure established over water held at 20°C will be 17.5 mm Hg even if the air is warmer and therefore could hold more water vapor. Water, we see, can properly be said to have an equilibrium or saturation vapor pressure that varies with its own temperature. Henceforth we shall simply call this the vapor pressure of the water. The vapor pressure difference between an animal and the surrounding air is defined as follows:

$$\text{vapor pressure difference} = \text{vapor pressure of body fluids} - \text{vapor pressure of air}$$

That is, it is the difference between (1) the equilibrium vapor pressure that would be established in air by evaporation from the body fluids in the absence of contravening factors and (2) the actual ambient vapor pressure. When this quantity is positive, net evaporation will occur; water moves from regions of higher vapor pressure to regions of lower vapor pressure. Furthermore, evaporation can be expected to take place at a rate directly related to the magnitude of the vapor pressure difference.

Assuming for the moment that the solutes in the body fluids exert no effect on evaporation, we note that if the animal and the air are at the same temperature, then the body fluid vapor pressure is the same as the saturation vapor pressure of the air, and the vapor pressure difference is identical to the saturation deficit. Thus it is in systems of uniform temperature that the saturation deficit can be properly related to the rate of evaporation. If, on the other hand, the body fluids are cooler than the air, their vapor pressure will be lower than the saturation pressure of the air, and the vapor pressure difference will be lower than the saturation deficit. Evaporation will occur more slowly than the saturation deficit would indicate. (This may be exemplified as follows: If the air temperature is 20°C, the saturation vapor pressure of the air is 17.5 mm Hg. If the body fluids are at 15°C, their vapor pressure is the same as the saturation vapor pressure of air at 15°C, or 12.8 mm Hg. Supposing the actual ambient vapor pressure to be 10 mm Hg, the saturation deficit is 7.5 mm Hg, but the vapor pressure difference is only 2.8 mm Hg.) When the body fluids are warmer than the air, the vapor pressure difference will be greater than the saturation deficit. This has what might be a surprising implication for animals in saturated air. Suppose, for example, that the body fluid temperature of an animal is 32°C while the surrounding air is saturated and at a temperature of 30°C. The air, being saturated, has a vapor pressure of 31.8 mm Hg, and its saturation vapor pressure is identical; the vapor pressure of the body fluids is the same as the saturation vapor pressure of air at 32°C, or 35.7 mm Hg. The saturation deficit is zero, and this might lead one to expect no net evaporation. (This would be true if the animal were of the same temperature as the air.) However, the vapor pressure difference is 3.9 mm Hg, and net evaporation will occur. The air, being saturated at its temperature, cannot hold more water vapor, and the added vapor from evaporation will rapidly condense. Insofar as it condenses directly onto the animal, no net water (or heat) loss will be realized by evaporation. But some condensation will occur as droplet formation in the air, and some will occur on nearby objects such as soil or rocks. To this extent, the animal will realize a net water (and heat) loss. Poikilotherms (as well as homeotherms) are often warmer than their surroundings when in saturated or nearly saturated air because of the continual internal production of metabolic heat and the low rate of heat dissipation through evaporation under such conditions. Thus in an experiment on frogs in saturated air at 20°C, the body temperature was sufficiently elevated so that about 20% of the metabolic heat production was lost in evaporation.

An additional factor that must be considered in computing the vapor pressure difference is the effect of solutes in the body fluids. You will recall that if pure water is placed in contact with air in a closed system at uniform temperature, the air becomes saturated. Because the vapor pressure of the water is the same as the saturation pressure of the air, the vapor pressure difference does not fall to zero until full saturation has been realized. If solutions of various concentrations are placed in this system, it is found that the maximum, equilibrium vapor pressure decreases with increases in solute concentration. (The meaning of *concentration* here is a special one and will be developed fully in the section on osmosis.) Solutes depress the vapor pressure of solutions below the saturation vapor pressure of air of the same temperature. Thus the vapor pressure difference falls to zero and net evaporation ceases before the air in a system of uniform temperature is fully saturated. The magnitude of this effect is small in the physiological range of concentrations. The vapor pressure of seawater is about 2% lower than that of pure water of the same temperature. This means that in a closed system of uniform temperature, the equilibrium relative humidity in air over seawater will be about 98%. Mammalian plasmas are much less concentrated than seawater, and the depression of vapor pressure caused by solutes is only about 0.5%. Regardless of other factors, solutes always reduce the vapor pressure difference between fluids and air and thus reduce the rate of evaporation. Under strictly comparable conditions, animals with relatively high body fluid concentrations will tend to lose water by evaporation slightly more slowly than animals with lower concentrations. Inasmuch as this effect is small, however, it is generally masked by other, more important influences on evaporation, and it has received little quantitative attention in comparative studies. It may be significant in some special cases.

Determinants of the rate of evaporation

The rate of evaporation from animal body surfaces is dependent not only on the vapor pressure difference but also on other factors, and computation of the vapor pressure difference itself in natural situations is often not practical. The relationships are sufficiently complex that rate of evaporation is generally measured empirically rather than being calculated from knowledge of the individual contributing factors. It is important, none-theless, for the physiologist to have at least a qualitative understanding of the parameters involved. The following list of factors affecting the rate of evaporation is in part a summary of what has been discussed above.

1. Evaporative rate is strongly and directly dependent on the vapor pressure difference. This has the following corollaries:
   a. Evaporative rate increases with decreases in ambient vapor pressure.
   b. Evaporative rate increases with increases in the temperature of the body fluids that are acting as the source of water vapor. Conversely, the rate decreases with decreases in temperature. The relevant temperature here is not necessarily equivalent to the deep body temperature. It is, again, the temperature of the body fluids from which evaporation is occurring. Thus the crit-ical temperature is the surface or subsurface temperature of the animal.
   c. Evaporative rate decreases slightly with increasing solute con-centration in the body fluids from which evaporation is oc-curring.
2. Rate of evaporation is dependent on the extent of convective move-ment of the air near the body surface. When the air near the sur-face is relatively stagnant, evaporation causes a buildup in the vapor pressure of the air near the surface, such that the vapor pres-sure near the surface may be significantly higher than that in the ambient air at large. Because rate of evaporation is dependent on the vapor pressure of air next to the body surface, evaporation will then be slower than one would expect from the gross ambi-ent humidity. Convective movement of the air near the body sur-face will replace water-laden air with fresh air and thus hasten evaporation. We all know how welcome a breeze is on a hot, humid summer day.
3. Rate of evaporation depends on the area and permeability of the body surfaces involved. When water is evaporating across the sur-face tissues, the tissues present a potential barrier to evaporation that is not present when water is directly exposed to air. The per-meability of the body surface is, in this context, a measure of the facility with which water passes across the tissues in response to evaporative effects. For given vapor pressure difference, permeabil-ity can be expressed as volume of water lost per unit area of surface per unit time. Some animal body surfaces are so permeable that the rate of evaporation is essentially the same as that from free-standing body fluids. This is true, for example, of many amphibians. Other animals have poorly permeable body surfaces. Low perme-ability is vital to the success of such groups as insects and mam-mals in environments where the vapor pressure difference is often high. Permeability varies over the body surfaces of animals, and the properties of each part of the surface must be considered individually. The cellular covering of respiratory surfaces must be

thin to allow rapid exchange of respiratory gases, and such surfaces are typically highly permeable to water, thus constituting potentially outstanding sites of evaporative water loss.

UPTAKE OF WATER FROM THE AIR

Evaporation is exclusively an avenue of water loss. One may wonder whether animals can take up water vapor from the air in net fashion, especially when the air is saturated or nearly so. In most animals this does not appear to occur to any appreciable extent. The solutes of animal body fluids do depress the body fluid vapor pressure below the saturation pressure of air at the same temperature. In the vast majority of terrestrial animals, however, the extent of the depression is 1% or less. This means that if the body surfaces are at the same temperature as the air, the vapor pressure difference will favor loss of water by evaporation whenever the ambient relative humidity is below about 99%. At humidities above 99%, when uptake of water vapor would be favored in a system of uniform temperature, animals are generally, as we have noted earlier, warmer than the surrounding air. This raises the body fluid vapor pressure sufficiently to establish, again, a vapor pressure difference favoring evaporation, not uptake. If the animal should be cooler than the surrounding air, clearly the vapor pressure difference could favor uptake of water. (We see this in the form of water condensation on the sides of a glass containing a cold drink on a warm, humid day.) It is reported that water condenses on the body surfaces of lizards emerging from cool burrows into warm air. The physiological significance of this is unknown, but it is possible that the moisture could be licked off or even absorbed to some extent through the skin, though lizard skin is poorly permeable to water. This situation in the lizard cannot persist for long because the warm air and the condensation itself will warm the lizard and ultimately reverse the vapor pressure difference to the direction of evaporation.

The important known cases of water uptake from the air are reported among arthropods. Certain insects, mites, and ticks are able to take up water vapor *against* vapor pressure differences favoring evaporation. The mechanism is unknown. Absorption of water at ambient relative humidities of 90–95% is not uncommon. Nymphs and adult females of a desert cockroach (*Arenivaga*) can take up water at relative humidities of 82% and above; the firebrat (*Thermobia*) can absorb water down to humidities at least as low as 63%; and the prepupal stage of the rat flea (*Xenopsylla brasiliensis*) can take up water at relative humidities as low as 50%. Such water uptake continues over long periods and can result in a significant increase in body water volume and weight. There is circumstantial evidence that absorption occurs across general body surfaces in some forms, but in the firebrat it seems clear that absorption is across the rectum, for blockage of the anus prevents water uptake.

PASSIVE MOVEMENTS OF SOLUTES

Having reviewed two mechanisms by which water moves between its liquid and gaseous phases, we shall now turn to several mechanisms of solute and water movement across membranes when both sides of the mem-

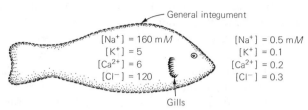

Figure 4–1. A schematic representation of a freshwater fish showing, inside the fish, the concentrations (mM) of certain ions that might prevail in the blood plasma and, outside the fish, the concentrations (mM) that might prevail in the environment.

brane are surrounded by aqueous solutions. These mechanisms are relevant to direct transfers between the animal and environment in aquatic organisms and are ubiquitous to solute and water movements across membranes within the animal. We shall start by considering passive movements of solutes.

Solutes move across biological membranes according to a good many mechanisms. Various classifications of these mechanisms have been proposed, and none has received universal acceptance. In part we are hampered by ignorance of mechanisms at the molecular level, and in part there is the problem that any gross scheme of classification will fail to recognize all the fine distinctions among the known mechanisms of movement. Here we shall subdivide the mechanisms into passive and active. The distinction between the two will be elaborated after development of some background information.

The ionic relationship between a freshwater fish and its environment is diagrammed in Figure 4–1. In all aquatic animals the internal body fluids are separated from the environmental fluids by a cellular body covering. Also in all animals that have been studied, at least some solutes differ in concentration on the inside and outside of the organism. The latter point is exemplified by the fish. Some solutes are able to pass through the body covering, and to them the body covering is said to be permeable. The covering is impermeable to other solutes.

When, for a particular permeating solute, there is a concentration difference across the body covering, the solute will tend to diffuse in net fashion from its area of greater concentration to its area of lesser concentration. The concentration difference will tend to be degraded. In the example of the freshwater fish, all four ions will tend to diffuse from the body fluids into the surrounding water. Given that the volume of water is much greater than the volume of the fish's body fluids, diffusion will not appreciably raise the concentrations of the ions in the water, and the ultimate result of diffusion in the absence of contravening factors would be to lower concentrations in the body fluids to those in the environment.

In simple diffusion, the primary motive force is the continuous random movement displayed by all atoms and molecules at temperatures above absolute zero (molecular kinetic energy). The average kinetic energy of random movement increases proportionally with absolute temperature. When a concentration difference exists for a solute, random motions tend to carry more particles of the solute away from areas of high concentration than are carried into such areas. Thus the distribution of the solute particles becomes progressively more even. The presence of the complex body covering between the internal and external fluids may alter this simple picture of diffusion to a greater or lesser extent. When there are pores through the covering that are sizable by comparison to the solute particles concerned, diffusion may proceed largely according to the classical picture. Narrow pores may somewhat alter the picture, and special carrier molecules in the cells of the covering that pick up solute molecules on one side and carry them to the other may alter it radically. (The latter is a form of facilitated diffusion.) Here it is sufficient to recognize that the ultimate result is a tendency toward degradation of concentration gradients.

Electrical potential differences across part or all of the body covering may also alter the simple picture of diffusion in the case of ions and other electrically charged solutes. Thus, for example, if the inside is positive relative to the outside, there will be a tendency for negatively charged solutes to be attracted inward and for positively charged solutes to be repelled outward. The movement of a particular solute will depend both on its concentration gradient and on the prevailing electrical gradient. Thus we speak of the *electrochemical gradient* as being the governing factor in passive solute movement. Electrical potential differences may augment or oppose movements resulting from concentration differences and are of predominant importance in certain situations. For example, consider a membrane bathed on the outside by a solution that is much less concentrated in Cl^- than the solution bathing the inside. Strictly on the basis of the concentration gradient, we would expect Cl^- to diffuse from inside to outside, but if the inside of the membrane were sufficiently positive relative to the outside, Cl^- would passively move inward in net fashion despite the opposing concentration gradient.

We are now in a position to define passive solute movements as *those that occur only in the direction of prevailing electrochemical gradients*. Passive movements, when unopposed, lead to an equilibrium situation in which electrochemical gradients no longer dictate net movement of solute across the membrane in one direction or the other. (If such an equilibrium is achieved, movements of solute across the membrane continue in both directions, but net movements cease.)

The rate of net passive movement is, in general, influenced by a variety of factors that interact complexly:

1. The rate typically increases with the magnitude of the electrochemical gradient, at least within broad limits. The prevailing electrical gradient affects all charged solutes, but each solute responds only to its own concentration gradient. The concentration gradient is defined as

$$\frac{C_1 - C_2}{L}$$

where C_1 is the concentration on one side of the membrane, C_2 is the concentration on the other side, and L is the thickness of the membrane or, more generally, the distance separating concentration C_1 and concentration C_2. The electrical gradient is defined analogously. You will note that for a given concentration difference,

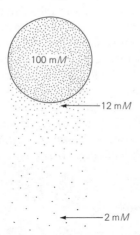

Figure 4–2. Diffusional concentration of sodium in a boundary layer next to the organism, schematic representation.

a solute that is not under electrical influence will diffuse more rapidly from one region to another, the smaller the distance L separating the two regions of differing concentration. Diffusion is a notoriously slow process in the macroscopic world, but when only the thickness of a cell or cell membrane separates two regions of differing concentration, exchange by diffusion may be rapid. Often the determination of the effective concentration gradient is far from simple. Suppose, to take an example, that the concentration of Na^+ in the body fluids of an animal is 100 mM and that the concentration in the external environment at large is 2 mM. Na^+ is lost from the animal by diffusion, and this steady loss may elevate the concentration in the environmental water right next to the animal's integument. As illustrated in Figure 4–2, the concentration may be 12 mM in the boundary layer next to the skin and fall progressively to 2 mM further away from the skin. The boundary layer of elevated concentration may be much less than 1 mm thick, but still the effective concentration difference governing Na^+ diffusion from the animal will be only $100 - 12 = 88$ mM, rather than the 98 mM that a more superficial analysis would indicate. This situation is analogous to the development of an elevated water vapor pressure next to the body surface during evaporation, as discussed earlier, and again, convective movement of the medium across the body surface will tend to wash out the elevated concentration and hasten diffusion. It has been suggested that the mucous films secreted by many aquatic animals may act to trap boundary layers next to the body surface and thus reduce the rate of diffusional transfer of solutes.

2. The rate of passive movement typically increases with temperature. For normal diffusion, the Q_{10} is typically less than 1.5, but it may be as great as 2 or 3.

3. The rate depends directly on the area of the surfaces across which

movement is occurring and on the permeability of the membranes to the solute in question. As emphasized in the discussion of evaporation, different parts of the body may exhibit different permeabilities to the same substance. Typically, the respiratory membranes are especially permeable to many solutes and, additionally, represent a considerable surface area; they thus are particularly important sites of passive solute exchange with the outside world. Any given membrane will exhibit different permeabilities to different solutes, and permeabilities often change with the physiological status of the membrane. When a membrane is impermeable to a solute, no passive exchange will occur, regardless of the electrochemical gradient.

ACTIVE TRANSPORT

Animals are capable of transporting many solutes *against their electrochemical gradients at the expense of metabolic energy.* The mechanisms involved are called active-transport mechanisms or pumps. It is generally hypothesized that energy-driven carrier molecules pick up solute molecules on one side of a cellular membrane and release them on the other. The various proposals that have been made regarding the exact nature of transport are reviewed in texts of cellular physiology. Active-transport mechanisms are always highly specific for certain solutes; many transport only one or two ordinary physiological solutes.

Active-transport mechanisms are important to such diverse functions as maintenance of electrical polarity in nerve and muscle cells, absorption of nutrients from the gut, and regulation of buoyancy in cuttlefish, as well as being intimately involved in kidney function and osmotic-ionic regulation in general. The hagfish *Eptatretus* and marine invertebrates of many phyla are reported to take up dissolved amino acids and some other small organic compounds from the water against gross concentration gradients of a million to one, or more. The direct active exchanges between organism and environment with which we will be mostly concerned are exchanges of ions.

As discussed earlier, the freshwater fish in Figure 4–1 is confronted with a continuous passive loss of sodium and chloride from its body fluids. These losses are replaced in large part by active uptake of sodium and chloride from the water across the gills. Net uptake of chloride from waters containing as little as 0.02–0.3 mM Cl^- has been reported in various species. Frogs, which face much the same problems as freshwater fish when in water, can take up Na^+ across their skin from water as dilute as 0.01 mM (concentration in the body fluids is about 110 mM). Marine bony fish have body fluids more dilute than seawater; plasma chloride concentration may be about 150 mM, whereas the concentration in standard seawater is 458 mM. These fish confront a continuous passive uptake of chloride and compensate by actively secreting Cl^- across their gills. Such active secretion of salts into the environment by a structure other than the kidneys is termed *extrarenal salt excretion.*

Active-transport mechanisms are known for a great variety of solutes, but there has yet been no conclusive demonstration of active transport of water. Interest in the possibility of active water transport remains high, and there are certain biological situations in which it is regarded as an attractive hypothesis (for example, uptake of water vapor from the atmosphere against the vapor pressure gradient by certain arthropods, as discussed earlier).

Suppose (unrealistically) that Na^+ were actively absorbed across a membrane without any contemporaneous movements of other ions. The net movement of Na^+ would then be self-limiting, for the accumulation of positive charges (Na^+ ions) on the inside of the membrane would ultimately establish such a large electrical potential across the membrane (inside positive) that the active-transport pump would be unable to overcome the electrical forces tending to repel Na^+ from the inside. This does not occur in real situations because active transport of one ion is accompanied by contemporaneous movements (active or passive) of other ions. The electrical potential gradient established by transport of a given ion across a membrane can be limited if ions of opposite charge move across in the same direction or if ions of the same charge move in the opposite direction. Thus transport of Na^+ across a membrane might be accompanied by movement of Cl^- in the same direction or movement of NH_4^+ in the opposite direction; either of these contemporaneous movements of other ions would act to limit the disequilibrium of electrical charge developed across the membrane. Often two contemporaneous ion movements are both active. In the squid giant axon, for example, Na^+ is actively transported out of the cell, and K^+ is actively transported inward. In other cases one movement is active and the other passive. To illustrate, we may consider a membrane in which Na^+ is actively transported inward and Cl^- follows passively in response to the electrical gradient established by Na^+ transport. The steady-state situation is perhaps best explained by imagining that we could suddenly "turn on" the Na^+ pump. At first Na^+ alone would be absorbed, but the accumulation of positive charges on the inside of the membrane would quickly establish an electrical gradient sufficient to draw Cl^- in passively even if Cl^- were less concentrated on the outside of the membrane than the inside. Once the electrical gradient became large enough for Cl^- to be drawn inward at the same rate as Na^+ was being actively transported, a steady state would be established. Each Na^+ would be accompanied by a Cl^-, and the electrical gradient across the membrane would stabilize at a fixed value. In many situations of environmental interest, we do not yet definitely know which ion movements are active and which are passive. The reader must thus recognize that sometimes when we speak subsequently of NaCl transport, for example, it may be sodium or chloride or both that are moved directly by active transport.

The rate of active transport typically increases with increasing temperature in the physiological range, Q_{10}'s of 2 to 3 being common. Unlike passive movements, active movements are not related to the concentration gradient in any simple way, though they do depend on the concentrations of solute on the two sides of the membrane. Figure 4–3 shows the relation of rate of sodium uptake to environmental sodium concentration in a sodium-depleted crayfish, *Astacus pallipes.* The rate is highly sensitive to concentration at low sodium levels, whereas it is essentially independent of concentration at higher sodium levels. This suggests that uptake is mediated by carrier molecules which require only a certain minimal environmental concentration for full operation. You will recognize the similarity to enzyme saturation kinetics,

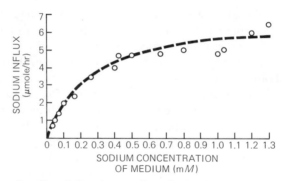

Figure 4–3. Rate of sodium influx in a sodium-depleted crayfish (*Astacus pallipes*) as a function of environmental sodium concentration at 12°–13°C. (From Shaw, J. 1959. J. Exp. Biol. 36: 126–144.)

as discussed in Chapter 3. Permeability of the membrane is only indirectly relevant to active transport, as transport mechanisms are adapted to carry solutes across even poorly permeable membranes. It is characteristically of advantage, in fact, for the membrane to be poorly permeable to actively transported solutes because the active-transport mechanism must operate against any tendency for solute to move passively in the direction opposite to transport. When an ion is moving passively according to the electrical gradient established by active transport of another ion, the membrane's permeability to the passively moving solute will be of obvious significance. If, for instance, a positive ion is being actively transported, membrane permeability to the various available negative ions can be an important factor in determining which anions accompany the cations.

Active-transport mechanisms establish and maintain nonequilibrium distributions of solutes and thus demand expenditure of metabolic energy. The coupling of energy sources to the transport mechanism is not as yet understood and is a subject of much current interest. Metabolic poisons and inhibitors characteristically interfere with active transport, and elimination of the oxygen supply to preparations of active-transport tissues usually decreases the rate of transport.

ADDITIONAL ASPECTS OF SOLUTE MOVEMENT

All animals exhibit internal concentrations of certain solutes that differ from environmental concentrations. There is a tendency for these differences to be degraded by passive solute movements, and the animal opposes this tendency by such energy-requiring activities as active transport. The active involvement of the animal is more widespread than production of active transport, however, and in this sense the word *passive* is unfortunate as applied to many solute movements. The rates of passive solute movements depend, for example, on the permeability properties of membranes, and because animals construct and maintain membranes with particular properties, they thereby exert "active" influences on these "passive" movements. Another excellent example of the active involvement of the animal in affecting passive movements is provided by cases in which electrical gradients established by active transport influence the passive movements of

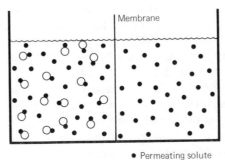

Numbers of molecules of the permeating solute

Free: 20   Free: 31
Bound: 15   Bound: 0
TOTAL: 35   TOTAL: 31

← – – – – – – – – – – –
Direction of net diffusion
of permeating solute

Membrane

• Permeating solute
○ Nonpermeating solute

Figure 4–4. A vessel divided into two compartments by a membrane. The nonpermeating solute is restricted to the left compartment. The total (bound plus free) concentration of the permeating solute is higher in the left compartment than in the right, but the permeating solute will diffuse from the right to the left because the free concentration is higher in the right compartment. See text for further explanation.

charged solutes; if Cl^-, for example, should be moving in tandem with Na^+ in response to the electrical gradient established by active transport of Na^+, the Cl^- movement, although passive in the strict sense, would obviously be only a small step removed from the active process of Na^+ movement. The distinction that is drawn between active and passive movement is generally useful, but the physiologist must maintain a certain perspective and not let semantics get too much in the way of a proper appreciation of the living beast.

The maintenance of a concentration difference across a membrane does not always signify even an indirect involvement of active transport, though some energy-demanding process is obviously necessary. Animals have in their body fluids a variety of high-molecular-weight solutes, such as respiratory pigments and other proteins, to which the body covering is typically impermeable. Concentration gradients between the organism and outside world for these substances are established and persist because the organism maintains the impermeable body covering and manufactures the solutes within the covering, both of these being energy-requiring activities.

A permeating solute may become bound to nonpermeating solutes, giving rise to a situation like that in Figure 4–4. In this circumstance the free concentration of the permeating material in the body fluids is lower than the total (bound plus free) concentration, and it is the free internal concentration that determines the concentration gradient for passive movement. Thus, as in the figure, there can be passive inward movement of the permeating solute despite what is superficially an opposing concentration gradient. There is good evidence that calcium, magnesium, and possibly some other divalent and polyvalent ions do become bound to a substantial extent to proteins within cells and body fluids in such diverse groups as crustaceans, molluscs, and mammals. Concentration gradients for these ions may thus be maintained passively, at least in part. Binding of monovalent ions and water seems, in general, to be of minor significance.

A more complex mechanism that very commonly results in "passive" maintenance of concentration gradients of permeating charged solutes is *Donnan equilibrium*. To understand the basis for this type of equilibrium, it is first necessary to recognize that animal body fluids contain solutes, such as proteins, that are charged and nonpermeating (or poorly permeating). Consider now that the *net* charge on all these nonpermeating solutes taken together may very well not be zero; that is, there may well be an excess of either positive or negative charges on the nonpermeating solutes. With this background we can turn to the simplified example shown in Figure 4–5. At the top we have two fluid compartments separated by a membrane that is permeable to Na^+ and Cl^- but impermeable to protein. The membrane separates two equimolar solutions (see "initial concentrations"), one composed of Na^+Cl^- and the other of Na^+ and protein$^-$ (the protein is presumed to be monovalently anionic). The situation on side 1 simulates that already discussed for body fluids in that the net charge on the nonpermeating solutes (protein) is not zero but negative. Now the system as we have described it is not at equilibrium. The concentrations of protein and Cl^- are strongly unequal on the two sides of the membrane, but diffusion cannot distribute both of these solutes evenly on both sides because the protein cannot cross to side 2. The equilibrium that develops thus is not one in which all solutes are equal in concentration on both sides but, instead, is a Donnan equilibrium. We can best understand the Donnan equilibrium by analyzing its development, starting with the system at the top of Figure 4–5. Cl^-, being able to permeate the membrane and initially being far more concentrated on side 2 than side 1, will tend to diffuse toward side 1. This movement of Cl^- in response to its concentration gradient, in turn, will cause side 1 to become negatively charged relative to side 2, thus establishing an electrical potential gradient that will attract Na^+ to side 1. Significantly, the movement of Na^+ to side 1 in response to the electrical gradient will establish a concentration gradient for Na^+, with the concentration on side 1 becoming increasingly greater than that on side 2. As the Na^+ concentration gradient increases, the tendency for Na^+ to return to side 2 in response to its concentration gradient will increase. Ultimately, this tendency will become great enough to oppose exactly the tendency for Na^+ to move toward side 1 in response to the electrical gradient established by Cl^- movement. Net Na^+ movement will then cease. Net Cl^- movement will also cease because the accumulation on side 1 of Cl^- ions that are not neutralized by accompanying Na^+ ions will establish an electrical gradient (side 1 negative) that will exactly oppose movement of Cl^- in response to the Cl^- concentration gradient. At this point the system will have arrived at a stable Donnan equilibrium, with concentrations on either side of the membrane being as shown at the bottom of Figure 4–5. The concentrations of Na^+ and Cl^- at equilibrium are described by the equation

$$\frac{[Na^+]_1}{[Na^+]_2} = \frac{[Cl^-]_2}{[Cl^-]_1}$$

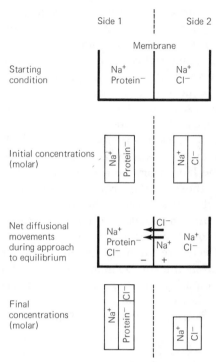

Figure 4–5. Development of a simple Donnan equilibrium. The membrane is permeable to Na+ and Cl− but is impermeable to the protein solute. The height of the concentration histograms is proportional to concentration. See text for explanation.

Note from the "final" concentration histograms that not only is each solute not distributed at concentration equilibrium, but the total molar concentration of all solutes taken together is greater on side 1 than on side 2. As we shall see in the next section, this implies a difference in osmotic pressure.

What we have seen with this example is that the presence of the non-permeating charged solute on one side of the membrane results in an equilibrium in which each permeating charged solute is more concentrated on one side than the other. Clearly, if the protein could cross the membrane, then, starting with the initial concentrations, protein and Cl− would diffuse in opposite directions to reach concentration equilibrium on the two sides of the membrane; electrical neutrality would be maintained because each Cl− that left side 2 would simply be replaced by a protein−, and, in the end, Na+ as well as the two anions would all be at concentration equilibrium. Instead, the nonpermeating protein constitutes a reservoir of negative charges that cannot be distributed equally on the two sides of the membrane, and this in turn prevents all other charged species from coming to concentration equilibrium.

The example in Figure 4–5 is analogous to the situation often pertaining across the outer body covering of aquatic animals in that nonpermeating protein solutes are present on one side (in the body fluids) and absent, or virtually so, on the other (in the ambient water). Donnan equilibria also develop across membranes within the animal. In vertebrates, for example, proteins are more concentrated in the blood plasma than in the tissue fluids, and there is a charge disequilibrium between the plasma proteins and tissue fluid proteins; this in turn produces a Donnan equilibrium in which permeating ionic species such as Na^+ and Cl^- are at unequal concentrations on either side of the blood capillary membranes.

In many marine invertebrates it is found that there are relatively small concentration gradients for many permeating ions across the outer integument; that is, concentrations of such ions are somewhat different in the body fluids than in seawater. The presence of nonpermeating proteins in the body fluids suggested that the ionic concentration gradients might largely be maintained by Donnan effects. This hypothesis has been tested by dialysis of body fluids against seawater: the body fluids are separated from seawater in the laboratory by a membrane whose permeability characteristics resemble those of the ordinary body covering. In this way the influence of active solute transport by the animal is eliminated. Echinoderm, annelid, and lamellibranch bloods are low in protein, and it is found that concentration gradients for permeating ions are essentially eliminated by dialysis; Donnan effects are virtually undetectable. Crustacean bloods are high in protein, and significant Donnan effects are found. These are sufficient to account for only a small part of the difference between normal blood and seawater, however, and active processes are thus predominant in maintaining concentration gradients of permeating salts even in the Crustacea.

TEST #1

OSMOSIS

The basic concept of osmotic movement of water

Consider a membrane separating two glucose solutions of different concentrations. If the membrane is permeable to glucose, we have seen that glucose will diffuse from the side with the higher glucose concentration to the side with the lower glucose concentration. It is also true that if the membrane is permeable to water, water will move passively from the side with the higher water concentration (lower glucose concentration) to the side with the lower water concentration (higher glucose concentration). These net movements of solute and solvent will continue, if unopposed, until the concentrations of glucose (and water) are the same on both sides of the membrane. The water is said to move across the membrane by *osmosis*. In the case of the freshwater fish in Figure 4–1, because solutes are more concentrated in the body fluids than in the ambient water, water will tend to enter the fish by osmosis. Accordingly, the fish faces a dual problem of diffusional loss of solutes to the medium and osmotic uptake of water, both processes tending to dilute its body fluids if unopposed.

Osmotic pressure, its definition and direct measurement

A membrane that is permeable only to water is known as a *semipermeable membrane*. No known biological membrane is strictly semipermeable, but some artificial membranes do meet this condition. Suppose that we separate a solution from pure water by a semipermeable membrane as in Figure 4–6A, with the membrane being mounted as the end of an ideal,

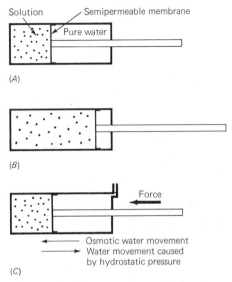

Figure 4–6. (*A*) A cylinder containing an aqueous solution separated from pure water by a semipermeable membrane. The membrane is mounted as the end of an ideal, frictionless piston. (*B*) A later condition of the system if the piston is left free to move. (*C*) Stable condition of the system if the increased hydrostatic pressure produced in the solution compartment by force on the piston is equal to the osmotic pressure of the solution. The open side arm on the pure-water compartment permits hydrostatic pressure in that compartment to remain constant as the hydrostatic pressure in the solution compartment is increased.

frictionless piston within a cylinder. If the piston is free to move, water will travel by osmosis from the side filled with pure water to the side filled with solution, and the resulting changes in volume of the two fluid compartments will cause the semipermeable membrane to move to the right as in Figure 4–6*B*. The solution will progressively become diluted. Because the solutes in the solution are unable to diffuse across a semipermeable membrane, we have a situation in which we can see the effects of osmosis alone in tending to bring the two fluid compartments into equilibrium.

Now if we return to the starting condition and exert a force on the piston as in Figure 4–6*C*, an increased hydrostatic pressure will be produced in the solution compartment. The difference in hydrostatic pressure induced between the two compartments will tend to drive water molecules through the pores of the membrane by ultrafiltration in the direction opposite to osmotic movement. Clearly, we can adjust the hydrostatic pressure so that there will be no net movement of water across the membrane, that is, so that osmotic movements will be exactly counterbalanced by movements resulting from the hydrostatic pressure difference. The increase in hydrostatic pressure required to achieve this is equal to the *osmotic pressure* of the solution, for the osmotic pressure of a solution is defined to be *equal to the increased hydrostatic pressure needed to prevent net movement of water when the solution is separated from pure water by a semipermeable membrane.* To illustrate, if the solution were 0.1*M* glucose, we would have to increase the hydrostatic pressure in the solution compartment by about 2.2 atm to prevent net movement of water. Then we would say that the osmotic pressure of the solution is 2.2 atm. If, by contrast, the solution were 0.2*M* glucose, we would have to increase the hydrostatic pressure in the solution

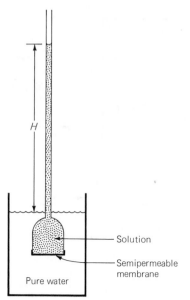

H

Solution

Semipermeable
membrane

Pure water

Figure 4–7. An apparatus for determining the osmotic pressure of a solution. See text for explanation.

compartment by about twice as much to prevent net water movement, and the osmotic pressure would be about twice as great. Clearly, the osmotic pressure of a solution provides a measure of the tendency for water to enter the solution by osmosis from pure water. The greater the osmotic pressure—that is, the greater the hydrostatic pressure needed to counteract net movement of water by osmosis—the greater must be the tendency for water to enter the solution by osmosis.

Figure 4–7 illustrates another apparatus for the determination of osmotic pressure. Here, as water moves from the pure water into the solution by osmosis, the increase in volume in the solution compartment results in an increase in the height *H* of the fluid column. The fluid column exerts a hydrostatic pressure on the solution compartment, and this pressure increases as the column rises. Ultimately, an equilibrium is reached such that water movement out of the solution compartment due to the hydrostatic pressure difference is exactly equal to inward movement due to osmosis. The pressure exerted by the fluid column (its height *H* multiplied by the specific gravity of the fluid) is then equal to the osmotic pressure of the solution.

The osmotic pressures of solutions are highly significant in the analysis of osmosis, for they determine the direction of osmosis between solutions and also affect the rate of osmosis. If two solutions are separated by a membrane that is permeable to water, osmosis will always occur from the solution of lower osmotic pressure into the solution of higher osmotic pressure. Further, if all other factors are held constant, the rate of osmosis will tend to increase as the difference in osmotic pressure between the solutions is increased. We shall return to these principles after looking further at the concept of osmotic pressure itself.

We have seen that the osmotic pressure of a solution is measured in relation to opposing hydrostatic pressures and is often expressed in units

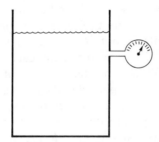

Figure 4–8. A beaker of water with a standard pressure gauge attached so as to measure the pressure at a given depth in the water. See text for further explanation.

of hydrostatic pressure such as atmospheres. Further, if a solution is separated from water by a semipermeable membrane as in Figure 4–7, we have seen that hydrostatic pressures are developed as water moves into the solution compartment and forces the fluid column upward. In fact, hydrostatic pressures developed through osmosis may be prodigious. If a 1.0*M* solution of glucose is placed in the solution compartment of Figure 4–7, for example, the fluid column will be over 700 ft high and exert a pressure of 22.4 atm at equilibrium. It is vital to recognize, however, that *osmotic pressures, in and of themselves, are not hydrostatic pressures.* Put another way, a solution by itself does not exert a hydrostatic pressure as a consequence of its osmotic pressure. Thus if we attach a standard pressure gauge to the wall of a beaker as in Figure 4–8, and, starting with pure water, we add glucose to the contents of the beaker so as to create a 1*M* solution, we know that the pressure gauge will register only the small changes in pressure that result from changes in the height and specific gravity of the solution in the beaker—certainly nothing resembling the 22.4 atm developed in the apparatus of Figure 4–7. Indeed, if the solutions in animal cells exerted hydrostatic pressures equal to their osmotic pressures, the cells would quickly burst. Hydrostatic pressures are developed only under certain circumstances: a solution must be separated from another solution of differing osmotic pressure by a water-permeable membrane, and there must be resistance to expansion in the compartment into which water moves by osmosis. Then the movement of water, tending to increase the fluid volume in a compartment that is not freely expansible, will increase the hydrostatic pressure in that compartment. The apparatuses in Figures 4–6*C* and 4–7 simply provide special cases of this type of situation, in which a solution is separated from pure water by a semipermeable membrane; in both cases there is resistance to volume expansion in the solution compartment that receives the osmotic influx of water.

   In summary, although solutions in and of themselves do not exert hydrostatic pressures by virtue of their osmotic pressures, hydrostatic pressures can be developed if solutions of differing osmotic pressure are placed in circumstances where osmosis occurs and the compartment receiving the increased fluid volume resists expansion.

Determinants of the osmotic pressure of a solution

   The osmotic pressure of a solution is determined basically by the total concentration of dissolved solutes, but this is a superficial statement that requires considerable refinement. It is to be emphasized from the start that only *dissolved* solutes contribute to the osmotic pressure; red blood cells or other suspended materials do not. We can consider a solution to consist of solute "particles" dissolved in water, with each separate dissolved entity constituting a particle. Thus if a glucose molecule is placed in solution, it constitutes a single dissolved particle. By contrast, if a NaCl molecule is placed in solution, it will dissociate into two ions, Na^+ and Cl^-, and each of these will constitute a separate dissolved particle. With this introductory background, we can state that the *osmotic pressure of a solution is proportional to the effective concentration of dissolved particles, regardless of the size or chemical nature of the particles*. The meaning of *effective concentration* will be elaborated below. As a first approximation, we can emphasize here that each dissolved particle—be it a glucose molecule, a large protein molecule, or a Na^+ ion—will make a roughly equal contribution to the osmotic pressure of a solution.

When nonelectrolytes are dissolved in water, the individual molecules go into solution separately, and each constitutes a dissolved particle. Because a mole of solute contains the same number of molecules (Avogadro's number) regardless of the solute in question, equimolar solutions of nonelectrolytes have equivalent particle concentrations and equivalent osmotic pressures. Thus a $0.1M$ solution of glucose has the same dissolved particle concentration as a $0.1M$ solution of urea, and both have virtually the same osmotic pressures. Similarly, a $0.1M$ solution of a large, nondissociating protein will also have essentially the same osmotic pressure. (You may have recognized that solutions of nonelectrolytes have the same ratio of dissolved particles to water molecules when they are of the same molality, not molarity. In point of fact, it is solutions of identical molality that exhibit the same osmotic properties, and the physical chemist uses molal concentrations in the study of osmosis. In biological work we are routinely confronted with "preconstructed" solutions such as body fluids and environmental media for which it is easier to determine concentrations in molar terms than in molal terms. At physiological concentrations, the errors involved in working with molar expressions are small, and this discussion will generally use such expressions.)

When electrolytes are dissolved in water, the individual molecules dissociate into two or more ions, each of which constitutes a dissolved particle. Ideally, then, a $0.1M$ solution of NaCl will exhibit twice the osmotic pressure of a $0.1M$ glucose solution, and a $0.1M$ solution of Na_2SO_4 will exhibit three times the osmotic pressure of the glucose solution. Electrolytes, however, do not show this ideal behavior. Dissociation may not be complete; and even when it is complete, as with strong electrolytes, the dissolved anions and cations tend to be attracted to each other, and the solution behaves as if it had a lower concentration of dissolved particles than would be surmised from simple dissociation kinetics. A $0.1M$ solution of NaCl, for example, exhibits about 1.9 times the osmotic pressure of a $0.1M$ glucose solution, not twice the osmotic pressure. The particle concentration of a solution as judged from its actual osmotic behavior is variously termed the *effective particle concentration*, the *osmotic concentration*, or the *osmotic activity*. It is important to recognize that the effective particle concentration is, in reality, a fictional concentration for strong electrolytes. A mole of NaCl in

Table 4–1. Cryoscopic coefficients for simple solutions of four electrolytes.

Concentration (molal)	NaCl	KCl	MgSO₄	CaCl₂
0.05	—	1.89	1.30	2.63
0.1	1.87	1.86	1.21	2.60
0.2	1.84	1.83	1.13	2.57
0.5	—	1.78	—	2.68
0.7	1.81	—	—	—

SOURCE: Heilbrunn, L. V. 1952. *An Outline of General Physiology.* 3rd ed. Saunders, Philadelphia.

solution does dissociate into 2 moles of particles, but the solution behaves as if there were fewer free particles because of electrostatic attractions among the ions.

The factor by which the molar (strictly, molal) concentration of an electrolyte is multiplied to get the effective osmotic concentration is termed the *activity coefficient* or *cryoscopic coefficient.* A 0.1 molal (*m*) solution of NaCl, for example, has a cryoscopic coefficient of 1.87 and an effective particle concentration of 0.187*m*. The cryoscopic coefficient varies with concentration in simple solutions of electrolytes (Table 4–1).

The physical chemist van't Hoff showed that the gas laws have application to the osmotic pressures of dilute solutions. The fundamental gas law may be written $PV = nRT$, where P is the pressure of the gas in atmospheres, V is the volume in liters, n is the number of moles of gas, R is the gas constant (0.082 liter • atm/°K/mole), and T is the absolute temperature. By rearrangement, we get

$$P = \frac{n}{V} RT$$

and it should be clear that n/V is the concentration. As applied to simple solutions of one solute, n/V is taken to represent the molar concentration C, and, substituting the osmotic pressure π for P, we get for solutions of *nonelectrolytes*

$$\pi = CRT$$

From this equation one can readily calculate that the predicted osmotic pressure of a 1*M* solution of nonelectrolyte at 0°C (273°K) is 22.4 atm—the same as the pressure exerted by 1 mole of ideal gas that has been compressed to a volume of 1 liter at 0°C. The early observation by Wilhelm Pfeffer that molar solutions of nonelectrolytes exert an osmotic pressure of approximately 22.4 atm at 0°C was, in fact, important in stimulating van't Hoff's application of the gas law to problems of osmotic pressure. The equation $\pi = CRT$ is fully accurate only for very dilute solutions, for in more concentrated solutions interactions among the solute and solvent molecules cause significant deviations from ideal behavior. Thus even for solutions of nonelectrolytes one must multiply the result by an osmotic coefficient G to get the true osmotic pressure: $\pi = GCRT$. The coefficient is 1.024 for a 0.1*M* solution of sucrose, for example, indicating a slight but significant departure from ideal behavior. (Because of such departures, equimolar solutions of various nonelectrolytes do differ slightly in osmotic pressure.) For simple solutions of *electrolytes*, G is the cryoscopic coefficient discussed earlier, and osmotic pressure is computed from the equation $\pi = GCRT$ using the cryoscopic coefficient.

The concept of osmolarity, an empirical measure of effective particle concentration

Biological solutions and environmental media are complex mixtures of many solutes, all of which interact with each other. It is impractical to compute their osmotic pressure using the equations for simple solutions of one solute presented above. Thus osmotic pressure is typically simply measured, and the effective particle concentration is expressed in empirical units of *osmolarity*. A 1-osmolar solution is one having an effective particle concentration of one Avogadro's number of particles per liter—or, in other words, is one having the osmotic pressure of a $1M$ solution of ideal nonelectrolyte (22.4 atm at 0°C). An osmole is an Avogadro's number of effective particles, and we speak analogously of milliosmolarity and milliosmoles. These are, it is to be emphasized, empirical units. Recalling our earlier example of a $0.1M$ solution of NaCl, the actual particle concentration is $0.2M$, but the solution behaves as if the particle concentration were $0.19M$ and thus has an osmotic concentration of 0.19 osmolar.

Colligative properties of solutions and the indirect measurement of osmotic pressure

Properties of solutions that depend on the effective particle concentration, regardless of the chemical nature of the particles, are termed *colligative properties*. There are four such properties: osmotic pressure, freezing-point depression, boiling-point elevation, and vapor pressure. An increase in effective particle concentration increases the osmotic pressure, increases the freezing-point depression (decreases the freezing point), increases the boiling-point elevation (increases the boiling point), and decreases the vapor pressure. The latter three effects are easily remembered by recognizing that increases in the effective particle concentration of a solution tend to impair any change in state of the water. Thus as the particle concentration is increased, lower temperatures are required for water to change from a liquid to a solid (the freezing point is lowered), higher temperatures are required for boiling, and the equilibrium vapor pressure established through evaporation in a closed system is reduced. The effect on osmotic movement of water is analogous. As the effective particle concentration is increased, the tendency for water molecules to leave a solution through a membrane is reduced. Thus if two solutions of different particle concentrations are separated by a membrane, water molecules move from the less concentrated to the more concentrated more readily than in the opposite direction, and net movement (osmosis) is toward the solution of higher concentration.

All four colligative properties vary predictably with each other as the effective particle concentration of a solution is varied. Accordingly, if any colligative property of a solution has been measured, all the others can be easily computed. This is the fundamental basis for the indirect measurement of osmotic pressure. For various technical and practical reasons, physiologists interested in osmotic pressures usually measure either the freezing-point depression or the vapor pressure of a solution and then calculate its osmotic pressure, rather than determining osmotic pressure directly. Both freezing-point depression and vapor pressure can be determined on very small volumes of fluid; the methods are thus adaptable to studies of even small animals. Again for technical and practical reasons, measures of freezing-point depression are much more commonly used than measures of vapor pressure, and we shall accordingly give special attention to the use of freezing-point depression.

The freezing-point depression of a solution is the amount by which its freezing point differs from 0°C. It is symbolized by ΔFP or simply by Δ and can be measured using substantially less than a drop of body fluid or environmental medium. The freezing point of a molar solution of ideal nonelectrolyte is $-1.86°C$. Thus the ΔFP of a 1-osmolar solution is 1.86°C, and for a solution of known ΔFP we get

$$\text{osmolarity} = \frac{\Delta FP}{1.86°C}$$

and

$$\pi \text{ at } 0°C = (22.4 \text{ atm}) \ \frac{\Delta FP}{1.86°C}$$

Often the osmotic pressure is simply expressed as ΔFP. This saves the effort of performing a mathematical conversion and is satisfactory because osmotic pressure and ΔFP are strictly proportional at a given temperature. Because we shall generally express osmotic pressure as ΔFP in this text, it is worth emphasizing again (and remembering) that a ΔFP of 1.86°C represents a 1-osmolar solution having an osmotic pressure of 22.4 atm (at 0°C). All three measures vary proportionally with each other; thus a solution with a ΔFP of 0.93°C is a 0.5-osmolar solution having an osmotic pressure of 11.2 atm, and so forth.

Basic principles of
water movement in biological systems

In situations of biological interest we are confronted with solutions separated by membranes, and we are concerned with movements of solutes and water across these membranes. Net movements of water by osmosis are governed by the osmotic pressure difference across a membrane and by the permeability of the membrane to water. Two solutions of the same osmotic pressure are termed *isosmotic*. If a solution A is of greater osmotic pressure than a solution B, A is said to be *hyperosmotic* to B, and B is said to be *hyposmotic* to A. When isosmotic solutions are separated by a semipermeable membrane, there is no net movement of water by osmosis. When solutions of different osmotic pressure are separated by such a membrane, water will move by osmosis toward the hyperosmotic side at a rate directly dependent on the osmotic gradient and membrane permeability.

Osmosis is not the only mechanism by which water moves across membranes. As we have seen earlier, water can also be driven across a membrane by ultrafiltration when there is a differential of hydrostatic pressure across the membrane. Accordingly, the net movement of water in a given situation is dependent on both osmosis and hydrostatic pressure effects. Such pressure effects are of widespread significance, but we shall introduce them later, as appropriate, and concentrate for the moment on osmotic movements in the absence of hydrostatic pressure differences.

The movements of water by osmosis are nicely illustrated by placing sea urchin (*Arbacia*) eggs in different osmotic environments. About 90% of the total volume of such eggs is composed of a solution of organic molecules and salts to which the cell membrane is only slightly permeable. Because sea urchin eggs are essentially isosmotic with seawater, an egg placed in normal seawater will exhibit no net uptake or loss of water. If an egg is exposed to artificially concentrated seawater, it will lose water by osmosis until the internal osmotic pressure is elevated to the osmotic pressure of the concentrated seawater. In the process, the cell volume will decrease. If an egg is exposed to diluted seawater, it will take up water by osmosis until isosmoticity is established. The cell volume will increase, and if the seawater is sufficiently dilute, the membrane will be unable to accommodate the volume necessary for isosmoticity, and the cell will burst.

Solutes and osmosis

The osmotic pressure of a solution is a collective effect of all solutes present, and it should be helpful to enunciate certain principles regarding the effects of particular solutes on osmotic water movements. First it must be recognized that biological membranes are typically permeable to at least some solutes as well as to water, and the permeability of a membrane to the particular solutes present on either side of the membrane is important in understanding both the movements of solutes and the movements of water. We can analyze some of the effects of solutes on osmosis in terms of solute effects on osmotic pressure differentials across membranes.

A solute, permeating or nonpermeating, that is equal in its osmotically effective concentration on the two sides of a membrane will not contribute to an osmotic pressure difference across the membrane. For example, if a membrane separates two complex solutions that both contain glucose and if the glucose concentration is equal on the two sides, the contribution of glucose to the osmotic pressure of each solution will be the same, and any differences in osmotic pressure between the solutions will have to be due to other solutes. This principle has the interesting corollary that if a membrane is permeable to a solute and the solute comes to equal concentration on each side by diffusion, the solute will no longer contribute to the osmotic pressure differential even though it may initially have been more concentrated on one side than the other.

If a solute is nonpermeating and more concentrated on one side of a membrane than the other, it will contribute a component to the osmotic pressure differential across the membrane that cannot be eliminated by passive movement of the solute, for the solute cannot diffuse to concentration equilibrium across the membrane. This is a significant principle in

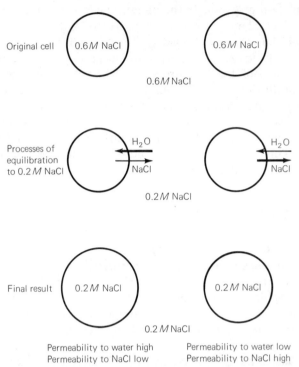

Original cell | 0.6M NaCl | 0.6M NaCl

0.6M NaCl

Processes of equilibration to 0.2M NaCl | H_2O / NaCl | H_2O / NaCl

0.2M NaCl

Final result | 0.2M NaCl | 0.2M NaCl

0.2M NaCl

Permeability to water high   Permeability to water low
Permeability to NaCl low   Permeability to NaCl high

Figure 4–9. Behavior of a simple cell containing 0.6M NaCl when placed in 0.2M NaCl. It is assumed that the volume of the solution outside the cell is sufficiently great that exchanges of NaCl and water do not alter the external concentration significantly. See text for explanation.

many situations. For example, the blood plasma of vertebrates contains higher concentrations of nonpermeating (or poorly permeating) proteins than the tissue fluids. Thus these proteins contribute to a differential of osmotic pressure between plasma and tissue fluids that cannot be eliminated by solute diffusion. This persistent osmotic pressure differential caused by proteins is termed the *colloid osmotic pressure* and will receive repeated attention in several subsequent chapters.

A permeating solute that initially is not distributed according to its electrochemical equilibrium across a membrane will move passively toward such equilibrium, and this solute movement may increase or decrease the prevailing osmotic pressure differential. Whereas uncharged permeating solutes will tend to diffuse to concentration equilibrium, charged solutes may diffuse so as to increase or decrease their difference in osmotically effective concentration across the membrane, depending on the initial concentrations on either side and the electrical effects that are present. The approach of a system to an equilibrium state will often involve a complex interplay among both water and solute movements.

Certain interactions between passive solute and water movements are illustrated by considering the simplified model cell in Figure 4–9. The cell is filled initially with 0.6M NaCl, and the membrane is presumed to be permeable to both NaCl and water. We shall assume that there is no electrical gradient across the cell membrane, so that NaCl diffuses strictly according to its concentration gradient. If the membrane is much more permeable to water than to NaCl and the cell is placed in 0.2M NaCl, the approach to equilibrium will be largely by osmosis. The cell will take up water rapidly, at the same time losing salts slowly, until isosmoticity has been established by the combination of the two processes. The cell volume will be greatly increased at equilibrium. If the membrane is much more permeable to NaCl than to water, the approach to equilibrium will be largely by diffusion of NaCl. The cell will lose NaCl rapidly, taking up only a small amount of water, and at equilibrium will be only slightly expanded relative to its initial condition. This shows, then, that solute movement can be an important factor in influencing water movement.

URINE AND FECES

The significance of urinary osmotic pressure in osmotic regulation

It is unfortunate that urine has gained the reputation among many of being simply a vehicle for voiding nitrogenous wastes. In fact, it often carries a great variety of salts and organic compounds and plays an exceedingly important role in the regulation of the body's content of these solutes. In many animals it plays a minor role in voiding nitrogenous wastes, and this serves to emphasize its more pervasive significance. The water in urine does act as an essential vehicle for transporting solutes from the excretory organs to the outside world, but it is, again, far too narrow to view water as simply a vehicle. Urinary water output is frequently intimately involved in regulating body water content. Thus we must view the kidneys as regulatory organs that exert control over all the constituents of the urine produced. Chapters 6 and 7 will deal with the structure and physiology of the kidneys and the voiding of nitrogenous wastes. Here we shall emphasize the characteristics of the urine as a factor in osmotic and ionic regulation.

Urine may be isosmotic, hyperosmotic, or hyposmotic relative to the body fluids. The relative osmotic pressure of the urine is expressed as the U/P ratio, or ratio of urinary osmotic pressure to plasma osmotic pressure. (Plasma is the solution that remains after blood cells and other osmotically inactive solid constituents have been removed from the blood, generally by centrifugation.) If U/P = 1.0, the urine is isosmotic to the plasma. If U/P is greater than or less than unity, the urine is, respectively, hyperosmotic or hyposmotic to the plasma.

Osmotic regulation is defined as maintenance of a stable osmotic pressure within the body fluids of the animal, and the relative osmoticity of the urine (U/P) directly reflects whether the urine can play a role in osmotic regulation. Suppose, to illustrate, that a freshwater fish (e.g., Figure 4–1), being hyperosmotic to the ambient water, takes up a quantity of water into its body fluids by osmosis from the medium. This will act to dilute the body fluids and reduce their osmotic pressure. Now the question is, can the fish help to restore its original osmotic pressure (osmotically regulate) through production of urine? To do this it must produce a urine more dilute than its body fluids, that is, a hyposmotic urine (U/P < 1.0). Such urine will preferentially void water over solutes, for it will contain more water in relation to its osmotically effective solute content than the body fluids. By this preferential loss of water—and the converse, preferential retention of solutes—the urine will serve to raise the osmotic pressure of the body fluids. Contrast this with the situation if the urine were always isosmotic to the body fluids. Then the fish, having suffered dilution of its body fluids through osmotic influx of water, would only void water in the same relation to solutes as pertained in its body fluids, and the urine would not serve to raise the osmotic pressure of the body fluids back to its original level. This introduces the general principle that production of isosmotic urine cannot serve directly in osmotic regulation. Production of hyposmotic urine, on the other hand, does serve osmotic regulation in animals hyperosmotic to their medium, and most freshwater animals have evolved the capacity to render their urine hyposmotic to the body fluids.

In animals hyposmotic to their medium, the passive movements of water and solutes between the body fluids and environment tend to raise the osmotic pressure of the body fluids. If the organism has suffered an increase in its internal osmotic pressure, production of isosmotic urine cannot serve osmotic regulation, that is, cannot lower the internal osmotic pressure back to its original level. However, production of hyperosmotic urine (U/P > 1.0) would serve osmotic regulation by preferentially voiding solutes in relation to water, thus favorably altering the ratio of solutes to water in the body fluids. Although some aquatic animals that are hyposmotic to their environment do have a capacity to produce hyperosmotic urine, others cannot raise their urinary osmotic pressure above isosmoticity with the body fluids. In the latter, the urine cannot directly aid osmoregulation.

Terrestrial animals are frequently faced with problems of water shortage and then can benefit by voiding solutes in as little water as possible. It is noteworthy in this context that the greatest capacities to concentrate the urine have evolved in terrestrial groups. Insects, birds, and mammals can all produce hyperosmotic urine, some of the mammals being able to raise urinary osmotic pressure to as much as 20 times the osmotic pressure of their body fluids. Other terrestrial groups, such as the reptiles, are unable to raise their urinary osmotic pressure above isosmoticity with the body fluids. Terrestrial animals may also sometimes have an excess of body water. Then production of hyposmotic urine is of advantage, for it voids the water with a minimal loss of body solutes. Many terrestrial animals are able to render their urine either hyperosmotic or hyposmotic as circumstances dictate.

In addition to whatever role it may play in osmotic regulation, urine also serves in the regulation of the body's water content (*volume regulation*) by voiding greater or lesser amounts of water as required. Even isosmotic urine can function in this capacity. Certain crabs that live in fresh water, for example, are strongly hyperosmotic to their environment and suffer an osmotic influx of water, yet produce only isosmotic urine. Although their urine production does not directly serve to keep their internal osmotic pressure high, it performs a vital role in ridding the body of the excess water gained osmotically. These two functions, regulation of osmotic pressure and regulation of body water content, must be kept clearly distinct.

Urine and ionic regulation

Ionic regulation is the maintenance of stable concentrations of ions in the body fluids, and urine often plays a role in this type of regulation. We can express the concentration of each ion in the urine as a ratio of its concentration in the plasma, that is, as a U/P ionic ratio. Taking Na^+ as an example, if the U/P Na^+ ratio is greater than unity at a given time, then the urine contains more Na^+ in relation to water than the plasma and is acting to lower the Na^+ concentration in the plasma. Converse arguments pertain if the U/P Na^+ ratio is less than unity. The functions of ionic regulation and osmotic regulation, though intimately interrelated, are hardly identical and must be clearly distinguished. In the case of osmotic regulation we are looking at control of the total osmotically effective concentration of all solutes combined, whereas in ionic regulation we are concerned with control of the concentrations of particular solutes. The distinction is illustrated nicely by the example of marine teleost fish, which are hyposmotic to sea-water, yet produce only isosmotic urine (osmotic U/P $\simeq$ 1.0). Because the total osmotically effective solute concentration in their urine is the same as that in the plasma, the urine cannot serve directly to regulate the osmotic pressure of the plasma. However, the solute *composition* of the urine is quite different from that of the plasma, and the urine does serve in ionic regulation. In particular, the urinary concentrations of Mg^{2+} and SO_4^{2-} are higher than those in the plasma. The fish are faced with excesses of these ions, which they pick up by diffusion from seawater, and the urine helps to maintain relatively low plasma concentrations.

Feces

The feces may be an important site of water and solute loss; in terrestrial animals faced with problems of water conservation, fecal water losses assume particular significance in water balance. Basically, two factors affect the amount of water lost in their feces: the amount of fecal matter produced and the water content of the fecal matter. There is evidence on mammals indicating that both of these factors may be adaptively modified in species occupying arid habitats. Desert kangaroo rats (*Dipodomys*), for example, were compared to laboratory rats in a setting where both were given the same diet, pearled barley. Per gram of barley consumed and digested, the kangaroo rats produced only about half as much fecal matter, measured in dry weight, as the laboratory rats. Further, the feces of the kangaroo rats were much drier, containing only about 0.83 g of water per gram of dry matter as compared to 2.25 g/g in the laboratory rats. Taking the two factors together, it turned out that, per gram of barley consumed, the kangaroo rats lost only about 19% as much water in their feces as the laboratory rats. A dromedary camel, another desert inhabitant, was found to void only 0.76 g of water per gram of dry fecal matter when on a diet of hay and dates with no drinking water, whereas a grazing cow voided 5.7 g/g. The feces of the camel became more moist (1.1 g/g) when drinking water was provided, illustrating the more general principle that fecal water content in mammals often varies with the diet and the animal's state of hydration. Some insects produce rather fluid feces, but in others the feces are dry, sometimes to the point of being described as a bone-dry powder. These scattered examples serve to illustrate that fecal water losses do vary to a considerable extent among species and indicate some of the measurements that should be made in taking fecal losses into account.

FOOD AND DRINKING WATER

The *overall* composition of available food and drinking water is an important consideration in water and salt relationships. In this section we shall consider a series of examples that will illustrate various dimensions of this principle.

Relative osmoticity of predator and prey in aquatic habitats

Each species of aquatic animal maintains certain osmotic and ionic concentrations in its tissues and body fluids. When an animal is consumed, its salt and water content, as well as its nutrient content, may have significant implications for the predator. Marine teleost fish and mammals are markedly hyposmotic to seawater, whereas most marine plants and invertebrates are isosmotic with seawater or nearly so. Consider a fish or mammal that consumes a meal of invertebrates. The tissues and body fluids of the invertebrate prey, being markedly hyperosmotic to those of the predator, impose a salt load on the predator that, if uncompensated, would tend to raise the osmotic pressure of the predator's body fluids. The predator, therefore, is obligated to eliminate the salt load in order to maintain osmotic-ionic homeostasis. By contrast, consider a fish or mammal that consumes a meal of fish. In this case the tissues and body fluids of the prey will be much closer to isosmoticity with those of the predator, and any salt load that may be incurred will be substantially less than that imposed by the invertebrate meal. In a very real sense, the fish-eating predator benefits from the osmotic work the prey performed in maintaining its body fluids hyposmotic to seawater—an intriguing lesson in ecological energetics. A similar situation is found among freshwater animals. All freshwater plants and animals are hyperosmotic to the medium, to a greater or lesser extent. Predators consume what are, in essence, packets of relatively concentrated solutes and benefit from the osmotic-ionic work done by their prey.

The water content of air-dried foods

Desert animals often face precarious problems of water balance. Many consume air-dried seeds and other dry plant matter, and the water content of the food, as well as its nutrient content, may be critical to survival. It thus becomes significant to recognize that these foods, even though ostensibly dry, equilibrate with air moisture and vary in their water content as humidity varies. At 25°C, "dry" pearled barley, for example, contains 3.7 g of water per 100 g dry weight at 10% relative humidity and 18.1 g per 100 g at 76% relative humidity. Ambient relative humidity in the desert is often as low as 10–15% during the day, but is higher at night, and hygroscopic plant material equilibrates rapidly with air humidity. Thus nocturnal feeders may receive considerable amounts of free water in their food. Humidities are often higher below ground than in the daytime air. Insects and other animals may dig for food buried in the sand, and burrowing rodents are known to store food in their burrows, where it may take up water.

Foods of high water content in the desert

Some desert herbivores live on succulent foods. Here the amount of water is high, but other problems may be presented. Cacti contain as much as 90% water, but besides the problem of negotiating the spiny surface (which the pack rat, for example, accomplishes with remarkable skill), the animal must handle the fine silicaceous needles and oxalic acid present in the tissues of some species. Oxalic acid, found in certain other succulent desert plants as well, is exceedingly toxic to some animals, such as sheep, which can be killed by grazing on plants with high concentrations. Pack rats live almost exclusively on cacti at certain times of year. By contrast, white rats given cholla cactus from which the spines had been removed died sooner than they would have without any food at all, possibly because of the oxalic acid. Why the pack rat is not adversely affected by this chemical is unknown.

Some succulent desert plants, known as halophytic or salt-loving species, live in saline soils, and many have very high salt concentrations in their tissues. Total salt concentration may exceed that of seawater by as much as 50%, and some species contain high amounts of oxalic acid as well. These plants constitute a major part of the diet of the sand rat (*Psammomys obesus*), for example, and are consumed at times by dromedary camels. Whereas their fleshy parts are juicy and can provide needed water to a predator, the sodium concentration of their tissue fluids—to take the case of just one particular ion—may be five times that in mammalian plasma. This means that the predator must be able to produce urine that is at least five times as concentrated in sodium as its plasma, for otherwise it would have to void more water in getting rid of the sodium than it took in with the sodium and would suffer a net water loss in processing the sodium load. Many mammals cannot achieve such a high urinary sodium concentration and thus could not gain water by eating these plants. Because the fleshy parts of the plants may be 80–90% water, an animal such as the sand rat that depends on them for nutrition must eat large quantities and incur a correspondingly large salt load. This example again emphasizes that the total composition of the food must be considered in any effort to appraise its osmotic-ionic significance.

Desert animals that eat animals or succulent plants (many of which have only moderate salt concentrations) are consuming packets of water that have been accumulated from the arid environment through the physiological efforts of the consumed organisms. The predator may thus sometimes take advantage of specialized abilities of those organisms, as may be illustrated by several examples. Plants, with their deep root systems, draw water from places that many of their predators do not enter; predators capitalize on this attribute when they consume the water-laden tissues of succulent plants. Animals of relatively low mobility may consume ones of greater mobility, which can collect food and water over a larger area. Animals with modest capabilities to conserve water may prey on ones with highly developed capabilities. For example, the kangaroo rat can maintain its body water content when eating air-dried seeds and drinking no water. Raptorial desert birds require considerable free water in their diets and get it when they consume a kangaroo rat. Thus, though the birds cannot survive with as little free-water intake as their prey, they are, in an indirect sense, able to do so, and they can live in arid regions. The adaptation of certain species to exploit the adaptations of other species, graphically illustrated in the desert, is a principle of the most widespread application.

The special significance of protein foods for water balance in terrestrial animals

Carbohydrates and lipids are composed primarily of carbon, hydrogen, and oxygen, and their oxidative catabolism results in formation of CO_2 and water. The CO_2 is voided by the respiratory organs, and the water contributes to the organism's overall water resources (see later section on metabolic water). In contrast, proteins contain large amounts of nitrogen as well as carbon, hydrogen, and oxygen, and their catabolism results in nitrogenous wastes as well as CO_2 and water (see Chapter 6 for a detailed discussion). The nitrogenous wastes are usually voided in the urine of terrestrial animals. The principal nitrogenous waste of mammals, for example, is a highly soluble compound, urea. Because urea is necessarily voided in solution, a given protein intake, implying production of a given amount of urea in the steady state, obligates the animal to a certain loss of urinary water, depending on the concentrating ability of the kidney. Thus the protein content of the food has direct implications for urinary water loss; if, for example, a mammal is concentrating urea maximally in its urine, high-protein foods will demand greater urinary water loss than low-protein foods. (The relationship is not always so simple, because mammals can vary the urea concentration of their urine.) Birds, lizards, and many insects, in contrast to the mammals, excrete waste nitrogen largely in the form of poorly soluble uric acid and urates, compounds which can be voided as a thick sludge or solid mass rather than in solution. The protein content of their diet thus does not have the immediate implications for water loss that it does in species producing mostly soluble nitrogenous wastes, though many animals that predominantly excrete uric acid are apparently obligated to loss of sufficient water to move the uric acid as a sludge through their excretory passages.

Animals and saline drinking water

Some animals can gain water from drinking saline waters such as seawater, whereas others cannot. A critical factor is whether the salts in the water can be eliminated in less water than was taken in with them. We shall see that some animals (e.g., marine teleost fish and some desert mammals) can achieve this when drinking seawater. However, others cannot and are therefore, somewhat paradoxically, dehydrated by drinking seawater. A good example is provided by man. We have all heard Coleridge's famous line from "The Rime of the Ancient Mariner," "Water, water, everywhere, nor any drop to drink." Sailors discovered long ago that drinking seawater was worse than drinking no water at all. We can now understand this in terms of fundamental physiological principles. The maximum osmotic concentration of human urine ($\Delta FP = 2.6°C$) exceeds the osmotic concentration of seawater ($\Delta FP = 1.9°C$). Some of the osmotic pressure of the urine, however, is contributed by urea and other organic excretions, rather than salts, and, what is more to the point, the maximum concentration of chloride in human urine is less than the concentration of chloride in seawater. Thus the predominant anion in seawater can be eliminated only at the cost of more water than was taken in, meaning that to excrete the chloride a man drinking seawater not only would have to use all the water ingested, but

also would need to draw on other bodily reserves of water. Here again we
see the importance of considering the concentrating ability for each solute
in the urine in addition to the overall osmotic U/P ratio. The problem of
water balance in man is further aggravated by the fact that the $MgSO_4$ in
seawater induces diarrhea, thus increasing fecal water loss. $MgSO_4$ under
the guise of Epsom salts is a well-known home laxative.

Concluding comment

One of the major points of this discussion of food and water has been
that total composition must be considered and that the animal, in the
process of taking in needed nutrients, salts, or water, may take in an excess of other materials. It is well to remember that although the physiological machinery can operate to correct such excesses, deficits must inevitably
be replaced by uptake from the outside world, often in food or drink, and
the animal must adapt to handle the total composition of the foodstuffs and
waters available in its environment.

METABOLIC WATER

When foodstuffs are aerobically catabolized, water and CO_2 are produced, as illustrated by the equation for glucose catabolism:

$$C_6H_{12}O_6 + 6O_2 \rightarrow 6CO_2 + 6H_2O$$

The water produced in catabolism is known as *metabolic water* or *oxidation
water*, in contrast to *preformed water*, which is taken in as such from the
environment. Metabolic water is produced in all animals.

Table 4–2 gives the amount of water formed per gram of foodstuff
oxidized. The yield of water from protein catabolism is dependent on the
disposition of the waste nitrogen. Urea has a higher hydrogen-to-nitrogen
ratio than uric acid (Figure 6–3); thus there are fewer hydrogens available
for water formation when urea is the nitrogenous end product.

The following discussion of metabolic water will emphasize terrestrial
animals inasmuch as it is in them that metabolic water has received the
most attention as a factor in water balance.

The calculation of net gain
or loss of water in catabolism

There are certain obligatory water losses the animal sustains directly
in the catabolism of foodstuffs; these may be classed as respiratory, urinary,
and fecal. To assess the net impact of catabolism of food on water balance,
the obligatory water losses must be subtracted from the gain of metabolic
water. This will be exemplified after briefly discussing the nature of the
obligated losses:

1. *Respiratory losses* of water are obligated in the oxidation of all
    foodstuffs. The animal must take in oxygen to meet the demands
    of cellular oxidation and in the process loses water evaporatively
    across its respiratory tract.
2. *Urinary losses* of water are obligated in the oxidation of proteins

but not in that of carbohydrates or lipids. The catabolism of proteins results in nitrogenous wastes that often must be excreted in the urine at a cost in water.

3. *Fecal losses* must be taken into account in the case of all foodstuffs when considering the catabolism of ingested, rather than stored, foods. Ingested foods usually contain at least some *preformed water*. If the animal loses more water in its feces than it took in with its ingested food, it then suffers a net fecal loss in processing the food, and the fecal loss must be subtracted from metabolic water production in assessing the overall effect of catabolizing the ingested food. On the other hand, if the feces contain less water than was taken in with the food, then the animal realizes a net gain of *preformed water*, which is ignored in calculating the net gain of metabolic water.

To give these concepts more specific meaning, we may now consider an example. Desert kangaroo rats (*Dipodomys*) were studied at a temperature of about 25°C and a relative humidity of about 33% and were fed pearled barley at hygroscopic equilibrium with the prevailing humidity. In the box below it is shown that they realized a net gain of water in the catabolism of the barley under these conditions. The net gain is computed by first determining the yield of metabolic water in the cellular oxidation of a gram of barley and then determining and subtracting the water losses obligated in the catabolism of a gram of barley.

METABOLIC WATER PRODUCTION (0.54 g H_2O/g barley)
By knowing the nutrient composition of pearled barley and the amount of water produced in the catabolism of the major foodstuffs (Table 4–2), it can be calculated that the nutrients derived by the kangaroo rat from each gram of barley (dry weight) will yield about 0.54 g of metabolic water during cellular oxidation.

OBLIGATED WATER LOSSES (total: 0.47 g H_2O/g barley)
1. RESPIRATORY (0.33 g H_2O/g barley). Oxidation of the nutrients derived from a gram of barley requires consumption of about 810 cc of oxygen; this entails an evaporative loss of about 0.33 g of water across the lungs at the conditions of the experiments.
2. URINARY (0.14 g H_2O/g barley). The protein in a gram of barley yields about 0.03 g of urea. Urea can be concentrated sufficiently in the urine that this amount can be voided in about 0.14 g of water.
3. FECAL (0 g H_2O/g barley). Each gram (dry weight) of barley ingested contains about 0.1 g of preformed water, and the feces resulting from the digestion of a gram of barley contain only about 0.03 g of water. Thus there is a net gain of *preformed* water in the digestion of the barley; this is ignored in calculating the net gain of metabolic water.

NET GAIN OF METABOLIC WATER (0.07 g H_2O/g barley)
Because about 0.54 g water is produced in oxidation of a gram of barley and a total of only about 0.33 + 0.14 = 0.47 g is obligatorily lost, the kangaroo rat realizes a net gain of about 0.07 g water in the metabolism of each gram of barley. This can be used to offset other water losses, or, if it represents an excess, can be excreted.

Table 4–2. Average amount of metabolic water formed in the oxidation of the basic foodstuffs.

	Grams of Water Formed per Gram of Food
Starch	0.56
Lipid	1.07
Protein when urea is the end product of nitrogen metabolism	0.40
Protein when uric acid is the end product of nitrogen metabolism	0.50

SOURCE: Schmidt-Nielsen, K. 1964. Terrestrial animals in dry heat: desert rodents. *In*: D. B. Dill (ed.), *Handbook of Physiology. Section 4: Adaptation to the Environment.* American Physiological Society, Washington, D.C.

It is instructive to compare the performance of the kangaroo rat with that of the white laboratory rat studied under comparable conditions (25°C, 33% relative humidity). Because white rats obtain about the same nutrients from each gram of barley as kangaroo rats, their production of metabolic water is similar. However, in contrast to the kangaroo rats, which suffer total obligated losses of about 0.47 g H_2O/g barley, the laboratory rats are obligated to a total loss of about 0.60 g H_2O/g barley, partitioned as follows: a respiratory loss of about 0.33 g, a urinary loss of about 0.24 g, and a net fecal loss of about 0.03 g. The higher urinary loss and the net fecal loss reflect the fact that the white rat, by comparison to the kangaroo rat, can achieve only a lower urea concentration in the urine and, for reasons discussed earlier, loses more fecal water per gram of barley ingested. Given a comparable metabolic water production, the white rat will thus sustain a net *loss* of about 0.06 g water in the catabolism of a gram of barley. That is, the white rat not only must void all the water produced metabolically from the oxidation of the nutrients in the barley, but also must draw on other water resources to meet its total obligated water losses. This result, taken in the context that the kangaroo rat realizes a net gain of water, shows clearly the importance of considering the water losses obligated by metabolism, as well as the metabolic water production, in determining the overall effect of metabolism on water balance. Even though the white rat (as with many other mammals) suffers a net loss of water under the conditions of these experiments, it is essential to recognize that this does not mean that metabolic water production imposes a water stress on the animal. Quite to the contrary, the rat must ingest and metabolize food to obtain energy and would have the same total obligated water loss, 0.60 g/g barley, even if no metabolic water were produced. The metabolic water production offsets 90% of this loss and thus plays a vital and positive role in the water economy.

The potential role of carbohydrate and lipid stores as sources of water

Animals store nutrients in the form of both carbohydrates and lipids, and it has frequently been suggested that they could oxidize these stores to obtain water in times of water shortage. This concept has been applied to lipids in particular, for you will note in Table 4–2 that oxidation of a gram of lipid actually yields somewhat more than a gram of water. From this observation it has sometimes been hypothesized that the abundant fat stores of some desert animals are, in a sense, water stores in disguise. It was claimed, for instance, that the hump of the camel, although not filled with water, is a type of water-storage site inasmuch as it is filled largely with adipose tissue. As we shall see shortly, this argument cannot be supported in the case of the camel, but the student who delves into the literature of physiological ecology will find continuing debate regarding the role of fat storage in the water economies of other animals.

To assess the potential role of stored carbohydrate and lipid as sources of water, we must apply the same type of analysis as introduced in the previous section. Obligatory water losses must be subtracted from metabolic water production to determine whether oxidation would result in a net water gain. The analysis is simplified by two considerations; first, oxidation of carbohydrates and lipids does not obligate urinary water losses, and second, in considering oxidation of stored nutrients we need not be concerned with fecal losses. Thus the problem resolves to considering metabolic water production in relation to the respiratory water loss obligated to supply oxygen for the oxidation. From the outset we may note that although lipids yield about 90% more metabolic water per gram than carbohydrates, they also require about 140% more oxygen per gram for oxidation. Thus under given conditions lipid metabolism entails a greater respiratory water loss per gram than carbohydrate metabolism, signifying, at the least, that the purported advantages of stored lipids as sources of water are not so straightforward as might be suggested by Table 4–2.

Two of the important parameters that affect water balance in the oxidation of stored compounds are fixed by biochemical considerations: the amount of metabolic water yielded per gram and the amount of oxygen required per gram. The amount of oxygen required, however, is only one factor affecting the obligatory respiratory water loss, the other being the amount of water lost per unit of oxygen consumed. It can be shown by simple calculations that if the respiratory water loss per unit of oxygen consumed exceeds about 0.7 mg H_2O/cc O_2, the catabolism of both lipid and carbohydrate results in a net water *loss*, though the loss of water is less for carbohydrate than for lipid. If respiratory water loss is in the range of about 0.44–0.7 mg H_2O/cc O_2, there is a net gain of water in carbohydrate metabolism that exceeds any gain realized in lipid metabolism; that is, carbohydrate provides a richer source of metabolic water. Below a respiratory water loss of about 0.44 mg H_2O/cc O_2, the net gain of water in lipid metabolism exceeds that in carbohydrate metabolism, and lipid is the richer source of metabolic water.

The rate of water loss per unit of oxygen consumption depends on the respiratory physiology of a species, and within a species it depends on conditions of temperature and humidity. Thus we can readily understand why different authors have made different claims regarding the absolute and relative merits of carbohydrate and lipid as water sources: The answer depends on both the species under study and the ambient conditions. Because the rate of water loss exceeds 0.7 mg H_2O/cc O_2 in camels at the ambient conditions of deserts, camels sustain a net *loss* of water in the catabolism of both lipids and carbohydrates, and it is clear that neither group of compounds can be viewed as a water store. Kangaroo rats at 25°C and at relative humidities of 0% to about 30% have a respiratory loss in the range of 0.44–0.54 mg H_2O/cc O_2 and thus realize a greater water gain in the catabolism of carbohydrates than lipids; at higher ambient humidities their water loss falls below 0.44 mg H_2O/cc O_2, and lipids are the richer source of metabolic water. It is possible that some insects realize a sufficiently small loss of respiratory water that lipids are the richer water source even at very low humidities. In any case it seems best at present to view storage of foodstuffs as primarily governed by the need for nutrients, recognizing, of course, that catabolism will yield water that may be of great importance to water economy, especially in arid environments.

Factors that affect the role of metabolic water in meeting water needs

Metabolic water may constitute anywhere from a dominant to a minor share of the total water supply, depending on the species and environmental conditions. At moderate temperatures and even at low relative humidities (as low as around 20%), the kangaroo rat and certain other desert rodents survive only on metabolic water and the small amounts of preformed water found in air-dried seeds. Some refuse drinking water even when it is available. In these animals under arid conditions, metabolic water may constitute over 90% of the total water supply necessary for life. Many rodents from more-mesic habitats can survive under these environmental conditions only when they receive more preformed water than is contained in air-dried foods. They require drinking water or succulent foods, and metabolic water accounts for a correspondingly smaller percentage of the total supply needed. Comparisons such as this have occasionally led to the misconception that animals such as the kangaroo rat produce especially large amounts of metabolic water. This is not the proper explanation.

It cannot be emphasized too much that the amount of water formed per gram of foodstuff metabolized is fixed by biochemical considerations and thus is the same for all animals, except that in protein metabolism the amount is dependent on the nitrogenous end product. Kangaroo rats have metabolic rates similar to those of more-mesic rodents of their size, and empirical evidence indicates that metabolic water production in kangaroo rats is comparable to that in other rodents. What, then, is the explanation for the greater role of metabolic water in supplying water needs in the kangaroo rat? The answer is that kangaroo rats *conserve* water more effectively than more-mesic rodents; we have seen certain evidence for this in earlier discussions. Accordingly, in the arid environment of the desert, kangaroo rats suffer less water loss than would mesic rodents and therefore require less water input. Metabolic water production need be supplemented with less preformed water, and it is thus that metabolic water meets a high proportion of total water needs.

In general, the role of metabolic water in water economy depends not merely on the absolute rate of metabolic water production, but also on the other factors affecting water gain and loss. The latter include physiological capacities to conserve water, ambient conditions such as temperature

and humidity that affect rates of water loss, and the availability of preformed water in food and drink. The importance of ambient conditions may
be illustrated, again, with examples from the rodents. In humid situations
at moderate temperatures, evaporative water losses are reduced by comparison to arid conditions. This, in turn, reduces water needs and thus
diminishes dependence on preformed, as opposed to metabolic, water. At a
relative humidity of 70% at 25°C, metabolic water production in the kangaroo rat, for example, may exceed absolute water need by over 35%; the rat
could get along with no preformed water at all. More-mesic rodents also
can survive with less preformed water at higher humidities, or even with
none at all. The availability of preformed water is also significant. Animals
that routinely drink or eat water-rich foods may receive so much preformed
water that metabolic water becomes rather incidental.

WATER AND SOLUTE
MOVEMENTS WITHIN THE BODY

For the most part we are concerned in this chapter with exchanges
between the animal and the outside world. It is often pertinent, nonetheless,
to consider exchanges of water and solutes within the body proper.

The aqueous solutions of the body are said to be distributed in several
compartments. The intracellular compartment comprises the contents of
the cells, and the extracellular compartment comprises fluids outside of
cells. Both compartments are further divisible. We may recognize the individual intracellular compartments of various tissues, such as liver and
skeletal muscle. In vertebrates and other groups with closed circulatory
systems, we distinguish the blood compartment, or contents of the circulatory system, from the lymph compartment, or fluids that immediately
bathe the cells and move through the lymphatic system if one is present.
Body water is distributed variously among the compartments, depending
on species. Intracellular water, for example, constitutes about 80% of the
total body water in man but only about 15% in the mollusc *Aplysia*. Concentrations of particular solutes may differ greatly from one compartment
to another. The intracellular concentration of sodium, for instance, is often
much lower than the concentration in the interstitial fluids bathing the cells.

Movements of water and solutes among various compartments have
been studied carefully in only a few species. It is clear that these movements may involve very substantial quantities of material, so much so that
they may exceed by a good measure the contemporaneous movements between the organism and its environment. In man, for example, the salivary
glands, gastric mucosa, and other glands secrete as much as 5–10 liters of
solution into the gut per day. Most of the water and salts are resorbed,
so that only about 100 cc is lost in the feces. In the human kidney, fluids
are driven by ultrafiltration from the blood into the kidney tubules. Nearly
200 liters per day may enter the tubules, and 99% of this volume is resorbed.

Solutes often are not distributed in electrochemical equilibrium across
the membranes separating body fluid compartments. Whenever the membranes are permeable to a particular solute, passive movements toward
electrochemical equilibrium will occur, and it is typical to find active transport mechanisms acting to maintain the disequilibria necessary for life. The

magnitude of the problem is illustrated by noting that the surface area of all the cells in a human may be 15,000 m^2 or more. The contents of each cell differ in composition from the surrounding interstitial fluids. In human erythrocytes the energy required to transport sodium and potassium against their electrochemical gradients so as to maintain proper intracellular concentrations amounts to some 10–20% of the total cellular metabolism. This appears to be a good approximation of the relative cost of ionic regulation in other mammalian cell types as well. The overall picture that emerges is one of considerable movement of water and ions within the body and of considerable energy expenditure to accomplish or counteract these movements.

When the organism, in its relationship to the environment, experiences a shortage or overload of water or salts, the distributions of these materials among the body compartments may prove to be critical in understanding the overall response. We have referred earlier to the principle that a decrease in blood volume in mammals exposed to dehydration and heat may be an important factor in causing death. As its water content decreases, the blood becomes more concentrated and viscous; circulation becomes more arduous; and convective movement of heat from the body core to the skin by the blood may slow to the point that core temperature rises to lethal levels. The dromedary camel can survive and recover without assistance from a water loss of over 25% of its body weight, as compared to only around 13% in many other large mammals, when exposed to the heat of the summer desert. At least part of the explanation of the camel's tolerance to dehydration appears to lie in the relative distribution of the total water loss among the fluid compartments of the body. Water is not lost from the various compartments in proportion to their volumes in the normally hydrated individual. In man, plasma volume decreases to a disproportionately large extent, whereas in the camel it decreases to a disproportionately small extent. Men, for example, who had lost about 10% of their body water exhibited about a 25% reduction in plasma volume; but camels that had lost about 30% of their body water had only about a 20% reduction in plasma volume. Because of these important differences in the distribution of the water loss, the camel can tolerate more total dehydration than man before blood viscosity rises to threatening levels.

Calcium is an essential mineral participating in such diverse processes as muscle contraction and coagulation of the blood. It is well known that mammals placed on a low calcium diet remove calcium from the skeleton to support proper plasma and tissue calcium levels, thus accommodating to an environmental stress through redistribution of material among the body compartments. This response, as in other tetrapod vertebrates, is mediated by the release of parathyroid hormone. It is also of interest to note the hormonally controlled shifts in calcium that occur during molting in many crustaceans. Prior to the molt, calcium is withdrawn from the old exoskeleton and deposited in the hepatopancreas or in concretions in the stomach called gastroliths. After molting, these calcium stores are directed to calcification of the new exoskeleton. Through these movements among the body compartments, the animal limits loss of calcium with shedding of the old exoskeleton, and its needs for calcium intake during calcification of the new exoskeleton are correspondingly reduced.

SELECTED READINGS

Davson, H. 1970. *A Textbook of General Physiology.* 4th ed. J. & A. Churchill, London.

Giese, A. C. 1973. *Cell Physiology.* 4th ed. Saunders, Philadelphia.

Lowry, W. P. 1969. *Weather and Life.* Academic, New York.

Potts, W. T. W. and G. Parry. 1964. *Osmotic and Ionic Regulation in Animals.* Macmillan, New York.

Schmidt-Nielsen, K. 1964. *Desert Animals.* Oxford University Press, London.

See also references in Appendix.

5 EXCHANGES OF SALTS AND WATER: INTEGRATION

Having reviewed the mechanisms operative in water and salt relationships, we shall now proceed to integrated study of the animal groups. The environmental challenges faced by organisms and their responses are exceedingly diverse, making for a rather lengthy treatment. Thus it may be helpful to provide a few words on the organization of the chapter at this point. The basic organization is by habitat rather than phylogeny in the belief that this provides a more straightforward synthesis. The first three sections will cover animals in fresh water, the ocean, and waters of intermediate salinity. Within each of these sections, the organization will largely follow mechanistic lines. However, a summary of the relations of aquatic organisms following these sections is organized along phyletic lines. The summary should help to crystallize the diverse responses of the various animal groups, and you may want to refer to it as you read the earlier sections. The material on terrestrial organisms toward the end of the chapter is approached largely phyletically throughout.

ANIMALS IN FRESH WATER

Introductory comments on the characteristics of natural waters

The total concentration of salts in natural waters is generally expressed as the *salinity*, in units of grams of solid matter per kilogram of water, or parts per thousand (‰). Seawater has a mean salinity of approximately 35‰, corresponding to a freezing-point depression of about 1.9°C. Fresh waters have salinities as low as 0.015‰, and salt lakes may have salinities in excess of 200‰. Fresh waters mix with more salty waters in many situations, such as estuaries, and there is, in fact, a rather continuous spectrum of salinities between the extremes given. Thus designation of an upper salinity limit for fresh waters is in part arbitrary. A commonly used upper limit is 0.5‰. An approximate average for lakes and rivers is 0.1‰.

"Fresh water," it should be clear, is not a single, invariant commodity. Total salinity and ionic composition vary with the source. Concentrations of individual common ions such as sodium, potassium, calcium, and chloride may vary from a few millimoles per liter down to a few tenths, or less, of a millimole per liter. Osmotic pressure is always very low, on the order of $\Delta 0.001°$–$\Delta 0.02°C$ (where Δ signifies freezing-point depression). The freshwater organism must adapt to living in this dilute medium. By comparison to animals in the open ocean, the freshwater organism is frequently also confronted with greater variation in temperature and pH.

Osmotic pressures and salt concentrations of the body fluids and other factors affecting passive exchanges of salts and water

The osmotic concentrations of the body fluids of some representative freshwater animals are given in Table 5–1. You will note that all are hyperosmotic to the medium. The mussel *Anodonta* has one of the lowest osmotic concentrations known, but even that is well above the concentration of fresh water. Apparently, body fluids as dilute as fresh water are incompatible with life. Freshwater animals have various evolutionary heritages, and the compositions of their body fluids have been influenced by both their ancestry and their adaptation to existence in the freshwater medium.

Being hyperosmotic to the medium, freshwater animals face the prob-

Table 5–1. Osmotic concentration and concentrations of some solutes in the blood plasma of some freshwater animals.

	Osmotic Concentration (ΔFP, °C)	Solute Concentrations (mM)					
		Na$^+$	K$^+$	Ca^{2+}	Mg^{2+}	Cl$^-$	HCO$_3^-$
Chlorohydra viridissima, hydrozoan coelenterate, tissue fluids	0.08						
Anodonta cygnaea, freshwater mussel	0.08	15.6	0.5	6	0.2	11.7	12
Lymnaea peregra, snail	0.23						
Astacus fluviatilis, crayfish	0.81	212	4.1	15.8	1.5	199	15
Aedes aegypti, mosquito larva	0.50						
Salmo trutta, brown trout	0.61	161	5.3	6.3	0.9	119	
Rana esculenta, frog	0.44	109	2.6	2.1	1.3	78	26.6

SOURCE: Potts, W. T. W. and G. Parry. 1964. *Osmotic and Ionic Regulation in Animals*. Pergamon, Oxford. Used by permission of Pergamon Press Ltd.

lem of continuous osmotic uptake of water, which, unopposed, would lead to dilution of their body fluids. As shown in Table 5–1, many ionic solutes in their body fluids are more concentrated than in fresh water. Thus these animals are also confronted with tendencies to lose these solutes to the medium by diffusion—a factor that, again, would lead to dilution of the body fluids if unopposed. In short, freshwater animals face two basic passive processes, osmotic gain of water and diffusional loss of salts, that both tend to bring their body fluids into equilibrium with their surroundings. Homeostatic maintenance of osmotic and ionic disequilibria between the body fluids and medium requires the active intervention of the animal.

In a broad sense the energetic demands of osmotic and ionic regulation are expected to vary directly with rates of passive water and salt exchange. The more rapidly water is taken up by osmosis and the more rapidly salts are lost by diffusion, the more metabolic energy the animal will have to invest in counteracting these tendencies so as to maintain homeostasis. Three factors are important in determining passive rates of exchange: (1) the osmotic and ionic gradients between the body fluids and surrounding medium, (2) the permeability of the body wall to water and ions, and (3) the surface area for exchange. Each of these deserves brief discussion.

Freshwater animals typically have less-concentrated body fluids than related marine forms. Most marine decapod crustaceans, for example, are virtually isosmotic with seawater ($\Delta FP = 1.9°C$), but freshwater decapods, which are believed to be descended from marine forms, have lower osmotic pressures (e.g., $\Delta 0.8°C$ for the crayfish *Astacus* in Table 5–1). Similarly, freshwater mussels have very low osmotic pressures but are derived from marine groups that are nearly isosmotic with seawater. The lower body fluid concentrations seen in freshwater forms result in lower osmotic and ionic gradients between the body fluids and environment than would otherwise be the case and can be viewed as adaptive to lowering the energy demands of osmotic-ionic regulation.

The permeabilities of freshwater forms are generally relatively low, this again being a factor in reducing rates of passive solute and water exchange and thus reducing the energy cost of maintaining homeostasis. Crayfish, for example, have been shown to be at least 10 times less permeable to both water and NaCl than certain strictly marine decapod crustaceans; the latter, being isosmotic with seawater, do not face problems of osmotic dilution or concentration of their body fluids. Typically the gill membranes of freshwater animals are particularly significant sites of passive water and salt exchange, not only because they tend to be more permeable than other parts of the body but also because they present a large surface area to the medium.

Comparing morphologically similar forms of different body size, surface-to-volume ratio can be a significant factor affecting passive water and salt exchanges. Other things being equal, small animals with relatively high surface-to-weight ratios will experience greater passive losses of salts and gains of water per unit of weight than large animals. This implies a greater weight-specific energy cost for maintenance of homeostasis in the small forms unless concentration gradients, permeabilities, or other factors are different.

Given that rates of passive exchange are influenced by a variety of factors, we might expect compensatory relationships among the factors. Relatively high permeability, for example, could be compensated by having relatively low internal osmotic and ionic concentrations, the latter acting to reduce osmotic and ionic gradients between the body fluids and medium. Various authors have pointed out possible examples of such compensatory interrelationships. The freshwater mussels, for example, have relatively high permeabilities; and, as we noted earlier, they have among the lowest internal osmotic and ionic concentrations known. Maintenance of high internal concentrations in the face of the rapid passive exchanges dictated by high permeabilities would be energetically demanding. The low metabolic rates of the mussels have also been cited as a potential contributing factor to their low internal concentrations inasmuch as their low tempo of energy utilization is probably incompatible with a highly demanding form of osmotic-ionic regulation. *Hydra* and other freshwater hydroids also have low body fluid concentrations. This property may be functionally correlated with their small size and low metabolic rates. Those freshwater animals with relatively high internal concentrations also tend to have low permeabilities.

It is interesting that among freshwater crustaceans, smaller species tend to have lower internal concentrations than larger species. Small size would tend (other things being equal) to increase weight-specific rates of passive exchange with the environment, and the reduced concentration gradient between animal and environment in small forms may be compensatory.

It is easy to speak rather glibly about evolutionary changes in the internal salt concentrations of animals under the influence of osmotic-ionic exigencies. Because intracellular and extracellular osmotic pressures are typically similar in multicellular animals, however, one must remember that any alteration in the osmotic concentration of the extracellular body fluids implies a change in that of the cells, which are the functional units of the organism. If extracellular osmotic pressure is reduced, for example, at least some intracellular solutes must be reduced in concentration. The biochemical machinery must adapt successfully to such changes. One cannot help but be impressed with some groups, such as the bivalve molluscs, which, having evolved in isosmoticity with the sea, have adapted to the low internal osmotic concentrations seen in freshwater forms. It is possible that certain other taxa, for whatever reason, have been less plastic, have been more tied to maintaining the internal concentrations of their ancestors.

Common mechanisms of osmotic-ionic regulation

There is a fundamental similarity in the mechanisms of osmotic-ionic regulation in many freshwater animals. This common pattern, to be discussed in this section, has been reported in such diverse groups as the freshwater mussels, crayfish, earthworms, leeches, mosquito larvae, teleost fish, and frogs.

As we have seen, these animals are faced with a continuous osmotic influx of water. The water is voided in a copious urine. Mussels (*Anodonta*), for example, have been reported to excrete water equivalent to 45% of their body weight per day; frogs (*Rana esculenta*), 30%/day; and crayfish (*Asta-*

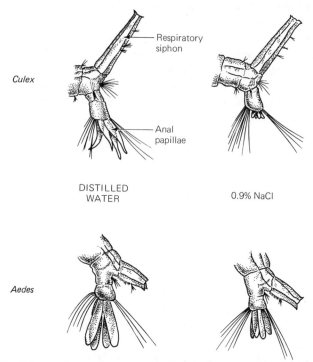

Figure 5–1. Posterior ends of larvae of *Culex pipiens* and *Aedes aegypti*, showing the difference in size of the anal papillae in animals reared in distilled water and in 0.9% NaCl. (From Wigglesworth, V. B. 1938. J. Exp. Biol. 15: 235–247.)

cus), 8%/day. These excretory rates provide some measure of the rate of osmotic influx. Typically, the urine of freshwater animals is strongly hyposmotic to their blood, thus conserving salts and contributing to osmotic regulation. The urine of the crayfish *Astacus fluviatilis* and some other animals can be almost as dilute as fresh water. *Astacus*, for example, with a blood sodium concentration of about 200 mM, can produce urine with as little as 1 mM Na$^+$. The capacities of most freshwater organisms to render their urine hyposmotic, though usually considerable, are more modest.

Freshwater animals also face problems of passive loss of salts by diffusion across their body surfaces and lose some salts in their urine. The extrarenal losses generally exceed the urinary losses, but in some animals, such as certain frogs, the opposite may be true. Salts are replaced through the food and by direct active uptake from the water across some parts of the body surface. Sodium and chloride, generally being by far the most-concentrated ions in the extracellular body fluids (Table 5–1), tend to be lost most rapidly by diffusion and are the predominant ions absorbed from the water by active transport. Generally, direct uptake from the water, as opposed to intake in food, is the main source of these ions.

The capacities of most freshwater organisms for active uptake of NaCl from the medium are remarkable. Various fish, with plasma Cl$^-$ concentrations of 100 mM or more, have been reported to remove Cl$^-$ from waters at least as dilute as 0.3 mM, for example; and frogs, having plasma Na$^+$ concentrations of over 100 mM, can absorb Na$^+$ from waters as dilute as 0.01 mM. The site of active transport varies among animal groups and in some taxa has yet to be positively identified. Teleost fish and decapod

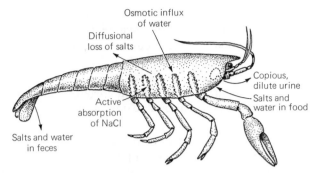

Figure 5–2. Summary of salt and water exchanges in crayfish. The renal organ, or green gland, is shown as a dashed, saccular structure opening at the base of the second antenna. The gills, which are covered by the carapace and not actually visible externally, are shown at the bases of the legs.

crustaceans absorb NaCl across the gill epithelium. Among frogs, transport occurs across the gills in tadpoles and across the general integument in adults. Midge and mosquito larvae absorb ions across their anal papillae; it is of interest that when larvae are reared in 0.9% NaCl, their internal osmotic pressure rises to that of the medium, the need for active uptake is essentially abolished, and the anal papillae are small by comparison to those of larvae reared in fresh or distilled water (Figure 5–1). The soft-shelled turtle *Trionyx spinifer* actively absorbs sodium across pharyngeal villi, which also serve as sites of aquatic respiration.

Freshwater organisms would not be expected to drink their medium because they already face the problem of osmotic water uptake and they are typically unable to produce a sufficiently dilute urine to realize a net gain of salts from drinking (in fact, they would usually sustain a net loss). Experiments on a number of species have indicated that drinking does not occur.

Diagrams summarizing the usual pattern of salt and water relationships in freshwater animals are provided in Figures 5–2 and 5–3. Although the pattern is qualitatively similar in the majority of these animals, there are many differences in quantitative detail, depending not only on species but also on environmental conditions. Data on crayfish (*Astacus*) under one set of conditions will provide a specific quantitative example. At 20°C in water containing 0.5 mM Na^+, a 29-g fasting crayfish excretes about 0.1 cc of urine per hour. This provides a measure of the rate of uptake of water by osmosis, probably principally across the gills. The urine is very dilute (1 mM Na^+) relative to the body fluids (204 mM Na^+). Only about 0.1 μmole of sodium is therefore lost per hour in the urine. Diffusional loss of sodium is much greater, approximately 10 μmole/hr. The urinary and diffusional losses of sodium are replaced by active uptake across the gills at a rate near 10 μmole/hr.

The advantage of
forming hyposmotic urine

Urine is formed from body fluids in the tubules of the kidneys or renal organs. Because hyposmotic urine is of lower osmotic pressure than the body fluids and contains reduced concentrations of various ions, its production involves establishment of osmotic-ionic disequilibria. This demands investment of metabolic energy. It is clear on analytical grounds that production of hyposmotic urine is not a necessity of freshwater life, and we shall see subsequently that some freshwater animals produce isosmotic urine. Urinary volume must be sufficient to void the osmotic water load, but the urine can be isosmotic rather than hyposmotic provided the increased rate of salt loss entailed by producing isosmotic urine can be compensated by uptake of salts in the food and through active transport from the medium. Why, then, do most freshwater animals produce hyposmotic urine?

The answer to this question is not entirely established, but there is good reason to believe that production of hyposmotic urine has energetic advantages. As we shall see in Chapter 7, the fluid that first enters the renal tubules of such diverse animals as the freshwater mussels, crayfish, and fish is similar in ionic composition and osmotic pressure to the blood plasma. Hyposmoticity is established as the fluid flows through the renal tubules by active resorption of ions across parts of the tubules that are poorly permeable to water. As the urine is rendered ionically and osmotically less concentrated than the body fluids surrounding the renal tubules, gradients favoring diffusion of salts into the urine from the body fluids and favoring osmotic loss of water from the urine to the body fluids increase. The active-transport pumps that establish hyposmoticity in the urine by removal of salts must operate against the passive tendencies for the urine to remain similar in composition to the body fluids. Suppose now that, starting with fluid similar in composition to the body fluids, an animal renders the urine similar in composition to fresh water before excretion. At first the osmotic and ionic gradients between the urine and surrounding body fluids will be small. Gradually the gradients will become larger, but it will not be until the process of establishing hyposmoticity is virtually complete that the absorption pumps will have to operate against gradients equivalent to those between the body fluids and fresh water itself. In other words, the pumps for the most part will be extracting ions from fluid that is more concentrated than fresh water. Through absorption of ions from the renal fluids, the animal producing hyposmotic urine conserves certain quantities of ions by comparison to an animal producing isosmotic urine. The animal producing isosmotic urine, by not resorbing these quantities from the urine, will instead have to replace them by direct absorption from the freshwater medium; and the pumps that accomplish this will have to operate entirely against the osmotic and ionic gradients between body fluids and fresh water. Because it is generally believed that the metabolic energy demanded for active accumulation of solutes increases with the gradients against which the pumps are working, we can see at this point why production of hyposmotic urine may have energetic advantages. The average gradients against which the pumps in the renal tubules must operate are smaller than the gradients against which the pumps at the body surface must operate. It is therefore probably less costly in energetic terms to resorb ions from the urine than to replace them by direct absorption from fresh water.

The energetics of osmotic regulation

As we have seen earlier, not only the osmoticity of the urine but also the ionic and osmotic concentrations maintained in the body fluids have implications for the energetics of regulation; other things being equal, the freshwater animal will have to expend more energy to maintain internal homeostasis, the greater the concentrations in its body fluids. Potts has performed interesting calculations to estimate the energetic significance of both the internal osmotic concentration and the production of hyposmotic urine in freshwater animals. He computed the cost of osmoregulation in fasting animals at 15°C under various assumptions of internal and urinary osmotic pressure. For two reasons his calculated costs should be considered minimal. First, he assumed the body surface of the animal to be impermeable to salts; extrarenal salt losses were ignored. Second, he computed the minimal cost dictated by thermodynamic considerations. The true cost will be greater insofar as the osmoregulatory mechanisms of the animal are energetically inefficient, that is, insofar as the metabolic energy invested exceeds the osmotic work accomplished.

Potts computed that a 60-g crayfish (*Astacus fluviatilis*) in water of Δ0.01°C with an internal osmotic pressure equivalent to that of its marine ancestors, Δ1.9°C, and producing urine isosmotic to the blood would have to expend about 1.3 cal/hr in osmotic regulation. Reduction of the internal osmotic pressure to the values seen in actual crayfish, about Δ0.8°C, while still assuming production of urine isosmotic to the blood, is predicted to reduce the energy requirement by a factor of about 13, to about 0.1 cal/hr. Production of urine as hyposmotic as seen in actual crayfish will further reduce the cost by a factor of about 3, to 0.037 cal/hr. This cost is about 0.3% of the total resting metabolism of *Astacus* at 15°C. If the energetic efficiency of the osmoregulatory mechanisms were only 20%, the cost in metabolic energy would be five times greater—still only a small fraction of total metabolism (1.5%). If a 60-g freshwater mussel (*Anodonta cygnaea*), which is much more permeable to water than the crayfish, had blood isosmotic with seawater and produced isosmotic urine, its minimal energy expenditure for osmoregulation would approximate 60 cal/hr, which is about 50 times greater than the total resting metabolism at 15°C. Reduction of internal osmotic pressure to the actual levels seen in these mussels (Δ0.08°C) cuts the energetic cost by a factor of about 2500, still assuming urine isosmotic to the blood. If, in addition, the urine is as hyposmotic as actually seen in the mussels (Δ0.04°C), the cost is further reduced almost twofold so that the actual minimal cost in these mussels is only about 0.015 cal/hr— or 1.2% of total resting metabolism at 15°C. Assuming an energetic efficiency of just 20%, the cost for osmoregulation would be 6% of total metabolism.

These calculations, although somewhat theoretical, indicate something of the magnitude of the energetic savings realized by production of hyposmotic urine and reduction of internal osmotic pressure from levels seen in the marine relatives of these freshwater animals.

Examples of species that do not adhere to the common pattern of osmotic-ionic regulation

The general pattern of regulation we have discussed does not apply to all freshwater animals, and it is instructive to examine some of the exceptional cases.

Though it is not commonly known to residents of North America, a good many species of true crabs live in fresh waters in other parts of the world. Two of these are *Potomon niloticus*, a freshwater form of Africa, and *Eriocheir sinensis*, the wool-handed crab of northern Europe, which spends much of its adult life in rivers but migrates to the sea for breeding. These crabs maintain relatively high internal osmotic concentrations in fresh water ($\Delta 1.2°C$ in *Eriocheir*, $\Delta 0.95°C$ in *Potomon*), yet both produce urine that is only isosmotic to their body fluids. Their permeabilities to water are exceptionally low. *Eriocheir* excretes water equivalent to about 3.6% of its body weight per day, and *Potomon* excretes less than 0.6% of its body weight per day. These values should be compared to those given earlier for some other freshwater forms (e.g., 8%/day in a crayfish and 30%/day in a frog). Because the crabs have relatively low permeabilities to water, they need to produce only relatively small amounts of urine to void their osmotic water load, and urinary losses of salts are thus limited despite the isosmoticity of their urine to the body fluids. Urinary salt losses are compensated by active uptake of salts from the water.

Lacewing larvae, *Sialis lutaria*, do not appear to take up ions from fresh water under normal physiological conditions. Their urine ($\Delta 0.4°C$) is only modestly hyposmotic to the blood ($\Delta 0.6°C$), but 90% of the urinary osmotic pressure is due to ammonium and bicarbonate ions produced in metabolism. Urinary concentrations of sodium and chloride are very low relative to those in the blood, indicating efficient conservation of these ions. The larvae are among the most impermeable of freshwater animals that have been studied. Osmotic uptake of water amounts to only about 4% of the body weight per day, remarkable for such small animals. Urinary volume and salt losses are correspondingly reduced. Extrarenal salt losses are probably also small, and it appears that these larvae can maintain homeostasis with only dietary inputs of salts.

These examples serve to emphasize the important interactions of the parameters involved in osmotic-ionic relationships.

ANIMALS IN THE OCEAN AND MORE-CONCENTRATED SALINE WATERS

The total salt concentration in the open oceans varies from 32‰ to 41‰ and is in the range of 34‰ to 37‰ in most waters. The freezing-point depression of ocean water of a salinity of 34.5‰ is 1.88°C, representing an approximately 1-osmolar solution. The concentrations of the principal inorganic ions in seawater are given in Table 5–2. Note that the most abundant monovalent ions are sodium and chloride and the most abundant divalent ions are magnesium and sulfate. Although the concentrations given in Table 5–2 pertain specifically to seawater of a salinity of 34.3‰, the *ratios* of the concentrations of the various ions are virtually the same in natural ocean waters regardless of salinity.

Inland saline waters, such as those of the Dead Sea or Great Salt Lake of Utah, have various salt compositions depending on their geological origin. Sodium and chloride are typically the most abundant ions, but the ratios of ionic concentrations may differ markedly from those in seawater. Salinities in excess of 200‰ have been reported.

For the most part we shall discuss in this section the animals living in the oceans. Life is thought to have originated in the sea, and to this day

Table 5–2. Concentrations of major inorganic ions in seawater of salinity 34.3‰.

Ion	Concentration (mM)
Sodium	470
Chloride	548
Magnesium	54
Sulfate	28
Calcium	10
Potassium	10
Bicarbonate	2

SOURCE: Barnes, H. 1954. J. Exp. Biol. 31: 582–588.

the fauna of the sea is more phylogenetically diverse than that in the other major habitats. All phyla and most classes of animals have marine representatives. Many groups have invaded fresh water and the land. In turn, there have been many reinvasions of the oceans. Thus, whereas some of our modern marine animals have a continuously marine ancestry, others trace their history to forms that occupied other habitats. The sea is a relatively stable and protective environment. Temperature, salt composition, and pH vary to a relatively small extent in any one place, and changes tend to occur slowly.

The condition of most marine invertebrates

Most marine invertebrates are isosmotic, or nearly so, to seawater. Included are such diverse forms as sponges, coelenterates, annelids, molluscs, echinoderms, and most crustaceans. These animals do not face problems of osmotic regulation. Sometimes their internal osmotic pressure is somewhat higher than that of the water, but this represents a passive osmotic equilibrium resulting from Donnan effects (see Chapter 4). The body fluids of all species that have received study differ in ionic composition from seawater, as illustrated in Table 5–3. The differences may be minor, as in the echinoderms (note *Echinus*), or may be more substantial, as in the squid *Loligo* and the crab *Carcinus*. Any given ion can be found relatively concentrated in some forms but relatively dilute in others (for example, note magnesium in *Loligo* and in *Carcinus*). The adaptive significance of the differences in ionic composition among these animals remains generally obscure.

The differences in ionic composition between the extracellular body fluids and seawater in isosmotic marine invertebrates imply ionic regulation. As discussed in Chapter 4, Donnan effects may produce significant differences in ionic composition in animals such as the crustaceans which have relatively high blood protein concentrations. But these effects do not account for more than a small part of the total difference between internal and external salt composition, even in crustaceans. The animals under discussion are typically relatively permeable to both water and many ions, and we must look to active processes of salt uptake or excretion to explain the regulation of internal ionic composition. Active uptake of various ions from the medium at the body surface or from ingested seawater is widely indicated. Active accumulation of ions can serve to elevate their concentra-

Table 5–3. Concentrations of certain ions in the body fluids of some marine animals. Figures are recalculated to apply to animals living in seawater of the composition given.

	Concentration (mmole/kg)					
	Na$^+$	K$^+$	Ca^{2+}	Mg^{2+}	Cl$^-$	SO$_4^{2-}$
Seawater	478	10.13	10.48	54.5	558	28.77
Aurelia mesogleal fluid, coelenterate	474	10.72	10.03	53.0	580	15.77
Echinus coelomic fluid, echinoderm	474	10.13	10.62	53.5	557	28.70
Mytilus blood, mussel	474	12.00	11.90	52.6	553	28.90
Loligo blood, squid	456	22.20	10.60	55.4	578	8.14
Carcinus blood, crab	531	12.26	13.32	19.5	557	16.46
Myxine blood, hagfish	537	9.12	5.87	18.0	542	6.33

SOURCE: Potts, W. T. W. and G. Parry. 1964. *Osmotic and Ionic Regulation in Animals.* Pergamon, Oxford. Used by permission of Pergamon Press Ltd.

tions in the body fluids above concentrations dictated by Donnan equilibrium with seawater. The urine of the crustaceans and molluscs is approximately isosmotic to the body fluids but differs significantly in ionic composition. The renal organs, by preferentially excreting certain ions and conserving others, contribute to ionic regulation. In most decapod crustaceans, for example, the urine contains magnesium and sulfate at higher concentrations than in the blood. This is important to maintaining reduced blood concentrations of these ions (see *Carcinus*, Table 5–3).

Hagfish: the only vertebrates with blood salt concentrations that render them isosmotic to seawater

The hagfish, an exclusively marine group of primitive jawless fishes, resemble the mass of marine invertebrates in having body fluids that are approximately isosmotic with seawater. Most of the blood osmotic pressure is due to dissolved salts, principally sodium and chloride (Table 5–3). Urine composition is known to differ from plasma composition, implicating the kidneys in ionic regulation. It seems likely that active uptake mechanisms are also involved. These fish appear to be the only modern vertebrates that might possibly claim a continuously marine ancestry.

Teleost fish: animals hyposmotic to seawater

A good many marine animals are hyposmotic to seawater. We shall consider the marine teleosts first, as they are well studied and illustrate many of the broad patterns of hyposmotic regulation (maintenance of internal osmotic pressures below those of the medium). The ancestral jawed fish likely evolved in fresh water, and it is therefore believed that the present teleost fauna of the oceans is descended from freshwater ancestors. The osmotic concentrations of modern marine teleosts range from about Δ0.6°C to Δ1.1°C, with most being lower than Δ0.9°C. This range is higher than that for freshwater teleosts (about Δ0.45°C to Δ0.66°C), probably reflecting the adaptations of the two groups to maintaining osmotic-ionic homeostasis efficiently in their respective habitats. Nonetheless, the concentrations of the marine fish are only about one third to one half those of seawater. This is taken as evidence of their freshwater ancestry.

Marine teleosts are confronted with a continuous osmotic loss of water. For the hyposmotic animal the sea resembles the land in being a desiccating environment. Because the concentrations of sodium, chloride, and many other ions are lower in the body fluids of marine teleosts than in the sea, these animals are also faced with a tendency to gain salts by diffusion. The passive movements of salts and water are opposite to those seen in fresh-water teleosts and, if unopposed, would tend to raise ionic concentrations and osmotic pressure in the body fluids to those of seawater. Most of the passive losses of water and gains of salts occur across the thin, vascular membranes of the pharynx and gills. The permeability of most of the body surface is low, as expected for animals that maintain large differences in composition between their body fluids and the environmental medium.

The fish must replace its osmotic and urinary water losses by drinking seawater and absorbing water from the gut. Experiments on some species have shown that death caused by dehydration occurs rapidly if drinking is prevented. When seawater is ingested, it is, of course, hyperosmotic to the body fluids, and the initial tendency would be for water to move by osmosis from the body fluids into the gut fluids rather than in the opposite direction. There is evidence from eels that this actually occurs. Salts diffuse from the gut fluids into the body fluids, and there is accumulating evidence of active absorption of sodium chloride from the gut fluids. This active absorption is believed to be central to the ultimate absorption of water. Once the gut fluids have become isosmotic to the body fluids by osmotic influx of water and loss of salts, further active absorption of salts, by tending to establish hyposmoticity in the gut fluids, will result in osmotic absorption of water. The gut fluids as a whole may or may not become measurably hyposmotic to the body fluids, depending on the extent to which osmotic water absorption lags behind ion uptake. Various investigators have found the fluids in the posterior gut to be approximately isosmotic or hyposmotic to the blood. It is important to recognize that because the seawater initially ingested is hyperosmotic to the blood, the condition of the fluids in the posterior gut reflects the fact that relatively more salt than water is taken up. That is, for the fish to absorb a certain proportion of the *water* ingested, it must absorb an even greater proportion of the *salts* ingested (otherwise the fluids in the posterior gut would still be as concentrated as seawater). This salt load incurred in meeting water needs adds to the diffusional salt load from the medium across the gills and other body surfaces. Measures on various fish have shown that the amount of sodium chloride absorbed from the gut per day can be so great as to equal or exceed the total steady-state quantity in the tissues and body fluids of the fish. In all, some 50–80% of the water in ingested seawater may be absorbed in its passage through the gut, and even greater proportions of the sodium and chloride are absorbed. The principal divalent ions, magnesium and sulfate, tend to be left behind in the gut though there is some diffusion into the body fluids. The residuum of water, divalent ions, and other ions is voided in the feces.

We now come to the question of how the fish eliminates excess ions that enter the body fluids from the gut and from the medium across the gills and other body surfaces. Excess divalent ions are voided in the urine at concentrations exceeding their plasma levels. Most of the excess sodium

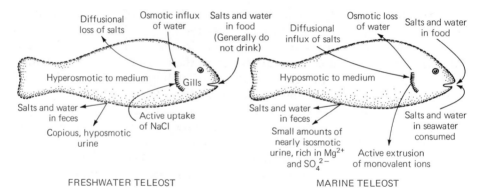

Figure 5–3. Summary of salt and water exchanges in freshwater and marine teleosts.

and chloride ions are excreted by active transport across the gills—the same site at which we find active uptake of ions in freshwater fish. Extrarenal excretion of ions is dictated by the fact that fish cannot produce urine that is hyperosmotic to the blood. Marine fish form essentially isosmotic urine. Such urine can serve as a route of preferential loss for *certain* solutes. This is what we see with the divalent ions. But the *total* osmotically effective concentration of solutes does not exceed that in the blood, and the preferential loss of solutes over water that is necessary to hyposmotic regulation must occur elsewhere. Present evidence indicates that the active excretion of monovalent ions by the gills is, in fact, accomplished without concomitant excretion of any water at all.

Because urine production involves loss of valuable water and does not immediately serve osmotic regulation, we would expect the volume to be limited to that necessary for excretion of solutes that cannot be voided by other routes. The principal ions, sodium and chloride, are excreted effectively by the gills, and nitrogenous end products are lost effectively by diffusion across the gills. The role of the kidneys is thus largely limited to excretion of divalent ions, and urinary volume can be, and is, small.

The routes of water and salt exchange in marine teleosts are contrasted with those in freshwater teleosts in Figure 5–3.

Movements of salts and water have been quantified in only a relatively few species of marine fish. Homer Smith's original data on water exchanges in eels (*Anguilla*) and sculpins (*Myoxocephalus*) are summarized in Table 5–4. You will note that 45–85% of the water swallowed is absorbed, that extrarenal, osmotic losses of water generally exceed urinary losses and may represent as much as 90% of total loss, that urine production amounts to only 0.5–4% of the body weight per day, and that the turnover of water is 3–17% of the body weight per day. It is well known that many fish are susceptible to a phenomenon called "laboratory" diuresis, tending to produce a greater volume of urine after handling. It is thus not unlikely that fish in nature excrete less urine than the fish in these experiments, and, correspondingly, urinary loss of water in nature may well represent a smaller proportion of total loss.

Hyposmotic regulation in certain arthropods of saline waters

Certain arthropods regulate hyposmotically in seawater. Among these are many grapsoid and ocypodid crabs, some isopods, and the palaemonid

Table 5–4. Water exchanges in eels and sculpins in seawater. Different lines represent results of different experiments. Amounts of water swallowed and absorbed and amounts excreted in the urine were determined experimentally. Extrarenal losses were determined by difference.

Water Swallowed (ml/kg/day)	Water Absorbed from Gut (ml/kg/day)	Water Excreted in Urine (ml/kg/day)	Extrarenal Water Loss (ml/kg/day)	Percent of Water Swallowed That Was Absorbed	Percent of Absorbed Water Lost Extrarenally
Anguilla rostrata, eel					
50.0	37.0	16.4	20.6	74	56
135.5	113.0	9.9	103.1	83	91
52.0	39.0	7.8	31.2	75	80
40.0	30.0	10.0	20.0	75	67
54.4	46.0	4.6	41.4	85	90
Myoxocephalus octodecimspinosus, sculpin					
225.0	169.0	41.6	127.4	75	76
89.5	44.1	16.8	27.3	63	62
109.0	58.3	25.1	33.2	55	57
59.3	44.5	37.6	6.9	75	16
96.8	43.0	28.0	15.0	45	35

SOURCE: Smith, H. W. 1930. Amer. J. Physiol. 93: 480–505.

prawns. The prawns *Palaemon serratus* and *Palaemonetes varians*, for example, maintain internal osmotic pressures of about $\Delta 1.3°C$; the fiddler crab *Uca pugnax*, about $\Delta 1.6°C$. The mechanisms of regulation are not well understood. The prawns and some crabs produce urine that differs in composition from the blood but is essentially isosmotic, implying a role in ionic, but not osmotic, regulation. Some crabs (such as *Uca pugnax* and *Ocypode albicans*) can produce somewhat hyperosmotic urine, meaning that the renal organs can serve osmotic regulation. Extrarenal salt excretion is also indicated.

The brine shrimp (*Artemia salina*) is remarkable for its ability to survive over an exceedingly wide range of salinities, including saturated solutions of sodium chloride (about 300‰ salinity). Blood osmotic pressure is about $\Delta 0.7°C$ in seawater. It rises with increasing environmental salinity but does not exceed $\Delta 1.9°C$ even in saturated NaCl. The mechanisms of hyposmotic regulation are broadly similar to those in marine teleosts, involving low permeability, oral and anal uptake of seawater, absorption of ions (particularly Na^+ and Cl^-) and water from the gut contents, and extrarenal excretion of Na^+ and Cl^- across the gills. It is notable that despite active absorption of sodium and chloride from the gut and despite the fact that the gut fluids of animals living in saline media become hyposmotic to the medium, the fluids remain hyperosmotic to the blood. The absorption of water from the gut is thus against the gross osmotic gradient between the gut fluids and blood. This is unexplained. Active transport of water is a possibility, but it is also possible that active absorption of ions from the gut estabishes localized osmotic gradients along the gut lining that are favorable to passive, osmotic absorption of water. To illustrate with a simple conjecture, active ion absorption could render gut fluids immediately next to the gut wall (perhaps in small diverticula of the wall) hyposmotic to the gut fluids at large and hyposmotic to the blood. Water would then be absorbed passively from the gut into the blood despite the opposing gross osmotic gradient.

The larvae of certain insects, such as the mosquito *Aedes detritus*, the chironomid *Cricotopus vitripinnis*, and the fly *Ephydra riparia*, can live in seawater while maintaining internal osmotic pressures around $\Delta0.7°C$. As we shall see in Chapter 7 (Figure 7–18), the excretory, or Malpighian, tubules of insects empty into the intestine near the junction of the midgut and hindgut. Thus both the gut contents per se and the urine pass through the hindgut and rectum, where their composition is altered before excretion. The fluids exiting the Malpighian tubules are typically isosmotic or slightly hyposmotic to the blood. In the saline-water larvae under discussion, these fluids and the other gut fluids are rendered strongly hyperosmotic to the blood in the rectum, and although many aspects of hyposmotic regulation await careful study, this facet is undoubtedly of much importance. *Aedes detritus* larvae, to illustrate, can produce excretions that are, osmotically, about three times as concentrated as the blood, and larvae of the fly *Ephydra riparia* can achieve a U/P ratio as high as 10. These animals are thus able to excrete solutes in great disproportion to their concentration in the blood. Some insect larvae are able to live in salinities much higher than those in seawater with only relatively modest increases in internal osmotic pressure. *Aedes detritus*, for example, is found in some of the more saline inland waters known, and *Ephydra cinerea* survives in $10M$ NaCl. *E. cinerea* and the brine shrimp, *Artemia salina*, are the only metazoans known to occupy the Great Salt Lake of Utah.

The insect larvae of saline waters not only are descended from terrestrial forms but develop into terrestrial adults. Their internal osmotic concentrations in seawater or other saline waters of like salinity are generally similar to those of insects in general, and their hyposmoticity seems clearly to be a reflection of their heritage. The capacity to produce hyperosmotic urine is widespread among insects, generally representing an adaptation to conservation of water on land; the production of hyperosmotic urine in larvae of saline waters would thus appear to be a manifestation of a more general capability of insects rather than a novel development in response to the problems of osmotic-ionic regulation in saline waters.

There is continuing debate over why the crustaceans discussed earlier (certain crabs, prawns, and *Artemia*) regulate hyposmotically in seawater. It seems likely that *Artemia* is derived from freshwater forms, and it is possible that the palaemonid prawns, and perhaps some of the crabs, trace a brackish water ancestry. In these cases, hyposmoticity may have historical roots, their internal concentrations being reflective of the concentrations maintained by ancestors in more dilute media. More will be said of hyposmotic regulation in these forms subsequently.

Marine reptiles, birds, and mammals: additional hyposmotic vertebrates

Marine reptiles, birds, and mammals, in common with the teleosts, are hyposmotic to the medium. All are descended from terrestrial forms, and their internal osmotic pressures, although often being slightly higher than those in terrestrial and freshwater relatives, reflect this heritage. The marine turtle *Caretta caretta* has an internal concentration of about $\Delta 0.8°C$; the marine fulmar *Fulmarus glacialis*, about $\Delta 0.7°C$; and the seal *Phoca foetida*, about $\Delta 0.7°C$.

Permeabilities in these animals are relatively low, in part because they have inherited integuments adapted to limiting water loss in the dehydrating terrestrial environment and in part because, being air breathers, they do not expose relatively permeable respiratory membranes to the medium. They nonetheless confront problems of water loss and salt loading. Water is lost through evaporation across the respiratory tract and is lost, to some extent, across the skin not only in seawater but also in air in the case of those forms that spend time in air. In addition, these animals do not depend on simple diffusion into the medium to rid themselves of nitrogenous wastes (see next chapter). Thus urinary water loss is obligated not only for salt excretion, but also for elimination of nitrogenous end products; the birds and many reptiles excrete most nitrogen in the form of precipitated uric acid and urates, but the mammals void most nitrogen as urea in solution. A salt load is imposed by the diet; organisms taken as food are often isosmotic with seawater (e.g., most invertebrates) and therefore strongly hyperosmotic to the body fluids of the predator, and quantities of seawater are probably often taken in with the food. You will note that such salt loading is a problem for animals like sea gulls and herons when they are consuming marine plants and animals, even though most or all of their body is in the air most of the time. The majority of marine reptiles, birds, and mammals do not appear to drink seawater. They thus avoid the salt load associated with drinking but must make up their water losses from the water in their food, the seawater that may be taken in during eating, and metabolic water.

There are relatively few marine reptiles, but they include representatives of all the major groups: snakes, lizards, turtles, and crocodiles. Reptiles are unable to produce urine hyperosmotic to their blood. There are a great many birds associated with the marine habitat, representing several orders. The capabilities of birds to concentrate their urine are incompletely understood, but for most the maximum urinary concentration appears to be isosmotic to the blood or only modestly hyperosmotic (up to on the order of twice the blood osmotic pressure). A central role in osmotic-ionic regulation in most or all of the truly marine reptiles and birds is assumed by *salt glands* that are located in the head and that empty into the nasal passages in birds and into the orbits or nasal passages in reptiles (Figure 5–4). These glands produce concentrated salt solutions composed principally of sodium and chloride, but also containing other ions such as potassium, calcium, and bicarbonate. Table 5–5 summarizes information on the sodium-concentrating abilities of some representative glands; chloride concentrations are typically about the same as sodium concentrations. Not only are the salt gland secretions very hyperosmotic to blood (by a factor of 4–5 in many forms), but the concentrations of sodium, chloride, and, in at least some cases, potassium exceed those in seawater. This means, for example, that these animals could drink seawater and void the major monovalent ions while still realizing a net gain in water. The kidneys alone do not provide such a capability. In one experiment on a double-crested cormorant fed fish and sodium chloride, about half the total NaCl load was voided by the

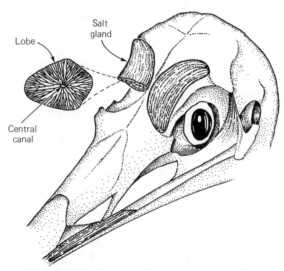

Figure 5–4. Salt glands of the herring gull. Each gland lies in a shallow, crescent-shaped depression in the skull above the eye. The cross section shows that the gland consists of many longitudinal lobes. Each lobe contains a great many branching, radially arranged secretory tubules that discharge into a central canal. The glands empty into the nasal passages via ducts, and their secretion flows out through the external nares. (From Schmidt–Nielsen, K. 1960. Circulation 21: 955–967.)

salt glands, and about half was voided via the cloaca in urine and feces; the accompanying cloacal water losses were over twice as great as those from the salt glands, however. The rate of secretion from salt glands is varied according to needs to excrete excess salt. In some species the glands increase in size when the animal experiences increased salt intake. The tears observed in marine turtles when they emerge onto land to lay their eggs are of some renown. We now understand that they are secretions of orbital salt glands. If you have ever watched a herring gull standing by the ocean, you may have observed the steady dripping of salt gland secretions from the end of its bill.

Functional salt glands have been reported in 13 orders of birds, including most of those that have truly marine representatives. They have not

Table 5–5. Sodium concentrations reported in the salt gland secretions of various reptiles and birds.

	Sodium Concentration (mM)
Standard seawater	470
Marine iguana (*Amblyrhynchus cristatus*), nasal secretion	840
Loggerhead turtle (*Caretta caretta*), orbital secretion	732–878
Leach's petrel (*Oceanodroma leucorhoa*)	900–1100
Herring gull (*Larus argentatus*)	600–800
Brown pelican (*Pelecanus occidentalis*)	600–750
Humboldt penguin (*Spheniscus humboldti*)	725–850

SOURCE: Schmidt-Nielsen, K. 1960. Circulation 21: 955–967; Schmidt-Nielsen, K. and R. Fange. 1958. Nature 182: 783–785.

been reported in passerines. Some passerines live in salt marshes, and one of these, a subspecies of the savannah sparrow, *Passerculus sandwichensis beldingi,* has attracted considerable interest. It can drink 75% seawater or 0.6*M* NaCl (osmotic pressure about Δ2.1°C) with impunity and may well consume water from the salt marshes. It does not possess the extrarenal route of voiding salts provided by salt glands but, compared to most birds, has kidneys with extraordinary concentrating capabilities. Mean maximum chloride concentration is reported as 960 m*M*, and the mean maximum osmotic pressure is Δ3.7°C. Another subspecies of this sparrow, *P. s. brooksi,* breeds in freshwater marshes and often overwinters in salt marshes. Its mean maximum urinary chloride concentration and osmotic pressure are 527 m*M* and Δ1.9°C, respectively. These values are lower than those in *P. s. beldingi* but still high relative to most birds. These birds provide an interesting case of physiological divergence at the subspecies level.

Mammals as a group have commonly evolved the capacity to produce highly concentrated urine. This has been important to species occupying arid terrestrial environments and appears to be central to hyposmotic regulation in the marine forms. Salt glands or other routes for extrarenal salt excretion have not been identified. The mechanisms of osmotic-ionic regulation in marine mammals are incompletely understood. Total osmotic pressures reported for whale and seal urines are greater than those of seawater, but urea accounts for a sizable portion of the solute composition, and measured concentrations of sodium and chloride have generally been below those in seawater. Calculations for seals (*Phoca vitulina*) eating herring indicate that the measured concentrating capacities of the kidneys are adequate to eliminating the dietary salt load in the amount of water available for urine formation; the tissues and body fluids of herring are hyperosmotic to those of seals but still quite hyposmotic to seawater. When eating invertebrates isosmotic with seawater, marine mammals incur a considerably greater salt load than when eating fish, and calculations and experiments have not conclusively demonstrated that the kidneys are adequate to eliminating this salt load in concentrations compatible with maintenance of water balance. The possibility remains that extrarenal mechanisms for salt excretion may be present. It must be remembered, however, that these animals are difficult to maintain and have been studied under a variety of unnatural conditions. The kidneys may be capable of producing greater salt concentrations than we have discovered. Concentrations that would be clearly adequate to the problems of marine mammals have been found in other mammals.

Elasmobranch fish: animals that have blood salt concentrations like those of teleosts but that are hyperosmotic to seawater

The elasmobranch fish (sharks, skates, and rays) have adopted a rather novel solution to the osmotic problems of living in the sea. Blood salt concentrations are similar to those of marine teleosts and well below those in seawater. The osmotic pressure of the blood, however, is slightly higher than that of seawater because of high concentrations of urea and, to a lesser extent, trimethylamine oxide (TMO). The hyperosmoticity established by the maintenance of high urea and TMO concentrations means that these fish experience a small osmotic influx of water rather than being faced with the potential problems of osmotic desiccation seen in teleosts.

Urea is the principal nitrogenous end product of protein catabolism in elasmobranchs. The gills of marine forms are reportedly less permeable to urea than those of marine teleosts, and urea is resorbed from the urine forming in the kidneys. These are important adaptations to maintaining the high plasma levels. Urea in high concentrations alters the structure of many proteins, and its concentration is kept fairly low in most vertebrates (around 10–40 mg/100 ml in human plasma). By contrast, in the spiny dogfish (*Squalus acanthias*), to cite one example, mean plasma urea concentration is about 360 mM, or 2160 mg/100 ml. (TMO concentration averages about 85 mM.) We do not understand how the tissues have adapted to functioning under these high concentrations of urea.

In most aquatic animals, blood osmotic pressure is due largely to dissolved salts. Thus problems of osmotic and ionic regulation are related in particular ways. If the animal is hyperosmotic to its medium, it confronts passive losses of salts and gains of water. If hyposmotic, the converse is true. In the elasmobranchs these relationships are to some extent uncoupled. Being slightly hyperosmotic to seawater, they tend to gain water by osmosis, but because blood salt concentrations are below those in seawater, they also tend to gain excess salts by diffusion and by consuming foods isosmotic with their medium. The osmotic inputs of water, water gained in the food, and water produced in metabolism are adequate to meet total water losses. Thus, unlike teleost fish, the marine elasmobranchs do not need to drink seawater to meet water needs. You will recall that teleosts incur a heavy salt load in extracting needed water from ingested seawater. The fact that elasmobranchs avoid this added salt load must be seen as one of the chief advantages of maintaining hyperosmoticity through retention of urea and TMO. Excess salts are removed in elasmobranchs by the kidneys and, extrarenally, by rectal glands. These glands void into the rectum a secretion that is isosmotic with the body fluids but that contains only traces of urea and approximates or slightly exceeds the concentrations of sodium and chloride in seawater. This is a significant route for elimination of these monovalent ions. Divalent ions in food tend to be retained in the gut. Excess divalent ions taken up by diffusion from the medium or absorbed from the gut are voided largely by the kidneys along with significant quantities of monovalent ions. The rate of urine production is low, and the urine is moderately hyposmotic to the blood. Whether active ion elimination occurs across the gills remains controversial.

The modern marine elasmobranchs are believed to be descended from freshwater ancestors. The moderate concentration of salts in their blood supports this argument. Retention of urea and TMO can be viewed as an adaptation to colonization of the sea. Numerous marine elasmobranchs penetrate into brackish and fresh waters. Probably, these are not primitively freshwater species, but rather have reinvaded fresh waters from the seas. In some cases, it may be that certain freshwater populations are landlocked and have been separated from conspecific marine populations for some period of time. It has been shown in some species that routinely enter brackish waters that blood urea concentration and osmoticity fall with decreasing ambient salinity. Numerous species in fresh water have been studied. Blood urea concentrations are only 25–35% of those in animals from seawater but are still well above the levels in most vertebrates and contribute an important component to the total osmotic pressure. You will note that the persistence of elevated urea concentrations in fresh water does not provide advantages in osmotic-ionic relationships but, to the contrary, aggravates problems of osmotic flooding. The high urea concentrations are taken as evidence of relatively recent reinvasion from the sea and are interpreted as indicating either a physiological dependence on high urea levels or some lack of physiological or evolutionary plasticity in the tissues and organs responsible for urea production and retention. Some elasmobranchs have exclusively freshwater distributions. Recently there has been study of two species of river stingrays (genus *Potamotrygon*) found thousands of kilometers from the sea in rivers of South America. Unlike all the other freshwater elasmobranchs studied, they have very low blood urea concentrations. This might possibly reflect a continuously freshwater ancestry or may indicate a longer, or at least more complete, adjustment to the freshwater habitat.

Latimeria: another marine fish with high urea concentrations

The crossopterygian fish, presumed ancestors of the terrestrial vertebrates, were thought to be completely extinct until, in 1939, a coelacanth, *Latimeria chalumnae*, was captured off Africa. Work on recent specimens has revealed that, like the elasmobranchs, their blood osmotic pressure approximates that of seawater, and a considerable part of the osmotic pressure is due to urea.

RESPONSES OF AQUATIC ANIMALS TO CHANGES IN SALINITY; BRACKISH WATERS

Up to this point we have primarily considered the physiology of aquatic animals living under relatively stable conditions of salinity in fresh waters or the ocean. Many species are exposed to ranges of salinity, and in this section we shall examine their responses. Emphasis will be placed on animals in the brackish waters of intermediate salinity that occur where ocean waters mix with fresh water.

Waters of salinities between about 0.5‰ and 30‰ are generally considered brackish, and when formed by the mixture of sea and fresh water, they typically have ion concentration *ratios* similar to those in seawater (see Table 5–2). Thus, for example, brackish waters of one fifth the salinity of open ocean water would have one fifth the sodium concentration, one fifth the chloride concentration, and so forth. Brackish waters cover less than 1% of the earth's surface but are of considerable physiological and ecological interest. They are found in estuaries, salt marshes, some inland seas (such as the Caspian Sea, salinity 13‰), and rain-washed tide pools. Animals in such habitats may be exposed to various salinities either by their own movements or by movements or changes of the medium. There are, for instance, gradients of salinity in estuaries, and these shift in position with the tides. An animal in a tide pool may be exposed to seawater one hour, diluted seawater after a heavy rain the next hour, and seawater again when the incoming tide again swirls about its habitat.

Among the animals found in consistently brackish habitats such as estuaries and brackish seas, four groups may be recognized: those with both marine and brackish distributions, those with both freshwater and brackish distributions, those generally found only in brackish waters (the brackish-water animals, sensu stricto), and those that range from fresh water to seawater on either a continuous or a migratory basis. Collectively, the animals in brackish waters have a distinctly more-limited species composition than those in the freshwater or marine environments.

Basic types of responses to changes in salinity

Some animals are able to survive only over narrow ranges of salinity, whereas others survive broad ranges. These forms are termed *stenohaline* and *euryhaline,* respectively. The terms have usefulness, but a two-part classificatory scheme can hardly hope to describe the full diversity of response that is seen.

Animals that maintain a relatively constant internal osmotic pressure over a range of environmental salinities are termed *osmoregulators* or *homeosmotic.* Those whose internal osmotic pressure approximately equals the external osmotic pressure over a range of the latter are termed *osmoconformers* or *poikilosmotic.* These characterizations represent extremes, and many animals lie somewhere in between.

Figure 5–5 illustrates the responses of three invertebrates to changes in external osmotic pressure. The dashed line is called the line of isosmoticity and is generally entered on graphs of this sort as a reference; the extent of departure of the response of an animal from the isosmotic line is a reflection of the degree of osmoregulation. Note that *Mytilus* is a strict osmoconformer. *Palaemonetes* is an osmoregulator over a wide range of external osmotic pressure. It maintains its internal osmotic pressure below that of the medium at ambient osmotic concentrations above $\Delta1.1°C$ but maintains its internal osmotic pressure above that of the medium at lower ambient concentrations. This shows that *Palaemonetes* possesses mechanisms for both hyposmotic and hyperosmotic regulation. *Carcinus* displays an intermediate response. It conforms at ambient osmotic concentrations of $\Delta2°C$ and above, being incapable of hyposmotic regulation. At concentrations below $\Delta2°C$, its internal osmotic pressure falls with external osmotic pressure but remains well above isosmoticity, indicating a substantial capability for hyperosmotic regulation.

Some of the physiological implications of regulation and conformity are analogous to those discussed under temperature relationships. The conformer does not incur any sizable energy demands for osmoregulation, but to survive over a range of external osmotic pressures, its tissues must be able to function over a similar range of internal solute concentrations. The regulator insulates its tissues from large changes in solute concentration but does so at a metabolic cost.

The responses of most marine invertebrates

Most of the invertebrates characteristic of the oceans are osmoconformers and when placed in diluted seawater soon become isosmotic with

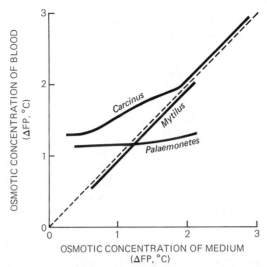

Figure 5–5. Osmotic concentration of the blood as a function of the osmotic concentration of the medium in three invertebrates exposed to various concentrations of sea-water. *Mytilus edulis,* a common mussel, is a strict osmoconformer. The green crab, *Carcinus maenas,* regulates hyperosmotically in brackish waters but is an osmoconformer at higher salinities. The shrimp *Palaemonetes varians* shows both hyperosmotic and hyposmotic regulation. The dashed line is the line of isosmoticity. For clarity, the curve for *Mytilus* has been displaced slightly below the isosmotic line, and the upper part of the curve for *Carcinus* has been displaced slightly above the line. In fact, these curves lie directly on the isosmotic line. (Sources of data—*Carcinus:* Duval, M. 1925. Ann. Inst. Oceanogr. 2: 232–407; *Mytilus:* Potts, W. T. W. and G. Parry. 1964. *Osmotic and Ionic Regulation in Animals.* Pergamon, Oxford; *Palaemonetes:* Panikkar, N. K. 1939. Nature 144: 866–867.)

the medium (see, for example, *Mytilus* in Figure 5–5); as we have seen before, marine invertebrates are usually relatively permeable and are adapted to ionic, not osmotic, regulation in seawater. Generally, there is a considerable osmotic increase in body water upon transfer to a dilute medium, manifested as an increase in overt volume and weight in soft-bodied organisms but more as an increase in weight alone in hard-bodied forms.

The implications of these changes depend, among other things, on the species, concentration of the medium, and duration of exposure. The great majority of species exhibit decreased vigor at salinities below approximately 30‰, and the marine fauna diminishes rapidly in natural waters as salinity falls below that level. Relatively few of the marine coelenterates, annelids, molluscs, or echinoderms, for example, are to be found in brackish waters. Only 25% of the marine invertebrates of the North Sea (salinity 34‰) are found in the mildly brackish waters (30‰) of the Skagerrak (the arm of the North Sea extending between Denmark and Norway).

Ecological as well as strictly physiological factors can be involved in responses to brackish waters. Animals that can survive in waters of a certain dilution in a protective laboratory setting may nonetheless be sufficiently impaired physiologically that they are unable to survive and reproduce at similar salinities in the ecological setting of natural waters. A species that is not physiologically impaired at all may be hindered from living in diluted waters if the organisms it uses for food are limited by osmotic stress. Duration of exposure is also important. Organisms that cannot tolerate continuous exposure to, say, 25‰ in an estuary may nonetheless not be harmed by a few hours of exposure to even lower salinities in rain-washed sands or tidal pools. Fluctuations of salinity at certain points in estuaries are sufficiently rapid that animals may not come to equilibrium with the extreme high and low salinities to which they are exposed, implying greater internal osmotic stability than simplistic extrapolation from environmental variability might suggest. Such considerations demand more experimental attention than they have received and must be kept in mind in interpreting laboratory results. Many molluscs and some other groups can insulate themselves from changes in their osmotic environment by closing themselves off. Certain bivalves, for instance, are able to maintain their body fluids near isosmoticity with seawater over at least several days in diluted waters by tightly closing their shells. Eventually the shells must be opened to permit respiration and feeding, and the approach to isosmoticity with the medium is then rapid.

There are some well-known examples of osmotically conforming animals that are routinely found down to quite low salinities. The common mussel *Mytilus edulis* (see Figure 5–5) is reported from 5‰ to full seawater, and the large burrowing polychaete *Arenicola marina* is found at salinities of 12‰ and above. The starfish *Asterias rubens* lives at 15‰ in the Baltic Sea and can survive 8‰. There is evidence that some of these forms are under stress at the lowest salinities in which they live. *Mytilus* from dilute waters, for example, are smaller than those from seawater and exhibit decreased metabolism, decreased rate of beating by the cilia of the gills, and decreased heat tolerance. *Asterias rubens* apparently cannot breed in waters of 15‰ in the Baltic; breeding individuals are found in the neighboring, more-saline waters of the North Sea.

The ability to restore original volume and weight after transfer to a dilute medium is known as *volume regulation* and has received considerable study. As discussed earlier, there is an initial increase in body water concomitant with achievement of isosmoticity in dilute waters. The isosmotic animal can restore its original volume (still remaining isosmotic) by net loss of body solutes. This can be done, for example, through production of isosmotic urine provided solutes are not replaced through food or other routes of uptake at a rate equivalent to their loss in the urine. *Mytilus* exhibits some volume regulation. *Asterias*, *Arenicola*, and the sipunculid *Golfingia* exhibit little volume regulation for at least several days. The latter two, both soft-bodied worms, remain noticeably swollen when in diluted seawater.

Osmoregulation in crustaceans that occur over broad ranges of salinity

Among crustaceans, osmoconforming marine species are generally limited to the oceans or only slightly dilute brackish waters. The spider crab *Maia squinado*, for example, is poikilosmotic, dies in a matter of a few hours at salinities much below 28‰, and has an exclusively marine distribution. The crustaceans that extend into dilute waters exhibit powers of osmotic regulation.

Two physiological groups of osmoregulating crustaceans may be recognized: (1) those that exhibit hyperosmotic regulation at reduced salinities but lack powers of hyposmotic regulation and conform at higher salinities and (2) those that regulate hyperosmotically at reduced salinities and hyposmotically at higher salinities.

Carcinus maenas (see Figure 5–5) is representative of the first group. It

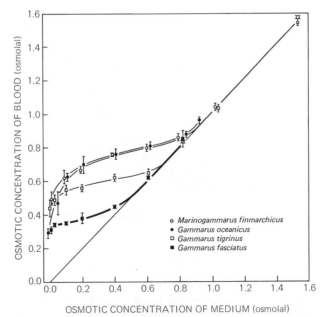

Figure 5–6. Osmotic concentration of the blood as a function of osmotic concentration of the medium in four species of gammarid amphipods exposed to various concentrations of seawater. *Marinogammarus finmarchicus* and *Gammarus oceanicus* are found in marine and brackish situations. *G. tigrinus* is largely brackish in distribution, and *G. fasciatus* is found in fresh waters. Note that the marine species regulate their blood concentrations at the highest levels and show hyperosmotic regulation at the highest salinities. *G. fasciatus* regulates its blood concentration at the lowest levels and begins to regulate hyperosmotically at much lower salinities than the marine species. *G. tigrinus* is intermediate in these respects. Blood concentration of the marine species falls sharply at salinities below about 4‰, whereas *G. fasciatus* and *G. tigrinus* exhibit only small drops in blood concentration even in fresh water. *M. finmarchicus* and *G. oceanicus* do not survive in fresh water. *G. tigrinus* survives over the entire salinity range from fresh water to full seawater. *G. fasciatus* shows poor survival in salinities exceeding 28‰. The urine of *G. oceanicus* is nearly isosmotic to the blood in all media. *G. fasciatus*, on the other hand, produces hyposmotic urine. [From Werntz, H. O. 1963. Biol. Bull. (Woods Hole) 124: 225–239.]

maintains its body fluids hyperosmotic to salinities below those of seawater but is essentially isosmotic with seawater and higher salinities. It is a common shore crab, often found in tidal pools, and penetrates in estuaries to salinities of about 10‰. *Carcinus* exhibits well-developed volume regulation as well as osmotic regulation when transferred from seawater to brackish water.

Certain gammarid amphipods display a qualitatively similar pattern of osmoregulation to that seen in *Carcinus,* as shown in Figure 5–6. *Gammarus oceanicus* and *Marinogammarus finmarchicus* remain hyperosmotic at salinities below about 29‰ and are distributed in marine and brackish waters. *G. tigrinus* and *G. fasciatus* are hyperosmotic below about 20‰ and 14‰, respectively, and are found in brackish and fresh waters.

Hyper-hyposmotic regulation is seen, for example, in *Pachygrapsus crassipes,* a common shore crab of the Pacific coast; many fiddler crabs (genus *Uca*), which spend long periods in air on sand or mud flats; many palaemonid shrimps found in shore and estuarine waters; the brine shrimp (*Artemia salina*), found from dilute brackish waters to the most saline waters known; and the migratory crab *Eriocheir sinensis,* one of the few known crustaceans that can live in both fresh water and seawater (Figures

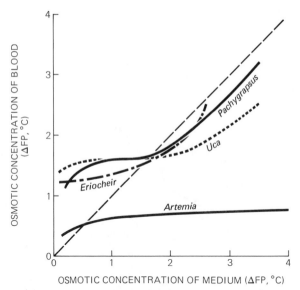

Figure 5-7. Osmotic concentration of the blood as a function of the osmotic concentration of the medium in the Pacific shore crab, *Pachygrapsus crassipes;* the fiddler crab, *Uca crenulata;* the wool-handed crab, *Eriocheir sinensis;* and the brine shrimp, *Artemia salina.* (Sources of data—*Pachygrapsus* and *Uca:* Jones, L. L. 1941. J. Cell. Comp. Physiol. 18: 79–92; *Artemia:* Croghan, P. C. 1958. J. Exp. Biol. 35: 219–233; *Eriocheir:* Scholles, W. 1933. Z. vergl. Physiol. 19: 522–554, and Conklin, R. and A. Krogh. 1938. Z. vergl. Physiol. 26: 239–241.)

5–5 and 5–7). Many of these animals are slightly hyposmotic in seawater. *Palaemonetes varians* and other palaemonid shrimps and *Artemia* are very much so.

The hyper-hyposmotic regulators not only are able to maintain relatively high internal concentrations in dilute waters, but often regulate hyposmotically in waters substantially more concentrated than seawater. The adaptive advantage of an extensive range of hyposmotic regulation is not always apparent. It seems clearly important in such species as *Artemia salina* and *Palaemonetes varians*, which can be found in inland saline waters of relatively high concentration. On the other hand, some species, such as *Eriocheir* and certain palaemonids, are probably not exposed to concentrated waters in nature. Their capacities for hyposmotic regulation in such waters are probably best viewed as an extension of their abilities for regulation in seawater and lower concentrations. Shore crabs, such as *Pachygrapsus, Uca,* and the ghost crab *Ocypode quadrata,* may be exposed to waters that have been concentrated by evaporation in pools. The frequency of such exposure deserves experimental attention. It has also been argued, but not altogether convincingly, that hyposmotic regulation may be an adaptation to coping with evaporative concentration of seawater carried in the gill chambers of forms such as *Uca* and *Ocypode* when on land.

We saw earlier that in hyperosmotic freshwater animals, the energetic demands of osmotic-ionic regulation may be reduced by lowering the osmotic gradient between organism and environment and by limiting permeabilities to water and salts. The same physicochemical principles apply to hyperosmotic regulators in brackish waters. Crustaceans that penetrate into brackish waters typically have lower permeabilities than strictly marine species, and permeabilities in freshwater forms may be lower yet. To illustrate, osmoregulating decapods that occupy shores and estuaries have considerably lower permeabilities than osmoconforming marine species; and crayfish and *Eriocheir*, which occupy fresh waters, have somewhat lower permeabilities than crabs of brackish waters. Similarly, the amphipod *Gammarus fasciatus*, which lives in fresh and brackish waters, is about half as permeable to water as *G. oceanicus*, which lives in brackish and marine waters. You will note from Figures 5–6 and 5–7 that internal osmotic pressure often decreases with external osmotic pressure in osmoregulating crustaceans. This implies the need for adaptation to various solute concentrations at the cellular level but also indicates that osmotic-ionic gradients, and thus energy expenditure for regulation, increase less with decreasing external concentration than if internal concentration were maintained at a constant high level.

The urine of hyperosmotic decapods is often approximately isosmotic with the body fluids at all ambient concentrations. The rate of production of urine increases with the osmotic gradient between animal and environment, thus serving to void the increasing osmotic water load but also aggravating the loss of salts. Diffusional losses of salts also increase as the difference between internal and external concentrations increases. Sodium and chloride are replaced and hyperosmoticity maintained by active uptake from the medium. Production of hyposmotic urine has been reported in the green crab (*Carcinus*) and in certain gammarid amphipods in dilute waters. In these cases the urine can also subserve hyperosmotic regulation. We have earlier reviewed the mechanisms of hyposmotic regulation in marine crustaceans. In hyper-hyposmotic regulators we would expect active ion transport to reverse from inward to outward upon tranfer from a medium in which hyperosmotic regulation occurs to one in which hyposmotic regulation pertains. Such reversal has been reported in *Palaemonetes varians* and *Pachygrapsus crassipes*, for example.

Osmoregulation in nereid polychaetes that occur in brackish and fresh waters

Among nereid polychaete worms, as in crustaceans, there is a good correlation between distribution in dilute waters and capacity for hyperosmotic regulation. *Nereis vexillosa* and *N. pelagica*, for example, are largely limited to marine waters, and both are essentially poikilosmotic at all salinities (Figure 5–8). *N. succinea* and *N. diversicolor* are found far up in estuaries (*diversicolor*, for example, to at least 4‰). These species, although poikilosmotic at high salinities, display hyperosmotic regulation at low salinities. *N. succinea* regulates hyperosmotically from about 15‰ down to very dilute brackish waters, but not in fresh water (Figure 5–8). *N. diversicolor*, which penetrates into lower salinities than *succinea*, remains strongly hyperosmotic even in fresh water. *N. limnicola* is distributed in fresh waters as well as brackish waters and regulates hyperosmotically in these habitats (Figure 5–8).

Active ion uptake and production of hyposmotic urine have been implicated as mechanisms of hyperosmotic regulation. There is evidence that the permeability of the body wall to salts is lower in species that occupy dilute waters than in species limited to waters of high salinity. When nereids

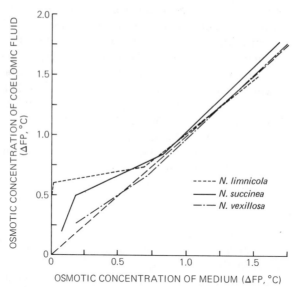

Figure 5–8. Osmotic concentration of the coelomic fluid as a function of the osmotic concentration of the medium in three species of nereid polychaetes collected in the region of San Francisco Bay. All species are osmoconformers at high salinities. *Nereis vexillosa* is found in marine habitats and is essentially an osmoconformer at low salinities also. *N. limnicola* is found in fresh waters and dilute brackish waters. It exhibits well-developed powers of hyperosmotic regulation at low salinities, remaining strongly hyperosmotic even in fresh water. *N. succinea* is widely distributed in brackish waters but does not penetrate fresh waters. Its capacities for hyperosmotic regulation are intermediate between those of *N. vexillosa* and *N. limnicola*. (From Oglesby, L. C. 1965. Comp. Biochem. Physiol. 14: 621–640. Used by permission of Pergamon Press.)

are transferred rapidly from seawater to diluted water, they initially lose salts and gain water. The degree of osmotic swelling is greater in marine species than in brackish-water species. The marine species show poor volume regulation. *N. vexillosa*, for example, remains permanently swollen after transfer to dilute waters. This condition interferes with bodily movements and probably increases vulnerability to predation. Species distributed in dilute waters typically show good volume regulation. *N. diversicolor*, for example, swells to about 160% of its original weight within a few hours when transferred from seawater to 20% seawater but approximately restores its original volume within a few days.

Marine and migratory fish

Many fish that live in the oceans are limited, physiologically or ecologically, to essentially marine salinities. A good many also have marine and brackish distributions. These include, for example, menhaden (*Brevoortia tyrannus*) and winter flounder (*Pseudopleuronectes americanus*), both of which penetrate well into estuaries. Some species, such as the killifish (*Fundulus heteroclitus*), occur routinely in fresh, brackish, and marine waters. Of particular interest are the fish that migrate between fresh waters and the oceans, typically breeding in one habitat and undergoing much of their growth and maturation in the other. Those that ascend rivers and streams from the sea to breed are termed *anadromous* ("running upward"), whereas those that grow in fresh waters and descend to the sea for breeding are termed *catadromous* ("running downward"). The former group includes species of salmon, smelt, shad, and lampreys (primitive jawless fish). Fresh-water eels of Europe (*Anguilla vulgaris*) and North America (*A. rostrata*) belong to the latter group.

The migratory fish are characteristically excellent osmoregulators, generally showing only a small drop in internal osmotic pressure in fresh water as compared to seawater. The chinook salmon (*Oncorhynchus tshawytscha*), for example, has an internal osmotic concentration of $\Delta 0.76°C$ when in the ocean and $\Delta 0.67°C$ when at its freshwater spawning grounds. Average values for *Anguilla vulgaris* in seawater and fresh water are $\Delta 0.71°C$ and $\Delta 0.61°C$, respectively. In nature, movements between fresh water and seawater can be accomplished more or less gradually because of the interposition of a gradient of brackish waters. Some species seem to require a gradual transition to their new physiological situation, whereas others (at least at certain stages) can be shifted directly from one extreme environment to the other without apparent harm. In many cases physiological capacities for living in the extreme environments are gained or lost during the life cycle: The species is not fully euryhaline at all stages of its development. Salmon eggs, for example, do not develop normally in saline waters. Young coho salmon (*O. kisutch*) require several months to develop full salinity tolerance and do not migrate to the oceans for a year or more. Chum salmon fry (*O. keta*) develop salinity tolerance more quickly and migrate sooner. Adult lampreys (*Petromyzon marinus* and *Lampetra fluviatilis*) progressively lose their ability to osmoregulate in highly saline waters as they migrate up rivers for spawning. They become stenohaline freshwater fish, and most die soon after breeding. Many adult fish remain euryhaline during their migrations. Atlantic salmon (*Salmo salar*), for example, may make several trips up rivers for spawning over their lifetime.

Migratory and other euryhaline teleosts and lampreys regulate hyper-osmotically at low salinities and hyposmotically at high salinities. The mechanisms of regulation are like those discussed for freshwater and marine teleosts, respectively. Upon movement from seawater to fresh water, for example, the direction of active ion transport across the gills reverses from outward to inward, drinking of the medium decreases, urine production increases, and the urine becomes hyposmotic to the blood. The mechanisms of control of osmoregulation in changing media are now beginning to receive the close attention that their interest level merits.

Common responses of freshwater animals in brackish waters

Most characteristically freshwater animals are capable only of hyper-osmotic regulation and do not tolerate substantial elevation of their internal solute concentration. If a leopard frog, crayfish, carp, or freshwater mussel is placed in progressively more concentrated brackish water, internal osmotic pressure will rise and ultimately become isosmotic to the medium (Figure 5–9). As a rule of thumb, freshwater animals do not survive well at salinities above their normal internal concentration in fresh water. Many species, whether for physiological or ecological reasons, do not enter even dilute brackish waters. Others, such as carp (*Cyprinus carpio*) and channel catfish (*Ictalurus punctatus*), venture well into estuaries. The characteristically freshwater fauna has virtually disappeared at salinities above 10‰.

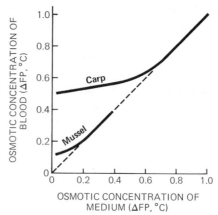

Figure 5–9. Osmotic concentration of the blood as a function of the osmotic concentration of the medium in a carp and freshwater mussel (*Anodonta*). (Data from Duval, M. 1925. Ann. Inst. Oceanogr. 2: 232–407.)

Euryhaline amphibians: some special cases

Most amphibians are virtually limited to fresh waters, but there are some exceptional species that penetrate into concentrated brackish waters. *Rana cancrivora*, the crab-eating frog of southeast Asia, occurs in coastal mangrove swamps up to salinities of 29‰. The green toad, *Bufo viridis*, of Europe, the Middle East, and western Asia may be found at nearly as high concentrations, though salinity tolerance varies seasonally and between populations.

In both species, internal osmotic pressure rises only slightly in waters up to 10‰ but then increases in parallel with external concentration, remaining somewhat hyperosmotic (Figure 5–10). The adaptation of *R. cancrivora* to high salinities is reminiscent of marine elasmobranchs. As depicted in the figure, about 40% of its increase in internal osmotic pressure is due to inorganic ions (mostly NaCl), but 60% is due to retention of urea. The frog in high salinities, being hyperosmotic, takes up water by osmosis. Because internal ion concentrations are below those of the medium, salts are also taken up passively—the reverse of the situation in waters of low salinities. Excess water is voided in the urine, which always remains hyposmotic to the blood. Urinary excretion also seems adequate to void the salt load, though extrarenal excretion remains a possibility. In *Bufo viridis*, most (85%) of the increase in internal osmotic pressure at high salinities is due to NaCl, though there is again some increase in blood urea concentration.

The tadpoles of *R. cancrivora* live at even higher salinities than the adults and exhibit a very different pattern of osmoregulation. They remain hyposmotic to the medium above about 10‰; in full seawater, their internal concentration is around $\Delta 1.0°C$. Mechanisms of regulation include drinking of the medium and extrarenal salt excretion. This species presents a particularly striking example of change in osmotic-ionic relationships during development.

Relationships between salinity and metabolic rate

Osmoconformers such as the spider crab *Maia*, the starfish *Asterias*, and the sea anemone *Metridium* generally exhibit a reduction in metabolic

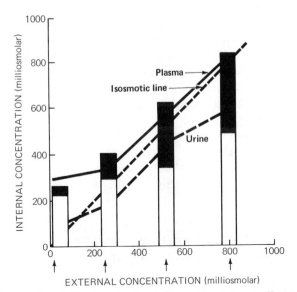

Figure 5–10. Properties of blood and urine in *Rana cancrivora* acclimated to fresh water and various concentrations of seawater for at least 48 hours. The solid and long-dashed lines portray the osmotic concentrations of blood plasma and urine, respectively, as functions of environmental osmotic concentration. The bars show the osmotic pressure caused by ions (open part) and urea (solid part) in the plasma at the external concentrations indicated by arrows. Most of the ionic osmotic pressure was due to NaCl. (From Gordon, M. S., K. Schmidt–Nielsen, and H. M. Kelly. 1961. J. Exp. Biol. 38: 659–678.)

rate as salinity decreases from its level in seawater. Many osmoregulators that have been studied have shown an increase in metabolism roughly in relation to increases in the osmotic-ionic gradient between animal and environment. In *Carcinus maenas*, a hyperosmotic regulator in brackish waters, for example, oxygen consumption increases with decreases in salinity and is elevated by almost 50% at 5‰ in comparison to its level at 32‰. The euryhaline stickleback *Gasterosteus* (a teleost) has a higher respiratory rate in fresh water than in isosmotic brackish water, and the crayfish *Astacus* exhibits a 40% increase in metabolism upon transfer from brackish water (15‰) to fresh water. Some osmoregulators show little or no consistent change in metabolism with salinity. This is true, for example, of *Eriocheir sinensis* when compared in fresh water and seawater. *Artemia* has about the same metabolic rate between 35‰ and 140‰, and the killifish, *Fundulus heteroclitus*, between fresh water and an isosmotic solution. In a few osmoregulators there is a decline in metabolic rate with increase in the osmotic-ionic gradient. The euryhaline starry flounder (*Platichthys stellatus*), for instance, has a lower rate of metabolism in fresh water than in isosmotic brackish water.

The interpretation of salinity-related changes in metabolism has been an area of uncertainty and controversy. The fact that many osmoregulators increase their metabolism in dilute waters whereas many osmoconformers do not (in fact, they typically decrease their metabolic rate) has suggested that the relatively large increases in metabolism seen in many osmoregulators are directly reflective of the increased work demanded for osmotic-ionic regulation in dilute waters. This conclusion is contraindicated by the responses of species such as *Eriocheir* and *Artemia*, which, although they are very effective regulators, show little change in metabolic rate over wide ranges of salinity. It is also contraindicated by calculations of the energetic demands of osmotic-ionic regulation, which indicate that regulation is expected to demand only a few percent of the total metabolism, even when gradients between organism and environment are large; this would imply that changes in the energetic demands of regulation would not be a dominant influence on overall metabolism and, in particular, would account for only a small part of any large increase in metabolism resulting from transfer to dilute waters. Although there is little doubt that osmoregulators must invest more energy in osmotic-ionic regulation as gradients between their body fluids and the environment are increased, there are thus reasons to question whether the large changes in metabolism observed in some forms are to be attributed entirely (or even primarily) to the demands of maintaining osmotic-ionic homeostasis.

A PHYLOGENETIC SUMMARY OF THE RELATIONSHIPS OF AQUATIC ANIMALS

All freshwater invertebrates regulate hyperosmotically. Ambient osmotic pressures in fresh water are $\Delta 0.001°$–$\Delta 0.02°C$, and the internal osmotic pressures of freshwater invertebrates range at least from $\Delta 0.08°C$ (in the clam, *Anodonta*) to $\Delta 1.2°C$ (in the migratory crab, *Eriocheir*). The mechanisms of osmotic regulation generally include active uptake of sodium and chloride from the medium and production of a copious urine that is hyposmotic to the blood (see Figure 5–2). Some forms deviate from this common pattern. *Eriocheir*, for example, produces only isosmotic urine.

Most marine invertebrates are isosmotic, or nearly so, with seawater ($\Delta 1.9°C$), but the ionic composition of their body fluids differs from that of seawater, indicating ionic regulation. Some marine invertebrates are hyposmotic to seawater, the known examples being arthropods such as prawns (*Palaemonetes*), fiddler crabs (*Uca*), the brine shrimp (*Artemia*), and insect larvae (*Aedes* and *Ephydra*). The mechanisms of hyposmotic regulation are generally incompletely known. The most thoroughly studied example, *Artemia*, uses mechanisms resembling those of teleost fish (discussed later). Some crabs (e.g., *Uca*) can produce urine that is slightly hyperosmotic to the blood, and the insect larvae produce strongly hyperosmotic urine; but only isosmotic urine is known in the prawns and other crabs.

Most marine invertebrates respond poikilosmotically when placed in brackish waters, lacking powers of osmoregulation. Most of these forms do not survive well at low salinities, but some, such as the mussel *Mytilus edulis* and the starfish *Asterias rubens*, are found in dilute waters despite their poikilosmotic condition. The majority of invertebrates that range from the sea into brackish waters display powers of osmoregulation. Among both crustaceans and nereid polychaetes, there are clear correlations between the ability to osmoregulate and the extent of penetration into brackish waters. Spider crabs (*Maia*), for example, are poikilosmotic and die at salinities below about 28‰. By contrast, such well-known animals of brackish waters as green crabs (*Carcinus*) and prawns (*Palaemonetes*) regulate hyperosmotically in dilute waters. The poikilosmotic nereid *Nereis vexillosa* is largely limited to marine waters, but *N. limnicola* lives in fresh and brackish waters and regulates hyperosmotically. Some nereids, the green crab (*Carcinus*), and some gammarid amphipods are examples of species that regulate hyper-osmotically in dilute waters but lack powers of hyposmotic regulation and conform osmotically at high salinities. By contrast, some crustaceans, such as certain crabs (*Pachygrapsus, Eriocheir,* and *Uca*) and prawns (*Palae-monetes*), have powers of both hyperosmotic and hyposmotic regulation. Some of these are forms that are hyposmotic when in normal seawater.

Invertebrates characteristic of fresh water typically lack powers of hyposmotic regulation. Placed in increasingly saline brackish waters, their internal osmotic pressure rises to that of the medium, and usually survival is limited to ambient osmotic concentrations below those found in the blood in fresh water.

In reviewing fish, we shall take the common view of their evolution, recognizing that it is still open to debate. Fish are thought to have first evolved in the ocean, and the hagfish may trace a continuously marine an-cestry. Hagfish have sufficiently high salt concentrations in their blood to render them isosmotic with seawater. Thus they resemble the mass of ma-rine invertebrates in showing ionic but not osmotic regulation.

Early fish penetrated fresh waters, and it is there that the ancestral jawed fish are believed to have evolved. Modern freshwater teleost fish are hyperosmotic to their medium, with internal osmotic pressures of about $\Delta 0.45°–\Delta 0.65°C$. The lowering of osmotic pressures from those presumably characteristic of their marine ancestors is believed to represent an adapta-tion to reducing the energetic cost of osmotic-ionic regulation. The mecha-nisms of regulation include active absorption of sodium and chloride across the gills and production of a copious hyposmotic urine (see Figure 5–3).

The marine teleosts are believed to be descended from freshwater forms. Though their internal osmotic pressures ($\Delta 0.6°–\Delta 1.1°C$) tend to be somewhat higher than those of freshwater teleosts, they are hyposmotic to the sea. They replace osmotic water losses by drinking seawater and absorb-ing water, along with a considerable salt load, across their gut. The load of sodium and chloride gained from ingested water as well as that gained by direct diffusion from the ocean is voided largely by extrarenal salt excretion across the gills. The urine is scant and essentially isosmotic to the body fluids, serving mainly to void divalent ions (especially magnesium and sul-fate). See Figure 5–3.

The marine elasmobranchs (sharks, skates, and rays) are also believed to be descended from freshwater forms and have blood salt concentrations resembling those of marine teleosts. However, they maintain high concen-trations of urea and trimethylamine oxide in their blood and are slightly hyperosmotic to seawater. As a consequence they do not need to drink to replace osmotic water losses and thus avoid the high salt load imposed on teleosts by ingestion and absorption of seawater. Rectal glands play a signifi-cant role in voiding excess monovalent salts. Elasmobranchs have reinvaded fresh waters and usually retain elevated concentrations of urea, though re-cently certain river stingrays have been shown to have low urea concen-trations.

Teleosts, such as the migratory forms, which occur in both fresh water and the ocean have powers of both hyposmotic and hyperosmotic regulation and maintain remarkably stable internal osmotic concentrations regardless of habitat. Their mechanisms of osmotic-ionic regulation in fresh and marine waters are like those of the freshwater and marine teleosts discussed earlier.

Other freshwater vertebrates, such as frogs, are hyperosmotic to the water and utilize mechanisms for osmotic-ionic regulation analogous to those seen in freshwater teleosts. Most fish and amphibians characteristic of fresh water, when placed in increasingly saline brackish water, ultimately become isosmotic to the medium and survive poorly at ambient osmotic concentrations exceeding their normal internal concentration in fresh water. Certain frogs are exceptional in being able to invade highly saline waters. The crab-eating frog, *Rana cancrivora*, regulates hyposmotically in high salinities as a tadpole. When adult, it maintains slight hyperosmoticity to the medium by increasing blood urea concentration as salinity rises, showing just a limited increase in blood salt concentration (see Figure 5–10). The green toad, *Bufo viridis*, also remains slightly hyperosmotic at high salinities but exhibits greater increases in salt concentration than *R. cancrivora* and smaller increases in urea.

Marine reptiles, like other reptiles, cannot raise urinary osmotic pressure above that of the blood, and in most marine birds the maximum osmotic U/P ratio is on the order of 2 or less. These groups possess salt glands that discharge highly concentrated solutions of sodium, chloride, and some other ions into the nasal passages or orbits. Certain savannah sparrows (*Passerculus sandwichensis*) that live in salt marshes and lack salt glands are able to produce especially concentrated urine. The salt and water relations of marine mammals are incompletely understood, but these animals are believed to rely considerably on the general mammalian ability to produce strongly hyperosmotic urine. Mammals are not known to possess extrarenal salt glands.

OSMOREGULATION AT VARIOUS STAGES OF THE LIFE CYCLE IN AQUATIC ANIMALS

This section and the next two will deal briefly with three topics relevant to aquatic animals: (1) osmoregulation at various stages of the life cycle, (2) effects of temperature and other environmental factors on water and salt relationships, and (3) responses to drying of the habitat. These topics are all relatively poorly understood, and emphasis will necessarily have to be placed on examples.

The animal must accommodate to its environment at all stages of its life cycle. Eggs and other early stages do not possess the differentiated organs and tissues employed by adults in osmotic-ionic regulation; these structures must undergo a gradual anatomical and physiological development. Young stages have high surface-to-volume ratios; and whereas eggs, cleavage stages, and larvae may be encased in relatively impermeable membranes, the young after hatching are frequently characterized by thin and delicate integuments. These factors would be expected to aggravate problems of passive water and solute exchange.

The eggs of most marine invertebrates are isosmotic with the medium and relatively permeable. Eggs of marine teleosts are hyposmotic to seawater, and those of teleosts and other animals in fresh water are hyperosmotic to the medium; in these groups the permeability of the eggs typically becomes very low soon after laying.

Eggs of freshwater trout and frogs take up some net water after they are laid, resulting in swelling and reduction in the internal osmotic pressure. Permeability soon falls, but because it is not reduced to zero, there is continued osmotic uptake. At least for a while, however, there is little further swelling; this may be due to buildup of internal hydrostatic pressure owing to resistance of the egg membranes to expansion or may be due to some as yet unknown mechanism of water excretion. Active uptake of ions has been reported in some freshwater fish and amphibian eggs.

Many freshwater invertebrates, by comparison to their marine relatives, produce larger eggs and undergo greater development within the egg membranes (often, through the early larval stages). This may be related to the demands for osmoregulation that are encountered after the protective egg membranes have been left behind.

Larvae of many euryhaline fish and invertebrates (e.g., *Nereis diversicolor* and the oyster *Ostrea edulis*) survive well only over narrower ranges of salinity than adults. We have previously discussed the development of salinity tolerance in eggs and larvae of salmon. Regulatory abilities of young are occasionally greater than those of adults. There is evidence, for example, that larvae of the spider crab *Libinia emarginata* and the calico crab *Hepatus epheliticus* osmoregulate. The adults of both species are stenohaline osmoconformers. The capacities of the larvae may be important in permitting dispersal through estuarine waters and in accommodating to the variable salinity of surface waters where much of their planktonic existence is spent. The osmoregulatory abilities of *Rana cancrivora* tadpoles have been discussed earlier.

Decapod crustaceans take up water before and during their molt. The increase in body volume helps to split the old exoskeleton and expand the new exoskeleton before it hardens. The amount of uptake of water varies from a few percent of the original body weight (as in *Gecarcinus*) to 100% (as in *Maia*). The mechanism of water uptake is not fully understood. *Carcinus* and other osmoregulating marine species that have been studied become slightly hyperosmotic to seawater during the molt, but this can account for only a small portion of the water uptake. Osmoconforming species that have been studied remain essentially isosmotic with seawater. Drinking of seawater, followed by water absorption from the gut, has been demonstrated in several species and accounts for a minor (*Pugettia producta*) to sizable (*Carcinus maenas*) proportion of the total water uptake. Changes in water relationships during molting are at least partly under hormonal control.

EFFECTS OF TEMPERATURE AND OTHER ENVIRONMENTAL FACTORS ON WATER AND SALT RELATIONSHIPS IN AQUATIC ANIMALS

Temperature has important influences on many of the individual processes involved in water and salt relations: rates of diffusion and osmosis, rates of active transport, rates of urine production, and permeabilities of membranes, among others. Thus we would expect temperature to affect regulation of body fluid composition. It is not unusual to find changes in total osmotic pressure and in concentrations of particular ions with changes in temperature. Range of salinity tolerance varies with temperature in some species, and in some cases certain temperatures cause a potentially lethal breakdown of regulation in the prevailing osmotic-ionic environment. Our knowledge of these effects is meager. Examples will serve to illustrate some of the possibilities.

Gammarus duebeni is an amphipod that regulates hyperosmotically in brackish waters. Internal osmotic pressure is maintained about $\Delta0.1°C$ higher at 7°C than at 20°C in the salinity range of 5–20‰. The difference is over $\Delta0.2°C$ in fresh water. The range of salinity tolerance varies with temperature, being maximal in the temperature range 4°–16°C. The shrimp *Crangon crangon* regulates hyperosmotically at low salinities. In Holland it occupies dilute estuarine waters in the summer. Regulation in such waters breaks down at temperatures below 3°C, however, and the shrimp migrate into higher salinities during the winter. The cod *Gadus morhua* cannot osmoregulate at temperatures below 2°C, and its distribution is limited to warmer waters in the north Atlantic.

The lamprey *Lampetra fluviatilis* in fresh water exhibits decreased integumentary permeability to water, decreased urine production, and greatly inhibited active ion uptake from the water at temperatures near 0°C. The decreased active uptake of ions is apparently not fully compensated by reduced rates of diffusional and urinary loss. Thus it appears that internal concentrations must fall during the winter. In the freshwater isopod *Asellus aquaticus*, the rate of active sodium uptake is halved when the temperature drops from 11°C to 1°C, but the rate of sodium loss is unaffected. Sodium concentration in the body fluids declines with temperature down to 5°C but inexplicably rises again at lower temperatures. Declines in blood concentration with temperature, possibly caused by decreased active ion uptake, have also been reported in the crayfish *Astacus*, the amphipod *Gammarus pulex*, and the salmon *Salmo trutta*, to cite some examples.

Frogs in water at temperatures below about 8°C (as in winter) exhibit water retention, largely resulting from decreased urine production. Body weight increases by up to 20% over its level at 20°C, and blood solute concentrations are correspondingly reduced. Because concentrations typically remain stable at their diluted values in cold waters, there is a balance between ion losses and gains despite a reduced rate of active uptake across the skin. The condition of hydration persists until temperatures rise again.

Factors in addition to temperature that may influence salt and water relationships include oxygen content of the water, pH, photoperiod, and concentrations of specific ions. Our knowledge of the effects of these factors is even more limited than that for temperature. In an interesting study of the American lobster, animals were acclimated to various combinations of temperature, salinity, and oxygen concentration. Then, while temperature and oxygen concentration were held at their acclimation levels, salinity was varied acutely to determine the level at which 50% of the animals would die within 48 hours. The lower lethal salinity was found to be dependent on the acclimation levels of all three parameters. Animals survived to lower salinities, the lower the salinity to which they had been acclimated, the lower the prevailing temperature, and the higher the prevailing oxygen concentration.

Effects of acclimation or acclimatization have been reported in a number of other studies as well. Populations of mussels (*Mytilus*) from the

North Sea (30‰) and Baltic (15‰), for example, were found to differ in the lower and, especially, upper salinity limits at which their gills continued to show ciliary activity. When animals from each habitat were acclimated to the salinity characteristic of the other habitat, the salinity limits for ciliary activity gradually came to approximate those shown by animals native to each habitat. Green crabs (*Carcinus*) from the Baltic maintain higher internal osmotic pressures at ambient salinities of 5–10‰ than crabs from the North Sea. The differences were reduced, but not eliminated, after crabs from each habitat had been acclimated to salinities characteristic of the other, indicating at least a partial basis in acclimatization.

Calcium is important to the maintenance of cell walls and intercellular cements. Low calcium levels in the water, implying decreased physiological availability of calcium, can lead to increased permeability of integumentary membranes. This may be a factor in the response of the polychaete *Nereis diversicolor* to calcium-free brackish waters. Worms placed from seawater into ordinary brackish water swell but quickly recover their original volume. Worms placed in calcium-free brackish water show continued swelling and do not commence volume recovery until the calcium is replaced. Many tropical marine teleosts that ordinarily do not penetrate very far into estuaries invade fresh waters that receive drainage from limestone regions. This phenomenon awaits careful study, but decreased permeability associated with the relatively high calcium levels may be involved.

RESPONSES TO DRYING
OF THE HABITAT IN AQUATIC ANIMALS

Residents of puddles, ditches, small ponds, intermittent streams, and the like, are often confronted with the drying of their habitat. It is hypothesized that this was an important factor in the emergence of vertebrates onto land. In the Devonian, two groups of related bony fish with lungs and fleshy fins appeared: the Crossopterygii and the Dipnoi. It is believed that these fish occupied transient bodies of fresh water and that their lungs evolved under a selective pressure to resort to air breathing when waters became stagnant. The lungs and lobed fins also probably permitted some species to migrate across land from drying streams or ponds to seek other bodies of water. The crossopterygians are thought to have given rise to the amphibians. They are, to our present knowledge, represented today by only one marine species, *Latimeria chalumnae*. The Dipnoi are represented by three genera of lungfish, one each in the tropical regions of Australia, Africa, and South America. These fish are found in transient bodies of fresh water where their lungs provide for respiration during periods of stagnation. The Australian genus, *Neoceratodus*, has sufficiently well-developed gills that it can live continually under water, but *Protopterus*, of Africa, and *Lepidosiren*, of South America, must utilize air breathing even when the waters are well aerated.

Protopterus and *Lepidosiren* are able to survive periods of drought by entering a resting condition within a chamber of mud dug in the bed of a lake or stream. This has been well studied in *P. aethiopicus*. When the water level falls to a low point, the animal digs into the mud and positions itself head upward. It makes occasional trips to the surface for breathing so long as water remains. Once the water has dried, the fish curls up, and mucus it has secreted hardens into a cocoon that opens only into the mouth. The mouth hole is used for breathing, and the cocoon serves to decrease evaporative water losses. Metabolism ultimately drops to about 10% of the ordinary resting level; this reduces respiratory water losses. The fish does not feed, and urine production virtually ceases. Urea becomes the major nitrogenous end product and builds up in the blood to levels approaching those of marine elasmobranchs. The fish can survive in this condition for more than a year.

Many other freshwater animals burrow into the substrate and enter a resting condition during times of drought. This has been reported, for example, in some leeches, snails, water mites, and amphibians. Mucoid coverings are common.

Many forms enter special dormant life stages that are resistant to desiccation in air, and often to other environmental extremes as well. Examples of such desiccation-resistant stages include the eggs of some nematodes, rotifers, ostracods, fairy shrimp, water fleas (cladocerans), and other crustaceans; embryonic cysts of coelenterates; and gemmules of sponges. Certain fish (mostly cyprinodonts) also produce desiccation-resistant eggs. Adult tardigrades ("water bears") and bdelloid rotifers are able to enter desiccated dormant conditions. A tardigrade exposed to drying retracts its head, posterior end, and appendages, loses considerable water, and assumes a rounded, shriveled appearance. Its metabolism becomes greatly depressed. The animal can survive in this condition for several years, reviving quickly when water again becomes available. Many of the other desiccation-resistant stages noted earlier also can endure impressively long periods of drought. Frequently they can be carried about by wind, thus leading to dispersal across land.

Many freshwater animals, according to present knowledge, have relatively little or no ability to cope with drying of their habitat other than by migration, if possible, away from the drying area. Many fish, amphipods, and isopods, for example, are in this category, though some do receive temporary protection by burrowing into mud. Animals in which a particular life stage is desiccation-resistant are typically vulnerable at other stages.

Numerous intertidal marine animals are exposed to air at low tides and commonly respond behaviorally. Small crustaceans and starfish may stay in among seaweed. Anemones contract into a compact mass, withdrawing their tentacles and closing their oral opening. Burrowing animals such as many bivalves and polychaetes withdraw from the surface.

Barnacles, snails, mussels, and the like, can seal themselves off by closing their shells at low tide. Maximum sealing, although it minimizes desiccation, may interfere with exchange of oxygen and carbon dioxide. Whereas some of these animals (e.g., certain bivalves) resort to anaerobic metabolism during low tide (see Chapter 11), others ordinarily do not. Intertidal barnacles expel water from their mantle cavity when the tide recedes and maintain a small opening between their opercular valves, allowing some evaporation of water but also providing for proper gas exchange. The opercular valves can be closed more tightly than they actually are. Although some intertidal bivalves appear to remain tightly closed throughout their exposure to air, others gape occasionally; and at least one, *Modiolus demis-*

sus, the ribbed mussel common to pilings in the high intertidal zone, leaves its valves partially open throughout its exposure to air.

Many intertidal animals are able to tolerate considerable desiccation: on the order of 15–30% of the body weight in certain sipunculids, mussels, and barnacles, for example. Some barnacles of the high intertidal zone have lower metabolic rates than ones of the low and middle zones. This may be an adaptation to lowering their needs for gas exchange, and correspondingly reducing water loss, during their longer periods of exposure to the air. Large individuals of a species have lower surface-to-volume ratios than small, and it is reported in certain snails that larger animals survive longer in dry air. This factor has been related to distribution in several species of snails, smaller individuals tending to be found lower in the intertidal zone, where air exposure is shorter.

ANIMALS ON LAND

Animal life originated and spent much of its evolutionary history in water. The land and its plant life represented a vast ecological reservoir within which animals would ultimately establish themselves in an immense variety of ecological niches. The earliest animals that ventured to spend time on land, to consume the productivity of the land, and, ultimately, to develop on land were able to escape competitors and predators in their primordial habitat. In these regards positive selective pressures for terrestriality must have been great. But early animal life was adapted to developing and living in an abundance of water, and evaporative losses of water constituted a physiological problem of paramount importance for all stages of the life cycle. Today we see groups, such as most amphibians and all terrestrial crabs, that still return to water for breeding. Many terrestrial animals remain tied, to a greater or lesser extent, to microhabitats in which the humidity is high. Only a few groups have made the full transition to relative independence of protected, humid environments—notably the insects, arachnids, reptiles, birds, and mammals.

Most terrestrial animals are able to replace salt losses rather readily through intake in their food. Their problems of water balance generally, and properly, receive emphasis. Evaporative losses add to urinary and fecal losses—all of which must be compensated through intake of preformed water and production of metabolic water.

The problems of evaporative water loss are intimately interconnected with thermal relationships. Elevated temperatures are often associated with increased vapor pressure differences between the animal and the air, leading to increased rates of evaporation. Many poikilotherms survive threateningly high ambient temperatures by virtue of some passive evaporative cooling of their bodies, and in many thermoregulators evaporative losses are actively increased at high temperatures. In general, high temperatures aggravate problems of water balance. High humidities, on the other hand, often act to limit the range of temperature tolerance by interfering with evaporative cooling.

Exchange of carbon dioxide and oxygen requires the animal to expose moist, thin membranes to the environment, and these are sites of relatively high evaporative water loss. Animals with fully exposed respiratory surfaces (e.g., the skin of earthworms and frogs) can suffer relatively large water losses; movement of air across such surfaces can greatly exceed that necessary for respiratory exchange, meaning that the animal will lose more water than that to which it is obligated to support its metabolism. Many terrestrial animals have invaginated respiratory structures that communicate to the outside via small openings (note the lungs of vertebrates and pulmonate snails and the tracheae of insects). In some cases the openings remain continuously open, and gas exchange with the atmosphere occurs by diffusion. Here the anatomical arrangement may limit exchange rates and, thus, water loss, while still providing for proper uptake of oxygen and loss of carbon dioxide. In many animals access of air to the respiratory surfaces is controlled actively and rather precisely according to needs for respiratory exchange, limiting water loss to the level obligated by metabolism. This is accomplished, for example, in most insects by opening and closing the tracheal orifices and in mammals by regulating ventilation of the lungs. These topics will be covered in more depth in Chapter 8. Here we emphasize that the structure and physiology of respiration have been influenced in important ways by the problems of evaporative water loss. In a more immediate sense, increases in metabolism augment the rate of ventilation in many animals and thus increase the rate of evaporative water loss.

There are a variety of additional adaptations to reducing water losses in terrestrial animals. These tend to be more highly developed in the more fully terrestrial species. Evaporative losses are reduced by decreased permeability of the general integument. Fecal losses can be diminished by removing water from the feces prior to elimination. Urinary losses can be reduced by decreasing the solute load and by increasing the concentration of the urine. A common mechanism for decreasing the solute load is elimination of nitrogenous wastes in the form of poorly soluble compounds (see Chapter 6 for a full discussion). Many insects, many birds, and mammals can produce urine that is hyperosmotic to the blood. Most other terrestrial animals cannot increase urinary concentration above isosmoticity with the blood. Behavioral mechanisms for reducing water loss are ubiquitous. For example, animals can occupy microhabitats of high humidity and limit their activity to cooler times of day. All land animals are tolerant of modest desiccation. Groups that have limited ability to restrict water losses or that live in arid regions tend to be especially so.

Often water intake is sufficient that animals do not have to exercise their capacities for retention to a maximal extent. Water outputs are adjusted to maintain internal homeostasis. Sometimes, as during flooding of the habitat, adequate water loss, rather than retention, may become the problem.

Earthworms

Earthworms have a highly vascularized, moist integument that serves respiratory gas exchange but that also poses little barrier to evaporative water loss. Because they desiccate rapidly in dry air, they are limited to the soil and other humid microhabitats. *Lumbricus terrestris*, a common nightcrawler, can produce urine that is essentially isosmotic to the blood or highly hyposmotic; we would presume that the concentration is adjusted according to needs for water balance. Earthworms avoid dry soils, digging deeper when upper layers begin to dry out. They can tolerate a remarkable degree of desiccation: 50–80% of body water in *Lumbricus*. Internal osmotic pressure is around Δ0.3°–Δ0.5°C in well-hydrated worms, but it varies widely according to the state of hydration. This lack of close regulation is something we shall see in many other highly permeable terrestrial animals. During periods of drought, earthworms burrow deep into the soil and assume a desiccated, quiescent condition ("estivation"). A few species surround themselves with a tough mucoid cyst that protects them from further water loss.

Earthworms live well in aerated fresh water, though they seldom venture into water in nature. They are effective hyperosmotic regulators, producing hyposmotic urine and actively absorbing ions. It seems unlikely that immersion is directly related to the death of worms that may be found on sidewalks and streets after heavy rains. It has been suggested that exposure to the ultraviolet rays of the sun is the lethal factor.

Terrestrial isopods

The terrestrial isopod crustaceans (wood lice, pill bugs) have received considerable attention and provide interesting comparisons. All make extensive use of secluded, humid microhabitats. Various species occupy and wander in a diversity of macroenvironments. *Ligia oceanica*, for example, is littoral, occurring in among rocks at the high-tide level and above. *Oniscus asellus* and *Porcellio scaber* are found in mesic terrestrial areas and seldom venture out except at night. *Armadillidium vulgare* occurs in relatively xeric areas, where its activity is chiefly nocturnal, and in mesic areas, where it may be found abroad during the day as well. *Hemilepistus reaumuri*, the "desert wood louse," is found in arid regions of northern Africa. It spends much of its time in vertical holes that it digs in the sand but may be found active on the surface during the morning and late afternoon as well as at night.

Overall permeability is assessed by measuring the total evaporative water loss under given conditions of temperature and humidity and expressing this as water loss per unit of surface area per unit of time (and often per unit of saturation deficit). This measure is essentially an expression of average surface-specific permeability. Parts of the body surface (e.g., the respiratory surfaces) may be more permeable, and other parts may be less permeable. The average value is useful in that it provides a measure of the total response of the animal. Terrestrial isopods are substantially less permeable than earthworms. Most isopods are considerably more permeable than most insects, but the least permeable isopods lose less water per unit surface area than the most permeable insects.

Among the species of isopods, there is a general correlation between permeability and dryness of the habitat. Under comparable conditions at temperatures around 30°C, the littoral *Ligia* loses water about twice as fast per unit surface area as the mesic *Porcellio*, about 2.7 times as fast as the mesic-xeric *Armadillidium*, and about 10 times as fast as the desert-dwelling *Venezillo arizonicus*. Reduction in permeability among the more xeric forms results from changes in the general integument and probably also from changes in respiratory anatomy. The main sites of oxygen uptake are flattened posterior legs called pleopods. In *Ligia* the pleopods are unmodified, and their entire respiratory surfaces are relatively exposed to the air. In many of the more terrestrial groups there are invaginated respiratory surfaces (pseudotracheae) within the pleopods, opening to the outside via small orifices (see Figure 8–14). This arrangement may serve to reduce respiratory water losses and is demonstrably important to proper oxygen uptake in dry air. In such air the surfaces of the general integument and the exposed surfaces of the pleopods may become so dry that little gas exchange can occur across them, but the respiratory surfaces of the invaginated pseudotracheae remain moist and functional. Oxygen consumption of *Ligia*, which lacks pseudotracheae, is reduced by almost 90% in dry air by comparison to moist air. The reduction in *Porcellio* and *Armadillidium*, both of which have well-developed pseudotracheae, is only about 10% and 6%, respectively. *Ligia* can suffocate to death in dry air. The general integument serves a respiratory function in terrestrial isopods. It is more important in *Ligia* than in *Armadillidium*, probably in correlation with the reduced water permeability seen in more terrestrial species.

The isopods are tolerant to considerable desiccation (up to about 40% of the body water in *Ligia*, for example). Blood osmotic pressure varies with the degree of hydration. On exposure to dry air, the percent reduction in body water per unit time depends not only on surface-specific permeability but also on body size; if two animals of the same permeability but different sizes are exposed to the same conditions, both will lose the same amount of water per unit surface area, but the larger will suffer a smaller percent drop in body water owing to its lower surface-to-volume ratio. Thus although the decreased permeabilities seen in more-xeric isopods do provide protection, it does not necessarily follow that any given species of low permeability will suffer less dehydration than one of higher permeability under the same conditions. *Ligia*, for example, has about twice the surface-specific permeability of *Porcellio* but is about 10 times larger and can live longer in dry air. The desert wood louse (*Hemilepistus*) combines the advantages of relatively large size and low permeability.

Whereas the terrestrial isopods do differ significantly in their rates of dehydration in dry air, they are, as a group, small and relatively permeable. Most dehydrate rapidly in dry, warm air and survive for only a few hours. Even *Hemilepistus*, one of the most resistant species, lost about 5.6% of its body water per hour when exposed to the sun in the northern Sahara Desert in April. The terrestrial isopods must avoid lengthy exposure to desiccating microclimates, and behavior assumes a vital role in their water economy. They spend much time in humid places; nearly everyone has observed them upon turning over a rock or rotting log. The relative humidity in the holes dug by *Hemilepistus* is reported to remain above 80% even when that at the desert surface is only 20%. Many species limit their outside wanderings to nighttime hours. Close aggregation of individuals is common and acts to reduce water loss by decreasing the collective surface-to-volume ratio. Individuals of many species, such as *Oniscus* and *Armadillidium*, can curl up into a ball, dorsal side outward—thus their common name, pill bug. This behavior can reduce water losses not only by decreasing effective surface-to-volume ratio, but also by limiting exposure of the relatively permeable pleopodal surfaces.

Isopods replace water losses by drinking and by anal uptake. In addition, it has been demonstrated that a number of species can recover water from their food, voiding less in the feces than was taken in.

Our knowledge of the role of humidity in the life of isopods has been extended by two recent experiments. Over short durations of exposure, isopods survive somewhat higher ambient temperatures in 50% relative humidity than in saturated air because evaporative water losses at the lower humidity cool the body below ambient temperature. We can thus expect isopods to move out of highly humid microhabitats when temperature becomes critically elevated. This response has been seen in *Ligia*, which emerges from seclusion when the sun has warmed the stones among which it lives (Figure 1-1). Because isopods dehydrate so rapidly, however, such behavior can provide only temporary protection. The animal must find a place in which conditions of both humidity and temperature are suitable.

Experiments on *Porcellio scaber* have indicated that animals in their secluded shelters absorb water from wet surfaces and from saturated or nearly saturated air. The excretory organs seem unable to void sufficient water for maintenance of homeostasis under prolonged exposure to saturated conditions. *Porcellio* emerges and climbs on trees at night, thus exposing itself to decreased humidities. The extent of this behavior appears to be governed primarily by needs to dissipate excess water through evaporation.

Snails and slugs

The terrestrial molluscs—snails and slugs—have not received extensive study. Their fleshy body surfaces are quite permeable. When extended from its shell, for example, the snail *Helix aspersa* loses water at nearly the rate from an open surface of water of the same area as its body. All the slugs and most snails respire primarily via a lung formed by the mantle and opening to the outside by a small, closable pore (see Chapter 8). This arrangement probably reduces respiratory water losses.

In the few species that have been studied, tolerance to desiccation is high, and internal osmotic pressure varies widely according to state of hydration. Behavior again plays an important role in the water economy. Slugs and many snails seldom venture from humid microhabitats, and nocturnality is common. When conditions become dry, terrestrial molluscs retreat deep into leaf litter or other protected places. Snails close up in their shells. Some (the prosobranchs) have a hard plate (the operculum) on their foot that covers the aperture of the shell when they retract. Others (the pulmonates) lack this plate and draw the edges of the fleshy mantle across the aperture; it has recently been shown that the surfaces of the mantle that remain exposed have a very low permeability by comparison to other parts of the body. In times of particular stress, many snails enter a quiescent, estivating condition. The aperture of the shell may then be sealed more thoroughly by secretion of a mucoid or calcareous membrane. Estivation may last for many months or even years; a specimen of *Helix desertorum* kept in the British Museum crawled away after four years of quiescence.

Many terrestrial snails routinely excrete a substantial proportion of their waste nitrogen as poorly soluble uric acid, and there is evidence that this proportion increases in some species during periods of water stress. Uric acid is stored in the kidneys in at least some forms during estivation, thus eliminating altogether the loss of water in nitrogen excretion.

Certain snails, when desiccated, are able to rehydrate very rapidly by crawling over a wet surface. The mechanism of water uptake is unknown. Some species carry water in their mantle cavity and apparently utilize this supply when in dehydrating environments.

Decapod crustaceans

The decapod crustaceans have radiated onto land, but all remain tied to water for breeding. The semiterrestrial crayfish present interesting problems but have received little study, and we shall concentrate here on the semiterrestrial and terrestrial crabs. They occupy a diversity of habitats, as will be illustrated by example. The ghost crabs (*Ocypode*) dig deep burrows on beaches near or above the high-tide line. The fiddler crabs (*Uca*) live on tidal sand or mud flats and in salt marshes, also digging burrows. Like these groups, many other terrestrial crabs stay in very close proximity to water. Some venture further inland. Land crabs of the genera *Gecarcinus* and *Cardiosoma* burrow in grassy areas and forests far from the sea. One of the most terrestrial crabs is the coconut or robber crab (*Birgus latro*) of the South Pacific. It climbs high in trees and has been reported at land elevations of 300 ft in the Solomon Islands.

All terrestrial crabs lose water by evaporation at a substantial rate in dry, warm air. The rate of loss tends to be lower in more fully terrestrial species. Two species of *Uca* lost 2.7% and 5%, respectively, of their original body weight per hour in still, dry air at 25°C. At 30°C and 78% relative humidity, *Ocypode quadrata* lost 0.74 %/hr; *Cardiosoma guanhumi*, 0.32%/hr; and *Gecarcinus lateralis*, 0.23%/hr. In these latter experiments, *Gecarcinus* tolerated a greater extent of desiccation than the other two species. On the average, *Gecarcinus* died after 89 hours, having lost 21% of its original body weight. Death occurred in *Cardiosoma* and *Ocypode* at 16% and 14% desiccation, respectively. Because *Cardiosoma* lost water less rapidly, it survived for an average of 53 hours as compared to 20 hours in *Ocypode*. It is clear that none of these animals can survive for very long in dehydrating conditions without access to water, and *Ocypode* is especially vulnerable.

The respiratory structures of semiterrestrial crabs are located in enclosed, ventilated branchial chambers. All have gills, but these tend to be reduced in size and number by comparison to marine species, and a respiratory role is assumed by the epithelial lining of the chambers (see Chapter 8). It remains controversial whether the respiratory structures are sites of especial water loss. Because they are enclosed and ventilated, losses can be controlled according to need for respiratory gas exchange, and some recent evidence indicates that most evaporative water loss occurs across the general exoskeleton.

Evaporation cools the crab, and this can be of critical importance when air temperatures are high or when the animal is exposed to intense solar radiation. *Cardiosoma guanhumi*, for example, has a body temperature of around 38°C at an air temperature of 42.5°C when the relative humidity is 36%. Because death occurs when body temperature rises to 39°C, we see that evaporative cooling is essential under such environmental conditions. Half of a sample of *Uca pugilator* died at an ambient temperature of 40.7°C in saturated air, whereas death occurred at 45.1°C in dry air.

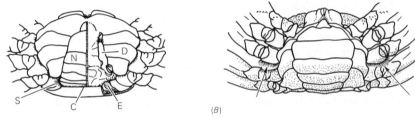

Figure 5–11. (*A*) Ventral view of male *Gecarcinus lateralis*. Portions of the abdomen and carapace are cut away on the right to reveal the protruding end of the pericardial sac on that side (E). Water is drawn into the branchial chamber from the substrate by capillarity among the setae at the end of the protrusion and along convolutions on the surface of the pericardial sac. Water is also drawn up into the branchial chambers among other tufts of setae (such as S and C). N: abdomen, D: pleopod. (*B*) Ventral view of female *Ocypode quadrata* showing setae on the second and third walking legs. The legs have been separated on the left to reveal the two rows of setae. Ordinarily the rows are close together, as on the right. Water is drawn into the branchial chambers from the substrate by capillarity among the setae. (From Bliss, D. E. 1968. Amer. Zool. 8: 355–392.)

Species such as *Uca pugilator* and *Ocypode quadrata* immerse themselves in seawater regularly. More inland species drink from pools of fresh and brackish water. It has been shown that several species of crabs can take up water from a damp substrate. *Ocypode cordimana*, for example, can live for many months on sand moistened with tap water; it is found further inland than species of the same genus that require periodic immersion in the sea. Desiccated *Cardiosoma carnifex* can rehydrate on sand dampened with seawater or fresh water. A proposed mechanism for water uptake from damp substrates will be discussed shortly.

The gills and branchial epithelium of terrestrial crabs must be kept moist for proper respiration. Amphibious fiddler crabs and ghost crabs routinely carry a small amount of water in their branchial chambers when on land, and this water moistens the gills. Many more-terrestrial crabs, such as *Gecarcinus lateralis*, carry only air in their branchial chambers. Recent studies have revealed mechanisms by which the gills and branchial epithelium of such species can be moistened by capillary uptake of water from the substrate. As a group, land crabs have enlarged pericardial sacs. In *Gecarcinus lateralis* these fill much of the ventral portions of the branchial chambers and protrude posteriorly from the exoskeleton (Figure 5–11). Water touching the protrusions is reported to move by capillary action along convolutions of the surfaces of the sacs, being drawn into the branchial chambers and into contact with gills that touch the sacs. *Gecarcinus* also has rows of closely spaced setae that project ventrally from the exoskeleton near the margins of the branchial chambers. Water is drawn up into the branchial chambers by capillarity in the spaces among these setae. A similar role for setae has been reported in several species of *Ocypode* (Figure 5–11), but the potential role of the pericardial sacs in *Ocypode* and *Cardiosoma* has not been investigated. As discussed earlier, several of these crabs can gain internal water from a damp substrate. It seems likely that this is accomplished by absorbing, across the gills or pericardial sacs, water that has been drawn into the branchial chambers by capillary action. Such absorption will occur completely passively if the water is of lower osmotic pressure than the body fluids. Active uptake of ions from the water could also act to establish a favorable osmotic gradient. These proposals require investigation. It remains obscure how species like *Cardiosoma carnifex* can remain hyposmotic to seawater when living on sand dampened with seawater.

Behavior plays an important role in the water and salt balance of terrestrial crabs. Burrowing and nocturnality are common. Many species enter the sea or pools of water as needed. The coconut crab drinks from waters of various salinities according to its relative needs for water and salts.

Insects, arachnids, centipedes, and millipedes

The insects have made the full transition to terrestrial life. They are the dominant land animals in terms of numbers of individuals and species and occupy all the major habitats on land, including deserts and such places as grain stores, which are often as dry as deserts. Insects and arachnids are the dominant invertebrate faunas of the driest places on earth.

Permeability of the exoskeleton is much reduced in most insects by the presence of a thin outer coating of greasy or waxy material. This property is central to the terrestrial existence of these small animals, with their high surface-to-volume ratios. Because it is the epicuticular lipoid layer that is primarily responsible for low permeability, soft-bodied insects, such as the larvae of the clothes moth (*Tineola*), may have permeabilities as low as hard-bodied forms. Species and life stages differ widely in permeability, and this is thought to be, in large part, a function of the chemical composition, molecular orientation, and thickness of the waxy or greasy layer. To illustrate with two extremes, insects with waxy, long-chain molecules oriented in ordered, tightly spaced arrays are expected to be less permeable than those with greasy, short-chain molecules oriented randomly. Removal of the lipoid layers by chemical solvents or abrasion may increase the rate of water loss by a factor of over 100.

If an insect is progressively warmed, there is typically a gradual increase in the rate of evaporative water loss per unit saturation deficit up to a critical temperature and thereafter a marked increase (Figure 5–12). High temperatures bring about a breakdown in the molecular organization of the lipoid layers, leading to increased permeability. The critical, or transition, temperature of many insects is above the thermal death point. In others, it is below, and temperatures that are not directly lethal can cause death by inducing rapid desiccation. The critical temperature varies to some degree with life stage within a species and tends to be higher in species that live in drier, warmer habitats. It is around 50°C in meal beetle larvae, *Tenebrio* (this is above the thermal death point); about 30°C in the cockroach *Periplaneta*; and about 24°C in the water beetle *Dytiscus*.

The greasy or waxy epicuticle of insects has proved to be a preadaptation to living in aquatic habitats, for just as it reduces evaporative water losses on land, it reduces osmotic transfers in water. Whereas many aquatic larvae are relatively permeable, many adults are as waterproof as the more permeable of terrestrial insects. If the water in which the water beetle *Dytiscus* lives is heated above the critical temperature of the lipoid layers, the animal is osmotically flooded with water and dies.

Typically, during molting the waxy layer of the new exoskeleton is laid down before shedding of the old, thus assuring continuous protection.

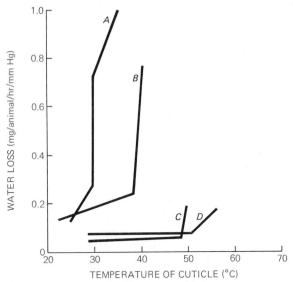

Figure 5–12. Evaporative water loss of dead insects exposed to rising temperature in nearly dry air. (*A*) large nymphal cockroach, *Periplaneta americana;* (*B*) larval butterfly, *Pieris brassicae;* (*C*) adult locust, *Schistocerca gregaria;* (*D*) mealworm, larval *Tenebrio molitor.* Water loss is expressed per unit of saturation deficit in the air. Temperatures were measured on the surface of the cuticle. (Data from Beament, J. W. L. 1958. J. Exp. Biol. 35: 494–519; Beament, J. W. L. 1959. J. Exp. Biol. 36: 391–422.)

Most insects respire by a system of invaginated tubes (tracheae) that penetrate to every region of the body (see Figure 8–28). These tubes open to the outside through small pores, or spiracles, which in the vast majority of species are closable. The spiracles are kept closed except as necessary for respiratory exchange, and water losses are correspondingly reduced to a minimal level. Some species regulate spiracular opening more stringently according to respiratory needs when they are dehydrated or the ambient humidity is low. Carbon dioxide acts as a stimulus for opening of the spiracles, and the importance of keeping them closed is illustrated by placing insects in an atmosphere of air plus a few percent carbon dioxide. Total evaporative water losses may then be increased by 2 to 12 times in resting animals; the factorial increase in respiratory losses is even greater, for some of the total loss is across the general exoskeleton. Evidence on the few species that have been analyzed indicates that spiracular water losses can account for more than half of total evaporative losses in the normal resting state: 66% in the pupa of the silk moth (*Bombyx mori*) and 70% in adult grasshoppers (*Gastrimargus*), but only about 20% in desert locusts (*Schistocerca*).

As discussed earlier, the Malpighian tubules of insects empty their isosmotic or slightly hyposmotic product into the intestine near the junction of the midgut and hindgut. The hindgut and rectum modify the composition of both the urine and feces before excretion. Generalizations concerning excretion are difficult not only because there is considerable variation among species, but also because only a relatively few species have received careful study. It is clear that many insects are very effective in resorbing water from their feces and urine before elimination. The principal site of absorption is apparently the rectal epithelium. Many beetles, locusts, grasshoppers, cockroaches, and larval butterflies and moths can produce quite dry excrement; in *Tenebrio* and some other species, it has been described as a bone-dry powder. Adult flies, moths, butterflies, hymenopterans, and fleas, on the other hand, always produce relatively fluid excretions. The ability to render the urinary fluid hyperosmotic to the body fluids appears to be widespread among insects from dry environments. Isosmotic and hyposmotic urine production are possibilities in many species. To cite some examples, the bloodsucking bug *Rhodnius*, after taking a meal, produces an abundance of crystal-clear urine that is isosmotic to the body fluids. It may void up to 75% of the water in the ingested blood over the first two or three hours. If it goes without eating for a period, however, the urine reaches twice the osmotic concentration of the body fluids, is much reduced in volume, and appears as a thick mixture of uric acid in yellowish fluid. The brackish-water mosquito larva *Aedes detritus* produces hyperosmotic urine (U/P about 3) when in seawater but hyposmotic urine in fresh water. The freshwater *Aedes aegypti*, on the other hand, although making hyposmotic urine in fresh water, is unable to raise the concentration above isosmoticity with the body fluids in saline waters. The osmotic U/P ratio can be raised to over 2 in the stick insect (*Dixippus morosus*) and to about 3 in the desert locust (*Schistocerca gregaria*). The U/P ratio can be as high as 10 in aquatic larvae of the fly *Ephydra riparia*, which live in salt marshes. There is accumulating knowledge of hormonal control of excretion in insects. In *Rhodnius*, for example, distention of the gut after a meal stimulates release of a diuretic neurohormone that controls the production of the copious, dilute urine. An antidiuretic hormone, producing opposite effects during dehydration, has been identified in the cockroach *Periplaneta*.

Terrestrial insects benefit in their water relations by the elimination of a high percentage of their waste nitrogen in the form of uric acid, urates, and other poorly soluble compounds. These appear as a whitish sludge or powder in the excrement.

The sources of water for terrestrial insects are the usual ones—food, drinking water, and metabolic water—plus, in some species, absorption of water vapor from humid air, as discussed in Chapter 4. Some soil insects can take up liquid water across special cuticular structures or extrusible rectal lobes.

Some insects from dry habitats are so effective in limiting water losses that they can maintain water balance at low humidities using only metabolic water and the small amounts of preformed water found in air-dried food. Examples include grain beetles, clothes moths, and wax moth larvae. The mealworm (*Tenebrio*), for instance, can maintain its body fluids at a nearly constant level over a month's fast in dry air, apparently replacing its small water losses through metabolic water production. Many other insects require some substantial amount of water in their diet at low humidities. Many drink and many routinely consume plant or animal foods with a relatively high water content.

Most insects that have been studied tolerate dehydration on the order of 20–25% of their body water. Larvae of the African midge *Polypedilum vanderplanki* are unusual in being able to survive for years in an almost completely desiccated condition. They develop in water holes in rocks, which are liable to dry up.

Insects often show strong behavioral reactions in humidity gradients. Hygroreceptors that mediate such responses have been identified in some groups. Certain insects—such as wireworms (*Agriotes*), many collembola, and many ants—routinely select high humidities. Others, when well hydrated, move to reduced humidities, though preference for higher humidities often sets in if dehydration should occur. Well-hydrated locusts (*Schistocerca*) and fruit flies (*Drosophila*), for example, select intermediate humidities; flour beetles (*Tribolium*) and locusts (*Locusta*) move to low humidities. Potential advantages of the selected humidity are not always clear. Wireworms live in the soil, where they suffer loss of their protective epicuticular lipoid layers through abrasion; being highly permeable, they require high humidities. The selected humidity for *Schistocerca* is near the optimum for development. Species that select reduced humidities may, in some cases, require such humidities for proper dissipation of water when feeding.

Much of the work on humidity preferences has not given close attention to interactions with other environmental factors. The careful work of Wellington on spruce budworm larvae (*Choristoneura fumiferana*) provides an example of the potential significance of such interactions. Well-hydrated larvae were exposed to a temperature gradient on which were superimposed various gradients of relative humidity. That is, the relative humidity coincident with any given temperature was varied. The larvae selected neither a particular temperature nor a particular relative humidity. Wellington placed water-filled micropipets in his gradients and measured the rate of evaporation from them. He found that the rate of evaporation was virtually identical at the various combinations of temperature and humidity selected by the larvae. Thus within certain limits, the larvae seemed to be seeking a particular evaporative situation regardless of the combination of temperature and relative humidity at which it occurred. The data on the larvae showed that the relative humidity was lower, the lower the temperature at the temperature-humidity preferendum. Recent calculations from Wellington's data indicate that, assuming evaporative cooling of the larvae, the vapor pressure difference between animal and air was virtually constant over the range of preferenda. Why the larvae should select a particular rate of evaporation is unclear. We may also note two other interactive factors. The larvae selected lower rates of evaporation when they became somewhat dehydrated, and the rate selected by well-hydrated larvae varied with the larval instar.

The arachnids—spiders, scorpions, ticks, and mites—are the other major invertebrate group to attain a fully terrestrial existence. As in the insects, permeability of the exoskeleton is much reduced by epicuticular layers of greasy or waxy material. The respiratory structures—tracheae or book lungs—open to the outside through small, closable pores. In spiders at least, much nitrogen is excreted in the form of guanine, which is even less soluble in water than uric acid. Certain ticks can take up water vapor from the air. The physiology of the arachnids has been much neglected.

It is interesting to note that the centipedes and millipedes lack the waxy cuticular coating characteristic of insects and arachnids and are generally confined to humid microhabitats. This observation emphasizes the importance of the waxy coating, for even though the exoskeleton of these arthropods looks similar to that of insects, it provides much less protection against desiccation. The exoskeletons of many millipedes are heavily calcified, a factor that could provide some added protection.

Amphibians

The amphibians have invaded a wide variety of terrestrial habitats, from the shores of ponds, to mesic forests, to deserts. With a few notable exceptions, they confront the problems associated with a highly permeable integument. Many species—including a variety of salamanders, frogs, and toads—lose water by evaporation in still air at about the same rate as evaporation from a free surface of water of equivalent surface area. It is noteworthy that some desert amphibians are no more resistant to water loss than mesic forms. Thus *Scaphiopus couchii* and *Bufo cognatus*, both fossorial desert species, lose water at about the same rate as the similarly sized semiaquatic frog *Rana temporaria* when placed in a stream of dry air at 26°C: on the order of 1.2–1.4% of body weight per hour (29–34% per day).

Urinary losses are reduced in amphibians exposed to water stress by limitation of volume and by increasing the osmotic pressure to near isosmoticity with the plasma. Amphibians cannot produce hyperosmotic urine. Soluble urea is the major nitrogenous end product in the majority of largely terrestrial amphibians, including most species from arid regions that have received study. Urine production may cease altogether under severe conditions of dehydration, and urea may then accumulate to considerable concentrations in the blood.

Tolerance of dehydration is often remarkable. In some groups it tends to be greater in species occupying more arid habitats, but this is a generalization with exceptions. The semiaquatic frog *Rana pipiens* tolerates 36% loss of body weight; the terrestrial toad *Bufo terrestris*, 43%; and the terrestrial-fossorial spadefoot toad *Scaphiopus holbrooki*, 48%.

Amphibians seldom, if ever, drink, and water losses are replaced by ingestion in the food, production of metabolic water, and osmotic absorption across the skin. The latter mechanism is of prime importance. Absorption does not necessarily require immersion in water. At least some frogs and toads can gain water from moist paper, moss, or soil. A recent study has shown that conditions are often favorable for spadefoot toads (*Scaphiopus*) to absorb sufficient water for proper maintenance of water balance from the sand in their burrows in the Arizona desert. In the genus of frogs *Neobatrachus*, species from drier habitats, when dehydrated, rehydrate by absorption more rapidly than those from mesic areas. A similar trend has been reported in other groups, but again there are exceptions. Water held in the bladder can be resorbed, and many species from dry areas use the bladder as a water store. The bladder volume is especially large in some amphibians. The filled bladder accounts for 30% of the body weight in the desert toad *Bufo cognatus*, for example.

A neurohypophyseal hormone that is structurally and functionally similar to mammalian antidiuretic hormones is involved in controlling the integrated response of amphibians to dehydration: the so-called water-balance response. This hormone increases retention and absorption of water by increasing permeability to water in three important structures: (1) the distal kidney tubules, leading to limitation of urine volume and production of more nearly isosmotic urine, as discussed in Chapter 7; (2) the urinary bladder, resulting in increased osmotic resorption from hyposmotic bladder contents; and (3) the skin, thus increasing the rate of osmotic water absorption. These effects tend to be more pronounced in more terrestrial species and are essentially absent from some strictly aquatic amphibians, such as the clawed toad, *Xenopus*. The normal stimulus for release of the hormone is apparently an increase in the osmotic pressure of the body fluids resulting from dehydration.

As in other relatively permeable terrestrial animals, behavior plays a central role in the water balance of amphibians. Many stay in close proximity to standing water, where they may retire to restore water losses incurred on land. Salamanders are largely restricted to secluded, humid microhabitats. Species that live in dry areas far from water also spend considerable time in such habitats and are often strictly nocturnal. They may occupy burrows dug by other animals, and some, such as the spadefoot toads, dig their own. The desert spadefoot toad, *S. couchii*, digs deep into the sand during the dry months and enters a state of dormancy; plasma urea levels may rise to nearly twice those of *Rana cancrivora* living in seawater. Many frogs are also known to burrow and enter dormancy in periods of water stress. Recently it has been found that several species of frogs from arid regions of Australia are surrounded by a thin cocoon in their burrows. The cocoon completely encloses the animal except for small openings into the nares and sometimes the mouth. It consists of a single layer of flattened cells and is formed by shedding the superficial layer of cells of the integument, in analogy to the familiar shedding seen in snakes. The cocoon is of much lower permeability than the living skin and provides protection against desiccation. The spadefoot toad seems to form a similar cocoon, though it consists of several cell layers.

It is remarkable that many amphibians of arid regions, when active, have been able to adapt to the stresses of their habitat by virtue of only relatively minor, quantitative improvements on mechanisms of water balance found in semiaquatic forms. They may show increased tolerance of dehydration, increased rapidity of rehydration, more stringent behavioral control of water loss, and so forth. Recently it has come to light that at least two species of frogs display remarkable and unexpected adaptations to the arid environment. One is a tree frog from semiarid regions of South America, *Phyllomedusa sauvagii*, the other, a south African arboreal frog that apparently spends long dry seasons in exposed locations, *Chiromantis xerampelina*. It is hypothesized that their habit of living in trees has demanded physiological adaptations unnecessary for desert species that remain in more protective microhabitats. Both of these frogs can excrete most of their nitrogenous wastes as insoluble uric acid, an ability unknown in other amphibians. In addition, in both these frogs total evaporative losses are dramatically lower than in other amphibians; in fact, their losses are comparable to those seen in desert lizards of similar size. Four *P. sauvagii* exposed to a stream of dry air at 26°C lost only 0.03–0.16% of

Table 5–6. Mean respiratory and cutaneous water loss in four reptiles at 23°C in dry air.

	Respiratory Water Loss (mg/ml O₂)	Cutaneous Water Loss (mg/cm²/day)	Total Water Loss (percent body weight/day)	Cutaneous Loss as Percent of Total
Pseudemys scripta, slider turtle	4.2	12.2	2.0	78
Terrapene carolina, Eastern box turtle	4.2	5.3	0.9	76
Iguana iguana, iguana	0.9	4.8	0.8	72
Sauromalus obesus, chuckwalla	0.5	1.3	0.3	66

SOURCE: Bentley, P. J. and K. Schmidt-Nielsen. 1966. Science 151: 1547–1549. Copyright 1966 by the American Association for the Advancement of Science.

their body weight per hour. This may be compared to losses of 0.03–0.04% per hour in desert iguanas (*Dipsosaurus dorsalis*) and about 1.4% per hour in fossorial desert spadefoot toads (*S. couchii*).

Reptiles

The skin of most reptiles is of low permeability by comparison to most amphibians. This has been a major factor in reducing water losses and permitting exploitation of terrestrial habitats. Recent work has shown that permeability is not always low in reptiles, despite their appearance. The mostly aquatic crocodile *Caiman schlerops*, for example, is 30–50% as permeable as the majority of amphibians. Nonetheless, permeability in terrestrial lizards and turtles is very much reduced. *Iguana iguana*, for example, loses cutaneously only about 5 mg/cm²/day at 23°C in dry air, by comparison to 33 mg/cm²/day in *Caiman*.

Table 5–6 presents data on evaporative water loss in two species of turtles and two species of lizards from environments of increasing aridity. The slider is an amphibious turtle. The box turtle is commonly found in mesic forests and swamps, and the iguana lives in tropical forests. The chuckwalla is a desert lizard. A number of points deserve emphasis. Total water losses, expressed as percentage of body weight, are very low by comparison to most amphibians. The reptiles studied are larger than amphibians, and there is thus some influence of surface-to-volume effects, but if total losses are expressed in surface-specific terms, the same relationship is in evidence. *Sauromalus*, for example, lost about 1.8 mg/cm²/day, as compared to about 95 mg/cm²/day in the desert toad *Bufo cognatus* under roughly similar conditions (26°C in dry, moving air). You will note that total water loss, expressed as percentage of body weight, decreases with the aridity of the habitat of the reptiles studied; this relationship is also clear when water losses are expressed in surface-specific terms. There is a clear inverse correlation between cutaneous permeability and aridity, this being an important factor in the trend seen in total evaporative water loss. The lizards lose much less respiratory water per unit of oxygen consumed than the turtles, and the chuckwalla loses 45% less than the iguana. Rep-

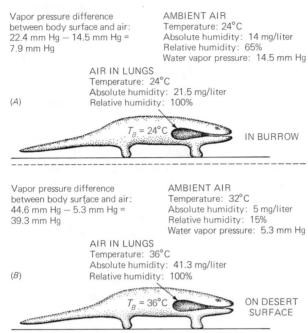

Vapor pressure difference
between body surface and air:
22.4 mm Hg — 14.5 mm Hg =
7.9 mm Hg

AMBIENT AIR
Temperature: 24°C
Absolute humidity: 14 mg/liter
Relative humidity: 65%
Water vapor pressure: 14.5 mm Hg

AIR IN LUNGS
Temperature: 24°C
Absolute humidity: 21.5 mg/liter
Relative humidity: 100%

(A)

$T_B = 24°C$ IN BURROW

Vapor pressure difference
between body surface and air:
44.6 mm Hg — 5.3 mm Hg =
39.3 mm Hg

AMBIENT AIR
Temperature: 32°C
Absolute humidity: 5 mg/liter
Relative humidity: 15%
Water vapor pressure: 5.3 mm Hg

AIR IN LUNGS
Temperature: 36°C
Absolute humidity: 41.3 mg/liter
Relative humidity: 100%

(B)

$T_B = 36°C$ ON DESERT
SURFACE

Figure 5–13. Conditions that might prevail for a desert lizard (*A*) in its burrow and (*B*) on the desert surface. See text for discussion. Effects of solutes on the saturation vapor pressure of body fluids have been ignored. It is assumed that the effective temperature for evaporation at the body surface is the same as the deep body temperature (T_B). At 24°C the saturation vapor pressure is 22.4 mm Hg. At 36°C it is 44.6 mm Hg. Absolute humidities are expressed in mg H_2O/liter air.

tilian skins were long assumed to be so impervious that cutaneous evaporation would be of only relatively minor significance. One of the striking results of these experiments was the discovery that cutaneous water losses can account for over half of total evaporation.

When chuckwallas were placed in dry air at 40°C, rather than at 23°C, both respiratory and cutaneous water losses increased, and the total rate of dehydration was elevated from 0.3% to over 1.1% of body weight per day. Cutaneous losses increased proportionately less than respiratory losses and accounted for 44%, rather than 66%, of total evaporation at 40°C.

It is instructive to analyze in physical terms some of the factors that influence evaporation in two major microhabitats of desert lizards: the burrow and the desert surface. Conditions that might prevail are illustrated in Figure 5–13. Air temperature in the burrow is lower than that on the desert surface. We shall assume that the lizard in the burrow has a body temperature the same as air temperature but that the lizard on the surface has warmed itself above air temperature by basking. The absolute humidity in the burrow is much higher than that on the surface. Cutaneous evaporative losses are in part dependent on the vapor pressure difference between animal and air. The vapor pressure difference is much lower in the burrow than on the surface not only because the ambient vapor pressure is higher but also because the animal is cooler; accordingly, cutaneous losses will be much lower in the burrow. Air inhaled by the animal is brought to saturation at the body temperature. In the burrow, the respired air is not warmed in the lungs, and exhaled air will carry 7.5 mg/liter more water than inhaled air. On the surface not only does the inhaled air contain less water but it is saturated at a higher temperature in the lungs, and air holds more water vapor as its temperature increases; these two factors combine to produce a net respiratory loss of 36.3 mg/liter, almost five times greater than the loss in the burrow. In addition, because metabolic rate increases with body temperature, the lizard on the surface will have to breathe more air to obtain the oxygen it needs. Activity will also increase ventilation on the surface. We shall see later that some lizards can reduce the temperature of exhaled air below their body temperature, thus achieving a saving in water. The basic contrasts in respiratory water loss between the burrow and surface still hold, however, because exhaled air is warmer when the lizard is on the surface than when it is in its burrow. It is apparent that a variety of factors can combine to render the surface environment much more rigorous than the burrow.

Snakes, lizards, and some turtles excrete much of their nitrogenous wastes as uric acid and urates, thus realizing considerable water economies. Reptiles cannot produce hyperosmotic urine, but can increase the concentration to isosmoticity with the blood and reduce urinary volume to very low levels. Some desert lizards possess salt glands like those discussed for marine reptiles and birds. These permit elimination of salts in concentrated solution.

By a variety of behavioral and physiological adaptations, many reptiles can reduce their total water losses to a low level. It seems unlikely, though, that metabolic water production alone can suffice to replace these losses in xeric habitats. Desert reptiles are mostly carnivorous and probably gain sufficient water from their food to achieve balance. Reptiles usually drink when water is available, but drinking would not be a reliable source of water in many arid regions.

Mammals and birds: evaporative water losses

Many aspects of the water relations of mammals and birds have been discussed in Chapters 3 and 4. The attempt here will be to provide an integrated overview without detailed review of topics considered previously.

Our direct knowledge of the permeabilities of avian and mammalian integuments is surprisingly limited, but all available evidence indicates that transpirational water losses are relatively low. Permeability differs among species, and transpirational cutaneous losses can constitute quite different percentages of total evaporative losses. In man and the laboratory rat at temperatures around 25°C in dry air, about half of the evaporative losses are across the skin. Kangaroo rats (*Dipodomys*), on the other hand, lose only about 5% across their skin, and many other rodents from arid regions also appear to have very low cutaneous losses. Cutaneous evaporation accounts for less than 2% of total evaporative water loss in the ostrich, a desert bird, but a much higher percentage in some other birds (e.g., nearly 60% in painted quail).

Respiratory losses in the absence of thermal stress can be viewed as dependent on the rate of oxygen utilization and the amount of water lost per unit of oxygen consumed:

loss (mg/hr) = oxygen consumption (cc/hr)
$$\times \text{ water loss per unit of oxygen consumed (mg/cc)}$$

We shall first discuss some factors that influence the latter parameter.

We noted earlier in our discussion of reptiles that, at given ambient conditions, if the body temperature is elevated above ambient temperature and the air warmed before exhalation, there will be a greater water loss in each volume of respired air than if the air is not warmed, owing to the fact that air becomes saturated in the respiratory tract and the saturation vapor pressure increases with temperature. This factor, tending to increase the water loss per unit of oxygen consumed, is of much significance for birds and mammals because of their high, regulated body temperatures. A mammal in its burrow, even though the ambient air may be saturated, will lose water through respiration insofar as it exhales air at a higher temperature than the burrow temperature. This principle is manifest when we "see our breath" in the winter. Clearly, our exhaled air contains more water vapor than the ambient air can hold at saturation. One of the frequent advantages of hypothermia is the reduced water loss.

At least two possibilities exist for reducing the water loss per unit of oxygen consumed. One is to increase the amount of oxygen extracted from each volume of respired air, thus reducing the amount of air that must pass through the respiratory tract. Another is to reduce the temperature of air before it is exhaled.

Differences in oxygen extraction among terrestrial vertebrates have received little systematic study. A preliminary analysis comparing mammals and nonpasserine birds has indicated that birds extract significantly more oxygen from each unit of air respired than mammals. Within each group differences in oxygen extraction among species appear to be relatively small, but they undoubtedly do account for some differences in loss of water per unit of oxygen consumption. At present it does not appear that birds and mammals from arid regions achieve increased oxygen extraction per unit of air respired in comparison to their mesic relatives.

It is becoming increasingly evident that a type of countercurrent heat exchange in the nasal passages acts to reduce the temperature of exhaled air below the body core temperature in many animals. Suppose, to illustrate the mechanism, that a mammal with a body temperature of 37°C inhales ambient air at 25°C. The air is progressively warmed to about 37°C as it moves up the nasal passages and, even if saturated at inhalation, takes up additional water vapor as its temperature is elevated, ultimately achieving saturation at 37°C. The heat that warms the air and the latent heat of vaporization for the added water vapor are drawn from the walls of the nasal passages; accordingly, the walls are cooled. At the end of the inspiration, the outer ends of the nasal passages may be left as cool as 25°C if the ambient air is saturated. They may be even cooler than 25°C if the ambient air is not saturated, because of evaporative cooling. There will be a gradient of increasing temperature along the walls of the nasal passages from the nostrils to the upper passageways. Now exhaled air reaching the upper nasal passages from the lungs will be saturated at about 37°C. As the air moves down the passages, it will encounter the increasingly cooler surfaces established during inspiration and lose heat to them. Cooling will lower the saturation vapor pressure, and water vapor will condense onto the passage walls, liberating the heat of vaporization. The air, when it leaves the nostrils, will be saturated but at a temperature below 37°C. Both water and heat are conserved by this mechanism. The water that condenses out of the exhaled air as it moves down the nasal passages may be absorbed osmotically across the nasal mucosa or may remain on the surface, where it will serve to saturate the incoming air on the next inhalation. Analogous statements may be made for the heat. The actual temperature of the exhaled air will depend on the dynamics of heat transfer along the nasal passages. Increased surface area, decreased diameter of the passageways, and decreased velocity of air flow all tend to facilitate heat exchange and will result in lower exhaled temperatures at given ambient conditions. Circulatory parameters are also important. If the nasal passages are quickly warmed by the blood, the thermal gradient established on inspiration will not persist fully, and cooling during expiration will be reduced. The similarity of countercurrent exchange in the nasal passages to that illustrated in Figure 3–18 is obvious. The difference is that the opposing flows are separated in time rather than space. Accordingly, the process in the nasal passages has been termed *intermittent countercurrent exchange.*

Kangaroo rats (*Dipodomys merriami*) in dry air at 28°C exhale air at 23°C, about 13°C cooler than body temperature; in saturated air they exhale at about 28°C, the same as ambient temperature. The effectiveness of countercurrent cooling in the laboratory rat is approximately the same. Recent tests on seven species of small and moderate-sized birds revealed cooling of the exhaled air in all. At ambient temperatures of 12°–30°C and relative humidities of 10–30%, exhaled air was always somewhat warmer than ambient temperature, but below body temperature. In the cactus wren (*Campylorhynchus brunneicapillum*), for example, exhaled air was at about 19°C at an ambient temperature of 12°C and at about 31°C in air at 30°C. Even man shows modest cooling of exhaled air, presumably because of countercurrent exchange. At ambient temperatures around 20°C in dry air, his exhalations are at about 32°–34°C. A recent report has extended our knowledge of countercurrent exchange to lizards. The desert iguana (*Dipsosaurus dorsalis*) exhales air at 35°C when its deep body temperature is 42°C in air at 30°C and 25% relative humidity.

The water savings realized by countercurrent cooling can be considerable. In the range of temperatures involved, a 12°C drop in temperature approximately halves the amount of water vapor contained in saturated air. A convenient way to express the water savings is as percent recovery of the water vapor added to the air during inhalation. A cactus wren inspiring air at 15°C and 25% relative humidity adds about 59 mg of water to each liter (at STP); this would all be lost on expiration in the absence of cooling, but the wren expires air at about 20°C, resulting in recovery of about 44 mg/liter, or about 74%. At ambient conditions of 30°C and 25% relative humidity, recovery amounts to about 49%. A kangaroo rat breathing saturated air at 28°C cools exhaled air to about the ambient temperature and thus recovers essentially all the water vapor added on inspiration. Percent heat recovery is about the same as percent water recovery.

A recent study on panting dogs has suggested that evaporative cooling may be controlled by an intriguing use of the nasal countercurrent mechanism. We have previously discussed the advantages of panting at the resonant frequency of the respiratory system. The question arises of how the amount of evaporative cooling can be varied according to the severity of heat loading if the panting frequency is kept rather constant, at or near the resonant frequency. One mechanism that can be used by animals that pant at more or less fixed frequencies is to modulate the amounts of time spent in panting and normal breathing. Another potential mechanism is now known in dogs. It turns out that under many conditions panting dogs inhale mostly through their nose, thus establishing a thermal gradient along the nasal passages on each inhalation. The air can be exhaled through either the nose or mouth. Exiting by the nose, it is cooled by the thermal gradient established on inhalation; but exiting through the mouth, it is not cooled. At ambient conditions of 23°C and 30% relative humidity, air exiting the nose is saturated at 29°C and carries about 15 added calories of heat per liter, whereas air exiting the mouth is saturated at 38°C and carries about 28 added calories per liter. It is hypothesized that the amount of cooling provided by panting can be modulated by varying the amount of air that exits by the two available routes.

Considering the number of factors, identified and not identified, that can influence the amount of water lost per unit of oxygen consumed, it is not surprising to find considerable variation in this parameter among both mammals and birds. Differences are largely unexplained, except that countercurrent cooling is undoubtedly significant in permitting the low values seen in some species. Values reported for birds at moderate temperatures and low humidities range from about 0.6 mg/cc O_2 (for example, in the budgerigar) to 3.7 mg/cc O_2 (in a poorwill). Many small rodents have losses as low as 0.5–0.6 mg/cc O_2 in dry air at 25°–28°C. In man at moderate temperatures and humidities, respiratory loss is around 0.9 mg/cc O_2. Note that this parameter is strongly dependent on ambient humidity, and comparisons among species should be performed under similar conditions.

Turning to the other factor that influences respiratory water loss, the amount of oxygen consumed, we note immediately that the basal metabolic rates of mammals and birds are much higher than the metabolic rates of resting reptiles at avian and mammalian body temperatures. Furthermore, resting metabolism rises as ambient temperature falls below the thermoneutral zone in homeotherms, but it falls with decreasing body temperature in reptiles. These factors are important in raising respiratory losses in mammals and birds above those of reptiles of equivalent size. The *total* evaporative losses of collared lizards (*Crotaphytus collaris*) at mammalian body temperature are only about 20% of those of white laboratory mice near thermoneutrality.

Because the weight-specific metabolic rates of small mammals and birds are greater than those of their larger relatives, we would expect a tendency for the weight-specific rate of respiratory water loss to be greater in smaller species. Small species are also characterized by greater ratios of skin area to body weight than large species, a factor that would tend to cause greater weight-specific rates of cutaneous evaporative loss in smaller species. From these considerations we would expect small species to suffer a greater percent loss of body water than large species under comparable conditions. Two obvious factors can intervene in the application of this simple analysis to the real world: (1) species differ widely in the rate of respiratory water loss per unit of oxygen consumed, and (2) there are also appreciable differences in cutaneous permeability to water. Nonetheless, there is a distinct tendency for total percent water loss per unit time to increase with decreasing size in resting mammals and birds under conditions of moderate temperature and humidity. Thus under comparable conditions, a 9-g house wren may lose 36% of its body weight per day by evaporation; a 23-g white-crowned sparrow, 13% per day; a 43-g cardinal, 6% per day; and a 100-g screech owl, 4% per day. A good many species deviate significantly from what is "expected" for their size. Notable, for example, are quite a few small desert rodents that, because of low cutaneous permeabilities and low rates of loss per unit of oxygen consumption, have much lower total evaporative losses than mesic relatives of similar size.

Activity increases respiratory losses above the resting level. Budgerigars, for example, lose about the same amount of water per unit of oxygen consumption when flying as when at rest, but their rate of oxygen consumption, and rate of respiratory water loss, can be 13 to 20 times greater in flight than under basal conditions.

Actively augmented evaporative cooling in hot environments can severely stress the water supplies of mammals and birds. Small animals tend to be in a considerably less advantageous position than larger species, as we have discussed in connection with Figure 3–19. Some examples will illustrate the extent to which evaporative water losses may be increased. Several finches evaporated water 2.5 to 5 times faster at 40°–44°C than at 34°C, the upper limit of thermoneutrality. Dogs can increase respiratory water losses tenfold when panting. Humans can maintain sweat outputs of 2 liters per hour for several hours; this rate represents 60 to 100 times the rate of transpirational cutaneous water loss at moderate temperatures and humidities.

As discussed in Chapter 3, mammals and birds exploit many possibilities for reducing demands for evaporative cooling in hot environments. These are often central to achieving water balance, especially in arid regions. Many species seek out protective microhabitats such as burrows or simply shade. Other common behavioral attributes include nocturnality (especially among mammals) and reduction of activity during the hottest parts of the day. Many birds and some mammals, such as Grant's gazelle (*Gazella granti*), become markedly hyperthermic, thus reducing thermal gradients between themselves and the environment. Some species undergo fluctuations in body temperature, absorbing heat and then dissipating it by radiation, convection, and conduction in favorable environments. These include the dromedary camel and the eland (*Taurotragus oryx*), which void excess heat each night, and the antelope ground squirrel, which cools on each return to its burrow.

Birds and mammals:
urinary and fecal characteristics

The urine and feces are other important routes of water loss. As discussed in Chapter 4, there are considerable differences among mammals in the extent of water resorption from the feces and in the extent of utilization of ingested food, which in turn affects fecal bulk. Within a species, utilization varies with diet, and water resorption may vary according to state of hydration. Capacities for limiting fecal water loss are especially developed in certain species from hot, arid habitats. The function of the gut in water retention in birds is little known, in part because of difficulties in experimental approach. It remains a possibility that the cloaca, which receives both urine and feces, may act to resorb water in some species. There is recent evidence that cloacal contents can be forced back into the rectum, large intestine, and intestinal caeca of domestic chickens, ducks, and roadrunners (*Geococcyx californianus*). Experiments on chickens indicate that substantial quantities of sodium and water can be resorbed from the *urine* in the intestine, thus providing a mechanism for added conservation of water and salts after the urine has left the ureters.

There are considerable differences in renal physiology between birds and mammals. Birds excrete much of their nitrogenous waste in the form of poorly soluble uric acid dihydrate and possibly also urates, permitting the water economies to which we have often referred. The capacities of bird kidneys to concentrate solutes have been difficult to ascertain because handling birds tends to induce diuresis (production of relatively copious, diluted urine). It does seem likely that the majority of species are limited, at most, to producing urine of around twice the osmotic pressure of their plasma. We have previously discussed the relatively high concentrating abilities of certain subspecies of the savannah sparrow (U/P can be nearly 6 in *Passerculus sandwichensis beldingi*). Such capacities probably evolved under the exigencies of living in salt marshes but can be regarded as preadaptive to occupancy of arid habitats; it is interesting that this species has established populations on desert islands off northwestern Mexico. The black-throated sparrow (*Amphispiza bilineata*), characteristic of American deserts, can also produce urine approaching six times the osmotic concentration of its plasma, and it seems likely that relatively concentrated urine can be produced by certain other seed-eating birds from arid regions, such as the Australian zebra finch (*Taeniophygia castanotis*). Some birds that occupy deserts have more ordinary concentrating abilities. For example, the mean maximum U/P ratio recorded for the house finch (*Capodacus mexicanus*) is 2.3.

The extent to which terrestrial birds utilize extrarenal salt excretion is of considerable current interest. Nasal salt glands producing secretions hyperosmotic to the blood have been reported, for example, in the ostrich, a North African partridge (*Ammoperdix heyi*), and several falconiforms, but not as yet in any passerines. In the falconiforms the nasal secretions are rich in sodium and chloride. A Gaber goshawk, for example, produced nasal fluid about three times as concentrated in sodium as the plasma. We have previously emphasized the importance of extrarenal excretion to eliminating salts while conserving water in marine birds.

Nitrogenous wastes are excreted virtually entirely in soluble form among mammals, principally as urea. This means, for example, that if a bird and a mammal were to produce urines of the same maximum total osmotic pressure and if they both were to have the same urinary load of nonnitrogenous solutes, the mammal would have to void more urine, and lose more water, because of its greater load of soluble nitrogenous material. Many mammals can produce very concentrated urine. Maximum reported osmotic U/P ratios for a number of species are 4.2 in man, 8.9 in the laboratory rat, 8.0 in the dromedary camel, 9.9 in the domestic cat, 14 in Merriam's kangaroo rat and the common gerbil, 17 in the sand rat (*Psammomys obesus*), and 22 in the Australian hopping mouse *Notomys alexis*. As a basis for comparison, the osmotic concentration of seawater is 2.3–3.3 times higher than the concentrations of mammalian plasmas. Production of highly concentrated urine can compensate appreciably in the water relations of mammals for the excretion of nitrogenous wastes in soluble form. There is a general correlation between concentrating ability and life habit among terrestrial mammals; those that live in arid regions or that consume saline waters or foods tend to have particular concentrating capacities. This trend may be illustrated by example. The beaver, which lives in an abundance of fresh water, can produce urine approaching only twice the plasma osmotic concentration. The dromedary camel, kangaroo rat, and gerbil inhabit deserts and have relatively high concentrating abilities, as cited earlier. The sand rat not only lives in deserts but subsists routinely on succulent, salty plants; it has one of the highest concentrating capacities known. Extrarenal paths of salt excretion have not been reported in mammals.

Birds and mammals: integrated study of species from arid regions

We have reviewed in the previous sections the processes that affect rates of water loss—or the converse, water conservation—in birds and mammals. Water is gained through production of metabolic water and ingestion of preformed water in food and drink. The balancing of gains and losses has received particular attention in species of arid and semiarid regions, and the remainder of this discussion will dwell largely on them. These animals provide particular insight into avian and mammalian capabilities, for from the point of view of water relations, they occupy the most demanding of terrestrial environments.

Some desert mammals, by virtue of well-developed capabilities to conserve water, are able to subsist in their arid environment on only metabolic water and the relatively small amounts of preformed water contained in air-dried seeds or other dry plant matter. A classic example is provided by the kangaroo rats (*Dipodomys*) of the American deserts. They can reduce their rate of water loss to a very low level by a combination of very low cutaneous permeability, cooling of exhaled air by a highly effective nasal countercurrent mechanism, production of small quantities of very hyperosmotic urine, and low fecal losses resulting from high utilization of ingested food and relatively great water resorption in the gut. Figure 5–14 depicts the *minimal* water losses attainable over a range of humidities at 25°C. You will note that evaporative, urinary, and fecal losses are superimposed on each other in the figure so that the heavy solid line represents total losses. Evaporative losses decrease with increasing humidity, but the minimal attainable fecal and urinary losses are independent of humidity. Total water losses reflect the decreasing evaporative losses as humidity increases. Water inputs are also shown in Figure 5–14 as dashed lines. Metabolic water production per 100 g of metabolized pearled barley is, of course, independent of humidity. The preformed water contained in air-dried barley increases with humidity because the barley comes to equilibrium with the water vapor in the air. It is assumed in Figure 5–14 that no drinking water is available. Metabolic and preformed water inputs are superimposed so that the heavy dashed line represents total water intake. The animal will be in water balance when water inputs equal water losses, and the figure indicates that balance is possible at

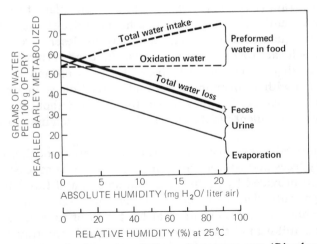

Figure 5–14. Parameters in the water relations of kangaroo rats (*Dipodomys merriami*) at 25°C. See text for explanation and discussion. Animals normally consume 100 g of pearled barley in about a month. (From Schmidt-Nielsen, B. and K. Schmidt-Nielsen. 1951. J. Cell. Comp. Physiol. 38: 165–181.)

humidities of around 10% and higher. Remember that the fecal and urinary losses depicted are minimal. The animal will not accumulate water at humidities above 10% but rather will increase urinary and fecal losses above their minimal levels so as to maintain proper balance.

Figure 5–14 was constructed from information on the individual parameters of the water balance equation. Experiments on kangaroo rats have shown that they can maintain body weight on a diet of pearled barley with no drinking water at 25°C and 24% relative humidity. Some individuals could also maintain their weight at 10% and 15% relative humidity, though others showed a small decrease. The tested animals could not maintain water balance at 5% relative humidity. These results basically confirm the abilities indicated by the analysis in Figure 5–14. In nature, of course, kangaroo rats do not live continuously under the conditions of these experiments. Nonetheless, ecological evidence indicates that they do get along for long periods eating air-dried foods without access to drinking water. As Knut Schmidt-Nielsen has aptly stated, the laboratory results, even though gathered under somewhat artificial conditions, show that there is "nothing mysterious" about the capacities exhibited by these remarkable animals in nature.

The analysis of water relations in the kangaroo rat remains one of the most complete. It is clear, though, from physiological and ecological evidence that a good many other rodents from desert regions survive in their natural habitats on air-dried foods without drinking. Included are at least certain gerbils and jerboas of the Old World and pocket mice and kangaroo mice of the New World. These are nocturnal, burrowing forms that produce relatively concentrated urines.

Many mammals of arid regions, although unable to survive on a dry diet without water, have sufficient abilities to conserve water behaviorally or physiologically that succulent plant or animal foods can supply their needs for preformed water. Like the mammals discussed above, they have achieved independence of standing water for drinking. Rodents in this category include the pack rats (*Neotoma*), which eat succulent plant foods, including considerable amounts of cacti; certain grasshopper mice (*Onychomys*), which consume large quantities of insects; and several desert ground squirrels (*Citellus*), which eat growing plants, insects, and dead vertebrates. The pack rats are interesting in that they dehydrate as rapidly as laboratory rats on a diet of dry food without drinking water and, in fact, die sooner because they tolerate less desiccation. Yet they live in the desert without drinking water by eating highly succulent plants such as the cacti, which may contain nearly 90% water.

The camel achieves considerable water economy through production of relatively dry feces and concentrated urine (U/P can be at least 8), through the protection provided by its fur as a heat shield, and through the advantages of large daily changes in body temperature (see Chapter 3). It can tolerate dehydration to the extent of at least 25% of its body weight when exposed to the full heat of the summer desert. Camels can travel for days or weeks, depending on conditions, without water but do, however, have to drink occasionally, at least in the warm months. Their drinking feats are of some note. Camels dehydrated by 15% of their body weight may replace their water losses completely within 10 minutes; if dehydrated more than 20%, they generally replace their losses in two drinking bouts. A lactating female that had been grazing for six days among the sand dunes first drank 107 liters and several hours later took another 60 liters. These prodigious drinking feats helped to give rise to the notion that camels store extra water to tide them over during long journeys in the desert. Schmidt-Nielsen has shown, however, that camels drink only enough to replace previous water losses.

Camels and other large desert animals are limited in their ability to escape to protective microhabitats. Shade is about the only possibility, and that may not be easy to find. Recent studies have shown, nonetheless, that several large, herbivorous ungulates inhabiting deserts and semideserts of East Africa do not require drinking water at all. Patterns of thermoregulation are intimately involved in reducing their rates of water loss.

The oryx (*Oryx beisa*) and Grant's gazelle (*Gazella granti*) can live continuously in dry, hot deserts and do not seek shade even at the hottest times of day. Both can become markedly hyperthermic. In air at 45°C moderately dehydrated animals allow their body temperature to remain 0.5°–2.0°C *above* air temperature for many hours, with no ill effects. Such hyperthermia reduces or eliminates needs for evaporative cooling. C. R. Taylor and his colleagues have demonstrated that the oryx can survive without drinking in an artificial desert (12 hours per day at the average desert daytime temperature, 40°C, and 12 hours per day at the average desert nighttime temperature, 22°C) provided that its food contains an average of about 30% water. The oryx grazes on grasses and shrubs, particularly a shrub of the genus *Disperma*. The leaves of *Disperma* vary in water content according to ambient conditions. They may contain as little as 1% water during the desert day. Leaves exposed to average nighttime temperature and humidity for 10 hours, however, contained 42% water. It seems likely that the oryx can obtain food averaging 30% in water by eating *Disperma* late at night, as well as more succulent plants at other times. A question of considerable interest is how the oryx and Grant's gazelle can survive body temperatures that would be rapidly fatal for most mammals. One dimension of this problem has received attention. The brain may be expected to be particularly sensitive to elevated temperatures, and there is evidence that the brain is kept cooler than the rest of the body through countercurrent cooling of blood flowing toward the brain in the external carotid arteries. Each artery divides into a plexus of small vessels in the cavernous sinus just below the brain. Veins carrying blood that has been cooled in the nasal passages also form a plexus in the cavernous sinus, the venous vessels intermingling with the arterial. The arterial blood is apparently cooled by this arrangement. A similar system has been documented in domesticated sheep and goats. In exercising gazelles, the brain may be as much as 2.9°C cooler than arterial blood leaving the heart.

The eland (*Taurotragus oryx*) is a large antelope that occurs in dry, semidesert regions of East Africa but, unlike the oryx and Grant's gazelle, does not extend its range into full deserts. It also survives without drinking water and provides interesting comparisons. Its body temperature rises several degrees during the day and falls at night. Such thermal cycling serves to dissipate some of the daytime heat load nonevaporatively, as is clearly the case in the oryx and Grant's gazelle as well. The eland, however, does not allow its body temperature to rise above 40°–41°C even when the air temperature is as high as 45°C. At an air temperature of 40°C, the body temperature remains slightly lower than 40°C. The eland does not have to cope with the effects of marked hyperthermia but pays a price in the form of increased water loss for evaporative cooling. Trees tall enough to provide shade are common in the eland's habitats, and the eland seeks shade at the hottest times of day. Taylor determined that elands in his artificial desert could achieve water balance without drinking if their food contained an average of about 60% water. Elands eat great amounts of acacia tree leaves, and the average water content of such leaves proved to be about 58% even during a severe drought.

Desert mammals with carnivorous habits, although gaining considerable preformed water in their food, must void large quantities of urea, at a cost in water, because of the relatively high protein content of their food. Unfortunately, most have not been studied in detail. Insectivorous grasshopper mice, as discussed earlier, do not require drinking water. They can produce highly concentrated urine (U/P = 12), and they live in burrows during the day, thus avoiding the environmental extremes of the surface. Ecological evidence indicates that the fennec (*Fennecus zerda*), a foxlike animal the size of a small cat from the Sahara Desert, survives on a diet of rodents, lizards, insects, and plant material without having to drink water. It is nocturnal and digs deep burrows, where it spends the day. It probably needs to expend little water in thermoregulation. Our knowledge of larger, nonburrowing desert carnivores such as jackals, coyotes, and foxes is meager. Although they may reduce problems of heat loading during midday by seeking favorable microhabitats and remaining relatively inactive, they undoubtedly must cope with greater water losses for evaporative cooling than the fennec. Data on domestic cats and dogs would indicate that carnivores can just maintain water balance without drinking if on a diet of fresh meat under *moderate* conditions of temperature and humidity. The stresses of exposure to the daytime desert probably impose a need for drinking to supplement preformed water in food, and it is a common observation that large desert carnivores tend to roam in the vicinity of reliable watering places. They have the advantage over small rodents of being able to travel long distances.

Desert birds in general do not appear to attain the high level of water conservation seen in many small mammals, in good measure because they are mostly diurnal and nonfossorial. Our knowledge of their water relations in nature is limited. To date, a number of desert species have been found to survive for long periods on air-dried food without drinking water under moderate conditions of temperature and humidity in the laboratory. Included are the black-throated sparrow (*Amphispiza bilineata*) of North American deserts, some larks of African deserts, and the zebra finch (*Taenopygia castanotis*) and budgerigar (*Melopsittacus undulatus*) of Australian deserts. It seems unlikely, however, that even these species can survive on air-dried foods alone during the warm seasons in nature. Not only do high temperatures and low humidities place added stresses on their water economy, but flight and other activities in the natural environment increase water losses by comparison to the laboratory situation. Indeed, zebra finches and budgerigars in nature are observed to visit water holes regularly in large flocks, and desert travelers have on occasion utilized this behavior to find water. Birds have the advantage of being able to fly long distances to watering places, and some species that require drinking water routinely penetrate the desert far from the nearest standing water. As discussed in Chapter 3, marked hyperthermia is a common response to hot conditions and serves to reduce demands for evaporative cooling.

Many birds of hot, arid regions appear either not to have regular access to standing water or not to visit watering places very frequently even when they are available. It is probable that most, if not all, of these species subsist on liberal amounts of preformed water obtained in succulent foods in addition to their metabolic water. Included are some species that are usually thought of as eating mostly seeds or other air-dried foods but that, on closer analysis, have proved to eat succulent vegetation or insects as well. Gambel's quail (*Lophortyx gambelii*), for example, dehydrate rapidly under moderate ambient conditions if denied drinking water on a diet of air-dried seed. Yet they are able to live and reproduce beyond their flying range from water in deserts of Mexico and the southwestern United States by subsisting on such succulent vegetation as leaves, buds, berries, and fruits of cacti.

Water relations
of young terrestrial animals

A brief comment on the water relations of young terrestrial animals is appropriate before concluding this chapter. Unfortunately, our quantitative knowledge is very limited.

Prior to birth or hatching, the young are protected from desiccation in a variety of ways. Terrestrial decapod crustaceans and most amphibians lay their eggs in water, and after hatching, the young undergo considerable development there. A variety of terrestrial animals retain their eggs or young within the reproductive tract of the female (e.g., most mammals, some reptiles, and some insects) or within protective brood pouches (e.g., some amphibians and the isopods) until the young have reached a more or less advanced stage of development. Direct physiological connections between the young and mother for transfer of nutrients, water, waste products, and so forth, are sometimes present (e.g., most mammals, some snakes, and some insects).

Many terrestrial animals lay eggs on land. Weeks or months may pass before the young hatch, and the need for protection against desiccation is obvious. In many cases, the young animal developing within the egg appears to get along with the metabolic water formed in the catabolism of foodstuffs and the preformed water enclosed in the egg at the time of its formation. The eggs of many insects, in addition, absorb water from moist surroundings, such as soil or the tissues of plant or animal hosts; specialized structures for water uptake are often present, and it is thought that active absorption processes are involved in some cases. Water absorption is also reported in certain turtle and snake eggs. Eggs are commonly enclosed in desiccation-resistant shells of one type or another and are often deposited in humid microhabitats. Earthworms enclose their egg masses in a tough cocoon secreted by the clitellum, and terrestrial snail eggs are usually enveloped in calcified shells. Some insect eggs are relatively permeable and require a humid environment, whereas others are quite resistant to desiccation. Many possess a thin, waxy layer laid down by the developing oocyte on the inside of the egg shell. This layer, in its properties resembling the waxy coating of the adult exoskeleton, provides protection against water loss.

In desiccation-resistant insect eggs, as in other such eggs, there has had to be reconciliation of two superficially conflicting demands: reducing permeability to water while at the same time providing for proper exchange of oxygen and carbon dioxide with the outside world. Many well-protected insect eggs are relatively impervious except for small, localized pores that communicate between the inside of the egg shell and the surface; probably this arrangement restricts the exchange surface to the minimum necessary for respiration and thus minimizes diffusional water loss. Less-protected insect eggs generally have pores all over their surface. Among vertebrates, birds provide striking examples of eggs that develop in open, desiccating situations. The special porous properties of the shell provide for proper diffusion of respiratory gases while at the same time limiting evaporative water losses. The production of desiccation-resistant eggs has been critical to the success of such groups as the insects, reptiles, and birds in arid regions.

The young animal after birth or hatching can be expected to be in a considerably less advantageous physiological position than the adult with regard to limiting water losses. Its greater surface-to-volume ratio will imply greater weight-specific cutaneous losses of water under comparable conditions. Often the young have greater weight-specific metabolic rates and therefore greater weight-specific respiratory water losses. To these considerations must be added the potential immaturity of systems for conserving water and potential ineptitude at finding suitable foods and sources of drinking water. These are all much neglected areas of study. In some cases the parent assumes responsibility for supplying food and water needs for a period after birth. Insofar as the young may have special problems in maintaining water balance, these are transferred to the parent—often several times over, as there may be more than one young.

The magnitude of changes in rate of evaporative water loss during ontogeny has recently been assessed in painted quail (*Excalfactoria chinensis*). Hatchlings (weighing about 4 g) enclosed in a chamber through which dry air was pumped lost an average of 10.9 mg H_2O/g body weight/hr at 35°C. This represented a cutaneous loss of about 7.0 mg H_2O/g/hr and a respiratory loss of about 3.9 mg H_2O/g/hr. Adults (weighing about 43 g), on the other hand, lost only an average of 4.3 mg H_2O/g/hr: 2.5 mg H_2O/g/hr cutaneously and 1.8 mg H_2O/g/hr across the respiratory tract. The lower weight-specific cutaneous loss of adults was largely a reflection of their lower surface-to-volume ratio. The lower weight-specific respiratory loss of adults resulted from their having both a lower weight-specific rate of oxygen consumption and a lower rate of water loss per unit of oxygen consumed. These data indicate that hatchlings suffer about 2.5 times the weight-specific evaporative water losses of adults. At 25°C, the contrast is even more dramatic, with hatchlings dehydrating about 3.4 times faster than adults. The greater rates of evaporation in hatchlings are significant to temperature regulation as well as water economy. At 25°C, a temperature that induces hypothermia in hatchlings and is below thermoneutrality for adults, evaporation was sufficient to dissipate all the metabolic heat production of young. The high rate of water loss is thus a significant factor in limiting capacities for homeothermy. Evaporation was responsible for dissipating only about 20% of the metabolic heat production in adults.

It is to be hoped that the present paucity of knowledge concerning water relations in young terrestrial animals will be remedied in the near future by more studies of this type. We cannot fully appreciate the animal's relationship to its environment until we understand the properties and special challenges of each step in the continuity of life.

SELECTED READINGS

Adaptations of Intertidal Organisms. (A Symposium.) 1969. Amer. Zool. 9: 269–426.
Bentley, P. J. 1966. Adaptations of Amphibia to arid environments. Science 152: 619–623.
Bentley, P. J. 1971. *Endocrines and Osmoregulation.* Springer-Verlag, New York.
Bliss, D. E. 1968. Transition from water to land in decapod crustaceans. Amer. Zool. 8: 355–392.
Brown, G. W., Jr. (ed.). 1968. *Desert Biology.* Academic, New York.
Dill, D. B. (ed.). 1964. *Handbook of Physiology. Section 4: Adaptation to the Environment.* American Physiological Society, Washington, D.C.
Edney, E. B. 1967. Water balance in desert arthropods. Science 156: 1059–1066.
Edney, E. B. 1968. Transition from water to land in isopod crustaceans. Amer. Zool. 8: 309–326.
Krogh, A. 1939. *Osmotic Regulation in Aquatic Animals.* Cambridge University Press, London.
Lockwood, A. P. M. 1966. *Animal Body Fluids and Their Regulation.* Harvard University Press, Cambridge, Mass.
Maloiy, G. M. O. (ed.). 1972. *Comparative Physiology of Desert Animals.* Symposia of the Zoological Society of London, No. 31. Academic, New York.
Potts, W. T. W. 1968. Osmotic and ionic regulation. Ann. Rev. Physiol. 30: 73–104.
Potts, W. T. W. and G. Parry. 1964. *Osmotic and Ionic Regulation in Animals.* Macmillan, New York.
Prosser, C. L. 1973. Water: osmotic balance; hormonal regulation. *In*: C. L. Prosser (ed.), *Comparative Animal Physiology.* 3rd ed. Vol. I. Saunders, Philadelphia.
Schmidt-Nielsen, K. 1964. *Desert Animals.* Oxford University Press, London.
Schmidt-Nielsen, K. 1972. *How Animals Work.* Cambridge University Press, London.
Taylor, C. R. 1969. The eland and the oryx. Sci. Amer. 220: 88–95.

See also references in Appendix.

6
NITROGEN EXCRETION AND OTHER ASPECTS OF NITROGEN METABOLISM

When animals catabolize organic molecules for release of chemical energy, the atoms of the molecules appear in various catabolic end products. Carbon and oxygen atoms, for example, often appear in carbon dioxide, hydrogen atoms in water, sulfur atoms in sulfate ions, and phosphorus atoms in phosphate ions. Carbon, hydrogen, and oxygen atoms are the most plentiful, occurring in all the major foodstuffs. The fourth most plentiful atom is nitrogen, which is a characteristic member of amino acids and proteins. Nitrogen atoms yielded in catabolism appear in a variety of end products, and their disposition is the topic of this chapter.

The animal receives a steady supply of amino acids in its diet and uses amino acids in the synthesis of proteins and other nitrogenous compounds, such as the nucleic acids. Amino acids in excess of those needed for synthesis of nitrogenous compounds are generally utilized in three ways. The first is the one alluded to above—they may be catabolized as energy sources. Alternatively, they may be used as general sources of carbon atoms for synthesis of nonnitrogenous compounds, or they may be converted to metabolically available storage compounds, often ultimately to be used as energy sources. Inasmuch as the metabolically available storage compounds of animals are typically lipids or carbohydrates (e.g., glycogen), the synthesis of storage compounds is often actually a particular type of synthesis of nonnitrogenous compounds. When amino acids are utilized as energy sources or for synthesis of nonnitrogenous compounds, reactions that remove the nitrogen atoms are among the first steps.

To give these concepts more specific meaning, Figure 6–1 depicts

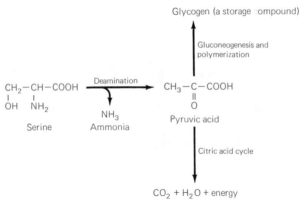

Figure 6–1. Some routes of serine catabolism showing the initial deamination reaction and two possible fates of the resulting pyruvic acid. The disposition of the nitrogen liberated in such reactions is the subject of this chapter.

some possible fates of serine. The nitrogen-containing amino group is first removed by a deamination reaction, forming ammonia and pyruvic acid. The pyruvic acid can be directed into the citric acid cycle and oxidized to carbon dioxide and water, with consequent release of chemical energy that is partially incorporated into ATP molecules. Alternatively, the pyruvic acid can be utilized in the synthesis of the storage compound glycogen. Although the route followed here by serine is especially direct, all deaminated amino acids can be channeled into the usual anabolic and catabolic pathways of the organism.

In the case of most amino acids, removal of the amino group is commonly accomplished not by direct deamination, as in the case of serine, but by transfer of the amino group to another compound—that is, by a transamination reaction. As shown at the top of Figure 6–2, for example, the amino group of aspartic acid can be transferred under the catalysis of a transaminating enzyme to α-ketoglutaric acid, forming oxaloacetic acid and glutamic acid; or, similarly, alanine can undergo transamination with α-ketoglutaric acid to form pyruvic acid and glutamic acid. Note that the effect of transamination is to remove the amino groups from the initial amino acids. The pyruvic acid arising from alanine can react as indicated in Figure 6–1. The oxaloacetic acid arising from aspartic acid can also enter the citric acid cycle or participate in glycogen formation through gluconeogenic reactions. Unlike deamination reactions, transamination reactions do not result in formation of ammonia but, as illustrated, generate other amino acids—in this case glutamic acid. In mammals, as well as in a number of other taxa, amino groups removed from most amino acids tend to be channeled, directly or indirectly, to the formation of glutamic acid by transamination reactions. As shown at the bottom of Figure 6–2, glutamic acid can then be deaminated to yield ammonia. When this occurs the ultimate effect of transamination to form glutamic acid and the subsequent deamination of glutamic acid is to remove amino groups as ammonia. The entire process has accordingly been termed *transdeamination*. Note in the figure, for example, that transdeamination ultimately liberates the amino groups of alanine and aspartic acid as ammonia.

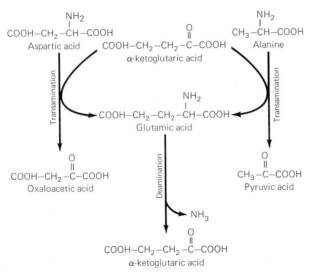

Figure 6–2. Removal of amino groups as ammonia by transdeamination. See text for explanation.

As suggested above, ammonia is a common end product of amino acid catabolism. Many animals excrete nitrogen in the form of ammonia. Ammonia, however, is highly toxic, and this militates against the final disposition of nitrogen in ammonia in many animals. Often the nitrogen removed from amino acids is incorporated into less-toxic compounds, such as urea or uric acid (Figure 6–3). Such compounds are commonly termed "detoxification compounds of ammonia." This is a convenient shorthand expression and has phylogenetic validity, for ammonia seems clearly to be the most primitive nitrogenous end product and has been replaced by other compounds in the evolution of some groups. However, the expression is biochemically somewhat misleading, for it suggests that an animal that synthesizes, say, urea or uric acid first forms free ammonia and then synthesizes the other compounds from ammonia. Actually, the production of free ammonia is often partly bypassed. Urea, for example, contains two amino groups (Figure 6–3). In the usual synthetic pathway in vertebrates, one of these amino groups is derived from free ammonia released in the deamination of glutamic acid (see Figure 6–2). The other amino group, however, is derived directly from the amino acid aspartic acid, rather than being released first as ammonia. Aspartic acid, in turn, can be formed from most other amino acids by transamination reactions. Thus amino groups from most amino acids can make their way into one of the amino positions of urea by transamination and subsequent direct incorporation from aspartic acid, meaning that the amino groups need never appear as free ammonia.

Animals consume and are composed of a great variety of types of nitrogenous compounds. When different types of compounds are catabolized, the nitrogen may appear in different end products. Nitrogen from one type of compound may also be incorporated into more than one end product. It is often important to know the ultimate source of the nitrogen atoms found in the various end products produced. Man, for example, produces large amounts of urea and small amounts of ammonia and uric acid. The urea and ammonia arise largely from catabolism of amino acids and proteins, whereas the uric acid is derived from catabolism of the nucleic acid purines adenine and guanine.

Because proteins are typically depolymerized to their component amino acids, protein and amino acid nitrogen enter a common pool and appear in the same end products. In turn, because proteins and amino acids are the dominant nitrogenous compounds of animals, their nitrogenous end products predominate among the various end products produced. Nucleic acid catabolism accounts for roughly 5% of catabolic nitrogen. Pyrimidines may be excreted as such, but usually pyrimidine nitrogen appears in ammonia or urea. The purines adenine and guanine give rise to a variety of nitrogenous end products in different animals. Their routes of catabolism will be outlined subsequently. Suffice it to say here that purine catabolism often results in nitrogenous end products different from those of protein and amino acid catabolism. We have seen this earlier in the example of man. The other nitrogenous compounds of animals are numerous; their nitrogen may appear in the same end products as the nitrogen of proteins and amino acids, or it may appear in special compounds.

The preceding considerations would lead us to expect that a given animal will produce a variety of nitrogenous end products, and this is borne out by the evidence. The relative dominance of any given end product is generally expressed as the percent of total catabolic nitrogen represented in that end product. Suppose, to illustrate, that an animal produces 1 mole of ammonia and 0.08 mole of urea over a period of time. The mole of ammonia carries a mole of nitrogen, whereas the 0.08 mole of urea carries 0.16 mole of nitrogen (2 atoms of nitrogen per molecule of urea). There is a total of 1.16 moles of catabolic nitrogen. Of this, 1/1.16, or 86%, is carried in ammonia, and 14% is carried in urea. Generally one or a few end products account for a large proportion of the total nitrogen, and these are characteristically the products of protein and amino acid catabolism. In many animals the major nitrogenous end product is ammonia, urea, or uric acid. Organisms excreting a preponderance of their nitrogen in one of these forms are termed, respectively, *ammonotelic, ureotelic,* and *uricotelic.* It is important to recognize that these terms are relative. Mammals, for example, are ureotelic. This does not mean that they produce only urea, nor does it necessarily mean that urea constitutes an overwhelming fraction of the catabolic nitrogen. The percentage of excreted nitrogen that is carried in urea is as low as 64% in some mammals and as high as 88% in others. The terms *ammonotelic, ureotelic,* and *uricotelic* are very useful in facilitating discussion, but it is to be stressed that no one term can fully characterize nitrogen excretion in an animal when, in fact, it involves production of many compounds in varying proportions. As a broad generalization, the predominant nitrogenous end product, whatever it may be, rarely constitutes over 90% of the total catabolic nitrogen, often accounts for 60–90%, and commonly constitutes only 40–60%. Sometimes excretion is so mixed that no one compound can properly be called major.

The products of catabolism have been termed end products, rather than "wastes," in this discussion for a biologically significant reason. It is true that catabolic end products—be they carbon dioxide, water, or nitro-

genous compounds—are generally not used as sources of elements for anabolic, or synthetic, processes (though, for example, ammonia can be used in
the synthesis of glutamine from glutamic acid in mammals, and other synthetic reactions using nitrogen from nitrogenous end products have been
demonstrated or are hypothetical possibilities in various taxa). It is also
true that catabolic end products generally cannot be, or are not, oxidized
further for release of metabolic energy. These considerations, however, do
not mean, in and of themselves, that these end products are wastes.
End products often perform important functions. We need only recall the
great significance of metabolic water in the water relations of many animals.
Carbon dioxide production renews bodily reserves of the various compounds
of the carbonate buffer system, which is often central to acid-base regulation. In the case of nitrogenous end products, we have seen in Chapter 5
that urea plays an important role in the osmotic relations of marine elasmobranchs and the frog *Rana cancrivora*. Other functional roles of nitrogenous
end products will be mentioned in this chapter. Substances become wastes
insofar as they have no useful function or are present in excess of needs for
maintenance of homeostasis. They are then generally excreted, though some
animals store excess nitrogenous end products in their bodies for short or
long periods of time.

Nitrogenous end products have received extensive attention from comparative physiologists because there are clear adaptive correlations between
the modes of life of animals and their forms of disposition of catabolic nitrogen. These correlations will be discussed after a review of the nature and
biochemical origins of the major nitrogenous end products.

THE MAJOR NITROGENOUS END PRODUCTS

The list of compounds known to play some role in nitrogen excretion
in various species is long, including allantoin, allantoic acid, amino acids,
ammonia, creatinine, guanine, hippuric acid, pyrimidines, trimethylamine oxide, urea, uric acid, and urates. Some of these compounds are illustrated in
Figure 6–3.

Ammonia

Deamination of amino acids yields ammonia, and ammonia can also
arise in the degradation of purines and pyrimidines. Ammonia is highly soluble in water and diffuses readily through both aqueous solutions and biological membranes. It also is rather toxic to animals, and blood concentrations are kept low: under 0.5 mg/100 ml in vertebrates and under 10 mg/
100 ml even in the more tolerant of invertebrates. Because of its toxicity, it
cannot be permitted to accumulate. Animals that excrete nitrogen in the
form of ammonia must remove the ammonia from their body fluids as
rapidly as it is formed.

Ammonia reacts with hydrogen ions to form ammonium ions:

$$NH_3 + H^+ \rightleftharpoons NH_4^+$$

This reaction is shifted strongly to the right at the ordinary pH's of animal

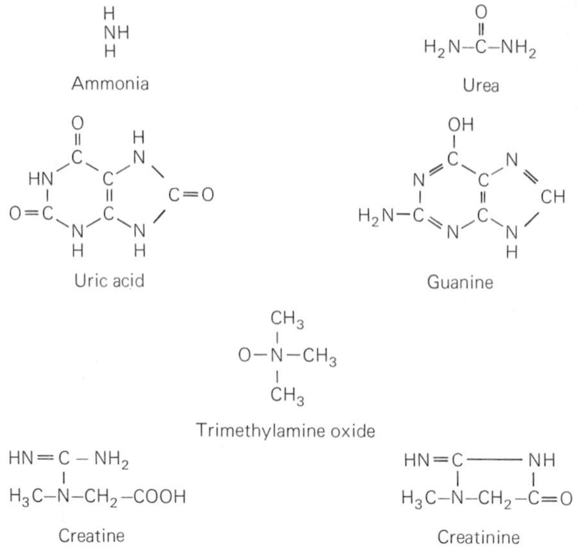

H
NH
H
Ammonia

$$O$$
$$\|$$
$$H_2N-C-NH_2$$
Urea

Uric acid

Guanine

$$CH_3$$
$$|$$
$$O-N-CH_3$$
$$|$$
$$CH_3$$
Trimethylamine oxide

$$HN=C-NH_2$$
$$|$$
$$H_3C-N-CH_2-COOH$$
Creatine

$$HN=C\text{——}NH$$
$$|\qquad\quad|$$
$$H_3C-N-CH_2-C=O$$
Creatinine

Figure 6–3. Some nitrogenous compounds excreted by animals.

body fluids; thus ammonia typically exists in animals primarily as the ammonium ion and often is dissipated to the environment as such. You will note that the preceding reaction is a buffer reaction, meaning that changes in the concentrations of NH_3 and NH_4^+ affect the pH of the body fluids. The reaction plays a significant role in acid-base balance in a number of situations, being important, for example, in the physiology of urinary acid excretion in mammals.

Urea

Urea is synthesized from amino acid nitrogen in many animals. It is slightly more soluble in water than ammonia but does not diffuse as readily. It is also much less toxic than ammonia. In man, blood concentrations are normally in the range of 15–40 mg/100 ml, and much higher concentrations, although abnormal, can be tolerated. We have previously discussed the high plasma concentrations found in marine elasmobranchs (around 2000 mg/100 ml), estivating lungfish, and *Rana cancrivora* in saline water.

The most thoroughly known route of synthesis of urea, the one used by ureotelic vertebrates, is the *ornithine-urea cycle*. Other routes for synthesis from general amino acid nitrogen have been proposed and may be used by some vertebrates or invertebrates. The ornithine-urea cycle is diagramed in Figure 6–4. Note that one of the nitrogens appearing in urea originates from free ammonia; in turn, it is typically derived from deamination of glutamic acid. The second nitrogen appearing in urea comes from the amino group of aspartic acid. As described earlier, amino groups from most amino acids can make their way to glutamic and aspartic acids by transamination reactions. It is noteworthy that the compound from which urea is immediately derived in the ornithine-urea cycle is the amino acid arginine; cleavage of arginine by the enzyme arginase yields urea and ornithine, the latter being recycled to again form citrulline and ultimately arginine. It is currently

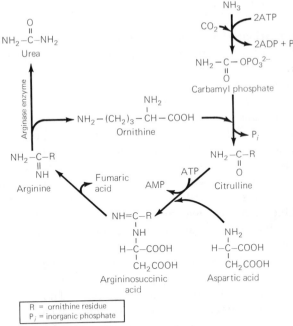

$$NH_2-C-NH_2$$
with O double bond above, Urea

$$NH_3$$
$$CO_2$$ 2ATP
2ADP + P_i

$$NH_2-C-OPO_3^{2-}$$ with O double bond below
Carbamyl phosphate

Arginase enzyme

$$NH_2-(CH_2)_3-CH-COOH$$ with NH_2 above
Ornithine

P_i

$$NH_2-C-R$$ with NH below
Arginine

Fumaric acid AMP ATP

$$NH_2-C-R$$ with O double bond below
Citrulline

$$NH=C-R$$
$$NH$$
$$H-C-COOH$$
$$CH_2COOH$$
Argininosuccinic acid

$$NH_2$$
$$H-C-COOH$$
$$CH_2COOH$$
Aspartic acid

R = ornithine residue
P_i = inorganic phosphate

Figure 6–4. An outline of the ornithine-urea cycle.

hypothesized that some of the enzymes of the ornithine-urea cycle originally evolved to serve a nutritive function, that of synthesizing arginine. This nutritive pathway, it is suggested, was then appropriated to the function of producing urea by the evolutionary addition of arginase and certain other enzymes. If this hypothesis is correct, the evolution of the ornithine-urea cycle itself would not have been so complex as might otherwise appear, for some of the critical reaction sequences and enzymes would have already been present in their nutritive role. This cycle is known today to supply arginine for protein synthesis in a number of ureotelic forms.

The synthesis of urea requires energy; in the ornithine-urea cycle, four high-energy phosphate bonds are utilized in the structuring of each molecule of urea (two high-energy bonds per nitrogen atom). Production of ammonia does not require this investment of energy. Thus ureotelic animals gain the advantage of producing an end product of relatively low toxicity but pay an energetic price in comparison to ammonotelic animals.

Urea synthesis occurs primarily in the liver in mammals and, apparently, in most other ureotelic vertebrates as well. There is controversial evidence of more pervasive ureogenesis in elasmobranch fish.

Some animals lack the ornithine-urea cycle as a whole but possess the enzyme arginase. They cannot channel nitrogen from general amino acid catabolism into urea but can hydrolyze dietary arginine to produce urea. Such hydrolysis is believed to be one of the sources of urea in these animals. They produce much smaller quantities of urea than animals with the ornithine-urea cycle. Urea is also formed in the degradation of purines and pyrimidines in various animals. At least some pyrimidine nitrogen can be channeled through the ornithine cycle, but purine nitrogen follows a different route (discussed later).

Purines and purine derivatives

The most important purines from the point of view of nitrogen excretion are uric acid and its derivatives such as urate salts and uric acid dihydrate (a hydrated form of uric acid). Animals excreting a preponderance of their nitrogen in any of these forms, or a mixture of them, are properly termed uricotelic. Uric acid is poorly soluble in water (6.5 mg/100 ml at 37°C). Its potassium and sodium salts are more soluble but still of very low solubility by comparison to urea or ammonia. We have often referred to the fact that uric acid can be excreted as a semisolid paste, with considerable savings in water by comparison to nitrogenous end products that are highly soluble and thus must be voided in solution. Uric acid and urates are of low toxicity. Like urea, uric acid requires energy for its synthesis. Uric acid carries more available chemical energy per nitrogen atom with it upon excretion than urea.

Uricotelic animals synthesize uric acid from the nitrogen of amino acids and proteins. The route of synthesis has been analyzed in detail in some birds and is presented in biochemistry texts. The nitrogen atoms are derived directly from certain amino acids (glycine, aspartic acid, and glutamine) but can be derived ultimately from other amino acids by transamination or from free ammonia. In some birds all the enzymes for uric acid synthesis are found in the liver. In others an intermediate purine is formed in the liver and then converted to uric acid in the kidneys. Pathways of purine synthesis are phylogenetically ancient and ubiquitous in modern animals, serving the nutritive function of producing purines for nucleic acids, ATP, and other vital compounds. There is little question that these nutritive pathways have been appropriated to the function of nitrogen excretion in uricotelic animals.

The catabolism of the nucleic acid purines adenine and guanine has received considerable attention. Many animals, regardless of their disposition of protein nitrogen, first convert adenine and guanine to uric acid. Species that excrete protein nitrogen predominantly as uric acid generally excrete nitrogen from nucleic acid purines in this form as well. Interestingly, purine nitrogen is also excreted as uric acid in man and other primates, though they are ureotelic. Most animals that excrete protein nitrogen in another form than uric acid convert the uric acid arising from purine catabolism to some other compound prior to excretion. The end product is dependent on which of a series of *uricolytic enzymes* are present. Figure 6–5 summarizes the enzymes and the reactions they catalyze. Most mammals, some turtles, and some gastropods possess only uricase, which oxidizes uric acid, splitting the pyrimidine ring to form *allantoin*. Purine nitrogen is then excreted in this form. Allantoin is a slightly more soluble compound than uric acid. Some teleost fish possess one enzyme in addition to uricase: allantoinase. Allantoinase converts allantoin to *allantoic acid*, splitting the imidazole ring. Allantoic acid is still only a slightly soluble compound. In addition to the preceding enzymes, elasmobranchs, terrestrial amphibians, and most teleosts possess allantoicase, which degrades allantoic acid to urea. Many invertebrates, either in their own tissues or in their intestinal

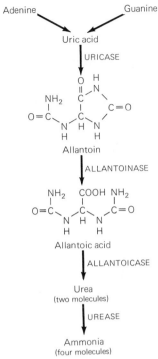

Figure 6–5. Uricolytic reactions and the enzymes that catalyze them. The structures of guanine, uric acid, urea, and ammonia are given in Figure 6–3.

microorganisms, have, in addition to all the above enzymes, a urease that degrades urea to ammonia.

Figure 6–6 provides a summary of patterns of purine and amino nitrogen excretion in a variety of animals. You will note that there has often been evolutionary convergence to excretion of purine and amino nitrogen in a common form (e.g., elasmobranchs, birds, many invertebrates, and so on). Equally, there are prominent examples of lack of convergence. Mammals excrete amino nitrogen largely as urea but purine nitrogen as uric acid or allantoin. Many teleosts excrete much of their amino nitrogen as ammonia but their purine nitrogen as urea or allantoic acid.

Allantoin and allantoic acid usually arise solely from the catabolism of nucleic acid purines and thus appear in only small quantities in the excreta. This is not always so. Terrestrial insects generally convert both amino and purine nitrogen to uric acid, and many excrete the nitrogen predominantly in this form. Many others (e.g., certain heteropteran bugs) exhibit high uricase activity and convert much uric acid to allantoin. In these, allantoin can be the primary nitrogenous end product, for it includes the amino nitrogen as well as the purine nitrogen. Still other insects (e.g., certain lepidopterans) possess both uricase and allantoinase and excrete the preponderance of their nitrogen as allantoic acid.

The purine guanine (Figure 6–3), an important component of nucleic acids, is the major nitrogenous end product of spiders. It is even less soluble in water than uric acid and contains more nitrogen per unit weight than uric acid.

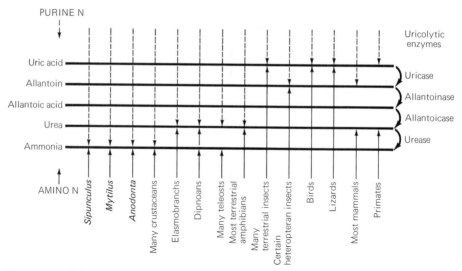

Figure 6–6. Principal nitrogenous end products of purine and amino catabolism in certain groups of animals. *Dashed lines:* purine catabolism; *solid lines:* amino catabolism. (Modified after Florkin, M. and G. Duchateau. 1943. Arch. Int. Physiol. 53: 267–307.)

Trimethylamine oxide

Trimethylamine (TMA) and trimethylamine oxide (TMO) are highly soluble and diffusible and of low toxicity. They are found in the tissues of many marine invertebrates and fish, and the oxide especially appears as a nitrogenous excretory product in some animals. Interestingly, these compounds are present, at most, in only small quantities in freshwater animals. (This is unexplained.) TMA and TMO have received the greatest attention in marine teleosts and elasmobranchs. Some teleosts excrete substantial amounts of nitrogen as TMO. The elasmobranchs produce considerable TMO, and TMO in the plasma accounts for 15–25% of the osmotic pressure due to organic constituents (the rest being contributed mostly by urea). There has been considerable debate concerning the origin of TMO in fish, and the issue has yet to be resolved. The favored hypothesis is that TMO is generally derived from TMA and TMO in the tissues of invertebrate foods. A recent study has indicated that nurse sharks (*Ginglymostoma cirratum*) can synthesize TMO de novo, probably ultimately deriving the nitrogen atom from amino acids. Similar studies on dogfish (*Squalus acanthias*), however, have failed to show de novo synthesis. Salmon in seawater do not "produce" TMO until they consume foods containing TMA and TMO, indicating origin in the diet.

The tissues of marine fish contain mostly TMO rather than TMA. After death, however, bacterial action converts TMO to TMA, and it is the TMA that is responsible for the characteristic odor of dead marine fish.

Creatine and creatinine

Creatine is a rather soluble compound, and its internal anhydride, creatinine, is considerably more soluble. Both are of low toxicity. They are found in the tissues of vertebrates and a few invertebrate groups, creatine phosphate serving as an important store of high-energy phosphate bonds for muscular contraction (see Chapter 11). As excretory compounds, they are similarly limited to vertebrates and a few invertebrate taxa. Mammals may excrete from less than 1% to over 10% of their nitrogen in these forms, mostly as creatinine. Certain fish also excrete significant amounts of nitrogen as creatinine. It appears at present that most excreted creatine and creatinine probably come directly from creatine in the muscle of the animal and its food. Because creatine is synthesized from amino acids, it and its anhydride could potentially serve as excretory products for amino nitrogen, but the possibility has not received much study.

Amino acids

Most animals lose only modest quantities of amino acids, but in some, such as certain crustaceans, insects, echinoderms, and turtles, amino acids may represent 10% or more of the total nitrogen loss. Amino acids are reported to represent over 20% of the total in a starfish (*Asterias*), for example. It is not clear whether the amino acids should be considered true excretory compounds. They may simply be lost because of their ability to permeate the integument or because of incomplete resorption from the urine forming in the excretory tubules.

In the carpet beetle *Attagenus* and clothes moth larva *Tineola*, cystine has been identified as a major compound of sulfur elimination. The diet of these insects is especially high in sulfur.

A NOTE ON ENERGETICS

In Chapter 2 we discussed the use of the bomb calorimeter in determining the energy available from foodstuffs. We noted that the heat released in the animal oxidation of a quantity of foodstuff will equal the heat released during combustion in a calorimeter provided the animal oxidation yields the same end products as combustion. This follows from the fact that the yield of heat depends only on the overall reaction, not on the particular pathway by which the reaction is accomplished. In oxidizing carbohydrates and lipids, animals produce the same fully oxidized products as are produced in combustion—carbon dioxide and water. Thus heats of combustion measured in a calorimeter may be applied straightforwardly to the animal oxidation of these foodstuffs. The same does not hold true for proteins. Combustion in a calorimeter fully oxidizes the nitrogen of proteins, but animals excrete nitrogen in a reduced form and frequently in compounds, such as urea and uric acid, that contain still additional chemical energy that would be released on full oxidation. The result is that heats of combustion measured for proteins do not apply directly to animal catabolism but must be corrected for the difference in nitrogenous end products. Mixed proteins characteristic of the human diet, for example, yield about 5.4 kcal of heat per gram on complete oxidation in a calorimeter but yield only about 4.3 kcal/g in human catabolism because of the fact that the urea produced in the human contains some of the energy that would be released as heat during complete oxidation. Comparatively, the proportion of the total heat of combustion of proteins that is released in animal catabolism is substantially less when uric acid is the end product than when urea is produced but is somewhat greater when ammonia, rather than urea, is the end product. In other words, the proportion of the available energy lost in the nitrogenous end product increases in the order: ammonia, urea, uric acid.

BASIC CONCEPTS OF THE INTERRELATIONSHIPS BETWEEN HABITAT, WATER RELATIONS, AND THE FORM OF NITROGEN EXCRETION

Turning now to a comparative examination of nitrogen excretion in the animal groups, it should first be clearly recognized that there are many gaps in our knowledge. Numerous groups have received little or no attention; and in many species that have received study, the chemical form of a substantial fraction of the excreted nitrogen remains unknown. We do not understand the role in adaptation to the environment, if in fact any exists, for some compounds, such as allantoin, that in some species comprise much of the excreted nitrogen. The nitrogenous excretions of a species are often very mixed. In emphasizing the predominant end products, as we often will, we achieve a degree of simplification that aids understanding in some respects but hinders it in others.

Early in this century it had become clear that the form of nitrogen excretion varies with habitat among phyletically related animals. In particular, aquatic animals tend to be ammonotelic, whereas terrestrial forms tend to be either ureotelic or uricotelic. These observations gave rise to certain classic concepts of the interrelationships between habitat and nitrogen excretion. It is the purpose of this section to outline these concepts as well as certain other basic principles. To a substantial extent the classic concepts have stood the test of time, though we shall see in subsequent sections that they have been subject to both challenge and refinement.

Ammonotelism was probably the most primitive form of nitrogen excretion. Today it is found in most aquatic groups but is uncommon in terrestrial animals. The correlation between ammonotelism and the aquatic habitat is understandable in terms of the chemical properties of ammonia. Because it is toxic it must be voided from the body fluids rapidly. As we shall see shortly, this would generally imply a high rate of urine production were renal excretion necessary and could then tax the animal's water resources. The aquatic animal, however, need not excrete ammonia renally, for it can simply allow ammonia to pass across its body surfaces into the surrounding water at no cost in body water whatsoever. Here two other properties of ammonia come into play, its high solubility and diffusibility. It readily passes through biological membranes and thus can diffuse away into the environment across the gills or other relatively permeable parts of the organism.

The fact that terrestrial animals are typically ureotelic or uricotelic suggests the validity of either or both of two propositions: first, that ammonotelism has disadvantages in terrestrial life and, second, that the other modes of nitrogen excretion have advantages. Both propositions can be rationalized in terms of the chemical properties of the nitrogenous end products. The contrasting implications of ureotelism and ammonotelism can be exemplified by looking at vertebrates. Many fish are ammonotelic, but terrestrial amphibians are largely ureotelic, suggesting a basic transition in nitrogen excretion with the emergence of vertebrates onto land. A critical factor for the terrestrial amphibian is that unlike its aquatic forebears, it cannot lose its nitrogenous end products by diffusion into surrounding water but must excrete them in urine at a cost in body water. To see the implications of the type of end product for urinary water loss, consider first that, in molar terms, steady-state blood concentrations of urea in ureotelic vertebrates are at least 50 times those of ammonia in ammonotelic vertebrates. This is a conservative estimate of the difference and reflects the relative toxicities of the two compounds: *urea, being far less toxic than ammonia, can be allowed to accumulate in the body fluids to a far higher level.* Remembering that one molecule of urea carries twice the nitrogen of a molecule of ammonia, we see that an animal excreting only ammonia would have to achieve a urine-plasma concentration ratio at least 100 times that of an animal excreting only urea in order to void the same amount of nitrogen in the same amount of water. Put another way, if both animals had the same urine-plasma concentration ratio for their respective end products, the ammonotelic animal would have to void 100 times more water to excrete the same amount of nitrogen. With their emergence onto land, animals confronted problems of conserving water. Reasoning such as the preceding indicates the important conclusion that excretion of a compound of low toxicity is more compatible with maintenance of water balance than excretion of ammonia in animals dependent on urinary nitrogen excretion.

Another factor of importance in analyzing the advantages of ureotelism for terrestrial life is that the animal on land can sometimes be faced with problems of water scarcity. In modern amphibians that have received study, a common response to such conditions is severe restriction of urine production. This conserves water but also means that urea is not excreted at the rate it is produced and therefore accumulates in the body fluids. The low toxicity of urea plays a critical role in permitting water balance to take transitory supremacy over nitrogen excretion. If ammonia were the nitrogenous end product, the need for its elimination would be of such urgency that a considerable urinary volume would have to be maintained despite its implications for water balance.

The accumulation of urea during periods of water stress elevates the osmotic pressure of the body fluids. It has been postulated that, within limits, this in itself could aid the water relations of animals such as toads and frogs by establishing osmotic gradients more favorable to absorption of water across the skin. If the moisture tension of the soil in the burrow of a dormant toad, for example, should be equivalent to 400 milliosmolar (mOs), a toad with an ordinary body fluid osmotic pressure of around 300 mOs would lose water to the soil; but if accumulation of urea raised the osmotic pressure of the body fluids to 450 mOs, the toad would absorb water in net fashion from the soil. Again, it is because urea is of relatively low toxicity that it can potentially serve this role; ammonia could not.

Uricotelism is more common than ureotelism among terrestrial animals. The distinctive advantages of uricotelism for terrestrial life rest primarily on the fact that uric acid is so poorly soluble that it is excreted mostly in precipitated form, thus requiring no more water than that needed to move it as a sludge or paste through the excretory passageways. Some animals, in fact, can void uric acid with virtually no water.

The low toxicity and poor solubility of uric acid also give it certain advantages over urea when circumstances favor curtailment of excretion. Because urea is so soluble, its concentration in the tissues and body fluids builds up steadily so long as it is not being excreted at the rate it is produced. This cannot continue indefinitely, for urea does become toxic at high concentrations (the actual concentration depending on species). Storage of urea is characteristically temporary. By contrast, uric acid and other poorly soluble compounds, when accumulated, are deposited in precipitated form within the body, and their concentrations in the body fluids cannot increase above low, saturating levels regardless of how much material is stored. These saturating concentrations typically are not toxic. Accordingly, such compounds are suited to indefinite storage, and as we shall see, some uricotelic animals are known to accumulate, rather than excrete, a portion of their nitrogenous end products for prolonged periods, possibly throughout life in certain cases.

INVERTEBRATE NITROGEN EXCRETION

With the background presented above, we shall now review patterns of nitrogen excretion from a phylogenetic perspective. In covering the invertebrates, we shall first take a general overview and then look in particular at two groups, the molluscs and insects.

The majority of primitively aquatic invertebrates are ammonotelic and dissipate ammonia primarily as NH_4^+ across their outer body surfaces. In freshwater forms, which produce a copious urine, appreciable amounts of nitrogen may be voided renally, but available evidence indicates that the volume of urine production is governed by needs for osmotic regulation rather than nitrogen excretion. Basically, the abundant osmotic flow of water through the body seems to act simply as another reservoir for dissipation of ammonia, just as the water of the environment serves this purpose in a more general sense.

So far as is now known, ammonotelism is rare among terrestrial animals, but it does occur in some invertebrates of humid environments. The terrestrial isopods that have received study are all ammonotelic and, interestingly, dissipate appreciable quantities of nitrogen as ammonia gas (NH_3) across their general integument, a mechanism that obviates the need for water loss in excretion. In some isopods it is reported that virtually all nitrogen is lost in this manner; in others less than half is volatilized as NH_3, the remainder presumably being dissipated as NH_3 or NH_4^+ in urine. Diffusional loss of gaseous ammonia is unusual so far as we now know; it would appear to require special adaptations of acid-base physiology inasmuch as relatively alkaline conditions are required to convert NH_4^+ to NH_3, but the mechanisms have not received study. Recently it has come to light that certain pulmonate snails not only dissipate considerable ammonia in gaseous form, but apparently do so mostly across their shell. It appears, though, that volatilization of ammonia is not the major route of nitrogen excretion in these animals. Earthworms (*Lumbricus*) are ammonotelic when well fed.

Whereas urea production is common among invertebrates, ureotelism is not. Most invertebrates excrete at most only small amounts of urea, and that probably originates largely from hydrolysis of dietary arginine and,

possibly in some cases, from purine breakdown. Some terrestrial invertebrates are ureotelic, and they provide most interesting insights into the
phylogenetic history of ureotelism. Among flatworms, at least one common
terrestrial planarian, *Bipalium kewense*, is ureotelic, excreting as much as
70% of its nitrogen as urea. Among earthworms, *Lumbricus* is ammonotelic
when well fed but becomes ureotelic when starved, and production of
considerable amounts of urea has been reported in some other genera,
at least under certain conditions. The evidence is strong that in both
Bipalium and earthworms, urea is generated by the ornithine-urea cycle,
thus suggesting that this biochemical mechanism may well be phylogenetically ancient.

Whereas ureotelism is not widespread among modern terrestrial invertebrates, excretion of nitrogen predominantly as uric acid or other poorly
soluble compounds is a dominant feature in the adaptation of several major
groups to life on land. Both the insects and terrestrial molluscs excrete
nitrogen mainly as uric acid or other purines or purine derivatives. The
independent evolution of this excretory mode in these two groups, as well
as in certain vertebrates (reptiles and birds), is a testimony to its distinct
advantages for terrestrial life. Uricotelism is found also in the myriapodous
arthropods (centipedes and millipedes), and spiders void nitrogen largely
as guanine. Recently it has come to light that at least one land crab, *Cardiosoma guanhumi*, sometimes stores large quantities of uric acid within its
body, possibly indicating the evolution of uricotelism, at least as a facultative response, in still another terrestrial group.

Molluscs

In this section and the next we shall look in more detail at two major
groups that occupy a diversity of habitats, the molluscs and insects. There
are large areas of uncertainty in the study of molluscan excretion, but recently there has been a revival of active research interest in this group, and
clarifications can be hoped for in the near future.

Looking first at aquatic groups, the cephalopods (e.g., squids and
octopods), which are all marine, seem clearly to be ammonotelic. Studies of
bivalves and primitively aquatic gastropods have indicated considerable
ammonia production in some, but little in others. There is a distinct possibility that some of the latter will prove to be exceptions to the rule that
primitively aquatic invertebrates are usually ammonotelic. Substantial dissipation of nitrogen in both urea and amino acids has been reported in
some forms under at least certain conditions. In the mussel *Modiolus
demissus*, for example, nitrogen is lost both in ammonia and amino acids,
and amino acid nitrogen can account for a quarter or a third of total
nitrogen.

Most terrestrial gastropods are uricotelic. Recent studies have shown
that although uric acid is their predominant nitrogenous end product, many
species also produce considerable quantities of guanine and xanthine—
other purines. Thus it has been suggested that they be termed *purinotelic*
rather than uricotelic. Terrestrial snails, as discussed in Chapter 5, commonly enter a state of dormancy when confronted with problems of water
balance. In this state excretion ceases and uric acid, guanine, and other
purines are stored, typically in the kidneys. It has been suggested that

some land snails, many of which live only a year or so, may store much, or even all, of their waste nitrogen throughout their lives, thus obviating to an especial degree the need for water loss in nitrogen excretion. This hypothesis requires experimental confirmation.

In the 1930s Needham gathered pioneering data on the content of uric acid in the renal organs and tissues of snails, taking this as an indicator of their mode of nitrogen excretion. He found that marine species, believed to be primitively aquatic, tended to contain low amounts of uric acid; terrestrial species contained relatively large amounts; and certain freshwater forms also had relatively large amounts. It has generally been postulated that such freshwater snails are secondarily aquatic, having been derived from terrestrial ancestors; their relatively high levels of uricogenesis have thus been viewed as a carry-over from their terrestrial heritage. This view has been challenged, however, and it may be that high levels of uric acid production evolved in some primitively aquatic lines, perhaps as an adaptation to periodic drying of their habitat. Recently another comparative study of the uric acid content of the renal organs and tissues was carried out, and it was found that even some marine species contained amounts of uric acid comparable to those in terrestrial species. Although data on the uric acid *content* of the animal do not provide direct information on nitrogen *excretion*, this finding gives reason to believe that production of uric acid in amino acid catabolism may be a fairly general property of the gastropods, perhaps having simply been elaborated into true uricotelism in groups confronted with the possibility of water shortage. No firm conclusion can be reached until further comparative studies of excretion itself are undertaken.

Enzymes of the ornithine-urea cycle are widely distributed among terrestrial gastropods, but the significance of urea production remains uncertain. Most forms appear to excrete little urea. In fact, they commonly possess a urease enzyme that degrades urea to ammonia and carbon dioxide in the tissues; thus urea is produced, starting partly with ammonia, and then degraded to yield ammonia. The function of this turnover of urea is unclear, but it may have to do with acid-base regulation, particularly in the context of depositing carbonates in the shell. In any case, production of urea does not generally appear to function primarily in nitrogen excretion. Earlier we mentioned that some terrestrial snails liberate considerable ammonia as a gas; this is at least partly derived from the degradation of urea by urease. There are some reports of ureotelism in gastropods. Two species of slugs, for example, have been claimed to be ureotelic, but the data are subject to controversy. A freshwater snail (*Lanistes*) has recently been reported to be ureotelic during some seasons and ammonotelic during others. There are a number of examples of terrestrial snails that accumulate urea during estivation, though apparently not being ureotelic during active phases. In at least one species, urea accumulates to high concentrations comparable to those seen in the body fluids of marine elasmobranchs; it then is voided upon arousal from estivation.

To summarize, patterns of nitrogen excretion in molluscs are highly diverse and not well understood. Certain aquatic groups are indisputably ammonotelic, and the terrestrial forms are predominantly purinotelic, but there are groups in which too little is known to draw synthetic conclusions. Ureogenesis is common in terrestrial snails, but ureotelism appears to be unusual. Storage of purines during inactive states is common in terrestrial snails, and accumulation of urea has been reported in a number of species under such conditions.

Insects

In the majority of terrestrial insects, the predominant nitrogenous excretion is uric acid, allantoin, or allantoic acid. The latter two compounds, you will recall, are products of uricolysis; being poorly soluble, they confer much the same advantages for terrestrial life as uric acid. As a group, terrestrial insects are usually termed uricotelic; it must be recognized that this description is accurate only if interpreted in the broad sense of including excretion of a preponderance of allantoin or allantoic acid as well as uric acid itself. Within a number of orders (such as the Lepidoptera, Heteroptera, and Orthoptera) it is known that some species excrete mostly uric acid, whereas others excrete mainly allantoin or allantoic acid. The adaptive significance of these interspecific differences is unknown; in particular, the pattern of nitrogen excretion does not appear to correlate with either diet or habitat. The excretion of nitrogen as poorly soluble uric acid or uricolytic products in insects and its excretion as guanine in spiders have undoubtedly been important factors in the evolution of these groups to be the dominant invertebrates of arid environments.

Storage of uric acid is widely reported in insects under a variety of conditions. Commonly, the site of storage is the fat body, either in the fat body cells generally or in specialized "urate cells." Storage is common in the Orthoptera, where deposits of uric acid in the fat body may account for as much as 10% of body weight in certain cockroaches. To cite some further examples, uric acid accumulates in fat body cells of starved mosquito larvae (*Aedes*) and is excreted when the larvae are fed. Insect pupae often store their nitrogenous end products either in the gut or elsewhere, to be excreted around the end of the pupal stage. We can venture plausible hypotheses regarding the adaptive value of storage in some cases. Pupae do not feed; having no input of preformed water, it is logical that they should conserve water resources to a maximum by excreting nothing. Other life stages that store uric acid may also face special problems of water balance. This may be true, for example, of the larvae of some desert flies and lacewings, in which no excretion can occur because the hindgut is blocked. In some cases storage may be related to other factors than water balance. There is evidence, for example, that at least some insects can use stored uric acid as a source of amino acids for synthesis of proteins or for other biochemical processes. Cockroaches on a protein-free diet are known to utilize uric acid deposited in their fat body.

Insects do not possess the ornithine-urea cycle, and in fact there is evidence that the cycle is lacking in arthropods as a group. Urea is widely found in the excreta of insects, and some mosquitoes void as much as 15% of their nitrogen as urea. Usually, however, the amount of urea is small. The source of urea is uncertain but seems most likely to be hydrolysis of dietary arginine; insects are not known to possess the enzymes required for production of urea in uricolysis.

Production of ammonia in substantial amounts is known in a few terrestrial insects. Certain meat-eating fly larvae—which occupy moist microhabitats—void large amounts of ammonia and ammonium in their urine. These compounds may constitute 90% of the excreted nitrogen in some blow-fly larvae, for example, but this is not an accurate reflection of their role in total nitrogen catabolism, for end-product nitrogen is also stored as uric acid in the body during larval life. The collective concentration of ammonia and ammonium in the excreta of these larvae can be nearly 60 times that in the body fluids, a factor that reduces the quantity of water obligated for nitrogen excretion. Interestingly, the pupae and adults of these species are uricotelic. Thus there is a transition in the pattern of nitrogen excretion correlated with the change from a moist larval habitat to the aerial adult habitat.

In contrast to most terrestrial insects, some aquatic insects, such as adult water beetles (*Dytiscus* and *Acilius*) and adult back-swimmer bugs (*Notonecta*), are ammonotelic. Similarly, the freshwater larvae or nymphs of certain species with terrestrial adults are ammonotelic. Examples include nymphs of the dragon fly *Aeschna* and larvae of the caddis fly *Phryganea* and the lacewing *Sialis*. *Sialis* larvae, for example, excrete some 90% of their nitrogen as ammonia. The aquatic larvae and nymphs undergo a transition to uricotelism at the time of metamorphosis into terrestrial adults, providing a notable example of a change in nitrogen excretion with change in habitat. At least some of the aquatic insects are so poorly permeable that we would expect greatly impeded loss of ammonia by diffusion across the integument, and studies on some species have shown that urinary excretion is predominant. Although water conservation is no problem in the freshwater habitat, the poorly permeable forms do not experience the osmotic flood of water found in many freshwater animals. Urinary volumes are relatively small, and ammonia and ammonium, collectively, may be concentrated in the excreta by as much as 300 times relative to the body fluids. The evidence is that ammonotelic insects produce ammonia directly rather than through uricolysis.

Amino acids and nitrogen excretion in aquatic invertebrates exposed to changing salinities

In Chapter 5 we saw that many euryhaline aquatic invertebrates, whether osmoregulating or osmoconforming, undergo changes in the osmotic pressure of their body fluids as the ambient salinity is changed, and we pointed out that the intracellular fluids come more or less to osmotic equilibrium with the extracellular fluids. These changes in osmotic pressure imply changes in the concentration of at least some solutes in the extracellular and intracellular fluids. Recently it has become increasingly evident that in many euryhaline invertebrates, changes in the concentrations of amino acids play an important role in the adjustment of total solute concentration, especially in the intracellular fluids. Thus in a variety of crustaceans and molluscs it has been shown that increases in osmotic pressure are accompanied by increases in intracellular amino acid concentrations, and vice versa. In the wool-handed crab (*Eriocheir*), for example, muscle amino acid concentration decreased by about 40% when animals were transferred from seawater to fresh water, concomitant with some decrease in the osmotic pressure of the body fluids. When prawns (*Leander*) were compared at salinities of about 10‰ and 35‰, their osmotic pressure proved to be about 30% higher at the higher salinity, and muscle amino acid concentration was about 23% higher.

It is interesting in this context that in a number of euryhaline invertebrates, ammonia dissipation is known to increase when they are transferred to a dilute medium and decrease when the salinity is raised. Thus it appears that the decrease in amino acid concentration at lowered salinities is at least partly mediated by an increased rate of degradation, and, conversely, the increase in concentration at higher salinities is at least partly mediated by a decreased rate of degradation.

VERTEBRATE NITROGEN EXCRETION

Fish other than lungfish

Freshwater teleost fish are strongly ammonotelic, though they do excrete some urea and other end products. Ammonia is dissipated primarily across the gills; there is some loss also in their copious urine, but this constitutes less than half of the total. There appear to be two significant mechanisms of ammonia loss across the gills. First, ammonia diffuses away into the water according to its concentration gradient. Second, there is now good evidence that ammonia, in the form of NH_4^+, is exchanged for Na^+ in the process of active Na^+ absorption. To elaborate, freshwater fish actively remove both Na^+ and Cl^- from the water, and because the Na^+ pump and Cl^- pump operate independently, the absorption of each of these ions must be accompanied by the simultaneous extrusion of ions of like charge to prevent the development of large electrical disequilibria. It is presently thought that HCO_3^- is exchanged for Cl^-. That is, the active absorption of each Cl^- ion is accompanied by the loss of a HCO_3^- ion, resulting in electrical balance; HCO_3^- is formed from metabolically produced CO_2. Apparently, either H^+ or NH_4^+ can be exchanged for Na^+. According to one model, the events in the gill epithelium can be viewed as follows. CO_2 enters the epithelial cells and reacts to form bicarbonate: $CO_2 + H_2O \rightarrow HCO_3^- + H^+$. The HCO_3^- is exchanged for Cl^-, and the H^+ either can be exchanged directly for Na^+ or can combine with ammonia to form NH_4^+, which is exchanged for Na^+. Similar processes are also indicated in the gills of crayfish.

Marine teleosts, like the freshwater forms, are ammonotelic, though other end products can again be present in significant amounts. Urea is reported, and in contrast to the freshwater teleosts, there can be considerable production of trimethylamine oxide. Urine volume in marine teleosts is small, and nearly all nitrogen, regardless of its chemical form, is apparently lost by diffusion across the gills.

There has been continuing controversy over the site of production of the ammonia lost across the gills of teleost fish. Is the ammonia formed centrally and thus preformed in the blood reaching the gills, or is it formed in the gills themselves—for example, by deamination of circulating amino acids? There might be some advantage to production of ammonia in situ in the gills, for concentrations of this toxic compound could then be kept lower in the general circulation than if it were formed in other tissues and carried to the gills as such. Information available at present is limited to a few species. Studies on carp indicate that ammonia is not produced in appreciable quantities in the gills. Rather, virtually all the ammonia lost across the gills is preformed in the blood supply. In the sculpin *Myoxocephalus scorpius*, preformed ammonia accounts for about 60% of that excreted; the remaining 40% is formed in the gills themselves, apparently by deamination of circulating amino acids. Information on more species is needed to clarify this issue further. The liver and, to a lesser extent, the kidneys have been identified as the sites of central ammonia production.

Though some teleosts do appear to have a functional ornithine-urea cycle, it seems usually to function at a low level; other teleosts, lacking certain of the necessary enzymes, do not have a functional ornithine-urea cycle. Many teleosts possess the enzymes required for generation of urea through uricolysis; this mechanism and perhaps hydrolysis of dietary arginine appear usually to be the sources of the relatively small quantities of urea formed.

In contrast to teleosts, marine elasmobranchs and many freshwater elasmobranchs possess an active ornithine-urea cycle and are ureotelic. As discussed in the previous chapter, urea retention plays a central role in maintaining the body fluids of marine forms slightly hyperosmotic to the medium, thus promoting osmotic influx of water. Excess urea is voided almost entirely by diffusion across the gills in marine elasmobranchs.

We noted in Chapter 5 that the coelacanth *Latimeria*, like the elasmobranchs, contains highly elevated concentrations of urea in its body fluids. This and enzymatic evidence suggest ureotelism. Ureotelism is also reported in holocephalan fish (chimaeras).

The evolution of nitrogen excretion in fish has been subject to considerable debate. According to one hypothesis, fish were primitively ammonotelic, and ureotelism evolved in elasmobranchs and some other groups as a response to the osmoregulatory problems arising from living in the sea and having blood salt concentrations much lower than those in seawater (see Chapter 5). According to another view, fish were primitively ureotelic. The ureotelism of some modern groups is then seen as a carry-over from the primitive condition, and it is argued that teleosts abandoned ureotelism for ammonotelism, perhaps partly because production of NH_4^+ provided for effective Na^+ absorption in fresh water through cation exchange. Neither hypothesis can at present be rejected.

Lungfish

The lungfish have received much attention for the insight that they might provide in understanding the transition of vertebrate life onto land. You will recall from Chapter 5 that certain lungfish, such as *Protopterus*, enter a dormant state in the mud when the bodies of water in which they live dry up. They then undergo most significant changes in nitrogen excretion. *Protopterus* in water are ammonotelic, though they do produce considerable amounts of urea. Both ammonia and urea are voided largely across the gills. When *Protopterus* is dormant in its cocoon, metabolism becomes depressed, urine production ceases, and though there is considerable protein catabolism, nitrogen is not excreted. The fish produces urea virtually exclusively, and urea accumulates to high levels in the body fluids. Clearly, the change from ammonotelism to ureotelism is adaptive; ammonia could not be allowed to accumulate, and its excretion in the urine would quickly drain the water resources of the dormant animal. Recent work has shown that urea is synthesized via the ornithine cycle in *Protopterus* and, furthermore, that the activities of the cycle enzymes do not increase during dormancy. The rate of urea production during estivation is about the same as that during life in fresh water, but release of free ammonia is suppressed almost completely. This reflects a considerable decrease in production of total nitrogenous end products during dormancy, as is permitted by the reduction in metabolism. Interestingly, the activities of the ornithine cycle enzymes are much lower in another lungfish, *Neoceratodus*, than in *Protopterus*. *Neoceratodus*, you will recall, does not estivate. Apparently, *Protopterus* continually maintains a relatively active ornithine cycle as protection against a nonrainy day.

The presence of an active ornithine cycle in *Protopterus* suggests either its evolution or its maintenance from ancestral fish as a response to the exigencies of living in intermittent bodies of water. The progenitors of amphibians are believed to have occupied such waters, and it seems likely that they were at least facultatively ureotelic. Ureotelism may well have originally become a feature of the line leading to amphibians owing to the advantages of urea as a storage compound during periods of water stress.

Amphibians

Adult amphibians occupy a diversity of habitats and provide interesting comparisons. Adult semiaquatic and terrestrial frogs and toads (anurans) are almost all ureotelic. This is true of several fossorial desert species, but, as discussed in Chapter 5, two arboreal frogs from arid regions are uricotelic. We may hypothesize that these arboreal frogs face greater problems of water conservation in their exposed habitat than do fossorial species, and uricotelism would then have strong advantages over ureotelism. The South African toad *Xenopus laevis* is normally fully aquatic as an adult and, in contrast to semiaquatic and terrestrial anurans, is ammonotelic. About 75% of its total nitrogen excretion is ammonia; the remainder is urea. The aquatic mudpuppy *Necturus* is also ammonotelic. *Xenopus* can live for long periods in humid conditions out of water if forced to do so by changes in its habitat. It then becomes ureotelic, and urea accumulates in its body fluids. In contrast to *Protopterus*, the rate of urea synthesis in *Xenopus* increases under such conditions; this is correlated with increases in the activity of presumably rate-limiting enzymes of the ornithine cycle in the liver. Ammonia production continues, but ammonia does not accumulate. Rather, the ammonia and some urea are excreted.

In summary, certain aquatic amphibians are ammonotelic when in water; most terrestrial and semiterrestrial amphibians are ureotelic; but two arboreal frogs of arid regions are uricotelic. *Xenopus* becomes ureotelic when out of water.

As noted earlier, it is known that many terrestrial frogs and toads enter a dormant condition during periods of water stress and then curtail urine production and accumulate urea.

Changes in nitrogen excretion over the life cycle have been studied in several semiterrestrial and terrestrial amphibians having aquatic tadpoles. In all cases the tadpoles are ammonotelic, but ammonotelism is abandoned for ureotelism at metamorphosis. Data for the bullfrog (*Rana catesbeiana*) are presented in Figure 6–7. The numbered stages on the abscissa refer to morphological steps in development. Animals of stages X to XVII are tadpoles. Those of stages XX to XXIV are undergoing metamorphosis. Metamorphosis is in good measure complete at stage XXV. You will note not only a great increase in urea excretion during metamorphosis but sudden increases in the activities of ornithine cycle enzymes in the liver. The transition from ammonotelism to ureotelism at metamorphosis provides a striking example of correlation between nitrogen excretion and habitat.

Tadpoles of the aquatic toad *Xenopus* are ammonotelic. During early metamorphosis the proportion of urea excretion increases to the point that urea and ammonia each constitute about half of the total nitrogen excretion. Then, as metamorphosis proceeds, the proportion of urea excretion declines to adult levels.

Reptiles

Collectively, reptiles exhibit a broad range of patterns of nitrogen excretion, and interesting correlations with habitat are again evident. The snakes and lizards, which commonly occupy xeric habitats, are predominantly uricotelic. The American alligator can be ammonotelic or uricotelic, depending on conditions; in its natural environment it excretes about half of its total nitrogen as ammonia and half as uric acid, but when overly hydrated it excretes predominantly ammonia and when dehydrated it excretes mostly uric acid. Urea production is slight, and it has been suggested that alligators are descended from terrestrial forms that were uricotelic and secondarily re-established a capability for excretion of nitrogen as ammonia. Ammonia is dissipated as NH_4^+ in the urine.

The turtles that have received study excrete ammonia, urea, and uric acid in various proportions. Recognizing that these compounds collectively may represent from 85% to as little as 40% of total nitrogen excretion, we may note certain correlations with habitat. In a comparison of eight species Moyle found the aquatic forms to excrete approximately equal amounts of ammonia and urea, and little uric acid; semiaquatic species were distinctly ureotelic; and species common to relatively dry terrestrial habitats were strongly uricotelic, although still excreting 10–20% of their nitrogen as urea. Studies on turtles have tended to come up with variable results, and we now understand that this arises in part from the fact that, at least in certain species, the pattern of nitrogen excretion can change markedly in a short period of time. Thus one individual tortoise, *Testudo leithii*, at one time excreted 80% of its total nitrogen as uric acid and only 2% as urea. Later it excreted only 3% as uric acid, 65% as urea. The possibility exists that terrestrial species capable of producing both urea and uric acid in large amounts vary their pattern of excretion according to availability of water.

Birds and mammals

Birds are uricotelic. For the most part we shall consider here the interesting situation in mammals.

We have seen throughout this discussion that nitrogen excretion is not only an adaptive character but, commonly, an exceedingly labile one as well; closely related species and groups often exhibit quite different excretory

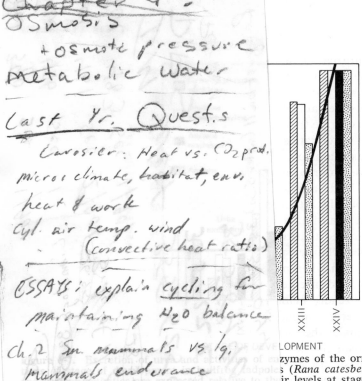

Chapter 4?
Osmosis
+ osmotic pressure
metabolic water

Last Yr. Quest.s
Lavosier: Heat vs. CO₂ prod.
micros climate, habitat, env,
heat & work
Cyl. air temp. wind
(convective heat ratio)

ESSAYS: explain cycling for
maintaining H₂O balance
Ch. 2 Sm. mammals vs. lg.
mammals endurance
under H₂O

...LOPMENT

...zymes of the ornithine cycle as func-
...s (*Rana catesbeiana*). Urea excretion
...ir levels at stage XXIV. See text for
...l P. P. Cohen. 1958. Biosynthesis of
...by and B. Glass (eds.), *A Symposium
...hns Hopkins University Press, Balti-

...rom this point of view the con-
dition in mammals is somewhat surprising. All mammals, regardless of habitat, are ureotelic. An informed biologist knowing the patterns of nitrogen
excretion in other animal groups but knowing nothing about this character
in mammals would probably predict the appearance of uricotelism, at least
in species facing severe problems of water balance. All the other terrestrial
classes of vertebrates have evolved uricotelism, and the mammals themselves
have the capacity to synthesize purines from amino nitrogen and to form
uric acid from other purines. Nonetheless, uric acid typically appears in
only small quantities in the urine as a specific product of purine catabolism.

Many mammals are able to restrict their urinary water losses to relatively low levels, while excreting their highly soluble nitrogenous end product, by virtue of kidneys that are uniquely capable among the vertebrates
of concentrating the urine. The urea U/P ratio is typically higher than that
for any other solute and can greatly exceed the total osmotic U/P ratio—
meaning that urea can be concentrated to a much greater extent than solutes
as a whole. In man, for example, the maximal osmotic U/P ratio is about
4.2, the maximal chloride ratio is about 3.5, and the maximal urea ratio is
about 170. In kangaroo rats the maximal osmotic ratio is about 14, but the
maximal urea ratio is about 350. These figures do not reflect the entire difference between man and the kangaroo rat. The concentration of urea in
human plasma is typically less than half of that in kangaroo rat plasma, and
the kangaroo rat can excrete a given amount of urea in about a fifth as
much water as man would require. A number of desert rodents, which possess some of the most effective of mammalian kidneys, can achieve urinary
urea concentrations of 2.8–4.3M, corresponding to 8–12 g of nitrogen per 100

cc. Comparative information on birds is limited, but based on what we know now, the attainable urinary nitrogen-water ratio in the desert rodents is the same as, or only slightly lower than, that realized by some desert birds. Thus although it is almost certainly true that most mammals require more water than most birds to excrete the same nitrogen and although the birds achieve their high nitrogen-water ratios without having to concentrate their urine to the same degree as mammals, the fact remains that some mammals have been able to realize water economies similar to those of birds despite their ureotelism. Some birds and reptiles, reported to excrete uric acid as a "solid pellet," probably attain substantially greater economies than the desert rodents, but quantitative information is needed.

It has been argued that the basic design of the mammalian kidney, as described in the next chapter, provided such great potential for concentrating the urine that even when mammals invaded deserts, there was little selective pressure toward uricotelism. The basic design of the kidney, so this argument goes, was readily elaborated to meet the increased needs for water conservation. The reverse argument is that mammals for some reason have not been able to make the biochemical and physiological adjustments required for uricotelism and have thus remained tied to ureotelism. This then would provide a basis for great selective pressures toward evolution of a kidney with pronounced concentrating abilities. We shall need much more knowledge of nitrogen metabolism and renal physiology before these hypotheses, or others, can be evaluated.

Mammals are not known to store urea in the sense of accumulating it. Through the activities of bacteria in the gut, however, ruminants and some other groups of mammals are able to derive nitrogen for protein synthesis from urea; and at least in ruminants, urea is retained especially well for this purpose—that is, urea excretion is especially curtailed—during periods of low dietary protein intake. Urea is introduced into the rumen fluid of ruminants, and there bacteria accomplish what the mammal cannot—synthesis of protein using urea as a nitrogen source. When the bacteria are carried through the rest of the gut, the bacterial protein is digested, and the resulting amino acids are absorbed and utilized in protein synthesis by the animal. In this way nitrogen can be cycled over and over within the confines of the ruminant's body.

Embryonic nitrogen catabolism in terrestrial animals

The problems of nitrogen disposition in the embryos of terrestrial animals have been of considerable interest. Reptiles and birds, as well as insects and terrestrial gastropods, produce eggs encased in tough shells that limit evaporative water loss. Embryonic nitrogenous wastes are stored within the egg, and it has been logically satisfying to find that the embryos of these groups are, to our present knowledge, predominantly uricotelic, though there is evidence that embryos of some reptiles and birds pass through periods of ammonotelism and ureotelism early in development. Urea, when accumulated in early development, may later be converted to uric acid. The embryos of placental mammals are ureotelic. The urea is carried into the maternal circulation and voided by the mother. Needham suggested that ureotelism in adult mammals is a carry-over from the fetal situation. From his point of view the different patterns of adult nitrogen excretion in mammals, on the one hand, and uricotelic birds and reptiles, on the other, are a direct reflection of the conditions of their embryonic life. As an explanation of ureotelism in mammals, this is not entirely satisfying in view of the general lability of nitrogen excretion in the animal kingdom and the numerous examples of changes in nitrogenous end products within the life cycles of other animals.

SELECTED READINGS

Bursell, E. 1967. The excretion of nitrogen in insects. Advances in Insect Physiology 4: 33–67.

Campbell, J. W. (ed.). 1970. *Comparative Biochemistry of Nitrogen Metabolism.* Academic, New York.

Campbell, J. W. 1973. Nitrogen excretion. *In:* C. L. Prosser (ed.), *Comparative Animal Physiology.* 3rd ed. Vol. I. Saunders, Philadelphia.

Campbell, J. W. and L. Goldstein (eds.). 1972. *Nitrogen Metabolism and the Environment.* Academic, New York.

Potts, W. T. W. 1967. Excretion in the molluscs. Biol. Rev. 42: 1–41.

See also references in Appendix.

7 RENAL ORGANS AND EXCRETION

The excretions of animals are numerous: carbon dioxide, water, nitrogenous end products, diverse inorganic ions, and a great variety of organic compounds. The animal must excrete each substance in a manner commensurate with overall homeostasis. In particular, as we have often emphasized, the excretion of one substance, say, chloride, must be accomplished without overly taxing the animal's resources of another, in this case, water.

Excretions leave the animal through a variety of structures and mechanisms. In most aquatic animals all excretions are carried away in aqueous solution. Carbon dioxide and ammonia diffuse across the membranes of the gills or other permeable parts of the body surface. Ions are sometimes actively excreted extrarenally, as observed in the gills of marine teleosts and the rectal glands of marine elasmobranchs. The renal organs assume diverse functions in various aquatic animals. In marine teleosts, for example, the kidneys function primarily in the excretion of divalent ions; water loss is minimized, and virtually none of the nitrogenous waste is carried in the urine. In freshwater teleosts, by comparison, the kidneys assume the task of excreting excess water, and some of the nitrogenous wastes are voided in the urine. Terrestrial animals and air-breathing aquatic animals excrete some wastes in gaseous form into the air as well as excreting others in aqueous solution. Obviously, the potential for voiding a substance into the air depends on whether the substance exists as a gas at physiological temperatures. Carbon dioxide is the principal gaseous excretion of animals and is voided across the respiratory organs and sometimes across other permeable body surfaces. A few terrestrial animals lose ammonia as a gas, and loss of water vapor can be considered an excretion when the animal faces problems of overhydration. Most excretions do not exist as gases at physiological temperatures and must be voided by the renal organs in solution, or by other structures—notably salt glands—that produce aqueous solutions.

The integration of excretion, involving not only diverse substances but a variety of structures, is understood in quantitative detail in only a few animals, particularly man and other mammals used in medical research. One of the more elegant integrated systems now appreciated is the bicarbonate buffer system in mammals, central to acid-base regulation of the body fluids. The basics of this system can be outlined with reference to the following equation, familiar to all students of chemistry:

$$CO_2 + H_2O \rightleftharpoons H^+ + HCO_3^-$$

The pH of the body fluids is a function of the hydrogen ion concentration. An increase in the CO_2 concentration of the body fluids, driving this reaction to the right, will increase the H^+ concentration and decrease pH. Contrariwise, a decrease in CO_2 concentration will raise the pH. Increases or decreases in the bicarbonate concentration will, respectively, raise and lower the pH. Here then is a buffer system with three components, namely CO_2, H^+, and HCO_3^-; and we see that changes in the concentration of any one of these will exert a direct influence on the acid-base status of the body fluids. The tissues continually produce CO_2, which in turn can react to form HCO_3^-. Working on these continually renewed buffer resources, the excretory organs function to maintain body fluid concentrations such that pH is held within narrow limits. The lungs regulate the concentration of the volatile component of the buffer system, carbon dioxide, whereas the kidneys regulate the concentration of bicarbonate and can also directly act on the hydrogen ion concentration by excreting greater or lesser amounts of free or combined hydrogen ions. If, for example, there is an increase in H^+ concentration producing overacidity, this might be compensated (1) by increasing the rate of pulmonary ventilation, thus blowing off CO_2 at a greater rate and decreasing the CO_2 concentration of the body fluids; (2) by increasing the rate of renal excretion of H^+; and (3) by decreasing the rate of renal excretion of HCO_3^-, thus elevating the concentration of HCO_3^- in the body fluids. The actual response of a mammal to an injected dose of acid can be understood in terms of the need for rapid acid-base compensation plus the need for ultimate restoration of normal blood levels of carbon dioxide and bicarbonate. Simply stated, the lungs assume the role of immediate compensation. There is a rapid increase in ventilation, and, essentially, the excess hydrogen ions combine with normal blood reserves of bicarbonate, producing carbon dioxide, which is excreted. This, however, depletes the bicarbonate reserves, leaving a subnormal bicarbonate concentration. The kidneys assume the role of restoring the bicarbonate reserves. CO_2 produced in metabolism reacts to form H^+ and HCO_3^-. The kidneys preferentially excrete H^+ and retain HCO_3^- until HCO_3^- levels return to normal. It is only then that the disturbance caused by the added acid has been completely overcome.

This discussion, though it may seem complex, is in fact oversimplified.

It does illustrate several important principles: (1) Excretory organs often act in highly integrated tandem. (2) Excretion is involved not only in the regulation of concentrations of particular solutes, but also in maintaining proper concentration relationships among solutes. Such relationships are, among other things, central to acid-base status. (3) The excretion of one substance often has important implications for the excretion of others. We have seen the latter point in the preceding example and, to illustrate further, may mention certain additional aspects of the acidification of the urine in mammals (as occurs in response to the acid load discussed). In acidification, an active transport mechanism pumps hydrogen ions into the urine forming in the kidney tubules, and this interferes with the addition of potassium ions to the urine. Thus when the urine is being acidified, that is, when hydrogen ions are being extruded, potassium excretion decreases, and there is a tendency toward accumulation of potassium in the body fluids. Another well-known concomitant of acidification of the urine is increased ammonia excretion. Ammonia added to the urine reacts with excreted hydrogen ions to form ammonium ions and permits the urine to carry away many more hydrogen ions at a given urinary pH than would otherwise be possible. The extra ammonia excreted during acidification of the urine reduces slightly the excretion of nitrogen as urea. When we consider that the mammalian kidney simultaneously plays vital roles in excreting wastes, in regulating the concentrations of myriad body solutes, and in regulating osmotic pressure and pH, we see that it must be classed as one of the marvels of animal evolution.

The remainder of this chapter will be devoted to a comparative review of the physiology and morphology of the structures responsible for producing urine, herein termed *renal organs.*

From the first we must emphasize what is meant, and what is not meant, by the designation *renal organ.* The renal organs of such diverse groups as the vertebrates, molluscs, and insects are not biologically homologous. The functions of various renal organs, although they may be said to be analogous, are often only roughly so. Thus, for example, the kidneys of marine teleosts and those of mammals are, in total, performing very different tasks. Whereas renal organs are uniformly excretory organs, it is misleading to think of them as *the* excretory organs. This designation would imply that the renal organs assume all the excretory functions of the animal, which, as we have seen, is never true. In fact, they sometimes assume a relatively minor, albeit essential, role in excretion. It is also improper to think of renal organs simply as the organs of nitrogenous excretion. Many in fact assume little or no role in excreting nitrogenous end products.

Recognizing, then, the need for keeping certain perspectives, we may ask what unites these organs. They are all tubular structures which communicate directly or indirectly with the outside world. They produce and eliminate aqueous solutions. Most importantly, their function is regulation of the composition of the body fluids through controlled excretion of solutes and water.

BASIC MECHANISMS OF RENAL FUNCTION

It will be helpful first to review certain basic mechanisms of renal function. Often it is appropriate to view urine formation as occurring in

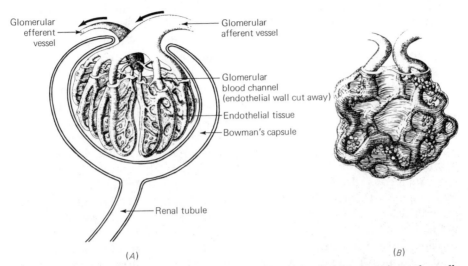

Glomerular efferent vessel

Glomerular afferent vessel

Glomerular blood channel (endothelial wall cut away)

Endothelial tissue

Bowman's capsule

Renal tubule

(A)

(B)

Figure 7–1. (*A*) Renal corpuscle of man. The endothelial cells that constitute the walls of the glomerular blood channels form continuous sheets of endothelial tissue. The tissue is cut away at the top of the glomerulus so that its laminar orientation can be seen. Arrows show direction of blood flow. The Bowman's capsule is depicted diagrammatically. The inner membrane of the capsule, constructed of specialized cells termed *podocytes*, actually interdigitates with the sheets of endothelial tissue so that there is intimate juxtaposition of all blood channels and the capsular membrane. The lumen of the blood channels is separated from the lumen of the Bowman's capsule by only a single layer of endothelial cells, a thin basement membrane, and a single layer of podocytes. (*B*) Glomerulus of a frog (*Rana pipiens*). Note that it also consists of blood channels running through continuous sheets of endothelial tissue but is of simpler construction than the human glomerulus. (Glomeruli from Elias, H., A. Hossman, I. B. Barth, and A. Solmor. 1960. J. Urol. 83: 790–798. © 1960 The Williams & Wilkins Company, Baltimore.)

two steps, though these "steps" may sometimes be in part contemporaneous. First, an aqueous solution is introduced into the renal tubules, and, second, this solution is modified as it moves through the renal tubules and other excretory passages, ultimately resulting in the definitive urine that is eliminated.

Ultrafiltration

One widespread mechanism by which fluids are introduced into the renal tubules is ultrafiltration. This mechanism may be illustrated with reference to the vertebrate renal corpuscle (Malpighian corpuscle). The vertebrate kidney consists of many tubules, or *nephrons*, each of which begins blindly with a hemispherical, invaginated *Bowman's capsule* (Figures 7–1 and 7–3). The Bowman's capsule surrounds a small, anastomosing cluster of capillaries, the *glomerulus* (Figure 7–1), which is supplied with blood at relatively high pressure (about 60–70 mm Hg in man) by the renal artery. The glomerulus and Bowman's capsule together constitute the renal corpuscle. (Sometimes the entire renal corpuscle is called a glomerulus.) The glomerular capillaries are closely applied to the walls of the Bowman's capsule, and the lumen of the capillaries is separated from the lumen of the Bowman's capsule by only two cell layers and a thin basement membrane. The latter structures act as a fine-pored filter. Fluid is driven through this filter from the blood plasma into the lumen of the Bowman's capsule by the hydrostatic pressure of the blood. The fluid moves through pores in the filtering membranes in the manner of small streams (bulk flow), and solutes are carried along with the water much in the way that they are carried through any fine-pored filter. Clearly, the size of the pores will determine which solutes will be allowed to pass. Inorganic ions and such small organic molecules as glucose, urea, and amino acids pass freely through the pores. Their concentrations are virtually the same in the Bowman's capsule as in the blood plasma. Plasma proteins, such as albumins, are too large to pass in appreciable quantities through the pores. Thus the fluid introduced into the Bowman's capsule is a true filtrate of the plasma, lacking the high-molecular-weight protein solutes. Donnan effects between the filtrate and plasma can slightly alter the concentrations of charged solutes in the filtrate from their values in the plasma.

Analogous ultrafiltration arrangements are found elsewhere in the animal kingdom. Clearly, the properties of the filtering membranes will determine the exact nature of the filtrate in any particular case. To assess whether in fact ultrafiltration occurs in a structure that anatomically would appear to provide for ultrafiltration, it is necessary to analyze certain parameters of fluid dynamics. Two types of forces can oppose ultrafiltration promoted by the hydrostatic pressure of the blood. One is the hydrostatic pressure in the renal tubule at the site of filtration (in the case of vertebrates, this is the pressure of the fluids in the Bowman's capsule). The other is an osmotic force produced by the proteins left behind in the capillaries. As fluids are driven into the pores of the filtering membranes, they are depleted of their protein content. The total solute concentration is thus reduced, and the osmotic pressure is correspondingly reduced relative to the plasma. A difference in osmotic pressure between the plasma and filtrate is established and persists because the plasma proteins cannot come to concentration equilibrium across the filtering membranes; this difference in osmotic pressure is known as the *colloid osmotic pressure* of the plasma (see Chapter 10 for further refinement of these ideas). Because the osmotic pressure of the plasma exceeds that of the filtrate, it is clear that osmotic movement of water will oppose movement caused by the plasma hydrostatic pressure. Ultrafiltration will occur whenever the *filtration pressure*, defined as follows, is positive:

$$\begin{aligned}
\text{filtration pressure} = {}& \text{hydrostatic pressure of the plasma} \\
& - \text{hydrostatic pressure of fluid} \\
& \quad\text{in the renal tubules} \\
& \quad\quad - \text{colloid osmotic pressure} \\
& \quad\quad\quad\text{of the plasma}
\end{aligned}$$

To illustrate, in man the blood pressure in the glomerular capillaries is 60–70 mm Hg; the hydrostatic pressure of fluids in the Bowman's capsules is variously estimated at 5–20 mm Hg; and the colloid osmotic pressure in the glomerular plasma is 20–30 mm Hg. Using mean values we get

$$\text{filtration pressure} = 65 - 12 - 25 = 28 \text{ mm Hg}$$

This shows, then, that the hydrostatic pressure of the blood is sufficient to overcome the opposing forces and cause ultrafiltration. In general, the rate of filtration will depend on the magnitude of the filtration pressure and the

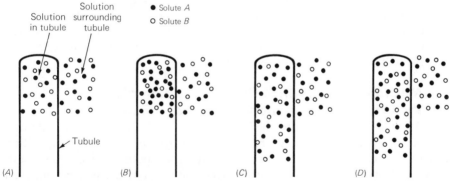

Figure 7–2. A highly simplified illustration of a secretory system. There are two uncharged solutes, symbolized by open and closed circles. The renal tubule is assumed to be completely surrounded by the outside solution, but only a small sample of the outside solution is shown at the upper right of the tubule in each step. Movement of water into the tubule is represented by an increase in the length of the tubule filled with solution. See text for explanation.

resistance of the filtering membranes. In young humans, on the average, some 630 ml of plasma passes through the glomeruli (collectively) per minute, and of this some 120 ml is filtered into the Bowman's capsules. The preceding analysis of filtration pressure will indicate why ultrafiltration systems are always associated with arterial blood supplies. Venous blood pressures are generally too low to overcome the colloid osmotic pressure of the plasma.

Solute secretion as a mechanism of introducing fluid into the renal tubules

Another method by which fluid can be moved into renal tubules is based on active solute secretion. This is the presumptive mechanism in insects, for example. Certain essentials of a secretory system are illustrated in artificial, stepwise fashion in Figure 7–2. We assume just two uncharged solutes, and we start, in Figure 7–2A, with a system in which the osmotic pressure and concentrations of both solutes are equal on the inside and outside of the renal tubule. We shall also assume for simplicity that the fluid outside of the tubule is sufficiently abundant that movements of solutes and water into the tubule do not appreciably alter the outside concentrations. In Figure 7–2B, an active-transport mechanism secretes a quantity of solute A into the lumen of the renal tubule, increasing the inside concentration of A and also increasing the inside osmotic pressure. In Figure 7–2C, water moves inward by osmosis according to the osmotic gradient set up by secretion, and the volume of fluid in the tubule increases. Because of this increase in volume, the inside concentration of solute B, initially the same as the outside concentration, is reduced. In Figure 7–2D, solute B diffuses inward such as to neutralize its concentration gradient. The system is still not at equilibrium, and the analysis presented here is simplified in many respects. Nonetheless, this treatment is sufficient to illustrate the important principle that active secretion of one solute into the renal tubule can lead to passive influx of water and other solutes. As in ultrafiltration, a complex solution of many body solutes is introduced into the renal tubule. The animal must expend energy to accomplish this task. In ultrafiltration systems energy is expended in maintaining a suitably high blood pressure, whereas in secretory systems energy is required for active transport. In the case of secretion, the membranes of the renal tubules again act as something of a filter. Permeability to the various solutes that might enter passively will determine which solutes do, in fact, enter the renal tubule. Additional aspects of secretory systems will be considered subsequently.

Mechanisms by which the composition of the urine is altered along the renal tubules

As fluid introduced into the renal tubule moves down the tubule and through other parts of the excretory system, its volume and composition may be altered extensively before elimination as the definitive urine. Both active and passive processes are involved. Some of these will be outlined here to illustrate certain widespread possibilities.

Solutes may be added to or extracted from the tubular fluid by active transport. Insofar as these active processes produce a net increase or decrease in total osmotically effective solute concentration in the tubules, the osmotic pressure of the tubular fluid will be altered. Water will tend to move by osmosis such as to neutralize osmotic gradients set up between the tubular fluid and the surrounding body fluids. The extent of water movement will obviously depend on the permeability of the tubular membranes to water. Suppose, for example, that the tubular fluids are initially isosmotic with the surrounding body fluids and that some solutes are actively resorbed from the tubular fluid. If the membranes are freely permeable to water, the reduction of tubular osmotic pressure will result in osmotic loss of water from the tubules and a reduction in the volume of tubular fluid. If, on the other hand, the membranes are relatively impermeable to water, resorption of solutes will result in persistent hyposmoticity of the tubular fluids relative to the body fluids. Resorption of solutes across parts of the renal tubule that are poorly permeable to water is, in fact, a very widespread, if not universal, mechanism of producing hyposmotic urine.

Active transport of certain solutes will also influence passive distributions of other solutes. Recall that passive solute movements depend on concentration gradients and also, in the case of electrically charged solutes, on electrical gradients. Passive movements are in the direction of electrochemical equilibrium. Active solute movements can influence both the concentration and electrical gradients affecting passive movements. When the actively transported solute is charged, transport will often tend to establish an electrical potential difference across the tubular membranes. Active sodium resorption, for example, will tend to render the outside of the tubule positive with respect to the inside, and this will tend to facilitate passive resorption of negatively charged solutes and hinder resorption of other positively charged solutes. Active solute movements, when they lead to osmotic changes in the volume of tubular fluid, will influence the tubular concentrations of all solutes. Concentration gradients thus established between the tubular fluids and surrounding body fluids will affect passive solute exchanges. To illustrate, if the volume of tubular fluid is decreased by active solute resorption across membranes that are freely permeable to water, concentrations of passively moving solutes in the tubular fluid will be increased, often establishing concentration gradients favorable to net diffusional loss of such solutes from the tubular fluid. Solutes to which the tubular membranes are permeable will move in response to their concentration gradients, but others, to which the membranes are poorly permeable, will persist at elevated concentrations in the tubular fluid. We have seen earlier that osmotic increase in volume of the tubular fluid can result in passive influx of solutes.

Clearly, the active and passive movements of solutes and water across the renal tubules can interact in complex ways. In appraising their effect on the urine that is ultimately eliminated, it is important to maintain a careful distinction between *concentration* and *mass*. To illustrate, when particular solutes are actively or passively resorbed, contemporaneous water resorption may be such that tubular *concentrations* of those solutes remain little changed. Nonetheless, there will have been net loss of solutes and water from the tubular contents, and the *masses* of water and solutes eliminated in the urine will be reduced. Measures of concentration provide important information concerning the relative proportions of urinary constituents and often provide insight into the dynamics of urine formation, but measures of mass are the appropriate expressions of actual excretion.

THE AMPHIBIAN NEPHRON

The physiology of the amphibian nephron has been carefully studied in certain species and provides a good base of comparison for the study of other vertebrate systems. The nephron (Figure 7–3) consists of (1) a Bowman's capsule; (2) a convoluted segment known as the *proximal convoluted tubule;* (3) a short, relatively straight segment of small diameter, the *intermediate segment;* (4) a second convoluted segment known as the *distal convoluted tubule;* and (5) a relatively straight segment, the *initial collecting duct.* The nephrons are microscopically small, and there are many thousands in each kidney. The initial collecting ducts of many nephrons feed into a common *collecting duct,* and all the collecting ducts of each kidney connect to the single *ureter,* which carries fluid from the kidney to the bladder.

Ultrafiltration and events along the proximal convoluted tubule

Measurements of glomerular blood pressure have demonstrated that it is adequate for ultrafiltration, and analysis of fluid taken from the Bowman's capsules by micropuncture has confirmed that the fluid is an ultrafiltrate of the plasma. Thus the fluid entering the proximal convoluted tubule resembles plasma in composition except that it lacks the plasma proteins. It is approximately isosmotic with the plasma. (The difference in osmotic pressure between the tubular fluid and plasma, due to the plasma proteins, is very important in analyzing ultrafiltration but represents less than 1% of the total osmotic pressure of either the plasma or tubular fluid.)

Reference to Figure 7–4 will aid in the discussion of what happens to the tubular fluid as it passes along the nephron. Fluid exiting the proximal convoluted tubule is still approximately isosmotic to the plasma. Its chloride concentration is also still about the same as the plasma chloride concentration. Nonetheless, it can be shown that the proximal convoluted tubule has altered the fluid in certain important ways. Ions are resorbed in the proxi-

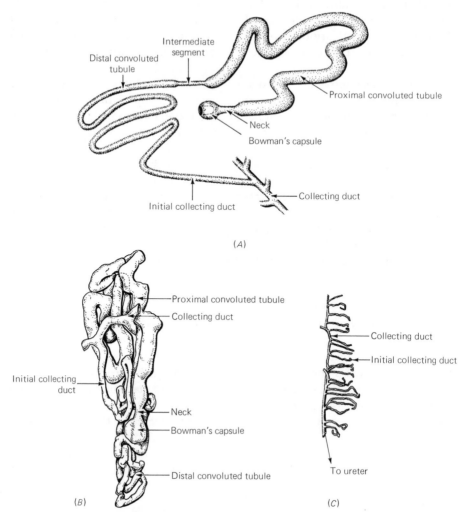

Figure 7-3. (*A*) Schematic representation of an amphibian nephron. (*B*) Nephron of a bullfrog (*Rana catesbeiana*), magnified 54 times. The nephron was reconstructed from analysis of serial sections of the kidney, and the natural orientation of the parts is illustrated. The transition from proximal convoluted tubule to intermediate segment and distal convoluted tubule occurs where the proximal convoluted tubule has finally doubled back to about the level of the Bowman's capsule and is hidden from view. (*C*) A collecting duct of *Rana catesbeiana* showing the connections of the initial collecting ducts of many nephrons. The many collecting ducts of the kidney connect to a common ureter. [*B* and *C* from Huber, G. C. 1928. Renal tubules. *In:* E. V. Cowdry (ed.), *Special Cytology*. Vol. I. Medical Department of Harper & Row, New York.]

mal convoluted tubule. Recent evidence indicates that the principal cation, sodium, is actively resorbed and that the principal anion, chloride, follows passively in response to the electrical gradient established by sodium resorption. The proximal convoluted tubule is freely permeable to water, and the resorption of ions is accompanied by osmotic resorption of water that acts to maintain the tubular fluids near isosmoticity with the surrounding body fluids. Thus although there is a reduction in the volume of fluid and in the amounts of sodium and chloride in the tubule, the tubular concentrations of these ions and the osmotic pressure remain essentially unchanged.

That there has, in fact, been a net reduction in volume by the time fluid exits the proximal tubule is demonstrated clearly by studies using

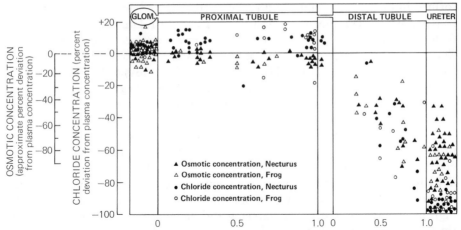

Figure 7-4. Chloride and osmotic concentrations in the renal tubular fluid of *Necturus maculosus* and the frog *Rana pipiens*. Fluid was sampled by micropuncture. The scales on the abscissa indicate the point of sampling as a fraction of the total length of the proximal tubule and distal tubule, respectively. Chloride concentration is expressed as percent deviation from the chloride concentration in the plasma. (For example, a value of −40 in the distal tubule of a frog would indicate a chloride concentration 40% lower than that in the plasma of the same frog.) Osmotic concentration is also expressed *approximately* as percent deviation from the osmotic concentration in the plasma; for the exact meaning of measured osmotic concentrations, see original research report. (From Walker, A. M., C. L. Hudson, T. Findley, Jr., and A. N. Richards. 1937. Amer. J. Physiol. 118: 121–129; percent deviation of urinary osmotic concentration from plasma osmotic concentration estimated from calibration data by linear regression.)

inulin. Inulin is a fructose polysaccharide derived from Jerusalem artichokes that is widely used in the study of renal function in animals with ultrafiltration systems. When administered to amphibians it is filtered into the Bowman's capsule at concentrations equivalent to its plasma concentration. It is not secreted into the renal tubule, nor is it actively or passively resorbed. Thus the mass of inulin exiting in the definitive urine is the same as the mass introduced into the Bowman's capsules by filtration. In one study of amphibians, the concentration of inulin in fluid exiting the proximal convoluted tubule proved to be elevated by some 30% over the prevailing plasma concentration. This result shows that there is a reduction in volume during the passage of fluid through the proximal convoluted tubule, and it is easy to compute that the volume reduction in the experiments was on the order of 25%.

Another important function of the proximal convoluted tubule is active glucose resorption. The preponderance of glucose in the ultrafiltrate has been removed by the time the fluid leaves the proximal convoluted tubule. Glucose is a valuable metabolite that, because of its small size, cannot be withheld from the urine at the site of ultrafiltration, but as we shall see, many animals, like the amphibians, promptly reclaim it. In a similar vein, amino acids filtered across the glomeruli are resorbed in the proximal tubule.

Events in the distal convoluted tubule; antidiuretic hormone

The action of the distal convoluted tubule depends strongly on the permeability of the tubular membranes to water. Permeability is under control of an octopolypeptide hormone, identified as arginine vasotocin, which is formed in the hypothalamus and released from the posterior pituitary (neurohypophysis). This hormone and certain closely related octopolypeptides formed in other tetrapod vertebrates are known as *antidiuretic hormones*. The antidiuretic hormone (ADH) of reptiles and birds is also arginine vasotocin. In most mammals the ADH is arginine vasopressin, but in pigs, hippopotamuses, and some other mammals it is lysine vasopressin. The most clearly demonstrated effect of ADH in the renal tubules of amphibians is to increase the permeability of the distal convoluted tubule to water.

Ions are actively resorbed across the distal convoluted tubule, as in the proximal tubule. When ADH levels are low, the permeability of the distal convoluted tubule to water is relatively low, and although there is considerable osmotic resorption of water concomitant to ion resorption, water resorption is impeded, with two important consequences. First, a relatively high proportion of the water entering the distal convoluted tubule passes through to be excreted in the urine; and, second, ion resorption results in persistent hyposmoticity of the tubular fluid. These were the conditions prevailing in the experiments depicted in Figure 7–4. You will note that the osmotic pressure of the tubular fluid falls in the distal convoluted tubule and that chloride concentrations are reduced to very low levels by comparison with the plasma. The animal produces a copious, dilute urine and is said to be in *diuresis*.

Now if ADH levels are high, the distal convoluted tubule becomes quite permeable to water. More water is resorbed in response to the osmotic gradient set up by ion resorption, and the tubular fluid may remain virtually isosmotic with the plasma. The animal produces a relatively small amount of relatively concentrated urine, this being the reason that the hormone is termed antidiuretic hormone.

It should be immediately apparent that the effects of ADH provide for adaptive response to the osmotic situation of the animal. It is adaptive for amphibians in fresh water to produce a copious, dilute urine. Likewise, it is adaptive for animals faced with problems of dehydration to produce a small amount of relatively concentrated urine. The rise in plasma osmotic pressure resulting from dehydration appears to be the primary stimulus for release of ADH. The renal response to ADH is strongly developed in semi-terrestrial and terrestrial amphibians but is reported to be essentially absent in the aquatic toad *Xenopus*.

There are reports that in some species ADH has other renal effects that contribute to water retention. ADH may reduce the glomerular filtration rate, thus decreasing the amount of fluid introduced to the renal tubules to begin with, and ADH may stimulate active ion resorption across the renal tubules, thus also increasing the extent of osmotic water resorption.

Additional considerations

The disposition of urea in the amphibian kidney has been of some particular interest. The nephron seems generally to be poorly permeable to urea. Thus as water is resorbed, the concentration of the urea introduced by filtration increases. This mechanism appears to be adequate to produce the urea concentrations found in *Necturus* urine. It is not adequate to explain those in bullfrog urine. In fact, the amount of urea in bullfrog urine may exceed the amount filtered across the glomeruli by severalfold. This and other evidence indicates that urea is actively secreted into the renal tubules.

Active secretion is also reported in a number of other semiterrestrial frogs.

The composition of amphibian urine may be modified in the bladder. Again, effects of ADH are critical. ADH increases the permeability of the bladder wall to water, leading to osmotic withdrawal of water from hyposmotic contents. ADH is also reported to stimulate active sodium resorption from the bladder, which will again augment water resorption. Many amphibians utilize the bladder as a water reservoir. The response of the bladder to ADH tends to be stronger in more terrestrial species and is apparently absent in *Xenopus*. Analogous effects of ADH on the bladder have been reported in certain turtles. ADH increases permeability to water and stimulates active ion uptake across the skin of amphibians, but such cutaneous effects have not been reported in other vertebrate classes.

FISH

Among fish we find great diversity of nephron structure and function. Glomeruli and distal convoluted tubules are lacking in various forms, whereas others possess nephrons much like those in amphibians. This diversity of structure has provided fertile ground for speculation about the evolution of the nephron in early vertebrates and the subsequent adaptation of the nephron according to demands for occupying the freshwater and marine environments.

The ultrafiltration system of glomerulus and Bowman's capsule provides a ready means for introducing relatively large quantities of fluid into the nephron. In this sense the system is well adapted to the needs of freshwater forms, which must excrete a copious urine. Various authors have argued that the glomerular nephron evolved under the exigencies of living in fresh water. Noting that most modern fish have glomerular nephrons and that the aglomerular condition of some species can be readily explained as secondary rather than primitive, they have argued that nephron structure provides support for the hypothesis that the earliest vertebrates evolved in fresh water. A recent review of paleontological and phylogenetic evidence has provided compelling (but not conclusive) support for the contrary hypothesis, that the earliest vertebrates were marine. If this is true, then the modern hagfish may well trace a continuously marine ancestry, and because they have glomerular nephrons, we would conclude that the glomerular condition evolved in the context of the sea. As we shall see later, certain invertebrates that are indisputably marine in origin have ultrafiltration systems. Thus the presence or absence of such a system cannot be accepted as an indicator of ancestral habitat, and the issue must be resolved on the basis of more direct (preferably paleontological) evidence. If fish evolved in the sea, it is clear that they migrated rather early into fresh water, where the glomerular kidney, if it evolved in the sea, undoubtedly proved to be a valuable preadaptation. The evidence is strong that modern marine teleosts are derived from freshwater ancestors, and it seems likely that the same is true for modern marine elasmobranchs.

Cyclostomes

The renal corpuscles of hagfish are very large. In each kidney many Bowman's capsules connect rather directly with a common duct, the archinephric duct or ureter, which empties to the cloaca. On cytological grounds the archinephric duct appears to be homologous with the proximal convoluted tubule of amphibians and other vertebrates. Thus we may state that the renal tubules of hagfish consist of a renal corpuscle plus proximal tubule, though the organization of the kidney differs from that of other vertebrates in that one proximal tubule serves many Bowman's capsules. Inasmuch as the proximal tubule leads directly to the cloaca, there are no distal convoluted tubules, initial collecting ducts, or collecting ducts. Probably the lack of these parts represents the primitive vertebrate condition. Hagfish, you will recall, are isosmotic with seawater and produce urine that is isosmotic to their plasma. In regard to the osmoticity of the urine, the output of the hagfish proximal tubule resembles that of the amphibian proximal tubule. In fact, throughout the vertebrates that have received study, the osmotic pressure of the tubular fluid exiting the proximal tubule is close to that of the plasma. Solutes may be resorbed or secreted along the tubule, leading to significant changes in composition of the fluid, but the tubule is sufficiently permeable to water that water moves toward establishing isosmoticity with the surrounding body fluids.

The hagfish are exclusively marine, but the other group of cyclostomes, the lampreys, are found in fresh water as well as in the sea. In fresh water they regulate hyperosmotically. Their urine then is strongly hyposmotic to the plasma and contains much-reduced concentrations of sodium and chloride. Unlike the condition in hagfish, the Bowman's capsules of lampreys do not connect directly to a common duct but lead to separate renal tubules. The tubules of each kidney, in turn, ultimately discharge into a common ureter. This basic design is reminiscent of that in amphibians and is found in the other groups of fish. The tubules of lampreys consist of a proximal convoluted segment homologous to that of amphibians and a distal convoluted segment. The homologies of the distal segment are not completely clear. Some authors have believed it to be homologous to the distal convoluted tubule of amphibians, whereas others believe it to be homologous with the initial collecting duct or collecting duct. Cytologically it resembles the collecting ducts, and we shall call it the collecting tubule. You will recall that in amphibians production of hyposmotic urine involves resorption of salts across parts of the nephron that are poorly permeable to water. The proximal tubule, being freely permeable, does not permit this. In the absence of direct experimentation on lamprey nephrons, we may presume that their proximal tubule has similar properties to the proximal tubule of amphibians and other vertebrates that have received study and hypothesize that it is in the collecting tubule that hyposmoticity is established. Thus it seems that the lampreys, in comparison with the hagfish, have added an important new capability through the evolution of this additional segment of the renal tubule.

Teleosts

The collecting tubule is a constant feature of the nephrons of jawed fish. In nearly all freshwater teleosts, a segment that is indisputably homologous with the distal convoluted tubule of amphibians is interposed between the proximal tubule and collecting tubule. It is generally presumed, in the absence of direct evidence, that the urine is rendered hyposmotic in the distal tubule, as it is in amphibians, and that the distal tubule enhances the capacity to produce hyposmotic urine over that provided by the apparently more-primitive collecting tubule. Almost certainly, the distal tubule evolved under the pressures of occupying the freshwater habitat.

Marine teleosts commonly lack the distal convoluted tubule. Inasmuch as they are believed to be descended from freshwater ancestors, the absence of the distal tubule may well represent a secondary loss rather than a primitive condition. This loss can be understood in terms of the osmotic relations of marine fish and the presumptive function of the distal tubule in producing hyposmotic urine. Marine teleosts are hyposmotic to their medium and produce small quantities of nearly isosmotic urine. They have no need of a nephron segment specialized for the establishment of hyposmoticity.

Teleosts that migrate between fresh water and seawater or that are routinely distributed in both habitats generally have the distal tubule.

The glomeruli of many marine teleosts are degenerate by comparison with those of freshwater forms, and in some marine fish some or all of the nephrons are entirely aglomerular. In some fish that have glomeruli the neck connecting the Bowman's capsule to the proximal convoluted tubule is so constricted that fluid may not be able to pass in appreciable quantities. These fish may be functionally aglomerular. The loss or reduction of the ultrafiltration system is probably related to the fact that marine teleosts produce only small amounts of urine. Fluid and solutes are introduced into the aglomerular nephron by secretion.

To summarize this review of renal anatomy in teleosts, there is a very clear correlation between the presence of the distal convoluted tubule and habitat. The tubule is virtually universal among species that live in fresh water or enter fresh water and is very commonly absent in strictly marine forms. Nearly all freshwater fish have well-developed glomerular nephrons, but glomeruli are lacking or reduced in some marine species. Our understanding of the function of the distal tubule and renal corpuscle permits a physiological interpretation of these correlations between morphology and habitat.

Some teleosts lack the usual morphological correlates for their habitat, and these examples should serve to caution against an overly facile interpretation of renal morphology as a critical adaptive factor. The freshwater pipefish of Malaysia, *Microphis boaja*, is aglomerular, and it seems likely that certain aglomerular marine pipefish enter fresh waters or dilute brackish waters. *Microphis boaja* is believed to be a recent immigrant from the sea, and its aglomerular condition can be understood in this light. Nonetheless, it demonstrates that a freshwater existence is possible without glomeruli and apparently without ultrafiltration as a means of introducing large quantities of fluid into the nephrons. *Microphis* also lacks the distal convoluted tubule. The euryhaline killifish *Fundulus heteroclitus* and three-spined stickleback *Gasterosteus aculeatus* are glomerular but lack the distal tubule. Urinary hyposmoticity in fresh water might be established in the collecting tubules of these fish.

Comparisons of teleosts with glomerular nephrons have revealed a clear tendency for total filtration surface and total rate of filtration to be substantially lower in marine forms than in freshwater forms. This may be correlated with the differences in volume of urine produced by fish in the two habitats. In both groups the volume of fluid excreted is generally considerably lower than the volume filtered. This is similar to the condition in amphibians and is indicative of active salt resorption and concomitant water resorption along the renal tubule. In mammals the glomerular filtration rate tends to remain stable, and the rate of urine flow is adjusted mainly by varying the rate of water resorption along the nephrons rather than by varying the rate at which fluid is introduced into the nephrons at the glomeruli. Studies on fish reveal quite a different situation. The glomerular filtration rate tends to vary over a wide range, and evidence indicates that the proportion of filtered water that is resorbed remains rather stable in some species. Their rate of urine flow thus varies closely with the rate of filtration. In experiments on one white sucker (*Catostomus commersonii*), for example, glomerular filtration rate varied between 1 cc/hr/kg body weight and 8 cc/hr/kg. The proportion of the filtered volume excreted as urine remained close to 35% over this entire range, and urine output varied from about 0.3 cc/hr/kg to 2.9 cc/hr/kg.

We discussed earlier the important role of the urine in voiding magnesium and sulfate in marine teleosts. Evidence from both glomerular and aglomerular species shows that these ions are actively secreted into the renal tubules.

Glucose is actively resorbed in the renal tubules of marine and freshwater teleosts.

Elasmobranchs

The nephrons of marine and freshwater elasmobranchs include a glomerulus, proximal convoluted tubule, distal convoluted tubule, and collecting tubule. The marine forms produce small amounts of slightly hyposmotic urine, whereas elasmobranchs in fresh water void a copious, dilute urine. The total filtration rate of *Pristis microdon* in fresh water is much higher than that of elasmobranchs in seawater. In both groups there is again a net resorption of water along the renal tubules. The elasmobranch nephron is noteworthy in its apparent ability to resorb urea and trimethylamine oxide actively. This ability is important for maintenance of high plasma levels of these compounds.

Control of renal function in fish

There is considerable evidence, mostly from studies of euryhaline species, that fish exert sensitive control over parameters of nephron function. When rainbow trout, for example, are transferred from fresh water to seawater, glomerular filtration rate and urine flow may drop by over 90%. Similar changes, adaptive to the marine environment, are reported in other euryhaline teleosts. In some species, such as the Japanese eel (*Anguilla japonica*) and southern flounder (*Paralicthys lethostigma*), the glomerular filtration rate returns close to its freshwater level after a period in seawater, but urine flow remains low because an increase in the permeability of the renal tubules to water results in increased water resorption along the nephrons. It seems in these species that decrease in glomerular filtration rate provides for rapid adjustment to the marine environment, but long-term adjustment is provided by changes in tubular permeability. When southern flounder are transferred into fresh water from seawater, it takes 12 to 24 hours for tubular water permeability to decline. At first the urine is essentially isosmotic to the blood, but urinary osmotic pressure declines steadily to a low level.

Plasma urea concentrations of elasmobranchs are reduced in diluted water by comparison to seawater. This results in part from increased urinary excretion of urea. In *Squalus acanthias* accommodated to diluted seawater, tubular resorption of urea is less than in seawater. In the skate *Raja erinacea*, the glomerular filtration rate and rate of urine production increase considerably in dilute waters, with a concomitant increase in urea excretion.

Alterations of renal function in fish are assumed to be under hormonal control, as they are in other vertebrates, but although a variety of hormones have been implicated, the situation remains very unclear. Fish pituitary extracts can exert an antidiuretic effect in tetrapods but do not clearly do so in the fish themselves. There is no evidence of a functional ADH of the same chemical family as those of tetrapods. Hormones of the interrenal gland (homologue of the adrenal) have been implicated in salt balance, and the counterpart of prolactin in fish, secreted by the anterior pituitary, appears to be important in some way to production of copious, dilute urine in euryhaline fish in fresh water.

MAMMALS

The freshwater progenitors of amphibians probably had nephrons like those seen in modern amphibians and most modern freshwater teleosts. This basic structure—including glomerulus, proximal convoluted tubule, and distal convoluted tubule—has been carried over essentially unmodified in the reptiles and persists with the addition of new features in the birds and mammals. Thus the tetrapod vertebrates have uniformly retained the proximal and distal convoluted segments that evolved in their piscine ancestors. We shall examine the mammalian kidney in this section and the avian kidney in the next. The mammalian kidney is the most thoroughly understood renal organ in the animal kingdom.

Anatomy of the mammalian kidney

The outstanding morphological difference between the mammalian nephron and the amphibian or reptilian nephron is the presence of an elongated hairpin loop between the proximal and distal convoluted tubules (Figure 7–5). This loop, known as the *loop of Henle*, consists of two parallel limbs, the *descending limb* leading from the proximal convoluted tubule to the hairpin bend and the *ascending limb* leading from the bend to the distal convoluted tubule. The descending limb begins with a *thick segment* that probably is derived evolutionarily from the proximal tubule, and the ascending limb terminates with a thick segment that is probably derived from the distal tubule. Interposed between these thick segments at various positions and for various lengths is a segment of very small diameter, the *thin segment*. The thin segment differs cytologically from the intermediate segment discussed earlier (Figure 7–3) and occurs only in mammals and birds. The loop of Henle varies considerably in length among species and among the nephrons within the kidneys of one species. The thin segment tends to be longer and of more prominence, the longer the loop of Henle.

Another outstanding morphological difference between the mammalian

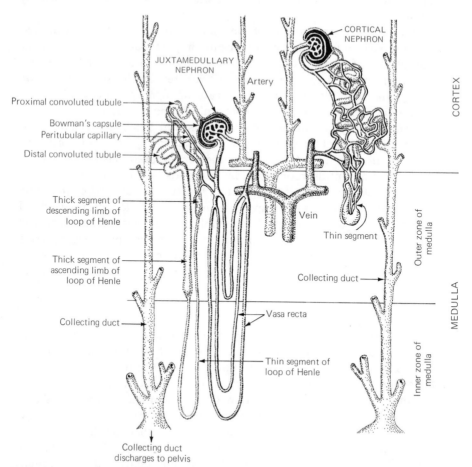

Figure 7–5. Diagram of human nephrons and their circulatory supply. See Figure 7–6 for relationships of the cortex, medulla, and pelvis in whole kidneys. The collecting ducts discharge into the renal pelvis at the inner surface of the medulla. Juxtamedullary and cortical nephrons are discussed subsequently in the text. (From Smith, H. W. 1951. *The Kidney. Structure and Function in Health and Disease.* Oxford University Press, New York.)

kidney and the amphibian or reptilian kidney is the highly ordered arrangement of the nephrons and collecting ducts. As depicted in Figure 7–5, the glomeruli and convoluted tubules are aggregated toward the outer surface of the kidney, whereas the loops of Henle and collecting ducts project inward toward the renal pelvis. This arrangement is in good part responsible for the zonation apparent on even gross histological examination of the kidney. In section the kidney is seen to consist of a morphologically distinct outer "layer," the *cortex*, which surrounds an inner "layer," the *medulla*. The cortex consists of renal corpuscles, convoluted tubules, the beginnings of collecting ducts, and associated vasculature. The medulla consists of loops of Henle and collecting ducts as well as the associated vasculature. The medulla contains a great many loops of Henle, and it is highly significant that the loops tend to run parallel to each other and that the collecting ducts run parallel to the loops.

We may trace the path of fluid through the renal tubules with reference to the nephron with a long loop of Henle to the left in Figure 7–5. After filtration in the renal corpuscle, fluid moves first through the proximal convoluted tubule and then descends into the medulla in the loop of Henle. After rounding the bend of the loop, the fluid returns to the cortex, passes through the distal convoluted tubule, and enters a collecting duct. The fluid then again passes through the medulla, ultimately being discharged from the collecting duct into the renal pelvis on the inner surface of the medulla. The renal pelvis is a tubular structure representing the expanded inner end of the ureter that drains the kidney.

As noted earlier, various nephrons in the kidney of a species may have loops of Henle of quite different lengths. In man and many other mammals, nephrons with their renal corpuscle located toward the outer surface of the cortex, termed *cortical nephrons*, have relatively short loops of Henle that penetrate only a short distance into the medulla or do not enter the medulla at all (Figure 7–5). Nephrons with their renal corpuscle situated deep in the cortex, termed *juxtamedullary nephrons*, have longer loops of Henle that penetrate far into the medulla. Some of these loops reach all the way to the inner surface of the medulla.

The total number of nephrons has been estimated for a variety of species and varies widely. The laboratory rat has about 30,000 nephrons in each kidney; the dog, about 400,000; and man, about 1.1 million.

Comparative anatomy of the kidney in relation to concentrating ability and habitat

We have seen that the mammalian nephron differs structurally from the amphibian or reptilian nephron and furthermore that the arrangement of the nephrons in the mammalian kidney is highly ordered, whereas nephrons in the amphibian or reptilian kidney are arranged in a distinctly more random fashion. It has seemed logical to hypothesize that these morphological differences are related to functional differences and, in particular, to suggest that the loops of Henle and arrangement of the nephrons are involved in the capacity of the mammalian kidney to synthesize urine of greater osmotic pressure than the body fluids. The importance of the loops of Henle to the production of hyperosmotic urine has been indicated by several types of comparative morphological data on mammals.

The mountain beaver, *Aplodontia rufa*, an inhabitant of cool, mesic areas along the Pacific coast, achieves a maximal urinary osmotic concentration (expressed as freezing-point depression) of only $\Delta 1.1°C$, and all of its nephrons are of the cortical type with short loops of Henle. The sand rat, *Psammomys obesus*, on the other hand, produces sizable quantities of highly concentrated urine (up to $\Delta 9.2°C$), and its nephrons all possess long loops of Henle. Comparisons such as this indicate a relationship between the possession of long loops of Henle and the ability to concentrate the urine. Many mammals possess both long and short loops. There is a tendency for species having a very low percentage of long loops to exhibit modest concentrating abilities by comparison to those having 15–25% or more of long loops, but among the latter forms there is little or no correlation between percentage of long loops and concentrating ability.

Commonly, the medulla of the kidney has a somewhat pyramidal shape,

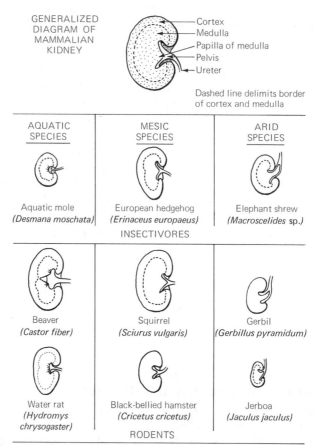

Figure 7–6. Diagrams of the kidneys of nine mammals. Aquatic species show little or no development of the papilla. (*Desmana* and *Hydromys* lack the papilla. *Castor* has two very shallow papillae.) Mesic species have papillae. The papilla is especially developed in arid species, so much so that it often penetrates well into the ureter. (From Sperber, I. 1944. Zool. Bidr. Upps. 22: 249–432.)

projecting into the renal pelvis as the *renal papilla* (Figure 7–6). Often there is just one pyramid and papilla, but in some mammals, such as man, the medulla consists of two or more pyramids that project in an equal or lesser number of papillae. Inasmuch as the medulla is composed in good part of loops of Henle, the prominence of the papilla provides some indication of the number and length of long loops. In 1944 Sperber reported observations on the medullary structure of about 140 species of mammals living in diverse environments and representing several orders. He found that the papilla was uniformly lacking or poorly developed in species inhabiting wet or aquatic environments. The papilla was present in species from mesic environments and was most developed in species from arid environments (Figure 7–6). Insofar as habitat may be taken as an indicator of demands for urinary concentration, these results indicate that there is a greater development of the long loops of Henle in species that produce relatively concentrated urine.

Figure 7–7 depicts the renal papillae of two rodents from arid environments. *Dipodomys* has about 25% long loops of Henle, whereas *Psammomys* has all long loops. This difference is reflected in the larger papilla of

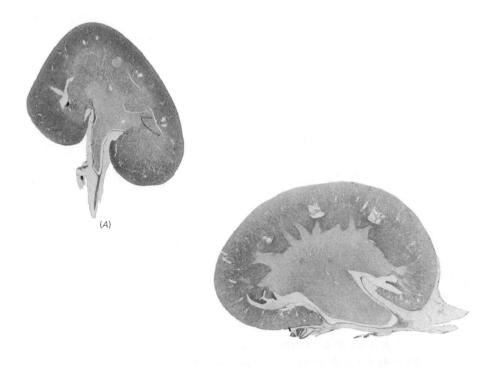

(A)

(B)

Figure 7-7. Cross sections of the kidneys of (*A*) a kangaroo rat (*Dipodomys*) and (*B*) the sand rat (*Psammomys obesus*). [From Schmidt-Nielsen, B. 1964. Organ systems in adaptation: the excretory system. *In:* D. B. Dill (ed.), *Handbook of Physiology. Section 4: Adaptation to the Environment.* American Physiological Society, Washington, D.C.]

Psammomys. You will recall that *Dipodomys*, the kangaroo rat, is a grain eater. *Psammomys*, the sand rat, subsists in large part on succulent plants with high salt content. By comparison to the kangaroo rat, the sand rat has a much greater urinary solute load owing to the high concentrations of electrolytes in its diet. The sand rat can produce urine that is somewhat more concentrated relative to plasma (maximal osmotic U/P ratio is 17, by comparison to 14 in the kangaroo rat), and the sand rat voids a much greater volume of urine (perhaps 20 times more per gram of body weight). The greater size of the papilla in the sand rat may thus be correlated with important differences in renal function.

Various authors have compared medullary size to the concentrating ability of the kidney. The kidneys of various species differ widely in size, and to correct for differences in overall size of the kidney, the size of the medulla has been expressed relative to total kidney size or to the size of the cortex. One set of results is shown in Figure 7-8, and it can be seen that there is a clear positive correlation between relative medullary thickness and maximum concentrating ability. If transverse sections are made of the kidney and the thickness of the medulla and cortex measured, it is found, for example, that the two zones are of about equal thickness in the mountain beaver (*Aplodontia*), whereas the medulla is about five times thicker than the cortex in the kangaroo rat. These results again indicate a relationship between the development of the long loops of Henle and the concentrating ability of the kidney.

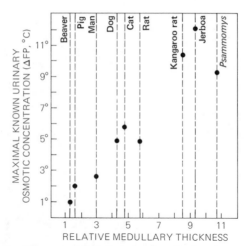

Figure 7–8. Maximal urinary osmotic concentration (in freezing-point depression) measured on nine mammals as a function of relative medullary thickness. Relative medullary thickness was determined as 10 times the ratio of medullary thickness to kidney size, kidney size being expressed as the cube root of the product of the three primary dimensions of the whole kidney. (From Schmidt-Nielsen, B. and R. O'Dell. 1961. Amer. J. Physiol. 200: 1119–1124.)

The mechanism of concentrating the urine: countercurrent multiplication

Many of the anatomical observations just discussed were made before the role of the loop of Henle in urinary concentration was elucidated and served to center attention on the physiology of the loop. The mechanism of urinary concentration defied understanding, however, until a radical shift in our thinking about the kidney was engendered by the seminal papers of Kuhn, Hargitay, Ramel, Martin, and Ryffel in the 1940s and 1950s. A complex countercurrent mechanism is involved, and it will be important to understand the distinction between the *countercurrent multiplier* in the loop of Henle and a *countercurrent exchanger*, such as that in Figure 3–18*B*. You will recall that in the countercurrent heat exchanger discussed earlier, thermal gradients exist between the two opposing bloodstreams along the entire length of the exchanger. These gradients are not established by energy expenditure in the countercurrent system itself. Rather, heat is generated in other tissues, and the exchanger simply acts on the thermal difference between the body core and the surrounding air (Figure 3–18*B*). This type of system is known as a *passive* countercurrent system. By contrast, it is possible to have a countercurrent system in which gradients between the two opposing fluid currents are established by energy expenditure in the system itself. This *active* type of system is what is found in the loop of Henle. We shall have to explore the behavior of the countercurrent system in the loop of Henle before important differences between active and passive systems can be fully appreciated.

The descending limb of the loop of Henle is permeable to both salts and water, whereas the ascending limb is permeable to salts but relatively impermeable to water. The descending limb does not transport ions, but in the ascending limb sodium and chloride are resorbed from the tubular fluid by active transport. Traditionally it has been believed that sodium is actively

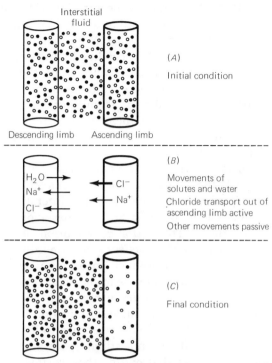

Interstitial
fluid

Descending limb   Ascending limb

(A)
Initial condition

(B)
Movements of
solutes and water
Chloride transport out of
ascending limb active

Other movements passive

(C)
Final condition

Figure 7–9. Schematic representation of the development of an osmotic gradient between adjacent parts of the descending and ascending limbs in the loop of Henle. The walls of the ascending limb are drawn thick to represent their relative impermeability to water. For simplicity, sodium and chloride are the only solutes represented, sodium as closed circles, chloride as open circles. In A sodium and chloride concentrations are uniform throughout. In B the chloride pump acts, accompanied by passive movements as discussed in the text. The consequences of the movements occurring in B are illustrated in C. The osmotic concentration of the fluid in the ascending limb is lowered from its original level, whereas the osmotic concentrations of the interstitial fluid and fluid in the descending limb are raised. Active chloride transport is assumed here on the basis of recent research findings. Movement of sodium out of the ascending limb may be entirely passive or may include an active component; this issue is not resolved.

removed and that chloride follows passively in response to the electrical gradient established by sodium transport. Recent studies indicate, however, that the opposite is likely true; chloride appears to be actively removed, with sodium following passively. In any case, both sodium and chloride are resorbed from the ascending limb. If we imagine for a moment two tubes, corresponding to the descending and ascending limbs, in which fluid is not flowing (Figure 7–9), we can see the consequences of the properties of the two limbs. Suppose that initially the solutions in the two limbs and in the interstitial space are identical in composition (and osmotic pressure). Now sodium and chloride are actively removed from the fluid of the ascending limb. This raises the sodium and chloride concentrations and the osmotic concentration of the interstitial fluid and lowers the ionic and osmotic concentrations of the fluid in the ascending limb. An osmotic gradient is established and persists because the ascending limb is poorly permeable to water. The fluid in the descending limb comes to equilibrium with that in the interstitial space by passive osmotic loss of water and passive uptake of sodium and chloride. Thus an osmotic gradient is established between the ascending limb, on the one hand, and the interstitial fluid and descending limb, on the other. More particularly, the osmotic concentration of the ascending limb is lowered from its initial value, whereas that of the descending limb and interstitial fluid is raised from its initial value. Note that this gradient between adjacent parts of the ascending and descending limbs is maintained all along the loop of Henle and is established actively within the loop. In man producing concentrated urine the osmotic difference maintained at any one point along the loop is on the order of 200 milliosmolar (mOs). This difference is known as the *single effect* of the countercurrent system. If countercurrent flow did not enter the picture, this would be the maximum osmotic difference established in the system.

It will be important in the following discussion to keep clearly in mind that the single effect is the osmotic gradient developed between the ascending limb and the *adjacent* descending limb and interstitial fluids. Now we shall turn to considering the osmotic gradient developed between the two *ends* of the loop of Henle. The major contribution of Kuhn and his coworkers was to show that the single effect can be multiplied in the presence of countercurrent flow so as to produce an osmotic gradient between the ends of the loop that is much greater than the single effect. For this reason the loop of Henle is known as a *countercurrent multiplier*.

The mechanism of countercurrent multiplication in the loop of Henle can be outlined in simplified form with reference to Figure 7–10. In part *A* the entire loop of Henle and the interstitial space are filled with fluid of the same osmotic pressure as that exiting the proximal convoluted tubule —approximately isosmotic with the body fluids. In part *B* active transport establishes a single-effect osmotic gradient of 200 mOs all along the loop. In part *C* fluid moves through the loop in countercurrent fashion. Fluid that was concentrated in the descending limb during step *B* is thus brought around opposite to the descending limb in the ascending limb so that both limbs and the interstitial space are filled with concentrated fluid at the medullary end of the loop. Now when, in *D*, the single-effect osmotic gradient is again established between the ascending limb and interstitial fluid, the interstitial fluid is elevated to 500 mOs at the medullary end, rather than the 400 mOs developed in step *B*, and the fluid in the descending limb also reaches this higher osmotic concentration. Steps *E* and *F*, and *G* and *H*, repeat this process. Fluid concentrated in the descending limb moves around into the ascending limb, providing a basis upon which the single effect can produce an ever-increasing osmotic concentration in the interstitial fluid and descending limb at the medullary end. The result is that the osmotic gradient between the cortical and medullary ends of the loop becomes greater and greater. We have discussed in detail here the action of the multiplier at the medullary end, but the reader should carefully examine Figure 7–10 to see how a gradient of osmotic concentration is established from end to end along the loop as concentrated fluid moves up the ascending limb and the single effect is exerted between the fluid in the ascending limb and that in the interstitial space and descending limb. You will note the importance of the single effect, countercurrent flow, and the fact that the two opposing flows are connected at the hairpin bend so that fluid concentrated in the descending limb in turn moves into the ascending limb.

We may now return to the distinction between active and passive countercurrent systems. In the passive countercurrent heat exchanger of Figure

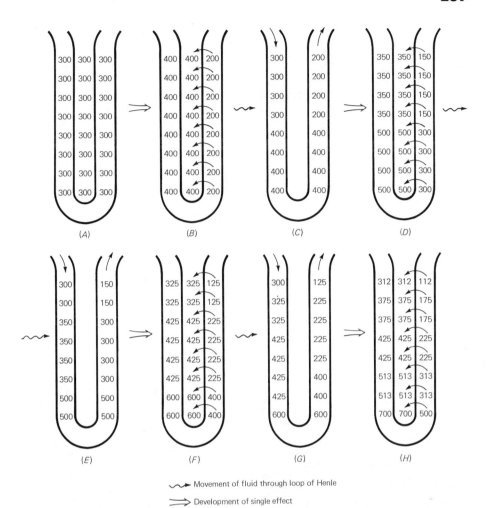

Movement of fluid through loop of Henle

Development of single effect

Figure 7–10. Simplified illustration of the operation of the countercurrent multiplier in the loop of Henle. Numbers are osmotic concentration in milliosmolarity. The operation of the multiplier is represented in a series of alternating steps. In *A* the entire system is at 300 m*Os*. In *B* a single-effect osmotic gradient of 200 m*Os* is developed all along the loop, and in *C* fluid moves through the loop. These steps are repeated in *D* through *H*. The amount of fluid movement through the loop decreases progressively in steps *C*, *E*, and *G*. Note that fluid entering the descending limb is always at 300 m*Os*. This produces a tendency for the osmotic concentration at the cortical end of the descending limb and interstitial space to remain near 300 m*Os*. For simplicity, osmotic pressures developed by the single effect have been computed by raising the osmotic pressure in the descending limb and lowering that in the adjacent ascending limb by equivalent amounts so as to produce a difference of 200 m*Os* between the two limbs. (Adapted from Pitts, R. F. 1974. *Physiology of the Kidney and Body Fluids*. 3rd ed. Year Book Medical Publishers, Inc., Chicago. Original figure copyright © 1974 by Year Book Medical Publishers, Inc. Adapted by permission.)

3–18*B*, there is a large thermal gradient between the two ends of the system, and heat is exchanged all along the two adjacent fluid streams. The exchanger, however, does not create the thermal difference between the body core and the air but rather acts in its presence, enhancing the thermal difference between the tissues at the two ends of the limb and conserving body heat. In the active system of the loop of Henle, countercurrent flow in conjunction with the single effect *establishes* a large osmotic gradient. This is an important distinction. If the heat exchanger were turned off, the thermal

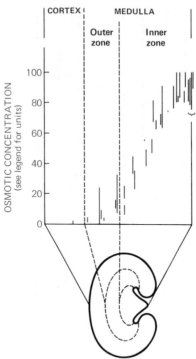

Figure 7–11. Osmotic concentrations determined along the axis of the papilla in kidneys of dehydrated rats. The value 0 corresponds to the osmotic concentration of plasma. The highest concentration measured (about 1000 mOs greater than plasma at the tip of the papilla) is set equal to 100, and other increases of concentration above the plasma concentration are expressed as a percentage of this maximal deviation from the plasma concentration. [That is, the ordinate is equal to: 100 (measured concentration − plasma concentration)/(maximal concentration − plasma concentration).] Throughout the cortex the osmotic concentration is equivalent to the plasma concentration. The osmotic concentration increases steadily with depth in the medulla. (From Wirz, H., B. Hargitay, and W. Kuhn. 1951. Helv. Physiol. Pharmacol. Acta 9: 196–207.)

difference between the body core and the air would persist, but if the loop of Henle were turned off, the osmotic gradient established between the two ends of the loop would disappear.

In the medulla of the kidney, you will recall, there are a great many loops of Henle that run in parallel, their hairpin bends all being directed toward the inner (pelvic) side of the medulla. These act in tandem to produce a gradient of increasing osmotic pressure from the cortical to the pelvic side within the medulla. Note, as stressed earlier, that this gradient is established within the interstitial fluids of the medulla as well as in the limbs of the loops of Henle. In pioneering experiments reported in 1951, Wirz, Hargitay, and Kuhn demonstrated the existence of a steep osmotic gradient in the medulla of rats that had been deprived of water for two days. They froze the kidney, sectioned it, and allowed the sections to thaw, determining the osmotic concentration at various levels in the kidney from the temperature at which ice crystals disappeared. Their results, illustrated in Figure 7–11, and other similar results have provided strong support for the countercurrent multiplier hypothesis.

Though we have discussed the operation of the countercurrent multiplier in terms of osmotic concentrations, it is important to recognize that

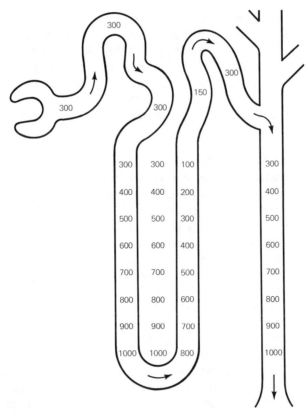

Figure 7–12. Summary of changes in osmotic pressure (in milliosmolarity) as fluid traverses the nephron and collecting duct in the concentrating kidney. Osmotic pressures given are meant only to illustrate general trends as they are now understood. It is assumed that the distal convoluted tubule is permeable to water and that the tubular fluid comes to osmotic equilibrium with surrounding body fluids in the distal tubule. This is discussed subsequently in the text.

both the single-effect osmotic gradient and the osmotic gradient established from end to end along the loops of Henle are reflective of gradients of salt concentration. The countercurrent mechanism produces a high medullary concentration of sodium chloride, which gives rise to the high medullary osmotic pressure.

You will probably have noticed that we have not yet completely explained the production of concentrated urine. Fluid in fact leaves the loop of Henle at a *lower* osmotic concentration than that at which it enters. Recall, though, that after passing through the distal convoluted tubule, the fluid makes a one-way passage through the medullary osmotic gradient in the collecting ducts before exiting the kidney as definitive urine. It is in this passage that final concentration occurs. The walls of the collecting ducts in the concentrating kidney are permeable to water. Thus as the fluid moves down through the increasing osmotic concentrations of the medulla, it comes to osmotic equilibrium with the interstitial spaces by losing water and exits the collecting duct at about the osmotic concentration of the inner end of the medulla. The changes in osmotic pressure of the tubular fluid in the concentrating kidney are summarized in Figure 7–12.

It can be shown that the magnitude of the osmotic gradient established in the medulla depends, among other things, on three parameters of the loops of Henle. The magnitude increases with the magnitude of the single effect, increases with the length of the loops, and decreases with increases in the rate of fluid flow through the loops. The increase in the osmotic gradient with increase in the length of the loops allows a physiological understanding of the observation discussed earlier, that the thickness of the renal medulla and prominence of the papilla tend to increase with increasing ability to concentrate the urine.

The circulation of the medulla and its role in maintaining the medullary concentration gradient

Because the final concentration of the urine involves osmotic loss of water from the collecting ducts to the medullary spaces, we must wonder why this would not dilute the osmotic gradient established by the loops of Henle. Clearly, if water were allowed to accumulate in the medulla, the concentrating capacity of the kidney would be degraded. This does not happen because water is carried away by the medullary blood supply.

As depicted in Figure 7–5, the blood vessels of the medulla have a special geometry, forming long hairpin loops (*vasa recta*) that parallel the loops of Henle and collecting ducts. This in itself is significant in the maintenance of the medullary osmotic gradient. As the blood flows down into progressively more-concentrated levels of the medulla, it tends to equilibrate osmotically and ionically with the interstitial fluids by osmotic loss of water and diffusional uptake of ions. If the blood simply left the medulla at the renal pelvis, it is easy to see that the blood flow would degrade the medullary osmotic gradient by leaving water behind and carrying solutes away. Instead, the blood reverses direction and flows back up through the medulla, tending again to equilibrate with the interstitial fluids, this time by the reverse processes of gaining water and losing salts. The vasa recta are known as *countercurrent diffusion exchangers*.

Remember that the blood vessel walls are poorly permeable to the large protein solutes of the blood. These solutes cannot reach concentration equilibrium across the vessel walls and create a colloid osmotic pressure in the blood. If all permeable solutes (such as ions) were to reach concentration equilibrium across the vessel walls, the blood would still not be at osmotic equilibrium with its surroundings. The colloid osmotic pressure would continue to cause osmotic uptake of water into the blood. This effect can be opposed or even overridden by hydrostatic pressure differences between the blood and surrounding tissues, as we have seen in the glomerulus. In the medullary vasculature the blood pressure is sufficiently reduced that the colloid osmotic pressure exceeds the hydrostatic pressure differential across the vessel walls, meaning that the net effect of the interplay of these forces is to cause osmotic uptake of water. As indicated above, the blood in the vasa recta goes through a dynamic series of osmotic exchanges with the interstitial fluids of the medulla, losing water as it flows down into the medulla and gaining water as it flows out again. The colloid osmotic pressure introduces a bias for the blood to gain more water than it loses over this whole series of events, and the volume of blood emerging from the vasa recta exceeds the volume entering. It is through this process that the blood takes up and carries away water added to the medulla from the collecting ducts.

An integrated review of events in the concentrating kidney

It is appropriate now to review the sequence of events in the nephrons and collecting ducts of the kidney producing concentrated urine with particular reference to the kidneys of man (see again Figure 7–12 for an overview). The fluid introduced into the Bowman's capsule by ultrafiltration is approximately isosmotic to the blood plasma and contains similar concentrations of inorganic ions, glucose, amino acids, and other solutes of low molecular weight. Along the proximal convoluted tubule, some 65–85% of the sodium is actively resorbed, chloride following in response to the electrical potential gradient. The proximal convoluted tubule is relatively permeable to water, and water is resorbed osmotically so that the tubular fluid remains essentially isosmotic with the plasma. Glucose, many amino acids, and potassium are virtually completely resorbed by active mechanisms in the proximal convoluted tubule. The water and solutes resorbed from the tubule are returned to the blood in the peritubular capillaries.

By the time the tubular fluid reaches the beginning of the loop of Henle, it has been reduced in volume by some 50–80% but remains approximately isosmotic to the plasma (about $300\,mOs$). The fluid then moves through the loop of Henle, reaching a maximal concentration at the bend of the loop. There is some net loss of sodium, chloride, and water during the passage through the loop of Henle. Approximately 5–20% of the originally filtered volume of water is lost in the loop, and the fluid emerges from the ascending limb hyposmotic to the blood (perhaps 100–200 mOs).

The distal convoluted tubule of the concentrating kidney is relatively permeable to water in some mammals, and the tubular fluid again reaches isosmoticity with the plasma through osmotic loss of water during its passage through the distal tubule. There is a consequent reduction of volume so that, considering the volume reductions throughout the nephron, only about 5–15% of the volume originally filtered reaches the collecting duct. There is a relatively modest active resorption of sodium in the distal tubule. Potassium is added to the fluid in the distal tubule. Because the tubular fluid is virtually free of potassium when it enters the distal tubule, this process in large part controls the amount of potassium eliminated in the urine.

As the tubular fluid moves down the collecting duct through the medulla, it is progressively concentrated by osmotic loss of water, with consequent reduction in volume. Sodium is actively resorbed along the collecting duct, and chloride follows passively. This further reduces the amounts of these ions excreted and enhances the osmotic reduction in volume by reducing the solute load of the tubular fluid. The effect of the osmotic equilibration of the tubular fluid with the medullary interstitial fluid is to concentrate solutes in the tubular fluid—primarily inorganic ions—so that their *collective* osmotic concentration equals the osmotic concentration of salts established in the interstitial fluids by the loops of Henle. The tubular fluid contains a diversity of ions (e.g., Na^+, K^+, NH_4^+, Ca^{2+}, Cl^-, and SO_4^{2-}), and its particular ionic composition is determined by those renal tubular mechanisms that adjust the amount of each ion excreted according to demands for maintaining proper concentrations in the body fluids. It is important to keep the dimensions of ionic and osmotic regulation distinct. The amounts of ions excreted must be compatible with ionic homeostasis. The countercurrent system in the concentrating kidney is essentially a mechanism for minimizing the water loss associated with the necessary excretion of ions.

In the end, 1% or less of the originally filtered volume is excreted. It may at first seem strange that the kidney should filter a considerable quantity of fluid only to resorb 99% or more of it, but this assures that the nephrons will have adequate minute-by-minute access to the body fluids to perform their vital functions of regulating body fluid composition.

Against the background of the events described here, a high urea concentration is typically established in the urine produced by the concentrating kidney. The mechanism of urea concentration is discussed in a later section.

Regulation of urinary concentration: the concentrating and diluting kidney

Thus far we have discussed the kidney producing concentrated urine. Mammals can also produce urine hyposmotic to the plasma. The osmotic concentration of human urine, for example, can be as low as 15% of the plasma concentration. In producing hyposmotic urine, the amounts of various solutes excreted are again governed by the need to maintain a proper level of each particular solute in the body fluids, but the water loss is increased. On a steady diet a person will excrete much the same quantities of solutes each day. If he has taken in little water, the solutes will be excreted in a relatively small amount of concentrated urine, but if he has taken in an abundance of water, the solutes will be excreted in a copious, dilute urine.

Antidiuretic hormone from the posterior pituitary plays a major role in regulating the concentration and volume of urine. The principal demonstrated effect of ADH is to increase the permeability of the collecting ducts to water. It has a similar effect on the distal convoluted tubules in some species (such as the laboratory rat), but not in others (such as the dog).

In the concentrating kidney, as discussed earlier, the collecting ducts, and in some cases the distal convoluted tubules, are relatively permeable to water. This permeability to water results from high blood levels of ADH and permits the renal tubular fluid to come to osmotic equilibrium with surrounding interstitial fluids, resulting in concentration of the tubular fluid and osmotic reduction in volume.

When blood levels of ADH are low, the collecting ducts and distal convoluted tubules are poorly permeable to water. This has the important result that the tubular fluid does not come to osmotic equilibrium with the surrounding interstitial fluids. Recall that fluid exiting the ascending limb of the loop of Henle is hyposmotic to the plasma. When, in the absence of ADH, the walls of the distal convoluted tubule are poorly permeable to water, this hyposmoticity is maintained through the distal convoluted tubule, and there is relatively little osmotic reduction in volume. Much the same can be said for the collecting ducts. In addition, you will remember that salts are actively resorbed from the distal convoluted tubule and, especially, the collecting ducts. When ADH levels are low, this resorption effectively lowers the solute concentration of the tubular fluid, and the fluid becomes more and more hyposmotic as it approaches the renal pelvis. In the absence of ADH, then, the tubular fluid is relatively isolated osmotically from its surroundings, with the important consequences that osmotic reduction in volume is limited and hyposmoticity can persist and be enhanced. The urine is dilute and copious. In man output may reach 15% of the filtered volume.

The principal factor controlling the release of ADH by the posterior pituitary is the osmotic concentration of the plasma, apparently sensed by osmoreceptors in the hypothalamus. An increase in plasma osmotic pressure leads to increased release of ADH, which in turn effects water conservation by the kidney.

When the kidney converts from producing concentrated urine to producing dilute urine, an additional important change is that the osmotic gradient established in the renal medulla by the loops of Henle is diminished. Conversely, in the transition from dilution to concentration, the medullary gradient must be amplified again. In an experiment on dogs administered excess water by stomach tube, the osmotic concentration of the inner medulla dropped from about 2400 m*Os* to about 500 m*Os* as the animals entered a diuretic state. Osmotic gradients across the collecting ducts in the diluting kidney are thus smaller than they would be in the absence of any change in the medullary gradient. Remember also that in the concentrating kidney the urine can only attain the osmotic concentration of the medullary interstitial spaces. Thus, in the dogs cited, if the collecting ducts of diuretic animals were rendered permeable to water, the maximal urine concentration would be only 500 m*Os* until the medullary osmotic gradient increased again.

The mechanisms of control of the medullary osmotic gradient are poorly understood. Considerable attention has been directed to medullary blood flow. An increase in the rate of flow through the medullary vessels (vasa recta) would tend to "wash out" the osmotic gradient by reducing the effectiveness of the countercurrent diffusion exchange which is believed to be important in preserving the gradient. To illustrate one dimension of this, the faster the blood flows down into the medulla, the less chance it has to come to osmotic equilibrium with each successive level of the medulla. This could mean that under conditions of rapid flow, blood reaching the deeper levels of the medulla would be more dilute than under conditions of slower flow. Then during rapid flow the osmotic gradient between the blood and medullary interstitium would be enhanced, and more water would be lost osmotically into the deep levels of the medulla. The efficiency of all countercurrent mechanisms tends to be reduced as rate of flow increases. There is still considerable debate over whether medullary blood flow is altered during transitions between diuresis and antidiuresis. Inasmuch as total renal blood flow is not altered, changes in medullary flow would have to result from redistribution of flow within the kidney. ADH is known to cause vasoconstriction in some vessels in mammals, and it has been suggested that increased blood levels of ADH during antidiuresis could effect a decrease in medullary flow. There is also some evidence that stimulation of renal nerves can cause a decrease in cortical flow and increase in medullary flow. This might occur during diuresis.

Another hypothesis to explain changes in the medullary osmotic gradient is that there may be a redistribution of blood flow to the glomeruli of the various types of nephrons. According to this suggestion, the cortical nephrons predominate during diuresis whereas the juxtamedullary nephrons predominate during antidiuresis. You will recall that the flow of fluid through the long loops of the juxtamedullary nephrons is essential to the establishment and maintenance of steep osmotic gradients in the medulla. If the mechanism of redistribution of blood flow to various types of nephrons should be important, it could help to explain why rodents with virtually all long-looped nephrons have difficulty in achieving diuresis sufficient to eliminate excess water loads.

Finally, it has been suggested that changes in the rate of active salt transport might be involved. Decreased transport during diuresis could reduce the single-effect osmotic gradient in the loops of Henle and thereby impair the maintenance of the medullary osmotic gradient. It has been suggested that ADH, perhaps acting through cyclic adenosine monophosphate, might stimulate active salt transport in the loops. Clearly, much remains to be learned about factors that control the magnitude of the medullary osmotic gradient.

Effects of adrenal hormones

Adrenal cortical steroids, particularly aldosterone, are known to affect active ion transport in the mammalian kidney. Increased hormone levels stimulate increased sodium resorption and increased potassium secretion. Sodium transport, of course, has important effects on chloride and water movements in many parts of the renal tubules. Thus, for example, a deficiency of aldosterone leads not only to decreased sodium resorption, but also to decreased chloride and water resorption from the renal tubular fluid; urinary losses of sodium, chloride, and water are then increased. Whereas the adrenal corticosteroids exert important controls on renal function and body fluid composition, they act on a fairly long time scale and do not appear to be involved, for example, in routine transitions between diuresis and antidiuresis.

An interesting recent study has shown that rabbits living in a sodium-poor region of Australia have much lower urinary sodium levels than rabbits living in nearby sodium-rich areas. This can be related to differences in dietary input of sodium. Probably aldosterone is involved in regulating urinary sodium output. The adrenals of the rabbits from the sodium-poor region are larger than those of the other rabbits, and circulating aldosterone levels are three to six times greater.

Correlations between habitat and storage and secretion of ADH

It is of interest that some evidence on rodents from arid and mesic habitats indicates greater capacities to synthesize and store ADH in the arid species. This evidence takes three forms. First, the posterior pituitary, a site of ADH storage, is larger relative to total pituitary size in the arid species. Second, the amount of ADH stored in the posterior pituitary is also larger (expressed relative to body weight). Third, under conditions that elicit production of concentrated urine, the arid species are less likely to deplete their stores of ADH than the mesic species. The greater capacities of arid species to synthesize and store ADH are probably related to the fact that their kidneys are maintained in an antidiuretic state more of the time. The arid species may also tend to maintain higher circulating levels of ADH. In laboratory rats and a number of other rodents, there are increases in hypothalamic neurosecretory activity and in the size of the posterior pituitary when animals are exposed to dehydrating circumstances.

Basic similarities between the mammalian and amphibian nephron

It is fascinating to note that the mammalian nephron has been able to accomplish important new tasks while retaining some of the physiological attributes that were laid down early in vertebrate evolution. You will recall that sodium is actively resorbed all along the amphibian nephron. Mammals differ from amphibians in being able to produce hyperosmotic urine. Hyperosmoticity could theoretically be established by actively secreting solutes into the renal tubules across membranes that are poorly permeable to water. This would involve a reversal of the direction of active sodium transport from the amphibian condition. Instead, mammals have achieved hyperosmoticity while retaining the primitive condition by throwing the nephrons into arrays of long, parallel loops in which countercurrent multiplication can occur. Sodium, where it is actively transported (or moves in tandem with active chloride transport), is still resorbed all along the nephron.

We may also note the basic similarity of ADH action in the mammalian and amphibian kidney. In both cases ADH increases permeability to water across tubular membranes that are otherwise poorly permeable. This increase in permeability has the important result that the tubular fluids come to osmotic equilibrium with the fluids surrounding the tubules. In amphibians the fluids surrounding the distal tubules are osmotically similar to the general body fluids, and in the presence of ADH, the urine approaches isosmoticity with the body fluids. Mammals have retained the same basic action of ADH but have surrounded the collecting ducts with fluids that are hyperosmotic to the general body fluids. Thus the urine is rendered hyperosmotic in the presence of ADH.

Concentration of urea in the concentrating kidney

The operation of the countercurrent multiplier system in the medulla of the concentrating kidney, discussed heretofore, can be understood in principle without taking urea into account. We must now, however, add an important new dimension to the operation of the concentrating kidney, that of urea concentration. Urea is typically highly concentrated in the urine of antidiuretic mammals and can contribute half or more of the total urinary osmotic pressure (see Chapter 6). Urea is concentrated in the urine by a different mechanism than salts. Urea is present at high concentration in the medullary interstitial spaces of the concentrating kidney, and the medullary collecting ducts are permeable to urea. Basically, then, high urea concentrations are established in the incipient urine by diffusion of urea to concentration equilibrium across the walls of the collecting ducts.

The countercurrent multiplier in the loops of Henle establishes high salt concentrations in the medullary interstitial spaces and thus also establishes a high osmotic pressure due to salts. The high medullary urea concentration also contributes to medullary osmotic pressure, so that the total medullary osmotic pressure is far higher than that explained by high salt concentrations. The incipient urine in the collecting ducts comes to osmotic equilibrium with the total osmotic pressure of the medullary interstitial spaces. Accordingly, urea plays a most important role in attaining high urinary osmotic pressures, and, in fact, mammals characteristically attain their highest urinary osmotic pressures when they are excreting urea at high concentration. Salts alone cannot be concentrated in the urine to establish osmotic pressures as high as those seen when the urine contains high loads of both salts and urea. You will recall that tubular fluid arriving in the medullary collecting ducts of the concentrating kidney is of lower osmotic pressure than the medullary interstitium and thus loses water to the interstitium by osmosis, a process that acts to reduce the amount of water excreted. The extent of osmotic water resorption is governed by the osmotic gradient between the tubular fluid and interstitial fluid. As noted above, the evidence is that urea diffuses to concentration equilibrium across the walls of the collecting ducts. We discussed in Chapter 4 the fact that once a particular solute has reached concentration equilibrium across a membrane, it will no longer contribute to a gradient of osmotic pressure across the membrane. Accordingly, though urea contributes substantially to the *absolute* osmotic pressures of both the tubular fluid and interstitial fluid, it does not, except transiently, contribute to an osmotic gradient across the membranes of the collecting ducts. Salts, by contrast, do not diffuse to concentration equilibrium across the walls of the collecting ducts. The fact that the total osmotic concentration of ions is initially lower in the tubular fluid than in the medullary interstitium thus creates a gradient of osmotic pressure that cannot be abolished by solute diffusion but instead is abolished by osmotic loss of water from the tubular fluid. Accordingly, it is the high concentration of salts in the medullary interstitium that exerts predominant influence on the reduction of tubular volume in the medullary collecting ducts, and it is thus that we have previously discussed the adjustment of water content in the context of the countercurrent multiplier system without reference to urea. The osmotic loss of water from the tubular fluid essentially acts to bring the total osmotically effective concentration of salts in the incipient urine up to that in the medullary interstitium, so that, considering both salts and urea, the total osmotic pressure established in the urine approximately equals that in the medulla. An important outcome of the processes outlined here is that salts can be concentrated (collectively) in the urine to much the same extent regardless of the urea concentration. This is a consideration of much importance to water economy. The animal can minimize the urinary volume needed to excrete salts by concentrating salts maximally, and that same volume of urine can, in addition, carry a high load of urea.

The question of how high urea concentrations are established and maintained in the medulla is not fully answered at present, and it would be beyond the scope of this text to go into the issue in detail. Suffice it to say that something around twice as much urea is filtered into the Bowman's capsules as is excreted per unit time, and the excess filtered urea is lost by diffusion from the tubular fluid along parts of the renal tubules, concentration gradients favorable to outward diffusion being established by osmotic loss of water from the tubular fluid and consequent concentration of urea contained in the fluid. Some of the urea that is lost enters the medullary spaces and tends to accumulate there, thus establishing the relatively high medullary concentrations. Tubular fluid arriving in the medulla in the collecting ducts actually has a higher urea concentration than the medulla owing in part to concentration of tubular urea through osmotic withdrawal of water; the tubular fluid loses urea to the medullary interstitial spaces by diffusion, but because the medullary concentration is high, the equilibrium concentration established in the incipient urine is also high. The accumulation of urea in the medulla can be explained satisfactorily without invoking active urea transport, and although there is some evidence that the mammalian kidney can transport urea actively, prevailing opinion holds that only passive urea movements are usually involved.

BIRDS

The avian kidney has received much less study than the mammalian, but it appears that countercurrent multiplication is the mechanism of urinary concentration. Most nephrons in the avian kidney are of the reptilian or amphibian type, lacking a loop of Henle. Other nephrons have loops of Henle, which penetrate inward in the kidney, running roughly parallel to each other and to the collecting ducts. The loops possess a thin segment as in the mammalian kidney, but all the loops are relatively short, resembling the cortical nephrons of mammals. We have noted earlier that birds cannot achieve urinary concentrations as high as those seen in many mammals. This can be related to the small proportion of nephrons that have loops of Henle and to the shortness of the loops. In two subspecies of Savannah sparrow, *Passerculus sandwichensis beldingi* and *P. s. brooksi*, and in the house finch, *Carpodacus mexicanus*, mean maximal osmotic U/P ratios have been reported, respectively, as 5.8, 3.2, and 2.3. A recent study has indicated that *P. s. beldingi* has about twice as many loops of Henle as *P. s. brooksi*, and *P. s. brooksi*, in turn, has about 50% more than the house finch. The correlation between concentrating ability and the number of loops supports the hypothesis that countercurrent multiplication is the mechanism of urinary concentration in birds.

Uric acid, the principal nitrogenous end product of birds, is actively secreted into the renal tubules from the peritubular capillaries. The amount of uric acid excreted thus exceeds the amount filtered. Active secretion of uric acid is also reported among reptiles. Uric acid precipitates as the volume of tubular fluid is reduced by water resorption. The ureters of birds are peristaltic, and it is postulated that this is important to moving the uric acid sludge down into the cloaca.

CRUSTACEANS

In moving on to consider the invertebrates, we shall first examine two groups—the crustaceans and molluscs—that are believed to possess ultrafiltration systems.

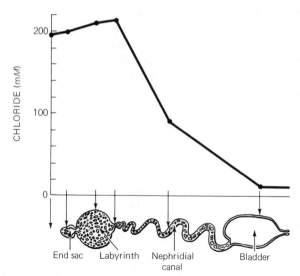

Figure 7–13. Mean chloride concentrations along the antennary gland of a crayfish (*Astacus fluviatilis*). The arrows pointing downward from the abscissa indicate where samples for chloride determination were obtained. The concentration recorded directly on the ordinate is for blood. Other concentrations were measured on fluid drawn from the renal duct. Studies on *Orconectes* have shown that chloride, sodium, and potassium are resorbed all along the duct. The diagram of the antennary gland is distorted. In life, the nephridial canal is tightly convoluted and closely juxtaposed to the labyrinth. [From Parry, G. 1960. Excretion. *In:* T. H. Waterman (ed.), *The Physiology of Crustacea.* Vol. I. Academic, New York; based on data of Peters, H. 1935. Z. Morphol. Oekol. Tiere 30: 355–381.]

The renal organs of adult decapod crustaceans are a pair of *antennary glands* or *green glands* located in the head and opening to the outside independently near the bases of the second antennae (see Figure 5–2). The basic structure of the antennary gland in freshwater crayfish is illustrated at the bottom in Figure 7–13. The renal duct begins with a closed end sac lying to the side of the esophagus. Following the end sac is the labyrinth, or green body, a highly convoluted canal. The nephridial canal leads from the labyrinth to an expanded bladder, and the bladder empties to the outside. The nephridial canal is found in only certain freshwater crustaceans. The modern marine forms, lacking the nephridial canal, are believed to represent the primitive crustacean condition, and it is thought that the canal made its appearance when crustaceans migrated from the sea into fresh water. The various parts of the renal duct vary morphologically among species. The labyrinth is lacking in some shrimps, and in hermit crabs the bladder ramifies into a complex tubular network that penetrates much of the thorax and sends diverticula into the abdomen. The physiological significance of much of this morphological diversity is unknown.

The walls of the end sac are thin. Arteries coming fairly directly from the heart supply a network of small vessels or lacunae on the outer surface of the end sac, though a true capillary bed has not been demonstrated. This morphological evidence has long suggested that fluid enters the end sac by ultrafiltration. There is physiological evidence to support this hypothesis in both marine and freshwater forms, and it is generally thought that ultrafiltration is sufficient to account for the observed rates of entry of fluid into the end sac.

The composition and volume of the tubular fluid may be altered during passage through the labyrinth, but all evidence indicates that the fluid remains essentially isosmotic to the body fluids. As shown in Figure 7–13, it is in the nephridial canal that osmotic pressure and salt concentration are lowered in crayfish, the mechanism apparently being active resorption of sodium or chloride, or both, across membranes that are poorly permeable to water. There is also considerable water resorption in the crayfish renal duct so that urinary volume is generally only 30–50% of filtration volume.

The nephridial canal is apparently necessary for production of hyposmotic urine. Marine crustaceans, lacking the nephridial canal, produce isosmotic urine. Similarly, freshwater forms, such as *Palaemonetes varians* and *Eriocheir sinensis*, that produce virtually isosmotic urine lack the nephridial canal. The crayfish and freshwater gammarids, known to void hyposmotic urine, have the canal. In certain respects the nephridial canal is functionally analogous to the amphibian distal convoluted tubule.

There are suggestions that sodium or chloride, or both, are actively resorbed along the renal duct of certain marine crabs, probably in the labyrinth. Isosmoticity persists because of concomitant osmotic water resorption, and fluid volume is reduced. There is evidence in various crabs for active resorption of calcium and glucose and for secretion of magnesium and sulfate. Interestingly, the bladder is the site of magnesium secretion and glucose resorption in the shore crab *Pachygrapsus*. The mechanism of producing slightly hyperosmotic urine in certain shore crabs is unknown. It has been suggested that parts of the labyrinth may be poorly permeable to water in these crabs and that hyperosmoticity is established by active ion secretion across those parts.

MOLLUSCS

The renal organs of molluscs are termed *kidneys* or *nephridia*. They are tubular or saccular structures that empty into the mantle cavity or directly to the outside. Each nephridium generally connects to the pericardial cavity via a renopericardial canal. The morphology of the nephridia is highly variable among species. In the bivalves, most cephalopods, and some gastropods, there are two nephridia, but in most gastropods there is only one.

Excretion in Octopus

The physiology of the molluscan kidney is generally poorly known. The most extensive work has been done on the octopus *Octopus dofleini*, and although the renal complex of the cephalopods includes several unusual features, we shall discuss this work first.

Cephalopods possess branchial hearts (generally two) in addition to the usual systemic heart. These receive venous blood from the body and pump it through the gills (see Chapter 10). The nephridia are associated with the branchial hearts, as diagramed schematically in Figure 7–14. Each branchial heart bears a relatively thin-walled protuberance, the branchial heart appendage, which communicates with the lumen of the heart. In *Octopus* the pericardium encloses only the side of the branchial heart bearing the heart appendage. A long renopericardial canal leads from the peri-

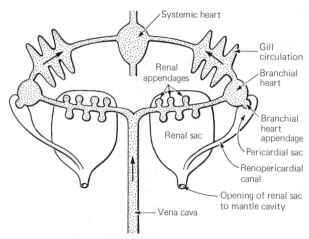

Figure 7–14. Schematic representation of the renal complex and associated circulatory system in *Octopus*. The parts of the circulatory system are stippled grey. Arrows show the direction of blood flow. Fluid is driven into the pericardial sacs by ultrafiltration across the branchial heart appendages and flows to the renal sacs through the renoperi-cardial canals. The renal sacs discharge into the mantle cavity. See text for further discussion. [From Martin, A. W. and F. M. Harrison. 1966. Excretion. *In:* K. M. Wilbur and C. M. Yonge (eds.), *Physiology of Mollusca*. Vol. II. Academic, New York.]

cardial cavity to the large renal sac. The sac, in turn, empties into the mantle cavity.

As noted, the connection of the nephridium with the pericardial cavity is a common molluscan feature. It has long been thought that fluid is introduced into the molluscan nephridium by ultrafiltration from the blood across the heart wall into the pericardial cavity. This hypothesis has been particularly appealing in view of the close association of the nephridia with the very source of blood pressure. The only clearly demonstrated example that such ultrafiltration occurs is in *Octopus*. Various evidence indicates that the pericardial fluid is an ultrafiltrate of the blood, the site of filtration being the branchial heart appendage and perhaps other parts of the heart wall enclosed in the pericardium. The filtrate flows to the renal sac via the renopericardial canal. There is good evidence of active resorption of glucose and amino acids along the canal. If inulin is administered to *Octopus*, the concentration in the final urine is about the same as that in the fluid introduced to the pericardial cavity, indicating little water resorption in the nephridium.

As depicted in Figure 7–14, blood approaches the branchial hearts in a large vein, the vena cava, which divides to form two lateral branches that enter the hearts. Blood leaving the hearts goes directly to the gills via the afferent branchial vessels. The lateral branches of the vena cava pass by the renal sacs and there bear many glandular diverticula, the so-called renal appendages, which are closely applied to the walls of the sacs. These appendages have long been thought to be sites of exchange between the blood and the fluid in the renal sacs. Cephalopods excrete considerable quantities of ammonia in their urine. Recent evidence on *Octopus* indicates that ammonia diffuses into the renal sacs from the blood across the renal appendages and effectively becomes trapped in the urine by reaction to form ammonium ion. Interestingly, much of this ammonia is chemically released

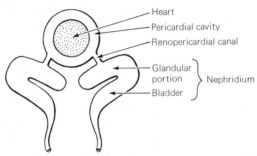

Figure 7-15. Schematic representation of the renal complex of bivalve molluscs. [From Martin, A. W. and F. M. Harrison. 1966. Excretion. *In:* K. M. Wilbur and C. M. Yonge (eds.), *Physiology of Mollusca.* Vol. II. Academic, New York; in turn, after Goodrich, E. S. 1945. Quart. J. Microsc. Sci. 86: 113–392.]

in the renal appendages themselves, at least in part through enzymatic breakdown of circulating glutamine by glutaminase. As a consequence, blood leaving the region of the renal appendages often has a higher ammonia concentration than blood entering, despite considerable diffusion of ammonia into the urine. It is significant that the blood travels directly to the gills from the branchial hearts, and there blood ammonia is lost by diffusion into the water in the mantle cavity. More ammonia is lost across the gills than in the urine.

Excretion in bivalves and gastropods

Bivalves and gastropods do not possess branchial hearts, and their nephridia open into the pericardial cavity surrounding the systemic heart (e.g., see Figure 7–15). Frequently, there are so-called pericardial glands on the wall of the heart or pericardium which are thought to play a role in producing the urine or altering its composition. These are of very variable morphology. Some, for example, are simple, thin-walled sacculations on the heart wall containing secretory cells, whereas others consist of a great complex of tubules opening into the pericardial cavity through the pericardium. Some types of pericardial glands may be homologous with the cephalopod branchial heart appendage. The function of pericardial glands is little known. Structures resembling the renal appendages of the vena cava in cephalopods are not present in other molluscs.

Formation of the urine by ultrafiltration in bivalves and gastropods has been indicated by studies on the composition of the pericardial and nephridial fluids and by data on inulin excretion. There is evidence that the heart wall is the site of ultrafiltration in bivalves and nonpulmonate gastropods. However, contradictory evidence has been provided recently in a study of five species of bivalves, in which it was found that the blood in the heart is somewhat hyperosmotic to the pericardial fluid and that the blood pressure is insufficient to cause ultrafiltration across the heart wall against the osmotic gradient. (Blood pressures in bivalves and gastropods are much lower than those in cephalopods.) Among the pulmonate snails there is convincing evidence that the renal sac, not the heart wall, is the site of introduction of fluid into the kidney; the renal sac is an enlarged chamber of the kidney into which the renopericardial canal (coming from the pericardium) leads. Though the fact that inulin appears in the fluid of the pulmonate renal sac at concentrations established in the blood would indicate that ultrafiltration is the mechanism by which fluid is introduced into the sac, the site of ultrafiltration has not been identified, and it remains possible that secretion, not ultrafiltration, is the mechanism involved. Clearly, there is much to be learned about the entry of fluid into the kidneys in these groups.

Freshwater mussels, *Anodonta*, have received considerable study. As in other bivalves, there are two nephridia that communicate with the pericardial cavity (Figure 7–15). The nephridia are U-shaped. The first limb, leading away from the pericardial cavity, is glandular; and the second, which doubles back and empties to the mantle cavity, forms a thin-walled bladder. Fluid in the pericardial cavity of *A. cygnea* has the composition expected of an ultrafiltrate of the blood. Concentrations of ions and total osmotic concentration are similar to blood values, whereas protein concentration is lower. If the pericardium is punctured, fluid drips out at a steady rate. These observations suggest ultrafiltration into the pericardial cavity, but ultrafiltration across the heart wall is contraindicated in another species, *A. imbecilis*, by evidence that cardiac blood pressure is inadequate. The composition of the urine is altered on its passage through the nephridia in *A. cygnea*. The osmotic concentration of bladder urine is lower than that of the blood or pericardial fluid by 45%, indicating establishment of hyposmoticity in the nephridia. Active resorption of ions is suggested. There is little resorption of water. Protein and nonprotein nitrogenous materials are added to the urine in the nephridia.

ANNELIDS

In polychaete and oligochaete annelids the usual arrangement is two renal tubules, or *nephridia*, per body segment—either in most or all of the body segments or in the segments of a limited part of the body. The morphology of the nephridia is highly variable among species. Typically, a nephridium is an unbranched, more or less convoluted tubule. The tubule may be differentiated, to a greater or lesser extent, into morphologically distinct regions. The nephridia project into the coelomic cavity, and usually each nephridium opens directly to the coelom at its coelomic end via a ciliated funnel (nephrostome). At the other end the nephridium opens directly or indirectly to the outside. Most commonly, each nephridium discharges via a pore through the body wall (nephridiopore), but in some earthworms the nephridia discharge into the gut.

The mode of introduction of fluid into the renal tubules awaits careful study. Ultrafiltration across nephridial blood vessels may occur but has not been investigated. In many species the tubular fluid is probably largely or completely derived from coelomic fluid entering at the nephrostome. Insofar as this is the case, the nephridium, ostensibly, is not an ultrafiltration system. It has been suggested, however, that the coelomic fluid itself may be formed in part by ultrafiltration across coelomic blood vessels. According to this view the coelomic cavity would assume a role roughly analogous to the Bowman's capsule of vertebrates or the end sac of crustaceans, and the coelomic circulation, coelomic cavity, and nephridium together would constitute an ultrafiltration system. Available evidence is insufficient to resolve this issue.

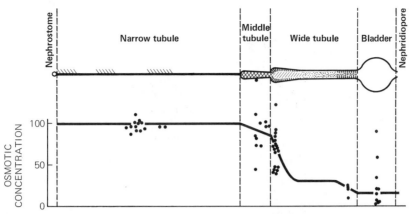

Figure 7–16. Osmotic concentration of the nephridial fluid in *Lumbricus terrestris.* Nephridia were dissected out in Ringer's solution and samples of nephridial fluid taken by micropuncture. Osmotic concentrations are expressed as percentages of the osmotic concentration of the surrounding Ringer's solution. The line drawn through the data represents the investigator's interpretation of the results. The diagonal dashed lines along the narrow tubule indicate ciliated portions of the tubule. (From Ramsay, J. A. 1949. J. Exp. Biol. 26: 65–75.)

Figure 7–16 depicts certain aspects of the anatomy and physiology of the nephridia in *Lumbricus terrestris,* a common nightcrawler. The nephridium consists of several morphologically distinct parts and is ciliated along much of its length. An expanded bladder is present in *Lumbricus* but is not a uniform feature of annelid nephridia. As indicated earlier, coelomic fluid probably enters the nephridium at the nephrostome. The osmotic pressure of the tubular fluid remains unchanged through the narrow tubule but falls in the middle and wide tubules, resulting in production of hyposmotic urine. The decline in osmotic pressure is due to resorption of sodium and chloride across membranes that are poorly permeable to water. Apparently sodium resorption is active, and chloride follows passively.

Interpretation of correlations between nephridial morphology and habitat in polychaetes and oligochaetes is hindered by lack of physiological information. Nephridia tend to be longer in freshwater and terrestrial forms than in marine forms. Perhaps the greater length of the nephridia in the former groups is functionally related to the capacity to form hyposmotic urine, as in crustaceans. As noted earlier, nephridia that discharge into the gut are found among certain earthworms. This may be an adaptation for water conservation inasmuch as the gut could resorb water that would be lost were the nephridia to discharge directly to the outside. An Indian earthworm, *Pheretima,* with nephridia opening into the gut is reported to be especially resistant to drought.

It is interesting to note here a basic similarity among the earthworm nephridium, the crayfish green gland, and the nephrons of many vertebrates. In all, the region in which hyposmoticity is established is near the end of the renal tubule which discharges to the outside. This may be fortuitous, but it may represent an adaptation for efficiency of renal function. Establishment of hyposmoticity demands transport of solutes against strong concentration gradients. Given an initial solution isosmotic to the body fluids, the *amount* of solute that must be transported to achieve a certain degree of hyposmoticity depends directly on the volume of fluid. In vertebrates it is known that the volume of tubular fluid can be greatly reduced along

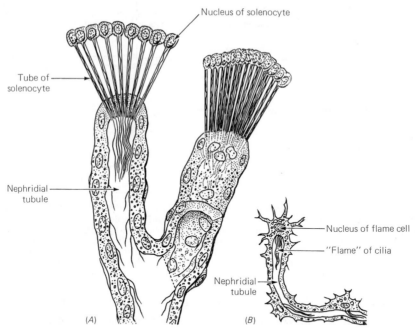

Figure 7–17. Inner ends of protonephridia. (*A*) Solenocytes as seen in a polychaete annelid (*Phyllodoce paretti*). (*B*) Flame cells as seen in a polyclad flatworm. (From Goodrich, E. S. 1945. Quart. J. Microsc. Sci. 86: 113–392.)

the early parts of the renal tubule (particularly in the proximal convoluted tubule), and there may be significant energetic advantages to postponing any effort at establishing hyposmoticity until this initial volume reduction has been achieved. Similar considerations may apply in the earthworm and crayfish.

SOLENOCYTES AND FLAME CELLS

In certain primitive groups of polychaete annelids, the nephridia do not open into the coelom at their inner ends but terminate blindly in clusters of specialized cells known as *solenocytes* (Figure 7–17). Each solenocyte consists of a rounded cell body with a long flagellum that projects into a very fine tubule. The tubule connects with the lumen of the nephridium. Nephridia ending blindly in solenocytes are also reported in the cephalochordate *Amphioxus* and in many larval annelids and molluscs. Among flatworms, rotifers, gastrotrichs, and several other groups the nephridia end blindly in *flame cells* or *flame bulbs* (Figure 7–17). The flame cell bears a tuft of cilia projecting toward the lumen of the nephridium; under the microscope the beating cilia give something of the appearance of a flickering candle, thus the name *flame cell*. Solenocytes are believed to be derived evolutionarily from flame cells, and nephridia ending in either type of cell are termed *protonephridia*.

The physiology of protonephridia is virtually unknown. There are several suggestions concerning how fluid could enter the nephridial tubule. Sometimes solenocytes or flame cells are in close association with blood vessels. Ultrafiltration caused by blood pressure is then a possibility. It

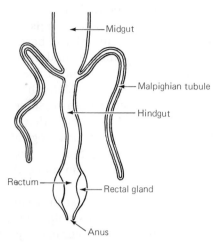

Figure 7–18. Generalized schematic representation of the posterior gut and Malpighian tubules in insects. The rectal glands are depicted as longitudinal thickenings of the wall of the rectum.

has also been suggested that the beating of the cilia or flagella could drive fluid away from the solenocytes or flame cells and create a negative pressure within the end of the nephridium, inducing ultrafiltration. Active secretion is another possibility for entry of solutes and water into the protonephridium.

There is evidence that in certain freshwater cercariae and rotifers, the rate of urine production by the protonephridia decreases as the salinity of the medium is increased. This indicates a role in regulation of body water content because osmotic influx of water will be reduced at higher salinities. Among rotifers and gastrotrichs, freshwater forms have long, coiled protonephridia, but marine forms lack the protonephridia altogether. In the euryhaline flatworm *Gyratrix hermaphroditus*, freshwater populations have more highly differentiated protonephridia than brackish-water populations, and protonephridia are absent in marine populations. These morphological correlations are reminiscent of those in crustaceans and annelids. Their physiological significance awaits study.

INSECTS

Anatomy

Most insects possess Malpighian tubules, and these are often referred to as the "excretory tubules." It is to be emphasized from the outset, however, that the lower gut plays such a central role in the formation of the urine that it must be included in any definition of the renal complex.

The Malpighian tubules typically arise from the junction of the midgut and hindgut (Figure 7–18). They are long and slender, end blindly, and number from two to several hundred, depending on species. There is a tendency for the tubules to be longer in species with fewer of them. The tubules project into the hemocoel and are bathed by the blood (hemolymph). The blind ends of the tubules may be free in the hemocoel, or they may be attached to the posterior hindgut or rectum. The walls of the tubules consist of a single layer of epithelial cells, supported on the outside by a prominent basement membrane. Bands of muscle cells often spiral around the tubules on the outside of the basement membrane, and muscular contractions cause the tubules to undergo writhing movements that may be important in moving fluid through the tubules or in mixing fluids inside or outside of the tubules. The Malpighian tubules show little histological differentiation along their length in some species but are differentiated into two to four distinct regions in others. In many insects the various tubules within an individual are morphologically similar. In others two or more types of tubules are present. For the most part the physiological significance of anatomical differentiation along the tubules or among the tubules is poorly understood.

The walls of the hindgut usually consist of a single layer of cuboidal epithelial cells. The rectum may present a similar appearance, but in many species there are groups of glandular, columnar epithelial cells termed *rectal glands* or *rectal pads* (Figure 7–18). The gross morphology of the rectal glands varies; in some species the columnar cells are oriented in longitudinal bands running along the rectum, whereas in others the cells form distinct papillae that project into the rectal cavity. In interspecific comparisons the ability to resorb large quantities of water from the excreta has been related to the presence of rectal glands.

Introduction of fluid into the Malpighian tubules

Ultrafiltration has long seemed unlikely as a mechanism of entry of fluid into the Malpighian tubules. The tubules are not supplied with blood vessels, and pressures in the hemocoel are low (see Chapter 10). Fluid collected from the Malpighian tubules does not have the composition expected of an ultrafiltrate of the blood.

Active secretion of potassium appears to play a central role in the formation of the tubular fluid. The exact sequence of events is not completely understood, but current thinking can be summarized as follows (refer also to Figure 7–2). Potassium is secreted from the blood into the tubular lumen. The concentration of potassium in the tubular fluid has been determined to be 6 to 30 times the blood concentration in various species. Sodium may also be actively secreted into the tubules, but tubular sodium concentrations generally remain below blood concentrations. Chloride probably accompanies the potassium and sodium passively. Water moves into the tubules so that the osmotic pressure of the tubular fluid remains close to that of the blood. In some forms, water appears clearly to enter the tubules osmotically, for the tubular fluid, owing to the active addition of solutes, remains somewhat hyperosmotic to the surrounding blood. In at least one species (*Carausius*), however, it has been shown that entry of water continues even when the tubular fluid is somewhat hyposmotic to the blood, meaning that, at least superficially, water is moving against an osmotic gradient; this remains unexplained. Amino acids, sugars, and many inorganic ions are believed to enter the Malpighian tubules passively according to concentration gradients established by water influx and electrical gradients arising from active ion transport. Large protein molecules are excluded, probably because of the permeability properties of the tubular epithelium. Uric acid, the principle nitrogenous end product of many insects, is actively secreted into the Malpighian tubules.

The fluid exiting the Malpighian tubules into the hindgut is, then, a complex solution, rich in potassium and similar to the blood in osmotic concentration. If this fluid were voided directly as urine, the insect's potassium reserves would be rapidly depleted, and the urine could play little or no role in osmotic regulation. Such observations serve to emphasize the importance of the lower gut in formation of the definitive urine.

Basic aspects of function
in the hindgut and rectum

The volume of fluid may be reduced to some extent in the hindgut, and there may be some changes in ionic composition; but it is in the rectum that osmotic and ionic composition are finely adjusted according to needs for regulating the composition of the body fluids. Urine exiting the rectum may bear little resemblance to the fluid that was introduced into the lower gut from the Malpighian tubules. Principal importance is attributed to the rectal glands.

Available evidence indicates that sodium, potassium, and chloride can all be actively resorbed in the rectum. Many insects can render the rectal fluids hyperosmotic to the body fluids. This phenomenon is a subject of considerable current interest to comparative physiologists. The establishment of hyperosmoticity cannot be explained by net addition of solutes to the rectal fluid. Rather, it seems clear that water is resorbed in excess of solute resorption. Hyperosmoticity can be established in the rectum of the locust *Schistocerca gregaria*, for example, simultaneously with *reductions* in the concentrations of sodium, potassium, and chloride. In one set of experiments the rectum of the locust was tied off from the hindgut and filled with a solution of trehalose, a carbohydrate that is neither resorbed nor secreted. Ion concentrations in the rectal fluid remained low during the experiments. The osmotic concentration of the rectal contents rose to be nearly three times the osmotic concentration of the blood, and the trehalose concentration apparently rose accordingly. Such experiments confirm that hyperosmoticity can be established by resorption of water against strong, opposing gross osmotic gradients between the rectal fluid and surrounding blood.

The possibility of active water resorption in the insect rectum has long been recognized, and the extensive studies on *Schistocerca* at first sight seem to confirm its occurrence. Historically, however, when active water transport has been claimed, intensive investigation has generally revealed the water movement to be passive, and at present it seems likely that the situation in the insect rectum will prove to be no exception.

Ultrastructural studies of the rectal glands in several species have revealed their anatomy to be complex. Though there are significant interspecific differences, we shall concentrate here on one example, the blowfly *Calliphora*. The rectal papilla of this species is diagrammed in simplified fashion in Figure 7–19. It can be seen that adjacent cells of the rectal epi-

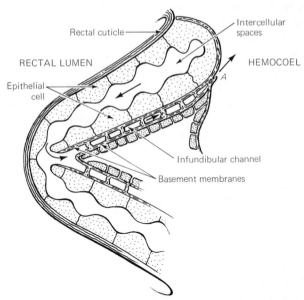

Figure 7-19. Highly diagrammatic representation of a rectal papilla of the blowfly *Calliphora erythrocephala*. The papilla is about 1 mm long and about 0.5 mm in diameter at its base. Each fly has four papillae. The epithelial cells run between the cuticle and outer basement membrane. The intercellular space, depicted for simplicity as a single broad cavity running through the epithelial cells, actually exists as a complex, interconnecting network of small channels and spaces *between* the epithelial cells. These channels and spaces ultimately communicate with the infundibular channels near the apex of the papilla. The infundibular channels, bounded by basement membranes (connective tissue), are traversed by bar-shaped bridges of basement material; these bridges do not form complete septa and do not prevent flow of fluid through the channels. Arrows depict the direction of hypothesized fluid movements. Fluid exits to general blood spaces (hemocoel) at the point marked *A*. There is evidence that blood cannot enter the infundibular channels, and it is hypothesized that the small flap of basement material at *A* acts as a valve, preventing backflow of blood. The anatomy of the rectal glands is basically similar in some other species that have received study but differs in detail. (After Gupta, B. L. and M. J. Berridge. 1966. J. Morphol. 120: 23–82.)

thelium are tightly joined on the side facing the rectal lumen and on the opposite side (basal side), but in between they are separated by fluid-filled intercellular channels and spaces. The intercellular spaces communicate with infundibular channels that run along the basal side of the epithelial cells and connect with general blood spaces. It has been hypothesized that active ion transport from the epithelial cells into the intercellular channels or spaces can render the fluid in the spaces hyperosmotic to the rectal fluid even when the blood is hyposmotic. Water would then tend to move from the rectal fluid into the intercellular spaces by osmosis, leaving the rectal fluid hyperosmotic to the blood. It is further hypothesized that the hyperosmotic fluid accumulated in the intercellular spaces flows into the infundibular channels and that active or passive ion resorption from the infundibular channels into the epithelial cells reduces the osmotic concentration of the fluid before it enters the blood spaces. Thus by recycling ions between the infundibular channels and intercellular spaces, water could be withdrawn from the rectal fluid by osmosis without introducing hyperosmotic fluid into the blood. The osmotic concentration of the rectal fluid would thus rise above that of the blood. There is some evidence that something like this actually occurs, though many details remain to be elucidated.

When water is being withdrawn from the rectal fluid of cockroaches (*Periplaneta*) or blowflies, the intercellular spaces swell, indicating that water enters them. More importantly, recent studies on dehydrated cockroaches have shown that the fluid in the intercellular spaces is hyperosmotic to that in the rectum, and the fluid in the analogues of the infundibular channels is hyposmotic. More work is needed before a definitive conclusion can be reached, and there are probably significant physiological differences among species. Such evidence, however, provides strong support for the hypothesis that hyperosmoticity is established in the rectal fluids of insects by passive water movements in conjunction with active solute movements rather than by active transport of water itself.

The function of the rectum in regulating excretion

We may now look briefly at some of the regulatory possibilities in the insect rectum. In general, the volume of fluid and the amounts of potassium, sodium, and chloride are reduced in the rectum. In the stick insect, *Dixippus morosus*, for example, approximately 95% of the sodium and 80% of the potassium exiting the Malpighian tubules are resorbed in the rectum. Considerable water resorption is indicated by the low water content of the excreta in many species. It has been calculated that in *Dixippus* the Malpighian tubules secrete the equivalent of all the body water each day and all the body potassium every three hours. The necessity of recovery in the rectum is clear. The resorption of potassium deserves emphasis. Potassium secretion provides the basis for entry of fluid into the Malpighian tubules, but the animal could ill afford to excrete the quantities of potassium that must be secreted. Potassium is recycled again and again through the system.

The urine can be rendered hyposmotic to the blood in some species. This occurs in the rectum and probably reflects active ion resorption in excess of water resorption. When the urine is rendered hyperosmotic, water resorption exceeds total solute resorption. That these processes are under control is indicated by adaptive adjustments in the osmotic concentration of the urine. In *Aedes detritus* larvae in seawater, for example, the urine is rendered hyperosmotic to the blood in the rectum. In larvae in fresh water, on the other hand, the urine becomes hyposmotic in the rectum. Other examples of rectal adjustment of osmotic pressure and evidence of hormonal control were discussed in Chapter 5.

Several studies have demonstrated rectal control of the ionic composition of the urine. Table 7–1 provides a comparison of the hindgut and rectal fluids of locusts (*Schistocerca gregaria*) deprived of food and supplied with either tap water or a saline solution containing sodium, potassium, chloride, and other ions. You will note that hindgut concentrations of ions were higher in the saline-fed animals. This was probably due, at least in part, to increased concentrations realized in the Malpighian tubules as a result of higher blood concentrations. The differences in composition between the rectal fluids of the two groups of animals are dramatic. Rectal fluid in the water-fed locusts was almost depleted of ions. It was hyperosmotic to the blood, probably because of high concentrations of organic solutes. Rectal fluid in the saline-fed locusts was much more concentrated in all ions than the blood or hindgut fluid, reflecting water resorption in

Table 7–1. Composition of the blood, hindgut fluid, and rectal fluid in *Schisto-cerca gregaria* deprived of food but provided with tap water or a saline solution. All values are means.

Experimental Treatment	Fluid	Osmotic Pressure ($\triangle$FP, °C)	Ion Concentrations (mM)		
			Cl⁻	Na⁺	K⁺
Water-fed	Blood	0.74	115	108	11
	Hindgut fluid	0.78	93	20	139
	Rectal fluid	1.52	5	1	22
Saline-fed	Blood	0.96	163	158	19
	Hindgut fluid	—	192	67	186
	Rectal fluid	3.47	569	405	241

SOURCE: Phillips, J. E. 1964. J. Exp. Biol. 41: 69–80.

excess of ion resorption. Its osmotic pressure was high. In addition, the saline-fed locusts accumulated much larger volumes of fluid in the rectum than the water-fed locusts. These results indicate that the ionic composition and volume of the urine are adjusted in the rectum according to environmental supplies of ions in a manner commensurate with maintenance of homeostasis in the body fluids.

Other groups with Malpighian tubules

Malpighian tubules are found among certain other terrestrial arthropods, namely, centipedes, millipedes, and spiders. There has been little study of the physiology of excretion in these groups.

SELECTED READINGS

Bentley, P. J. 1971. *Endocrines and Osmoregulation.* Springer-Verlag, New York.
Dicker, S. E. 1970. *Mechanisms of Urine Concentration and Dilution in Mammals.* Williams & Wilkins, Baltimore, Md.
Kirschner, L. B. 1967. Comparative physiology: invertebrate excretory organs. Ann. Rev. Physiol. 29: 169–196.
Kümmel, G. 1973. Filtration structures in excretory systems. A comparison. *In:* L. Bolis, K. Schmidt-Nielsen, and S. H. P. Maddrell (eds.), *Comparative Physiology.* North-Holland, Amsterdam.
Pitts, R. F. 1974. *Physiology of the Kidney and Body Fluids.* 3rd ed. Year Book Medical, Chicago.
Potts, W. T. W. 1967. Excretion in the molluscs. Biol. Rev. 42: 1–41.
Riegel, J. A. 1972. *Comparative Physiology of Renal Excretion.* Oliver and Boyd, Edinburgh.
Schmidt-Nielsen, B. 1964. Organ systems in adaptation: the excretory system. *In:* D. B. Dill (ed.), *Handbook of Physiology. Section 4: Adaptation to the Environment.* American Physiological Society, Washington, D.C.

See also references in Appendix.

8
EXCHANGES OF OXYGEN AND CARBON DIOXIDE: BASIC PRINCIPLES, RESPIRATORY ENVIRONMENTS, AND EXTERNAL RESPIRATION

We have seen that the organism is a self-sustaining organization that, far from being physically isolated from its environment, is in a continual state of exchange with the outside world. We have examined exchanges of energy, water, ions, and nitrogenous materials and now proceed to the exchanges of the respiratory gases. The uptake of oxygen and elimination of carbon dioxide are intimately related to the utilization of chemical energy contained in foodstuffs. Carbohydrates, lipids, and proteins are catabolized to compounds containing less chemical energy, and some of the energy thus released is incorporated into molecules of ATP, subsequently to be utilized in the various energy-demanding activities of the organism. Maximal energy yield from foodstuffs is realized by complete oxidation, which in turn entails operation of the electron transport system. The electron transport chain cannot function without oxygen, which serves as its final hydrogen receptor. The cells of most animals cannot survive for long without the operation of the electron transport system and for this reason demand a more or less steady supply of oxygen. The two major products of the complete oxidation of foodstuffs are carbon dioxide and water. This water is what we have previously described as metabolic water; it becomes part of the body's general water content, and its disposition can be understood as part of the overall water exchanges discussed in earlier chapters. Carbon dioxide is continually produced in the catabolism of foodstuffs but in animals, unlike plants, cannot be used in the synthesis of organic compounds. Carbon dioxide may be directed to formation of carbonates in skeletons or egg-shells, and in the body fluids certain concentrations of the members of the carbonate buffer system are often necessary for proper acid-base regulation. In general, however, carbon dioxide must be eliminated as it is produced, for accumulation in the body would disturb acid-base balance and exert other deleterious effects. For many animals the exchanges of the respiratory gases are the most urgent of all exchanges with the outside world. Man, for example, can live for many hours or days without exchanging nutrients, nitrogenous wastes, water, or ions, but he dies within minutes if denied oxygen.

CONCEPTS OF PARTIAL PRESSURE, CONCENTRATION, AND SOLUBILITY

You will recall from Chapter 4 that the partial pressure of a particular gas in a mixture of gases is the total pressure multiplied by the volume or mole fractional concentration of the gas in question. Ordinary dry air consists approximately of 20.95% oxygen, 78.1% nitrogen, 0.03% carbon dioxide, and small percentages of other gases such as argon, neon, and helium. The volume or mole fractional concentration of a gas is simply 1/100 of its percent concentration, and we find that

$$p = \frac{x}{100}\, P$$

where p is the partial pressure of the gas in mm Hg, x is its percent concentration, and P is the total pressure of the gas mixture, again in mm Hg. It is easy to compute that in ordinary dry air at 760 mm Hg pressure, the partial pressures of oxygen, nitrogen, and carbon dioxide are 159, 594, and 0.23 mm Hg, respectively.

When water vapor is added to air, the vapor pressure of the water must be taken into account if the partial pressures of the other components of the air are to be computed from knowledge of the composition of dry air. Suppose, for example, that air at 20°C and 760 mm Hg is saturated with water vapor. The water vapor pressure is 17.5 mm Hg, meaning that the other components together contribute only 742.5 mm Hg. Again, 20.95% of this latter pressure is due to oxygen, meaning that the partial pressure of oxygen in the saturated air is $(0.2095)(742.5) = 155.6$ mm Hg. In general, the partial pressure of a gas can be computed from its percent concentration in dry air (x) by the formula

$$p = \frac{x}{100}\, (P - vp)$$

where vp is the water vapor pressure. At a given total pressure, the partial pressures of oxygen, nitrogen, and other components of dry air decrease with increasing vapor pressure.

It is often important to know the concentration of a given gas. Concentration is the amount of the gas per unit volume. Expression of the concentration of oxygen in air, for example, in terms of weight of oxygen per liter or moles of oxygen per liter is straightforward. But if it is desired to express the concentration as the volume of oxygen per liter, the conditions of temperature and pressure under which the volume of oxygen is measured must be stated. Generally the volume is corrected to standard conditions of 0°C and 760 mm Hg (STP). This sometimes leads to some confusion. Consider, for illustration, dry air at 20°C and 740 mm Hg. This consists of 20.95% oxygen and therefore at the prevailing conditions contains about 210 cc of oxygen per liter. Correcting the volume of oxygen to standard conditions, we get $(273°K/293°K)(740 \text{ mm}/760 \text{ mm})(210 \text{ cc}) = 190 \text{ cc}$. Thus the air contains 190 cc O_2 at STP/liter. The two volumes, 210 cc and 190 cc, must be carefully distinguished. The 210 cc is the actual volume of oxygen that would be measured under prevailing conditions. However, if we are to compare the oxygen concentrations of various gas mixtures at various temperatures and pressures, we must express their concentrations of oxygen in comparable terms; and it is in this light that the expression at STP is indispensable. Concentrations expressed in weight/liter, mole/liter, and volume at STP/ liter are interconvertible by simple proportionalities.

Water vapor pressure, temperature, and barometric pressure all influence the concentration of oxygen, or any other gas, in air. Note, for example, that dry air at 0°C and 760 mm Hg contains 210 cc O_2 at STP/liter, whereas dry air at 20°C and 740 mm Hg contains only 190 cc O_2 at STP/liter, as computed earlier. The former gas contains more molecules of oxygen per liter than the latter. It is convenient to note that at any given temperature, the concentration of a gas is proportional to its partial pressure, regardless of the total pressure or the partial pressures of other gases. At given partial pressure the concentration of a gas decreases with increasing temperature.

Gases dissolve in aqueous solutions. Molecules of oxygen, for example, become distributed among water molecules in much the same way that the molecules of glucose or the ions of sodium chloride are incorporated among water molecules during dissolution of solids. The molecules of oxygen do not appear in the solution as tiny bubbles any more than sodium chloride in solution appears as tiny crystals. Bubbles of gas represent gas that is not in solution. The concentration of a gas in solution can be expressed in the usual units: weight/liter, moles/liter, and volume at STP/liter. Concentration is frequently expressed as milliliters of gas at STP per liter of solution. This is not to be taken as implying that the gas exists as a gas in solution. The gas does not occupy the STP volume stated in the expression ml/liter when it is in solution. Rather, dissolved gases can be extracted from solution into their gaseous phase by various means and then will occupy the stated volume at STP. If, for example, some water contains 2 ml O_2/liter, then 2 ml of gaseous oxygen (measured at STP) could be extracted from each liter of water.

If oxygen-depleted water is brought into contact with air containing oxygen at a partial pressure of 159 mm Hg, oxygen will dissolve in the water until a certain equilibrium concentration is achieved. The solution is then said to have a partial pressure of oxygen of 159 mm Hg. If this solution is brought into contact with air containing oxygen at 140 mm Hg, the solution will lose oxygen to the air until a new, lower equilibrium concentration is realized. The partial pressure of oxygen in the solution will then be 140 mm Hg. In general, *the partial pressure of any given gas in solution is precisely equal to the partial pressure of the same gas in a gas phase with which the solution is at equilibrium.* The partial pressure of a gas in solution is often called the gas *tension*. Some authors also apply the term *tension* to partial pressures of gases in gas mixtures.

As indicated above, there are limits to the solubility of gases in solutions. An aqueous solution exposed to a given gas-phase partial pressure will dissolve only so much gas. The solubilities of gases in solutions are expressed in a standardized way, namely, as the concentration, in volume of dissolved gas at STP/liter, when the solution is at equilibrium with a gas phase in which the partial pressure of the gas is 760 mm Hg (that is, when the partial pressure of the gas in solution is 760 mm Hg). This quantity is known as the *absorption coefficient*. The absorption coefficient of oxygen in distilled water at 0°C, for example, is about 49 ml/liter. This means that distilled water at 0°C will dissolve 49 ml of oxygen (at STP) per liter if allowed to equilibrate with an atmosphere in which the partial pressure of oxygen is 760 mm Hg. Absorption coefficients are tabulated in various standard reference works.

There are several important characteristics of gas solubility that can be elucidated by examination of absorption coefficients. The absorption coefficients of oxygen, nitrogen, and carbon dioxide in distilled water at 0°C are, respectively, 49, 24, and 1713 ml/liter. These values illustrate that the solubilities of various gases are different and, in particular, the solubility of carbon dioxide is far greater than the solubilities of oxygen and nitrogen in aqueous solutions. The absorption coefficients of oxygen in distilled water at 0°C, 20°C, and 40°C are, respectively, 49, 31, and 23 ml/liter. These values exemplify the important point that *solubilities of gases decrease strongly with increasing water temperature*. The absorption coefficients of oxygen at 0°C in waters of 0‰, 29‰, and 36‰ salinity are, respectively, 49, 40, and 38 ml/liter. Here we see that *increasing salinity decreases gas solubilities*.

The concentration of a gas in aqueous solution at any given partial pressure can be readily computed from the absorption coefficient according to the following formula:

$$C = p\,\frac{A}{760}$$

where C is the concentration in ml of gas at STP/liter, p is the partial pressure of the gas in solution in mm Hg, and A is the absorption coefficient in ml/liter. Clearly, the concentration of any given gas is related proportionally to its partial pressure in any given body of water. The absorption coefficient varies with the gas under consideration and with the temperature and salinity of the water. Thus the constant of proportionality, $A/760$, also varies with these parameters.

To illustrate the use of the preceding formula, consider distilled water at 0°C in equilibrium with ordinary dry air at 1 atm of pressure. As given earlier, the absorption coefficient of oxygen is 49 ml/liter, and the partial pressure of oxygen in the air and water is 159 mm Hg. We get that

$$\text{ml of } O_2 \text{ (at STP) per liter} = (159)\,\frac{49}{760} = 10.2$$

Similarly, we can compute the concentrations of nitrogen and carbon dioxide as 18.8 ml/liter and 0.5 ml/liter. Carbon dioxide, though it has the highest solubility, has the lowest concentration because of its low partial pressure.

Inasmuch as the absorption coefficient for a given gas varies with the temperature and salinity of the water, it should be clear that the concentration of the gas at a given partial pressure also varies with temperature and salinity. Consider, for example, solutions of oxygen in distilled water at 40°C, distilled water at 0°C, and 36‰ seawater at 0°C—all at oxygen tensions of 159 mm Hg. These solutions, respectively, will contain 4.8, 10.2, and 8.0 ml O_2/liter even though their oxygen tensions are identical. This example serves to emphasize that the partial pressure of a gas in solution does not in itself tell much about the concentration of the gas. At given temperature and salinity, concentration is proportional to partial pressure, indicating that a solution at higher partial pressure than another solution will also contain a higher gas concentration. The same does not necessarily hold true when solutions of various temperatures and salinities are being compared. It is possible for a solution that has a higher partial pressure than another solution to have a lower concentration. For example, distilled water at 40°C and 159 mm Hg oxygen tension contains 4.8 ml O_2/liter. Distilled water at 0°C and 130 mm Hg oxygen tension—a lower tension—contains 8.4 ml O_2/liter—a much higher concentration. This is a reflection of the higher solubility of oxygen in colder water.

We come now to the important point that *gases tend to diffuse from areas of higher partial pressure to areas of lower partial pressure*. This is true within aqueous solutions, within mixtures of gases, and across gas-water interfaces. Sometimes, but *only* sometimes, diffusion in the direction of the partial pressure gradient also means diffusion in the direction of the concentration gradient. In a body of water of uniform temperature and salinity, if the oxygen tension is greater in one region than another, the oxygen concentration will also be greater. Diffusion will occur from the region of higher tension to the region of lower tension, which in this case also happens to be from the region of higher concentration to the region of lower concentration. As shown earlier, it is possible for cold water to have a higher oxygen concentration but lower oxygen tension than warm water; diffusion will again occur according to the tension gradient, but now would be against the concentration gradient. Consider distilled water at 0°C with an oxygen tension of 200 mm Hg in contact with dry air at 0°C and 1 atm of pressure (oxygen partial pressure: 159 mm Hg). The oxygen concentration in the water is 12.9 ml/liter, much lower than that in the air, 209 ml/liter. Nonetheless, the water will lose oxygen to the air until its oxygen tension is reduced to equilibrium with the air. The importance of a knowledge of partial pressure to understanding the behavior of gases is obvious. In any system equilibrium is attained when the partial pressure is uniform throughout.

In aqueous solutions only gas molecules that are in physical solution as gas molecules contribute to the partial pressure. Conversely, knowing the partial pressure of a gas in solution, the concentration as computed from the absorption coefficient is the concentration of gas in solution as gas molecules. In natural waters these considerations are chiefly relevant to carbon dioxide, for, unlike nitrogen, oxygen, and other atmospheric gases, carbon dioxide may react with the water, forming bicarbonate and carbonate ions. These ions do not contribute to the partial pressure of carbon dioxide and are not included in the concentration computed from the absorption coefficient for carbon dioxide. The alkalinity of seawater favors the reaction of carbon dioxide to form bicarbonate and carbonate. At a given partial pressure and temperature, seawater does contain a lower concentration of dissolved carbon dioxide molecules than distilled water, as anticipated from its greater salinity. However, seawater can dissolve much more carbon dioxide than distilled water (or a neutral salt solution) before it comes to equilibrium with a given gas-phase partial pressure. This is because much of the carbon dioxide goes into solution not as carbon dioxide, but as the ionic reaction products. Similar considerations are relevant not only to carbon dioxide, but also to oxygen in animal bloods. Hemoglobin is one of a number of blood pigments that combine chemically with oxygen. Oxygen molecules that become bound to hemoglobin molecules do not contribute to the partial pressure of oxygen dissolved in the blood solution and therefore do not interfere with the capacity of the blood solution to take up more oxygen. Uptake ceases only when the concentration of oxygen dissolved as oxygen rises to the point that the partial pressure is equal to that of the oxygen source.

PROPERTIES OF AIR AND
WATER AS RESPIRATORY ENVIRONMENTS

The two major environments of animals, air and water, differ in several physical properties important to respiratory physiology. Table 8–1 gives the concentration of oxygen in several media at a partial pressure of 159 mm Hg. This partial pressure is, of course, the partial pressure in dry air at 1 atm of total pressure. The concentrations in the aqueous media are thus representative of the upper limits of concentration realized in natural waters in equilibrium with air. The values for distilled water are essentially the same as those found in natural fresh waters at the stated oxygen tension. It is immediately obvious that oxygen concentrations in air greatly exceed those in water, and concentrations in fresh water are higher than those in seawater at the same temperature. Air provides a richer source of oxygen than water. The importance of this factor is illustrated by the case of a hypothetical animal attempting to obtain a liter of oxygen by completely extracting the oxygen from a volume of its medium. At 0°C, with the medium at an oxygen tension of 159 mm Hg, the animal would have to pass over its respiratory surfaces 4.8 liters of air, 98 liters of fresh water, or 125 liters of seawater. The difference between air and water is even more marked at higher temperatures. The difference between oxygen concentrations in air and water is undoubtedly significant to the observation that the highest known metabolic rates are found in air-breathing animals: in insects among invertebrates and in birds and mammals among vertebrates. It is probable that the inherently low oxygen concentrations in natural waters have imposed definite limits on the metabolic rates of aquatic animals.

The oxygen concentration of air at constant oxygen tension falls about 8% as the air temperature rises from 0°C to 24°C because of the reduction in gas density (for example, see Table 8–1). Given the high concentrations usually found in air of whatever temperature, this decline is probably not

Table 8–1. Concentration of oxygen in air and water at a partial pressure of 159 mm Hg at three temperatures. Concentrations are expressed as cc O_2 at STP/liter.

Medium	Temperature		
	0°C	12°C	24°C
Air	209	200	192
Distilled water	10.2	7.7	6.2
Seawater (36‰)	8.0	6.1	4.9

generally of much significance to terrestrial animals. The oxygen concentration of fresh water or seawater at constant oxygen tension falls about 40% as the water is warmed from 0°C to 24°C, a reflection of decreasing oxygen solubility (again see Table 8–1). The percentage decline of oxygen concentration with increasing temperature in water is much greater than that in air, and given the much lower concentrations in water, it is clear that elevated temperatures can threaten the oxygen supplies of aquatic animals. You will recall that the metabolic rates of poikilotherms tend to rise with increasing temperature. At elevated temperatures the aquatic animal can be caught in a respiratory trap of increasing demands for oxygen coupled with much reduced availability of oxygen in the medium.

Rates of diffusion of gases are much greater in air than in water. At 20°C, for example, oxygen diffuses about 300,000 times faster in air. This means that if two regions of the medium differ by a given oxygen tension, are separated by a given distance, and are exposed across a given cross-sectional area, about 300,000 times as many oxygen molecules will diffuse into the region of lower tension per unit time in air than in water. Carbon dioxide diffuses somewhat more slowly than oxygen through air but about 25 times faster than oxygen in water. Rates of gas diffusion bear considerable import for the analysis of respiratory physiology, as will be seen later.

Water has a higher density and viscosity than air. The density of air at 760 mm Hg of pressure and a temperature of 17°C is 0. 0012 g/ml, whereas that of fresh water at the same temperature is essentially 1 g/ml—over 800 times higher. Viscosity is a measure of internal resistance to flow in fluids. When air or water is flowing through a tube or respiratory passage, all regions of the fluid stream do not move at the same velocity; because of cohesive forces between the fluid and walls of the passageway, the outer concentric layers of fluid move more slowly than the more central layers. A type of frictional resistance develops between adjacent layers of fluid that are moving at different velocities, and this is quantified as viscosity. The rate of flow through a passageway with a given pressure drop from end to end is inversely related to the viscosity of the fluid. The viscosity of water is 35 times higher than that of air at 40°C and over 100 times higher at 0°C.

The greater density and viscosity of water dictate that, within broad limits, aquatically respiring animals must expend more energy in moving a given volume of medium through their respiratory passages than air-breathing animals. The problem is compounded by the fact that each volume of water carries less oxygen than each volume of air. The aquatically respiring animal thus must often work harder than the air-breathing animal to obtain a given volume of oxygen, which, more interestingly stated, means that a greater percentage of the oxygen taken up must be directed to the metabolic effort of obtaining more oxygen. In resting man, 1–2% of metabolism is involved in ventilating the lungs, whereas in resting fish 10–20% is probably directed to ventilating the gills.

You will recall that an object immersed in a fluid is buoyed up by a force equal to the weight of the fluid displaced. The density of water is close to that of protoplasm. Gill filaments of aquatic animals are thus supported near neutral buoyancy by the water, and this is often vitally important in keeping the individual filaments suspended so that their entire surface area is in contact with the water. The buoyant force provided in air is negligible in comparison to the weight of protoplasm. We have all observed the gills of aquatic animals droop into a wet mass when in air. Respiratory structures of terrestrial animals that project into the medium must have sufficient structural rigidity to support themselves so that the individual members do not bend into contact with each other, with consequent reduction in the surface area exposed for respiration.

The fact of evaporative water loss in air has previously received much emphasis but deserves mention again because it has probably been a principal factor influencing the anatomy and physiology of respiration in terrestrial animals.

DETERMINANTS OF AMBIENT OXYGEN AND CARBON DIOXIDE TENSION

Oxygen and carbon dioxide tensions are known to assume a wide range of values within the habitats of animals. As we shall see, the prevailing tensions in any particular habitat have immediate relevance to the respiratory physiology of the animals living there. A fish, for example, that has no difficulty in meeting the oxygen demands of maximal activity in waters at an oxygen tension of 100 mm Hg may not be able to meet even the minimal requirements for continued life should the oxygen tension fall to 10 mm Hg. It is important to understand some of the factors that influence oxygen and carbon dioxide tensions in animal habitats. These factors are both physical and biological.

Figure 8–1 depicts the basic processes at work in a segment of an aquatic or terrestrial environment. The animals and plants living there exert strong influences on oxygen and carbon dioxide tensions. During the day, under adequate illumination, the net effect of photosynthetic plants is to add oxygen to the medium and extract carbon dioxide. Opposite effects are exerted by animals, by saprotrophic bacteria and fungi, and by photosynthetic plants at night. An important question is whether the oxygen produced in photosynthesis is less than, equal to, or greater than the oxygen demands of all organisms in the system. This will determine the net effect of the resident organisms on the oxygen tension. Similarly, the consumption of carbon dioxide in photosynthesis must be compared to total carbon dioxide production to determine the net effect of the resident organisms on the carbon dioxide tension. As illustrated in Figure 8–1, any given segment of the environment exchanges oxygen and carbon dioxide with neighboring

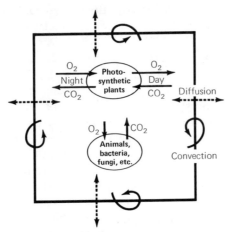

Figure 8–1. Schematic representation of processes affecting oxygen and carbon dioxide tensions in a segment of an aquatic or terrestrial environment. Oxygen and carbon dioxide are exchanged with adjacent segments of the environment by diffusion and convection.

segments through the physical processes of diffusion and convection. Diffusion will always tend to equalize gas tensions throughout, and convection very commonly exerts the same effect. The oxygen and carbon dioxide tensions prevailing in a given segment of the environment at a given time depend on the interplay of the biotic and physical processes involved. The net effect of the resident organisms, for example, may be to raise or lower the oxygen tension by comparison to that in neighboring parts of the environment. Diffusion and convection, on the other hand, will exert an equalizing effect. If the rate of exchange provided by diffusion and convection is high, very little tension difference may develop. But if the rate of exchange is low, the oxygen tension may rise well above or fall well below that in surrounding segments of the environment.

It is important to remember some of the important distinctions between diffusion and convection. The rate of gas transfer by diffusion between two parts of the medium depends in a well-defined manner on the tension difference and the distance: the rate decreases with decreasing tension difference and with increasing distance. Although diffusion provides for very rapid gas transfer over microscopically small distances, transfer over distances of inches or feet is quite slow. By contrast, convection, in the form of wind or water currents, operates independently of the tension difference and can mix air or water rapidly over even large distances.

With these introductory remarks, we may now proceed to examine in a more specific manner some of the major habitats of animals, starting with some terrestrial situations.

According to all available measures, the percentage composition of dry air in reasonably open habitats is remarkably uniform from place to place on the earth's surface both at sea level and over the altitudinal range occupied by animals. This uniformity of composition reflects the continual convective mixing of winds. Oxygen represents about 20.95% of the volume of dry air, and carbon dioxide averages about 0.03%. The level of carbon dioxide displays more variation than that of the other major components of dry air but generally does not exceed 0.06% even on city streets during peak activity or in forests when dissipation of carbon dioxide produced by resident animals and plants is limited by the effects of atmospheric thermal inversion.

Given that the percentage composition of dry air is highly stable and uniform in open terrestrial habitats, variations in the partial pressures of oxygen and carbon dioxide arise chiefly from changes in barometric pressure and water vapor pressure. A most important factor is the decrease in barometric pressure with altitude. The mean barometric pressure falls by about half for every increase of 18,000 ft above sea level. At 14,900 ft in the Peruvian Andes, for example, the average barometric pressure is about 446 mm Hg. Dry air still consists of 20.95% oxygen, but the oxygen tension is only about 93 mm Hg. The oxygen concentration is, of course, also reduced at altitude. At 24°C dry air at 14,900 ft averages only 112 ml O_2 at STP/liter, by comparison to 192 ml/liter at sea level.

The exchange of gases between secluded terrestrial microhabitats and the open atmosphere may be sufficiently restricted for the metabolic activities of resident organisms to lower the oxygen tension and raise the carbon dioxide tension of the air. Unfortunately, our knowledge of gas tensions in such microhabitats is limited, and generalizations are not possible. A few examples will illustrate possibilities. In studies of occupied rodent burrows, oxygen tensions have been reported to range from approximate equality with those in the outside air to as much as 35 mm Hg lower; carbon dioxide tensions have been reported to range from equality with the outside air to 35 mm Hg higher. The oxygen tension in a decaying beech trunk was measured as 133 mm Hg, probably a reflection of the metabolic activities of saprotrophs. Carbon dioxide tension in a group of anthills was reported to reach 14 mm Hg in the summer, a large increase over atmospheric tension.

Many air-breathing animals occupy soils. The interstitial air spaces in the soil provide paths for diffusional and microconvective gas exchange with the open atmosphere, and in porous soils the composition of air in the interstitial spaces may remain close to atmospheric down to some depth so long as the exchange paths with the atmosphere remain air-filled. During rains, the interstitial spaces can become filled with water. Because diffusion of gases is vastly slower in water than in air and because microconvective exchange is diminished by water's greater viscosity and density, gas tensions at depth in wet soil can deviate profoundly from those in the atmosphere. In one study oxygen tension at a depth of 30 cm in a field fell from 153 mm Hg to 46 mm Hg after a rain, and carbon dioxide tension rose from 1.5 mm Hg to 46 mm Hg. Such changes can strongly affect the organisms resident in the soil and may, for example, be responsible for the emergence of earthworms during heavy rain. The metabolism of saprotrophs that are relatively insensitive to low oxygen tensions can render permanently wet soils so oxygen-poor as to be uninhabitable by many animals.

Turning to consideration of bodies of water, we may first note some basic and important differences between the aquatic medium and air. We have seen that the atmosphere is so thoroughly mixed by convective currents that oxygen and carbon dioxide tensions in terrestrial habitats generally do not deviate significantly from those in open air unless there is some physical barrier to gas exchange. Convective mixing can be much less active in water, in part because of water's greater density and viscosity, and diffusion of gases through water is much slower than in air. As a result, there are often large differences in gas tension from one region of a body of water to another even without intervening physical barriers. The air above the ground in a forest or field will generally be of virtually uniform composition. In a lake of equivalent volume, gas tensions can vary significantly from place to place.

The physical processes of diffusion and convection tend to bring bodies of water into equilibrium with the atmosphere. The extent to which the partial pressure of oxygen or carbon dioxide deviates from that in the atmosphere depends on the rates of these equilibrating processes and on the metabolic activities of organisms living in the water. Waters in which organisms consume oxygen more rapidly than it is produced in photosynthesis may become essentially depleted of oxygen if exchange with the atmosphere is sufficiently slow. On the other hand, oxygen tensions of up to twice the equilibrium value with air have been reported in waters in which there is intense photosynthetic activity. Carbon dioxide tensions ranging from 0 to over 45 mm Hg have been reported, the latter in acid bogs and lakes.

Oxygen tensions in the open oceans are generally reasonably high, even at abyssal depths. Surface waters tend to be richest in oxygen, because of exchange with the atmosphere and planktonic photosynthesis. Light intensity falls off with depth, and in deeper waters there is little or no photosynthesis—and little or no local oxygen production. Thus oxygen supplies of the deeper waters must be replenished from the surface through circulation of the water. The fact that deep waters in the oceans generally have reasonably high oxygen tensions indicates that circulation is sufficient to bring oxygen into these waters at an adequate rate to replace the oxygen consumed by resident organisms. In some bodies of the sea various factors militate against adequate circulation, and deep waters may be virtually depleted of oxygen. Waters of the Black Sea below about 200 m, for example, contain so little oxygen that no animal life can survive, the only inhabitants being anaerobic bacteria. Poorly aerated bottom waters in certain Norwegian fiords are occasionally forced upward in great quantity, threatening the lives of the animals in the surface waters.

The oxygen tension in fast-flowing streams is maintained near equilibrium with the air by turbulence. In lakes, ponds, and slow-moving streams or rivers, lower turbulence provides the opportunity for oxygen tension to rise well above or fall well below the equilibrium value with air. In lakes, as in the oceans, decreasing light penetration with increasing depth results in a diminution of photosynthetic oxygen production. The depth at which oxygen production by resident plants is just sufficient to meet the oxygen demands of resident organisms is termed the *compensation level*. Above the compensation level there is an excess of oxygen production over oxygen demand, and below it oxygen production is not sufficient to meet oxygen demand. Waters below the compensation level must receive a convective input of oxygen from the surface waters, or their oxygen tension will gradually fall. Adequate convective input is impaired in many lakes during the summer by a phenomenon known as *thermal stratification*. Surface waters are warmed by the sun and, because of their reduced density, tend to float on top of the colder, deeper waters. There may then be active convective circulation *within* the warm and cold layers, respectively, but only restricted convective exchange across the interface between the two layers. It often happens that photosynthetic oxygen production within the cold, deep layers is insufficient to meet oxygen demands, and with effective input from the surface waters cut off, a severe state of oxygen depletion can develop. This is illustrated in Figure 8–2. Note that temperature and oxygen concentration are virtually uniform throughout the upper 10 m of water. This uniformity reflects good convective mixing within the upper, warm layers of the lake. The temperature drops off rapidly between 11 m and 15 m, and below 15 m it is again virtually uniform but over 10°C colder than in the surface waters. The oxygen concentration in the surface waters is near the level expected for equilibrium with the atmosphere. The deep waters, however, are almost devoid of oxygen. Sessile inhabitants of the deep waters are unable to migrate away when such conditions develop. Also, some of the mobile animals prefer or require low temperatures (e.g., some trout); these animals may be caught between the conflicting demands of needing to swim to upper waters to find satisfactory oxygen tensions, yet needing to avoid the upper waters owing to their elevated temperatures. Sometimes, massive die-offs of certain species ("summer kills") result.

In the fall, the upper waters of stratified lakes cool off, and as the density of the upper waters approaches that of the lower waters, convective mixing is restored between the lower waters and surface (fall overturn). High oxygen tensions are then re-established throughout the lake. A danger in winter arises from ice formation. Ice impairs aeration of the water from the air, and particularly when the ice is covered with snow, light penetration may be sufficiently limited that photosynthesis cannot compensate for oxygen utilization. Oxygen demand is lower than in summer because of the depressing effect of low temperatures on the metabolic rates of most organisms, but sometimes adverse photosynthetic circumstances persist sufficiently long that the entire lake becomes substantially oxygen-depleted. Winter kills can result.

Stratification caused by thermal effects can occur in a variety of waters, and in estuaries stratification can arise through salinity effects, less-saline waters tending to float on top of more-saline waters. Again oxygen stratification can develop as a consequence. In parts of the estuarine Chesapeake Bay, for example, the oxygen concentration in bottom waters falls to half of that in surface waters during the summer.

Fish, crayfish, snails, and other macroscopic aquatic organisms coexist with a great variety of microscopic organisms, many of which live on dead organic matter suspended in the water. It is important to recognize that these microscopic saprotrophs draw on the oxygen supplies of the water and add carbon dioxide to the water. The degree of saprotrophic activity is often an important determinant of the oxygen tension. Saprotrophic activity tends to increase with the concentration of suspended and dissolved organic matter, and it is not uncommon, for example, to find considerably lower oxygen tensions in rivers with a heavy organic load than in rivers with a relatively light organic load. In a thermally stratified lake, depletion of oxygen in the deep waters will occur more rapidly and will have a greater likelihood of reaching critical extremes if the waters support an abundance of saprotrophic activity than if they support only relatively little.

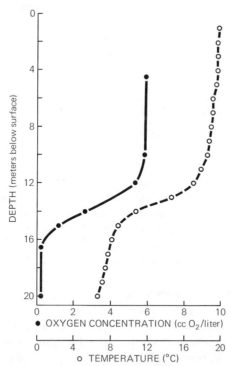

Figure 8–2. Dissolved oxygen concentration (●) and temperature (○) as functions of depth in Douglas Lake, Cheboygan County, Michigan, during July, 1969. (Data gathered by a class of physiology students.)

Man has caused an increase in the organic load of many bodies of water through two major processes. One is the straightforward addition of organic matter, such as sewage. The other is the addition of inorganic chemicals such as nitrate and phosphate that essentially fertilize the waters, promoting increased algal growth and eventually producing an increased load of decaying material when the algae and organisms that feed on the algae die. These forms of pollution often have demonstrable effects on oxygen tension. Polluted rivers commonly show reduced oxygen levels, and oxygen depletion in thermally stratified lakes often proceeds to a more extreme extent after increase in the organic load. Such alterations in oxygen relationships can strongly affect the resident fauna. Animals that require relatively high oxygen tensions, such as many game fish, are replaced by carp, catfish, and other species that can tolerate lowered oxygen levels. The faunas of swift-flowing waters are perhaps particularly sensitive because they have become adapted to an environment of assuredly high oxygen concentration over their evolutionary past.

Earlier we discussed the problems of gas exchange through wet soils. The bottom sediments of bodies of water present similar problems. Movement of oxygen by diffusion and convection from the open water through the interstitial spaces of sediments occurs slowly and can compensate only a small oxygen demand. When organic matter in the sediment supports any degree of saprotrophic activity, completely anaerobic conditions may prevail within millimeters of the substrate surface. Most animals that live or burrow in bottom sediments have adopted the expedient of drawing their oxygen directly from the open water above, rather than depending on interstitial supplies. Clams, for example, draw water over their gills through siphons that project to the water through the substrate surface, and burrowing *Tubifex* worms project their posterior end into the open water.

FUNDAMENTAL ASPECTS OF ANIMAL GAS EXCHANGE

The gas exchange systems of all animals can be schematized as in Figure 8-3. There are fundamentally three important steps: (1) oxygen and carbon dioxide must move between the medium and the respiratory exchange membranes, (2) they must traverse the exchange membranes, and (3) they must move between the exchange membranes and the internal tissues of the animal. All these processes must proceed at a sufficient rate to meet the oxygen demands of the tissues and to void carbon dioxide into the environment so that it does not accumulate in the tissues. The three "steps" will be discussed here individually.

The exchange membrane

There is always a membrane separating the internal tissues of the animal from the environmental medium. In a unicellular animal the membrane is simply the cell membrane. In a fish the membrane is the epithelial covering of the gills, which separates the water from the blood capillaries. Transmission of gas molecules across the membrane is always by diffusion. There are no demonstrated examples of active gas transport in the animal kingdom. Gases diffuse faster than ions, but the other factors that influence the rate of diffusion are identical to those elaborated in Chapter 4. The direction of diffusion of gases is governed by the partial-pressure gradient, and the larger the gradient, the faster diffusion occurs. Oxygen diffusion into the animal, for example, will not occur unless the oxygen tension on the animal side of the exchange membrane is lower than that on the environmental side and will occur more rapidly, the larger the difference in oxygen tension. The rate of diffusion increases with the area of the exchange surface, and the respiratory membranes of animals are often thrown into elaborate patterns of invagination or evagination that increase their surface area. Finally, the rate of diffusion varies inversely with the thickness of the exchange membrane or, more properly, with the distance separating the gas tensions that establish the tension gradient. For this reason the respiratory exchange membranes of animals must be thin. It has been emphasized before that this necessity can create problems in the area of water and salt relationships, for the thin membranes present a reduced barrier to movements of salts and water as well as to those of oxygen and carbon dioxide.

Movement of oxygen and carbon dioxide between the medium and respiratory exchange membrane

Oxygen in the medium may be brought to the environmental surface of the exchange membrane, and carbon dioxide carried away, by diffusion or convection or a combination of both. Exchange by diffusion alone is seen

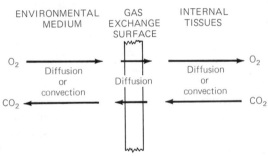

Figure 8–3. Schematic representation of the fundamental processes in animal gas exchange. See text for discussion.

in many air-breathing animals, but convective exchange of one type or another is probably universal among aquatically respiring animals. Diffusion in water is too slow to provide the sole means of transport of gases between the medium and respiratory surface. Convective movement of the medium across the respiratory membranes is termed *ventilation*. Animal respiratory surfaces may be washed by ambient air or water currents (passive ventilation). In such cases ventilation is realized without the need for energy input from the animal but may not be very reliable. Many animals create convective movement of the medium across their respiratory membranes by developing forces of suction or positive pressure at the expense of metabolic energy. This is termed active ventilation and is amenable to control by the animal according to respiratory circumstances. The exchange of gases between the medium and the respiratory surfaces is commonly termed *external respiration*.

Movement of oxygen and carbon dioxide between the respiratory exchange membrane and internal tissues

It is instructive from the outset in discussing this topic to review some calculations performed by the great physiologist August Krogh. Using an equation derived by E. N. Harvey, Krogh sought to define the maximum distance over which diffusion through the organism could supply the gas exchange requirements of tissues. Attention was centered on oxygen because it diffuses more slowly than carbon dioxide in aqueous media. Oxygen diffuses less rapidly through animal tissues than pure water, and Krogh assumed, on the basis of some evidence, a rate of one third the rate in water (at 20°C). He considered a spherical cell and, postulating a metabolic rate of 0.1 cc O_2/g/hr and an oxygen tension of about 159 mm Hg at the cell surface, computed that diffusion of oxygen through the cell will suffice to meet the oxygen demands of all parts of the cell only if the cell radius does not exceed 0.9 mm. Convective movement of oxygen through the cell is necessary if the cell has a larger diameter, if the metabolic rate is higher, or if the oxygen tension at the surface is lower. Conversely, diffusion would suffice at a higher metabolic rate if the cell were smaller and at a larger cell size if metabolism were lower. These calculations, though somewhat theoretical, indicate in any case that whether we are considering spherical cells or tissues of other shapes, movement of oxygen through the tissues by diffusion alone will generally suffice to meet oxygen demands only over short distances even when the oxygen tension of the source is high.

Two general possibilities for gas exchange between the respiratory surface and tissues become evident. Either the respiratory surface itself must be very near all cells, in which case exchange by diffusion could meet the requirements of all tissues; or the respiratory surface can be distant from some cells, in which case convective transport of gases within the organism will generally be essential. One of the major roles often assumed by circulatory systems is convective transport of gases; oxygen, for example, can be picked up by the circulating blood at the respiratory surface and carried convectively throughout the body. In many animals the requirements of gas transport have exerted a primary influence on the functional attributes of the circulatory system and the composition of the blood.

Protoplasmic circulation, or streaming, has been observed in many types of cells and also provides opportunity for convective gas transport. The influence of this process on the rate of gas exchange through cells and tissues has not been quantified, but at least in certain cases it is probably important. The protozoan *Paramecium*, for example, has a substantial metabolic rate, about 1 cc O_2/g/hr. Calculations suggest that oxygen diffusion through the protoplasm may not be adequate to supply needs, and attention has been focused on the potential role of protoplasmic streaming, which is very much in evidence. Protoplasmic circulation within the cells of multicellular animals could also enhance the rate of gas exchange through the tissues and thus increase the permissible distance between tissues and their oxygen source beyond that dictated by diffusion alone. The oxygen source in this case could be either the respiratory membrane itself or blood that has carried oxygen convectively from the respiratory membrane.

EXTERNAL RESPIRATION: SOME GENERAL CONCEPTS

In mammals it is easy to define the "organs of respiration." Most of the body surface is very poorly permeable to gases, and on a simple anatomical basis the lungs stand out as distinctively appropriate sites of gas exchange. Things are not so straightforward in many animals. It is important to remember that any thin membrane exposed to the medium can serve as a site of gas exchange with the environment. In those animals in which a circulatory system is important to gas transfer within the animal, we usually expect a membrane that serves an important respiratory role to be well vascularized. Within these limits any thin membrane should be suspect as a respiratory site. Characteristically, only one part of an animal is identified by name as respiratory, be it the lungs of a frog, the gills of a fish, or the branchial papulae of a starfish. This does not mean that other parts do not play important respiratory roles. The skin of frogs and the tube feet of echinoderms, for example, are important to gas exchange. In some cases the part having a respiratory name turns out not even to be the principal site of respiration. All this suggests that a certain degree of openmindedness is essential in the study of respiratory physiology.

Sometimes a particular structure of the body seems clearly to represent a special adaptation for respiration. This is still not to say that it is the only important site of respiration. The specialized respiratory structures are usually characterized by having thin membranes thrown into extensive patterns of invagination or evagination that enhance the surface area for gas exchange. Structures evaginated from the body and surrounded by the environmental medium are termed *gills*. Those invaginated into the body and containing the medium are termed *lungs*.

Convective ventilation of respiratory structures, as noted earlier, may be active or passive, depending on whether the animal creates ventilatory currents through investment of metabolic energy. Active ventilation may be nondirectional, unidirectional, or bidirectional. These three types may be illustrated by example. Mudpuppies ventilate their exposed gills by simply swishing them back and forth in the water; the flow of water across the gills probably has little directionality. Most teleost fish drive a unidirectional current of water across their gills; the water enters at the mouth and exits through the opercular openings behind the gills. Mammals undergo bidirectional, or tidal, ventilation. Air is drawn into the lungs, then driven out via the same passageways.

A factor of importance is the percentage of available oxygen extracted from the medium on its passage across the respiratory surface, termed the *oxygen utilization*. If, for example, water entering the mouth of a fish contains 6 ml O_2/liter and that exiting from the opercular openings contains 2 ml/liter, then the oxygen utilization across the gills is 4 ml out of every 6 ml available, or 67%. Clearly, this factor must be considered along with the oxygen concentration of the medium in determining the volume of medium that must pass the respiratory structures for the animal to obtain a given volume of oxygen. Given that water typically contains so much less oxygen than air, we might expect a higher premium to be placed on high utilization in aquatically respiring animals than in air-breathing animals. Utilization is most easily defined in animals with directional ventilation, for one can readily identify the medium going to and coming from the respiratory surface. This is difficult or impossible when ventilation lacks directionality.

Countercurrent exchange between the blood and medium occurs in some animals and has important implications for the physiology of respiration. This may be illustrated by comparing gas exchange with and without counterflow, as diagramed in Figure 8–4.

Part *A* depicts a situation in which the blood flows along the respiratory exchange surface in the same direction as the medium. Oxygen-depleted blood meets fresh medium of high oxygen tension at the left, and there is at first a large diffusion gradient between the two. As the blood and medium flow along together, the tension of the blood gradually rises and that of the medium gradually falls. Thus the tension gradient diminishes until the blood and medium reach equilibrium. Note that the tension of the efferent medium cannot be below that of the efferent blood. The final equilibration of the medium is with blood that has reached a substantial oxygen tension, and the final equilibration of the blood is with medium that has a considerably reduced tension.

Part *B* depicts a situation in which the blood and medium flow in oppo-

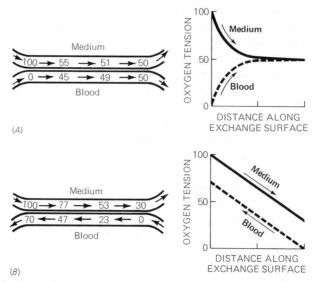

Figure 8–4. A schematic representation of oxygen transfer between blood and environmental medium (*A*) when the blood and medium flow in the same direction along the respiratory exchange surface and (*B*) when they flow in opposite directions (countercurrent exchange). Numerical values are oxygen tensions. It has been assumed for simplicity that the flow rates of blood and medium are the same and that oxygen content and tension are related proportionally and similarly in the blood and medium.

site directions. Here, at the right on the diagram, oxygen-depleted afferent blood meets medium that has already been substantially deoxygenated. The initial tension gradient between the afferent blood and medium is relatively small. As the blood flows along the exchange surface and picks up oxygen, however, it continually encounters medium of higher and higher tension. Thus a tension gradient is maintained all along the exchange surface. The final exchange of the blood is with fresh medium of high oxygen tension; as a result the tension of efferent blood is greater than in part *A*. The final exchange of the medium is with fully oxygen-depleted blood; thus, the tension of the efferent medium is lower than in part *A*. Clearly, the net result of countercurrent exchange is a more complete transfer of oxygen from the medium to the blood. Oxygen utilization is greater than in part *A*, and the blood leaves the exchange membrane at a higher tension, carrying more oxygen.

It should be stressed that the representation in Figure 8-4 is schematic. The actual benefit of counterflow over parallel flow in animals depends on a number of parameters, including the relative flow rates of the blood and water. There is evidence in certain animals, such as teleost fish, that counterflow provides considerable advantage.

EXTERNAL RESPIRATION
IN AQUATIC INVERTEBRATES
AND ALLIED GROUPS

We shall now proceed to review the mechanisms of external respiration in animals, treating first the aquatic invertebrates and their allied groups, then the vertebrates, and finally the insects and arachnids. Internal gas

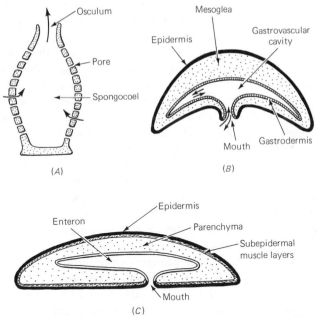

Figure 8–5. Highly diagrammatic representations of the body form of a sponge, coelenterate, and flatworm: (*A*) radial section of an asconoid sponge, (*B*) radial section of a coelenterate medusa, and (*C*) longitudinal section of a turbellarian flatworm. Arrows in *A* and *B* indicate water currents through body cavity.

transport in some groups will be covered in this chapter, but for the most part, discussion of internal transport is deferred to Chapters 9 and 10.

Sponges, coelenterates, and flatworms

Sponges, coelenterates, and flatworms (such as *Planaria*) lack a circulatory system. Once oxygen has penetrated the body proper, it probably moves only by diffusion or via the highly localized convective movements of protoplasm within cells and of pockets of body fluid between cells. This suggests that the rate of oxygen movement within the body proper would be insufficient to supply the needs of cells far removed from an exchange surface with the environment. It is not surprising, then, to find the bodies of these animals so constructed that the medium can get close to most cells.

Sponges are penetrated by many pores, and it is significant that most cells lie near the outer surface of the animal or along the lining of a pore or other water channel within the body. Water is drawn into the pores by flagellary action and exits via larger openings, the oscula, which receive water from many pores (Figure 8–5). There is, then, unidirectional ventilation. Reported oxygen utilizations vary from less than 20% to as much as 90%. The spaces between the cellular linings of adjacent pores are filled with a nonliving mesochyme containing gelatinous and skeletal material. Wandering amoebocytes in the mesochyme are the only cells that are at all distant from an exchange surface with the water. Oxygen supply to the amoebocytes has not been studied, but diffusion through the mesochyme may be sufficient given that the cells are present in relatively low density.

Coelenterates can be large. A good number of anemones and jellyfish reach diameters of many inches. The largest anemone can attain a diameter of over 1 yd, the largest jellyfish, over 2 yd. Coelenterates possess an internal body cavity, the gastrovascular cavity, which opens to the outside by a single orifice, or mouth (Figure 8–5). It is notable that most of the cells of the body are located near the outer surface of the animal or near the gastrovascular cavity. The region between the outer surface and gastrovascular cavity can be quite thick, as in many jellyfish, but it is filled largely with a nonliving membranous or gelatinous mesoglea. The mesoglea may be occupied by wandering amoebocytes. There is little direct physiological information on respiration in coelenterates. It seems probable that the oxygen needs of cells near the outer surface of the animal are met by diffusion directly from the water. Water currents are established in the gastrovascular cavity by the action of cilia and flagella and at least sometimes are highly directional. These may bring oxygen-laden water from the environment to all parts of the cavity at a sufficient rate to assure oxygen tensions adequate to diffusional supply of the cells near the cavity walls. Data on some species indicate that oxygen tensions in the gastrovascular fluid are maintained near ambient levels. The larger species have not been studied, however, and in these the distances over which water must travel from the mouth to the outer reaches of the cavity may be sufficiently great to pose problems. Oxygen supply to the amoebocytes has not been investigated and, again, may pose problems, especially in thick-bodied forms. It is often pointed out that the metabolic rates of coelenterates are low, 0.01 cc O_2/g/hr or less. This is sometimes taken to indicate either that diffusion could supply the needs of cell layers over a greater distance than in most animals or that the tension of the oxygen source could be relatively low and still supply cellular needs over more usual distances. Oxygen consumption expressed relative to the entire weight of the animal is deceptive, however, for much of the animal may consist of nonliving mesoglea. The critical parameter in assessing the adequacy of diffusion is the oxygen consumption of the cells themselves, and this will be higher than the figures given above indicate. It is interesting that nerve and muscle cells as well as epidermal and gastrodermal cells are localized near the internal or external surfaces of the animal. Probably requirements of gas exchange have imposed this restriction on the body plan of coelenterates.

The flatworms, as their name suggests, are flattened dorsoventrally. The present discussion will be limited to the free-living forms, or Turbellaria. The body is covered on the outside with an epidermis, and the major muscle layers lie just beneath the epidermis. Except in acoels, there is an internal body cavity, or enteron, opening to the outside via a single orifice, the mouth (Figure 8-5). The space between the gastrodermal cells lining the enteron and the subepidermal muscle layers is largely occupied by a population of cells termed the mesenchyme or parenchyma. Fluid-filled spaces are much in evidence among these parenchymal cells. The turbellarians are mostly aquatic, but there are some terrestrial species, and the terrestrial forms are the largest in the group; some reach over 20 in. in length. We are again faced with a marked paucity of information on respiration. Circulation of water in the enteron probably plays a minor role, if any, in supplying oxygen. Thus most oxygen enters the animal across the epidermis. Convective exchange with the water in aquatic forms is generally aided by epidermal cilia. We may note a number of features of turbellarians that are probably significant in gas exchange. First, the major muscle layers, which are probably important sites of oxygen uptake, are located just beneath the exchange surface with the medium. Second, the pockets of fluid among the parenchymal cells are squeezed during body movements. Convection within these pockets is probably important to gas transfer through the animal. Third, the highly flattened shape of turbellarians assures a relatively short distance between all cells and an exchange surface with the environment. It is generally thought that the limited capacities for gas transfer within the animal have been a significant factor in limiting body thickness in this group.

It is to be hoped that this review of respiration in some of the lower invertebrate phyla, although providing little concrete information, will demonstrate that they present interesting and significant questions worthy of physiological investigation.

Polychaete annelids

A fairly sophisticated circulatory system has made its appearance by the level of annelids. This not only permits cells to be further from the body surface but permits the meaningful development of specialized respiratory structures, for oxygen taken up in a specialized region of the body can be carried to cells throughout the body.

As in many other groups, respiratory physiology is so diverse in the annelids as to preclude even a semblance of comprehensive treatment, and physiological studies have been performed on only a relatively few species. Among polychaetes, the general integument is usually sufficiently thin and well vascularized to assume part of respiratory gas exchange. In some forms the general integument must assume the entire task, but in many there are localized evaginations that amplify the surface area for exchange and are often particularly heavily vascularized. Such evaginations are properly termed gills. The gills of the various species are extremely diverse in morphology and location. Frequently they are developments of the lateral appendages, or parapodia, which extend on either side of each body segment. They vary (Figure 8–6) from simple, flattened plates of tissue (as in *Nereis*) to elaborate, branching trees of filaments (as in *Arenicola*). Gills may also develop as head appendages or as extensions from the dorsal surfaces of body segments.

Epidermal cilia are found in many groups of polychaetes and often create water currents over the respiratory surfaces. Dorsal cilia may propel water along the dorsal integument; often the direction of flow is anterior to posterior, and inasmuch as blood in certain dorsal vessels flows in the opposite direction, there is a potential for countercurrent exchange. Cilia on the parapodia may circulate water over the parapodial respiratory surfaces. In swimming and crawling forms, water is driven across respiratory surfaces by general body movements through the water and movements of the parapodia in particular.

Many polychaetes live in tubes constructed of sand grains, hardened viscous secretions, or calcareous material. Parapodial gills are often reduced in these forms, and anterior gills that are presented to the water at the mouth of the tube are common. The most spectacular of these are the beautiful fans of pinnately divided tentacles found in sabellids and serpulids, the

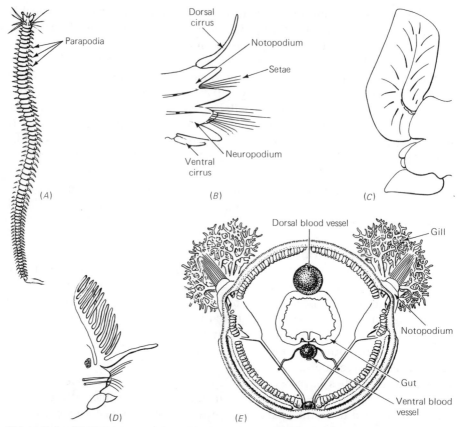

Figure 8–6. (*A*) Dorsal view of a nereid polychaete showing the position of the parapodia. (*B*) Lateral view of a parapodium of the nereid *Nereis pelagica*. The parapodium consists of a dorsal division, the notopodium, and a ventral division, the neuropodium. Each division bears a cluster of chitinous bristles, the setae, and a fleshy tentacular process, the cirrus. The parapodia consist of broad, well-vascularized lobes and constitute important sites of respiratory gas exchange.

The parapodia of *Nereis* are gills in the general sense of the word, but commonly the word *gill* is used in a more restrictive sense to refer to specialized developments of the parapodia (or other parts of the body). Specialized branchial developments are seen in *C*, *D*, and *E*. (*C*) Parapodium of *Phyllodoce groenlandica*. The dorsal and ventral cirri are developed into large, broad lamellae. (*D*) Parapodium of *Eunice harassii*. A filamentous branchial structure arises from the base of the dorsal cirrus. (*E*) Transverse section of a gill-bearing body segment in *Arenicola marina*, viewed from the anterior end. The gill consists of hollow, branching outgrowths of the body wall and is attached just behind the notopodium. Gills are found on only 13 of the body segments. (*A–D* from Fauvel, P. 1923. *Faune de France.* Vol. 5. *Polychètes errantes.* Fédération Française des Sociétés de Sciences Naturelles, Office Central de Faunistique, Paris; *E* from Wells, G. P. 1950. J. Mar. Biol. Ass. U. K. 29: 1–44.)

so-called fanworms (Figure 8–7). Water is driven along the tentacles by cilia, and the tentacles are used in feeding as well as respiration. Some tube-dwelling polychaetes (such as *Arenicola*) have well-developed parapodial gills, and in most, or all, species the general integument still presents possibilities for respiratory exchange. Water is commonly circulated through the tube by ciliary action or by undulatory or peristaltic body contractions. In some forms, well-developed parapodia act as paddles, driving water through the tube. Oxygen utilizations measured for tube- or burrow-dwelling polychaetes have ranged up to 50–70%.

A recent research report has pointed out some interesting and hitherto

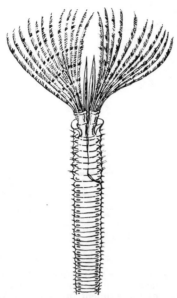

Figure 8–7. Dorsal view of the anterior end of the tube-dwelling sabellid polychaete *Sabella pavonina*. The array of pinnately divided tentacles is projected into the water at the mouth of the tube in which the worm lives. The sabellids and serpulids are often called fanworms in reference to their elaborate fans of tentacles. The tentacles function in both feeding and respiration. (From Fauvel, P. 1927. *Faune de France*. Vol. 16. *Polychètes sédentaires*. Fédération Française des Sociétés de Sciences Naturelles, Office Central de Faunistique, Paris.)

unrecognized possibilities for passive ventilation in animals, and because the principles involved may apply to some tube-dwelling polychaetes, this is an appropriate point at which to introduce the basic ideas. Some polychaete burrows are U-shaped (Figure 8–8*A*), opening to the surface at both ends. When this is the case, forces creating a difference in ambient pressure at the two openings will result in passive flow of water through the tube. When the movement of ambient water at the substrate surface is turbulent, it is immediately apparent that the pressure exerted at either opening may be transiently greater than that at the other, thus producing passive ventilation. What is not so apparent is that in certain circumstances passive ventilation can result when the flow of water along the substrate is laminar. When one end of the burrow or tube opens on an elevation of the substrate, as in Figure 8–8*A*, the fluid stream flowing along the substrate must accelerate as it passes over that orifice. According to the Bernoulli principle (familiar in the study of airfoils in physics), the lateral pressure exerted by a fluid stream decreases as the velocity of the stream increases. This effect is predicated by the laws of conservation of energy: simply stated, if more of the total energy is in the form of kinetic energy of flow, then less must be in the form of potential energy of pressure. According to these principles, then, the ambient pressure at the elevated orifice of the tube will be less than that at the lower orifice, and water will flow through the tube as indicated in Figure 8–8*A*. It is interesting that in this type of passive ventilation the direction of the ambient current along the substrate does not affect the direction of flow through the animal tube. Note, for example, that if the direction of the ambient current were opposite to that in Figure 8–8*A*, the pressure at the elevated orifice

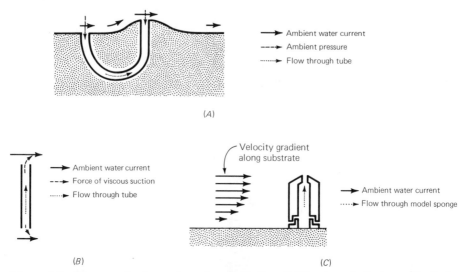

Figure 8-8. Some mechanisms of passive ventilation. Solid arrows indicate qualitatively, by their length, the velocity of the ambient current. Dashed arrows indicate qualitatively, by their length, the magnitude of the ambient pressure (in *A*) or the magnitude of the force of viscous suction (in *B*). (*A*) Passive ventilation of a U-shaped animal burrow or tube based on the Bernoulli principle. The velocity of the fluid stream along the substrate increases as the stream passes over the elevation of the substrate. The lateral pressure exerted at the elevated orifice of the tube is thus less than that at the lower orifice, and passive ventilation results. The physical principle is similar to that explaining the development of lift on an airfoil. (*B*) Passive ventilation of a tube based on the principle of viscous suction. Fluid tends to be drawn from the tube into a current flowing across an orifice of the tube. The force of suction is greater at the orifice exposed to the greater current velocity, and passive ventilation results. (*C*) A plastic model of a sponge, about 2 cm high, placed in a laminar water current. Passive ventilation was observed experimentally and can be explained according to the principles of viscous suction. Water flowing in laminar fashion along a substrate exhibits a velocity gradient near the substrate surface, and the higher and lower orifices of the model sponge are accordingly exposed to different velocities. (After Vogel, S. and W. L. Bretz. 1972. Science 175: 210–211. Original figures copyright 1972 by the American Association for the Advancement of Science.)

would still be lower than that at the lower orifice, and the direction of flow through the tube would be unaltered.

Another possibility for passive ventilation induced under conditions of laminar flow rests on the principle of viscous entrainment or viscous suction. When a current flows across the opening of a tube or other such structure, there is a tendency for fluid to be drawn from the tube into the current because of the viscosity, or resistance to shear forces, of the fluid. Other things being equal, this force of suction will be greater, the greater the velocity of the current. Thus if the tube is open at two places and the two orifices are exposed to different ambient current velocities (Figure 8–8*B*), fluid will tend to flow through the tube toward the end exposed to the greater velocity. This principle can lead to passive ventilation in situations where the considerations of the Bernoulli principle do not apply. When a fluid flows in laminar fashion along the substrate (Figure 8–8*C*), there is a gradient of fluid velocity within a short distance of the substrate, fluid further from the substrate flowing more rapidly than that near the substrate. In this simple and commonplace situation, there is no difference in pressure associated with the velocity gradient. Bernoulli's principle does not say that *any* rapidly moving stream will exert less pressure than *any* more slowly moving stream; rather, it applies only when a *particular* fluid stream, such

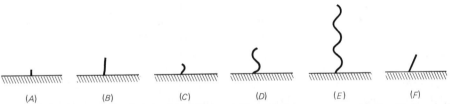

Figure 8–9. Responses of the tubificid worm *Limnodrilus* as oxygen tension is lowered from well above 160 mm Hg (*A*) to values approaching zero (*F*). At high tensions, no wriggling movements are evident, and the length of the worm extended from the substrate increases as tension is lowered (*A–B*). The length extended decreases when wriggling commences (*C*). Further reduction in tension (*C–E*) results in increased extension and increased frequency of wriggling. At very low tensions (*F*), worms stop wriggling and withdraw into the substrate. *Tubifex* shows similar responses except that as stage *E* is approached, the worms leave the substrate and migrate. The wriggling motions are believed to produce a current of water over the worm, drawing water from upper layers toward the substrate. (From Alsterberg, G. 1922. Lunds Univ. Aarssk. N. F., Avd. 2, 18: 1–176.)

as that near the substrate in Figure 8–8*A*, is accelerated or decelerated. The upper and lower orifices of the artificial sponge in Figure 8–8*C* may not be exposed to different ambient pressures, but they are exposed to different ambient velocities, and the principle of viscous suction can account for passive ventilation. Note again that the direction of the ambient current is immaterial to the direction of flow through the animal.

It is only recently that these ideas have been put forth, and their applicability and importance in nature have not been fully assessed. Clearly the principles involved could contribute to passive ventilation in many situations besides the worm tubes and sponge used as examples above, and in fact recent studies have shown their applicability to prairie-dog burrow systems. The mounds built at different openings of these systems differ in height and shape, and the evidence is that even light winds at the surface of the ground induce a significant flow of fresh air into and through the burrows, thus serving to renew oxygen supplies and wash out carbon dioxide.

Oligochaete annelids

Although the polychaetes are predominantly marine, the oligochaetes are found mostly in fresh water and on land (earthworms). Oligochaetes are generally smooth-bodied, lacking gills, and respire across the general body surface, which is often richly vascularized.

Tubifex and other tubificids live in mud at the bottoms of ponds, lakes, and pools of water. They bury their anterior end in the mud and leave the tail projecting from the surface for direct gas exchange with the water. The tail often wriggles about, ventilating the exposed integument. The tail is projected further, the lower the oxygen tension of the water (Figure 8–9). The muds in which these animals live can be very oxygen-poor, and the need for direct exchange with the water is obvious.

External and internal gills

Before proceeding to the other groups of aquatic invertebrates, we should recognize the distinction between external and internal gills and note some of the implications of having each type. Gills, such as those of many polychaetes, which project directly into the medium are termed external gills. By contrast, gills located within a body cavity of the animal are termed internal gills. External placement of the gills, by exposing them directly to the medium, probably enhances possibilities for passive ventilation by ambient water currents and thus may reduce ventilatory energy demands on the animal. However, external placement may also make the delicate gill surfaces especially vulnerable to injury. Location of the gills in a body cavity reduces the potential for injury but commonly also imposes the necessity of steady active ventilation. Protection of the gills is not the only potential advantage of having them in a body cavity. Internal placement can provide for improved efficiency of ventilation by allowing more rigorous control of the direction of water flow across the gills; this is so because water passing through the gill chamber can be channeled. As we shall see, many animals with internally placed gills ventilate unidirectionally; water is taken in at one point, passed through the gill chamber, and voided at another point. This assures that water that has been in contact with the respiratory surfaces and lost some of its oxygen content is steadily replaced by fresh water at or near ambient oxygen tension. The opportunity for admixture of oxygen-depleted water with incoming fresh water is reduced.

Polychaetes that live in tubes in effect realize internal placement of the gills. As we shall now see, the gills of most molluscs are located in a true external body cavity.

Gill-breathing molluscs

In molluscs, outfolding of the dorsal body wall produces a sheet of tissue, the mantle, that commonly overhangs part or all of the rest of the body. In shelled forms the mantle is responsible for generating the shell. Where the mantle overhangs the rest of the body, it encloses an external body cavity, the mantle cavity. The mantle and mantle cavity are diverse in morphology among the molluscan groups. The basic arrangement in some groups is illustrated in Figure 8–10.

The gills of molluscs are typically suspended in the mantle cavity. In many snails there is but a single gill, whereas in certain chitons there may be over 20 pairs. The more primitive gills (as seen in chitons and certain snails) are plumose, consisting of many pinnately arranged filaments along a central axis. Various modifications of this structure occur, and in many bivalves the filaments have become fused so that the gills appear as broad, thin plates or lamellae (Figure 8–11). The mantle and other body surfaces of molluscs are often sufficiently thin and well vascularized to perform some role in gas exchange.

The surfaces of both the mantle and the gills are generally heavily ciliated, and ventilation of the mantle cavity and gills is most commonly accomplished by ciliary action. Usually flow through the cavity is unidirectional, there being well-defined incurrent and excurrent openings. Some molluscs have replaced ciliary ventilation to a greater or lesser extent with muscular ventilation. Muscular ventilation reaches its zenith in the cephalopods. Most cephalopods swim by using muscular contractions of the mantle. Water is alternately sucked into the mantle cavity through incurrent openings and then driven forceably outward through the ventral funnel, producing a propulsive force. The gills in the mantle cavity are thus ventilated.

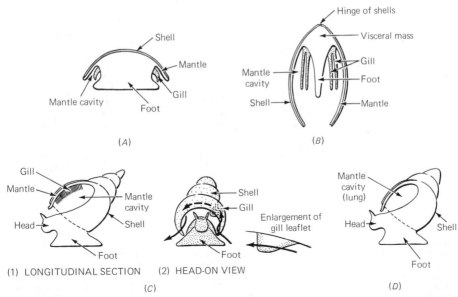

Figure 8–10. Schematic representations of the arrangement of the mantle, mantle cavity, and gills in several groups of molluscs. (*A*) Transverse section of a chiton. The mantle, extending laterally away from the foot, delimits a mantle trough (mantle cavity) running the length of the animal on each side. Many pairs of gills project into the mantle troughs in a serial arrangement from anterior to posterior. In life the lateral edges of the mantle are appressed to the substratum except in localized regions at the anterior and posterior ends of the animal (inhalant and exhalant openings). Ciliary action on the gills produces a ventilatory current. Water enters anteriorly, follows the channel between the gills and mantle, passes across the gills to enter the channel between the gills and foot, and then flows posteriorly in that channel to exit. (*B*) Transverse section of a lamellibranch clam. The mantle cavity is relatively capacious, and the gills are suspended in the cavity. There are just two gills, each being folded to produce two half-gills or demibranchs. Ventilation in eulamellibranchs is discussed subsequently in the text. (*C*) Diagrams of the condition in many prosobranch gastropods. There is only one gill, the left, and it has become modified and fused to the mantle, assuming the form of many triangular gill leaflets that hang into the mantle cavity something like the pages of a book. There is a broad anterior orifice into the mantle cavity (see *C*2). Ciliary action on the gill leaflets produces a ventilatory stream, water entering at the left of the animal, passing across the gill leaflets, and exiting at the right (indicated by arrows in *C*2). (*D*) Longitudinal section of a terrestrial pulmonate gastropod. Gills are lacking, and the walls of the mantle cavity have become richly vascularized, transforming the mantle cavity into an air-breathing lung. The mantle cavity opens to the outside only through a small pore-like orifice. When the mantle cavity is ventilated, air passes both in and out through this pore. See text for further discussion.

External respiration in the eulamellibranch clams is somewhat specialized and intriguing. One of the four gill lamellae in *Anodonta*, a freshwater species, is illustrated in Figure 8–11. The gill is perforated by many small pores, or ostia, which open into long water channels running dorsoventrally within the gill lamella. Cilia around the ostia drive water into the water channels, wherein it flows dorsally. Collectively, this ciliary action generates a water current through the mantle cavity. Water enters the mantle cavity at the inhalant siphon, travels into the ostia and up the water channels to a suprabranchial chamber, and then flows out through the exhalant siphon. *Anodonta* buries itself in the substrate with only the two siphons opening to the water above. In many clams the siphons are elongated into distinct tubes, so that the animal can be deeper in the substrate and still maintain communication with the water. The clams are unusual among molluscs in using their gills for feeding as well as respiration. Particulate organic

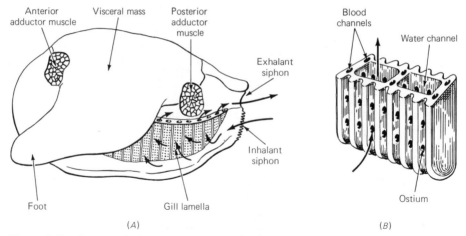

Figure 8–11. Aspects of gill structure and ventilation in eulamellibranch clams as exemplified by the freshwater mussel *Anodonta*. (*A*) A specimen of *Anodonta* with the left shell removed (semidiagrammatic). One of the four gill lamellae suspended in the mantle cavity is shown. It has been sectioned longitudinally near its dorsal attachment with the body to reveal the water channels within. (*B*) The structure of the lamella in greater detail (semidiagrammatic). Ostia on the surfaces of the lamella lead into the water channels, which run dorsoventrally within the lamella.

Ventilatory currents are indicated by arrows. Water from the mantle cavity is forced into the water channels through the ostia by the action of cilia surrounding the ostial openings. The water then passes dorsally through the water channels to enter a suprabranchial chamber above the gill (not shown). The water flows posteriorly in the suprabranchial chamber to exit through the exhalant siphon. The action of the ostial cilia pulls water into the mantle cavity through the inhalant siphon. [After Parker, T. J. and W. A. Haswell. 1940. *A Text-book of Zoology.* Vol. I. 6th ed. (as revised by O. Lowenstein) Macmillan and Company, Ltd., London. Used by permission of Macmillan London and Basingstoke.]

material in the water is filtered out as the water enters the ostia. This material then is carried across the outer gill surface by special cilia, ultimately to arrive at the mouth. It seems likely that the specialized morphology of the gills is more an adaptation to this filter-feeding mode of life than to any unique respiratory requirement in the clams.

The oxygen utilizations of bivalves have usually proved to be in the range of 5–15%. These values are much lower than those found in some other molluscs, and it has been argued that the flow of water through the gills for feeding in clams is so copious that oxygen extraction is subsidiary and not the governing factor for water flow.

Water flow through the internal water channels of the gill lamellae in eulamellibranch clams is, of course, highly directional, and it is counter to the direction of blood flow in the gill vessels most important to gas exchange. Countercurrent exchange is, in fact, common in molluscs. Cilia on the gill surfaces are responsible for creating the ventilatory current in many groups, and typically the cilia drive water across the surfaces in a direction opposite to that of blood flow in the branchial exchange vessels. As depicted schematically in Figure 8–10*C*, for example, the gill of many snails consists of numerous triangular leaflets suspended from the laterodorsal mantle wall. The individual leaflets are attached transversely in the mantle cavity, and they are arranged like pages of a book from anterior to posterior. Cilia on the leaflets produce a water current flowing from left to right across the leaflet surfaces, and blood flow through the exchange vessels of the leaflets is from right to left.

Pulmonate gastropods

The predominant land snails and the slugs are members of the Pulmonata. They probably evolved from aquatic snails with a single gill. The molluscan gill is a delicate structure and tends to collapse in the absence of the buoyant support of water. It is not surprising, then, to find that the gill has disappeared in the pulmonates. Respiration is subserved by the internal mantle wall, which is well vascularized. The mantle cavity (see Figure 8–10D) opens to the outside by a single, closable, porelike opening, the pneumostome, formed by the edge of the mantle. The mantle cavity in the pulmonates is a true lung. The internal surface area for exchange is sometimes enhanced by ridges, and in a few groups there are even long tubular invaginations of the mantle wall into surrounding tissues and blood spaces.

Various calculations indicate that simple diffusion between the ambient air and the mantle lining, across the pneumostome and mantle cavity, will suffice to meet requirements for respiratory gas exchange in the pulmonates. The process is termed *pore diffusion* in reference to the fact that the communicating passage between the animal and environment is a pore, and the lung may be termed a *diffusion lung* insofar as diffusion is the only mechanism of exchange. Diffusion is adequate to meet exchange requirements both because the metabolic rates of these animals are fairly low and because oxygen and carbon dioxide diffuse relatively rapidly in air. Diffusion could not suffice in water. Despite the fact that diffusion seems adequate to the needs of all pulmonates, some species ventilate the lung by movements of the floor of the mantle cavity. Air is forced out when upward movement of the floor compresses the air in the lung, thus raising its pressure above ambient pressure. When the floor is lowered, the pressure in the lung is reduced below ambient, and air flows in. The pneumostome is alternately closed and opened in both ventilating and nonventilating forms. Closing of the pneumostome probably serves to prevent evaporative water loss when open communication with the ambient air is unnecessary for respiratory gas exchange.

The pulmonates have radiated back into the aquatic environment (chiefly fresh water), and here three respiratory patterns are evident. Some species have retained the air-breathing habit and periodically come to the surface to refresh the air in their lung or, in certain cases, respire while submerged through a tubular elongation of the mantle edge that extends to the surface. Some are believed to ventilate the lung with water, probably tidally. Finally, gills have made a new appearance in some. These evaginated respiratory structures are not homologous to the primitive molluscan gill, which was lost in the original evolution of the pulmonates. In the freshwater limpets, for example, the mantle cavity is reduced, and the gill has developed as an evagination of the foot. Such cases emphasize the strong evolutionary bias toward evaginated respiratory structures in aquatically respiring animals.

The air-breathing aquatic pulmonates are, in fact, *dual breathers*, using both the water and air as sources of oxygen. When the water is well aerated, they probably obtain about as much oxygen from the water across the general body surfaces as they obtain from the lung. When the oxygen tension of the water is reduced, they become more dependent on oxygen uptake through the lung and surface more frequently.

Crustaceans: general features
and respiration in aquatic decapods

Some of the very small crustaceans, such as the copepods, lack specialized respiratory structures and apparently respire entirely across the general body surface. Their surface-to-volume ratio is high not only because of their small size, but also because they possess many elongated appendages such as antennae and legs. In most larger crustaceans and many of the smaller ones, gills are present. In these the general integument may still play some role in gas exchange.

Crustacean gills are nearly always closely associated with the thoracic or abdominal appendages. They may arise from the appendages themselves or close to the bases of the appendages. They vary in morphology from simple platelike processes (lamellae) to highly divided filamentous structures. The entire body surface of crustaceans is covered with a chitinous cuticle. The gills are no exception, but the covering over the gills is thin and permeable. External cilia are lacking. Thus ventilation is always accomplished by muscular contraction, typically by beating of the appendages. The gill-bearing appendages themselves may be responsible, or there may be certain appendages specialized for producing the ventilatory current.

Decapods have received the greatest amount of attention from physiologists and will be emphasized here. A well-defined carapace covers the head and thorax dorsally and overhangs the thorax laterally, fitting more or less closely around the bases of the thoracic legs (Figure 8–12A). The carapace delimits two lateral branchial chambers in which the gills lie. The gills, which are all thoracic, arise from the first segments of the thoracic legs, from the body wall above the articulations of the legs, and from the articulating membranes between the legs and the body wall. There are from 3 to 26 gills on each side. Each gill consists of a central axis to which are attached a great many lamellar plates, filaments, or dendritically branching tufts, depending on species. The gills are richly vascularized. Each branchial chamber is ventilated by a specialized appendage located toward its anterior end and known as the scaphognathite or gill bailer (Figure 8–12B). The gill bailer beats back and forth, generally driving water outward through an anterior exhalant opening and thus creating a negative pressure within the branchial chamber. Water is drawn in at a variety of places, depending on species. In some the edge of the carapace fits only loosely against the sides of the body, and water can enter all around the posterior and ventral margins of the carapace. In the crayfish and some other decapods water can enter only at the posterior margin and around the bases of the legs. Sometimes entry is limited to one orifice. In any case, ventilation is unidirectional. Recent studies on green crabs (*Carcinus maenas*) have indicated counterflow between the water and the blood in the gills.

Measured oxygen utilizations in various decapods have ranged from about 15% to 90%. In crayfish (*Astacus*) the rate of ventilation increases with decreasing ambient oxygen tension. In one set of experiments, for example, the volume of water passing through the branchial chambers more than doubled when the oxygen tension was reduced from 150 mm Hg to 50 mm Hg, whereas the oxygen utilization remained fairly constant at 60–70%. Increase in ventilation with decrease in oxygen tension has been

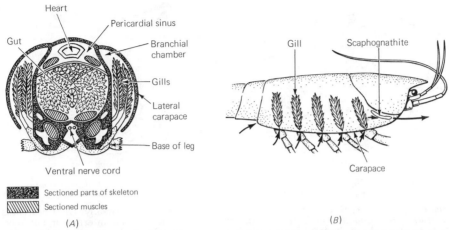

Figure 8–12. (*A*) Transverse section of a crayfish at the level of the heart (semidiagrammatic). The carapace overhangs the thorax laterally, delimiting an external body cavity, the branchial chamber, on each side. The gills arise near the bases of the legs and lie in the branchial chambers. (*B*) Water flow through the branchial chambers of a crayfish. The gills and scaphognathite are enclosed by the carapace and not visible externally. The beating of the scaphognathite drives water out of the branchial chamber anteriorly. Water is drawn in around the bases of the legs and the posterior margin of the carapace according to the pressure gradient established by the scaphognathite. Arrows indicate water flow. (*A* after Plateau, F. 1880. Arch. Biol. 1: 595–695.)

reported in several other decapods and in some isopods and amphipods. In some crustaceans, on the other hand, ventilation is relatively unaffected by oxygen tension. Lobsters (*Homarus*), for example, showed little change in ventilation when the ambient oxygen concentration was reduced from 5.8 ml/liter to 2.4 ml/liter; oxygen utilization increased from 30% to 55%, but this did not fully compensate for the decreased oxygen concentration, and oxygen consumption fell.

Semiterrestrial and terrestrial crabs

All the semiterrestrial and terrestrial crabs retain gills. The cuticular covering of crustacean gills is probably significant in this regard because it stiffens the gills by comparison, for example, to molluscan gills. In some groups of land crabs, the branchial lamellae are especially rigid and are held apart by various structural arrangements. The gills of terrestrial crabs tend to be reduced in size and number by comparison to those of marine crabs. The branchial chambers tend to be enlarged, and frequently some part of the epithelial lining of the branchial chamber has become highly vascularized and thus well suited to respiratory exchange. In some species the surface area of the epithelium is increased by evaginated folds or papillae (Figure 8–13). The trends toward reduction of the gills and development of a lunglike branchial chamber in terrestrial crabs provide a striking parallel to the situation in pulmonate gastropods.

The scaphognathite still assumes the role of ventilating the branchial chamber in terrestrial crabs. In some species the chamber is kept partially full of water when on land, but in the more terrestrial groups the chamber is filled entirely with air. The beating of the scaphognathite generally circulates air through the chamber in either case.

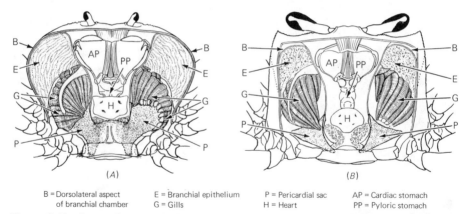

| B = Dorsolateral aspect | E = Branchial epithelium | P = Pericardial sac | AP = Cardiac stomach |
| of branchial chamber | G = Gills | H = Heart | PP = Pyloric stomach |

Figure 8–13. Internal anatomy of two terrestrial crabs viewed dorsally. (*A*) *Gecarcinus lateralis.* (*B*) *Ocypode quadrata.* The dorsal carapace and certain internal organs have been removed. Note the large branchial chambers (especially in *Gecarcinus*) and the thin, vascularized branchial epithelium lining the walls of the branchial chambers. The branchial epithelium of *Ocypode* (*B*) bears vascularized tufts or papillae, whereas that of *Gecarcinus* (*A*) does not. (From Bliss, D. E. 1968. Amer. Zool. 8: 355–392.)

Amphipod and isopod crustaceans

Amphipods have thoracic gills. Terrestrial forms are limited to humid microhabitats, and their gills remain well developed. In the isopods the gills are formed from the abdominal appendages, or pleopods. Each of these appendages has two branches, and in the isopods the branches take the form of broad lamellae that lie flat against the underside of the abdomen. Among terrestrial isopods the inner (dorsal) branch of each pleopod is delicate and vascular and acts as a gill, whereas the outer (ventral) branch serves as a gill cover, or operculum. The gills are protected under a series of overlapping opercula along the abdomen. Similar arrangements are found in many marine isopods. The striking feature of respiratory morphology in many terrestrial isopods is the development of invaginations of various types on the inner surfaces of the opercula (Figure 8–14). These *pseudotracheae* act as diffusion lungs and were discussed in Chapter 5. Some gas exchange occurs across the general body surfaces of isopods and amphipods.

Horseshoe crabs

The five species of horseshoe crabs are marine arthropods of the subphylum Chelicerata. Their gills are only one of a number of interesting features. The underside of the abdomen (Figure 8–15) bears six pairs of large, heavy, flaplike appendages, termed opercula, that overlap each other sequentially from anterior to posterior. Except for the most anterior pair of opercula, the underside of each operculum gives rise to about 100 broad, thin gill lamellae that are stacked on top of each other dorsoventrally under the operculum. These lamellae appear very much like the pages of a book, thus the name *book gill.* The opercula undergo rhythmic flapping motions. When they drop away from the abdomen, water is drawn in among the gills, largely laterally. When the opercula are pulled back up against the abdomen, the water is forced out, largely posteriorly.

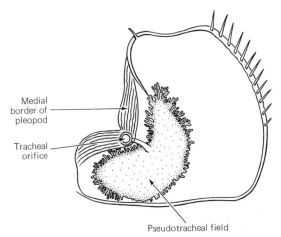

Medial
border of
pleopod

Tracheal
orifice

Pseudotracheal field

Figure 8–14. The outer branch (operculum or exopodite) of the first pleopod in the terrestrial isopod *Porcellio scaber*, showing the pseudotracheae. A branching, air-filled hollow within the operculum (pseudotracheal field) communicates with the outside through the tracheal orifice. The pseudotracheae are well supplied by the circulatory system and play an important role in respiratory gas exchange. (After a drawing by Verhoeff as represented in Vandel, A. 1960. *Faune de France.* Vol. 64. *Isopodes terrestres.* Fédération Française des Sociétés de Sciences Naturelles, Office Central de Faunistique, Paris.)

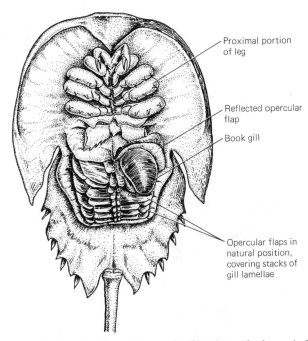

Proximal portion
of leg

Reflected opercular
flap

Book gill

Opercular flaps in
natural position,
covering stacks of
gill lamellae

Figure 8–15. A ventral view of a horseshoe crab (*Limulus polyphemus*). There are five pairs of overlapping gill opercula, each operculum covering a book gill. A sixth pair of opercula lies anterior to the gill opercula; these are the genital opercula and do not bear gills. One gill operculum has been reflected anteriorly, spreading the numerous, very thin lamellae of the book gill out like an accordion. In life the lamellae are stacked closely, like pages of a book, under the operculum. Only the proximal segments of the legs are shown. (From a drawing by Ralph Russell, Jr.)

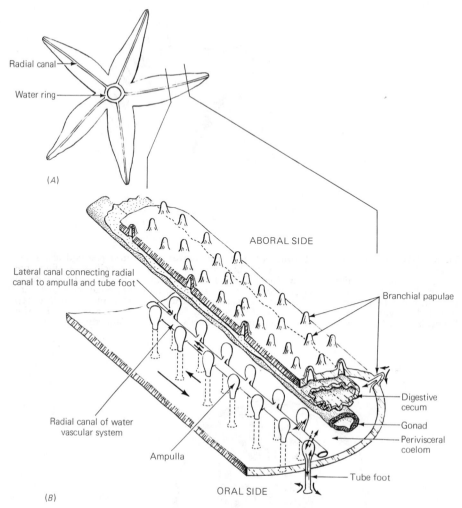

Figure 8–16. Semidiagrammatic representation of major structures involved in gas exchange in a starfish. (*A*) The general plan of the water vascular system. A circular canal, the water ring, sends out a radial canal along the length of each arm. The water ring and radial canals are situated on the oral side of the body cavity. (*B*) Diagram of part of an arm, with the aboral-lateral integument cut away on one side. The ampullae lie in the perivisceral coelom and connect with tube feet that project through the integument on the oral side. Each ampulla connects with the radial canal of the arm through a valved lateral canal. Two digestive (pyloric) ceca and two gonadal branches run along each arm in the perivisceral coelom; only one of each of these is shown. The branchial papulae are thin-walled evaginations of the perivisceral coelomic wall and appear externally as minute, fingerlike projections. Solid arrows show movements of ambient water, water vascular fluid, and coelomic fluid. Dashed arrows indicate diffusion of gases between the ampullar fluid and perivisceral coelomic fluid.

Echinoderms

The starfish have a limited hemal circulatory system, and most gas movement within the animal is accomplished by circulation of fluid in the perivisceral coelom and the water vascular system. As shown in Figure 8–16, the water vascular system sends a tube, the radial canal, along the length of each arm. Attached laterally to the radial canal are a great many tube feet, or podia, that project on the oral side of the animal. Associated with each tube foot is a muscular bulb, the ampulla. The most obvious function of the tube feet is in locomotion: each tube foot terminates in a sucker with which it can grip objects in the environment, and the animal moves about by co-ordinated activity of its many tube feet. The tube feet are retracted by the action of longitudinal muscles in their walls but are extended under forces of hydrostatic pressure developed by contraction of the ampullae. In addition to their role in locomotion, the tube feet are important sites of external respiration, for they present a considerable surface area to the environment, and gases diffuse readily across their thin walls. The walls are lined internally, like the rest of the water vascular system, with a ciliated epithelium, and the cilia circulate water vascular fluid between the tube feet and ampullae. Cilia or flagella on the external surfaces of the tube feet circulate ambient water across the surfaces. Convective movement of the internal and external fluids is also produced by ordinary locomotory activity. As shown in Figure 8–16, the ampullae and canals of the water vascular system lie in the perivisceral coelom. Oxygen taken up across the tube feet probably enters the perivisceral coelomic fluid by diffusion across the thin walls of the water vascular system. The perivisceral coelomic fluid is circulated by the action of cilia on the walls of the coelom, thus carrying oxygen throughout the animal and supplying the digestive organs, gonads, and other structures lying in the coelomic cavity.

The *branchial papulae* of starfish are fine, fingerlike evaginations of the coelomic wall found on the aboral surface of the animal and sometimes on the oral surface as well. They are ciliated internally and externally and, being thin-walled, provide sites for direct exchange of gases between the coelomic fluid and external medium. The branchial papulae play a significant role in external respiration and have been named for their respiratory function, but various measures indicate that the tube feet play at least as great a role, if not greater, in many species. The respiratory system of starfish seems simple by comparison to that of many other of the larger invertebrates, but the oxygen exchange requirements of these sluggish animals are not great.

Various respiratory modifications occur in other groups of echinoderms, but it is among the holothuroideans, or sea cucumbers (Figure 8–17), that a truly unusual respiratory mechanism makes its full appearance. Most sea cucumbers retain a well-developed water vascular system. Certain of the podia are elaborated into highly branched tentacles at the oral end, and at least some of the podia along the body wall retain their suckers and function in locomotion. The podia and water vascular system probably still play a significant role in respiration, and the tentacular podia may be particularly important. The remarkable feature of most sea cucumbers is the presence of well-developed, tidally ventilated water lungs. These *respiratory trees* arise as invaginations of the lower gut, or cloaca. They extend far up into the coelomic cavity and branch into many fine tubules. The trees are filled with water by a series of cloacal contractions. The anal sphincter closes, and water in the cloaca is driven up into the trees; the sphincter then opens, the cloaca dilates and refills from the environment, the sphincter closes, and water is again propelled into the trees. Six or 10 such cycles are required for one inhalation. The trees are emptied in a single step by contraction of the respiratory tubules and muscles of the body wall with the anus open. In *Holothuria tubulosa* the trees are ventilated once every one to

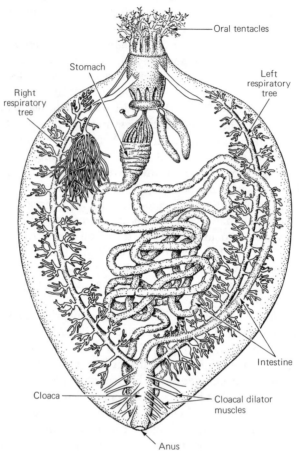

Figure 8–17. Internal anatomy of a sea cucumber (*Thyone*), viewed from the ventral side. The two respiratory trees, which actually branch more extensively than shown, arise from the anterior end of the cloaca and lie within the coelomic cavity. (After Coe, W. R. 1912. *Echinoderms of Connecticut.* Connecticut State Geological and Natural History Survey, Bull. 19: 1–152. Used by permission of the Connecticut State Geological and Natural History Survey.)

four minutes, and oxygen utilization varies from 20% to 50%. Available evidence indicates that the respiratory trees of sea cucumbers are responsible for about half of total oxygen uptake. Oxygen diffuses across the walls of the trees into the perivisceral coelomic cavity and is carried about in coelomic currents. There is also a possibility that oxygen is picked up and carried by the hemal circulatory system.

Tidally ventilated lungs: implications in water and air

The respiratory trees of sea cucumbers are remarkable, for invaginated respiratory structures that must be ventilated tidally are uncommon among aquatically respiring animals. The density of water has probably been an important factor militating against the development of such structures, for it implies a relatively high inertia that must be overcome by the animal in setting the medium into motion.

Unidirectional ventilation requires only a single acceleration of each volume of the medium that is passed over the respiratory surfaces. In tidal ventilation, on the other hand, the animal must accelerate the medium in one direction on inspiration and then, after its inward motion has stopped, accelerate it again in the opposite direction, overcoming the inertia of the medium in both instances and, in aquatically respiring animals, compounding the problems presented by water's high density. Simply analyzed, tidal ventilation implies a greater investment of metabolic energy to bring a given volume of the medium into contact with the respiratory surfaces. Also, more time should be required to exchange a volume of medium tidally than unidirectionally. Sea cucumbers have relatively low oxygen demands and can ventilate their respiratory trees at a fairly leisurely pace. It is difficult to see how some of the more active aquatic animals would be able to exchange water at a sufficient rate to meet their oxygen demands if ventilation were tidal. Both the energetic and temporal problems of tidal ventilation are exacerbated in aquatically respiring animals by the fact that natural waters have relatively low oxygen concentrations, meaning that the animal can extract only a comparatively small amount of oxygen from each volume of medium ventilated. Air is a much more favorable environment for tidal ventilation on these counts, and such ventilation is quite common in terrestrial animals. Because of its low density and inertia, air can be moved back and forth rapidly at relatively little metabolic cost, and each volume of air can yield a great deal of oxygen so that the temporal problems of tidal ventilation are of much more minor consequence than in water.

FISH

The anatomy of the respiratory system in teleosts

The buccal cavity of teleost fish communicates with the environment not only via the mouth but by lateral pharyngeal openings, the gill slits. The gills are arrayed across these openings and are covered by protective external flaps, the opercula. The structure of the gills is illustrated in Figure 8–18. On each side of the fish are four branchial arches that run dorsoventrally between the gill slits. Each arch bears two rows of gill filaments splayed out laterally in a V-shaped arrangement. Both the arches and the filaments are supported by skeletal elements. The tips of the posterior filaments on one arch lie very close to those of the anterior filaments on the next arch back. The filaments thus form a corrugated array separating the buccal cavity on the inside from the opercular cavity on the outside.

Each gill filament bears a series of folds, the secondary lamellae, on its upper and lower surfaces. These lamellae run perpendicular to the long axis of the filament. As depicted in Figure 8–18C, the lamellae divide the space between one filament and the next lower filament into rows of elongated pores. The entire array of filaments and their secondary lamellae thus forms a sievelike arrangement between the buccal and opercular cavities. The secondary lamellae are the major sites of gas exchange. They are richly vascularized and very thin-walled. The distance between blood and water across the walls of the lamellae is generally only 1–5 μ. During ventilation water passes through the gill sieve from the buccal to the opercular side. The arrangement of the gill elements assures that most water will pass very close to a secondary lamella. The lamellae of the tench, for instance,

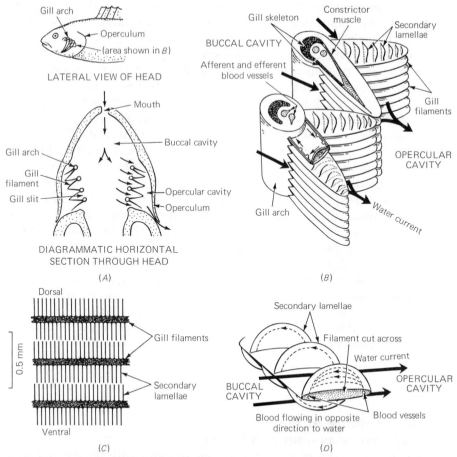

Figure 8–18. Major features of the branchial respiratory system in teleost fish. (*A*) The general arrangement of the gills. The lateral view shows the orientation of the gill arches under the operculum. The horizontal section shows the orientation of the buccal cavity, gill filaments, opercular cavity, and operculum. Arrows indicate direction of water flow. (*B*) Enlarged view of segments of two gill arches. Note the two rows of gill filaments on each arch and the secondary lamellae that project dorsally (and ventrally) from each gill filament. The secondary lamellae are the major sites of gas exchange between blood and water. Heavy arrows show direction of water flow. Light arrows within the upper left gill filament show direction of blood flow within the filament. (*C*) Diagrammatic longitudinal section through three gill filaments in a tench. Note that secondary lamellae on the upper and lower surfaces of the filaments divide the spaces between adjacent filaments into many minute channels, the arrangement of the respiratory surfaces being like that of a sieve. (*D*) Enlarged view of a gill filament showing that blood flow in the secondary lamellae (indicated by dashed lines) is counter to water flow across the lamellae. (Horizontal section in *A* from Bijtel, J. H. 1951. Arch. Neer. Zool. 8: 267–288. *B*, *C*, *D*, and lateral view in *A* from Hughes, G. M. 1961. New Sci. 11 (247): 346–348; these drawings first appeared in New Scientist London, the weekly review of science and technology.)

average only about 0.03 mm apart. Of all the gills studied, those of the fish appear to be the most elaborately organized. The surface area of the gills has been estimated to be 10 to 60 times that of the rest of the body.

Figure 8–18D illustrates that blood in the capillaries of the secondary lamellae flows from the opercular to the buccal side of the gill filament. This is counter to the direction of water flow through the spaces among the lamellae and provides for the high oxygen utilizations that can be realized by fish: up to 80–85%.

Active fish tend to have more gill surface area per unit of body weight than sluggish fish. They tend to have thinner lamellae, more closely spaced lamellae, and a shorter diffusion distance between the blood and water across the lamellar membranes. These features are conducive to meeting the greater gas exchange requirements of an active way of life.

Ventilation in teleosts

In general, water flow across the gills is maintained almost without interruption by the synchronization of two pumps: a buccal pressure pump, which forces water from the buccal cavity through the gills into the opercular cavity, and an opercular suction pump, which sucks water from the buccal cavity into the opercular cavity. The actions of these pumps will be discussed separately before reviewing their integration over the respiratory cycle. It is important to remember throughout that water will flow from regions of higher pressure to regions of lower pressure. The pumps work by establishing suitable pressure gradients.

The buccal cavity is filled with water when the floor of the cavity is depressed with the mouth open. The depression of the floor increases the volume of the cavity, thus decreasing buccal pressure below ambient pressure and resulting in influx of water. The mouth is then closed and the floor of the cavity raised. This increases buccal pressure and drives water from the buccal cavity through the gills into the opercular cavities. Thin flaps of tissue, which act as passive valves, project across the inside of the oral opening from the upper and lower jaws. During the refilling phase of the buccal cycle, when buccal pressure is below ambient, these valves are pushed inward and open by the influx of water through the mouth. During the positive-pressure phase, however, the valves are forced against the oral opening on the inside and help to prevent reflux of water from the buccal cavity through the mouth.

The opercular cavities can be expanded and contracted by lateral movements of the opercula and other muscular activities. Running around the rim of each operculum is a thin sheet of tissue that acts as a passive valve, capable of sealing the slitlike opening between the opercular cavity and the ambient water. When the opercular cavity is expanded, the pressure in the cavity falls below the pressures in the buccal cavity and ambient water. Water is thus sucked into the opercular cavity from the buccal cavity through the gill sieve and would be sucked in readily from the environment were it not for the action of the rim valve; because the pressure in the opercular cavity is lower than ambient during the sucking phase, the rim valve is pulled medially against the body wall, substantially sealing the opercular opening and preventing influx of ambient water. After the sucking phase the opercular pump enters its discharge phase. Contraction of the opercular cavity raises opercular pressure above ambient pressure, forcing the rim valve open and discharging water from the opercular cavity through the opercular opening.

The buccal and opercular pumps are synchronized in such a way that flow of water from the buccal cavity into the opercular cavity across the gill sieve is almost continuous. This is illustrated schematically in Figure 8–19. In step *A* the buccal cavity is being refilled. Expansion of the cavity produces a pressure below ambient; if the buccal pump were the only pump,

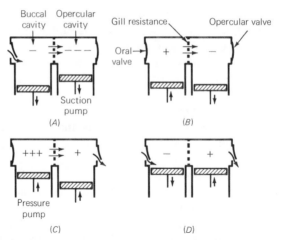

Figure 8–19. Schematic diagram of phases of the respiratory cycle in teleost fish. Plus (+) and minus (−) symbols indicate pressures in the buccal and opercular cavities relative to *ambient* pressure. The buccal and opercular pumps are represented by pistons. The gill sieve is interposed between the buccal and opercular cavities; arrows through the gill sieve indicate water flow. The oral valves are thin flaps of tissue that project across the inside of the oral opening from the upper and lower jaws; they are forced against the oral opening when pressure in the buccal cavity exceeds ambient. The opercular valve is a thin sheet of tissue running around the rim of the operculum; it is forced medially against the body wall when pressure in the opercular cavity is below ambient. Phases *B* and *D* are transitional and short in duration. See text for discussion. [From Hughes, G. M. 1961. New Sci. 11 (247): 346–348. This figure first appeared in New Scientist London, the weekly review of science and technology.]

flow through the gills from the buccal side would not occur at this point, and, in fact, there would be backflow through the gills into the buccal cavity because of the lowered pressure in the cavity. It is at this time, however, that the opercular pump is in its sucking phase. Opercular pressure is reduced well below buccal pressure, and water is drawn through the gills from the buccal cavity. Step *B* is a short transition stage in which the opercular pump is completing its sucking phase and the buccal pump is beginning its pressure phase. In step *C* the opercular pump is in its discharge phase. Pressure is elevated in the opercular cavity, but because the buccal pump is in its pressure phase, buccal pressure is elevated to an even greater degree, and water again flows through the gills from the buccal cavity. Only in step *D*, which occupies just a short part of the respiratory cycle, is the pressure gradient such as to cause backflow of water through the gills. In all, the two pumps are beautifully integrated to produce almost continuous, unidirectional flow across the gills. The opercular pump sucks while the buccal pump is being refilled, and the buccal pump exerts pressure while the opercular pump is being emptied. The velocity of flow across the gills varies over the respiratory cycle, but in many fish the flow is in the buccal-opercular direction 90–95% of the time.

The development of the buccal and opercular pumping mechanisms is not uniform among teleosts. In some species, such as many bottom-dwelling forms, the opercular pumping mechanism is particularly well developed, and opercular suction (stage *A* of Figure 8–19) accounts for most of the water movement across the gills. In other species, including some but not all pelagic forms, the buccal pump (stage *C*) predominates, or the two pumps are of similar importance.

Actively swimming fish may continue to ventilate their gills using the buccal and opercular pumps at all swimming speeds, or once they have reached a certain speed, they may cease pumping movements and allow their motion through the water to ventilate the gills, a phenomenon sometimes termed *ram-jet ventilation*. Sockeye salmon, for example, stop pumping while swimming and, leaving the mouth and opercula slightly open, allow water to flow "passively" into the mouth and across the gills. Ram-jet ventilation transfers the muscular effort of ventilation from the buccal and opercular mechanisms to the swimming muscles and may reduce the energetic cost of ventilation. Some fish, such as mackerel and tuna, swim continuously and ordinarily use ram-jet ventilation all the time. Mackerel commence buccal and opercular pumping if swimming is experimentally restricted, but oxygen exchange into the blood is then greatly impaired, and the fish can suffocate. Mackerel are dependent on continuous motion through the water for proper ventilation.

Teleosts increase the rate of ventilation of their gills in response to increased metabolic demands for oxygen (as during activity) and in response to lowered oxygen concentrations in the water. The ventilation rate may be increased by increasing the rate of buccal-opercular pumping, by increasing the volume of water moved on each pump cycle (stroke volume), or both. Tuna and probably other fish that utilize ram-jet ventilation increase the gape of the mouth as respiratory demands increase.

Elasmobranchs

Respiratory morphology and physiology in other groups of fish differ to a greater or lesser extent from the condition in teleosts. Elasmobranchs, in analogy to teleosts, ventilate their gills almost steadily in a single direction by a combination of a prebranchial pressure pump and a postbranchial suction pump. The anatomy of their ventilatory apparatus is quite different, as, for example, water is drawn in through valved spiracles as well as through the mouth and flows out on each side through several discrete gill slits not covered by an operculum. There is evidence on a number of species that postbranchial blood is at a higher oxygen tension than exhalant water, indicating countercurrent flow across the gills or another exchange arrangement that achieves a similar result. Some sharks rely heavily on swimming and ram-jet ventilation to produce their respiratory current.

The postbranchial suction pump appears to dominate in bottom-dwelling skates and rays. Interestingly, prebranchial pressure remains higher than postbranchial pressure throughout the respiratory cycle in skates and rays that have been studied. The absence of a reversed gradient in stage *D* (Figure 8–19) has also been observed in some teleosts, and in the bottom-dwelling elasmobranchs is hypothesized to be adaptive to preventing influx of sand through the gill slits. These animals take in water only through the dorsally positioned spiracular openings when buried in the sand but utilize both the mouth and spiracles when swimming.

Air breathing in fish

Low oxygen tensions occur commonly in sluggish bodies of fresh water, especially in swamps and other habitats supporting high levels of saprotrophic activity. Adaptations to utilize the air as a source of oxygen are much more widespread among freshwater fish than is commonly realized.

Some fish that come to the surface and gulp air have no particular morphological adaptations for air breathing. This is the case in the common goldfish, for example; unfortunately, the respiratory significance of air gulping in the goldfish remains unknown. The American eel (*Anguilla vulgaris*) is another species that utilizes air but exhibits no marked anatomical differences from the usual teleost pattern. The gills of the eel are typical. Eels are known to come out onto land in moist situations, and at low temperatures receive sufficient oxygen to support life by a combination of air gulping and cutaneous gas exchange. Air is taken into their buccal and opercular cavities and held until the oxygen tension falls to perhaps 100 mm Hg; then the air is released and another breath taken. Oxygen uptake from the inhaled air is thought to occur primarily across the gills. In eels on land at 7°C, about 60% of oxygen uptake occurs across the skin, 40% across the gills. At elevated temperatures aerial respiration becomes inadequate to meet the increased oxygen demands of the animal.

In most air-breathing fish part of the alimentary canal has become highly vascularized and serves as the principal site of gas exchange with the air. Sometimes it is the stomach or intestine. In such cases air is swallowed and then voided via either the mouth or anus. Armored catfish (*Plecostomus*), for example, are among those fish that use stomach breathing.

Often part of the buccal cavity has become adapted to air breathing. It may simply be highly vascularized, or it may in addition be thrown into evaginations or invaginations. In mudskippers (*Periophthalmus*), which spend much time on land near the water's edge, the inner walls of the opercula and adjacent parts of the gill chambers are vascularized and folded. Some catfish (e.g., *Saccobranchus*) have diverticula of the gill chambers that extend all the way to the tail region. Sometimes the presence of structures for aerial respiration is accompanied by reduction of the gills. This is shown in extreme degree by the electric eel (*Electrophorus electricus*). The walls of the buccal and pharyngeal cavities in this eel are thrown into highly vascularized papillae, with blood capillaries actually protruding above the surfaces of the papillae. The gills essentially lack secondary lamellae and are so reduced as to be of little respiratory significance. *Electrophorus* is an obligate air breather.

Most teleost fish possess a swimbladder, which develops as a dorsal evagination of the anterior gut. It generally functions as a buoyancy organ and may be filled with a very high concentration of oxygen. This oxygen is derived from the blood, and the high concentration is established and maintained by a vascular countercurrent multiplier system. Some fish call on the supply of oxygen in their swimbladder to support their metabolism in certain situations. When the American eel is out of water, for example, about 35% of the oxygen consumed over the first half hour is drawn from the swimbladder. In the eel the swimbladder cannot be ventilated with air. Thus when the animal emerges onto land, the oxygen available is limited to the reserve built up previously. This reserve becomes depleted, and after four hours only a negligible portion of oxygen consumption is supplied from the swimbladder.

Whereas many fish (termed physoclistous) lack an opening between the swimbladder and gut, many others (termed physostomous) have such an opening. In some of the physostomous species the swimbladder is used as an aerial respiratory organ. It is filled with air by the buccal pressure pump. The fish takes a gulp of air, closes its mouth, and forces the air into the swimbladder by compressing the buccal cavity.

Aerial respiration using the swimbladder has recently been studied in the bowfin, *Amia calva* (a holostean fish). The surface area of the lining of the swimbladder in *Amia* is greatly increased by comparison to that of most fish by a complex pattern of folding. *Amia* has well-developed gills and is a dual breather. The lower the oxygen tension of the water, the more frequently *Amia* breathes air at the surface. This pattern is observed in many other dual-breathing fish and is reminiscent of the behavior of air-breathing aquatic pulmonate snails. When the oxygen tension of the water is low, *Amia* may depend almost completely on air breathing to supply its oxygen needs. In well-aerated water *Amia* relies mostly on gill breathing at 10°C and visits the surface infrequently. When the temperature is raised to 30°C, however, oxygen demands increase, the bowfin comes to the surface every one or two minutes, and aerial respiration accounts for about two thirds of oxygen uptake. An interesting problem faced by air breathers with well-developed gills in oxygen-poor waters is that oxygen may be lost across the gills to the water. Oxygen will diffuse according to its tension gradient, and the tension of blood reaching the gills in an air breather may well be above the water tension when the water tension is low. When *Amia* is receiving most of its oxygen by aerial respiration, blood flow through the exchange vessels of the gills is substantially curtailed, thus limiting any loss of oxygen to the water that might occur.

The three genera of lungfish (dipnoans) have enlarged diverticula of the pharynx that are utilized in aerial respiration. These are situated dorsally to the gut but connect to the pharynx ventrally via a tube that curves around the right side of the esophagus. The diverticula arise embryologically as a ventral evagination of the pharynx. This is also the origin of the lungs of terrestrial vertebrates, and the diverticula of the lungfish are believed to be homologous to the lungs of higher vertebrates and are generally termed lungs for this reason. ("Lung" is here being used in a restrictive sense. According to the general definition of a lung as an invaginated respiratory structure, the diverticula of the lungfish are obviously lungs, as are the swimbladders of fish that use those structures for aerial respiration.) It remains debatable whether the lungs of lungfish are to be considered homologous with the swimbladders of teleosts. The swimbladder, you will recall, arises as a dorsal evagination of the gut.

The internal surfaces of the dipnoan lung are thrown into a complex pattern of ridges and septa (Figure 8–20), resembling the lining of many amphibian lungs. The Australian lungfish (*Neoceratodus*) has a single lung and well-developed gills. It cannot survive for long out of water and is a typical dual breather. It makes little or no use of its lung in well-aerated water but fills the lung regularly in oxygen-poor water. The African and South American lungfishes (*Protopterus* and *Lepidosiren*) have bilobed lungs and much-reduced gills that lack secondary lamellae. They are obli-

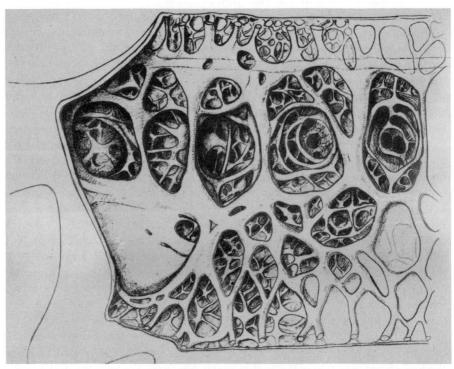

Figure 8–20. The ventral internal surface of the anterior portion of a lung of *Protopterus aethiopicus*. Respiratory surface area is greatly enhanced by a complex pattern of vascularized ridges and septa. Compartmentalization in other parts of the lung is similar but less elaborate. The side compartments in the wall of the lung open to a central cavity that runs the length of the lung and communicates anteriorly with a short pulmonary canal leading to the esophagus. (From Poll, M. 1962. Ann. Mus. Roy. Afr. Centr., Ser. 8, 108: 129–172.)

gate air breathers and drown if they are prevented from ventilating their lungs. As discussed earlier, they survive for long periods out of water in a state of dormancy during droughts.

AMPHIBIANS

Gills: anatomy and ventilation

The gills of aquatic amphibian larvae (tadpoles) are of different origin and structure than the gills of adult fish. They develop as outgrowths of the integument of the pharyngeal region and project directly into the medium from the body wall. They typically consist of branching, filamentous tufts. Each tuft has something of the appearance of a tree or brush, with the "trunk" or "handle" attached to the integument (Figure 8–21). The gills of all young amphibian larvae are external, but in the frogs and toads (Anura) an outgrowth of the integument, termed the operculum, soon encloses the gills in a chamber that opens to the outside posteriorly, usually via a single aperture. The operculum of salamander larvae appears only as a much-reduced integumentary fold and does not enclose the gills.

The gills are generally lost at metamorphosis, but external gills remain throughout life in certain salamanders, which gain reproductive maturity while retaining much of their larval morphology. An example is provided by the common mudpuppy (*Necturus*), an exclusively aquatic sala-

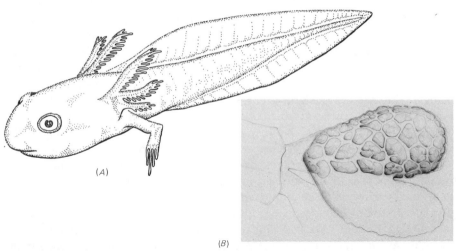

Figure 8–21. Respiratory structures of amphibians. (*A*) A three-week-old larva of the salamander *Ambystoma maculatum*, showing the external gills. (*B*) A lung of the frog *Rana temporaria*, showing the compartmentalization of the wall by ridges and septa. The lung has been sectioned and the dorsal aspect reflected to reveal the inner surface of the ventral wall. (*A* from a drawing by Ralph Russell, Jr.; *B* from Poll, M. 1962. Ann. Mus. Roy. Afr. Centr., Ser. 8, 108: 129–172.)

mander that retains its external gills in adulthood. At the time of metamorphosis it develops large but simple lungs as well.

The gills of the adult mudpuppy appear as large dendritic tufts, colored deep red by their rich blood supply. The tufts are provided with muscles and are waved back and forth in the water. These ventilatory movements increase in frequency with declining oxygen tension and increasing temperature. Muscular movements of the gills are also reported in many amphibian larvae with external gills, and cilia on their gill surfaces may aid in ventilation. In anuran larvae with the gills enclosed in the operculum, ventilation is accomplished by buccal pumping. Water is taken in through the mouth and nares and driven back through the pharyngeal gill slits into the opercular cavity containing the gills. The water then exits via the opercular aperture.

Lungs: anatomy and ventilation

Paired lungs develop from the ventral wall of the pharynx in most amphibians near the time of metamorphosis. The plethodontid, or lungless, salamanders lose their larval gills but fail altogether to develop lungs; and in some other salamanders the lungs are rudimentary. The lungs of many adult amphibians are simple, well-vascularized sacs. Internal surface area is increased little, if at all, by folding, and in this respect the lungs are less well developed than those of the modern lungfish. Particularly among frogs and toads, the walls of the lung may be thrown into a complex pattern of folds and septa, giving them something of a honeycombed appearance (Figure 8–21). Still, the lung retains its basic saclike form. The central cavity of the lung remains open and provides access to the various side compartments formed by the folding of the walls. The folding greatly increases the surface area for gas exchange.

It is instructive to compare the actual vascularized surface area in the lung with the area that would be realized by complete vascularization of the walls of a simple sac of the same gross dimensions as the lung. The ratio of these two quantities is less than 1 in amphibians with simple, unfolded lungs because the entire internal surface area is not vascularized. In *Rana esculenta*, which has a highly divided lung, the ratio is about 8—illustrating the extent to which folding can increase the surface for respiratory exchange. The highest surface areas realized in amphibians are, however, much lower than those attained in mammals for equivalent lung volume. Man has about 15 times the respiratory surface per unit volume as *R. esculenta*. This reflects the greater extent of folding in the mammalian lung.

Amphibians fill their lungs by buccopharyngeal pressure. This basic mechanism is presumably carried over from their piscine ancestors and, as mentioned earlier, is often employed by amphibian larvae to ventilate their gills. Most studies of pulmonary ventilation have been performed on frogs, and though several patterns differing in detail have been reported, the essentials of the buccopharyngeal pressure pump are quite uniform. Air is taken into the buccal cavity through the nares or mouth when the pressure in the cavity is reduced by lowering the floor of the cavity. When the floor of the cavity is raised with the mouth closed and the nares sealed by valves, the increase in pressure forces air down into the lungs. This stretches the lungs and elevates pulmonary pressure. The lungs would discharge upon opening of the mouth or nares were it not for the fact that the glottis, the slitlike opening of the lung passage into the pharynx, is closable. The glottis is closed by muscular contraction after inhalation. The nares are then opened, and the animal often pumps air in and out of the buccal cavity through the nares by lowering and raising the floor of the cavity. This so-called buccopharyngeal pumping is easily observed in common frogs and is to be distinguished from pulmonary ventilation. After a period of time, the glottis is opened, and air from the lungs is exhaled. Exhalation results in part from the elastic recoil of the expanded lungs and to this extent is passive, that is, does not involve muscular contraction. Active forces of exhalation may also be brought to bear in the form of contraction of muscle in the walls of the lungs and body wall.

As noted earlier, differences in detail in the sequence of ventilatory events have been reported. Recent data on the bullfrog (*Rana catesbeiana*) will illustrate one possibility (Figure 8–22). In this species filling of the buccal cavity in preparation for inflation of the lungs occurs before pulmonary exhalation. With the glottis closed, air is drawn into the buccal cavity through the nares (step 1, Figure 8–22). Much of the air comes to lie in a posterior depression of the buccal floor, situated ventrally to the opening of the glottis. Next the glottis is opened, and pulmonary exhalant air passes in a coherent stream across the dorsal part of the buccopharyngeal cavity to exit through the nares (step 2). There is little admixture of exhalant air with the fresh air located in the posterior depression of the buccal floor. This fresh air is then driven into the lungs when, in step 3, the buccal floor is raised with the nares closed. After inflation of the lungs, the frog commences buccopharyngeal pumping with the glottis closed and nares open. It is evident from the data on bullfrogs that this pumping plays an important respiratory role, for it acts to wash out residual pulmonary ex-

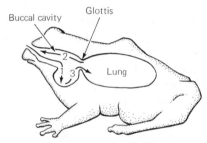

Figure 8–22. Diagrammatic representation of a bullfrog, showing three major steps in the pulmonary ventilatory cycle. See text for explanation. (From Gans, C. 1970. Evolution 24: 723–734.)

halant air from the buccal cavity so that when the next pulmonary ventilatory cycle begins, the cavity is filled with a relatively fresh mixture.

It is noteworthy that the lungs are not completely emptied of air on each exhalation, and air forced into the lungs on inhalation thus mixes with a residual volume of air left in the lungs. This is a typical consequence of tidal ventilation and means that, even immediately after inhalation, the partial pressure of oxygen in the lungs is below ambient, and the partial pressure of carbon dioxide is above ambient. More will be said about this important factor later.

The relative roles of various respiratory sites in gas exchange

Collectively, amphibians present at least four potentially important sites of gas exchange: (1) lungs, (2) gills, (3) the buccal cavity, and (4) the skin. Usually at least three of these are present in a given individual, and it is not surprising that we find considerable diversity in the emphasis placed on various routes of exchange. Not only are there interspecific differences, but the pattern of gas exchange can vary significantly in one individual depending on environmental conditions, activity state, and other factors. The present discussion will emphasize only a few salient possibilities.

The skin of larval and adult amphibians is characteristically permeable and well vascularized and typically plays an important role in gas exchange. Adult plethodontid salamanders, which lack lungs or gills, rely especially on cutaneous respiration. Some 75–95% of their oxygen and carbon dioxide exchange occurs across the skin; the remainder takes place across the buccal surfaces, which are ventilated by buccopharyngeal pumping. In adult frogs and toads and lunged salamanders, the skin is also an important site of respiration. Studies of these animals have revealed many interspecific differences in the pattern of gas exchange, but a basic similarity is apparent in many species and may be portrayed by discussing their average response. In air, around 75–80% of carbon dioxide is lost across the skin at temperatures from 5°C to 25°C. Carbon dioxide loss from the skin increases approximately proportionally with metabolic rate as temperature is raised within this range so that cutaneous loss accounts for a rather steady fraction of total loss. Oxygen uptake across the skin, on the other hand, does not increase proportionally with metabolic demand as temperature is elevated, with the result that the responsibility for oxygen uptake falls more and more to the lungs at increased temperatures. At 5°C about two thirds of oxygen uptake is cutaneous, with the lungs contributing only about one third. At 25°C, by contrast, these percentages are approximately reversed. Pulmonary oxygen uptake increases strongly between 5°C and 25°C concomitantly with increases in the tidal volume and rate of pulmonary ventilation, and pulmonary oxygen absorption comes to account for about two thirds of total uptake. For simplicity, "pulmonary" exchange has been used here to refer to what is actually the combination of exchange across the lungs and the buccopharyngeal surfaces. Experimentally it is not difficult to partition combined buccopulmonary respiration from cutaneous respiration, but the separation of buccal and pulmonary respiration has proved more challenging inasmuch as both occur through the mouth and nares. With some indirect evidence, it is generally presumed that the lungs account for a high proportion of the combined buccopulmonary exchange.

In winter, frogs often hibernate at the bottoms of ponds and lakes. Exchange of oxygen and carbon dioxide is then believed to be entirely cutaneous. An important factor in permitting cutaneous exchange to meet the entire respiratory requirement for long periods is the depressed metabolic rate of the animal, resulting in part from low temperatures and also, in at least some species, from a seasonal reduction in metabolism. In some frogs, resting demands for gas exchange can be met cutaneously during submergence even at relatively high temperatures. *Rana esculenta*, for example, is reported to survive for two or three weeks submerged in well-aerated water at 15°C. Experiments on leopard frogs, *Rana pipiens*, at 20°C have shown that resting oxygen consumption during submergence in aerated water is about the same as that in air, though it is highly unlikely that cutaneous exchange under water could support active metabolic rates equivalent to those supported by the combination of pulmonary and cutaneous respiration in air. Some leopard frogs survived 24 hours of submergence in aerated water at 20°C without ill effect. It is noteworthy that those that died were the largest specimens in the test group; this may have to do with the fact that larger animals have lower surface-to-volume ratios than smaller ones. In bullfrogs submerged at 20°C, cutaneous exchange is insufficient to prevent a rapid fall in blood oxygen tension, and large size may be a significant factor in this response.

We have earlier noted that the permeability of the skin of frogs presents problems of high evaporative water loss on land and relatively rapid passive exchange of salts and water during submergence. In terms of gas exchange, however, the permeability and vascularization of the skin give frogs a great deal of flexibility in their amphibious way of life. The skin provides a route of exchange when use of the lungs is denied during submergence, a route that can meet at least the resting demands of some smaller frogs at relatively high temperatures and that can apparently sustain many species at low temperatures.

Of those amphibians that retain gills at maturity, the mudpuppy (*Necturus*) has probably received the most attention from physiologists. As noted earlier, this species develops simple, saccular lungs at metamorphosis. The lungs, however, are involved in respiration to but a slight extent; perhaps they function primarily in control of buoyancy. The skin and gills are the prominent sites of gas exchange. When metabolic demands are low, as during rest at low temperatures, the gills may be relegated to a minor role, judging from the fact that they are frequently disengorged of blood and held motionless against the side of the head. At higher temperatures or during activity even at lower temperatures, the gills are brought into play and then can account for 50–60% of oxygen and carbon dioxide exchange.

Some salamanders and anurans that lead a chiefly aquatic existence as adults are typical dual breathers, using their lungs to respire air and exchanging with the water across their skin, and sometimes gills. *Xenopus*, the African clawed toad, for example, lives mostly in water and surfaces regularly to fill its lungs. It lacks gills but achieves an appreciable exchange across its skin, as indicated by the fact that it can survive for at least a day if denied access to the air even at a relatively high temperature of 20°C.

REPTILES

Anatomy of the lungs

Among reptiles the lungs of some species (such as the primitive tuatara) are rather simple and saccular, but in many species the lungs reach a stage of complexity well beyond that of amphibians. In these, folding of the walls may be so extensive and elaborate that the lung no longer appears as a sac with folded walls and an open central cavity. Rather, the lung is filled with septa and foldings and presents a rather solid, albeit spongy, appearance. In amphibians, air entering the anterior end of the lung reaches all parts via the central cavity. In the more complex of reptilian lungs, however, there is a well-defined tubular passageway (primary bronchus) running through each lung and giving rise to secondary tubes (secondary bronchi) that supply the terminal air spaces, or alveoli. Not uncommonly in reptiles the anterior part of the lung is more elaborately developed than the posterior part. The lungs of crocodilians are among the more complex of reptilian lungs. Perhaps the most highly developed lungs are found in monitor lizards, animals noted for their especially active way of life (see Chapter 11). Monitor lizards exhibit considerably more elaborate and extensive subdivision of the lung than most lizards and unlike most lizards have true cartilage-lined bronchi within the lung. Even the lungs of monitor lizards, however, are substantially less elaborate than those of most mammals.

Ventilation

In reptiles we find a basic transition in the mode of ventilation that is carried over into the birds and mammals: the lungs are filled by suction rather than by buccal pressure. That is, air is drawn into the lungs by an expansion of pulmonary volume that creates a subatmospheric pressure in the lungs.

As will be discussed in more detail subsequently, the lungs assume a certain volume, termed the relaxation volume, if there are no active muscular forces tending to expand or contract them. This volume is determined by an interplay of elastic forces in the lungs themselves and the surrounding thoracic structures. In some reptiles, such as lizards and crocodilians, the lungs are expanded beyond their relaxation volume by muscular activity during inspiration, and during expiration they contract at least in part under forces of elastic rebound. In other reptiles, such as at least some snakes, the lungs are compressed below their relaxation volume by muscular activity during expiration and expand elastically on inspiration. In either case, the basic process of inspiration is suction. The effort of ventilating the lungs is transferred from the buccal cavity to muscles of the thorax and abdomen, and respiratory constraints on the form and function of the buccal cavity are relaxed.

Whereas modern amphibians have short, poorly developed ribs, the thoracic cavity of reptiles other than turtles is enclosed in a flexible rib cage. Running over and between the ribs on each side of the body are sheets of muscle, the contractions of which can expand or contract the volume enclosed by the rib cage. These muscles, termed *costal muscles* (*costa* = "rib"), are believed to play a major role in ventilation in lizards but probably play a more secondary role in snakes and crocodilians. The ventilatory cycle in lizards may be outlined as follows. The lungs are filled by an expansion of the thoracic cavity with the glottis open. Certain costal muscles are active at this time, and it is thought that these muscles expand the rib cage, concomitantly expanding the lungs. This mechanism has been referred to as a costal suction pump. After inflation of the lungs, the glottis is closed, and the inspiratory muscles relax. Elastic forces in the lungs and thoracic wall tend to return pulmonary volume to its preinspiratory condition, but because air cannot exit, complete return is prevented and the pressure in the lungs rises somewhat above atmospheric. During the ensuing respiratory pause the buccal cavity is often ventilated by buccopharyngeal pumping. After a period of several seconds to several minutes, expiration occurs, followed quickly by another inspiration. Both passive and active forces are involved in expiration. The glottis is opened, elastic forces drive air out, and, in addition, contraction of thoracic and abdominal muscles compresses the thoracic cavity. Certain costal muscles are active in this process. The overall pattern is one of pulmonary ventilatory cycles, performed with the glottis open, separated by periods of apnea (no breathing) during which the glottis is closed and the lungs are filled. In this sense, ventilation in lizards resembles that of amphibians. A similar pattern, of ventilation interspersed with apnea, is observed in the other groups of reptiles as well, though the muscular mechanics of filling and emptying the lungs differ to a greater or lesser extent.

It was long believed that turtles, in contrast to other reptiles, would have to fill their lungs by buccopharyngeal pressure because the ribs are fused to the shell and the skeletal elements surrounding the lungs are more or less inflexible. However, it is now known that contractions of certain abdominal and thoracic muscles act to expand and contract the volume of the visceral cavity and that these forces are transmitted to the lungs, dilating and compressing the pulmonary cavity. As in other reptiles, the lungs are filled by suction.

Buccopharyngeal pumping during periods of apnea is conspicuous in many lizards and terrestrial turtles and is believed usually to be primarily related to olfaction. The nasal olfactory organs might otherwise not have adequate access to the external environment during apnea. Turtles under water may ventilate the buccal cavity with water. Soft-shelled turtles (*Tri-*

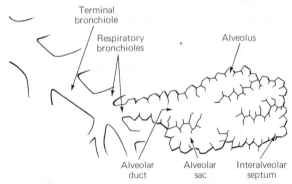

Figure 8–23. Diagrammatic cross section of the terminal airways in the mammalian lung. Terminal bronchioles, having diameters of 0.5 mm or less in man, represent the final branches of the purely conducting (nonrespiratory) bronchial-bronchiolar tree (see discussion of the anatomical dead space subsequently in the text). Each terminal bronchiole branches into two or more respiratory bronchioles. Respiratory bronchioles lead to alveolar ducts, which in turn terminate in alveolar sacs. The alveoli are minute out-pocketings of the alveolar ducts and sacs. Alveoli first appear along the respiratory bronchioles and form a continuous lining along the alveolar ducts and alveolar sacs. Adjacent alveoli are separated by thin interalveolar septa. The walls of the alveoli are richly invested with capillaries. [From Hildebrandt, J. and A. C. Young. 1965. Anatomy and physics of respiration. *In:* T. C. Ruch and H. D. Patton (eds.), *Physiology and Biophysics.* 19th ed. Saunders, Philadelphia.]

onyx), for example, possess many vascularized pharyngeal villi, and considerable oxygen uptake from the water occurs across the buccopharyngeal membranes, which are ventilated. Oxygen uptake also occurs across their skin and cloaca, the latter being ventilated. These sites of oxygen exchange are important during long dives and during winter hibernation in the mud at the bottoms of lakes and ponds.

Buccopharyngeal pressure, although not utilized to fill the lungs during normal ventilation, is used by some reptiles to cause overinflation in certain situations. Thus the buccal force pump, so prominent in amphibians, has not been entirely lost. When disturbed, for example, chuckwallas (*Sauromalus*) wedge themselves in among rocks by greatly inflating their lungs through a series of cycles in which air is driven into the lungs under positive pressure from the buccal cavity.

MAMMALS

Basic anatomy of the lungs

The lungs attain a very high degree of complexity in mammals. The bronchi branch dendritically into tubes of smaller and smaller diameter, the finer branches being termed bronchioles. The final branches lead into alveolar sacs composed of many semispherical alveoli (Figure 8–23). The alveolar clusters have something of the appearance of a bunch of grapes. The alveolar clusters of adult man are separated from the trachea by a mean of 23 branches of the bronchial-bronchiolar tree. There are some 300 million alveoli in the lungs of adult man, each measuring 150–300 μ across and having a collective surface area of about 75 m². This is greater than the floor area of a 25- by 30-ft room. These data demonstrate the elaborate complexity of the mammalian lung and the amplification of surface area that results.

Functional parameters
of ventilation and gas exchange

The functional parameters of ventilation and gas exchange have been studied exhaustively in man and certain other mammals and can be discussed in particularly specific and quantitative terms. The reader should recognize that many of the concepts developed below have clear applicability to many other vertebrate lungs despite variations in anatomical and physiological detail.

The trachea, bronchi, and all but the last few branches of bronchioles in the mammalian lung are not much involved in gas exchange and accordingly are termed the *anatomical dead space*. They are lined with a relatively thick epithelium and do not receive a particularly rich vascular supply. The last few branches of bronchioles and the alveoli are lined with thin, highly flattened epithelial cells and are richly supplied with blood capillaries. It is here that gas exchange occurs. The alveoli constitute most of the exchange surface, and some 75% of the alveolar surface is invested with capillaries. Gases must diffuse across the alveolar epithelium, the capillary endothelium, and a basement membrane separating the two. These structures are all very thin (Figure 8–24), and the total blood-to-gas distance is considerably less than 1 μ (0.2–0.6 μ in man). By comparison, this distance is about 4 μ in simple unfolded amphibian lungs and about 1 μ in turtle lungs.

During ventilation the volume of the air spaces in the lung alternately increases and decreases. In resting man the volume of air exhaled and inhaled in each ventilatory cycle—that is, the *resting tidal volume*—is about 500 cc. The volume of air left in the lungs at the end of expiration is about 2400 cc. This is termed the *resting expiratory volume* (or, more formally, the functional residual capacity). The volume of the anatomical dead space varies somewhat over the breathing cycle, but most of the change in lung volume occurs in the alveolar ducts and sacs and the alveoli (see Figure 8–23). The average volume of the dead space in man is about 150 cc.

With the preceding background information, we can appreciate certain important features of alveolar gas exchange. During expiration in resting man 150 cc of air left in the dead space from the previous inhalation is expired, along with 350 cc of air from the respiratory spaces. At the end of expiration the dead space is left full of air that has come up from the respiratory spaces, and the entire 2400 cc of air remaining in the lungs is therefore air that has been in the respiratory spaces. This air has a lowered oxygen tension and elevated carbon dioxide tension. On inspiration, when the respiratory spaces expand, the first air to reach the alveoli is that left at the end of expiration in the alveolar ducts and sacs, respiratory bronchioles, and dead space. This air—of lowered oxygen tension and elevated carbon dioxide tension—effectively fills the alveoli. Of the 500 cc of fresh air inspired, 350 cc passes through the dead space and enters the respiratory bronchioles and alveolar sacs and ducts. The alveoli themselves are not washed to any appreciable extent with the fresh air because they have previously been filled. The important result is that exchange between the fresh air and alveolar air is largely by diffusion. This is not any serious impediment to respiratory exchange because fresh air is carried convectively to the apertures of the alveoli, and diffusion across the minute distances to the alveolar membranes occurs rapidly in air. Note that the

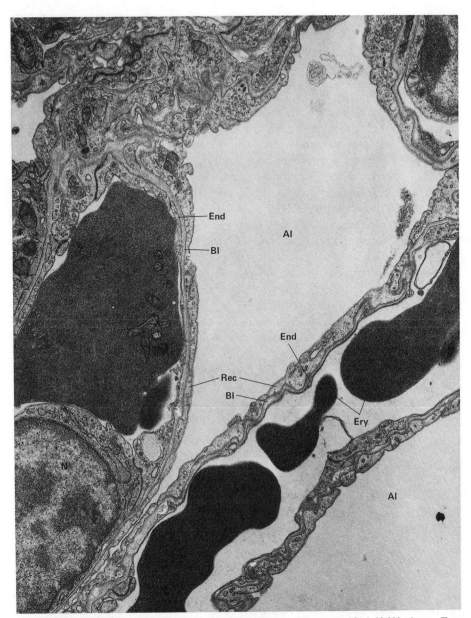

Figure 8–24. Electron micrograph of the lung of a mouse, magnified 32,000 times. Two alveoli (Al) are separated by an interalveolar septum containing a blood capillary in which erythrocytes (Ery) can be seen. The lumen of the alveolus is separated from the lumen of the capillary by: (1) respiratory epithelial cells (Rec) lining the alveolus, (2) a basement membrane (Bl), and (3) capillary endothelial cells (End) lining the capillary. The capillary endothelial cells are very attenuated except in the region of the cell nucleus (N). The respiratory epithelial cells are similarly attenuated. (Electron micrograph kindly supplied by K. R. Porter and M. A. Bonneville.)

fact that the alveoli are not washed convectively is a direct result of tidal ventilation in a lung whose passageways end blindly. This phenomenon is a potentially serious limitation for tidally ventilated lungs in aquatically respiring animals because diffusion is so much slower in water than in air. It is often pointed out that diffusional exchange with the alveoli in mammals enhances the stability of alveolar gas composition and thus may have advantages. The alveolar respiratory surfaces are not exposed to oxygen-depleted air one moment (at the end of expiration) and fresh air the next (at the end of inspiration), but are separated from the fresh air by a diffusional buffer.

When the respiratory organs are ventilated unidirectionally, as in teleost gills, medium exposed to the respiratory surfaces at one moment can be replaced completely with fresh medium at full ambient oxygen tension by a single ventilatory cycle. When the organs are ventilated tidally, however, the only way to accomplish this same result would be to empty and refill the respiratory structure completely on each ventilatory cycle. This is probably never realized in animals with tidally ventilated respiratory structures and certainly is not realized in vertebrate lungs. We have noted earlier that in resting man 2400 cc is left in the lungs at the end of expiration. Even when man undergoes maximal expiratory effort, 1200 cc remains. This has the important consequence that fresh inhaled medium is equilibrated with medium that has already been exposed to the respiratory surfaces. Alveolar oxygen tension is consequently always below ambient, and carbon dioxide tension is always above ambient.

These conclusions are dictated by simple physical principles. More information is needed to determine the exact levels of oxygen and carbon dioxide in the alveoli. Taking oxygen tension as an example, we must know the rate at which metabolic demands tend to reduce the tension and the rate at which ventilatory exchange tends to raise it. In other words, the alveolar oxygen tension is a dynamic property. It turns out that in mammals the rate of ventilatory exchange is regulated so that alveolar oxygen and carbon dioxide tensions remain remarkably constant over a broad range of rates of oxygen consumption. In man, for example, alveolar oxygen tension at rest is about 100 mm Hg, and carbon dioxide tension is about 40 mm Hg. These tensions are altered only slightly in all but strenuous exercise.

The rate of ventilatory exchange is commonly expressed as the *minute volume*—or volume of air exchanged per minute. This in turn is partitioned into two components, the tidal volume, or volume exchanged on each ventilatory cycle, and the frequency of ventilatory cycles:

minute volume = tidal volume × ventilatory frequency (breaths/min)

Given that resting man breathes about 12 times per minute, his minute volume is about 500 × 12 = 6000 cc/min.

When mammals exercise, the minute volume is increased according to increased metabolic demands so that alveolar gas tensions remain stable. The minute volume is increased by increases in both ventilatory frequency and tidal volume. The tidal volume can be increased relative to the resting condition both by increasing the extent of compression of the lungs on expiration and by increasing the extent of expansion on inspiration (Figure 8–25). The amount of air that can be expelled beyond the resting expiratory level is termed the *expiratory reserve volume*. Similarly, the amount that can be inhaled beyond the resting inspiratory level is termed the *inspiratory reserve volume*. As shown in Figure 8–25, the average inspiratory reserve in healthy young men is about 3100 cc, and the average expiratory reserve is about 1200 cc. The maximal possible tidal volume, termed the *vital ca-*

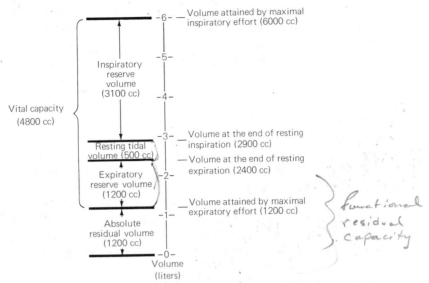

Figure 8–25. Average lung volumes in healthy young adult males. Volumes include the anatomical dead space as well as respiratory spaces.

pacity, is the sum of the resting tidal volume and the inspiratory and expiratory reserve volumes, or about 4800 cc in young men. Vital capacity tends to decrease with advancing age in man and is decreased by some disease states. Physical training tends to enhance the vital capacity.

During exercise it is not possible to maximize both tidal volume and ventilatory frequency inasmuch as the time needed for one ventilatory cycle tends to increase with the tidal volume. Trained athletes can achieve minute volumes of over 100 liters/min (compared to about 6 liters/min at rest), combining tidal volumes of around 3000 cc with respiratory frequencies in excess of 30/min. This performance represents a compromise between frequency and tidal volume. Usually most of the increase in tidal volume beyond the resting condition in man is achieved by enhancing inspiratory volume beyond the resting inspiratory level—that is, by utilizing the inspiratory reserve volume. The expiratory reserve volume is also utilized to some extent.

Oxygen utilizations in mammals are generally in the range of 20–30%. These utilizations are measured, as usual, by comparing the oxygen content of air entering the lungs with that of air exiting the lungs. Oxygen utilization typically varies with exercise state, but it should be recognized that for the most part this variation is not due to changes in the percent extraction of oxygen reaching the respiratory spaces. If the extraction in the respiratory spaces varied, alveolar oxygen tension would also vary. Rather, changes in oxygen utilization are largely attributable to changes in the percent of ventilatory volume relegated to the dead spaces. This may be illustrated by example. Assuming total air pressure to be 1 atm (760 mm Hg), the oxygen tension of inhaled air when it reaches the respiratory spaces of the lungs is about 149 mm Hg. This is lower than the tension in dry air, 159 mm Hg, because the air becomes rapidly saturated with water vapor at the body temperature of the animal when it is inhaled. Over a broad range of exercise states, air exiting the respiratory spaces in man has an oxygen tension of about 100 mm Hg. The percent extraction of oxygen in the respiratory spaces is thus relatively stable at approximately 33%, and if all inhaled air were to reach the respiratory spaces, this would be the oxygen utilization. At rest, however, only about 350 cc of every 500 cc inhaled reaches the respiratory spaces. Thus about two thirds of the inhaled air suffers an oxygen extraction of 33%, but the one third relegated to the dead space suffers virtually no oxygen extraction. Accordingly, the oxygen utilization measured for the entire inspired volume is close to 20%. When tidal volume is increased during exercise, a smaller fraction of the inspired air is relegated to the dead space, and accordingly, the oxygen utilization increases toward the percent extraction realized in the respiratory spaces. Thus at a tidal volume of 2000 cc, over 90% of the inspired volume enters the respiratory spaces and suffers an oxygen extraction of 33%, and the oxygen utilization measured for the entire inspired volume is close to 30%. In man oxygen utilization increases with exercise according to these principles except at high exertion, when reduced extraction in the respiratory spaces brings about a decrease in utilization.

You will note that the oxygen utilizations of mammals are much lower than those realized by teleost fish (often near 80%) despite the higher metabolic rates of the mammals. The oxygen concentrations of water and air help to explain this. Oxygen is so much less concentrated in water that a fish extracting 80% of the oxygen from fresh water at 0°C and an oxygen pressure of 149 mm Hg receives only 7.7 cc from each liter, whereas a resting man extracting 22% from air at 37°C and an oxygen pressure of 149 mm Hg receives 38 cc (at STP) from each liter.

Relationships between pulmonary parameters and body size

In mammals ranging in size from shrews to whales, lung volume constitutes a rather steady proportion of total body volume. Specifically, lung volume in liters averages about 6% of body weight in kilograms. A mammal weighing 10 kg, for example, would have a lung volume of about 0.6 liter. Vital capacity in liters is also close to 6% of body weight in kilograms, and resting tidal volume is about one tenth of vital capacity, or 0.6% of body weight. The resting weight-specific rate of oxygen consumption of mammals tends to increase with decreasing body size (Chapter 2), and as we would expect, weight-specific ventilatory minute volume tends to increase in parallel. Thus we arrive at the conclusion that minute volume per unit weight is greater in small animals than in large ones, but resting tidal volume per unit weight is about the same in the two groups. This implies the common observation that ventilatory frequency must be greater in the smaller animals. Because they require more oxygen per unit weight but obtain about the same amount per unit weight on each ventilatory cycle, they must breathe more frequently. A resting laboratory mouse breathes well over 100 times per minute, as compared to about 12 times per minute in resting man.

If the ratio of respiratory surface area to lung volume were to be the same in all mammals, it is clear from the proportionality of lung volume and body weight that all mammals would have about the same respiratory surface area per unit weight. This would mean that the rate of oxygen diffusion across each unit of surface area would have to be greater in small mammals to supply their greater weight-specific oxygen demands. There is some controversy over the relationship between respiratory surface area and body size, but one extensive study has indicated that surface area tends to be proportional to weight-specific metabolism rather than body weight, meaning that oxygen diffusion per unit of surface area tends to be similar across the range of mammalian body size. The relationship implies that animals with high weight-specific oxygen demands have more respiratory surface area per unit of lung volume than animals with lower weight-specific oxygen demands, a property resulting partly from the fact that alveolar diameter tends to decrease with decreasing body size. The average diameter of alveoli in the laboratory mouse, for example, is about 47 μ, whereas in man the average diameter is about 200 μ. This means that each volume of mouse lung is divided into more alveoli than each volume of human lung, and surface area in the mouse lung is amplified accordingly.

Mechanics of ventilation

Unlike other vertebrates, mammals have a true diaphragm that completely separates the thoracic and abdominal cavities and plays a central role in ventilation. This sheet of muscular and connective tissue is dome-shaped, projecting further into the thorax at its center than at its edges. The edges are attached to the body wall. Contraction of the diaphragm muscles tends to flatten the diaphragm, pulling the center away from the thorax toward the abdomen. This increases thoracic volume, resulting in expansion of the lungs and inflow of air by suction. It also reduces abdominal volume, as witnessed by the bulging of the abdominal wall on inspiration.

Among the other muscles important to ventilation are the external and internal intercostals. These are sheets of muscle that run obliquely between each pair of adjacent ribs. Contraction of the external intercostals rotates the ribs anteriorly and outward, expanding the thoracic cavity. The fibers of the internal intercostals run crossways to those of the externals, and, in general, contraction of the internal intercostals rotates the ribs posteriorly and inward, decreasing the volume of the thoracic cavity. You can easily demonstrate the action of these muscles on yourself by consciously expanding and contracting your rib cage while feeling the ribs with your fingers.

A significant development in mammals (also seen in birds) is that the glottis is no longer used to close off the lungs between inspiration and expiration, as it is used in reptiles and amphibians. Thus the lungs remain filled only so long as contraction of the inspiratory muscles maintains the thorax in its expanded condition.

Having noted these features, we may outline the major attributes of the ventilatory cycle with reference to man. Both the lungs and the thoracic wall are elastic structures, and as such they each have a relaxation volume, that is, a volume they will assume if isolated and neither stretched nor compressed by external forces. An analogy can be drawn with a rubber ball. If the ball is simply allowed to empty to the air, it will come to a certain volume, its relaxation volume. If we suck air out of the ball or pump air into it, we can alter this volume. The isolated thorax of man has a relaxation volume of about 5 liters, meaning that the thoracic cavity will contain about 5000 cc of air. The isolated lungs have a relaxation volume of about 1000 cc. In the intact body, if no muscular forces are applied, the lungs will tend to contract to their inherent relaxation volume, but in doing so they will tend to pull the thoracic walls in and reduce thoracic volume below the thoracic relaxation volume. Contrariwise, the thorax will tend to retain its inherent relaxation volume and thus will tend to maintain the lungs at a greater volume than the inherent pulmonary relaxation volume. The net result is that an equilibrium is reached wherein the contractive forces in the lungs are exactly counterbalanced by expansive forces in the thoracic wall. The volume of the lungs in this passive equilibrium state is the functional pulmonary relaxation volume. Any deviation from this elastic equilibrium volume requires muscular effort, and contrariwise, if lung volume is greater than or less than the functional relaxation volume, the relaxation volume will be restored by elastic forces in the absence of contravening muscular forces. The functional relaxation volume of the lungs in man is about 2400 cc. You will note that this is identical to the resting expiratory volume, and, in fact, the volume assumed by the lungs at the end of expiration in resting man is that dictated by the passive interplay of elastic forces in the lungs and thoracic wall.

At rest, inspiration is active but expiration is largely or completely passive. During inspiration the thoracic cavity is expanded by contraction of the diaphragm, external intercostal muscles, and anterior internal intercostal muscles. These muscles progressively relax at expiration, and elastic forces bring about a reduction in lung volume and thoracic volume to the elastic equilibrium condition.

During exercise, not only is the tidal volume increased, but also the respiratory frequency is increased. Additional muscular activity is required to amplify changes in lung volume over the ventilatory cycle and to hasten the inspiratory and expiratory processes. The external intercostals assume a greater role in inspiration than during rest. Expansion of the rib cage by these muscles during quiet breathing is of relatively minor importance by comparison to contraction of the diaphragm in bringing about expansion of the lungs, but during heavy exertion, expansion of the rib cage comes to account for about half of the change in lung volume during inspiration. In addition, active forces contribute to expiration during exercise. The most important muscles are the internal intercostals, which actively contract the rib cage, and muscles of the abdominal wall, which contract the abdominal cavity, forcing the diaphragm upward into the thoracic cavity. These muscles hasten expiration by supplementing the elastic rebound of the lungs and may also cause compression of the lungs beyond their functional relaxation volume, thus enhancing tidal volume through utilization of some of the expiratory reserve volume.

The same basic groups of muscles are utilized for ventilation in other mammals, but their relative importance varies. The forelimbs are suspended from the rib cage through the shoulder blades, and in larger quadrupeds especially, this places constraints on movements of the rib cage. The diaphragm then assumes greater overall importance in ventilation. Aquatic mammals are reported to rely heavily on the intercostal muscles.

The energetics of ventilation

The energetics of respiration have been studied in considerable detail in man. It has been estimated that the energy required for the necessary alveolar ventilation at rest is minimal at the usual respiratory frequency and tidal volume. The same alveolar ventilation could be attained by breathing more rapidly with a smaller tidal volume, but then a greater volume of air would have to be moved because a lesser fraction of the total minute volume would reach the respiratory spaces rather than the dead spaces. The energetic cost would be increased. On the other hand, the same alveolar ventilation could be attained by increasing the tidal volume and breathing more slowly. Again, though, the energetic cost would be increased because more energy would have to be expended in expanding the thorax to greater dimensions.

At rest, on the order of 1–3% of total oxygen consumption is due to the effort of ventilating the lungs. During exercise the cost of ventilation increases roughly in proportion to the rate of oxygen extraction across the lungs up to a minute volume of about 40 liters/min (about seven times the resting level). At higher minute volumes, however, the cost of ventilation increases out of proportion to the rate of oxygen extraction because the cost of moving a volume of air becomes greater when frequency and tidal volume are high. At a minute volume of 120 liters/min, ventilation demands 10–15% of oxygen consumption.

The control of ventilation

The control of ventilation has been more thoroughly studied in man and other mammals used in medical research than in any other animals. As we find when pursuing almost any subject in depth, the control of ventilation turns out to be far from simple, and many important questions remain unresolved. The present discussion will be limited to certain highlights.

The respiratory control center is located in the medulla of the brain and sends out motor nerves that activate the respiratory muscles. If the medulla is isolated experimentally from all neural inputs, an animal will continue to breathe rhythmically. This type of evidence has demonstrated that there is an endogenous respiratory rhythm in the medulla, but it is abundantly clear that this rhythm is modified, or modulated, by a variety of neural inputs in the intact animal. Another portion of the brain, the pons, is routinely involved in controlling the rhythm, and we all know that higher, conscious centers can intervene to modify the pattern of ventilation. (We can stop breathing, for example, when we want to.) Sensory nerves from the thorax relay information concerning the degree of expansion of the lungs. Some modulate the respiratory rhythm by inhibiting inspiration when the lungs are expanded, whereas others act to excite inspiration when the lungs are compressed.

Because the primary function of the lungs is supply of oxygen and removal of carbon dioxide, and because we know that ventilation is regulated so as to maintain stable alveolar tensions of these gases, it is clear that the parameters of ventilation must ultimately be dictated by information concerning gas exchange. This information is provided by chemoreceptors and chemosensitive areas that sense certain parameters of blood composition that, in turn, provide a reliable indication of respiratory status.

When the concentration of carbon dioxide in the blood is increased, the hydrogen ion concentration will also increase. One or both of these concentrations are sensed in the medulla, and deviations from normal levels exert a potent influence on respiration. Ventilation increases or decreases in such a way as to bring the concentrations back to normal—a negative feedback system. Thus if the carbon dioxide concentration of the blood is elevated, ventilation is increased, resulting in a greater rate of removal of carbon dioxide. On the other hand, a decrease in carbon dioxide concentration will tend to decrease the rate of ventilation, allowing carbon dioxide to accumulate. The potency of these effects is illustrated by the observation that an increase in arterial carbon dioxide tension from 40 to 44 mm Hg will result in a doubling of the respiratory minute volume in man. Oxygen tension is sensed in chemoreceptive bodies, termed the *aortic* and *carotid bodies*, located in the aorta and carotid arteries; information is relayed to the medulla via nerves. Respiratory sensitivity to oxygen tension is considerably lower than that to carbon dioxide tension and/or H^+ concentration within the usual range of blood composition, but at very low oxygen tensions, ventilation is stimulated strongly. Thus at constant carbon dioxide tension, ventilation in man increases by only about 25% with a reduction in arterial oxygen tension from 100 mm Hg to 60 mm Hg. Below 40–50 mm Hg of oxygen pressure, however, ventilation increases markedly. The carotid and aortic bodies become more sensitive to lowered oxygen tension as carbon dioxide tension is elevated. Although sensation of oxygen tension probably plays some subtle role in controlling ventilation under usual conditions, the paramount role is attributed to sensation of carbon dioxide tension or H^+ concentration in the medulla. Response to lowered blood oxygen tension becomes very apparent when the air has a low oxygen pressure, as at high altitude.

Many gaps remain in our knowledge of ventilatory control. The integration of various inputs to the respiratory control center in the medulla, for example, is not understood in quantitative terms. The control of ventilation during normal exercise presents many enigmas and deserves brief consideration.

Blood oxygen and carbon dioxide tensions and H^+ concentration remain relatively stable during exercise, along with and because of a great increase in minute volume. Deviations of these parameters from their usual levels are inadequate to explain maintenance of high minute volumes on the basis of the simple negative feedback systems discussed above. To illustrate, we noted earlier that an increase in carbon dioxide tension of 4 mm Hg will cause a doubling of minute volume, but carbon dioxide tension during exercise may not be elevated to this extent even when the minute volume is 10 or 15 times greater than the resting minute volume. It is clear that other factors are involved in elevating ventilation during exercise, though sensation of oxygen, carbon dioxide, and H^+ may exert fine control of ventilation against the background of these more immediate factors. What the other factors are remains largely obscure. It has been suggested that ventilation could be increased, for example, (1) by hormones such as adrenaline and noradrenaline, (2) by nervous input from receptors that sense the movements of the joints, (3) by some influence of motor activity in the higher centers of the central nervous system on the respiratory center, or (4) by accumulation of metabolites such as lactic acid. Some of these suggestions are supported by experimental evidence, but as yet the control of ventilation during exercise cannot be explained adequately.

Knowledge of respiratory control in the other vertebrate classes is fragmentary. It is clear that the motor impulses that activate the ventilatory muscles originate in the medulla in all groups. Representatives of all classes have been shown to respond to changes in carbon dioxide and oxygen tensions, but the receptors and their neural connections have been studied in only a few species. It is noteworthy that in teleost fish that have received study, the rate of ventilation is typically more sensitive to oxygen tension than to carbon dioxide tension, a situation that contrasts clearly with that in mammals.

BIRDS

The structure of the respiratory system

The avian respiratory system differs in many fundamental respects from that of mammals and reptiles. The structure of the system is complex and will be discussed here only in outline, with reference to the highly simplified diagram of Figure 8–26A. The trachea bifurcates to give rise to two primary bronchi that, respectively, enter the two lungs. The primary bronchus passes through the lung and within the lung is termed the *mesobronchus*. A number of *secondary bronchi* arise from the mesobronchus. These may be divided into anterior and posterior groups. For simplicity, each group is represented as a single passageway in Figure 8–26A, but remember that, in fact, there are several bronchi in each group. The anterior and posterior secondary bronchi are connected by a great many small tubes measuring a millimeter or less in diameter and termed *tertiary bronchi* or *parabronchi*. It is along the parabronchi that most gas exchange occurs. As depicted in Figure 8–27, each parabronchus gives off radially along its length an immense number of finely branching *air capillaries*. These are profusely surrounded by blood capillaries and are the sites of gas exchange. The air capillaries are on the order of only 3–10 μ in diameter, and the exchange surface between air capillaries and blood capillaries is enormous, 200–300 square millimeters per cubic millimeter of tissue in the parabronchial walls. Air flows through the parabronchi, but exchange between the parabronchial lumen and the surfaces of the air capillaries is probably largely by diffusion. The parabronchi, air capillaries, and associated vasculature constitute the bulk of the lung tissue. A recent study has indicated that the domestic duck and pigeon, both strong fliers, possess about four times as many parabronchi as the domestic chicken, a poor flier.

In mammals the bronchioles end blindly in clusters of alveoli, and air must move tidally to and from the respiratory surfaces. In birds, on the other hand, the parabronchi are open at both ends. This means that air might move unidirectionally past the openings of the air capillaries; as we shall see shortly, this appears to be the case.

Outside of the lungs, in the body cavity, are the air sacs. Usually there are nine, divisible into two groups. The anterior sacs (two cervical sacs, two anterior thoracic sacs, and a single interclavicular sac) open to various anterior secondary bronchi. The posterior sacs (two abdominal sacs and two posterior thoracic sacs) open to the posterior mesobronchi. (Each

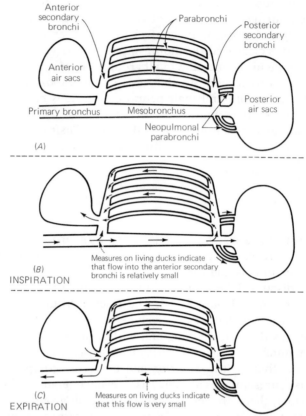

Figure 8–26. Aspects of anatomy and air flow in the respiratory system of birds. (*A*) A schematic representation of the anatomy of the bird lung and its connections with the air sacs. (For a detailed description, see Duncker, H.-R. 1972. Resp. Physiol. 14: 44–63.) (*B* and *C*) Known and presumptive directions of air flow in the lung and air sacs during inspiration (*B*) and expiration (*C*). (Based on Bretz, W. L. and K. Schmidt-Nielsen. 1971. J. Exp. Biol. 54: 103–118; Scheid, P., H. Slama, and J. Piiper. 1972. Resp. Physiol. 14: 83–95.)

mesobronchus terminates at its connection with an abdominal air sac.) In Figure 8–26*A* the anterior and posterior groups of air sacs are represented artificially as single structures. In fact, there are no direct connections among the various sacs. The air sacs are thin-walled, poorly vascularized structures that play little direct role in gas exchange with the blood. They take up a considerable fraction of the body cavity.

The structures of the lung described thus far (mesobronchus and anterior and posterior sets of secondary bronchi connected by parabronchi) are present in all birds, and their connections with the air sacs are similar in all birds. These pulmonary structures are collectively termed the *paleopulmonal system*, or simply *paleopulmo*. The paleopulmo is the only system found in some relatively primitive birds, such as penguins and emus. Most birds, in addition to the paleopulmonal system, also have a more or less extensively developed system of parabronchial tubes running directly between the posterior air sacs, on the one hand, and the posterior parts of the mesobronchi and posterior secondary bronchi, on the other. This system is called the *neopulmonal system*, or simply *neopulmo*. The tubes of the neopulmo, being invested with air capillaries, are respiratory and

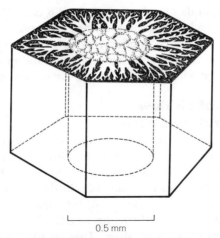

0.5 mm

Figure 8–27. Diagram of a segment of a parabronchus. The section at the top shows the air capillaries. The stippled network along the wall of the lumen of the parabronchus shows the pattern of air capillary orifices. (From Hazelhoff, E. H. 1943. Verslag van de gewone vergaderingen der Afdeling Natuurkunde van de Nederlanse Akademie van Wetenschappen 52: 391–400; English translation available in Poultry Sci. 30: 3–10, 1951.)

are termed neopulmonal parabronchi (see Figure 8–26A). The neopulmonal system may be relatively simple or may reach a high level of complexity in which direct connections are formed by neopulmonal parabronchi between the posterior and anterior sets of secondary bronchi (not shown in Figure 8–26), thus essentially paralleling the parabronchi of the paleopulmo. The paleopulmonal system always remains dominant, the neopulmo never representing more than 20–25% of the total lung volume even in its most highly developed form.

Lung weight in birds is roughly proportional to body weight and is similar to that of mammals of equivalent size. The lungs of birds, however, are more compact than those of mammals; that is, they occupy a smaller volume. When the volumes of the air sacs are considered, the total volume of the respiratory system in birds far exceeds that in mammals of the same size.

Ventilatory mechanics

Avian lungs are rigid by comparison to mammalian lungs, and the lung passages appear to undergo relatively little change in volume over the ventilatory cycle. The air sacs are expanded and contracted appreciably, and it seems clear that these expansions and contractions of the air sacs ventilate the lungs by a bellowslike action. To state this proposition in its extreme form, the lung passages would be viewed as rigid tubes, and air would be driven through the tubes over the ventilatory cycle as the air sacs are filled and emptied. There is evidence that the lung passages themselves expand and contract to some extent, but the role of these volume changes in ventilation is not understood.

As in mammals, air is drawn into the respiratory system from the environment by suction and expelled by positive pressure. The glottis remains open through the cycle.

The rib cage surrounding the lungs themselves is relatively rigid. During inspiration contraction of internal intercostals and several other thoracic muscles expands other parts of the rib cage (especially those posterior to the lungs), and the sternum swings downward and forward. These movements expand the thoracoabdominal cavity and enlarge the air sacs. Some of the external intercostals and various abdominal muscles act to compress the thoracoabdominal cavity and air sacs at expiration.

The pattern of air flow in the respiratory system

Of all the mysteries surrounding the avian respiratory system, the most profound has been the pattern of air flow. There is a complexity of interconnections among the various components of the system (Figure 8–26A) that, a priori, provides for many possibilities. The absence of obvious valves has severely limited the amount of information that can be deduced from morphological evidence, throwing nearly all the weight of investigation on physiological experimentation. This in turn has been difficult because the system is so complex and because experimental intervention has been limited by the usual need of not overly disrupting the system under study. Most investigations have been indirect. Deductions have been drawn from comparisons of gas composition in various parts of the respiratory system. Birds have breathed special gas mixtures or suspensions of fine carbon particles, and attempts have been made to trace the path of these substances in the lungs and air sacs. Only recently have investigators succeeded in placing flow detectors at various places in the mesobronchi and secondary bronchi of ducks. Their results provide direct confirmation of a basic hypothesis that has been supported by indirect evidence on a diversity of other species, including pigeons, geese, crows, finches, and the ostrich. Though many uncertainties remain and there is need for direct study of additional species, the basic pattern of flow now seems to be clear.

We shall first examine flow through the paleopulmo, that part of the lung that is dominant in all birds. During inspiration, both the anterior and posterior sets of air sacs are expanded and therefore must receive air. As depicted in Figure 8–26B, most air from the outside flows through the mesobronchus into the posterior air sacs. Some may also reach the anterior sacs via the anterior secondary bronchi. Some also enters the posterior secondary bronchi and passes anteriorly through the parabronchi. Air exiting the parabronchi enters the anterior air sacs. Three aspects of the events at inhalation deserve emphasis: (1) The posterior air sacs are filled with relatively fresh air coming directly from the environment. (2) The anterior air sacs are filled substantially with air that has passed across the respiratory surfaces in the parabronchi; this air has suffered a drop in oxygen content and rise in carbon dioxide content. (3) The direction of ventilation of the parabronchi in the paleopulmo is from posterior to anterior during inspiration.

During expiration the air sacs are compressed and discharge air. As shown in Figure 8–26C, air exiting the posterior air sacs predominantly enters the posterior secondary bronchi to pass anteriorly through the parabronchi. This air is relatively fresh, having entered the posterior sacs more or less directly from the environment during inspiration. Air exiting the parabronchi anteriorly, combined with air exiting the anterior air sacs, is directed into the mesobronchus via the anterior secondary bronchi and is exhaled. You will recall that the anterior air sacs took up air from the parabronchi during inhalation. Thus the exhaled air is mostly air that has passed across the respiratory surfaces, some of it having temporarily been stored in the anterior air sacs. Three aspects of the expiratory events deserve emphasis: (1) The relatively fresh air of the posterior air sacs is directed mostly to the parabronchi; direct measurements in the living Pekin duck have shown that there is very little flow from the posterior sacs through the length of the mesobronchus to the outside. (2) The exhaled air has largely passed the respiratory surfaces. (3) Again, as in inspiration, air flows through the paleopulmonal parabronchi from posterior to anterior; measures on the Pekin duck indicate that the flow through the parabronchi is stronger during exhalation than inhalation.

The ventilation of the neopulmonal airways is incompletely understood, but it seems probable, as indicated in Figure 8–26, that flow through the neopulmonal parabronchi connecting the posterior air sacs with the mesobronchus and posterior secondary bronchi is bidirectional. That is, air flows posteriorly into the air sacs during inspiration and anteriorly from the air sacs during expiration. This pattern of air flow is virtually mandated in some birds by the fact that the mesobronchi themselves are highly constricted at their connections with the abdominal air sacs, leaving the neopulmonal parabronchi as the main connecting passages between the abdominal sacs and the lungs. An important implication of the presumptive pattern of flow in the neopulmonal parabronchi is that air entering the posterior air sacs through the neopulmonal parabronchi on inhalation is exposed to respiratory surfaces. This is thought to explain why the air of the posterior sacs is commonly characterized by carbon dioxide and oxygen tensions that are, respectively, appreciably higher and lower than those that would be predicted if all air simply entered the sacs through the nonrespiratory mesobronchi.

As noted earlier, our present concept of the direction of air flow through the avian lung is supported by a variety of both direct and indirect evidence. Though direct measures of flow in the airways have predominantly been performed on ducks, the anatomical similarity of the paleopulmo in all birds suggests that the results have widespread applicability. It should be pointed out that direct measures of the *direction* of air flow in the lung passages have been obtained more commonly than measures of the *rate* of flow. Thus our understanding of the direction of flow is on far firmer ground than that of relative flow rates at various times over the ventilatory cycle.

One type of indirect evidence for the ventilatory pattern described above is the finding that air in the anterior air sacs is characterized by lower oxygen tensions and higher carbon dioxide tensions than that in the posterior air sacs. In the ostrich at rest, for example, air in the posterior sacs consists of 17–18% oxygen and about 4% carbon dioxide, whereas that in the anterior sacs consists of 13–15% oxygen and 6–7% carbon dioxide.

The mechanisms that control the direction of air flow through the lungs are not well understood. As noted earlier, the airways are not definitely known to contain valves (though there have been occasional reports of presumptive valves); thus the most obvious possible mechanism cannot at present be invoked. It has been hypothesized repeatedly that the complex anatomy of the lung passages is such as to direct air along its inspiratory and expiratory paths without the need for either passive or active valves. This concept has recently received interesting support from studies of excised duck lungs. The lungs, removed from the animal and disconnected from the air sacs, were ventilated by an alternating pump attached to the anterior mesobronchi; significantly, it was found that air entered the posterior secondary bronchi to pass anteriorly through the paleopulmonal parabronchi on both "inhalation" and "exhalation." No passive valves could be observed microscopically, and active (i.e., muscular) valving was impossible.

The differences between the avian respiratory system and the lungs of mammals and other vertebrates invite comparisons. In particular, it has been argued in various contexts that the avian system might provide advantages in respiratory gas exchange, perhaps being capable of meeting higher oxygen demands or providing more effective gas exchange at the high altitudes that birds sometimes enter during flight. The finding that flow in the paleopulmonal parabronchi is unidirectional (in the same direction on inhalation and exhalation) at first suggested an obvious mechanism by which gas exchange could be more efficient in the avian system; namely, the blood and air could exchange countercurrently. For this to occur, however, blood perfusing the air capillaries would have to flow longitudinally along the parabronchi, paralleling the parabronchial lumen and moving in the direction opposite to air flow. It cannot be adequately determined from the anatomy of the parabronchial vasculature whether this occurs, but physiological evidence indicates that, at least in the duck, it does not. As yet, then, the physiological significance of unidirectional air flow in the parabronchi remains unclear. It must also be said that even though the dynamics of gas exchange differ in the avian and mammalian lungs, there is no compelling evidence that either type of lung provides fundamentally distinct advantages in the overall result: uptake of oxygen from the environment and dissipation of carbon dioxide. At present perhaps the most reasonable conclusion in the debate concerning the relative advantages of the mammalian and avian systems is that voiced by Piiper and Scheid, two of the foremost investigators of the avian lung. They have pointed out that different types of respiratory structures may have been favored in the avian and mammalian lines because of differences in the general body plan of the two groups, the body plan of birds being in particular conditioned by their flying habit. Thus the differences between the groups may have evolved initially not as a response to different gas exchange requirements but as a response to different demands placed on the overall body plan for effective function in two dramatically different ways of life. It would then be most appropriate to state that the avian and mammalian lines both evolved efficient, albeit divergent, respiratory systems and that these systems have both evolved to a high level of performance in modern forms.

Comparative parameters of external respiration

Analyses of resting birds, other than passerines, permit the following comparisons with mammals. Rates of oxygen consumption are similar to those of mammals of equivalent size. Minute volume varies approximately proportionally with metabolic rate in the birds, but minute volumes are on the order of 10–25% lower than those of mammals of similar size. As these observations would suggest, oxygen utilization tends to be higher in the birds. Birds achieve their minute volumes by a combination of lower respiratory frequencies and higher tidal volumes than in mammals. Respiratory frequency tends to be about a third of that in mammals of equivalent size and increases approximately proportionally with weight-specific metabolic rate. Tidal volume tends to be about 70% higher than that in comparably sized mammals and varies approximately proportionally with body weight.

The respiratory system and evaporative cooling

Many birds increase evaporative cooling by panting in hot environments. Ventilation is then increased above the level demanded for respiratory gas exchange, and as noted in Chapter 3, alkalosis is a potential problem. It has long been conjectured that birds might obviate this problem partly or completely by passing some or all of the increased ventilatory volume exclusively through air sacs. In this manner evaporative cooling could be increased without a comparable increase in the ventilation of the respiratory exchange surfaces in the lungs. For example, some of the inspired air that enters the posterior air sacs could be exhaled directly through the mesobronchus rather than through the posterior secondary bronchi and parabronchi (Figure 8–26).

The hypothesis that some air might be ventilated exclusively to air sacs during panting received strong support in studies of the ostrich in the late 1960s. During periods of panting lasting up to eight hours, carbon dioxide tensions in the air sacs fell greatly, demonstrating increased ventilation of the sacs, but the birds did not develop alkalosis, indicating that ventilation of the parabronchi was not increased beyond demands for respiratory gas exchange. Studies on 10 other species of birds, representing several orders, performed at about the same time as those on the ostrich yielded quite different results, however. During panting at ambient temperatures of 43°–51°C, all species became alkalotic, most of them severely so. Pekin ducks, for example, had blood carbon dioxide tensions of about 30 mm Hg and pH's of about 7.5 at moderate temperatures, but when panting at 45°C mean carbon dioxide tension fell to about 13 mm Hg, and pH rose accordingly to about 7.7. These results indicated that bypass of the lungs, insofar as it occurred at all during panting, was insufficient to prevent pulmonary hyperventilation, and the ostrich stood as the exception rather than the rule. Recently, additional studies of Pekin ducks have provided new insights. In these studies the ducks were examined during panting at 30°–35°C, much lower temperatures than those employed in the earlier experiments. By comparison to nonpanting animals, the panting ducks exhibited greatly increased rates of ventilation and respiratory evaporative water loss. Yet they showed only slight tendencies toward alkalosis: a mean decrease in blood carbon dioxide tension of just 3 mm Hg and an increase in pH of only 0.02 unit. Calculations indicated that although overall ventilation increased sixfold during panting, parabronchial ventilation increased by only 50%. These results demonstrate that the ducks in some way were able to shunt ventilated air away from the respiratory exchange surfaces during panting, though the extent to which air was ventilated specifically to air sacs cannot be ascertained. At least in Pekin ducks, then, it now seems that panting under moderate heat loads (30°–35°C) is accomplished without excessive parabronchial ventilation, but problems of alkalosis develop under severe heat loads (e.g., 45°C). It is significant that temperatures of 30°–35°C are representative of those that ducks would encounter in shade on hot summer days in nature.

INSECTS

Early in this chapter we noted that oxygen and carbon dioxide must move between all cells of the body and the surfaces of exchange with the ambient medium and that transport by diffusion and microconvection within the cells and intercellular fluids will generally suffice only over short distances. We have seen that most of the reasonably large and active animals have localized respiratory exchange surfaces, remote from many cells, and thus depend on circulation of blood or other body fluids to move gases rapidly within the body. Among the insects, we find a truly remarkable respiratory system, one that brings the exchange surface itself close to all cells of the body.

The basic anatomy
of the insect respiratory system

Certain basic features of the insect respiratory system are illustrated in Figure 8–28. Access to the outside is provided by pores, termed *spiracles*, along the lateral body wall. Air-filled tubes, or *tracheae*, penetrate the body from each spiracle and branch repeatedly, reaching all parts of the animal. Figure 8–28 shows only the major branches. In a few insects, the tracheal trees arising from different spiracles remain independent, but usually they anastomose via large longitudinal and transverse connectives to form a fully interconnected system. Longitudinal connectives connect the tracheal trees along each side of the body and can be seen in Figure 8–28. Transverse connectives connect the trees on one side of the body with those on the other.

The spiracles number from a single functional pair to as many as 10 or 11 pairs. They are segmentally arranged and may occur on the thorax, abdomen, or both. In some insects they are not closable, but in most they can be closed by the activity of spiracular muscles.

The tracheae represent invaginations of the epidermis and are lined with a thin cuticle. Typically, the cuticle is thrown into folds that run a spiral course along the tracheae, providing resistance against collapse. The tracheae become finer with increasing distance from the spiracles and finally give rise to very fine end tubules termed *tracheoles*, believed to be the major sites of gas exchange with the tissues. The tracheolar membrane is very thin but still may possess a spiral ridge. Tracheoles are perhaps 200–350 μ long and are believed to end blindly. They generally taper from a diameter approximating 1 μ at their origin to a diameter of 0.1–0.2 μ at the end.

The layout of the tracheal system varies immensely among species, but the usual result is that all organs and tissues are thoroughly invested with fine tracheae and tracheoles. The number of tracheoles in silkworm larvae (*Bombyx*) has been estimated at 1.5 million. Collectively, the tracheoles provide a very great surface area for gas exchange. The degree

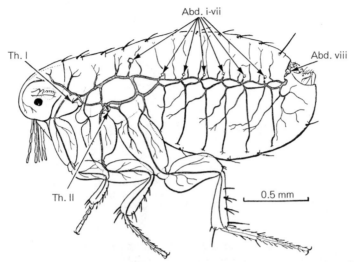

Figure 8–28. The main tracheae in the flea *Xenopsylla*. There are ten pairs of spiracles, two thoracic pairs and eight abdominal pairs, as labeled. The tracheae arising from the abdominal spiracles are narrowed for a short distance near the spiracular openings. All spiracular openings are closable. In the case of the abdominal spiracles, closing is realized by muscular compression of the narrow tracheal segment near the spiracular opening. (From Wigglesworth, V. B. 1935. Proc. Roy. Soc. London, Ser. B, 118: 397–419. Used by permission of The Royal Society.)

of tracheation of various organs and tissues appears generally to vary directly with metabolic requirements. In most tissues the tracheoles usually run between the cells, but in the flight muscles of many species, the tracheoles penetrate the muscle cells, indenting the cell membrane inward. Within the muscle cells, tracheoles run among the individual myofibrils, in close proximity to the longitudinal arrays of mitochondria between the myofibrils. The average distance between adjacent tracheoles within the flight muscles of strong fliers is commonly on the order of 3 μ. Other muscles, the nervous system, the rectal glands, and other active tissues also tend to be richly supplied by the tracheal system, though intracellular penetration of tracheoles is not nearly so common as in flight muscles. In the epidermis of the bug *Rhodnius*, tracheoles are much less densely distributed than in active flight muscles, but, still, cells are usually within 30 μ of a tracheole. In other words, no cell is separated from a tracheole by more than two or three other cells.

Movement of fluid within the tracheoles

Commonly, the terminal ends of the tracheoles are filled with fluid in resting insects. During exercise or when the animals are exposed to oxygen-deficient environments, the amount of fluid decreases, and gas penetrates further into the tracheoles. This is rather easily observed under the microscope because air-filled tubes stand out clearly from the surrounding tissues whereas fluid-filled tubes are much more difficult to visualize. Figure 8–29 depicts the extent of penetration of air in a tracheolar bed of a flea (*Xenopsylla*) exposed to diminishing ambient oxygen concentrations.

The extent of fluid penetration in the tracheoles of insects resting in normal air varies among species and among tissues within a species. In some the tracheoles contain little, if any, fluid. In general, the extent of the fluid column is believed to represent a balance between capillary forces,

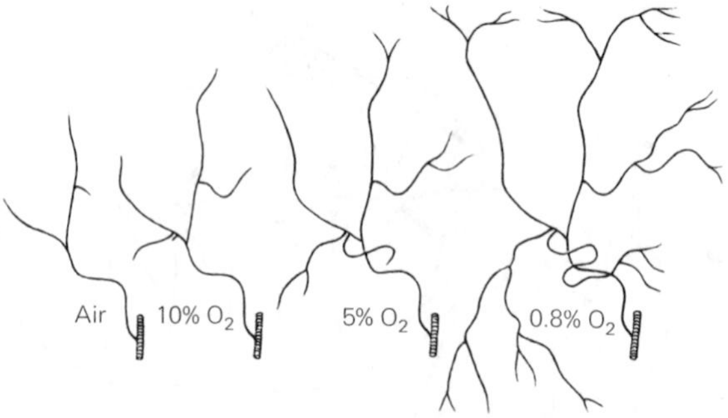

Figure 8–29. The extent of penetration of gas in a tracheolar bed of a resting flea (*Xenopsylla*) when exposed to air and to diminishing ambient concentrations of oxygen. Gas-filled tracheoles stand out from surrounding tissues and are readily visualized. (From Wigglesworth, V. B. 1935. Proc. Roy. Soc. London, Ser. B, 118: 397–419. Used by permission of The Royal Society.)

tending to draw fluid into the tracheoles, and the colloid osmotic pressure of the surrounding cells and body fluids, tending to restrain entry of fluid into the tracheoles. The mechanism of fluid withdrawal during activity or exposure to oxygen deficiency is incompletely understood. There is considerable support for the hypothesis that inadequate oxygen supply to the tissues results in accumulation of anaerobic end products and that the consequent increase in osmotic pressure of the cells and body fluids surrounding the tracheoles results in osmotic withdrawal of water. Whether this phenomenon is adequate to explain the withdrawal of water completely remains debatable. Whatever the mechanisms involved, increased penetration of air into the tracheoles is adaptive inasmuch as it will facilitate exchange of gases between the tissues and environment via the tracheal system.

Air sacs

Air sacs are a common feature of the insect respiratory system. Generally they are associated with major tracheal branches. Some air sacs occur as swellings along tracheae, and air can flow through. Others occur as blind endings of tracheae or as blind, lateral diverticula of tracheae. The degree of development and morphology of air sacs vary widely among species and among life stages with a species. Sacs can be found in almost any part of the body in one species or another. They tend to be particularly well developed in active insects and may occupy a considerable fraction of the body volume (Figure 8–30). The entire tracheal system has been found to occupy from as little as 5% to as much as 50% of the body in various insects, much of this variability being due to differences in the development of air sacs.

Diffusion as a mechanism of gas exchange through the tracheal system

Many insects, at least when relatively inactive, display no regular ventilatory movements. This observation suggests that gas exchange through the tracheal system might occur entirely by diffusion, and numerous analyses have confirmed that diffusion between the spiracles and tracheoles is often adequate to meet the respiratory requirements of insects. As in

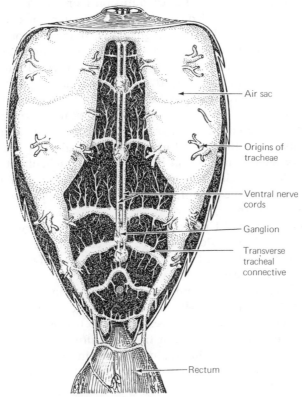

Figure 8–30. A dorsal view of the abdomen of a worker honeybee (*Apis*), showing the large air sacs and associated tracheae (including transverse connectives). The gut is reflected posteriorly, and the tracheal supply to the rectum is shown. Major air sacs are also found in the head and thorax. (From Dreher, K. 1936. Z. Morphol. Oekol. Tiere 31: 608–672.)

many pulmonate gastropods, this is an example of respiration through pore diffusion, the pores in this case being the spiracles. The tracheoles are usually sufficiently densely distributed in all tissues that diffusion can also account for exchange between the tracheolar membranes and all cells. There is thus little or no need for gas transport by the circulatory system.

Whether diffusion through the tracheal system will be adequate to meet the requirements of gas exchange depends on a number of factors, among which body size and metabolic rate deserve some discussion. The rate of exchange by diffusion, you will recall, is inversely related to the length of the diffusion path. Other things being equal, then, oxygen will arrive at the tracheoles at a greater rate in small insects than in large ones. In a given insect the rate of oxygen diffusion can be increased according to metabolic demands through an increase in the oxygen diffusion gradient. If we presume for simplicity that the diffusion path dimensions and the oxygen tension at the spiracles are constant, an increase in the diffusion gradient would have to be accomplished through a reduction in the tracheolar oxygen tension. With a fall in the tracheolar tension, the difference in tension along the tracheal system from spiracles to tracheoles will increase, and the rate of diffusion of oxygen to the tracheoles will rise accordingly. This process can be carried only so far. The tracheolar tension must remain high enough to support diffusion from the tracheoles to the tissues so as to maintain adequate tissue oxygen tensions. If oxygen demands in the tissues should reach high levels, it may become impossible for the two diffusional systems—the tracheal system and the tracheolar-tissue system—to meet oxygen needs. Clearly, the tracheal path length is a critical parameter in this context. At given tracheolar tension, the shorter the path length, the more rapidly diffusion will occur through the tracheae. Thus, other things being equal, small insects will realize more rapid oxygen delivery to the tracheoles at given tracheolar tension than large ones, and it is more probable in the small forms that tracheal diffusion sufficient to meeting high oxygen demands will be attained at tracheolar tensions compatible with adequate diffusion between the tracheoles and tissues. This analysis also indicates the advantages to be realized if an insect can shorten its tracheal path length.

The preceding physical principles suggest that gas transport through the tracheae by diffusion alone may be expected in even relatively large insects when metabolic rates are low but is likely only in small insects when metabolic rates are high. The biological evidence generally supports these conclusions. Diffusion appears to be the principal or only mode of transport in most larvae, all pupae, and most resting adults. During flight many insects are known to ventilate the tracheal airways actively. Calculations have demonstrated that tracheal diffusion should be adequate to meet oxygen demands during flight in small insects, such as fruit flies, and diffusion may be the principal mechanism of transport in these forms even during flight.

Ventilation of the tracheal system

Many insects ventilate the tracheal system. This is seen in some larger forms at rest and, as noted above, is common during activity, especially flight. It is important to recognize from the outset that only the major tracheae are ventilated. From these, exchange with the tracheoles via the smaller tracheae is still largely or completely by diffusion. Essentially, then, ventilation serves to reduce the path length for diffusion by moving air convectively to a certain depth in the tracheal system.

The mechanisms of ventilation are diverse. In the broadest terms, the airways are alternately compressed and expanded by muscular activity. The air sacs commonly act as bellows, forcing or sucking air through the tracheae. This role of the air sacs is reminiscent of the situation in birds and helps to explain the especial development of air sacs in active insects. Ventilation is sometimes tidal, with the ventilated air flowing in and out via the same tracheae and spiracles. In other cases it is unidirectional, with the air flowing in through certain spiracles and traveling along major tracheal connectives to exit at other spiracles. Pumping movements of the abdomen commonly provide the propulsive force for ventilation. During expiration, abdominal volume is decreased, generally by dorsoventral compression or by longitudinal telescoping of the abdominal segments. This phase is active, involving muscular activity. Inspiration is often passive, with the original abdominal volume being restored by elastic forces.

Ventilation in the locust *Schistocerca gregaria* has been studied in detail and will serve here as an example. These locusts ventilate their tracheal system at rest, the major pumping movements being alternate raising and lowering of the ventral abdominal wall. There are 10 pairs of spiracles. During inspiration all spiracles are closed except the two most anterior pairs, located on the thorax, and air enters via these spiracles. During expiration only the most posterior pair of spiracles is open, and air exits through them. The opening and closing of the spiracles thus establishes a unidirectional flow of air through the major tracheae; air enters at the thorax and flows posteriorly through the longitudinal connectives to exit near the posterior of the abdomen. As metabolic rate increases in the absence of flight, the intensity of ventilatory movements increases and more spiracles are brought into play. The anterior two to four pairs may serve as inspiratory openings and all the more posterior pairs as expiratory openings.

It has long been hypothesized that the contractions of the flight muscles of insects during flight might serve to ventilate the thoracic tracheal system that supplies those muscles. Experiments on *Schistocerca* have shown that this process, termed autoventilation, does occur. In flight the second and third most anterior pairs of spiracles remain open continuously, and air is pumped tidally through them by the action of the flight muscles and other thoracic muscles. The volume exchanged on each wing stroke is small, about 25 mm³, but at over 1000 wing strokes per minute, ventilation of 1.6 liters of air per hour is attained. This is more than adequate to meet the metabolic demands of the flight muscles, and it is a demonstrated fact that locusts can fly even though their abdomen has been removed, thus eliminating the abdominal component of ventilation. Abdominal ventilation in normal locusts is strong during flight. The most anterior pair of spiracles continues to open during inspiration, and all the abdominal spiracles open during expiration. Abdominal ventilation moves only about a fifth as much air as the thoracic ventilation.

Control of diffusional and ventilatory exchange

As emphasized in Chapter 5, the tracheal system is a potentially outstanding site of water loss in insects. Water vapor that evaporates into the tracheal airways across the tracheal membranes can readily diffuse out through the spiracles or be carried out in a ventilatory airstream. It will thus often be of advantage for insects to control access to the tracheal system rather strictly according to demands for respiratory gas exchange.

We shall first look at forms that respire exclusively by diffusion. In these, access to the tracheal airways is controlled by closing and opening the spiracles. Constriction of the spiracular openings has clear implications for restriction of water loss. Assuming the tracheal air to be generally saturated with water vapor, the water vapor tension gradient across the spiracular openings will be constant in a given atmosphere whether the spiracles are fully open or constricted. Because the rate of diffusion varies directly with the cross-sectional area of the spiracular openings, however, water loss will be reduced as the openings are constricted.

It is important to recognize that partial closing of the spiracles need not interfere with proper exchange of oxygen and carbon dioxide despite its effect on the rate of water vapor diffusion. Assume, for example, that the rate of oxygen diffusion to the tracheoles is sufficient to meet metabolic demands in an insect with spiracles fully open when the ambient oxygen tension is 150 mm Hg and the tracheolar tension is 140 mm Hg. That is, a total tension difference of 10 mm Hg is sufficient for an adequate rate of diffusion through the tracheal system. The tension drop across the spiracular openings alone will be very small. Now if the spiracles are partially closed, the reduction in pore dimensions will tend to reduce the rate of diffusion through the spiracles. This will result in oxygen depletion in the tracheae, but ultimately the tension on the inside of the spiracles will fall sufficiently to compensate for the reduced rate of diffusion imposed by decreased pore size. That is, an increased tension difference will develop across the spiracles, and because the rate of diffusion increases with the tension difference, this can counteract the negative effect of decreased pore size. Perhaps a tension drop across the spiracles of 20 mm Hg will bring about the same rate of oxygen diffusion through the spiracles as existed when the spiracles were fully open. We know that an additional drop of 10 mm Hg is needed for an adequate rate of diffusion through the tracheae to the tracheoles. Thus, all in all, oxygen supply will be sufficient at a tracheolar tension of 120 mm Hg. The insect, in short, can reduce water loss by allowing lower tracheolar and tissue oxygen tensions. The advantages are clear provided only that tissue oxygen tensions remain compatible with proper function; the theoretical ideal in terms of water conservation would be to constrict the spiracles to the maximum extent compatible with maintaining adequate tissue oxygen tensions.

When at rest, insects relying on diffusional exchange commonly either maintain their spiracles partially closed or periodically open and close them. That this behavior has advantages for water economy is demonstrated by the increased rates of water loss that result when the spiracles are forced open experimentally (see Chapter 5). During activity the spiracles often are observed to open more fully or more frequently. This response permits more rapid oxygen exchange at given tracheolar tension but also allows a greater rate of water loss.

The spiracles are at least partly controlled according to parameters that reflect the adequacy of respiratory gas exchange. The most potent stimulus for opening of the spiracles is an increase in carbon dioxide tension and/or acidity in the body fluids; decreased oxygen tension may stimulate spiracular opening but is far less influential. In these respects the control of the spiracles resembles the control of pulmonary ventilation in mammals. There is some evidence also that the spiracles may be controlled according to immediately prevailing evaporative conditions. In resting tsetse flies (*Glossina*), for example, the spiracles remain 30% open at 60% relative humidity but only 10% open in dry air. As a result, the rate of water loss is about the same in the two environments.

In insects that ventilate the tracheal system, analogous controls are known. Increased carbon dioxide tensions and decreased oxygen tensions stimulate ventilation, with sensitivity to carbon dioxide being generally greater than that to oxygen. Responses to evaporative conditions are also reported. For example, when locusts (*Locusta*) are held in dry air, the amplitude and frequency of ventilatory movements are decreased by comparison to the condition in moist air.

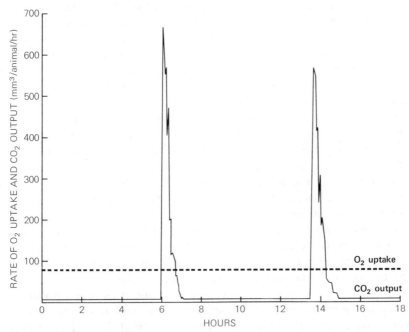

Figure 8–31. Rates of oxygen uptake and carbon dioxide release from a 5-g diapausing Cecropia pupa, illustrating discontinuous respiration. [From Schneiderman, H. A. and C. M. Williams. 1955. Biol. Bull. (Woods Hole) 109: 123–145.]

Discontinuous respiration

In many insects at rest both oxygen and carbon dioxide are exchanged at more or less continuous rates with the atmosphere. A different pattern, termed *discontinuous respiration*, has been observed in various pupae and quiescent larvae and adults. Whereas oxygen consumption proceeds continuously, the rate of carbon dioxide elimination displays strong, intermittent bursts (Figure 8–31). This pattern typically disappears when the insect emerges from its inactive or quiescent state.

Discontinuous respiration represents a special case of spiracular control of diffusional gas exchange. The most intensive experimental work has been performed on the large, diapausing pupae of Cecropia moths and will form the basis of this discussion. At 20°C bursts last around 15 to 30 minutes and are separated by 4 to 17 hours. During a burst the spiracles are fully open, but between bursts they are virtually closed.

At the end of a burst tracheal oxygen tension is near atmospheric, and carbon dioxide tension is about 25 mm Hg. Oxygen tension falls to 30–40 mm Hg within an hour once the spiracles are closed and remains at this level through the remainder of the interburst period. This low tracheal tension provides a strong tension gradient across the spiracles, and oxygen apparently diffuses inward at a steady rate compatible with oxygen demands. Carbon dioxide tension in the tracheae rises very slowly during the interburst period. About two thirds of the carbon dioxide produced between bursts is held in solution in the body fluids as bicarbonate. Most of the remainder diffuses outward across the spiracles. Some accumulates in the tracheae so that carbon dioxide tension reaches about 50 mm Hg by the end of the interburst. During the interburst, because much of the metabolic carbon dioxide is held in the body fluids and because the respiratory quotient is about 0.7 (see Chapter 2), the volume of oxygen removed from the tracheal system is not replaced by an equivalent volume of carbon dioxide. This tends to create a negative pressure in the tracheae, and air is sucked in across the spiracles during at least part of the interburst period. This bulk flow aids the supply of oxygen to the tissues, and it has been hypothesized that it hinders outward diffusion of water vapor.

Apparently the rising tracheal carbon dioxide tension ultimately serves as the stimulus for opening of the spiracles at the burst. The opened spiracles permit rapid outward diffusion of carbon dioxide. Tracheal carbon dioxide tension falls, and the large amount of carbon dioxide stored as bicarbonate during the interburst is released to the tracheae and diffuses out, thus the burst in carbon dioxide release. The open spiracles also permit greatly increased output of water vapor.

Discontinuous respiration is believed to represent an adaptation to water conservation, though the mechanisms by which water conservation is enhanced relative to more usual respiration are not well understood. Pupae do not consume water and thus have special requirements for conserving it. Cecropia pupae normally lose only about as much water as they produce in metabolism. If their spiracles are held open experimentally, however, water loss increases by a factor of about 25, and fatal dehydration occurs within a matter of weeks at moderate ambient humidities.

Respiration in aquatic insects

The respiratory adaptations of aquatic insects have attracted considerable interest. Nearly all retain an air-filled tracheal system. Perhaps the simplest adaptation has been limitation of the functional spiracular openings to one end of the body, which then is projected to the water surface. This is seen, for example, in common mosquito larvae. The spiracles open at the end of a posterior abdominal tube (Figure 5–1), and the larvae hang at the surface of the water, with the spiracles open to the air. One of the potential challenges confronted by insects with such arrangements is prevention of entry of water into the tracheal system. In all, the spiracular openings are rendered hydrophobic. Because water molecules have a greater attraction for each other than for the hydrophobic surfaces of the insect, the surface film of water pulls away from the spiracular openings, leaving them free access to the air. In some insects, including mosquito larvae, perispiracular glands produce a hydrophobic oily secretion around the spiracular openings. In others, the openings are surrounded by hydrofuge hairs (*hydrofuge* = "water-repelling"). Mosquito larvae, in common with many other aquatic insects with similar forms of respiration, have relatively capacious tracheal systems. This permits them to carry a relatively large air store when they break contact with the surface for underwater activity, sufficient to meet their oxygen needs for 5 or 10 minutes under usual circumstances.

Many insects carry bubbles or films of air on the outside of their bodies when under water. These air pockets communicate with the tracheal system via spiracles. Probably everyone has observed bubbles being carried under the wings of water beetles. Bubbles are also carried at the posterior tip of the abdomen or among the legs in various species. These bubbles are established and maintained because hydrofuge structures on the body surface render the affinity of the surface to air greater than that to water.

It is obvious that the bubbles contain a store of oxygen when they are established at the water surface. It is less obvious that this store can be replenished while the insect is under water. As oxygen is withdrawn from the bubble, its oxygen tension falls. When it has fallen below the oxygen tension of the surrounding water, dissolved oxygen from the water will diffuse into the bubble, renewing its oxygen supply. The bubble, then, acts as an "air gill." A complete analysis of the dynamics of the bubble requires attention to carbon dioxide and nitrogen as well as oxygen. Carbon dioxide is highly soluble in water, and carbon dioxide added to the bubble from metabolism diffuses away rapidly into the surrounding water. The carbon dioxide tension in the bubble thus remains low, and, effectively, oxygen withdrawn from the bubble in metabolism is not replaced by an equivalent volume of carbon dioxide. This means that the percent composition of nitrogen and nitrogen tension in the bubble will rise as the oxygen tension in the bubble falls. Thus there will be a tendency for nitrogen to diffuse out of the bubble into water of lower nitrogen tension, as well as a tendency for oxygen to diffuse into the bubble. The loss of nitrogen gradually reduces the size of the bubble. This has two detrimental effects on the functioning of the bubble as an oxygen supply. First, as the size decreases, the surface area of the air gill decreases, and the rate of diffusional replenishment of oxygen decreases accordingly. Second, at given oxygen tension, the smaller its size, the smaller the reserve of oxygen that the bubble contains. Eventually the insect must return to the surface and renew the bubble from the atmosphere. The bubble lasts longer under water and is more effective in extracting oxygen than might at first be expected because of the different properties of oxygen and nitrogen. At given tension difference, oxygen diffuses across an air-water interface about three times as rapidly as nitrogen. This means that when the composition of the bubble is altered by metabolic oxygen uptake, there is a physical bias toward re-establishing the original composition by inward diffusion of oxygen from the water rather than by outward diffusion of nitrogen.

It is instructive to examine here the results of an intriguing experiment on back-swimmer bugs (*Notonecta*), which carry a bubble of air on their ventral surface. Bugs in a system equilibrated with atmospheric air lived under water for six or seven hours. On the other hand, if the atmosphere was pure oxygen and the water was equilibrated with pure oxygen, the bugs survived for only 35 minutes. The bugs in pure oxygen carried about five times more oxygen in their bubble when they submerged as the bugs in air (given that air is about 20% oxygen). They could not survive as long when submerged, however, because their air gill was nonfunctional. As oxygen was withdrawn from their bubble, only oxygen remained, and little or no tension gradient was established between the bubble and water. Thus the animals could utilize the oxygen established in the bubble at the water's surface but could not gain appreciable amounts of oxygen by diffusion when submerged. This experiment illustrates the essential place of the inert gas nitrogen in the functioning of the air gill. It is the presence of nitrogen that permits the oxygen in the bubble to be diluted by metabolism, thus establishing a favorable tension gradient for diffusion. We also see that because bugs in the air-equilibrated system survived 10 to 12 times as long as bugs in pure oxygen even though they carried less oxygen in their bubble upon submergence, oxygen derived from the gill action of the bubble during submergence far exceeds that obtained at the water's surface.

In some aquatic insects, parts of the body are covered extremely densely with fine hydrofuge hairs (about 2 million hairs per square millimeter in the bug *Aphelocheirus*, for example). The layer of gas trapped among these hairs cannot be displaced. This thin, gaseous film, termed a *plastron*, remains constant in volume over a wide range of conditions, including submergence in water that is entirely free of dissolved gases (in which a substantial vacuum must develop in the plastron). Because the gas space is permanent in plastron breathers, it will serve indefinitely as an air gill, unlike the compressible type of air gill already discussed, in which gradual reduction in volume progressively impairs gill function. Some insects with plastron respiration are known to remain submerged for months in well-aerated water.

Thus far, we have considered aquatic insects in which the tracheal system remains open to the body surface via spiracles. By contrast, all the spiracles are obliterated or nonfunctional in many aquatic insects, and the immediate respiratory exchanges between animal and environment occur entirely by diffusion across the integument. The body surfaces are bathed with water, and exchange is directly with the dissolved gases in the water.

With only a few exceptions, the tracheal system in such forms is filled with gas and thus continues to function as the path of least resistance for diffusion of gases through the body from the body surface. Commonly, there is a dense proliferation of fine tracheae under the cuticle. Oxygen diffuses across the cuticle into this subcutaneous network from the water and then diffuses, as usual, through the tracheal system to the deeper tissues. In striking parallel to numerous other groups of aquatic animals, many of these insects have developed evaginations of the body surface that are covered with a thin cuticle and densely supplied with superficial tracheae. These *tracheal gills* may occur on the outer body surface or in the rectum, taking the form of papillae, lamellae, or fine feathery structures. Insects that possess a subcutaneous tracheal network but no tracheal gills include midge larvae (*Chironomus*) and blackfly larvae (*Simulium*). The larvae of caddis flies and the nymphs of stone flies, mayflies, and dragonflies possess tracheal gills in addition to subcutaneous tracheae.

ARACHNIDS

The scorpions, many spiders, and members of several other orders of arachnids (Palpigradi, Uropygi, and Amblypygi) possess a novel type of respiratory structure, the *book lung* (Figure 8–32). The number of book lungs varies from a single pair (as in certain spiders) to four pairs (in scorpions). Book lungs represent invaginations of the ventral abdomen and are lined with a thin cuticle. Each consists of a chamber, the atrium, opening to the outside by a ventral pore, or spiracle. The dorsal or anterior surface of the atrium is thrown into a great number of lamellar folds. Blood

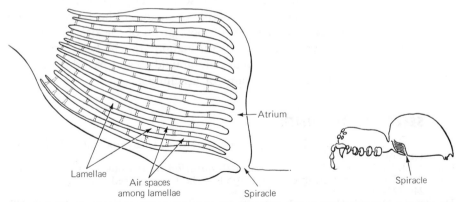

Figure 8–32. Section of a book lung of a spider (semidiagrammatic). Inset shows the position of the book lungs in the anteroventral abdomen of a two-lunged spider. The lamellae of the book lung are attached not only at the anterior wall of the lung cavity, but also along both lateral walls of the cavity. (From Comstock, J. H. 1912. *The Spider Book*. Doubleday, Garden City, N.Y.)

circulates within the lamellae, whereas the spaces among them are filled with air. In the South American tarantula, the lamellae are about 10 μ thick and are separated by air spaces measuring about 5 μ across. The blood-to-gas distance across the lamellar walls is only about 1 μ.

Book lungs bear a notable resemblance to the book gills of horseshoe crabs, and much discussion has centered on the possibility that the aquatic ancestors of modern arachnids had book gills and that these were internalized, with modification, upon emergence onto land. The issue remains unresolved.

Despite their widespread distribution and novel morphology, book lungs have received only the barest attention from physiologists. Calculations indicate that diffusion should be fully adequate to meet requirements for gas exchange between the lamellae and ambient air. Ventilation of the atrium is claimed in some spiders but appears not to be necessary. The size of the spiracular openings can be altered, and what little evidence there is indicates that the openings are kept small except during activity or other circumstances that demand decreased resistance to respiratory gas exchange. The spiracles thus appear to function analogously to those of insects in limiting water loss. Transport of oxygen and carbon dioxide between the book lung and the rest of the body is accomplished by circulation of the blood.

The scorpions have only book lungs, but tracheal systems are found in many spiders. Basically, spiders have two pairs of respiratory structures, always opening via ventral abdominal spiracles. In some spiders there are two pairs of book lungs. Others have a single pair of book lungs and a pair of tracheal trees, whereas still others have two pairs of tracheal trees and no book lungs. The tracheal systems take two forms. Sometimes the tracheal tree consists of a tuft of many relatively short tubes and functions analogously to a book lung; the tracheae do not permeate throughout the body but simply rest in abdominal blood spaces, and the blood must carry gases between the tracheae and the rest of the animal. This form of tracheal system has been termed a tracheal lung. In other cases the tracheal system is analogous to that of insects, though probably of independent evolutionary origin; tracheae run throughout the body and presumably supply the respiratory needs of most or all of the tissues directly.

Tracheae of the type that ramify throughout the body are found in daddy longlegs, mites, ticks, sun spiders (solpugids), and pseudoscorpions. Book lungs are absent in these groups of arachnids.

SELECTED READINGS

Comparative Physiology of Respiration in Vertebrates. (A Symposium.) 1972. Resp. Physiol. 14: 1–236.
Comroe, J. H., Jr. 1966. *The Physiology of Respiration.* Year Book Medical, Chicago.
Gans, C. 1970. Respiration in early tetrapods—The frog is a red herring. Evolution 24: 723–734.
Hughes, G. M. 1961. How a fish extracts oxygen from water. New Sci. 11: 346–348.
Hughes, G. M. 1965. *Comparative Physiology of Vertebrate Respiration.* Harvard University Press, Cambridge, Mass.
Johansen, K. 1970. Air breathing in fishes. *In*: W. S. Hoar and D. J. Randall (eds.), *Fish Physiology.* Vol. IV. Academic, New York.
Krogh, A. 1941. *The Comparative Physiology of Respiratory Mechanisms.* University of Pennsylvania Press, Philadelphia.
McCutcheon, F. H. 1964. Organ systems in adaptation: the respiratory system. *In*: D. B. Dill (ed.), *Handbook of Physiology. Section 4: Adaptation to the Environment.* American Physiological Society, Washington, D.C.
Mill, P. J. 1972. *Respiration in the Invertebrates.* Macmillan, London.
Piiper, J. and P. Scheid. 1973. Gas exchange in avian lungs: models and experimental evidence. *In*: L. Bolis, K. Schmidt-Nielsen, and S. H. P. Maddrell (eds.), *Comparative Physiology.* North-Holland, Amsterdam.
Schmidt-Nielsen, K. 1972. *How Animals Work.* Cambridge University Press, London.
Steen, J. B. 1971. *Comparative Physiology of Respiratory Mechanisms.* Academic, New York.

See also references in Appendix.

9
EXCHANGES OF OXYGEN AND CARBON DIOXIDE: TRANSPORT IN BODY FLUIDS

Oxygen that enters the animal at sites of exchange with the external environment must reach all cells of the body, typically at a rate commensurate with current metabolic demands. Similarly, carbon dioxide produced in the cells must reach regions of the body where it can be voided. In many animals all cells are sufficiently close to a site of external exchange that diffusion and microconvection within the cells and tissue fluids will suffice for internal gas transport. In many other animals this is not the case, and macroconvective movement of blood or other body fluids plays a vital role in achieving rates of internal transport compatible with the metabolic demands of the tissues. This chapter is chiefly concerned with the respiratory functions of these body fluids.

SECTION 1: Oxygen Transport

Suppose that a particular tissue has a total oxygen requirement of 5 cc/hr, that it receives all of its oxygen from the blood, and that it extracts all the oxygen from the blood that circulates through it. Clearly, the volume of blood that must pass through the tissue to supply the oxygen requirement depends on the amount of oxygen carried by each unit volume of blood. If the blood carries 0.2 cc O_2/100 cc, a circulation of 2500 cc of blood per hour will be required. If, on the other hand, the blood carries 20 cc O_2/100 cc, circulation of only 25 cc/hr will be necessary. In terms of demands placed on the circulatory system, then, there is a clear advantage to increasing the amount of oxygen carried by each unit volume of blood.

Oxygen dissolves in body fluids according to the physical laws outlined in Chapter 8. The amount of oxygen carried in solution per unit volume increases proportionally with the oxygen tension but is never very great at physiological tensions. In many animals that depend on circulation of body fluids to transport oxygen, the fluids contain compounds that combine reversibly with oxygen molecules and serve to increase the amount of oxygen carried per unit volume at given partial pressure. The *hemoglobins* are one such class of compounds. Human plasma at equilibrium with alveolar oxygen tension (100 mm Hg) carries approximately 0.3 cc O_2/100 cc, but whole blood, which includes the hemoglobin in the red blood cells, carries about 20 cc O_2/100 cc. In other words, only about 1.5% of the oxygen carried by blood leaving the alveoli is in physical solution as oxygen. The remainder is chemically combined with hemoglobin molecules.

Clearly, for a compound to perform any useful function in oxygen transfer within the body, its combination with oxygen must be reversible; that is, it must tend to take up oxygen at high partial pressures and to release the oxygen at lower partial pressures. Further, the range of partial pressures over which the compound "operates" must be within the physiological range of partial pressures for the animal concerned. The compounds that meet these criteria are chemically very diverse but are all pigments and fall into four basic groups: *hemoglobins, hemocyanins, hemerythrins,* and *chlorocruorins*. Whereas these respiratory pigments commonly act in the routine transport of oxygen in circulating body fluids, this is by no means always the case. Hemoglobins, for example, are commonly found in muscle or nerve cells, where they are believed to facilitate oxygen delivery from the circulating body fluids to these metabolically active tissues. The various functions of the respiratory pigments will be discussed in detail after review of their chemical properties and distribution in the animal kingdom.

RESPIRATORY PIGMENTS

Respiratory pigments are generally lacking in those animal groups that lack circulatory systems or other systems in which mass internal circulation can occur (e.g., a coelomic cavity). Whereas hemoglobins are reported in some Protozoa and flatworms, respiratory pigments are not found in most members of these groups or in the sponges or coelenterates. Circulating respiratory pigments are widespread in the major higher phyla and are virtually universal among relatively large and active animals— with the notable exception of the tracheate arthropods, which generally do not depend on circulation for internal gas transport. Undoubtedly, respiratory pigments have evolved independently many times, a testimony to their adaptive advantages. Pigments are lacking in some animals that depend on circulation for internal gas transport. In these, all oxygen must be carried in simple physical solution. Generally such animals have relatively low metabolic rates. Included, for example, are many of the echinoderms and some of the small or relatively inactive polychaetes.

It is to be emphasized from the outset that the hemoglobins, hemocyanins, hemerythrins, and chlorocruorins are *groups* of compounds. For

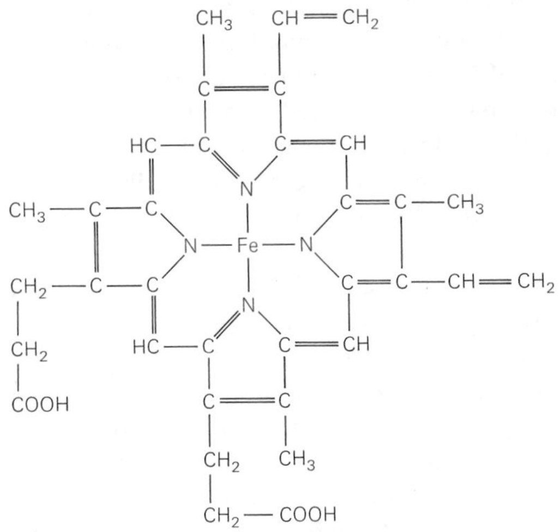

Figure 9–1. The chemical structure of heme. Ferrous iron is complexed with protoporphyrin. The positions assigned to double and single bonds in the porphyrin ring are arbitrary inasmuch as resonance occurs. Resonance also occurs in the central complex of iron with the four nitrogen atoms.

example, there are a great many different hemoglobins, all united by a common subunit of their chemical structure, the heme group. The four classes of respiratory pigments are all metalloproteins. The metal is either iron or copper and is bound in an organic complex. The combination with oxygen is at the site of this complex. The pigments combine with and release molecular oxygen (O_2).

Hemoglobins: basic properties

The basic molecular unit of hemoglobins consists of a heme group (Figure 9–1) bound to a protein, or globin, moiety. Heme is a particular metalloporphyrin, specifically ferrous protoporphyrin. (The complex ring structure surrounding the ferrous iron in Figure 9–1 is known as protoporphyrin.) So far as is known, the heme groups of all hemoglobins are identical. Hemoglobins of different species differ in their chemical and physical properties, and several different hemoglobins may occur normally in any one species. These differences are due to differences in the globin moieties conjugated with heme. Recent biochemical studies have revealed that only small differences in the globin moiety can cause highly significant alterations in the properties of hemoglobin. In man, for example, nearly 100 mutant hemoglobins have been identified, and most differ from the usual adult hemoglobin in only one of the over 140 amino acid residues of the protein conjugated with heme. In one mutant form, known as hemoglobin Rainier, a histidine residue occurs at a point in the protein structure where there is normally a tyrosine residue. This seemingly small change in the globin moiety greatly increases the affinity of the hemoglobin for oxygen in comparison to normal hemoglobin. Another mutant form, hemoglobin Kansas, in which there is also a single residue replacement, exhibits greatly reduced affinity for oxygen.

Oxygen molecules combine with the heme loci in hemoglobin molecules. One oxygen molecule can combine with each heme unit, or, in other words, the binding ratio of oxygen molecules to iron atoms is 1:1. Hemoglobin combined with oxygen is termed *oxyhemoglobin.* Hemoglobin that is not so combined is termed *deoxyhemoglobin* or sometimes *reduced hemoglobin.* The process of binding oxygen is termed *oxygenation,* whereas the process of releasing oxygen is termed *deoxygenation* or sometimes *reduction.* It should be recognized that the combination of heme with oxygen does not, in the simple sense of the word, oxidize the iron in the heme group to the ferric state. In fact, oxidation of the heme of hemoglobin in the laboratory to ferric protoporphyrin results in a compound termed methemoglobin which cannot combine with oxygen and is therefore useless as an oxygen transport substance. Oxygenation should not be equated with oxidation. *Reduced hemoglobin* is in common usage as a synonym for *deoxyhemoglobin,* and *reduction* is commonly used synonymously with *deoxygenation,* but these usages should not be taken to imply that heme is being reduced from an oxidized condition when it is deoxygenated. With these provisos we may note that recent evidence indicates that during oxygenation there is a partial transfer of an electron from the ferrous iron of heme to the combined oxygen molecule. The iron is thus shifted toward the ferric condition, and the O_2 molecule is shifted toward a superoxide, OO^-.

Porphyrins are colored compounds (pigments); that is, they absorb light over particular wavebands. Heme, like other porphyrins, exhibits a characteristic absorption spectrum that can be analyzed spectrophotometrically. The absorption spectrum is closely similar in all hemoglobins, though influenced to a small extent by the nature of the globin moiety. Oxyhemoglobins exhibit absorption bands peaking in the yellow, green, and far violet-ultraviolet regions of the spectrum. Often spectrophotometry has been used to evaluate whether a suspected oxygen transport compound is in fact a hemoglobin. It is a commonplace observation that oxygenated and deoxygenated hemoglobins are different in color; witness the fact that human arterial blood is bright red or scarlet, whereas venous blood is more purplish. Because the absorption spectrum of heme is different depending on whether the hemoglobin molecule is oxygenated or deoxygenated, spectrophotometric methods can be used to determine the relative proportions of oxygenated and deoxygenated heme groups in a sample of blood.

Unit molecules of hemoglobin, consisting of a heme group and associated globin, are frequently linked to form larger molecules. Circulating hemoglobins in vertebrates, for example, are usually four-unit molecules. The molecular weight of each unit molecule varies with the size and composition of the globin moiety but tends to be about 16,000–17,000 in vertebrates. The four-unit circulating hemoglobins thus have molecular weights of approximately 64,000–68,000. In the usual adult circulating hemoglobin of man, there are two types of unit molecules, termed α and β, which differ in the composition of their globin moieties. Each molecule of the circulating hemoglobin consists of two α units, each with 141 amino acid residues, and two β units, each with 146 amino acid residues. Because each molecule of hemoglobin includes four heme groups, it can bind four molecules of oxygen. Relatively huge hemoglobin molecules are found in some invertebrates. The circulating hemoglobin of the polychaete lugworm *Arenicola cristata,* for example, has a molecular weight of 2.85 million. This molecule includes 96 heme groups, or one heme group per

29,700 units of molecular weight. The linkage of hemoglobin units to form larger molecules has important consequences for the function of the molecules in respiratory transport, as we shall see later.

Hemoglobins found in different parts of the animal typically differ structurally and in their chemical and physical properties. Whereas the usual circulating hemoglobin of adult man, for example, is a four-unit molecule consisting of α and β chains, the hemoglobin found in human muscle tissue (termed *myoglobin*) is a single-unit molecule of molecular weight 17,500, consisting of one heme group and a globin moiety of different structure than the α or β chains in circulating hemoglobin. Not uncommonly the nature of hemoglobin changes over the life cycle. To again use man as an example, the circulating hemoglobin of the fetus is different from that of adults, being a four-unit molecule composed of two α chains (as in the adult) and two so-called γ chains, which differ in amino acid composition from the β chains of adult hemoglobin. Fetal hemoglobin is replaced by adult hemoglobin after birth, ordinarily being almost entirely gone within four to six months.

Although so far it may have been implied that circulating hemoglobin is of uniform composition in the stabilized adult condition, it is not uncommon to find more than one chemical form of hemoglobin in the blood, and evidence that has recently been accumulating at a rapid pace indicates that this may in fact be usual. Chemically different forms of hemoglobin exhibit different mobilities when analyzed by electrophoresis, and most evidence on the multiplicity of hemoglobin types comes from this form of analysis. In adult man about 97.5% of blood hemoglobin is typically of the form described heretofore, but about 2.5% is of a different chemical structure. Two forms of hemoglobin have been reported in the blood of cows, sheep, buffalo, and several species of birds. Some other animals have three or more forms; examples include chickens (three forms); the cyclostome *Petromyzon marinus* (six forms); herring, *Clupea harengus* (eight forms); and many other fish. Only one form of circulating hemoglobin has been reported in some animals, such as the domestic pig and a penguin. When a species has two or more forms of hemoglobin in its blood, it is possible for these hemoglobin types to exhibit significantly different functional attributes in oxygen transport. Such differences have been documented in some fish, for example, and, as will be discussed subsequently, there is reason to believe that the multiplicity of hemoglobins in these species assures adequacy of oxygen transport over a broader range of conditions than would be possible with only a single hemoglobin type. For the most part, however, potential differences in the transport properties of multiple hemoglobins have not been assayed. The vast majority of studies of oxygen transport have been performed on whole blood or solutions prepared from whole blood and thus indicate the combined properties of whatever hemoglobin types are present without discriminating the contributions of the individual chemical forms.

Hemoglobins: distribution in the animal kingdom

Hemoglobins are the most widely distributed of the respiratory pigments. They are the only respiratory pigments of vertebrates, and, with a few interesting exceptions to be discussed later, all vertebrates have hemoglobin in their blood. The circulating hemoglobin of vertebrates is always contained in blood cells, the *erythrocytes*. As noted earlier, vertebrate circulating hemoglobins are usually four-unit molecules; exceptions are found among the cyclostome fish. When oxygenated, for example, the hemoglobin of the large lamprey *Petromyzon marinus* is in the form of single-unit molecules. These molecules tend to aggregate into two-, three-, and four-unit molecules upon deoxygenation.

Hemoglobins are widespread in the muscles of vertebrates. Muscle hemoglobins are termed *myoglobins* and among vertebrates appear always to be single-unit molecules. Myoglobins tend to be especially concentrated in various active muscles, such as the heart, and then impart a distinctly reddish color to the tissue.

Circulating hemoglobins of invertebrates (occasionally termed *erythrocruorins*) may be contained in blood corpuscles or may be dissolved directly in the plasma. It is noteworthy that although hemoglobins contained in corpuscles are generally of relatively low molecular weight, as in vertebrates, plasma hemoglobins typically have molecular weights in excess of 1 million. We shall see subsequently that other types of respiratory pigments also tend to be combined into molecules of high molecular weight when dissolved in the plasma. Several arguments have been presented to explain this. If, for example, we assume a unit molecular weight of 17,000 for hemoglobin, then 100 molecules with a molecular weight of 1 million will contain the same number of oxygen-binding loci as 1470 molecules with a molecular weight of 68,000. Thus one obvious result of the polymerization of hemoglobin units into very large molecules is a great reduction in the number of molecules needed to carry a given number of heme units. Because the osmotic pressure of a solution depends on the number of dissolved molecules and not on their size, it has been suggested that the great polymerization of dissolved respiratory pigments may represent an adaptation to limiting the contribution of the pigments to the colloid osmotic pressure of the blood. Pigments contained in corpuscles, of course, do not influence this colloid osmotic pressure. Another hypothesis centers on the fact that biological membranes are typically less permeable to large molecules than to small ones. Respiratory pigments contained in corpuscles cannot diffuse out of the blood or be forced out by ultrafiltration, but dissolved pigments might do so, and their high molecular weights could be an adaptation to confining them to the circulatory system. These and other hypotheses demand rigorous experimental evaluation.

Circulating hemoglobins are widespread in annelid worms. Most commonly, they are found dissolved in the plasma. This is true in common earthworms, for example, and their blood, when held to the light, is wine red and clear—quite unlike vertebrate bloods, which are opaque because of the high concentration of red blood cells. Molecular weights of between 2.4 million and 3 million have been reported in the earthworm *Lumbricus* and in the marine polychaetes *Nereis* and *Arenicola*. In some polychaetes, hemoglobins of low molecular weight are found in coelomic corpuscles. Certain of these forms also have hemoglobins dissolved in their blood plasma, whereas others lack a functional circulatory system and possess only coelomic hemoglobin.

The distribution of circulating hemoglobins in other invertebrate phyla is sporadic. Hemoglobins are commonly reported among the small, entomostracan crustaceans (e.g., *Daphnia*, *Artemia*) but not at all in the malacostracans (crabs, crayfish, lobsters, shrimp, and so on). The water flea, *Daphnia*, displays a remarkable phenomenon, which is also observed in some of the other Entomostraca. *Daphnia* in well-aerated waters appear pale. When ambient oxygen tension is reduced to low levels, however, hemoglobin synthesis accelerates, and within a matter of days blood hemoglobin concentration is elevated and the animals appear pink or red. Respiratory pigments are generally lacking in insects, but hemoglobins occur in some species exposed to low oxygen tensions. Midge larvae, *Chironomus*, for example, exhibit a marked increase in circulating hemoglobin when placed in poorly aerated water; it is not uncommon to find these brilliantly red animals in stagnant pools. Their hemoglobin is noteworthy in that it is dissolved in the plasma but has a low molecular weight (16,000–32,000). Circulating hemoglobins, dissolved in the blood, have been reported in some aquatic pulmonate snails of the family Planorbidae but not elsewhere among gastropod molluscs. In *Planorbis corneus* the molecular weight of the plasma hemoglobin is in the neighborhood of 1.5 million. Most bivalves lack circulating oxygen transport pigments, but a few have hemoglobins, as evidenced by the unusual reddish color of their tissues. These hemoglobins are virtually always contained in blood corpuscles and then are of relatively low molecular weight. A number of holothuroidean echinoderms (sea cucumbers) have corpuscular hemoglobins in their blood and coelomic fluid, but respiratory pigments are otherwise lacking in the echinoderms. Circulating hemoglobins are also found sporadically among echiuroids, phoronids, nemerteans, and nematodes.

Hemoglobins are widely and sporadically distributed in muscle, nerve, and other tissues of invertebrates. These hemoglobins are generally of the single-unit type, as are muscle hemoglobins in vertebrates, but a few two- and four-unit molecules have been reported. Muscle hemoglobin is found commonly in the radular muscles of gastropods and chitons; the radula is a grinding or scraping organ utilized in feeding, and its muscles are particularly active. Muscle hemoglobins are also found in the pharyngeal muscles, stomach, and heart in various gastropods; in the heart, adductor muscles, and pedal muscles of certain bivalves; and in muscles of the body wall in various annelids and parasitic nematodes. Nerve hemoglobins, which sometimes impart a pinkish color to the ganglia or nerve tracts, are found in some annelids and molluscs. A few bivalves have hemoglobins in the tissue of their gills. In a few insects hemoglobins are found in the cells of certain structures termed tracheal organs, which are very richly supplied with tracheae. This is the case in certain backswimmer bugs that occupy poorly aerated waters and that utilize oxygen bound to the hemoglobin during dives. The oxygen is released to the tracheae when the tracheal oxygen tension falls and then can diffuse throughout the body.

We have emphasized the distinctly sporadic distribution of hemoglobins in the invertebrates. Hemoglobins may occur within certain subgroups of a phylum but not in others and even within certain species but not in other closely related species. This suggests not only that hemoglobins have evolved independently many times, but also that the evolution of hemoglobins is biochemically relatively "easy." In fact, the cytochromes of the electron transport system consist of protein groups conjugated with iron porphyrins similar to heme. The cytochromes are ubiquitous among aerobic organisms and undoubtedly evolved very early. Thus it can be suggested with some confidence that animals, in evolving hemoglobins, have capitalized on their ancient ability to synthesize iron porphyrins and porphyrin-protein conjugates.

Hemocyanins

Second to the hemoglobins, the hemocyanins are the most widely distributed respiratory pigments. At this point we encounter a minor problem common also to the other two groups of respiratory pigments: the names given to these compounds give no clue as to their chemical structure. Hemocyanins do not contain heme groups and, in fact, contain neither iron nor porphyrin. Their structure is poorly understood. The metal of hemocyanins is copper, and it appears to be directly bound to the protein. Oxygen binds in the ratio of one molecule of oxygen to two atoms of copper. Hemocyanins typically have high molecular weights—several hundred thousand to several million. Generally there are two copper atoms, corresponding to one oxygen-binding locus, for every 50,000–75,000 units of molecular weight. Hemocyanins are bluish when oxygenated and colorless when deoxygenated. The blood of squids and octopuses can be observed to turn blue as it traverses the gills, and when a horseshoe crab (*Limulus*) is bled, the blood takes on a strongly bluish color upon exposure to the air.

Hemocyanins are found among the arthropods and molluscs and are always dissolved in the plasma. Among molluscs they are found in cephalopods and in many chitons and gastropods. They have not been reported in bivalves, which generally lack circulating respiratory pigments. Molecular weights of molluscan pigments are in the millions, reaching 9 million or more in the whelk *Busycon*, for example. Hemocyanins are found in horseshoe crabs (e.g., *Limulus*) and very commonly in crayfish, crabs, shrimp, and other malacostracan crustaceans but are not found in the entomostracans. Crustacean pigments typically have molecular weights in the hundreds of thousands or low millions. A final phyletic position of the hemocyanins is among spiders and scorpions, which often depend on circulation of the blood to distribute oxygen through the body from book lungs. Hemocyanins are not reported in muscle, nerve, or other such tissues. In those molluscs that have muscular respiratory pigments, they are hemoglobins.

Chlorocruorins

The remaining groups of respiratory pigments, the chlorocruorins and hemerythrins, have considerably more limited distributions than the hemoglobins and hemocyanins. The chlorocruorins have close chemical similarities to the hemoglobins, the basic unit consisting of an iron porphyrin, very similar to heme, conjugated with protein. The porphyrin differs from heme only in that one of the vinyl chains ($-CH=CH_2$) on the periphery of the protoporphyrin ring (Figure 9–1) is replaced with a formyl group ($-CHO$). Among other effects this alters the absorption spectrum of chlorocruorins from that of hemoglobins. Oxygen binds in the ratio of one molecule per iron atom, as in hemoglobins. Chlorocruorins are always found in plasma solution and are always polymerized into large molecules. Their coloration is striking. In dilute solution they are greenish. In more concentrated solution, they are deep red when viewed by transmitted light but greenish when viewed by reflected light. Chlorocruorins are currently known from only four families of polychaete annelids, the fanworms—Sabellidae and Serpulidae—and the Ampharetidae and Flabelligeridae. The chemical similarity of chlorocruorins to hemoglobins is paralleled by a close phyletic juxtaposition of the two types of compounds. *Serpula* has both pigments in its blood. In the serpulid genus *Spirorbis* one species has chlorocruorin, whereas another has hemoglobin. The sabellid *Potamilla* has muscle hemoglobin but chlorocruorin in the blood. It is hypothesized that the chlorocruorins represent a parallel evolutionary development from cytochromelike precursors or that they were, in fact, derived from hemoglobins by mutation.

Hemerythrins

The hemerythrins contain iron, but not in a porphyrin complex. The iron is apparently combined directly with the protein, and oxygen binds in the ratio of one molecule to two iron atoms. Hemerythrins are reddish-violet when oxygenated but colorless when deoxygenated. They are always contained in corpuscles and occur in four phyla in a distribution that at present would suggest four independent evolutionary origins. They have been reported in both genera of the small phylum Priapulida; in a single genus of polychaete annelids, *Magelona*; and in one genus of brachiopods or lamp shells, *Lingula*. Their principal distribution is among sipunculid worms, where they occur in corpuscles in the coelomic fluid and tentacular circulatory system. Molecular weights as determined in sipunculids are relatively low: 66,000 in *Sipunculus* and 108,000 in *Golfingia*. In these examples there is one iron atom for every 6000 to 7000 units of molecular weight—two to three times the iron concentration of vertebrate hemoglobins.

THE OXYGEN-BINDING CHARACTERISTICS OF RESPIRATORY PIGMENTS

The combination of oxygen with each oxygen-binding locus of a respiratory pigment is stoichiometric. That is, one and only one molecule of oxygen will bind with each heme group of a hemoglobin or with each pair of copper atoms in a hemocyanin. In blood or other body fluids containing a respiratory pigment, there is a large population of oxygen-binding loci. Human blood, for example, contains about 5.4×10^{20} heme groups per 100 cc. The fraction of these loci that will be oxygenated depends, among other things, on the oxygen tension. If the tension is sufficiently high that all loci are oxygenated, the respiratory pigment is said to be *saturated*. The amount of oxygen carried by the pigment per unit volume of blood at saturation depends directly on the "concentration" of oxygen-binding loci or, in other words, on the concentration of pigment and the number of loci per molecule of pigment. Human blood, for example, has about twice as many hemoglobin molecules per 100 cc as blood of the frog *Rana esculenta*. In both there are four heme groups per molecule. Thus although human blood carries about 20 cc of oxygen per 100 cc of blood as oxyhemoglobin when saturated, blood of the frog carries only about 10 cc/100 cc. Of course, when we state, for example, that the blood carries 20 cc of oxygen per 100 cc, we do not mean that the oxygen is in its gaseous form in the blood and actually occupies 20 cc. Rather, it will occupy this volume (at STP) when extracted from the blood. Blood levels of oxygen are often expressed in units of *volumes percent* (vol %): the volume of oxygen (at STP) carried per 100 volumes of blood. Blood carrying 20 cc O_2/100 cc, for example, is said to have an oxygen content of 20 volumes percent.

When blood is equilibrated with oxygen tensions progressively lower than the tension required for saturation, the fraction of oxygenated binding sites falls and, with it, the amount of oxygen carried by the respiratory pigment. The degree of oxygenation at any given tension can be expressed as the percentage of binding sites oxygenated, as oxygen content in volumes percent, or as the percentage of saturation oxygen content (*percent saturation*). The first and third of these expressions are equal, and the second is proportional to them. For example, when 50% of the heme groups in human blood are oxygenated, the hemoglobin carries 50% as much oxygen as when saturated, or 10 vol %.

The oxygen-binding characteristics of a respiratory pigment are generally summarized in a graph of volumes percent or percent saturation as a function of the partial pressure of oxygen with which the pigment is at equilibrium. The sigmoid curve of Figure 9–2 presents this relationship for human arterial blood. Blood carries oxygen in simple solution as well as in the form of oxyhemoglobin. The curve in Figure 9–2 is a composite of these two functions. The line at the bottom depicts the amount of oxygen carried in plasma solution. It is small by comparison to the amount carried as oxyhemoglobin, and the binding characteristics of hemoglobin exert an overwhelming influence on the curve for whole blood. The curve for hemoglobin alone could be obtained by subtracting the amount of oxygen in plasma solution from the total oxygen content at each partial pressure and clearly would differ only slightly from the curve for whole blood. Note that the amount of oxygen in solution increases linearly with oxygen tension, as expected. In contrast, the amount of oxygen bound as oxyhemoglobin bears a strongly sigmoidal relationship to tension. This will be discussed in more detail subsequently. Curves such as the sigmoid curve in Figure 9–2 are termed *oxygen equilibrium curves* or *oxygen dissociation curves*. As a general principle, it is important to draw a clear distinction between curves that give the oxygen content due to the respiratory pigment alone and curves that give the total oxygen content of blood, including dissolved oxygen. Both types of presentations are common in the literature. In animals such as man that have high concentrations of respiratory pigment, the effect of the pigment is so overwhelmingly greater than that of simple oxygen dissolution that the distinction between these two types of curves is, for many purposes, virtually inconsequential. In those many animals that have low concentrations of respiratory pigment, the distinction can be of considerable importance.

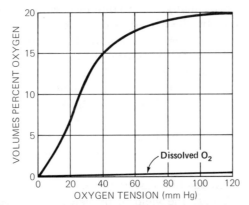

Figure 9–2. Typical oxygen equilibrium curve for human blood at arterial temperature, carbon dioxide tension, and pH. Curve includes oxygen bound to hemoglobin and dissolved oxygen. The portion of oxygen content due to dissolved oxygen is indicated at the bottom. As in other animals, there is significant individual variation: a recent study of 14 individuals revealed that their oxygen equilibrium curves varied in position along the abscissa such that the oxygen tension at which hemoglobin reached half-saturation ranged over a span of 6 mm Hg. [Constructed under the assumption of 20 vol % oxygen content at 120 mm Hg from data of Roughton, F. J. W. 1964. Transport of oxygen and carbon dioxide. *In:* W. O. Fenn and H. Rahn (eds.), *Handbook of Physiology. Section 3: Respiration.* Vol. I. American Physiological Society, Washington, D.C.]

The oxygen equilibrium curve of a respiratory pigment is influenced by many factors, such as temperature and pH, that we shall examine later in this section. First it will be important to understand how respiratory pigments function in the living animal and to appreciate the value of the oxygen equilibrium curve in interpreting the physiology of oxygen transport. These principles can be developed by looking at oxygen transport by the blood in man.

Oxygen transport in man: a case study

We have seen in Chapter 8 that an oxygen tension of around 100 mm Hg is maintained in the alveoli of the human lung. When blood arrives at the alveoli it is at a lower tension. Thus oxygen diffuses into the blood across the alveolar and capillary membranes, raising the blood oxygen tension; and hemoglobin takes up oxygen according to its oxygen equilibrium curve. If the blood oxygen tension were to rise to 100 mm Hg, we see from Figure 9–2 that the hemoglobin would be virtually saturated. For reasons beyond the scope of this text, mixed blood leaving the lungs is at a somewhat lower tension—perhaps 95 mm Hg. This makes little difference in its oxygen content because, as seen in Figure 9–2, oxygen content is rather insensitive to variations in tension at these high tensions. Accordingly, blood delivered to the systemic arteries is almost entirely saturated. Note that the relatively small variation in oxygen content with tension at high tensions signifies that alveolar tension can vary to some extent without greatly affecting oxygenation of the blood in the lungs. The close "matching" of alveolar oxygen tension and the saturation tension of hemoglobin reflects a notable degree of evolutionary coadaptation in pulmonary parameters, on the one hand, and the properties of the hemoglobin molecule, on the other.

Turning now to the events in the systemic tissues, we must first note that oxygen is continually being combined with electrons and hydrogen ions to form water by the activities of the electron transport chain in the mitochondria of the tissues. Essentially this means that oxygen molecules are being withdrawn from solution in the cells, and the effect is to lower oxygen tensions prevailing in the vicinity of the mitochondria. Blood arriving in the systemic capillaries from the lungs is at a high oxygen tension. Accordingly, oxygen diffuses from the blood to the mitochondria along its tension gradient, this being the primary motive force for the withdrawal of oxygen from the blood. As the oxygen tension of the blood falls, hemoglobin releases (unloads) oxygen according to its oxygen equilibrium curve.

Knowing that hemoglobin leaves the lungs in a virtually saturated condition, we can calculate the yield of oxygen in the passage through the tissues by obtaining a measure of the degree of saturation in the venous blood draining the tissues. The simplest way to do this is to determine the degree of saturation in blood drawn from the great veins leading back to the heart; this is termed *mixed venous blood*, for it represents a mixture of the venous blood coming from all parts of the body. In resting man the oxygen tension of mixed venous blood is about 40 mm Hg. From Figure 9–2, we can see that this blood contains about 15 vol % oxygen. Recalling that arterial blood contains about 20 vol %, we find, then, that the oxygen content falls by about 5 vol % during circulation through the systemic tissues in resting man, meaning that 5 cc of oxygen is released from each 100 cc of blood. Frequently the release of oxygen is expressed as a percentage of the arterial oxygen content, this expression being termed the *oxygen utilization coefficient*. In resting man, recognizing that arterial blood contains 20 vol % oxygen and that 5 vol % is released in the tissues, the oxygen utilization coefficient is 25%, meaning that only 25% of the oxygen brought to the tissues in arterial blood is actually utilized.

This analysis of the oxygen yield during circulation through the tissues is simplified because the oxygen equilibrium curve of venous blood is somewhat different from that of arterial blood (Figure 9–2) owing to changes in carbon dioxide tension and pH as the blood becomes venous. The basic conclusions are approximately correct, however, and we shall defer consideration of the effects of carbon dioxide and pH to a later section.

It is important to appreciate the significance of the oxygen tension of mixed venous blood. Basically it represents an average of the oxygen tensions of blood leaving the various systemic tissues. As such, it indicates the overall drop in oxygen tension during perfusion of all tissues combined, but it does not necessarily reflect the drop in tension across any particular tissue. For example, blood entering a given organ at a tension of 95 mm Hg might exit at a tension that is either higher or lower than the mixed venous tension; then the drop in oxygen content across that organ would differ from the mean value calculated from the mixed venous tension. The mixed venous tension is, in fact, a weighted average of the oxygen tensions of blood leaving the various tissues, weighted according to the rate of blood flow through each tissue. Tissues with high rates of circulation will influence the mixed venous tension more than those with low rates of blood flow.

It is also important to recognize that the oxygen tension to which the blood falls in its passage through a given tissue is not a static property of that tissue, but a dynamic function of the several factors which affect the rates of oxygen supply and utilization. These factors include the rate of blood flow to the tissue, the arterial oxygen tension, and the rate of oxygen consumption of the tissue. Clearly, if the rate of blood flow to a tissue were to decrease while the arterial tension and rate of oxygen consumption remained unchanged, the venous tension would be lowered, for each unit volume of blood would have to yield more oxygen in its passage through the tissue. Similarly, a decrease in arterial tension or increase in rate of oxygen consumption could cause decreases in venous tension. The venous tensions seen in resting man are thus conditioned by the prevailing arterial tension, circulatory rate, and rate of metabolism.

We may now examine oxygen delivery during exercise in man. As discussed earlier, only about 25% of the oxygen carried by arterial blood is utilized at rest. Thus the blood has a considerable reserve capacity to supply oxygen. This reserve is utilized during exercise. It is significant that the mixed venous tension during rest, being 40 mm Hg, is low enough to be below the plateau displayed by the oxygen equilibrium curve at high tensions (see Figure 9–2). Thus further drops in venous tension during exercise will result in relatively large increases in oxygen yield from the blood, for oxygen content varies to a relatively great extent with tension on the steep portion of the equilibrium curve below 40 mm Hg. To illustrate, we have seen that a decline in tension from the arterial value of near 95 mm Hg to the resting venous value of 40 mm Hg—a total drop of 55 mm Hg— causes release of about 5 vol % oxygen. A further drop of only 15 mm Hg to a venous tension of 25 mm Hg will double the oxygen yield, and decline of the venous tension to 15 mm Hg will triple the oxygen yield (Figure 9–2).

Over a wide range of exercise states, the tension of blood leaving the skeletal muscles is about 20 mm Hg in man and several other mammals on which experiments have been performed. This drop of about 20 mm Hg from the resting venous tension increases the amount of oxygen released from each volume of blood by a factor of about 2.5 and increases the oxygen utilization coefficient to about 65%. As noted, a tension of about 20 mm Hg prevails in the blood draining the muscles over a range of exercise states; that is, as the intensity of exercise increases and the oxygen demand of the muscles increases, the venous tension and oxygen utilization coefficient remain fairly stable. How can it be that oxygen delivery is increased to meet increasing oxygen demands and yet the yield of oxygen from each unit volume of blood does not increase? Again we must remember that the venous tension is a dynamic function not only of the rate of oxygen consumption by the tissue, but also of the rate of blood flow. As oxygen demand is increased, the rate of blood flow through the muscles increases, and it is this factor that permits venous tension to remain stable. The rate of blood flow cannot increase indefinitely. Once it is maximized, further increases in the intensity of exercise and rate of oxygen consumption result in an increased percent utilization of blood oxygen and a drop in venous tension. This occurs during strenuous exercise, and the tension of blood leaving some muscles may then fall close to zero, indicating virtually complete deoxygenation (100% oxygen utilization).

The minimal oxygen tensions that must be maintained at the mitochondria for full function of the electron transport system are not well defined but are probably on the order of 0.2–2 mm Hg. The actual tension prevailing at the mitochondria is a dynamic function of the rate of mitochondrial oxygen utilization and the rate at which oxygen diffuses to the mitochondria from the capillaries. In turn, the rate of diffusion depends on the gradient of oxygen tension between the blood and mitochondria. Various experiments indicate that when the tension of venous blood leaving the muscles falls below about 10 mm Hg, diffusion becomes inadequate to support full mitochondrial function. That is, the supply of oxygen becomes a limiting factor for aerobic catabolism, and oxidation of foodstuffs cannot proceed at its full rate. The venous tension at which this occurs, about 10 mm Hg, is termed the critical venous oxygen tension. We have seen that increases in blood flow assure that venous tension remains above the critical value over a range of exercise states, though it does fall below the critical value during strenuous exertion. It is noteworthy that the oxygen equilibrium curve is such that about 90% of the oxygen available in arterial blood can be released at tensions above the critical tension (see Figure 9–2). This provides another illustration of the close integration between the oxygen-binding properties of hemoglobin and other physiological features of the organism.

We may now look briefly at whole-body oxygen utilization during exercise. Though blood draining active muscles may be rather thoroughly deoxygenated during heavy exercise, the tension of mixed venous blood is not generally observed to fall below 16–20 mm Hg in man even during severe exertion, because blood from the muscles mixes in the great veins with blood from other parts of the body in which oxygen utilization is not so great. The whole-body oxygen utilization coefficient, computed using the mixed venous tension, thus can rise to a peak of about 60–75%—indicating that 2.5 to 3.0 times as much oxygen is extracted from each volume of blood as during rest. During heavy exercise in average young people, the amount of blood circulated by the heart per unit time may rise to 4 to 4.5 times the resting level. These figures, taken together, show that the total rate of oxygen delivery by the circulatory system can increase to 10 to 13 times the resting rate. Trained athletes can achieve higher delivery rates, largely because of an increase in the maximal rate of circulation.

This review of oxygen delivery in man, although simplified and specialized, should indicate that the oxygen equilibrium curve of respiratory pigments is a most important element in understanding the physiology of oxygen transport.

General principles
in the study of oxygen transport

Using the background developed in the preceding sections, we can recognize several principles that should be applied to the study of circulating respiratory pigments in animals in general. (1) We must inquire into the oxygen tensions prevailing at the sites of external respiration and, with reference to the oxygen equilibrium curve, assess the extent of loading that may be expected. (2) To appraise the extent of unloading, we must know something about oxygen tensions in the tissues. The mixed venous tension is a useful indicator but does not provide information on unloading in particular tissues when the tensions in the tissues differ. (3) Considerations such as the above are essential to deciding whether a pigment functions in oxygen transport at all and to assessing the magnitude of its contribution if it does function in transport. If oxygen tensions in the tissues should be too high for appreciable unloading or if those at the sites of external respiration should be too low for loading, the pigment can play little part in oxygen transport. When the oxygen content of blood flowing toward the systemic tissues exceeds that of blood returning from the tissues, circulatory oxygen transport is indicated. The yield of oxygen from simple solution in the plasma must be subtracted from the total oxygen yield to obtain the contribution of the respiratory pigment. (4) The rate of blood flow must be considered along with the yield of oxygen per unit volume of blood in computing circulatory oxygen delivery. Further, the rate of blood flow is a factor in determining venous tension and, thus, percent oxygen utilization. (5) Parameters of respiratory transport are influenced by such factors as exercise state and environmental oxygen tension. Full understanding of the function of the transport system demands study of the animal over the range of physiological and environmental conditions that it experiences in its natural habitat.

The shape of the oxygen equilibrium curve, its determinants and significance

We shall now turn to a more thoroughgoing analysis of oxygen equilibrium curves, looking first at the shapes of the curves and then at a number of other important factors.

Figure 9–3 depicts oxygen equilibrium curves for circulating respiratory pigments in a number of species, and Figure 9–4 presents the curve for human myoglobin. You will note that these curves vary in shape from hyperbolic to strongly sigmoidal. A hyperbolic curve, like that of human myoglobin, is predicted from the principles of mass action when oxygenation of some oxygen-binding sites does not influence the affinity of the remaining sites for oxygen. In this case the sites behave independently, and the reaction with oxygen behaves as any other simple equilibrium reaction:

$$Mb + O_2 \rightleftharpoons MbO_2$$

where Mb stands for myoglobin and MbO_2 for oxymyoglobin. The slope of the equilibrium curve decreases steadily with increasing tension, indicating, as expected, that less and less oxygenation occurs for a given increase in tension as tension rises and the reaction is shifted to the right. You will recall that vertebrate myoglobins and most of the tissue hemoglobins of invertebrates have only one heme group per molecule. The heme groups thus behave independently, and hyperbolic equilibrium curves result.

Most circulating respiratory pigments have multiple oxygen-binding sites per molecule. In these it is possible for oxygenation of some sites on a molecule to alter the molecular configuration in such a way as to change the oxygen affinity of the remaining sites on the same molecule. Such *interactions* among sites are very common and typically are facilitating. That is, once some sites on a molecule are oxygenated, the affinity of remain-

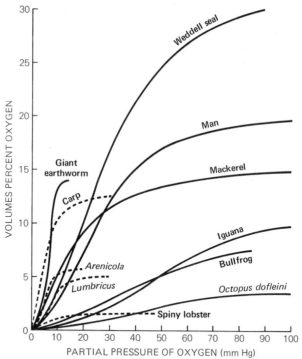

Figure 9–3. Typical oxygen equilibrium curves for 11 species: (1) Weddell seal (*Leptonychotes weddelli*), 37°C, 44 mm Hg CO_2; (2) man, 38°C, pH 7.4; (3) mackerel (*Scomber scombrus*), 20°C, 1 mm Hg CO_2; (4) iguana (*Iguana iguana*), 35°C, pH 7.4; (5) bullfrog (*Rana catesbeiana*), 22°C, 10 mm Hg CO_2; (6) *Octopus dofleini*, 11°C, 3.1 mm Hg CO_2; (7) a giant South American earthworm (*Glossoscolex giganteus*), 20°C, pH 7.50–7.58; (8) carp (*Cyprinus carpio*), 15°C, 1–2 mm Hg CO_2; (9) lugworm (*Arenicola* sp.), 20°C, pH 5.4–6.8; (10) nightcrawler (*Lumbricus terrestris*), 20°C, pH 7.3; (11) spiny lobster (*Panulirus interruptus*), 15°C, pH 7.53. [Sources of data: (1) Lenfant, C., R. Elsner, G. L. Kooyman, and C. M. Drabek. 1969. Amer. J. Physiol. 216: 1595–1597; (3) Root, R. W. 1931. Biol. Bull. (Woods Hole) 61: 427–456; (4) Wood, S. C. and W. R. Moberly. 1970. Resp. Physiol. 10: 20–29; (5) Lenfant, C. and K. Johansen. 1967. Resp. Physiol. 2: 247–260; (6) Lenfant, C. and K. Johansen. 1965. Amer. J. Physiol. 209: 991–998; (7) Johansen, K. and A. W. Martin. 1966. J. Exp. Biol. 45: 165–172; (8) Black, E. C. 1940. Biol. Bull. (Woods Hole) 79: 215–229; (9) data of C. Manwell as given in Prosser, C. L. and F. A. Brown, Jr. 1961. *Comparative Animal Physiology*. 2nd ed. Saunders, Philadelphia; (10) Haughton, T. M., G. A. Kerkut, and K. A. Munday. 1958. J. Exp. Biol. 35: 360–368; (11) Redmond, J. R. 1955. J. Cell. Comp. Physiol. 46: 209–247. Oxygen capacities for *Arenicola* and *Lumbricus* from Prosser, C. L. and F. A. Brown, Jr. 1961. *Comparative Animal Physiology*, 2nd ed., and from Prosser, C. L. (ed.). 1973. *Comparative Animal Physiology*, 3rd ed. Saunders, Philadelphia.]

ing sites is increased. A sigmoid equilibrium curve results (see, for example, line *A*, Figure 9–4). The facilitating interaction among sites is reflected in the fact that the slope of the curve *increases* with oxygen tension up to a certain point, indicating, in contrast to the hyperbolic curve, that more and more oxygenation occurs for a given increase in tension as tension rises and molecules become partially oxygenated. The degree of facilitation varies among species. At low facilitation, the equilibrium curve approaches the hyperbolic shape characteristic of zero interaction among oxygen-binding sites. When facilitation is strong, the equilibrium curve is strongly sigmoidal. At least some interaction is the rule in pigments with multiple oxygen-binding sites per molecule, but there are exceptions. In the case of hemo-

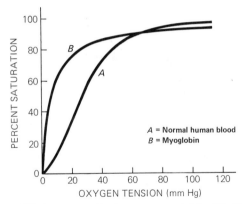

Figure 9–4. Oxygen equilibrium curve for human myoglobin (*B*). The equilibrium curve for hemoglobin in whole blood (*A*) is shown also for comparison. Both curves were determined under similar conditions: 38°C, pH 7.40. [From Roughton, F. J. W. 1964. Transport of oxygen and carbon dioxide. *In:* W. O. Fenn and H. Rahn (eds.), *Handbook of Physiology. Section 3: Respiration.* Vol. I. American Physiological Society, Washington, D.C.]

globins the interactions have generally been termed *heme-heme interactions.* This term may have misleading implications for the chemistry of the interactions, inasmuch as it suggests that the facilitating chemical effects are exerted directly between the heme groups of the molecule. Recent evidence indicates that oxygenation of some hemes does not alter the other hemes but instead alters the configuration of the globin moieties of the molecular subunits. If this is so, the interactions would better be called *subunit interactions.*

Certain implications of the shape of the equilibrium curve for oxygen transport can be illustrated with reference to Figure 9–5. We consider two hypothetical pigments having equal saturation tensions and carrying equal amounts of oxygen at saturation, but one has a hyperbolic equilibrium curve and the other, a sigmoid curve. Assuming that both become nearly saturated at the sites of external respiration, it is clear that the pigment with a sigmoid equilibrium curve will release more oxygen at any given tissue tension than the pigment with a hyperbolic curve. Put another way, the tissues of an animal with the hyperbolic curve would have to function at far lower tension than those of an animal with the sigmoid curve in order to achieve the same extent of oxygen release from the circulating pigment. Given that a certain minimal tissue tension is required for full operation of the electron transport chain, there are often clear advantages to the sigmoid curve. For example, at the critical venous oxygen tension of man (about 10 mm Hg), the pigment with a sigmoid curve will be about 94% desaturated, whereas the pigment with the hyperbolic curve will be only about 65% desaturated. The venous tension necessary for 94% desaturation of the latter pigment is near 1.5 mm Hg.

The concept of oxygen affinity

You will note from Figure 9–3 that various pigments become saturated at different oxygen tensions. That is, the functional range of oxygen tensions varies. The hemoglobin of the lugworm (*Arenicola*), for example, is virtually

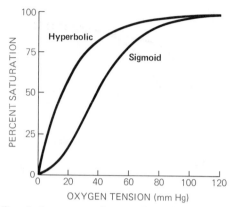

Figure 9–5. Oxygen dissociation curves for two hypothetical transport pigments. Both are assumed to reach saturation at about the same oxygen tension, but one exhibits a hyperbolic dissociation curve whereas the other exhibits a sigmoid dissociation curve. Assuming that both become virtually saturated at the respiratory organs and that both carry the same amount of oxygen at saturation, the release of oxygen at given tissue tension is greater for the pigment exhibiting sigmoid kinetics. See text for further discussion.

saturated at 10 mm Hg and loads and unloads predominantly between 0 and 10 mm Hg. Human hemoglobin, by contrast, is virtually saturated at 100 mm Hg and functions over the range from 0 to 100 mm Hg. This property of pigments is clearly of great importance, for it determines what tensions are necessary for relatively complete oxygenation at the sites of external respiration and determines the degree of unloading that will be realized at any particular tissue tension. *Arenicola* hemoglobin, for example, will not unload at all at the typical tissue tensions of resting man, whereas we have seen that human hemoglobin, about 98% saturated in the lungs, releases about 25% of its oxygen content in the tissues at rest.

Pigments that require relatively high oxygen tensions for full loading and that, conversely, unload appreciable amounts of oxygen at relatively high tensions are said to have a relatively *low affinity* for oxygen (for example, hemoglobin of man or the iguana in Figure 9–3). Pigments that load fully at low tensions and consequently also require low tensions for appreciable unloading are said to have a relatively *high affinity* for oxygen (for example, hemoglobin of *Arenicola* or the giant earthworm in Figure 9–3). Affinity *decreases* with displacement of the saturation level of the equilibrium curve to the *right* along the oxygen tension axis; note, for instance, in Figure 9–3 that the curve for human hemoglobin, which has a relatively low affinity, extends much further to the right before reaching saturation than the curve for *Arenicola* hemoglobin, which has a relatively high affinity. A convenient index of affinity is the *partial pressure of oxygen at which the pigment is 50% saturated*, symbolized as P_{50}. The P_{50} of hemoglobin in human arterial blood, for example, is typically about 26–27 mm Hg, whereas that of *Arenicola* hemoglobin is about 5 mm Hg (see Figure 9–3). P_{50} gives some idea of the positioning of the equilibrium curve along the oxygen tension axis; high P_{50}'s characterize curves that extend well to the right before reaching saturation, and low P_{50}'s characterize curves that reach saturation at low oxygen tensions toward the left. As indicated by the comparison between man and *Arenicola*, affinity decreases as P_{50} increases.

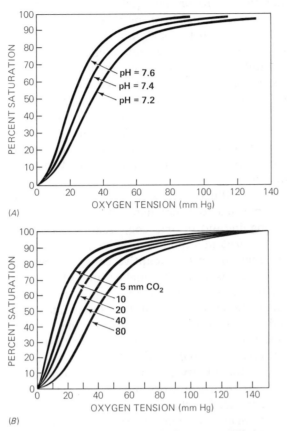

Figure 9–6. Illustrations of the Bohr effect. (*A*) Oxygen equilibrium curves for hemoglobin in human blood at 38°C. Affinity for oxygen decreases as pH decreases. Arterial blood typically has a pH of about 7.4. The pH of mixed venous blood is typically about 0.04 unit less than that of arterial blood at rest. (*B*) Oxygen equilibrium curves for hemoglobin in dog blood at 38°C. Affinity for oxygen decreases as carbon dioxide tension increases. Data are from the original work of Bohr and his coworkers in 1904. [*A* and *B* from Roughton, F. J. W. 1964. Transport of oxygen and carbon dioxide. *In:* W. O. Fenn and H. Rahn (eds.), *Handbook of Physiology. Section 3: Respiration.* Vol. I. American Physiological Society, Washington, D.C.]

The effect of carbon dioxide tension and pH on affinity: the Bohr effect

Respiratory pigments are frequently sensitive to such factors as carbon dioxide tension, pH, temperature, and salt concentration. We shall now consider these effects, starting with the influences of CO_2 tension and pH on affinity.

Commonly, an increase in CO_2 tension or decrease in pH causes decreased oxygen affinity, and the oxygen equilibrium curve is shifted to the right. This effect, illustrated for human and dog hemoglobin in Figure 9–6, is termed the *Bohr effect* or *Bohr shift* after its discoverer. Increased CO_2 tensions are typically accompanied by increased acidity and thus affect respiratory pigments by influencing pH, but there is evidence that increased CO_2 has, in addition, a direct effect. Bohr effects have been reported among all four classes of respiratory pigments. Sometimes the Bohr effect is absent, and sometimes an increase in CO_2 tension or decrease in pH causes an increase in affinity or shift of the equilibrium curve to the left. The latter phenomenon is termed a *negative,* or *reverse, Bohr effect* and, at physiological pH's, is known principally among certain animals that possess hemocyanins (e.g., the horseshoe crab *Limulus* and the snail *Fusitriton*).

The magnitude of the Bohr effect, when it occurs, varies considerably among species and is commonly expressed as the change in the common logarithm of P_{50} per unit change in pH, $\Delta\log P_{50}/\Delta pH$. This logarithmic expression is used because it typically remains constant over an appreciable range of pH; the simpler expression, $\Delta P_{50}/\Delta pH$, does not remain constant. Note that negative values of $\Delta\log P_{50}/\Delta pH$ indicate a positive or normal Bohr effect, for they reflect decreasing P_{50} (and log P_{50}) with increasing pH. The range of variation observed in the circulating hemoglobins of vertebrates may be illustrated with some examples of values for $\Delta\log P_{50}/\Delta pH$: 0 in some elasmobranch fish; −0.3 in the bullfrog; −0.5 in rainbow trout, Adelie penguin, and prairie dog; −0.6 in man and woodchuck; −0.7 in a sea lion; and −0.96 in a mouse.

The Bohr effect often has adaptive consequences for oxygen delivery. Because CO_2 tension is generally higher and pH, lower, in the systemic tissues than at the sites of external respiration, a respiratory pigment that displays a positive Bohr effect will shift to lower affinity as the blood enters the tissues and return to higher affinity when the blood returns to the respiratory organs. The shift to lower affinity in the tissues promotes release of oxygen, for deoxygenation will proceed further at any given tissue oxygen tension, the lower the affinity (see Figure 9–6). Conversely, the shift back to higher affinity at the respiratory organs promotes uptake of oxygen, for oxygenation will proceed further at a given oxygen tension in the respiratory organs, the higher the affinity. In essence, the respiratory pigment shifts back and forth between two equilibrium curves as it flows alternately to the tissues and respiratory organs. Looking at the schematic representation in Figure 9–7, we see that the consequence of this shift is that for given oxygen tensions in the respiratory organs and tissues, more oxygen is delivered to the tissues than would be if the pigment functioned along either curve alone. Put another way, a given oxygen delivery to the tissues can be realized at higher tissue tensions with the Bohr effect than without it. This is evident in Figure 9–7; with the arterial tension fixed at 70 mm Hg, a tissue tension lower than 30 mm Hg would be required for delivery of 7.2 cc O_2/100 cc of blood if the pigment functioned along either curve *A* or curve *B* alone instead of shifting between the two.

The magnitude of the contribution of the Bohr effect to oxygen delivery depends on the difference in CO_2 tension and pH between the tissues and sites of respiration, on the extent to which these parameters influence oxygen affinity, and on the rate of equilibration of the pigment to new conditions of pH and CO_2 tension. These factors must always be considered. It is not enough to know simply that a pigment exhibits the Bohr effect. If physiological differences in CO_2 tension and pH should be insufficient to cause a significant shift in affinity or if the shift should occur very slowly, the contribution of the Bohr effect to oxygen delivery may be slight.

During exercise, CO_2 tension in the tissues may rise above the resting level, and tissue pH may be depressed not only because of the increase in CO_2 tension, but also because of accumulation of acid metabolites such as

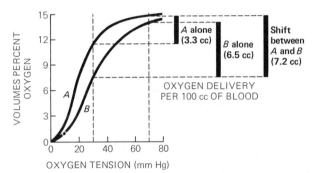

Figure 9-7. Diagrammatic representation of implications of the Bohr effect for oxygen delivery. Oxygen equilibrium curve *A* is that for arterial blood in a hypothetical animal. Curve *B* is that for venous blood, reflecting a decrease in oxygen affinity owing to increased carbon dioxide tension and H+ concentration. Assuming the oxygen tension established in arterial blood to be 70 mm Hg and that established in venous blood to be 30 mm Hg, oxygen release from the blood in the tissues is greater in the presence of the Bohr shift (shift from curve *A* to curve *B*) than it would be were the blood to follow either curve *A* or curve *B* alone throughout the circulatory cycle. The magnitude of the Bohr shift depicted is far greater than that observed under physiological conditions in man.

lactic acid. These changes may facilitate oxygen delivery to the active tissues by augmenting the Bohr shift.

Earlier we analyzed oxygen delivery in man with reference to the arterial equilibrium curve alone. We noted that the arterial curve is not fully adequate for analyzing the relationships between blood oxygen content and tension, and now we see one reason why this is so. At rest, the arterial CO_2 tension in man is near 40 mm Hg, whereas the mixed venous tension is about 46 mm Hg. Venous blood is slightly more acid (pH 7.36) than arterial blood (pH 7.40). From Figure 9–6 we see that these differences in CO_2 tension and pH are sufficient to cause a small but significant Bohr shift. The venous equilibrium curve is shifted somewhat to the right of the arterial curve, and venous oxygen content at given tension is thus somewhat lower than would be predicted from the arterial curve. Measurements on human red blood cells have shown that the Bohr shift occurs rapidly enough to be of physiological significance in the tissue capillaries. The half-time of the Bohr shift is about 0.12 sec. The residence time of red blood cells in the capillaries is on the order of 0.25–2 sec.

The Root effect

The Bohr effect influences oxygen affinity only; the oxygen tension required for full saturation is elevated by increased CO_2 tension or decreased pH, but oxygen content at saturation is not affected. Sometimes increased CO_2 tension or decreased pH not only reduces affinity, but also reduces the oxygen content at saturation (Figure 9–8). The reduction in saturation level is termed the *Root effect*, after its discoverer, and is always accompanied by a Bohr effect (shift of the equilibrium curve to the right). The Root effect is far less common than the simple Bohr effect, having been reported primarily from a variety of fish. The decrease in saturation level, insofar as it occurs under conditions in the tissues, will act in conjunction with the decrease in affinity to increase unloading at given tissue oxygen tension. There seems to be a general correlation between the occurrence of the Root effect and the presence of a swimbladder in fish, and current hy-

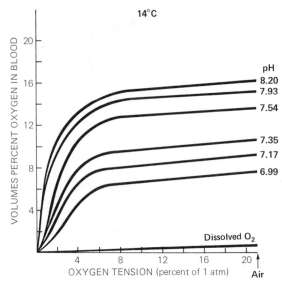

Figure 9–8. The Root effect in the eel *Anguilla vulgaris*. The pH was adjusted to six levels, indicated at the right, by addition of lactic acid. Oxygen equilibrium curves are for whole blood and include dissolved oxygen as well as oxygen bound to hemoglobin. Dissolved oxygen content of the blood is indicated at the bottom. Note that saturation of hemoglobin is indicated when the oxygen content of whole blood increases with oxygen tension in parallel with the line for dissolved oxygen. (From Steen, J. B. 1963. Acta Physiol. Scand. 58: 124–137.)

potheses attribute an important function to the Root effect in a circulatory countercurrent multiplier system that is believed to establish and maintain high oxygen tensions in the swimbladder. This may explain why the Root effect has evolved most commonly among fish.

Thermal effects

The oxygen affinity of respiratory pigments is commonly sensitive to temperature. Increases in temperature characteristically cause a decrease in affinity (shift to the right), and, conversely, decreases in temperature increase affinity (Figures 9–9 and 9–10). Thermal shifts as well as Bohr and Root shifts have considerable import for the physiological ecology of animals. These ecological considerations will be considered subsequently. Here we may note simply that when the temperature of actively metabolizing tissues exceeds that at the respiratory organs, thermal shifts in affinity have effects on oxygen delivery analogous to those discussed earlier for the Bohr effect. In man, for example, increased temperatures in the muscles during exercise help to increase the extent of deoxygenation at given oxygen tension (Figure 9–9).

Effects of pigment and ion concentrations

The shape of the equilibrium curve, oxygen affinity, and other important parameters may be influenced by factors such as salt concentration and pigment concentration. When vertebrate hemoglobins are released from red blood cells, their oxygen affinity commonly increases; this effect is often due largely to dissociation of the hemoglobin from red cell organophosphate compounds, as discussed later in this chapter. Studies of the liberated hemoglobin indicate that both dilution of the hemoglobin and removal of

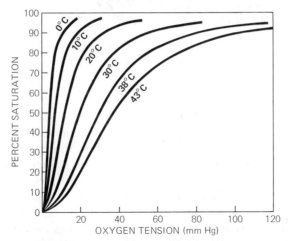

Figure 9–9. Oxygen equilibrium curves of human blood at six temperatures, with carbon dioxide tension held constant at 40 mm Hg. [From Roughton, F. J. W. 1964. Transport of oxygen and carbon dioxide. *In:* W. O. Fenn and H. Rahn (eds.), *Handbook of Physiology. Section 3: Respiration.* Vol. I. American Physiological Society, Washington, D.C.]

salts from solution frequently cause increases in affinity, and at high dilution the oxygen equilibrium curve may become hyperbolic. In many animals, pigment concentration and ion concentrations are regulated within sufficiently narrow limits that such effects probably have little bearing for normal physiology. The main point to remember is that the investigator should study the function of respiratory pigments in their normal milieu—whole blood. In some animals pigment concentration or ion concentrations may vary sufficiently under different physiological or environmental conditions that effects on oxygen affinity and other kinetic parameters could have important implications for respiratory transport. Unfortunately, this possibility has received little systematic experimental attention. It is becoming increasingly apparent that vertebrates do modulate affinity through alterations of red cell organophosphate concentration, a topic that, again, will be discussed subsequently.

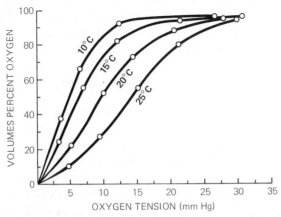

Figure 9–10. Oxygen equilibrium curves for hemocyanin in fresh blood of the spiny lobster (*Panulirus interruptus*) at four temperatures (pH 7.5). (From Redmond, J. R. 1955. J. Cell. Comp. Physiol. 46: 209–247.)

THE FUNCTION OF RESPIRATORY PIGMENTS IN OXYGEN TRANSPORT

Having reviewed some of the basic properties of respiratory pigments, we may now proceed to a more thoroughgoing, comparative study of their function in the living animal.

Oxygen-carrying capacity

A simple but important parameter is the amount of oxygen that can be carried by each unit volume of blood when the respiratory pigment is saturated, this being termed the *oxygen-carrying capacity* of the blood. The carrying capacity sets an upper limit on the oxygen delivery per circulatory cycle. Of course actual delivery depends on oxygen tensions in the systemic tissues and respiratory organs and on the oxygen equilibrium curve. Thus carrying capacities must be interpreted with some caution.

In practice, the oxygen-carrying capacity of blood is usually determined by equilibrating the blood with air and then measuring its oxygen content. Most pigments are saturated under these conditions. The oxygen capacity, being measured on whole blood, includes both dissolved oxygen and oxygen combined with the respiratory pigment. The amount of dissolved oxygen will be rather steady among bloods equilibrated with air, and variations in the oxygen capacity from animal to animal will therefore largely reflect variations in the amount of pigment present.

In Figure 9–3 all the pigments are not fully saturated at the highest oxygen tension shown, but a wide range of oxygen-carrying capacities is evident. You may wish to refer to the figure in the following discussion.

Among birds and mammals the oxygen capacity of the blood is generally in the range of 15–20 vol %. Some, but by no means all, diving birds and mammals have especially high oxygen capacities. The guillemot *Uria troile*, for example, has a capacity of about 26 vol %; the harbor seal (*Phoca vitulina*) and sperm whale (*Physeter catodon*) have capacities of about 29 vol %; and the Weddell seal (*Leptonychotes weddelli*) and bladdernose seal (*Cystophora*) exhibit capacities in excess of 31 vol %. These are among the highest oxygen capacities known. The high capacities of these diving animals provide for increased oxygen storage during dives, and this may be their primary adaptive significance (see Chapter 13).

Fish, amphibians, and reptiles generally have lower oxygen capacities than birds and mammals: in the range of 6–15 vol %. This is reflective of lower hemoglobin concentrations and is correlated with their lower metabolic demands. Active fish tend to have higher oxygen capacities than sluggish fish.

Among the invertebrates oxygen capacities tend to be relatively low and can be highly variable among individuals within a species. Perhaps the highest oxygen capacity reported in any invertebrate is 14 vol % for the hemoglobin-containing blood of the earthworm *Glossoscolex giganteus*, a huge species that reaches over a meter in length. Such a high capacity is quite exceptional. Oxygen capacities averaging 4–6 vol %, for example, are found in such smaller annelids as the nightcrawler *Lumbricus* and the polychaetes *Arenicola* and *Glycera* (the latter having hemoglobin in coelomic corpuscles). Capacities of 3–7 vol % have been reported among squids and octopuses. These values include the highest known capacities for animals with hemocyanin as their respiratory pigment. The cephalopods are large and active animals, and it is not surprising that they should have the highest capacities for their pigment type, but it is noteworthy that their capacities are toward the lower end of the range for fish. Some other molluscs with hemocyanin have capacities in the range of 1–3 vol %. Among decapod crustaceans capacities are usually below 3 vol %, and values around 1.5 vol % are common. As a base of comparison, seawater at 24°C dissolves about 0.5 vol % at an oxygen pressure of 159 mm Hg.

In some cases, oxygen capacity is known to vary in response to environmental factors. Most notable are the increases in capacity in some animals when exposed to decreased oxygen tensions. We noted earlier the striking increases of circulating hemoglobins in midge larvae and certain small crustaceans (such as *Daphnia*) when in oxygen-depleted waters. Some fish also show an increase in carrying capacity when the oxygen tension of their water is reduced. Among terrestrial animals a number of forms are known to increase their carrying capacity when exposed to the low oxygen tensions of high altitudes; included are man, sheep, dogs, and young chickens (see Chapter 12). Increased oxygen-carrying capacities under conditions of lowered ambient oxygen tension are adaptive insofar as they compensate for decreases in percent loading at the sites of external respiration.

Oxygen transport by respiratory pigments in the living animal: some case studies

In this section we shall review a number of case studies that will illustrate certain patterns of oxygen transport found in various groups of animals.

The example of man illustrates one widespread pattern of respiratory transport. The respiratory pigment becomes nearly saturated in the respiratory organs, and given the circulatory rate, oxygen consumption, and carrying capacity, it is far from completely desaturated in the tissues at rest. You will recall that the coefficient of utilization in resting man is about 25%. The limited resting utilization provides a significant reserve capacity for unloading during exercise or other states in which oxygen demand is increased.

The pattern in certain cephalopods is similar in that their hemocyanin is virtually fully saturated at the gills in aerated waters, but differs in that there is a much higher utilization under conditions of comparative rest, and there is a remarkably large Bohr effect, which contributes to the high utilization. Data for the squid *Loligo pealei* in well-aerated waters are presented in Figure 9–11. After passing the gills, blood exhibits a CO_2 tension of about 2 mm Hg and is virtually saturated with oxygen, at an oxygen tension of about 120 mm Hg. An increase in CO_2 tension to about 6 mm Hg in the tissues, accompanied by a fall in pH of about 0.13 unit, results in a very large Bohr shift. Even at the relatively high venous oxygen tension of about 50 mm Hg, the blood yields 90% of its oxygen content, or about 3.9 vol %. Note the implications of the Bohr shift. Without it, venous oxygen tension would have to be 25 mm Hg (rather than 50 mm Hg) to achieve the same amount of oxygen delivery per circulatory cycle. Alternatively

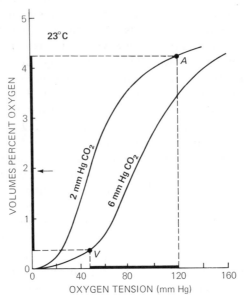

Figure 9–11. A summary of oxygen transport in the squid *Loligo pealei.* Arterial blood withdrawn from the systemic heart is at a carbon dioxide tension of about 2 mm Hg, whereas venous blood withdrawn from major venous channels is at a carbon dioxide tension of about 6 mm Hg. The blood exhibits a very large Bohr effect, and dissociation curves for the two carbon dioxide tensions are indicated. Oxygen tension falls from about 120 mm Hg to about 48 mm Hg as the blood traverses the systemic circulation (indicated on the abscissa). Arterial blood is characterized by point *A* and venous blood by point *V*, the release of oxygen being about 3.9 vol %, as indicated on the ordinate. The large Bohr shift is very important to unloading. The fall in oxygen tension to 48 mm Hg would result in much less unloading (see arrow on ordinate) if there were no Bohr effect and blood followed the curve for 2 mm Hg CO_2 throughout the circulatory cycle. Data represent the average response of a group of squids studied. (Constructed from data of Redfield, A. C. and R. Goodkind. 1929. J. Exp. Biol. 6: 340–349.)

stated, if venous tension were 50 mm Hg and there were no Bohr shift, the blood would yield just 2.3 vol % oxygen—only about 60% of the actual yield. Thus at the prevailing venous oxygen tension, the Bohr shift is responsible for over a third of the oxygen unloading observed, in contrast to the relatively small effect in man. Similar results are reported in the octopus *Octopus vulgaris.*

Recently oxygen transport has been studied in free-roaming octopuses, *Octopus dofleini,* in well-aerated waters at 11°C (the oxygen equilibrium curve at 3 mm Hg CO_2 is included in Figure 9–3). An interesting feature in these animals is that, though the mean oxygen tension in the experimental aquaria was 127 mm Hg, the mean tension in arterial blood was only 78 mm Hg, and the blood was only about 82% saturated as it left for the tissues. The mean oxygen tension of mixed venous blood was 10 mm Hg, sufficiently low to produce unloading of nearly 90% of the arterial oxygen. As in *Loligo,* then, the oxygen utilization coefficient under conditions of comparative rest is very high. *Octopus dofleini* exhibits a large, positive Bohr effect. In the experimental animals mean venous CO_2 tension (4.5 mm Hg) was greater than mean arterial tension (3.1 mm Hg), and venous pH (7.08) was lower than arterial (7.13); these changes are sufficient to produce a significant decrease in oxygen affinity. The Bohr shift probably acts to facilitate oxygen release during intermediary stages of unloading; but (unlike the case of *Loligo*) it is not critical to interpreting the oxygen content of mixed venous blood because at the low venous oxygen tensions observed, the curves for venous and arterial blood converge and are similar.

All studies of cephalopods have indicated the most interesting property that the oxygen utilization coefficient under conditions of relative inactivity is high. It appears that during heightened activity oxygen delivery must be increased largely by increasing circulatory rate, there being relatively little room for enhancement of the oxygen utilization coefficient. It would also appear that at lowered ambient oxygen tensions, decreased arterial loading in the gills could not be compensated to much extent by increasing the extent of unloading in the tissues and, thus, that adequate oxygen delivery in hypoxic conditions would depend largely on increased circulatory rate. These aspects of cephalopod physiology await systematic investigation.

It is a striking observation that the blood of cephalopods turns perceptibly blue on its passage through the gills, indicating clearly that the hemocyanin is undergoing oxygenation. A change of color is generally not observed in the gills of decapod crustaceans, and early investigators, noting also the relatively low hemocyanin concentrations in many decapods, questioned whether hemocyanin functions in routine oxygen transport in these animals. It has now been known for a long time that it does.

The approach taken in a number of early investigations of crustacean circulatory transport was the one that seemed most feasible at the time: blood was not sampled from animals living under water, but, rather, the animal was removed from water and holes were drilled through the exoskeleton as quickly as possible so that blood could be drawn from representative blood spaces upstream and downstream from the gills. A comparison of oxygen content in such samples provided information on the extent to which the blood was taking up oxygen in the gills and liberating it in the systemic tissues. Studies on a number of species revealed (or seemed to reveal) a common and most intriguing pattern, namely, that even though the hemocyanins of these animals were of relatively high affinity and even though the animals were living in well-aerated water, the hemocyanin was far from saturated in the gills. The hemocyanin of the spiny lobster, *Panulirus interruptus*, for example, becomes virtually saturated at only 25 mm Hg oxygen tension. Yet in aerated seawater, blood coming from the gills in this species was found not to be saturated; its oxygen tension was only 7 mm Hg, and the hemocyanin was carrying only 54% of the oxygen it could carry. Similar studies on other species generally revealed 50–70% saturation in postbranchial blood. The low percentage saturation suggested that the surfaces of the gills in decapods might present an unusually high barrier to oxygen diffusion between blood and water. These studies did reveal unequivocally that hemocyanin plays an important role in oxygen transport. Returning to the example of *Panulirus*, mixed blood draining the systemic tissues was found to have an oxygen tension of 3 mm Hg, and the hemocyanin in the blood was only about 20% saturated. Given a carrying capacity of about 1.5 vol %, the decline in hemocyanin saturation from about 54% to about 20% as the blood perfused the tissues indicated a yield (from the hemocyanin) of about 0.5 cc O_2/100 cc of blood. This yield is low by vertebrate or cephalopod standards. It indicates, nonetheless, that the hemocyanin plays a most important role in oxygen delivery, for blood lacking hemocyanin and carrying oxygen only in solution would yield but a miniscule amount of oxygen in going from an arterial tension of 7 mm Hg to a venous tension of 3 mm Hg.

Recently, study of several species of crabs has been carried out using indwelling catheters (small tubes) positioned in pre- and postbranchial blood spaces. These catheters are inserted surgically, and the animal, after recovery, is returned to its natural aquatic habitat. Blood can then be drawn when desired without disturbing the animal, leaving him free to roam about in an aquarium. These studies have yielded different results from those in the earlier investigations, as is exemplified for the large crab *Cancer magister* in Figure 9–12. At rest in well-aerated water (note "pre-exercise" points on equilibrium curve and "rest" bar on the ordinate), the blood is virtually saturated in the gills at a tension of about 105 mm Hg and yields about 40% of its oxygen during passage through the tissues, returning to the gills at near 25 mm Hg. There is a considerable reserve capacity to yield oxygen, which is utilized during exercise. When the animal is stimulated to continuous activity for five minutes (see "exercise" data), oxygen yield is more than doubled. The oxygen tension of venous blood is only 18 mm Hg less than at rest, but this greatly enhances unloading inasmuch as the equilibrium curve is very steep within the range of tensions involved. Interestingly, the tension established in arterial blood is much lower during exercise than during rest; this is unexplained. The degree of loading in the gills during exercise is not greatly impaired because the tension established remains high enough to be on the plateau of the equilibrium curve. During recovery from exercise, oxygen transport parameters return to their pre-exercise levels.

The studies on *Cancer* and other species using indwelling catheters present a very different picture of oxygen transport than the earlier studies which involved sampling of blood from animals removed from water. In fact there are many qualitative similarities between oxygen transport in *Cancer* and that in mammals. The investigators of *Cancer* (Johansen, Lenfant, and Mecklenburg) raise the clear possibility that the methods of sampling utilized in earlier studies of decapods introduced critical experimental artifacts in the data; the low arterial tensions and poor levels of saturation that were reported could, for example, have reflected impairment of external gas exchange during the sampling period. Although the arguments for methodological artifacts seem persuasive, it remains possible that the differences in results are indicative of interspecific differences. A definitive conclusion must await restudy of some of the same species utilized in the earlier experiments.

The hemoglobins of certain fish and lizards are not fully saturated in blood coming from the gills or lungs at rest. A recent study on eels (*Anguilla vulgaris*) has revealed an interesting phenomenon. In eels at rest in water with an oxygen tension of 150 mm Hg, the tension in the arterial blood was often as low as 30–40 mm Hg, sufficient for roughly 80–85% hemoglobin saturation at prevailing pH and temperature. During activity, however, the arterial tension often rose to 75–110 mm Hg, and percent saturation increased by about 10%. These data suggest that eels augment oxygen delivery during activity in part by increasing the available oxygen in arterial blood. The basis for increased arterial tension during activity is not known. It is hypothesized that some blood bypasses the exchange surfaces of the gills

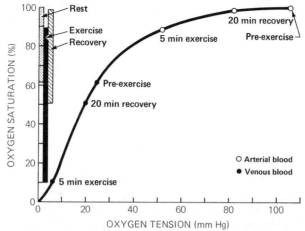

Figure 9–12. A representative example of results obtained on *Cancer magister* at 8°–10°C. Blood was sampled using indwelling catheters. The animal was studied while at rest in its aquarium and then was provoked by external stimulation to five minutes of continuous activity. Points on the oxygen equilibrium curve depict oxygen tension and percent saturation in arterial and venous blood during rest ("pre-exercise"), during exercise, and after 20 minutes of recovery from exercise. Bars on the ordinate depict oxygen utilization. The mean carrying capacity in crabs studied was 3.4 vol %. The hemocyanin of *Cancer* exhibits a positive Bohr effect, but arteriovenous changes in pH were insufficient to shift the oxygen equilibrium curve appreciably. Arterial blood was sampled from blood spaces immediately efferent to the gills. Venous blood came from sinuses directly afferent to the gills. (From Johansen, K., C. Lenfant, and T. A. Mecklenburg. 1970. Z. vergl. Physiol. 70: 1–19.)

during rest. The oxygen tension of blood that traverses the exchange surfaces may be the same during both rest and activity, but if at rest this blood mixes in the arteries with blood that has bypassed the surfaces, lowered arterial tensions will result. It is suggested that a greater fraction of the blood is delivered to the exchange surfaces in activity. What could be the benefit of having an appreciable fraction of the blood bypass the exchange surfaces at rest? Possibly it has to do with osmotic-ionic regulation. By shunting some blood away from these highly permeable surfaces, the fish might reduce passive losses and gains of salts and water. Another hypothesis is that vascular shunts around the gills pose a lower resistance to blood flow than the branchial capillary beds and that shunting helps to reduce the work load of the heart when, at rest, full perfusion of the respiratory exchange surfaces is unnecessary.

The ice fish:
vertebrates without hemoglobin

Most vertebrates have at least moderately high carrying capacities owing to the presence of hemoglobin in their blood. In this context considerable interest has centered on the fish of the family Chaenichthyidae, which occupy frigid antarctic seas and lack circulating hemoglobin altogether (or have negligibly small amounts). Without red blood, they are whitish and translucent—thus their common name, *ice fish*. They are not small fish, and their lack of hemoglobin is thus all the more remarkable. *Chaenocephalus aceratus*, for example, often weighs over 2 kg and reaches a length of about 2 ft. Without hemoglobin the oxygen-carrying capacity of the blood is limited to the amount that can dissolve in the plasma.

The habitat of these fish is undoubtedly important to understanding their respiratory physiology. The temperature remains near 0°C the year round. This tends to depress the metabolism of the fish and provides for relatively high solubility of oxygen in both the blood and the ambient water. Also, the antarctic seas are generally well aerated, thus assuring relatively high ambient oxygen tensions. The physiology of the ice fish is gradually being elucidated, and direct and indirect evidence on *C. aceratus* permits the following tentative conclusions. Resting oxygen demands are close to those of fish of the genus *Notothenia*, which occupy the same waters and have usual piscine levels of hemoglobin in their blood. *Notothenia* have carrying capacities of 6–7 vol %, whereas the carrying capacity of the plasma in *C. aceratus* is just 0.7 vol %. Thus at rest the ice fish are transporting oxygen at about the same rate as *Notothenia* but with only about one tenth the carrying capacity. This is probably to be explained in two ways. First, the ice fish likely circulate their blood considerably more rapidly than most other fish (their heart is three to five times larger); and second, the resting coefficient of oxygen utilization in the ice fish, being close to 65%, is probably two or three times the value in *Notothenia*. In short, the blood of ice fish leaves the gills and other sites of external exchange carrying much less oxygen than that of *Notothenia*, but a greater fraction of this oxygen is actually extracted, and blood flows more rapidly through the tissues.

What, then, are the disadvantages, if any, of lacking hemoglobin? Recent experiments have shown that *C. aceratus* dies when the ambient oxygen tension falls to 40–50 mm Hg, whereas *Notothenia* survives to much lower tensions, about 15 mm Hg. This difference probably, at least in part, reflects differences in the loading characteristics of blood with hemoglobin and blood lacking hemoglobin. The oxygen content of arterial ice fish blood must fall almost proportionally with ambient tension because oxygen is carried only in solution. By contrast, loading of blood containing hemoglobin may not be substantially affected over a considerable drop in tension if the affinity of the hemoglobin is sufficiently high. Data on a variety of antarctic fish indicate that nearly full loading of their hemoglobin may be expected at 50 mm Hg, a tension that can kill ice fish. The relative sensitivity of ice fish to lowered oxygen tension may not be of ecological significance, for it seems that tensions in the antarctic seas remain consistently high.

The high circulatory rates and high oxygen utilization coefficients ascribed to ice fish at rest suggest that these fish may not have as much reserve capacity to increase oxygen delivery as other fish. Their capacity for activity may thus be more limited, and they may be more sensitive to increased temperature (implying increased metabolic rate and decreased carrying capacity). These possibilities demand study.

The function of circulating respiratory pigments of very high affinity

In some animals the affinity of the circulating respiratory pigment is remarkably high. The P_{50} of hemoglobin in lugworms, *Arenicola*, for example, is variously estimated as 1.6 mm Hg to 5 mm Hg under physiological conditions of pH and temperature (Figure 9–3). Other examples include a P_{50} of 0.6 mm Hg in the aquatic oligochaete *Tubifex*; 3 mm Hg in *Daphnia*; 0.6 mm Hg or less in midge larvae, *Chironomus plumosus*; and

5 mm Hg in *Octopus vulgaris*. Such high-affinity pigments will load at low oxygen tensions and thus can help to assure adequate oxygen delivery in poorly aerated waters. On the other hand, low tissue tensions are required for unloading. The question arises: Are tissue tensions always low enough for unloading or do they only become low enough in certain circumstances, such as during activity or during exposure to low ambient oxygen tensions? In the latter case the pigment would function in oxygen transport only in certain situations.

In a number of these animals there is evidence that the respiratory pigment plays a significant role in oxygen transport even in well-aerated waters. The P_{50} in *Octopus vulgaris*, for example, is only 5 mm Hg; yet measures of venous oxygen content show that the blood is almost completely desaturated, as in other cephalopods, and tissue oxygen tensions must be very low. The aquatic oligochaete *Tubifex* exhibits reduced oxygen consumption when oxygen transport by its hemoglobin is blocked through administration of carbon monoxide. This indicates that transport by the hemoglobin is significant to oxygen delivery under normal circumstances.

The habitats of intertidal worms, such as the common burrow-dwelling lugworm *Arenicola*, are not washed by water over long periods at low tide. The high affinity of *Arenicola*'s hemoglobin suggested that it might not function during high tide but, rather, might serve largely as an oxygen store, loading during high tide and unloading as tissue tensions fell during low tide. To the question of whether *Arenicola*'s hemoglobin is involved in oxygen transport at high tide, the answer appears to be that it is. Recent studies have shown that even when the animals are in fully aerated water, meaning that the blood will carry a maximal load of dissolved oxygen upon leaving the gills, the hemoglobin goes from approximately full saturation to near 90% saturation in its passage through the body; this small decline in saturation level accounts for half or more of the oxygen liberated by the blood, the remainder being liberated from solution. Turning to the question of the function of hemoglobin at low tide, it must first be noted that calculations taking into account the carrying capacity of the blood and blood volume demonstrate that the amount of oxygen bound to hemoglobin at the start of low tide could not meet metabolic demands for more than 10 to 20 minutes in *Arenicola* and some other worms; this is inconsequential relative to the potential length of low tide. Some studies have indicated that oxygen tensions can remain high enough in *Arenicola*'s burrow during low tide (perhaps 7 mm Hg) for hemoglobin to continue to load at the body surface and carry oxygen to the tissues. Others have demonstrated anaerobic metabolism during low tide, indicating that, at least in certain circumstances, whatever oxygen transport occurs is inadequate to meet full demands. It is worthy of note that since the hemoglobin of *Arenicola* can hold sufficient oxygen to meet needs for 10 to 20 minutes, it can serve as an important reservoir of oxygen over short periods, as, for example, during intervals at high tide when the worm interrupts ventilation of its burrow. All respiratory pigments, by their very nature, can provide for some oxygen storage.

Chironomus larvae and *Daphnia* develop an abundance of hemoglobin only when placed in oxygen-depleted water, and there is evidence that their hemoglobins, which have high affinities, function in oxygen transport primarily at reduced ambient oxygen tensions. Poisoning of the hemo-globin of red *Chironomus* larvae by carbon monoxide, for example, does not affect oxygen consumption in well-aerated water but depresses it at mod-erate to low ambient oxygen tensions, indicating that the hemoglobin is not necessary to adequate oxygen delivery at high ambient tensions but is necessary at low tensions. *Chironomus* larvae live in tubes, and oxyhemo-globin likely serves as an oxygen store over periods of interrupted ventila-tion of the tube. The activity of water fleas, *Daphnia*, does nót appear to be affected by carbon monoxide poisoning at high ambient oxygen tensions. However, in experiments with no carbon monoxide, red *Daphnia* lived longer and produced more eggs than pale *Daphnia* at low oxygen tensions, indi-cating that oxygen transport by hemoglobin is of adaptive value under such conditions.

Correlations of oxygen affinity with habitat and other factors

Having reviewed some of the diversity of respiratory transport physiol-ogy, we may now examine additional comparative information that comes from study of oxygen equilibrium curves. Studies of transport in vivo are un-fortunately quite limited in number, and for many animals the only informa-tion available is the oxygen equilibrium curve. It should now be obvious that deductions about the function of a pigment in vivo from the oxygen equilibrium curve alone must be made with caution.

Among vertebrates there is a clear tendency for affinity to vary in-versely with the usual ambient oxygen tension. Species that inhabit environ-ments in which oxygen tension tends to be low generally have hemoglobins of higher affinity than those that inhabit environments in which the tension tends to be high. Hemoglobins of high affinity load more completely at low tensions, and the high affinities in species that regularly experience low tensions can be understood in this light. But why should affinity decrease with increasing ambient tension? Hemoglobins of high affinity will load just as well at high tensions as at lower tensions, and as affinity decreases, the animal becomes more vulnerable to inadequate loading should the ambient tension decrease from its usual high level. The advantages of low affinity are to be found in the processes of unloading. Low-affinity hemo-globins unload more completely at a given tissue tension than high-affinity hemoglobins. Thus a given percent utilization can be realized at higher tissue tensions with a low-affinity pigment, and, other things being equal, the low-affinity pigment will yield more of its oxygen before the tissue tension falls below the level that impairs aerobic metabolism. If we assume, then, that there are advantages to maintaining high tissue tensions, the evolu-tionary tendency would be toward lower affinity, within the limits of not impairing loading at prevailing ambient oxygen tensions. This is a sweeping and somewhat teleological statement. It may be helpful but is supported only tangentially by experimental evidence.

Fish that inhabit waters low in oxygen generally have hemoglobins of higher affinity than those that occupy waters high in oxygen. For exam-ple, at roughly equivalent CO_2 tensions and temperatures, the P_{50}'s of cat-fish and carp bloods are 1.4 mm Hg and 5 mm Hg, respectively, whereas those for rainbow trout and mackerel are 18 mm Hg and 16 mm Hg,

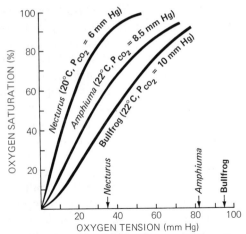

Figure 9–13. Oxygen equilibrium curves for three species of amphibians. For each species, carbon dioxide tension (P_{CO_2}) was held constant at a value about 2 mm Hg higher than the mean arterial carbon dioxide tension. Because none of the species showed a large Bohr effect, the differences in oxygen affinity of their hemoglobins are not to be explained by differences in experimental carbon dioxide tension. The mudpuppy, *Necturus maculosus*, though it possesses simple lungs, is chiefly an aquatically respiring animal, exchanging oxygen almost entirely across its gills and skin. *Amphiuma tridactylum* is a salamander that lives exclusively in water, like *Necturus*, but that lacks gills and depends strongly on air breathing. The bullfrog, *Rana catesbeiana*, is amphibious and spends much time out of water. There is a progressive drop in the oxygen affinity of the blood with increasing dependence on air breathing. The arrows on the abscissa indicate the mean oxygen tension of arterial blood of unrestrained animals living at 20°–22°C in environments of high oxygen tension (water: 125 mm Hg; air: 152 mm Hg). The blood of all three species is at least 90% saturated at prevailing arterial tension. Note that *Necturus*, though characterized by a far lower arterial tension than the other species, realizes a comparable percent loading owing to the greater affinity of its hemoglobin for oxygen. (From Lenfant, C. and K. Johansen. 1967. Resp. Physiol. 2: 247–260.)

respectively (see Figure 9–3). Air-breathing fish are commonly found in waters that become low in oxygen. They can resort to the atmospheric supply of oxygen and typically have hemoglobins of lower affinity than fish from similar habitats that exhibit strictly aquatic respiration; P_{50}'s of dual breathers and obligate air breathers are generally in the range of 12–20 mm Hg under physiological conditions. Among adult amphibians, semiterrestrial forms such as the bullfrog have hemoglobins of lower affinity than aquatic forms such as *Necturus* (Figure 9–13). Interestingly, tadpoles of semiterrestrial frogs have hemoglobins of higher affinity than the adults of their species. The synthesis of a new type of hemoglobin is a significant biochemical change accompanying metamorphosis. In *Rana esculenta*, for example, adult hemoglobin has a P_{50} of 13.2 mm Hg whereas tadpole hemoglobin has a P_{50} of 4.6 mm Hg under comparable conditions.

The P_{50}'s of mammalian and avian hemoglobins are typically in the range of 20–50 mm Hg. If mammalian bloods are brought to a CO_2 tension of 40 mm Hg, there is an evident correlation between P_{50} and body size, smaller animals tending to have higher P_{50}'s and lower oxygen affinities than larger animals. Under these conditions, for example, the elephant has a P_{50} of about 22 mm Hg; man, about 27 mm Hg; guinea pig, about 34 mm Hg; and a deer mouse (*Peromyscus*), about 49 mm Hg. Assuming that an arterial oxygen tension of 95–100 mm Hg is realized in the lungs, percent loading in the low-affinity hemoglobins of small mammals is not much lower than that in the higher-affinity hemoglobins of larger mammals. This is so because even the low-affinity hemoglobins have substantially entered their high-tension plateau phase by a tension of 95 mm Hg and are about 90% saturated. The low-affinity pigments achieve a given degree of unloading at a higher capillary tension in the systemic tissues than the high-affinity pigments. For example, at 40 mm Hg CO_2 tension, mouse hemoglobin is 50% desaturated at about 49 mm Hg O_2, whereas elephant hemoglobin is 50% desaturated only at a much lower oxygen tension, about 22 mm Hg. It is hypothesized that this is the adaptive significance of lower affinity in smaller mammals. Because small mammals have far higher weight-specific oxygen demands than large mammals, oxygen must diffuse from the capillaries into surrounding tissue cells at a greater rate. Simply analyzed, the rate of diffusion can be increased either by decreasing the capillary-to-tissue diffusion distance or by increasing the capillary-to-tissue oxygen tension difference. There is some evidence that the density of capillaries in muscle of small mammals is greater than that in medium-sized and large mammals. Though this aids diffusion in the small mammals by decreasing the diffusion distance, the effect does not appear to be nearly large enough to provide for their greater oxygen demands. By having relatively low-affinity hemoglobins, it is hypothesized that the small mammals maintain a higher capillary-to-tissue oxygen tension difference inasmuch as their blood unloads at higher capillary tensions. This is thought to be a significant factor in assuring diffusion at a sufficient rate to meet oxygen demands.

Effects of organic phosphates on affinity

Recently the effects of 2,3-diphosphoglycerate (2,3-DPG) and other organic phosphates (such as ATP) on the oxygen affinity of mammalian hemoglobins have come to light and aroused considerable interest. 2,3-DPG is formed from 1,3-DPG, a member of the glycolytic chain. It is the most abundant organic phosphate in the red cells of man and many other mammals. In human red cells it occurs at about the same molar concentration as hemoglobin. Its presence in large amounts in red cells has long constituted an enigma because it occurs in only trace amounts in other cell types and its production bypasses the ATP-generating reaction of 1,3-DPG to 3-phosphoglycerate (see Chapter 11). Now it seems likely that the primary function of 2,3-DPG in red cells is to reduce the affinity of hemoglobin for oxygen. The affinity of purified human hemoglobin in solution at physiological concentration is low (P_{50} less than 10 mm Hg). However, if 2,3-DPG is added in physiological concentration to a solution of hemoglobin that is also at physiological concentration, the affinity of the hemoglobin approaches that measured in whole blood. ATP at physiological concentrations exerts a similar but lesser effect. The shift to lower affinity is very significant to normal function. You will recall that human hemoglobin of normal affinity is about 25% desaturated at tissue tensions of 40 mm Hg. The affinity of hemoglobin in a purified solution without 2,3-DPG is so high that it is less than 10% desaturated at 40 mm Hg. A recent study has shown that 2,3-DPG does not always play an important role in reducing affinity in mammals. The red cells of adult man, horse, dog, rabbit, rat, and guinea pig have 2,3-DPG concentrations of 5–10 mM, and their hemoglobins are highly sensitive to 2,3-DPG concentration. However, the red cells of adult sheep, cow, goat, and cat contain far lower concentrations of 2,3-DPG

(<1 mM); their hemoglobins are virtually insensitive to 2,3-DPG concentration and have low affinities in the absence of 2,3-DPG.

In those animals with 2,3-DPG-sensitive hemoglobins, the most significant factor to the physiological ecologist is that changes in red cell 2,3-DPG levels can act to control affinity. This provides a mechanism for rapid shifts in affinity. If changes in affinity were entirely dependent on synthesis of red cells with different types of hemoglobin molecules, the changes would necessarily be slow. Instead the affinity of extant hemoglobins can be altered by changes in their cellular milieu. Many patients with anemia exhibit decreased oxygen affinity. The right shift of their equilibrium curve is not sufficient to jeopardize loading but increases unloading at given oxygen tension in the systemic tissues and is believed to be compensatory to the decreased carrying capacity of their blood. The decreased affinity of their hemoglobin is almost certainly due to an increase in 2,3-DPG levels; an increase in DPG concentration of 15% will shift the P_{50} from 26.5 to 29.5 mm Hg. As will be discussed in detail in Chapter 12, persons acclimatized to high altitude also exhibit a right shift of their oxygen equilibrium curve. This is believed to be due to increased 2,3-DPG. It is noteworthy that the shift can be half completed in just 6 hours and completed in 24 hours.

Other organic phosphates in the red cells exert effects similar to 2,3-DPG in other groups of vertebrates. Inositol hexaphosphate has been identified as the principal compound in birds and turtles, and ATP is believed to assume this role in teleost fish (see Chapter 11). Our knowledge of the control of affinity by organic phosphate compounds is in its infancy, and we can probably look forward to exciting developments as research progresses.

The functions of tissue hemoglobins

As discussed earlier, many animals have hemoglobins in muscle or other metabolically active tissues in addition to their circulating respiratory pigment. It is notable that these tissue hemoglobins are characteristically of higher affinity than the circulating pigment. In the large chiton *Cryptochiton stelleri*, for example, the P_{50} of the myoglobin of the radular muscle is about 3 mm Hg, whereas that of the circulating hemocyanin is about 17 mm Hg (Figure 9–14). Similarly, human circulating hemoglobin has a P_{50} of about 27 mm Hg, but human myoglobin has a P_{50} around 6 mm Hg (Figure 9–4). By virtue of their higher affinities, the myoglobins are more fully saturated at given oxygen tension than the circulating pigments. They tend to load at the expense of unloading of the circulating pigment and thus draw oxygen from the blood.

The function of myoglobins is incompletely understood. One hypothesis is that they provide an oxygen store, releasing oxygen whenever circulatory delivery is inadequate to maintain high tissue oxygen tensions. Another is that they facilitate diffusion of oxygen through the tissues. These proposals are not mutually exclusive.

If two gas mixtures of different oxygen tension are separated by a thin layer of saline solution, the rate of diffusion of oxygen through the solution is considerably greater if the solution contains hemoglobin or myoglobin than if it does not. In this situation, then, myoglobin facilitates oxygen diffusion. It was originally suggested that the myoglobin molecules

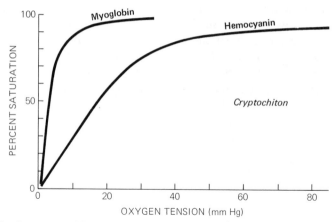

Figure 9–14. Oxygen equilibrium curves for circulating hemocyanin and radular myoglobin (hemoglobin) in a chiton, *Cryptochiton stelleri*. Both pigments were obtained from the same individual and were analyzed at 10°C. The radular muscles are colored deep red by their high content of myoglobin and are bathed by blood containing hemocyanin. (From Manwell, C. 1958. J. Cell. Comp. Physiol. 52: 341–352.)

acted as something of a "bucket brigade," enhancing the rate of diffusion by passing oxygen molecules, one to another, across the thickness of the saline solution. Now it seems clear, however, that the facilitation results from diffusion of oxymyoglobin molecules through the solution from the side of high oxygen tension to the side of low tension. Oxygen molecules diffuse through the solution as such whether myoglobin is present or not. With myoglobin present, the concentration of oxymyoglobin is higher on the high-tension side than on the other side, and the diffusion of oxymyoglobin according to this gradient provides a second "route" for oxygen movement, thus increasing the total rate of oxygen delivery. The laboratory experiments clearly suggest that myoglobin facilitates oxygen diffusion across muscle cells, enhancing the rate of oxygen movement from the surfaces of the cells to the mitochondria. As yet, however, there is no definitive evidence that this is so, and the question of facilitated diffusion in vivo remains unresolved.

The hypothesis that myoglobin acts as an oxygen store is on firmer ground. By its very nature myoglobin must release oxygen if tissue tensions fall sufficiently low. But in what situations does myoglobin release oxygen, and how much is released relative to the oxygen demands of the tissues? Unfortunately, present information is limited.

The reserve of oxygen in oxymyoglobin may potentially serve two "storage" functions. One is the traditional storage function of supplying oxygen demands during protracted periods of inadequate circulatory delivery. The other may be termed an oxygen buffer function. The oxygen reserve may serve to meet oxygen demands during transitory periods of inadequate circulatory delivery, thus providing a buffer against these transitory effects. At the start of exercise, for example, there may be a short lag before circulatory parameters adjust to the new state, and deoxygenation of myoglobin could supply some of the added demands for oxygen until a new circulatory steady state is established.

Human skeletal muscle contains about 0.5 mmole of myoglobin per kilogram. This concentration is near the high end of the range of values reported for terrestrial mammals. We can readily compute that the myoglobin holds about 0.01 cc O_2/g of muscle when saturated. How does this compare with oxygen demands? The oxygen consumption of human muscle is probably at least 0.003–0.03 cc O_2/g/min over the usual range of activity. Depending on the rate of oxygen consumption, the myoglobin stores will thus be sufficient to meet oxygen demands for anywhere from several minutes to less than 20 seconds. In terrestrial mammals, prolonged inadequacy of circulatory oxygen supply occurs primarily during heavy exercise, when metabolic demands are great. Under these conditions the myoglobin reserve will be adequate for only a short period, and the contribution of myoglobin as a storage compound in the traditional sense is seen to be limited. The reserve of oxygen in oxymyoglobin is sufficient to serve an important short-term buffer function, however, and there is some evidence that it does so. If the soleus muscle of the cat, for example, is stimulated to vigorous contraction, deoxygenation of myoglobin commences immediately and proceeds for a matter of 20 seconds until a low percentage of saturation is reached. Over this period circulatory parameters adjust so that circulatory supply of oxygen becomes adequate to meet the increased oxygen demand.

Myoglobin concentrations are much higher in many diving mammals and birds than in terrestrial forms, and the store of oxygen in oxymyoglobin is probably of considerable importance during dives. This will be discussed in Chapter 13.

Other cases in which oxygen is transferred from one pigment to another within the organism

In the exchange between circulating pigments and myoglobins we have seen one example of oxygen transfer from one respiratory pigment to another. Another example is the exchange between maternal and fetal hemoglobins in placental mammals. The fetus has its own circulatory system, which transports oxygen to its tissues from the exchange surfaces with the maternal circulation in the placenta. The fetal hemoglobin must load in the placenta at oxygen tensions that fall considerably lower than those in the adult lung; maternal arterial blood arrives in the placenta at high tension, but studies on a number of species have indicated that the tension of maternal blood can fall to 35–50 mm Hg before the blood leaves the placental exchange surfaces. Fetal hemoglobins typically have higher affinities than adult hemoglobins. For instance, the P_{50} of human fetal hemoglobin at a pH of 7.4 is about 22 mm Hg, whereas that of adult hemoglobin is about 27 mm Hg. The difference in some mammals is much greater. In rhesus monkeys fetal hemoglobin has a P_{50} of 19 mm Hg as compared to 32 mm Hg in the adult. The higher affinity of fetal hemoglobin permits more complete loading at the tensions in the placenta than would otherwise be the case. Fetal hemoglobins generally differ from adult hemoglobins in their protein structure. Sometimes, as in goats, these differences are immediately responsible for the differences in affinity. In man and some other species the fetal and adult hemoglobins differ in structure but do not differ in affinity when studied in the absence of 2,3-diphosphoglycerate; when 2,3-DPG is added in normal red cell concentrations, usual differences in affinity appear, indicating that the differences in protein structure give rise to differences in responsiveness to 2,3-DPG. In some species, such as the rhesus monkey, differences in affinity between adult and fetal hemoglobins arise partly from the immediate differences in protein structure and partly from differences in responsiveness to 2,3-DPG.

Placental-type attachments between the young and mother occur in various lower vertebrates. In two species that have received study, the garter snake (*Thamnophis*) and the spiny dogfish (*Squalus suckleyi*), fetal hemoglobin is also of higher affinity than maternal hemoglobin.

Sipunculid worms have a tentacular "circulatory system," diverticula of which are bathed by the coelomic fluid. In *Dendrostomum zostericulum*, the hemerythrin of the coelomic corpuscles has a higher affinity ($P_{50} = 3$ mm Hg) than that of the corpuscles of the "circulatory system" ($P_{50} = 15$ mm Hg), suggesting that oxygen taken up across the tentacles may be transferred to the coelomic circulation via the two pigments. Internal transport has not been investigated directly.

RESPIRATORY PIGMENTS IN RELATION TO ENVIRONMENTAL GAS TENSIONS AND TEMPERATURE

In this section we shall examine the properties of respiratory pigments as they influence the responses of animals to environmental temperatures and oxygen and carbon dioxide tensions.

Animals with circulating pigments of low affinity are generally more sensitive to lowered oxygen tensions than those with pigments of high affinity because lowered tensions threaten the adequacy of loading at the sites of respiration. Examples are numerous. Trout, which have pigments of relatively low affinity, die in poorly aerated waters in which carp thrive. Squid such as *Loligo* are very sensitive to reduced oxygen tension, octopuses such as *O. vulgaris*, less so. Worms such as the sabellids, which have chlorocruorins of low affinity, require higher oxygen levels than species such as *Arenicola* and *Tubifex*.

Elevated ambient CO_2 tensions, resulting in increased blood CO_2 tensions, can have considerable impact on oxygen transport when the respiratory pigment exhibits a Bohr or Root effect. Many fish that inhabit well-aerated waters, for example, have hemoglobins that are of relatively low affinity and show strong Bohr and Root effects. Increases in ambient CO_2 tension can reduce the affinity and carrying capacity of blood in their gills to such an extent as to seriously threaten the adequacy of oxygen loading at prevailing oxygen tensions. The fish may then be forced to seek higher oxygen tensions and in some cases may die even in thoroughly aerated water. The effect that increased ambient CO_2 can have on oxygen loading is illustrated by data on the South American paku, a species that normally inhabits well-aerated waters. In the paku an increase in CO_2 tension from 0 mm Hg to 25 mm Hg reduces the amount of oxygen carried by the blood at an oxygen tension of 80 mm Hg by nearly 40% owing to decreases in oxygen affinity (Bohr effect) and carrying capacity (Root effect). At a CO_2

tension of 0 mm Hg the blood carries as much oxygen at an oxygen tension of 17 mm Hg as it carries at an oxygen tension of 80 mm Hg when the CO_2 tension is elevated to 25 mm Hg.

Fish with high-affinity pigments may show a pronounced Bohr effect, but its potential for adverse consequences is typically much less than in fish with low-affinity pigments. This is so because a right shift in a high-affinity pigment is much less likely to place the oxygen equilibrium curve out of the useful range of oxygen tensions than a right shift in a low-affinity pigment. Catfish and carp, for example, have pigments with very low P_{50}'s at low CO_2 tensions. If the CO_2 tension is raised to 10 mm Hg, a Bohr shift occurs, but their P_{50}'s remain below 10 mm Hg. Thus even in the presence of elevated CO_2, the hemoglobins of these fish will load at relatively modest oxygen tensions (90% loading at 30–40 mm Hg). By contrast, a sucker (*Catostomus*) has a P_{50} of about 12 mm Hg at low CO_2 tension, and its P_{50} is raised to over 35 mm Hg by an increase in CO_2 to 10 mm Hg. The latter P_{50} is sufficiently high that relatively complete loading will require blood oxygen tensions that may not be attainable in common habitats of the sucker (90% loading requires over 80 mm Hg). Because hemoglobins of low affinity are commonly found in fish from well-aerated waters, it is these fish that are typically most threatened by elevated CO_2 tensions.

Root effects are commonly found in fish from well-aerated waters and, when present, add to the problems caused by the Bohr effect when CO_2 tensions rise. The Root effect, in fact, is probably often more threatening than the Bohr effect. A fish with only a Bohr effect will require higher oxygen tensions for loading when CO_2 tensions are elevated, but the full carrying capacity of its blood is available if oxygen tensions are high enough. The Root effect, on the other hand, reduces the carrying capacity; the blood cannot carry as much oxygen at high CO_2 tension as at low CO_2 tension no matter what the oxygen tension.

These considerations are exemplified in Figure 9–15 for two fish from Uganda. *Mormyrus* lives in lakes that are thermally stratified for long periods each year and feeds heavily on insect larvae found in the bottom deposits; it must enter waters of very low oxygen content when foraging near the bottom. *Mormyrus* possesses hemoglobin of high affinity. It does not show a Root effect, and the Bohr effect does not shift the equilibrium curve beyond low oxygen tensions. *Lates* is found in waters that usually have a consistently high oxygen content and low CO_2 content. It possesses hemoglobin of relatively low affinity, and elevated CO_2 tensions can seriously impair oxygen loading through Bohr and Root effects; at 70 mm Hg oxygen tension, for example, its hemoglobin binds only 70% as much oxygen at 25 mm Hg of CO_2 as at 0 mm Hg of CO_2. Large die-offs of *Lates* have been reported and appear to have been due to asphyxiation during periods when hydrological events produced reduced oxygen tensions and elevated CO_2 tensions.

The effects of CO_2 on affinity are minor in some fish, and the Root effect is minor or absent in many. Taking these observations as evidence that Root effects and large Bohr effects are not obligatory properties of piscine hemoglobins, we may ask why Root and Bohr effects are so prominent in some species, given that they have the potential for deleterious conse-

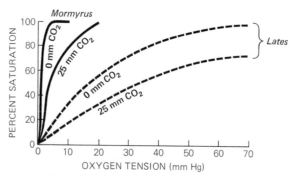

Figure 9-15. Oxygen equilibrium curves for solutions of hemoglobin from two species of fish from Uganda, *Lates albertianus* (dashed curves) and *Mormyrus kannume* (solid curves). Each species was studied at two carbon dioxide tensions: 0 mm Hg and 25 mm Hg. Hemoglobin solutions for both species were prepared identically; but because hemoglobin was removed from the erythrocytes and diluted to some extent, results may not apply, at least quantitatively, to the condition in vivo. Both species exhibit similar oxygen-carrying capacities. (From data of Fish, G. R. 1956. J. Exp. Biol. 33: 186–195.)

quences. As discussed earlier in this chapter, both effects aid unloading in the tissues, and the Root effect in particular is probably of great importance to the functioning of the countercurrent system which concentrates oxygen in the swimbladder. Thus Root and Bohr shifts in response to elevated CO_2 in the tissues provide advantages. It is when CO_2 becomes elevated at the gills that these shifts have adverse potentials.

A study on a series of tropical fish is interesting in this light. All species exhibited similar P_{50}'s at low CO_2 tensions. The paku, an inhabitant of turbulent waters, showed strong Bohr and Root shifts when CO_2 tension was elevated (as noted earlier). CO_2 is unlikely to rise to high levels in its habitat. The effects of elevated CO_2 on the oxygen equilibrium curve were smaller in species from rivers and were least in species from acid ponds. Recognizing that CO_2 tension is more likely to reach high levels in slow-moving river waters than in turbulent waters and that it is especially likely to do so in acid waters, these results suggest that Bohr and Root shifts have been limited according to the likelihood of exposure to elevated ambient CO_2 tensions in the natural habitats of the fish.

In active fish from well-aerated waters that have hemoglobins of low affinity and exhibit large Bohr shifts, it is claimed that accumulation of lactic acid during strenuous exertion can shift the oxygen equilibrium curve so far to the right as to threaten continued activity or even survival.

Recently studies on certain trout, salmon, and suckers have revealed an interesting property relevant to this discussion. Studies on whole blood show that there is a large Bohr shift. However, a more detailed analysis reveals that these fish have two types of hemoglobin in their blood, one that shows a Bohr shift and another that is insensitive to pH or CO_2 tension. In the sucker *Catostomus clarkii*, for example, about 20% of the hemoglobin molecules show no Bohr effect. This pH-insensitive component of the hemoglobin has a constant P_{50} of about 10 mm Hg over a pH range of 6.7 to 7.2, whereas the P_{50} of the pH-sensitive component increases from 13 mm Hg to 76 mm Hg with increasing acidity over this range. It is suggested that the pH-insensitive hemoglobins of these fish provide protection against the deleterious consequences both of elevated ambient CO_2 tension and of accumulated lactic acid during strenuous exercise.

The oxygen affinity of respiratory pigments commonly decreases with increasing temperature. Particularly in aquatic animals with pigments of relatively low affinity, thermal shifts, like CO_2-induced shifts, can displace the oxygen equilibrium curve sufficiently to threaten the adequacy of loading. In three species of trout, for example, P_{50} rises about 1 mm Hg for each increase of one degree centigrade. These animals can be caught in a threatening convergence of several factors at elevated temperatures. Oxygen demands increase while at the same time the solubility of oxygen in the ambient water decreases and the oxygen affinity of their hemoglobin falls. They may be compelled to seek cooler water.

Lowered temperatures often increase the oxygen affinity of respiratory pigments. This effect aids loading at decreased temperatures but means that tissue tensions must be lower to achieve a given degree of unloading. Generally, reduced tissue temperatures are accompanied by reduced oxygen demands. This may permit adequate oxygen supply at a lesser degree of unloading than at higher temperatures and to this extent will permit tissue oxygen tensions to remain more nearly like those at higher temperatures than would otherwise be the case. Temperatures in the hands and feet of man may drop to 10°C or 15°C in cold environments, and we see from Figure 9–9 that such temperatures are sufficient to cause a very pronounced shift to the left of the oxygen equilibrium curve from its position at the temperature of the body core. The shift is so great that hemoglobin at 10°C remains essentially fully loaded at 40 mm Hg O_2, the usual average venous tension of resting man in thermally moderate environments. Oxygen tensions in a hand or foot at 10°C must be below 10 mm Hg for significant unloading to occur. Such tensions are sufficiently low to act potentially as a limiting factor for aerobic metabolism.

In general, transport pigments seem to be adapted to function within the usual range of body temperatures for the animal concerned, though, as we have seen, extremes of temperature can cause potentially deleterious shifts in affinity. An interesting recent study on fish of the genus *Trematomus* from the antarctic seas serves to emphasize the adaptation of hemoglobin according to the usual regime of temperatures in the habitat. The waters in which these fish live are well aerated and remain near −1°C the year around. The P_{50} of the hemoglobin of *T. borchgrevinki* is about 21 mm Hg at a temperature of −1.5°C. This P_{50} is far higher than those of fish from temperate waters at the same temperature; mackerel, sea robin, and Atlantic salmon, for example, have similar affinities at temperatures of 15°C to 20°C. The hemoglobin of *Trematomus* is highly sensitive to increased temperature, showing reduced affinity and the fairly unusual property of a thermally induced reduction in carrying capacity. A relatively small rise in temperature from −1.5°C to 4.5°C, for example, is sufficient to reduce carrying capacity by 30%.

CONCLUDING COMMENTS

It is appropriate to end this discussion of oxygen transport pigments on a cautionary note. The supply of oxygen involves integration of several functions: external respiration, loading and unloading of transport pigments, and circulation. In studying one attribute of such an integrated system, we must not lose sight of the others. This is true both within the framework of evolutionary time and within the framework of the response of the individual animal to changes in its physiological or environmental condition. In an evolutionary perspective, oxygen transport pigments have developed along with and in the context of the attributes of the respiratory and circulatory systems. Comparing modern species, we find many interesting and significant correlations between the physiology of transport pigments and environmental or other physiological features. For example, aquatic species inhabiting well-aerated waters tend to have pigments of lower oxygen affinity than those from poorly aerated waters, and animals with high metabolic rates tend to have higher carrying capacities than those with lower metabolic rates. We should hardly expect such correlations to be perfect, however; in part this is true because the transport pigment can occupy a different role in the total oxygen transport system in different animals, depending on the properties of the circulatory and respiratory systems. To cite a simplistic example, an animal with a high circulatory rate may have a lower carrying capacity than another animal that has similar oxygen demands but has a lower circulatory rate. In considering the responses of an individual animal to changes in its environment, it is unlikely that alteration of any one component of the oxygen delivery system will exert a deterministic influence on oxygen transport, for compensations within other components are possible. Consider, for example, an aquatic animal that, upon exposure to elevated CO_2 tension, suffers a decrease in the oxygen affinity of its transport pigment sufficient to impair oxygen loading at the gills. A number of compensatory responses are possible. To cite two of them, the animal might allow tissue tensions to fall sufficiently so that each volume of blood would still yield the same amount of oxygen per circulatory cycle, or it might increase the rate of blood flow to offset the decrease in oxygen delivery per unit volume of blood. Life is threatened only when the entire, integrated system responsible for oxygen delivery cannot meet the animal's necessary demands.

SECTION 2: Carbon Dioxide Transport and Elements of Acid-Base Regulation

Carbon dioxide must ordinarily be voided into the environment as rapidly as it is produced metabolically. Its accumulation in the body can cause serious disturbances of acid-base balance and exert other deleterious effects, including disturbance of oxygen transport and, at high tensions, depression of nervous function. In many animals movement of carbon dioxide from the general tissues to sites of respiratory exchange with the environment at a sufficiently rapid rate is dependent on transport in circulating body fluids.

BASIC PRINCIPLES OF CARBON DIOXIDE TRANSPORT

In this section we shall develop the basic principles of carbon dioxide transport as they apply in particular to mammals. Many of the same principles, we shall see, are central to carbon dioxide transport in other animal groups as well.

The reactions of carbon dioxide in solution and the concept of total carbon dioxide content

When carbon dioxide is dissolved in aqueous solutions, it undergoes a series of reactions, an understanding of which is essential to the study of carbon dioxide transport. The first of these is hydration to form carbonic acid:

$$CO_2 + H_2O \rightleftharpoons H_2CO_3 \tag{1}$$

Carbonic acid, in turn, dissociates to hydrogen ions and bicarbonate ions,

$$H_2CO_3 \rightleftharpoons H^+ + HCO_3^- \tag{2}$$

and bicarbonate dissociates to hydrogen ions and carbonate ions,

$$HCO_3^- \rightleftharpoons H^+ + CO_3^{2-} \tag{3}$$

The entire series of reactions may be summarized as follows:

$$CO_2 + H_2O \rightleftharpoons H_2CO_3 \rightleftharpoons H^+ + HCO_3^- \rightleftharpoons 2H^+ + CO_3^{2-}$$

Suppose that we have an aqueous solution that is entirely devoid of carbon dioxide, carbonic acid, bicarbonate, and carbonate, and suppose that we expose this solution to gaseous carbon dioxide. Some carbon dioxide will dissolve in the solution; and depending on conditions in the solution, more or less of this carbon dioxide will react to appear as carbonic acid, bicarbonate, and carbonate. We can measure the amount of gaseous carbon dioxide that has dissolved in the solution, say 35 mmole/liter, and we can express this as the *total carbon dioxide content*, T, of the solution, recognizing, of course, that the carbon dioxide exists in solution as H_2CO_3, HCO_3^-, and CO_3^{2-} as well as CO_2 per se. Now the H_2CO_3, HCO_3^-, and CO_3^{2-} in the solution are all potentially available to form CO_2 through reversal of reactions (1), (2), and (3). Thus under appropriate conditions our solution could subsequently yield its entire total carbon dioxide content in the form of gaseous CO_2 (and would liberate 35 mmole of CO_2 per liter). From these considerations we see that the total carbon dioxide content of a solution can be viewed in two complementary ways. It is the amount of CO_2 that would have to be added to establish the concentrations of dissolved CO_2, H_2CO_3, HCO_3^-, and CO_3^{2-} present in the solution. And it is the amount of CO_2 the solution is capable of yielding, given its content of all these compounds.

In Chapter 8 we discussed the relationship between the concentration of dissolved carbon dioxide and the partial pressure of carbon dioxide in aqueous solutions. We pointed out that these quantities are related proportionally in any given body of water according to the formula

$$\text{concentration (ml at STP/liter)} = \frac{A}{760} \times \text{(partial pressure in mm Hg)} \quad (4)$$

where A is the absorption coefficient expressed as milliliters of gas (at STP) dissolved per liter at a partial pressure of 760 mm Hg. You will recall that A depends on the temperature and salinity of the solution. We stated in Chapter 8 that the concentration of dissolved CO_2 computed from partial pressure according to the absorption coefficient is the concentration of CO_2 per se. In fact, absorption coefficients for CO_2 are commonly calculated to refer to both dissolved CO_2 and H_2CO_3 (but *not* HCO_3^- or CO_3^{2-}). Thus if the absorption coefficient for a given solution is 800 ml/liter and the partial pressure of CO_2 is 19 mm Hg, we calculate the CO_2 concentration to be 20 ml/liter—meaning that the sum of the concentrations of dissolved CO_2 and H_2CO_3 is equivalent to 20 ml of gaseous CO_2 (at STP) per liter. In physiological solutions the great preponderance of the CO_2 content so calculated is represented by dissolved CO_2 per se. H_2CO_3 typically represents less than 0.5% of the total.

Earlier we developed the concept of the total carbon dioxide content, T, of a solution and indicated that it includes carbon dioxide represented as CO_2, H_2CO_3, HCO_3^-, and CO_3^{2-}. We can clearly write that T, in moles per liter, is equal to the sum of the molar concentrations of these solutes:

$$T = [CO_2] + [H_2CO_3] + [HCO_3^-] + [CO_3^{2-}]$$

We could also express T in milliliters of CO_2 at STP/liter by applying a simple conversion factor to the preceding equation. We can define another quantity S to be the sum of the molar concentrations of dissolved CO_2 and H_2CO_3:

$$S = [CO_2] + [H_2CO_3]$$

S could also be expressed in milliliters of CO_2 at STP/liter. Now it should be obvious that S can be computed from equation (4) and therefore bears a simple proportional relationship to the partial pressure of CO_2 in any given solution. It should also be clear that equation (4) does not necessarily provide a good estimate of the total carbon dioxide content, T. When the concentrations of HCO_3^- and CO_3^{2-} are significantly greater than zero, T will exceed S to a significant extent. This same principle was emphasized in Chapter 8. The relationship between T and the partial pressure of CO_2 is not at all so simple as that between S and the partial pressure of CO_2 and, in fact, can be quite complex. The reader who is interested in the calculation of T from partial pressure according to physicochemical principles is referred to advanced texts (such as that by Edsall and Wyman, cited in the selected readings).

Carbon dioxide equilibrium curves: introductory considerations

Earlier in this chapter we discussed oxygen equilibrium curves in body fluids. You will recall that the oxygen equilibrium curve can be represented as a graph of volumes percent oxygen as a function of the partial

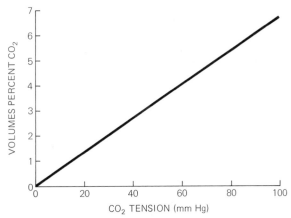

Figure 9-16. An approximate carbon dioxide equilibrium curve for a sodium chloride solution of the ionic strength of human plasma at 37°C.

pressure of oxygen. The curve often includes all oxygen present in the body fluid, both that in solution and that combined with a respiratory pigment. Analogously, we can define the *carbon dioxide equilibrium curve* or *dissociation curve* to be a graph of total carbon dioxide content as a function of the partial pressure of carbon dioxide. We can express the total carbon dioxide content in a variety of ways, including volumes percent.

When CO_2 is dissolved in distilled water or a neutral salt solution (such as a solution of NaCl), the amounts of HCO_3^- and CO_3^{2-} formed by reactions (2) and (3) are so small as to be virtually negligible. In the special case of such solutions, then, the total CO_2 content of the solution, T, is never very much greater than the amount of CO_2, S, carried as dissolved CO_2 and H_2CO_3. Thus a graph of S as a function of CO_2 tension will provide a good approximation of the CO_2 equilibrium curve. Suppose we make up a sodium chloride solution of the same total ionic strength as human blood plasma and place it at a temperature of 37°C, simulating human body temperature. The CO_2 absorption coefficient of such a solution is about 510 ml/liter, and according to the principles outlined earlier, we can construct a good approximation of the CO_2 equilibrium curve of the solution using this absorption coefficient and equation (4). The curve is depicted in Figure 9-16. Even at the highest CO_2 tension shown, 100 mm Hg, the actual total CO_2 content does not exceed the CO_2 content estimated from equation (4) by more than 0.1 vol %.

We may now inquire whether a body fluid described by Figure 9-16 would be adequate to the CO_2 transport requirements of resting man. Suppose that the rate of blood flow through the lungs of an individual is 5.5 liters/min and that his rate of CO_2 production is 210 cc/min. These values are well within normal limits for resting man. It is easy to calculate that each 100 cc of blood must unload 3.8 cc of CO_2 on its passage through the lungs. That is, the total CO_2 content of the blood must fall by 3.8 vol %. It is well known that the CO_2 tension of venous blood is near 46 mm Hg and that of arterial blood is near 40 mm Hg in resting man. We see that if the blood behaved as the solution in Figure 9-16, it would yield only 0.4 cc CO_2/ 100 cc with this drop in CO_2 tension of 6 mm Hg. This is much less than the requisite yield. To obtain the requisite yield, there would have to be a drop in tension of 57 mm Hg (i.e., if arterial blood were at 40 mm Hg, venous blood would have to be at 97 mm Hg).

Of course the solution described in Figure 9–16 is highly artificial. Blood plasma is not a simple solution of sodium chloride. However, the absorption coefficient of a solution depends only on its ionic strength and temperature. Thus the absorption coefficient of plasma is the same as that of our simple solution. In whole blood, unlike a sodium chloride solution, we cannot accurately estimate total CO_2 content from partial pressure and the absorption coefficient according to equation (4) because there is extensive reaction of CO_2 to form HCO_3^-. Equation (4), however, can, as always, be used to compute the sum S of dissolved CO_2 and H_2CO_3. When the CO_2 tension drops from 46 mm Hg to 40 mm Hg during the passage of blood through the lungs, the yield of gaseous CO_2 from the pool of dissolved CO_2 and H_2CO_3 is thus only around 0.4 cc/100 cc of blood.

We are faced, then, with a situation not unlike that discussed in connection with oxygen transport. Just as the quantities of oxygen dissolved in the blood in simple solution are insufficient for adequate oxygen transport, the quantities of CO_2 dissolved in simple solution are insufficient to meet CO_2 transport requirements. Mammals and many other animals have evolved transport pigments, such as hemoglobins, which carry oxygen in chemical combination and increase capacities for oxygen transport. Specialized transport substances for CO_2 are not known, but again, mechanisms have evolved that allow for the transport of CO_2 in chemically combined form and that similarly increase capacities for transport above those permitted by the simple solution of CO_2 in aqueous solutions.

Figure 9–17 depicts the CO_2 equilibrium curve for human arterial blood along with a curve, similar to that in Figure 9–16, for the amount of dissolved CO_2 (and H_2CO_3) as a function of partial pressure. You will note, first, that the blood carries much more CO_2 at any given partial pressure than that which is present as dissolved CO_2 and H_2CO_3. In this respect, blood differs markedly from the simple solution of sodium chloride we discussed earlier. Note also that although the curve for dissolved CO_2 and H_2CO_3 is again linear as predicated by equation (4), the relationship between total CO_2 content and partial pressure is strongly nonlinear. Especially note that the slope of the curve for total CO_2 content is considerably steeper in the range of 40–46 mm Hg than that of the curve for dissolved CO_2 and H_2CO_3 alone. In other words, blood releases much more CO_2 when its partial pressure drops from 46 mm Hg to 40 mm Hg than that which is released from the pool of dissolved CO_2 and H_2CO_3. Much of the CO_2 released comes from the blood's pool of bicarbonate.

Bicarbonate:
an important component of the total
carbon dioxide content of the blood

Some 90% of the total CO_2 content in the blood of resting man is carried as bicarbonate. To understand why this is so, it may be helpful to review a simple illustrative problem, which is not meant to apply specifically to the situation in blood but rather to develop intuition about what is going on in blood.

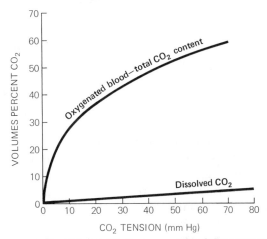

Figure 9–17. A representative CO_2 equilibrium curve for fully oxygenated human blood, indicating total CO_2 content as a function of CO_2 tension. The portion of total CO_2 content due to dissolved CO_2 and H_2CO_3 is indicated at the bottom. Most of the excess of total CO_2 content over dissolved CO_2 content is due to bicarbonate. (Equilibrium curve from Bock, A. V., H. Field, Jr., and G. S. Adair. 1924. J. Biol. Chem. 59: 353–378.)

The reactions involved have been reviewed earlier:

$$CO_2 + H_2O \rightleftharpoons H_2CO_3 \rightleftharpoons H^+ + HCO_3^- \qquad (5)$$

The dissociation of HCO_3^- to CO_3^{2-} has been omitted for simplicity inasmuch as it occurs to only a very slight extent at physiological conditions in mammals. At equilibrium, the reaction sequence in equation (5) can be described by the mass-action expression:

$$\frac{[H^+][HCO_3^-]}{[CO_2] + [H_2CO_3]} = K$$

The denominator of this expression is somewhat unusual in being the sum of the concentrations of CO_2 and H_2CO_3, but it is valid to use this sum inasmuch as the two concentrations vary directly together and bear a precisely quantifiable relationship to each other. The quantity in the denominator is S according to our earlier formulation and can thus be determined from partial pressure according to equation (4). As stated earlier, $[CO_2]$ accounts for better than 99.5% of the sum, $[CO_2] + [H_2CO_3]$. The equilibrium constant K is $10^{-6.1}$ in solutions of the ionic strength and temperature of human plasma.

Suppose now that we set up a CO_2-free sodium chloride solution of this ionic strength and temperature. As discussed earlier, such a solution will have a CO_2 absorption coefficient of 510 ml/liter. Its pH will be 6.8, the neutral pH at 37°C. We may now ask how much CO_2 we must add to this solution to raise its CO_2 tension from 0 to 40 mm Hg. This is another way of asking what the total CO_2 content will be at 40 mm Hg. From equation (4) and the absorption coefficient, we calculate that S at a tension of 40 mm Hg will be 2.7 vol %—or 1.2 mmole/liter. We must thus add 1.2 mmole of CO_2 to each liter of the solution to account for the CO_2 present as dissolved CO_2

and H_2CO_3 at 40 mm Hg. In a complete calculation we must also account for the CO_2 that reacts to form HCO_3^-. Note that the dissociation of H_2CO_3 releases H^+ as well as HCO_3^-. Thus H^+ will accumulate in the solution and the pH of the solution will fall as the tension is raised to 40 mm Hg. Knowing that S at 40 mm Hg is 1.2 mmole/liter, you can easily compute from the mass-action expression that the concentration of HCO_3^- at equilibrium will be 0.03 mmole/liter, and the pH will be 4.5. Thus we must add 1.23 mmole of CO_2 to each liter of solution to raise the CO_2 tension from 0 to 40 mm Hg. About 97.5% of this will be carried as dissolved CO_2 and H_2CO_3, and only about 2.5% will be carried as HCO_3^-.

Now suppose that we return to a CO_2-free solution and again raise its CO_2 tension to 40 mm Hg, but suppose that in this instance we are able to remove H^+ ions from solution as they are released in the dissociation of H_2CO_3 so that the pH remains at 6.8 and the dissociation results in the accumulation of HCO_3^- only. Again, S will be 1.2 mmole/liter at 40 mm Hg. From the mass-action expression we compute, however, that the concentration of HCO_3^- at equilibrium will be 6.0 mmole/liter. Thus we must add 7.2 mmole of CO_2 to each liter of solution to raise the CO_2 tension from 0 to 40 mm Hg. About 17% of this will be carried as dissolved CO_2 and H_2CO_3, and fully 83% will be carried as HCO_3^-.

It is not difficult to see why dissociation to HCO_3^- proceeds so much more extensively when the H^+ concentration is not allowed to increase. By not permitting one of the reaction products to accumulate, we allow the reactions summarized in equation (5) to proceed further to the right. Put another way, we know that the ratio in the mass-action expression will always be the same at equilibrium. When $[H^+]$ does not increase, $[HCO_3^-]$ will increase more before the product in the numerator, $[H^+][HCO_3^-]$, comes to bear its fixed relationship to the denominator.

These calculations, although somewhat artificial, show clearly that if the H^+ ions formed in the dissociation of H_2CO_3 are prevented from accumulating in solution and thus prevented from raising the H^+ concentration and lowering the pH, then much more of the CO_2 available to the system will be converted to HCO_3^-, and the total CO_2 content will be much greater at given partial pressure. In the real world, solutes that act as buffers can act to limit the buildup of H^+ concentration and thus limit changes in pH. Buffers are not perfect in the sense of preventing any change in H^+ concentration and pH, but they can exert a strongly limiting effect on these changes. Various solutes in mammalian blood act to buffer the blood against changes in pH incident to the dissociation of H_2CO_3. Among these are plasma proteins and phosphate compounds, but the most important blood buffer is hemoglobin itself. Various groups on the hemoglobin molecule serve a buffering function. Much of the buffering in the physiological range of pH is done by the imidazole groups associated with residues of the amino acid histidine in the protein structure of the hemoglobin molecule, and we shall concentrate on these groups. The buffering reaction of the imidazole group is illustrated in Figure 9–18. There are some 5×10^{19} imidazole groups per milliliter of human blood. At physiological pH some of these are in the dissociated form, Im, whereas others are in the form ImH^+. The kinetics of the imidazole reaction are such that the imidazole groups act as a very effective buffer at physiological pH. If H^+ ions are added to the blood, the imi-

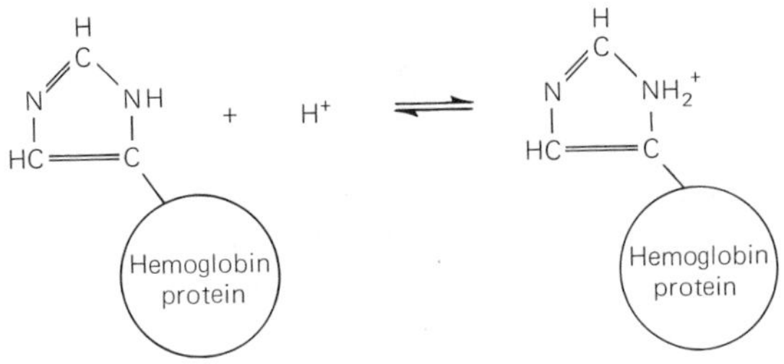

Shorthand expression:

$$Im + H^+ \rightleftharpoons ImH^+$$

Figure 9–18. The buffering reaction of the imidazole groups associated with residues of histidine in the protein structure of the hemoglobin molecule.

dazole reaction is shifted to the right, so that some of the added H^+ ions are combined to form ImH^+ rather than being left free in solution. As a result, the H^+ ion concentration is not elevated nearly so much as it would be if all the added H^+ ions were left free in solution, and the pH does not fall to nearly as great an extent. Contrariwise, if H^+ ions are removed from solution, the imidazole reaction is shifted to the left. Effectively, some of the removed H^+ ions are replaced from the pool of H^+ in ImH^+, tending to keep the concentration of free H^+ near its original level.

It is now possible to see why blood behaves much more like our simple system in which pH was held constant than like the system in which pH was uncontrolled. When CO_2 is added to the blood and the dissociation of H_2CO_3 proceeds, most of the H^+ ions released are immediately taken up by imidazole groups and other buffer groups. The H^+ ions are prevented from accumulating in free solution, and the dissociation of H_2CO_3 to HCO_3^- can proceed extensively. As a result, the HCO_3^- concentration at a given partial pressure can be high, and the total CO_2 content can be correspondingly high. In addition, the pH is only slightly depressed by the addition of CO_2.

We noted earlier in connection with Figure 9–17 that some 90% of the total CO_2 content of human blood at physiological partial pressures is represented as HCO_3^-. In fact, HCO_3^- is responsible for most of the total CO_2 content at all partial pressures depicted. Thus although a variety of factors interact in determining the shape of the CO_2 equilibrium curve, the kinetics of HCO_3^- formation dominate and impart a hyperbolic shape. Some brief comments about the kinetics are appropriate at this point, for they have bearing for understanding the CO_2 equilibrium curves of all animals.

We noted earlier that the following mass-action expression applies to solutions of the ionic strength and temperature of human plasma:

$$\frac{[H^+][HCO_3^-]}{[CO_2] + [H_2CO_3]} = 10^{-6.1}$$

This can be rewritten:

$$[HCO_3^-] = \frac{10^{-6.1}}{[H^+]} S$$

where S is the sum of the concentrations of dissolved CO_2 and H_2CO_3. The latter expression shows that the concentration of HCO_3^- in plasma is simply proportional to S at fixed H^+ concentration. We also know that S is proportional to CO_2 tension. Thus we conclude that the concentration of HCO_3^- is proportional to CO_2 tension at constant H^+ concentration. This allows us to draw the family of straight lines in Figure 9–19. The relationship between $[HCO_3^-]$ and CO_2 tension is proportional at any given pH, but the constant of proportionality changes with pH. Looking for the moment at the line for pH = 8, we see that if we could hold the pH of a solution at 8 while at the same time increasing the CO_2 tension, the concentration of HCO_3^- would increase very rapidly and linearly with tension. If instead we held the pH at 7.2, the concentration of HCO_3^- would again increase linearly with tension but considerably more slowly. If we held the pH at 5.8, $[HCO_3^-]$ would increase still more slowly with tension, and if we held the pH at 4.0 (not shown), $[HCO_3^-]$ would increase hardly at all. It is not difficult to see why the slope of the line relating the concentration of HCO_3^- to CO_2 tension decreases with pH. We have already seen that formation of HCO_3^- proceeds more extensively when the H^+ generated in the reactions of CO_2 is prevented from accumulating than when it accumulates freely, and it should be clear from equation (5) and the mass-action expression that the concentration of HCO_3^- in equilibrium with any given concentration of dissolved CO_2 will be greater, the lower the H^+ concentration.

So far, everything that has been said regarding Figure 9–19 has been based solely on knowledge of the reaction kinetics of CO_2. A solution of the ionic strength and temperature of human plasma must behave as described. We know, however, that the reactions of CO_2 to form HCO_3^- generate acid and that the buffers in blood cannot remove all of this acid from free solution. Thus as the CO_2 tension in the blood increases, the pH falls. This adds a new and critical dimension to the problem.

The extent to which pH decreases as the CO_2 tension of blood increases is a function of the buffer capacity of the blood. What would happen if the blood had no buffer capacity? CO_2-depleted blood has a pH in the vicinity of 8. Suppose, for argument's sake, that we add CO_2 to such blood with the assumption that no buffers are present. The first molecules of CO_2 to enter solution would react extensively to form HCO_3^-, for the slope of the line relating $[HCO_3^-]$ to CO_2 tension is very steep at high pH (Figure 9–19). However, the free H^+ concentration would increase rapidly with the formation of HCO_3^-, and we would rapidly shift to flatter and flatter pH lines. In fact, by the time the HCO_3^- concentration had risen to only 0.03 mmole/liter, the pH would be about 4.5, and the HCO_3^- concentration would be increasing hardly at all with further increase in CO_2 tension. We thus arrive at the same conclusion as earlier: an unbuffered solution carries little CO_2 as HCO_3^-.

In reality, blood has a high buffer capacity in the range of physiological pH. Thus considerable HCO_3^- can accumulate with only a relatively small drop in pH. To construct a HCO_3^- equilibrium curve for the blood using Figure 9–19, we need only measure the pH of the blood at a variety of CO_2 tensions and connect the appropriate points. Because a solution of given CO_2 tension and given pH *must* have a certain HCO_3^- concentration, a graph of HCO_3^- concentration against CO_2 tension results. This method was used to construct a typical HCO_3^- equilibrium curve for man on the figure. It is important to examine this curve to understand the kinetics of

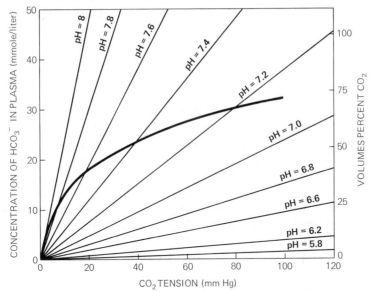

Figure 9–19. Relationships of bicarbonate concentration to carbon dioxide tension in human plasma. The straight lines radiating from the origin depict the concentration of HCO_3^- as a function of CO_2 tension at constant pH in a solution of the ionic strength and temperature of plasma. The heavy curved line depicts HCO_3^- concentration as a function of CO_2 tension in the plasma of oxygenated whole blood, taking account of the buffering properties of the blood. See text for further explanation. Content of HCO_3^- is expressed as mmole/liter on the ordinate to the left and as vol % CO_2 on the ordinate to the right.

bicarbonate formation in the blood. When CO_2 is first added to CO_2-depleted blood, the pH is high, and the blood buffers serve to keep the pH high despite considerable release of H+ in the process of HCO_3^- formation. Thus the concentration of HCO_3^- increases rapidly with CO_2 tension at low tensions. Gradually the pH does fall. This exerts a limiting effect on the further increase in HCO_3^- concentration with tension, and the slope of the HCO_3^- equilibrium curve declines. Thus we see that the equilibrium curve is the complex product of several factors: the simple reaction kinetics of HCO_3^- formation, the changing pH as the concentration of HCO_3^- increases, and the buffering action of the blood buffers—which determines the extent of the change in pH. The importance of the buffers in determining the shape of the equilibrium curve is illustrated by the obvious fact that if the pH decreased either more rapidly or more slowly with increasing CO_2 tension, the shape of the curve would be altered. More will be said about this later.

Several points concerning the bicarbonate equilibrium curve in Figure 9–19 deserve mention. First, the curve illustrates how effective the blood buffers really are. For the pH to decrease from 7.6 to 7.4, for example, the free H+ concentration must increase by only 0.000015 mmole/liter. We see from Figure 9–19 that the HCO_3^- concentration increases by about 6.4 mmole/liter between these two pH's. Each HCO_3^- is accompanied by one H+, and we see that over 99.999% of these H+ ions are removed from free solution by the buffers.

Note also that the bicarbonate equilibrium curve still has an appreciable slope in the physiological range of CO_2 tensions, around 40 mm Hg.

The slope of the bicarbonate curve for whole blood is slightly lower than that for the plasma component alone but still substantial. When the equilibrium curve for total CO_2 content of whole blood is constructed, the slope of the bicarbonate curve is added to the slope for dissolved CO_2, giving the total CO_2 curve a greater slope than the curve for dissolved CO_2 alone (see Figure 9–17).

Formation of carbamino compounds

At this point we have identified two significant components of the total CO_2 content of mammalian blood: (1) the pool of dissolved CO_2 and H_2CO_3, which, for simplicity, we shall henceforth term just "dissolved CO_2," and (2) the pool of CO_2 in the form of HCO_3^-. The third significant component of the total CO_2 content is the pool of chemically combined CO_2 in the form of *carbamino compounds*. CO_2 can react with amino groups of proteins to form carbamino compounds according to the following reaction:

$$P—NH_2 + CO_2 \rightleftharpoons P—NHCOO^- + H^+ \qquad (6)$$

where P represents protein. The CO_2 taken up in carbamino compounds is removed from the pool of dissolved CO_2 and no longer contributes directly to the partial pressure of CO_2. Thus the formation of carbamino compounds increases the total CO_2 content of the blood at given partial pressure. Carbamino compounds are formed to some extent with plasma proteins, but more than half of the carbamino formation in mammalian blood occurs with amino groups on hemoglobin. Carbamino-CO_2 accounts for some 4–5% of the total CO_2 content in human blood at physiological partial pressures.

The carbon dioxide equilibrium curve of oxygenated blood: a summary

We see now that there are three important pools of CO_2 in the blood: dissolved CO_2, HCO_3^-, and carbamino-CO_2. The amount of CO_2 held in each pool increases as CO_2 tension is increased. The total CO_2 equilibrium curve for oxygenated blood in Figure 9–17 represents the sum of the individual equilibrium curves for dissolved CO_2 (resembling Figure 9–16), HCO_3^- (resembling Figure 9–19), and carbamino-CO_2.

Effects of oxygenation and deoxygenation of hemoglobin

As discussed earlier, hemoglobin plays a central role in the formation of both major pools of chemically combined CO_2 in the blood. The buffering provided by hemoglobin is important to the formation of bicarbonate, and amino groups on hemoglobin react to form over half of the carbamino-CO_2 in the blood. It is to be stressed that the state of oxygenation of hemoglobin strongly affects both its buffering function and its reaction to form carbamino compounds. There are, as a result, striking interactions between oxygen transport and CO_2 transport.

Oxygenation of hemoglobin causes the important buffering groups of the hemoglobin molecule to become more strongly acidic. If we write the buffer reaction of the imidazole groups as follows,

$$ImH^+ \rightleftharpoons H^+ + Im$$

we are saying that oxygenation tends to shift the reaction to the right, increasing the extent of dissociation and tending to increase the H^+ concentration at equilibrium. Contrariwise, deoxygenation causes the buffer groups to become less acidic; the reaction tends to be shifted to the left, decreasing the extent of dissociation and tending to decrease the H^+ concentration at equilibrium. In short, oxyhemoglobin tends to buffer the blood at a lower pH than deoxyhemoglobin. We have seen earlier that the extent of HCO_3^- formation is strongly dependent on the buffering of H^+ concentration and, in turn, that the total CO_2 content of the blood is strongly affected by the extent to which HCO_3^- formation proceeds. It should be clear from equation (5) and the mass-action expression for HCO_3^- formation that more HCO_3^- will form at a given CO_2 tension, the lower the H^+ concentration maintained by available buffers. This can be seen readily by reference to Figure 9-19. The bicarbonate curve on the figure is that predicated by the buffering properties of oxygenated blood. In deoxygenated blood the pH is higher at a given CO_2 tension than in oxygenated blood, and accordingly the bicarbonate curve for deoxygenated blood lies above that for oxygenated blood. The fact that deoxyhemoglobin exhibits a more alkaline reaction than oxyhemoglobin thus increases the total CO_2 content of deoxygenated blood over that of oxygenated blood at given partial pressure by enhancing HCO_3^- formation.

The effects of oxygenation and deoxygenation on carbamino formation lead to a similar result. Deoxyhemoglobin binds more CO_2 as carbamino-CO_2 than oxyhemoglobin at given partial pressure, thus again tending to increase the total CO_2 content of deoxygenated blood over that of oxygenated blood.

As a consequence of the effects of oxygenation on the buffering function of hemoglobin and the reaction to form carbamino-CO_2, oxygenated blood and deoxygenated blood have different CO_2 equilibrium curves, as depicted in Figure 9-20. The shift in the CO_2 equilibrium curve with deoxygenation and oxygenation is termed the *Haldane effect*. We know that the partial pressure of CO_2 falls from 46 mm Hg in mixed venous blood to 40 mm Hg in arterial blood as the blood traverses the lungs in resting man. To determine the yield of CO_2 during this process, we must compare (1) the CO_2 content at 46 mm Hg of mixed venous blood, which is only about 70% oxygenated, with (2) the CO_2 content at 40 mm Hg of arterial blood, which is virtually fully oxygenated. The ellipsoid between points A and V in Figure 9-20 illustrates this functional relationship between total CO_2 content and partial pressure in resting man. The slope of the dashed line connecting A and V is the slope of the functional relationship and is greater than the slope of the CO_2 equilibrium curve for either oxygenated or deoxygenated blood alone. Earlier we computed that the total CO_2 content of the blood must fall by about 4 vol % during passage through the lungs in order for CO_2 removal by the circulatory and respiratory systems to keep pace with CO_2 production in resting man. We now see how this is achieved with a drop in partial pressure of only 6 mm Hg.

Two additional aspects of the interaction between oxygenation and the buffering function of hemoglobin deserve consideration: (1) its stabiliz-

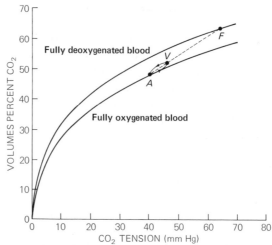

Figure 9–20. Representative CO_2 equilibrium curves for fully oxygenated and almost fully deoxygenated human blood. Because in the living organism changes in CO_2 content of the blood are accompanied by contemporaneous changes in state of oxygenation, neither equilibrium curve alone describes the true functional relationship between CO_2 content and CO_2 tension. The dashed line indicates this true functional relationship under the assumptions that fully oxygenated blood carries 20 vol % O_2 and the respiratory quotient is 0.80. If arterial blood carries 20 vol % O_2 and 48 vol % CO_2 (point A), complete utilization of the oxygen in the tissues would raise CO_2 content to 64 vol % on the curve for deoxygenated blood (point F). In resting man blood is not fully deoxygenated in the tissues. The ellipsoid between A and V describes the approximate functional relationship between CO_2 content and CO_2 tension as blood alternately becomes arterial (A) and venous (V) in resting man. Note that an increase in CO_2 content of 4 vol % as the blood becomes venous is accompanied by an increase in CO_2 tension of about 6 mm Hg. If deoxygenation did not affect the CO_2 equilibrium curve and blood followed the curve for oxygenated blood throughout the circulatory cycle, an increase of 4 vol % in CO_2 content would raise the CO_2 tension by about 9 mm Hg. (From Peters, J. P. and D. D. Van Slyke. 1932. *Quantitative Clinical Chemistry.* Vol. I. Williams & Wilkins, Baltimore, Md. Original figure © 1932 by The Williams & Wilkins Co., Baltimore.)

ing effect on the pH of the blood and (2) its import for the oxygen equilibrium curve.

If fully oxygenated human blood is equilibrated first with CO_2 at a tension of 40 mm Hg and then with CO_2 at 46 mm Hg, the pH declines according to the buffering properties of oxygenated blood from about 7.41 to about 7.32. This decrease in pH of almost 0.1 unit may sound so small as to be negligible, but remember that pH is expressed in logarithmic units; the H^+ concentration increases by almost 25%. In resting man the pH of arterial blood at a CO_2 tension of 40 mm Hg is about 7.41, but pH decreases to only about 7.37 in venous blood at 46 mm Hg. This drop in pH of about 0.04 unit (corresponding to an increase in H^+ concentration of only 10%) is much smaller than the drop observed in fully oxygenated blood over an equivalent increase in CO_2 tension. The explanation of the difference lies in the fact that, in vivo, the blood does not remain fully oxygenated as the CO_2 tension increases. As blood becomes partly deoxygenated in the tissues, the buffer groups on hemoglobin shift to a more alkaline reaction and tend to take up H^+ from solution. If other processes did not at the same time act to augment the H^+ concentration, this change in buffer function would, in fact, reduce the H^+ concentration and render venous blood more *alkaline* than arterial blood. Of course, processes that act to increase H^+ concentration do occur. CO_2 that is added to the blood as oxygen is extracted reacts to form HCO_3^-, with consequent liberation of H^+; and because of the increased carbamino formation of deoxyhemoglobin, CO_2 is additionally combined as carbamino-CO_2, a reaction that also liberates H^+ [equation (6)]. Because deoxyhemoglobin, by comparison to oxyhemoglobin, tends to extract H^+ ions from solution, it turns out that about 0.7 mmole H^+ can be added to the blood for each mmole O_2 extracted with *no change* in the free H^+ concentration or pH. In other words, if 0.7 mmole of CO_2, which will react to form about 0.7 mmole of H^+, were added to the blood for each mmole of O_2 removed, the pH of the blood would not change at all. In resting man the respiratory quotient (ratio of CO_2 production to O_2 consumption) is generally closer to 0.85 than 0.7. Thus there is addition of more H^+ than can be buffered without change in pH, and the pH does fall to some extent as the blood becomes venous. Nonetheless we see that the interaction between deoxygenation, the buffering function of hemoglobin, and the addition of CO_2 is a notably integrated system acting to limit pH changes incident to CO_2 transport.

In the section on oxygen transport we discussed the Bohr effect: a decrease in pH or increase in CO_2 tension acts to decrease the affinity of hemoglobin for oxygen, shifting the oxygen equilibrium curve to the right and decreasing the degree of oxygenation at given oxygen tension. The Bohr effect, at least in part, is the reciprocal of the effect of oxygenation and deoxygenation on the buffer function of hemoglobin (Haldane effect). We have noted that the oxygenation of heme groups on a hemoglobin molecule tends to increase dissociation of buffer groups on the molecule, driving reactions like the imidazole reaction to the right:

$$ImH^+ \rightleftharpoons Im + H^+$$

It is also true according to the principles of mass action that factors that drive these buffer reactions to the left, increasing the number of undissociated buffer groups on the hemoglobin molecule, tend to cause dissociation of the heme-oxygen complex. It is at least partly by this mechanism that an increase in H^+ concentration, as caused by the addition of CO_2 in the tissues, causes a decrease in the affinity of hemoglobin for oxygen. You will recall that the decrease of oxygen affinity as blood becomes venous aids the release of oxygen to the tissues. Again we have evidence of the evolution of a remarkably integrated system in the blood for meeting the dual responsibilities of supplying the tissues with oxygen and removing metabolic CO_2 from the tissues.

A summary of carbon dioxide transport in mammals

At this point we have reviewed the basic mechanisms involved in CO_2 transport in mammals. It should now be helpful to summarize what goes on in the blood during its circulation through the tissues and lungs, adding some further pertinent details. The broad outlines of these processes are believed to apply to mammals in general. The quantitative information given applies to a typical resting man.

As arterial blood approaches the tissues, it is highly oxygenated, has a CO_2 tension of about 40 mm Hg, and is at a pH of about 7.4. Its total CO_2

content is near 48 vol %. The CO_2 resides in three pools. Dissolved CO_2 accounts for about 5% of the total CO_2 content, HCO_3^- for about 90%, and carbamino-CO_2 for about 5%.

In the systemic capillaries CO_2 diffuses into the blood because the CO_2 tension in the blood is lower than that in the tissues. A substantial fraction of the added CO_2 is destined to be carried as HCO_3^-, but in order for the formation of HCO_3^- to proceed, the CO_2 must first be hydrated to H_2CO_3. The reaction of CO_2 and water to form H_2CO_3 is relatively slow—so slow that if uncatalyzed it would interfere with the blood's ability to take up adequate CO_2 during the short residence time in the capillaries. An enzyme catalyst for this reaction, *carbonic anhydrase*, is present in the red blood cells but not the plasma. (The hydration of CO_2 is the only catalyzed reaction in the entire reaction sequence involved in CO_2 transport.) Because carbonic anhydrase is absent in the plasma, little hydration of CO_2 occurs there, and, correspondingly, little formation of HCO_3^- occurs. Most CO_2 diffuses through the plasma into the red blood cells, and there hydration takes place rapidly. The H_2CO_3 so formed dissociates to H^+ and HCO_3^- in the red cells, where hemoglobin is immediately available to perform its important role as a buffer, acting to take up the H^+ ions released in the dissociation. The contemporaneous deoxygenation of hemoglobin increases its capacity to buffer the added H^+ ions and thus strongly limits the decrease in pH of the blood. The concentration of HCO_3^- in the red cells increases as a result of the dissociation of H_2CO_3, and HCO_3^- diffuses into the plasma according to the concentration gradient so established between the red cells and plasma. The plasma, in fact, ultimately carries most of the HCO_3^- added to the blood in the capillaries. Maintenance of electrical balance in the red cells demands that the outward diffusion of HCO_3^- be accompanied by the outward diffusion of positive ions or the inward diffusion of other negative ions. Because the erythrocyte membrane is poorly permeable to the major cations in the red cells but rather freely permeable to chloride, the outward diffusion of HCO_3^- ions is in fact accompanied by inward diffusion of Cl^- ions. The movement of chloride from the plasma into the red cells in the capillaries is known as the *chloride shift*. When hemoglobin is deoxygenated, it increases its formation of carbamino compounds. Some of the CO_2 added to the blood in the capillaries is thus taken up as carbamino-CO_2. The H^+ ions liberated in the formation of the carbamino groups are buffered by the buffering groups on hemoglobin. Figure 9–21 provides a schematic summary of the major processes of CO_2 uptake in the systemic capillaries.

When the blood leaves the capillaries, its CO_2 tension has increased to about 46 mm Hg, and its pH has fallen by about 0.04 unit. The total CO_2 content has risen by about 3.8 vol %. Dissolved CO_2 has increased by about 0.4 vol %; HCO_3^-, by about 2.4 vol %; and carbamino-CO_2, by about 1 vol %. The increased CO_2 content represents the CO_2 that is actually transported from the tissues to the lungs on one circulatory cycle. Note that about 10% of the transported CO_2 is carried as dissolved CO_2; about 65% is carried as HCO_3^-; and about 25% is carried as carbamino-CO_2.

When the blood arrives at the alveolar surfaces, it is exposed to a CO_2 tension of 40 mm Hg. Being at a higher tension, it loses CO_2 from the dis-

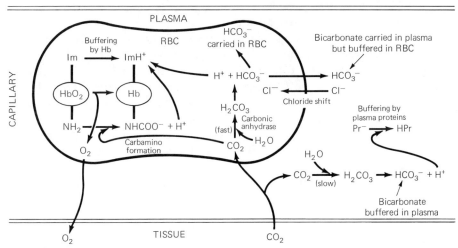

Figure 9–21. Schematic summary of processes of CO_2 uptake by the blood in the systemic capillaries of mammals. Processes occur in reverse in the lungs. RBC: red blood cell; Hb: hemoglobin; Im: imidazole groups; Pr: plasma proteins.

solved pool of CO_2 to the lungs. This reduces the dissolved CO_2 concentration and also the H_2CO_3 concentration according to equation (1). The reduction in H_2CO_3 concentration promotes the reaction of H^+ and HCO_3^- to form H_2CO_3 according to equation (2), and the H_2CO_3 formed is then dehydrated to form CO_2. The loss of dissolved CO_2 from the blood thus "pulls" the relevant reactions in the direction of forming more CO_2, which, of course, is then lost to the lungs so long as the blood CO_2 tension remains above 40 mm Hg. The reaction of HCO_3^- to form H_2CO_3 and, then, CO_2 occurs mostly in the red blood cells, where the hemoglobin buffer groups are available to provide H^+ and where carbonic anhydrase catalytically enhances the rate of dehydration of H_2CO_3. As the HCO_3^- concentration falls in the red cells, HCO_3^- diffuses back into the cells from the plasma, and chloride diffuses from the red cells into the plasma (reverse chloride shift). The reaction of HCO_3^- with H^+ is promoted by the fact that oxygenation of hemoglobin renders the hemoglobin buffer groups more acid, enhancing their dissociation to form free H^+. Oxygenation also promotes the dissociation of carbamino groups on the hemoglobin molecule to release CO_2. In summary, the processes of release of CO_2 in the lungs are the reverse of those shown in Figure 9–21.

COMPARATIVE PHYSIOLOGY
OF CARBON DIOXIDE TRANSPORT

With the review of CO_2 transport in mammals, we have seen the basic mechanisms known to be involved in CO_2 transport throughout the animal kingdom. Our detailed knowledge of CO_2 transport in groups other than the mammals is distinctly more limited. Most attention will be directed to transport of dissolved CO_2 and bicarbonate. The participation of carbamino formation in CO_2 transport in other groups is little understood and is contraindicated by data on some forms, such as certain fish.

Buffers and the carbon
dioxide equilibrium curve

We have seen that blood buffers are very important to the formation of bicarbonate and that the extent of bicarbonate formation is a major factor in determining the total CO_2 content of the blood at given partial pressure. The forms of the CO_2 equilibrium curves of various animals are diverse (see Figure 9–24), and it is important to recognize that the equilibrium curve for any particular species is in good part a reflection of the buffer groups present in the blood and their concentrations. This significant point will be examined further after a brief review of some pertinent details concerning buffers.

Buffer reactions can be represented by the following general equation:

$$HX \rightleftharpoons H^+ + X^-$$

where HX represents the undissociated form of the buffer compound, which can yield H^+, and X^- represents the dissociated form of the buffer compound, which can combine with H^+. Together, HX and X^- are termed a buffer pair. The mass-action expression for the buffer reaction is

$$\frac{[H^+][X^-]}{[HX]} = K'$$

K' is the apparent dissociation constant of the reaction; it is a constant in a solution of given ionic strength and temperature but assumes different values at different ionic strengths and temperatures. When the concentrations of the two members of the buffer pair, HX and X^-, are equal, it is clear from the mass-action expression that the concentration of H^+ is equal to K' and therefore that the pH is equal to the negative logarithm of K', symbolized pK'. These relationships are of considerable importance, for it is when the pH of a solution equals pK' (and the concentrations of HX and X^- are equal) that the buffer reaction acts most effectively to limit changes in pH caused by the addition or removal of H^+ from the solution. If H^+ is added to a solution of pH near pK', most of the H^+ combines with X^- to form HX, and only a small fraction of the added H^+ remains free in solution to increase the free H^+ concentration and decrease pH. If addition of H^+ is continued, driving the buffer reaction more and more to the left, the buffer gradually becomes less effective in taking up further H^+ as HX and thus becomes less effective in limiting changes in pH caused by the addition of acid. As a rule of thumb, by the time sufficient H^+ has been added to lower the pH by a unit from pK' (that is, increase the free H^+ concentration tenfold), the performance of the buffer will be seriously impaired. By that point, addition of H^+ will cause large changes in the free H^+ concentration and pH despite the presence of the buffer. Similar statements can be made if, starting at a pH near pK', H^+ is removed from solution. At first most of the H^+ removed will be replaced by dissociation of HX to yield H^+, and the pH will rise to only a slight extent; but by the time sufficient H^+ has been removed to elevate the pH by a unit from pK', the pH will no longer be effectively buffered against further re-

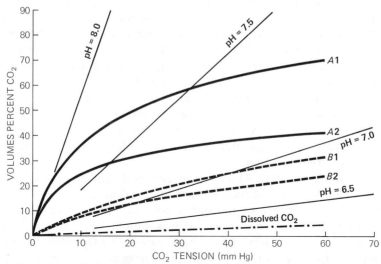

Figure 9–22. CO_2 equilibrium curves for four solutions ($A1$, $A2$, $B1$, $B2$), each being of the ionic strength and temperature of human plasma and each containing a single buffer. Buffer A (present in solutions $A1$ and $A2$) has a pK' of 8, whereas buffer B (present in solutions $B1$ and $B2$) has a pK' of 7. Buffer is supplied in equal molar concentrations in $A1$ and $B1$. Concentrations of buffer in $A2$ and $B2$ are identical to each other but half of those in $A1$ and $B1$. In each solution the concentration ratio of X^- to HX is 2:1 at a CO_2 tension of zero. CO_2 content, expressed in vol %, includes HCO_3^- and dissolved CO_2. Dissolved CO_2 alone is indicated at the bottom for all solutions. Straight lines radiating from the origin indicate the relationship of CO_2 content to CO_2 tension at constant pH.

moval of H^+. In summary, then, any particular buffer reaction acts effectively to limit changes in the free H^+ concentration caused by the addition or removal of acid over a span of about two pH units centered on the pK' of the reaction.

Body fluids may contain a great variety of potential buffer pairs, but the ones that are functioning importantly will generally be those with pK''s within a unit of the current pH of the body fluid. The imidazole groups of hemoglobin are very important buffers in CO_2 transport not only because they are present in great numbers but also because many of them have pK''s very near the pH of the blood.

The effects that the type of buffer and the buffer concentration can have on the CO_2 equilibrium curve are illustrated in simple fashion in Figure 9–22. Solutions of the ionic strength and temperature of human plasma were constructed with two different buffers, each buffer at two different concentrations. The four solutions exhibit very different CO_2 equilibrium curves. Because all the solutions carry the same amount of dissolved CO_2 at given tension, the differences in total CO_2 content (dissolved CO_2 plus HCO_3^-) are to be explained in terms of bicarbonate formation.

The influence of the particular buffer utilized can be seen by comparing curves $A1$ and $B1$. The two buffers were supplied in equal concentrations in these instances. Buffer A has a pK' of 8 and tends to hold the pH near 8, whereas buffer B has a pK' of 7 and tends to hold the pH near 7. Both buffers were established so that at zero CO_2 content, the ratio of the concentration of X^- to that of HX was 2:1. The pH is shifted to the alkaline side of pK' when this ratio exceeds 1:1; thus the pH of solution $A1$ was near

8.3 at zero CO_2 content, and the pH of solution $B1$ was near 7.3. Note that the slope of the CO_2 equilibrium curve for buffer A at low CO_2 tensions is much greater than that for buffer B. This results because of the pH difference between the solutions. Bicarbonate formation proceeds according to the steep line for pH's near 8 in the case of solution A and according to the much flatter line for pH's near 7 in the case of solution B. The equilibrium curves for both solutions decrease in slope as CO_2 tension increases. This is a consequence of decreasing HCO_3^- formation as the pH falls. Solution A always contains much more CO_2 at given tension than solution B because the H^+ concentration is maintained at a lower level in A, thus assuring a higher HCO_3^- concentration.

If we compare curves $A1$ and $A2$ or curves $B1$ and $B2$, we see the effect of buffer concentration. Solution $A2$ was identical to $A1$ at zero CO_2 content except that each member of the buffer pair was present in only half the concentration. Similarly, buffer pair B was supplied in solution $B2$ at half the concentration of solution $B1$. In both cases the CO_2 content is lower at given tension in the solution of lower buffer concentration. This follows from the fact that as H^+ is added to a solution of low buffer concentration, as compared to one of high concentration, less of the H^+ is removed as HX, and the pH accordingly falls more rapidly. The pH at given CO_2 tension is lower in the solution of lower buffer concentration, and the HCO_3^- concentration is correspondingly reduced.

Animal body fluids generally contain a variety of buffer groups with a variety of pK' values. For example, in man the imidazole groups of the plasma proteins alone have pK'''s ranging from 5.5 to 8.5, depending on their molecular milieu, and in addition there are other buffer groups on the proteins and nonprotein buffers such as the phosphate system ($H_2PO_4^-$ and HPO_4^{2-}). In this respect body fluids differ from the simple solutions containing only one buffer that are depicted in Figure 9–22. Nonetheless it should be clear from Figure 9–22 that both the nature of the buffers present in a body fluid and their concentrations influence the shape and amplitude of the CO_2 equilibrium curve.

Effects of temperature on the carbon dioxide equilibrium curve

As illustrated in Figure 9–23, decreases in blood temperature characteristically cause the total CO_2 content of the blood at given CO_2 tension to increase and impart a steeper slope to the equilibrium curve at low tensions. These effects are observed in human blood but are generally not important physiologically because man's body temperature is closely regulated. The influence of temperature is very significant among poikilotherms.

The thermal effect on the CO_2 equilibrium curve has a twofold explanation. First, decreases in temperature increase the solubility of CO_2, resulting in a greater concentration of dissolved CO_2 at given CO_2 tension. The second and quantitatively more important effect is on the blood buffers and HCO_3^- formation. Decreases in temperature generally cause buffers to become more weakly acidic, meaning that any particular buffer will tend to buffer the pH at a higher level as temperature declines; the pK' of the buffer is shifted upward. The effect of increasing pK' is clear from Figure 9–22. Formation of HCO_3^- occurs more extensively in the presence of buffers that

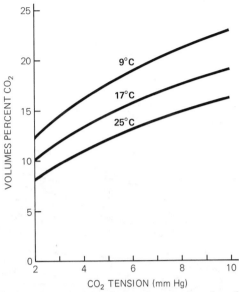

Figure 9–23. CO_2 equilibrium curves at three temperatures for the blood of a single specimen of the spotted dogfish, *Scyliorhinus stellaris*. The CO_2 equilibrium curve in this species is not perceptibly affected by state of oxygenation. (From data of Albers, C. and K. Pleschka. 1967. Resp. Physiol. 2: 261–273.)

tend to hold the pH at a more alkaline level, and the CO_2 equilibrium curve shifts upward.

The diversity of carbon dioxide equilibrium curves

We may now proceed to examine some of the details of CO_2 transport in various animal groups. Figure 9–24 depicts CO_2 equilibrium curves for oxygenated blood in a variety of species and illustrates the considerable diversity that is seen. Within the limits of available data, curves have been selected that were determined at similar temperatures, but the fact that all the curves were not determined at the same temperature should be noted.

The differences between the curves are largely to be explained in terms of the extent of HCO_3^- formation and thus in terms of the buffering properties of the bloods. Hemoglobin is an important buffer in HCO_3^- formation in all vertebrates, and differences in the "CO_2 combining power" of blood among vertebrates have been explained in part on the basis of differences in the hemoglobin content of the blood. Man, for example, has a higher CO_2 equilibrium curve than the dogfish *Scyliorhinus stellaris* (Figure 9–23), but the curves become quite similar if human blood is diluted to the same hemoglobin concentration as dogfish blood and analyzed at similar temperatures. In the experiments yielding the curves for bullfrogs and *Necturus* in Figure 9–24, the bullfrogs averaged 27% more hemoglobin per unit volume of blood than the mudpuppies. CO_2 equilibrium curves have been determined on a variety of fish, and, again, a rough correlation is evident between CO_2 combining power and hemoglobin concentration; mackerel, for example, were found to have more than twice the oxygen-carrying capacity of toadfish (Figure 9–24). In animals possessing hemocyanins as respiratory

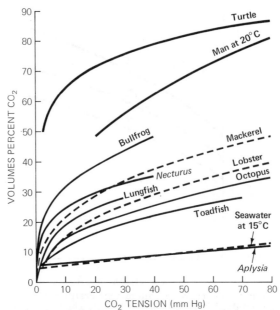

Figure 9–24. Typical CO_2 equilibrium curves of oxygenated blood in 10 species at 15°–25°C. There can be considerable variation in the curve among individuals of a single species. (1) River turtle (*Pseudemys floridana concinna*) at 25°C; (2) human blood at 20°C; (3) bullfrog (*Rana catesbeiana*) at 22°C; (4) mackerel (*Scomber scombrus*) at 20°C; (5) mudpuppy (*Necturus maculosus*) at 20°C; (6) lungfish (*Neoceratodus forsteri*) at 18°C; (7) lobster (*Palinurus vulgaris*) at 15°C; (8) octopus (*Octopus macropus*) at 15°C; (9) toadfish (*Opsanus tau*) at 20°C; and (10) sea hare (*Aplysia limacina*), a gastropod mollusc, at 15°C. [Sources of data: (1) Southworth, F. C., Jr. and A. C. Redfield. 1926. J. Gen. Physiol. 9: 387–403; (2) Harms, H. and H. Bartels. 1961. Pflügers Arch. Gesamte Physiol. Menschen Tiere 272: 384–392; (3) and (5) Lenfant, C. and K. Johansen. 1967. Resp. Physiol. 2: 247–260; (4) and (9) Root, R. W. 1931. Biol. Bull. (Woods Hole) 61: 427–456; (6) Lenfant, C., K. Johansen, and G. C. Grigg. 1966. Resp. Physiol. 2: 1–21; (7), (8), (10), and curve for seawater, Parsons, T. R. and W. Parsons. 1924. J. Gen. Physiol. 6: 153–166.]

pigments, the hemocyanins, like the hemoglobins, perform an important role as buffers in the formation of HCO_3^-. This is the case, for example, in *Octopus* and the lobster *Palinurus* (Figure 9–24). The blood of the sea hare *Aplysia* contains no hemocyanin and little protein of any kind; lacking appreciable concentrations of protein buffers, it exhibits a CO_2 equilibrium curve similar to that of seawater. Seawater does have a greater total CO_2 content than a neutral salt solution of equal salinity because it contains various inorganic buffers that promote HCO_3^- formation.

The functional range of carbon dioxide tensions

The range of CO_2 tensions over which the blood actually functions in an animal is influenced by a variety of factors. These may be analyzed by asking, first, what parameters affect the tension attained at sites of external respiration and, second, what parameters determine how much the tension must increase in the systemic tissues for the blood to transport the metabolic load of CO_2.

In analyzing exchange at the sites of external respiration, we must first recall certain basic aspects of metabolism, external gas exchange, and the properties of air and water as respiratory media. For simplicity we shall consider animals that extract most of their oxygen and void most of their CO_2 across discrete respiratory organs, be they gills or lungs. In such cases oxygen is extracted from and CO_2 is liberated to the same ventilatory stream. In steady state the properties of aerobic catabolism dictate that the amount of CO_2 added to a volume of the medium must bear a defined relationship to the amount of oxygen extracted. This relationship is indicated by the respiratory quotient, or ratio of CO_2 produced in metabolism to oxygen consumed. If the RQ is 1.0, for example, and the animal extracts 100 cc of oxygen from a volume of the medium, then the animal must also add 100 cc of CO_2 to the same volume of the medium. Now suppose that an aquatically respiring animal receives 100 cc of oxygen by extracting all the oxygen from the water passing across its gills. Assuming the oxygen tension of inhalant water to be 159 mm Hg, this will involve lowering the oxygen tension by 159 mm Hg and at 10°C in fresh water will demand ventilation of about 12.5 liters of water. Addition of 100 cc of CO_2 to this same 12.5 liters of water will raise the CO_2 tension of exhalant water to only about 5.1 mm Hg above that of inhalant water. The important point to note here is that even with a maximal decline in oxygen tension, the increase in CO_2 tension is small. This is a reflection of the differing solubilities of oxygen and CO_2 in water; CO_2 is so much more soluble than oxygen that even if water yields all its oxygen, the necessary contemporaneous addition of CO_2 will cause only a modest increase in CO_2 tension. Circumstances in air are very different, for in air the addition or removal of a given volume of gas will cause essentially the same change in partial pressure regardless of the gas in question. Thus if an air-breathing animal with an RQ of 1.0 were to reduce the oxygen tension of ventilated air by 159 mm Hg, it would also raise the CO_2 tension by 159 mm Hg. If respiratory oxygen utilization were only 20%, causing a drop in oxygen tension of 32 mm Hg, the CO_2 tension would still be elevated to a far greater extent than in the aquatically respiring animal, 32 mm Hg. In conclusion, then, the physical properties of water and air introduce a bias toward greater elevation of the CO_2 tension at the respiratory organs in air breathers than in water breathers.

The CO_2 tension established in the blood at sites of external respiration is affected by a number of additional factors. Clearly, the tension in inhalant (ambient) medium sets a lower limit on the tension that can be realized in arterial blood. When there is countercurrent exchange between blood and medium, as in fish gills, blood tension may approximate ambient tension because blood leaving the respiratory surfaces exchanges with fresh, incoming medium. In aquatic animals we have seen that the CO_2 tension of the medium cannot be greatly elevated above ambient even in the exhalant stream; so even if the final exchange of the blood is with exhalant medium, the blood tension may still be rather close to ambient. In a tidally ventilated air lung such as that of mammals, blood may exchange with air having a CO_2 tension that is very much above ambient.

The CO_2 tension reached in the systemic tissues depends on several factors: (1) the initial tension of the blood as it reaches the tissues, that is, the arterial tension; (2) the amount of CO_2 added to each volume of blood as it traverses the systemic tissues; and (3) the slope of the CO_2 equilibrium curve, which determines how much the tension must rise for the blood to carry the CO_2 added. The amount of CO_2 added to each volume of blood depends on the rate of CO_2 production and the rate of blood flow. CO_2 transport, like oxygen transport, involves integration of many physiological systems.

From the considerations developed earlier, we would expect arterial CO_2 tensions of air breathers to be generally higher than those of water breathers in well-aerated waters. Accumulating evidence indicates that this is a concept of considerable generality. Resting mammals and birds exhibit arterial CO_2 tensions of 25 mm Hg or higher. Fish in aerated waters, on the other hand, commonly have arterial tensions of only 1–3 mm Hg. It is instructive to look at groups of animals that have undergone a transition from aquatic to aerial respiration. Obligatorily air-breathing fish, such as *Electrophorus* and the lungfish *Protopterus*, have substantially higher arterial tensions than aquatically respiring fish. Bullfrogs in air have about twice the arterial tension of mudpuppies in aerated water, and land crabs such as *Gecarcinus* have higher tensions than aquatic crabs. A rise in arterial CO_2 tension appears, then, to be a significant accompaniment of the transition from aquatic to aerial respiration.

When arterial CO_2 tensions are high, as in mammals, the functional range of tensions in the blood is placed at relatively high tensions. We have seen, for example, that the blood of resting man shifts back and forth between about 46 mm Hg and 40 mm Hg as it alternately flows to the systemic tissues and lungs. The functional range is on the high, relatively flat portion of the CO_2 equilibrium curve, but the slope of the curve, given the Haldane effect, is still sufficient for the blood to carry the requisite CO_2 load with only a modest change in tension. In aquatically respiring vertebrates and invertebrates, the functional range of tensions is much lower; arterial *and venous* CO_2 tensions in aerated waters are generally within the span of 1–10 mm Hg. As is evident in Figure 9–24, CO_2 equilibrium curves are typically particularly steep in this range, and aquatic animals with such curves realize a considerable amount of CO_2 transport for a given change in blood tension. For example, CO_2 tension drops from about 6.5 mm Hg to about 3.6 mm Hg across the gills in the primarily water-breathing lungfish *Neoceratodus forsteri*. This drop of about 3 mm Hg is sufficient to cause unloading of somewhat over 4 vol % CO_2 according to the equilibrium curve for either arterial or venous blood alone (see Figure 9–24) and actually results in unloading of about 5 vol % because of a marked Haldane effect. This performance should be compared to that in man, where a drop of 6 mm Hg in tension causes unloading of about 4 vol %. In *Octopus dofleini* average CO_2 tension was found to drop from 4.5 mm Hg to 3.1 mm Hg across the gills. This species, like some others with hemocyanins, exhibits a Haldane effect, and with the contribution of the Haldane effect, the fall in CO_2 tension of 1.4 mm Hg resulted in release of 3.4 vol % CO_2.

Carbonic anhydrase

The importance of carbonic anhydrase as a catalyst for the dehydration of H_2CO_3 and the hydration of CO_2 was discussed earlier. This enzyme appears to be universally present in the red cells of vertebrates. Among invertebrates, it is reported in the blood of some annelid worms but is generally either absent or in low concentration in the bloods of crustaceans and molluscs. Carbonic anhydrase is found in a variety of tissues besides blood in both vertebrates and invertebrates. In the crustaceans and molluscs, it is commonly found at appreciable activity in the gills and is sometimes found in metabolically active tissues such as muscle. Given that these animals lack appreciable quantities of circulating enzyme, it has been suggested that the localized supply in the gills facilitates the removal of CO_2 from the bicarbonate pool of the blood and that localized supplies in muscle or other active systemic tissues facilitate the converse reaction. This hypothesis as yet requires rigorous evaluation. Carbonic anhydrase is also reported in the gills of fish and polychaete annelids and in the respiratory trees of sea cucumbers. These localized supplies may also play a role in the liberation of CO_2 from the bicarbonate pool to the environment.

The significance of the acid nature of carbon dioxide

Up to this point we have discussed CO_2 transport largely in terms of the elimination of a physiologically important gas. This is a limited perspective, for CO_2 is not only a gas that must be voided to the environment, but it is also a fundamentally important source of acid within the animal, and its level in the body fluids is an important factor in acid-base status.

The pH of the body fluids must ordinarily be controlled within fairly narrow limits. Changes in H^+ concentration can lead to marked and deleterious changes in the conformation and other properties of proteins; enzymes, for example, typically have well-defined pH optima, and their activity drops off markedly with deviations of pH from the optimum. The importance of control of pH is well known in clinical medicine. The normal arterial pH in man is near 7.4, and a rise in pH to 7.7 or fall to 6.8 will place a man near death. The usual blood pH differs among the various animal groups and species but is typically within a single unit of neutrality. In vertebrates blood pH is usually some 0.6–0.8 unit to the alkaline side of neutrality (the neutral pH being 6.8 at 37°C, 7.0 at 25°C, and 7.4 at 5°C).

Two dimensions of the role of CO_2 in acid-base physiology deserve note here: (1) the potential threat that the constant production of CO_2 poses to acid-base status and (2) the importance to acid-base physiology of maintaining a controlled tension of CO_2 in the blood.

Animals generate a variety of acid products in their metabolism, among which CO_2, in its reaction to form H_2CO_3, is one of the more important. Failure to void CO_2 can rapidly acidify the body fluids to a life-threatening extent. Not only must CO_2 be excreted to the environment, but the transport of CO_2 from the tissues to sites of external respiration must be accomplished in a manner that assures adequate stability of pH. This is an important dimension of the function of the blood buffers. An average man produces something near 8 mmole of CO_2 per *minute;* if the blood were entirely unbuffered, addition of this amount of CO_2 to the bloodstream would cause the pH to fall below 5.

Although accumulation of CO_2 can perturb pH deleteriously, it is also true that maintenance of a controlled tension of CO_2 in the blood is an important factor in normal acid-base regulation. From Figure 9–19 it is apparent that the maintenance of a normal arterial pH of about 7.4 in man is dependent on the control of CO_2 tension at near 40 mm Hg by the respiratory system. If ventilation should be inadequate to void CO_2 at the usual rate, causing CO_2 tension to rise, the pH will fall (this is termed *respiratory acidosis*). On the other hand, overventilation, as may occur during panting in many animals, will cause CO_2 tension to fall and result in a rise in pH (*respiratory alkalosis*). Alterations of pH resulting from failure of the respiratory system to maintain a proper CO_2 tension can be compensated in mammals by the kidneys, which are the other organs with a major involvement in acid-base regulation; but for normal function proper CO_2 tensions must ultimately be restored by the respiratory system. In general, organisms have evolved highly integrated systems that control acid-base status. The components of these systems include the buffers of the blood and other tissues, the processing of acid and base by the renal organs or other excretory structures, and—most importantly in the present context—the control of CO_2 tension by the respiratory organs.

The adaptive significance of the physiology of carbon dioxide transport

One of the important goals of comparative physiology is elucidation of the adaptive features of physiological systems. A natural question that arises is whether the diverse CO_2 equilibrium curves of animals exhibit features that appear to be adaptive to the peculiar requirements of the various groups and species. In the study of oxygen transport, attention is focused on blood compounds, the oxygen transport pigments, which to a substantial extent meet a discrete function, that of aiding the movement of oxygen within the animal; and it is often appropriate to consider the properties of the transport pigments rather directly within the context of adaptation to oxygen transport requirements. In the case of CO_2 transport, we find a more complex situation. The circulating compounds of interest are the blood buffers, which act to limit changes in pH caused by addition or removal of acid of whatever origin, not just those caused by changes in CO_2 content; the CO_2 equilibrium curve does not reflect the properties of compounds specifically adapted to transporting CO_2 but, rather, is a function of the blood buffer capacity. Further, CO_2, unlike oxygen, cannot be viewed simply as a gas but also represents a source of acid; and study of its transport must involve consideration of its acid properties. In short, the physiology of CO_2 transport is inextricably connected with the physiology of acid-base regulation, and attempts to analyze differences in CO_2 equilibrium curves in terms of gas transport alone are likely not to prove fruitful. Given the complexity of factors involved, it is not surprising to find that relatively little of a synthetic nature can be said.

One of the clearest generalizations that can be made from data now available is that in a great variety of animals, the amount of CO_2 that must be transported by each volume of blood, given the rate of CO_2 production and rate of circulation, is transported with a rather small arteriovenous change in CO_2 tension, less—often much less—than 10 mm Hg. This indicates that the slope of the CO_2 equilibrium curve in the functional range is sufficient that venous and tissue tensions need not be much above the tension established at the respiratory organs. In large part this relatively steep slope is a reflection of the action of the blood buffers, for they permit much of the CO_2 to be carried as HCO_3^-. The buffers are also responsible for another generalized attribute of CO_2 transport, namely, that shifts in blood pH incident to the removal and addition of CO_2 are limited. Given that the buffer systems that we see in modern animals are instrumental both in limiting the extent to which tissue CO_2 tension must exceed arterial tension and in holding blood pH within narrow limits over the cycle of CO_2 transport, it should be recognized that either or both of these effects could have been major selective determinants in the evolution of the buffers. In particular, the stabilization of pH may well have been the primary selective factor—leading to the interesting conclusion that the CO_2 equilibrium curve itself may well be simply a by-product of a more fundamental evolutionary development in the realm of acid-base regulation.

THE BICARBONATE AND NONBICARBONATE BUFFER SYSTEMS AND NONCARBONIC ACID

In closing this discussion of carbon dioxide in the body fluids, it is appropriate to recognize some additional features of blood buffering. Although we have not heretofore recognized it as such, the CO_2-bicarbonate system in the blood is in fact a buffer system. The reaction of HCO_3^- with H^+ to form H_2CO_3 and CO_2 can act to remove H^+ ions added to the blood, and the reverse reaction can act to replace H^+ ions removed from the blood. In the study of CO_2 transport, we have concentrated on the reactions of CO_2 as the *sources* of potential pH changes. When these reactions act as the sources of potential pH changes, it is the other buffer systems of the blood that must act to limit those changes by taking up or releasing H^+ ions. Thus we have concentrated on the *nonbicarbonate buffers* (mostly protein buffers) in the study of CO_2 transport. If, by contrast, H^+ should be added to the blood from a source other than carbonic acid, the bicarbonate buffer system as well as the nonbicarbonate systems can act to buffer the pH against the added H^+. That is, H^+ could be removed from solution by the same buffer groups that remove H^+ generated in the dissociation of carbonic acid or it could be removed by reaction with HCO_3^- to form carbonic acid and CO_2. In summary, H^+ added in the dissociation of carbonic acid must be buffered by nonbicarbonate buffers, but H^+ added in the dissociation of other acids (noncarbonic acids) can be buffered by both the nonbicarbonate systems and the bicarbonate system.

The metabolism of animals can generate an increased noncarbonic acid load under a variety of circumstances. Of these, the one of most general interest to physiological ecology is the production of lactic acid and other acid metabolites during anaerobic metabolism, as frequently occurs, for example, in exercise (see Chapter 11). The bicarbonate buffer system can act as an effective buffer for these acids; in fact, depending on the relative buffer capacities of nonbicarbonate and bicarbonate buffer, it can act as the principal buffer. If, for example, 10 mmole of noncarbonic H^+ is added to a liter of human arterial blood maintained at a constant CO_2 tension of 40 mm Hg, about 30% of the H^+ will be taken up by nonbicarbonate buffer and about 70% by bicarbonate buffer. The drop in pH caused by the addition of the acid will be about 0.14 unit, indicating that only about 0.00015% of the H^+ remains free in solution after buffering.

The reactions of noncarbonic acid with bicarbonate buffer are summarized as follows, using lactic acid as an example (Lac = "lactate"):

$$HLac + HCO_3^- \rightleftharpoons Lac^- + H_2CO_3$$
$$H_2CO_3 \rightleftharpoons H_2O + CO_2$$

H^+ from lactic acid reacts with bicarbonate, leaving the acid anion, lactate; and the CO_2 that is ultimately formed can be dissipated to the environment at the respiratory organs. One of the reasons that the bicarbonate buffer system can be a highly effective buffer system is the very fact that the ultimate product of the reaction with H^+, CO_2, can be readily voided from the body. In the nonbicarbonate buffer systems, such as the blood proteins, the product, HX, simply accumulates in the blood, and its increase in concentration tends to restrain further reaction of X^- with H^+ according to the principles of mass action.

After the blood has buffered a load of noncarbonic acid, it is found that the equilibrium bicarbonate concentration at given CO_2 tension is reduced (Figure 9–25). This effect may be understood in terms of our earlier discussion of the relationship between bicarbonate concentration and CO_2 tension as follows. The addition of noncarbonic H^+ to the blood essentially utilizes some of the acid buffering capacity of the nonbicarbonate buffers which are so important in bicarbonate formation, titrating them in an acid direction. As a result, the pH at given CO_2 tension is lower than usual (see Figure 9–25), and the equilibrium concentration of bicarbonate at given tension is accordingly diminished. The fact that the pH at given CO_2 tension is reduced after buffering of noncarbonic acid is significant for acid-base status. As depicted in Figure 9–25, when the blood of man has received an acid load of 5.4 mmole H^+/liter, the plasma pH of oxygenated blood at the usual arterial CO_2 tension of 40 mm Hg is no longer 7.4 but is near 7.3.

In a sense the nonbicarbonate and bicarbonate buffers provide the animal's first line of defense against alteration of blood pH when a noncarbonic acid load is imposed, and the capacity of these buffers to protect pH is therefore of importance. If a given amount of acid is added to the arterial blood of various animals, it is found that the drop in pH differs among species. This is a reflection of differences in buffer capacity, those species that exhibit a relatively small fall in pH having a higher buffer capacity than those that exhibit a relatively large fall in pH. Mammals, for example, exhibit considerably higher buffer capacities than fish, and there is substantial variation among species within both groups. There is evidence of an increase in buffer capacity with the transition from aquatic to aerial respiration; bullfrogs, for example, exhibit far higher buffer capacities than mudpuppies, and lungfish have high buffer capacities relative to other fish. It has been suggested that differences in buffer capacity among related species (as among teleost fish) might help to explain differences in their capacity for heavy exercise entailing a noncarbonic acid load, but available data permit no conclusion.

It is important to recognize that the primary depression of pH caused by an acid load can be compensated or aggravated by a number of mechanisms. A complete discussion is beyond the scope of this text, but within the present framework, examination of the effects of CO_2 tension is instruc-

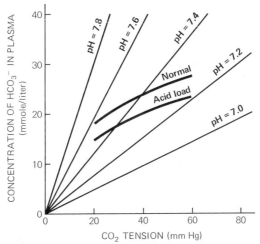

Figure 9–25. Relationship of HCO_3^- concentration to CO_2 tension in the plasma of oxygenated blood in man. "Normal" curve is identical to that in Figure 9–19. "Acid load" curve is that observed after addition of 5.4 mEq of strong, noncarbonic acid per liter of blood.

tive. As discussed earlier, pH decreases or increases, respectively, with rise or fall of the CO_2 tension along any given CO_2 equilibrium curve. If the pH at given CO_2 tension is reduced because of accommodation of a noncarbonic acid load, the pH can be returned toward its original level by decreasing the CO_2 tension. In essence, the decrease in pH caused by the noncarbonic acid load can be compensated by a reduction in the carbonic acid load. On the other hand, if CO_2 tension should increase along with the imposition of a noncarbonic acid load, the pH will be depressed by even more than the amount dictated by the noncarbonic acid. Respiratory regulation of CO_2 tension is thus an important element in the overall response of an animal to excess noncarbonic acid, just as it is intimately involved in acid-base status generally.

An example of respiratory compensation to a decrease in pH caused by lactic acid is provided by some recent experiments on men performing heavy exercise. The average response of 10 individuals will be described. Figure 9–25, although not pertaining quantitatively to these experiments, can be used as a guide in understanding the events observed. During the heavy exercise, sufficient lactic acid was added to the blood and buffered to cause a 7% reduction in arterial bicarbonate concentration and lower total CO_2 content by something over 3 vol % at 40 mm Hg (analogous to the shift from the "normal" to the "acid load" curve in Figure 9–25). Had arterial CO_2 tension remained at its usual resting level, the arterial pH would have been depressed by 0.04 unit. However, there was a respiratory compensation. Increased ventilation lowered the arterial CO_2 tension by about 3 mm Hg, and, as a result, no change in arterial pH from the resting level occurred. Note in Figure 9–25 that lowering the CO_2 tension along the "acid load" curve acts to return pH toward its usual level.

Regardless of what compensatory mechanisms may intervene, the animal which has incurred a load of lactic or other noncarbonic acid is left in an altered acid-base status. The added H^+ has titrated the blood buffers in an acid direction, and return to a normal acid-base status ultimately requires that an equal amount of H^+ be removed from the blood, thus titrating the buffers back in an alkaline direction. The return to normal acid-base status is characterized by restoration of normal pH and bicarbonate concentration at given CO_2 tension. The mechanisms involved depend on the disposition of the acid anions. We shall use lactate as an example. Mammals and many other animals retain the lactate ions generated during anaerobic metabolism and "destroy" them metabolically once adequate oxygen delivery to the tissues is re-established. The lactate either is oxidized to CO_2 and water via the citric acid cycle or is channeled into anabolic reactions leading to synthesis of glucose (see Chapter 11). In either case, the reactions essentially involve reconstitution of lactic acid from lactate:

$$H^+ + Lac^- \rightarrow HLac$$

In this manner, the H^+ ions added to the body fluids when lactic acid was generated in anaerobic metabolism are again withdrawn, restoring normal acid-base status. Some animals excrete lactate or other acid anions produced in anaerobic metabolism. When this occurs, the reconstitution and metabolic destruction of the acid is not possible and cannot serve as a route for restoration of normal acid-base status. Thus the animal must remove H^+ from the blood through enhanced excretion of H^+ to the environment, as via renal organs.

SELECTED READINGS

Davenport, H. W. 1969. *The ABC of Acid-Base Chemistry.* 5th ed. University of Chicago Press, Chicago.

Edsall, J. T. and J. Wyman. 1958. *Biophysical Chemistry.* Vol. I. Academic, New York.

Ghiretti, F. (ed.). 1968. *Physiology and Biochemistry of Haemocyanins.* Academic, New York.

Kagen, L. J. 1973. *Myoglobin. Biochemical, Physiological, and Clinical Aspects.* Columbia University Press, New York.

Masoro, E. J. and P. D. Siegel. 1971. *Acid-Base Regulation: Its Physiology and Pathophysiology.* Saunders, Philadelphia.

Prosser, C. L. 1973. Respiratory functions of blood. *In*: C. L. Prosser (ed.), *Comparative Animal Physiology.* 3rd ed. Vol. I. Saunders, Philadelphia.

Roughton, F. J. W. 1964. Transport of oxygen and carbon dioxide. *In*: W. O. Fenn and H. Rahn (eds.), *Handbook of Physiology. Section 3: Respiration.* Vol. I. American Physiological Society, Washington, D.C.

See also references in Appendix.

10 CIRCULATION

The macroconvective movement of body fluids through the organism that we term *circulation* accomplishes the transport of many commodities from one region of the body to another. Among these are oxygen, carbon dioxide, nutrients, metabolic wastes, hormones, and heat. The circulation often provides a vital link between specialized organs or tissues and the body in general. For example, cells that do not immediately receive ingested food (and that could not digest it even if they did receive it) often gain nutrient molecules from digestive organs via the circulation; and cells that lack immediate access to environmental oxygen often receive oxygen from sites of external respiration via the circulation. The circulation, then, is an essential intercommunicating system in many animals that have evolved toward a high degree of division of labor, in which specialized and often localized tissues assume certain functions and are in turn dependent on other tissues for the performance of other functions.

The anatomy and physiology of circulatory arrangements evolve under selective pressures to meet the organism's demands for internal convective transport. A demand for rapid transport imposes a selective pressure for a circulatory system with such a capability. The selective pressure is relaxed insofar as demands for internal transport are reduced, and an organism with modest demands may have a far more sluggish circulatory system than one with high demands. The demands imposed on the circulation are exceedingly diverse, as implied by the observation that the circulatory system often interconnects many different organs and tissues and transports many different commodities among them. The circulatory arrangement of an organism is thus susceptible to a multitude of evolutionary influences that affect both its general and particular features. Although the evolutionary response of the circulation to physiological demands deserves emphasis, the converse dimension of the problem is also important. At any point in evolution the transport capacity of the extant circulation can impose limits on the function of other systems, and the evolutionary development of the organism may be limited in important ways by limitations in the evolutionary plasticity of the circulation. To illustrate, recall the hypothesis that the thin, dorsoventrally flattened body form of turbellarian flatworms is at least partly predicated by their failure to develop a system for rapid internal convective transport of oxygen. In summary, the circulation and the structural and functional attributes of the organism that place demands on the circulation affect each other mutually and coevolve. At any point in evolution (such as the present), it is often a "chicken and the egg" problem to question whether circulatory development has imposed limits on the development of dependent functional attributes or whether the state of circulatory development has been limited according to the demands imposed on the circulation by the evolution of the dependent attributes.

As noted in Chapter 8 the exchanges of respiratory gases between animal and environment are typically the most immediately pressing of all exchanges. In turn, when the circulation is involved in the transport of respiratory gases between sites of external respiration and the general tissues, the demands for gas transport are often the most immediately pressing of demands on the circulation. There is good evidence to suggest that requirements for gas transport have been of primary importance in the evolution of the anatomy and physiology of circulatory systems, and there seems to be little doubt that many animals could function with a far more sluggish internal circulation than they actually have were the need for gas transport to be eliminated. Because a discussion of circulation must properly give emphasis to respiratory gas transport, the present chapter is in many respects a continuation of the treatment of oxygen and carbon dioxide exchange in Chapters 8 and 9.

Many textbooks and more advanced works in physiology provide lucid and thorough reviews of the anatomy and functional attributes of animal circulatory systems (see Selected Readings). It is not the purpose in this discussion to duplicate these treatments. Rather, aspects of circulatory systems that are of particular interest to environmental physiologists will be emphasized, and certain aspects that are of more purely physiological interest will be excluded or treated only briefly. Some features of circulatory systems that are important to the study of environmental physiology are discussed elsewhere. These include, for example, countercurrent heat exchangers (Chapter 3), the circulatory supply of renal organs (Chapter 7), and circulatory responses during diving (Chapter 13).

MAMMALS AND BIRDS

Research on the mammalian circulatory system has been intense and, historically, has provided many insights into fundamentals of circulatory dynamics. It is appropriate therefore to examine the mammalian system first. Many of its features are shared by birds. The metabolic demands of

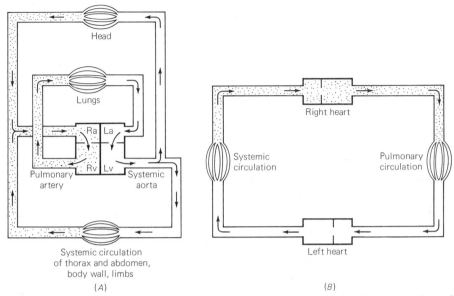

Figure 10–1. (*A*) Schematic representation of the circulatory plan in mammals and birds. Stippled parts carry relatively deoxygenated blood. Ra: right atrium of the heart; La: left atrium; Rv: right ventricle; Lv: left ventricle. (*B*) Another representation of the circulatory plan, serving to emphasize the arrangement of the pulmonary and systemic circuits in series with each other.

the tissues in mammals and birds are relatively great, and the circulation is accordingly adapted to rapid and consistent internal transport.

The basic circulatory plan

The basic circulatory plan of mammals and birds is illustrated in Figure 10–1. Oxygen-depleted blood draining the systemic tissues returns to the right heart via the great veins and is pumped by the right ventricle to the lungs, where oxygen is taken up and carbon dioxide released. The oxygenated blood from the lungs returns to the left heart and is then pumped by the left ventricle to the systemic aorta, which divides to supply all the systemic tissues. It is important to note that this circulatory plan places the respiratory organs in series with the systemic tissues. The series arrangement, emphasized in Figure 10–1*B*, maximizes the efficiency of oxygen delivery to the tissues. All the blood pumped to the tissues by the heart is freshly oxygenated, and the tissues receive blood at the full oxygen tension established in the lungs. This feature is an important component in the ability of the circulatory system to meet the high oxygen demand of the animal.

The closed nature of the circulatory system

Circulatory systems are commonly classed as *closed* or *open*, depending on whether the entire circulatory path is enclosed in discrete vessels. In an open system blood leaves discrete vessels to bathe at least some tissues directly, whereas in the ideal closed system there is always at least a thin vessel wall separating the blood from the tissues. The distinction between closed and open systems is relative, for there are many intergradations between systems that closely approximate the ideal closed condition and systems in which vessels invest only a small part of the circulatory path. Birds and mammals are among the animals that have essentially closed circulatory systems, though even in these groups there are vestiges of an open condition in some organs, as in the liver.

The systemic vascular system: its anatomy and functional attributes

The vessels at various points in the circulatory path differ anatomically and functionally in important ways. We shall discuss the major types of vessels in the order in which blood passes through them.

The great arteries are thick-walled, heavily invested with elastic and muscular elements. They are equipped to convey blood under considerable pressure from the heart to the peripheral circulation. Their elasticity allows for two important functions, the damping of pressure oscillations and the provision of a pressure reservoir. If the heart were to discharge into rigid, inelastic tubes, the head of pressure in the great arteries would oscillate violently upward and downward with each contraction and pause of the heart. By virtue of their elasticity, the great arteries in fact stretch when they receive blood discharged from the heart. Some of the energy developed by the heart is thus stored as elastic potential energy, and the increase in pressure is limited to some extent. The energy stored at the time of cardiac contraction is released as the arteries contract again during the intercontractile pause of the heart. In this way some of the energy generated during the cardiac contraction is used to maintain the head of pressure in the great arteries between contractions. The end result is that variations in pressure over the cardiac cycle are reduced (damping effect), and a substantial head of pressure is maintained in the arteries even when the heart is at rest (pressure reservoir effect). The two effects are obviously complementary.

The arteries become smaller as they branch outward toward the periphery. They also become more thinly walled, a fact that at first may appear paradoxical because the blood pressure is hardly diminished at all over a considerable distance into the periphery. According to the analysis provided by Laplace in 1820, however, the tension developed in the walls of a cylindrical vessel containing fluid under given pressure decreases proportionately with the radius of the vessel. Thus even though a small artery may be exposed to almost the same pressure as a large artery, the tension developed in the walls of the small artery will be substantially lower, and the walls accordingly need not be so well fortified to resist overexpansion under the force of the pressure. The same principle explains how capillaries can be exceedingly thin-walled yet resist an appreciable pressure.

Ultimately the systemic arteries deliver blood to the microcirculatory beds of the organs and tissues (Figure 10–2). The microcirculatory beds consist basically of arterioles, capillaries, and venules, all of which are microscopically small.

The walls of the arterioles contain important muscular and fibrous elements and are rather thick, given the small dimensions of the vessels themselves. Among the arterioles of man, to illustrate, the mean diameter of the lumen is about 30 μ (0.03 mm), and the mean thickness of the walls is about 20 μ. The smooth muscles in the walls of the arterioles are exceed-

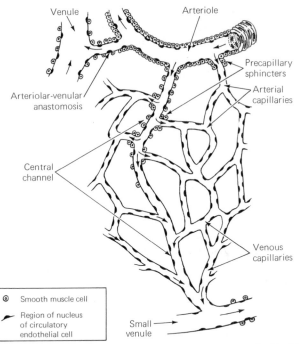

Venule

Arteriole

Precapillary sphincters

Arterial capillaries

Arteriolar-venular anastomosis

Central channel

Venous capillaries

◎   Smooth muscle cell

~   Region of nucleus of circulatory endothelial cell

Small venule

Figure 10–2. Diagram of a microcirculatory bed of a mammal. Capillaries form an anastomosing network between the arteriole and venule. The endothelial cells of the capillaries are thin and flat; they are thickened in the region of the cell nucleus. The arteriole is well invested with smooth muscle cells. The arteriolar end of the central thoroughfare channel contains dispersed smooth muscle cells and is known as a metarteriole. The venular end of the central channel lacks muscle cells and is structurally a capillary. Precapillary sphincters are located where capillaries arise from arterioles or metarterioles. "Arterial" and "venous" capillaries are distinguished by their position relative to arterioles and venules; they are not structurally different. (From Copenhaver, W. M., R. P. Bunge, and M. B. Bunge. 1971. *Bailey's Textbook of Histology.* 16th ed. The Williams & Wilkins Company, Baltimore. © 1971 The Williams & Wilkins Company, Baltimore.)

ingly important in the vasomotor control of blood distribution, for they mediate changes in the luminal diameter of the arterioles which strongly affect the rate of blood supply to the capillary beds serviced by the vessels. According to the Poiseuille equation, if we have a small vessel of fixed length with a fixed pressure difference from end to end, the volume of fluid passing through the vessel per unit time will vary with the *fourth* power of the luminal radius. This means that only modest changes in radius can cause large changes in the rate of flow. To illustrate, if the pressure gradient, length of the vessel, and viscosity of the blood remain constant, a reduction of one half in the radius of an arteriole will reduce the rate of flow through the arteriole to about one sixteenth of the original value. A reduction of just 16% in the radius will cut the flow by half. Contraction of the arteriolar muscles constricts the lumen, and relaxation of the muscles permits widening of the lumen under the force of the blood pressure. Control of the muscles is mediated by the autonomic nervous system, by circulating hormones, and by incompletely understood chemical and physical effects at the local tissue level. Through changes in the state of contraction of the arteriolar muscles, the organism can exert sensitive control over the perfusion of the tissues. To illustrate, you will recall from Chapter

3 that blood flow to superficial cutaneous tissues is often curtailed in mammals exposed to cold; this is brought about by vasoconstriction in the arterioles supplying the superficial vascular networks, the vasoconstriction being mediated by activation of sympathetic nerve fibers supplying the arteriolar muscles. In man, blood flow through skeletal muscles can increase tenfold or more during exercise; in part this results from arteriolar vasodilation, which appears to be mediated by a variety of factors, including activation of sympathetic vasodilator fibers. Where arterioles connect to capillaries, circular muscles termed precapillary sphincters are common. These sphincters and smooth muscles in the walls of small terminal arteries are also involved in controlling blood flow into the capillary beds.

The heart produces a head of pressure that is transmitted to all microcirculatory beds via the arteries. This driving force is always available and assures that vasodilation or vasoconstriction in the arterioles will result in virtually immediate changes in tissue perfusion. Each microcirculatory bed has its own arteriolar valves and thus is readily controlled independently of other vascular beds. These important features, which allow for highly sensitive temporal and spatial control of blood distribution, are obviously related to the fact that the entire circulatory system is enclosed in vessels. They are features that must be viewed as significant benefits of the closed type of circulation.

From the arterioles the blood enters the capillaries, fine vessels that are often barely wide enough to allow the passage of red blood cells. The walls of capillaries consist of only a single layer of endothelial cells and contain no muscular or fibrous elements. Because of the thinness of their walls (only about 1 μ), capillaries are the pre-eminent sites of exchange between the blood and tissues. Exchange of nutrients, oxygen, carbon dioxide, and other materials occurs by diffusion and ultrafiltration across vessel walls, and even the relatively thin walls of arterioles are usually thick enough to pose an effective barrier to these modes of exchange.

Capillary beds consist of many fine vessels that branch and anastomose among the tissue cells. The density of capillaries is different in different tissues. In man, for example, fat has a relatively low capillary density; skeletal and cardiac muscle and the brain have high densities; and the thyroid gland has a very high density. The gastrocnemius muscles of various mammals have been found to have from 300 to over 600 capillaries per square millimeter when studied in cross section. Studies on various mammalian skeletal and cardiac muscles indicate that the total capillary surface area ranges from 150 cm²/cc of tissue to over 1000 cm²/cc. Clearly, the capillary beds of these muscles provide an enormous exchange surface. In resting skeletal muscle only a fraction of the capillaries are open. A given capillary may be closed one moment and open the next, or vice versa—a phenomenon attributed to closing and opening of the precapillary sphincters. During exercise the entire capillary bed is brought into play to meet the increased metabolic demands of the muscle fibers.

The capillaries drain into small vessels, the walls of which contain fibrous and muscular elements but are nonetheless quite thin (2–5 μ in man). These vessels are termed venules, and at least the smaller ones are sufficiently thin-walled to permit a meaningful amount of exchange between blood and tissues.

449

The blood is led from the venules through a series of veins of increasing diameter, ultimately reaching the great veins that supply the heart. Because the blood pressure has declined precipitously by the time the blood leaves the capillary beds, the walls of the veins need not be capable of resisting the high tensions associated with the high pressures in the arterial circulation. The walls of the veins are thus thin by arterial standards. They contain important elastic elements but are comparatively sparsely invested with muscle. One of the significant functions assumed by the venous circulation is that of a capacity reservoir. At a given time the circulatory system contains an essentially fixed volume of blood, which must always be accommodated somewhere. We have seen that the microcirculatory beds of the tissues can vary considerably in their extent of openness and thus in the volume of blood they contain. If the capacity of the capillary beds is reduced, the veins enlarge, thus providing a compensatory increase in venous capacity. Conversely, a reduction in venous capacity accompanies increases in capillary capacity.

Cardiac output, heart rate, and stroke volume

The volume of blood pumped by the left ventricle into the systemic aorta per unit time is termed the *cardiac output* and corresponds to the rate of blood flow through the systemic vasculature. Because the right and left sides of the heart are connected in series across the pulmonary and systemic vascular circuits, the output of the right ventricle must, on the average, equal the output of the left ventricle. Cardiac output is the product of the heart rate and the volume of blood pumped per heart cycle, the latter being termed the *stroke volume*. Changes in both heart rate and stroke volume can participate in modulating cardiac output. Cardiac output is often expressed as *minute volume*, which is variously stated as the volume of blood pumped per minute (e.g., ml/min) or as the volume pumped per minute per unit of body weight (e.g., ml/kg/min).

Among mammals, resting cardiac output per unit body weight tends to increase with decreasing body size. This is not unexpected in view of the inverse relationship between weight-specific oxygen demand and body size. The heart represents somewhat the same proportion of body weight in large and small mammals, and small mammals have achieved their high weight-specific cardiac outputs more by an increase in heart rate than by an increase in weight-specific stroke volume. There is a clear inverse relationship between resting heart rate and body size: investigators have reported mean heart rates of about 30 beats/min in the Asiatic elephant, 70 beats/min in man, 250 beats/min in the laboratory rabbit, 500–600 beats/min in the house mouse, and 780 beats/min in the minute masked shrew (*Sorex cinereus*). An inverse relationship between heart rate and body size is also observed in birds.

Concepts of blood pressure, vascular resistance, and flow

The rate of blood flow through the circulatory system depends directly on the arteriovenous pressure difference and inversely on the resistance to flow presented by the circulatory bed. In rough translation this means that a higher head of arterial pressure is required to attain a certain rate of flow when the resistance posed by the vascular bed is high than when it is low. (This concept is analogous to Ohm's law: a higher voltage is required to produce a certain current in a high-resistance circuit than in a low-resistance circuit.) Earlier we stated the implication of the Poiseuille equation that in a small tube with fixed pressure drop from end to end, the rate of flow varies directly with the fourth power of the radius. Stated another way, this means that the resistance posed by the tube varies inversely with the fourth power of its radius; halving the radius increases resistance sixteenfold and results in a drop in flow to one sixteenth of the original value if the pressure gradient is unaltered. Large vessels such as the aorta present a minor resistance to blood flow, but small vessels such as the arterioles and capillaries present a considerable resistance. In the closed circulatory system of birds and mammals, each tissue is densely invested with minute vessels. This has advantages in assuring consistency of tissue perfusion and in allowing for exacting temporal and spatial control of blood distribution. It also results in a high vascular resistance and imposes the necessity of high arterial pressures to achieve adequate rates of blood flow.

The excess (or deficit) of blood pressure over ambient pressure is termed simply the *blood pressure* and is measured by using either a manometer or another device that has been calibrated against a manometer. Accordingly, blood pressures are expressed in manometric units, such as mm Hg. Blood pressure in the avian or mammalian systemic aorta varies over the cardiac cycle. The highest pressure attained during the contraction of the heart (*systole*) is termed the *systolic pressure*. The lowest pressure attained during the relaxation of the heart (*diastole*) is termed the *diastolic pressure*. In resting young people, for example, the average systolic pressure is about 120 mm Hg, and the average diastolic pressure is about 75 mm Hg. These pressures are often expressed as a pseudo-ratio, e.g., 120/75. The mean aortic pressure is obtained by averaging the pressure over the entire cardiac cycle and does not (except incidentally) correspond to the average of the two extreme pressures, systolic and diastolic. In resting young people the mean aortic pressure is 90–100 mm Hg. As a group, birds and mammals exhibit the highest pressures in the great systemic arteries of any animals. Mean resting systolic-diastolic pressures measured in several species are 98/64 mm Hg in male horses, 129/91 mm Hg in Norway rats, 135/105 mm Hg in common pigeons, and 180/130 mm Hg in starlings. As noted earlier, these high pressures are associated with the maintenance of rapid blood flow through high-resistance circulatory systems.

Interestingly, mean aortic pressure in a standing giraffe has been measured as about 220 mm Hg. Such a remarkably high aortic pressure is necessary to assure an adequate perfusion pressure within the vascular beds of the brain, which may be over 160 cm above the aorta in this long-necked animal. As in any fluid column, the excess of pressure over ambient pressure decreases with height in the large arteries supplying the head owing to simple hydrostatic effects (there are also small losses resulting from frictional effects as discussed later). When we say, for example, that the mean aortic blood pressure in a standing man is 100 mm Hg, we mean that the blood pressure is adequate to support a column of mercury 100 mm high or, given that mercury is about 12.9 times as dense as blood, a column of blood 129 cm high. If the carotid arteries were simply extended upward indefinitely, the blood would rise in them to a height of 129 cm, and at the top of this column of blood the blood pressure would be zero (equal to ambient). If man's brain were 129 cm above his aorta, there would be no head of pressure remaining to drive the blood through the vascular beds of the brain. In fact, man's brain is about 35 cm above his aorta, and at this height the blood pressure is still about 70 mm Hg above ambient. In essence, some 30 mm Hg of the aortic blood pressure is required to support the carotid blood column, leaving 70 mm Hg as perfusion pressure for the vascular beds of the brain. Now it is clear that the aortic pressure in man would be insufficient even to raise the blood to the level of the brain in a giraffe. With their high aortic pressure, however, giraffes not only support a carotid blood column that can be over 160 cm high but provide a perfusion pressure of some 90 mm Hg at the level of the brain. Correlated with their high aortic pressures, giraffes have an especially well-developed left ventricle.

The dissipation of energy over the circulatory cycle

The pressure developed in the systemic aorta by the heart represents a store of potential energy that is dissipated in driving the blood through the circulatory system. Broadly stated, the potential energy of pressure is converted to kinetic energy of flow, and in turn the kinetic energy of flow is degraded to heat through internal friction within the moving blood. The net result is that blood leaving the left ventricle ultimately makes its way back to the right atrium, but in the process the entire store of energy imparted by the left ventricle is lost as heat. The latter is the fundamental reason why the heart must keep beating. Without the continual conversion of chemical energy to mechanical energy by the heart muscle, the circulation of the blood would rapidly cease.

A comment on the frictional loss of energy as heat in the moving blood is appropriate. It is sometimes said that the friction develops between the moving blood and the walls of the blood vessels. This is erroneous, for all the friction develops within the blood itself, between adjacent layers of blood that are moving at different velocities. As we have noted previously, when a fluid moves through a tube, the fluid at the center of the tube moves more rapidly than that at the periphery of the tube. In cross section the fluid stream may be viewed as a series of thin concentric layers, with the velocity of flow in a given layer being greater the more centrally the layer is located in the blood vessel. A given layer does not slide effortlessly past its peripherally adjacent and more slowly moving layer. Rather, there is a resistance to their relative motion, which is quantified as viscosity, and some kinetic energy is lost as heat in overcoming this internal friction. It turns out that fluid immediately adjacent to the wall of the vessel does not flow at all. Thus the outermost moving layer of fluid must overcome frictional resistance between itself and a static layer of fluid, not between itself and the wall of the vessel. All the frictional resistance to flow is therefore internal to the moving fluid. For a given rate of flow this internal friction increases as the radius of the vessel decreases, thus the inverse relationship between vascular resistance and radius.

Pressure, flow, and dissipation
of energy in the systemic circuit

With the background presented earlier, we may now review the changes in pressure and flow and the dissipation of energy as the blood traverses the systemic vasculature, taking resting man as an example.

According to a simplified form of Bernoulli's equation, the total high-grade energy of the blood consists of the sum of its kinetic energy (energy of flow) and its potential energy of pressure. When the left ventricle ejects blood into the aorta, it not only increases its pressure but sets it in motion. Both factors must enter into the calculation of the total energy imparted to the blood by the ventricle. It turns out that the kinetic energy accounts for only 1–3% of the total energy in the aorta of man at rest, however, and the kinetic energy accounts for only a minor part of the total in other parts of the circulatory bed as well. This permits us to use pressure alone as an indicator of changes in energy content as the blood passes through the circulation. Kinetic energy is lost as frictional heat during circulation, thus requiring a steady conversion of potential energy of pressure to kinetic energy to keep the blood in motion. *The drop in pressure across any part of the circulation therefore indirectly reflects the loss of energy resulting from viscous (internal frictional) resistance to flow.*

The resistance posed in the large arteries is small. Thus, for example, blood arrives at the wrist in the radial artery of man at a mean pressure only 3–4 mm Hg lower than the mean pressure in the aorta. The small losses of energy in the large arteries assure that a large head of pressure remains for perfusion of the microcirculatory beds.

With the progressive branching of the circulatory system from the aorta to the capillaries, the total cross-sectional area increases. There is an abrupt increase as the blood approaches the arterioles, such that the total cross-sectional area of the myriad arterioles is over 100 times greater than that of the aorta. The cross-sectional area again increases abruptly in the capillaries, to over 500 times that of the aorta. Because the volume of blood flowing through the capillaries per unit time is the same as that flowing through the aorta, the velocity of flow in the capillaries must be very much slower. At rest, the mean aortic velocity might be 200 mm/sec and the mean capillary velocity less than 1 mm/sec. The volume of blood flow per unit time through individual capillaries is much lower relative to that in the aorta than these figures might suggest; a velocity of 1 mm/sec in a small-diameter capillary would represent a far lower flow rate than a like velocity in the large-diameter aorta. Because the capillaries are less than 1 mm long, a reasonably low velocity of flow in the capillaries is necessary to assure a sufficient residence time for exchange between the blood and tissues.

Even though the total cross-sectional area increases in the terminal arteries, arterioles, and capillaries and even though the rate of blood flow through individual vessels decreases in tandem, these vessels, because of their small diameters, are the major sites of resistance in the circulation. This is reflected in precipitous drops in blood pressure. Mean pressure declines from about 90 mm Hg to about 60 mm Hg across the terminal arteries, from about 60 mm Hg to 30 mm Hg across the arterioles, and from about 30 mm Hg to 15 mm Hg across the capillaries and initial venules. Whereas a pressure drop of only 2–3 mm Hg is sufficient to move blood at a relatively high flow rate from the shoulder to the wrist in the major arteries of the arm, a drop of perhaps 45 mm Hg is required to move blood at a relatively low flow rate over a distance of a few millimeters from the beginnings of arterioles to the initial venules in a microcirculatory bed. There is no more compelling evidence of the energy demand imposed by the type of closed circulatory system seen in birds and mammals.

The venous circulation is a low-pressure, low-resistance system. The average pressure at which blood is supplied to the venous vasculature (postcapillary pressure) in a resting, recumbent man is only around 10–15 mm Hg. The head of pressure developed by the left ventricle is reduced to less than 1 mm Hg by the time the blood has reached the entrance of the great veins into the right atrium.

Exchange of fluid across capillary walls

In Chapter 7 we discussed ultrafiltration in the renal glomerulus as the mechanism by which fluid is introduced into the nephrons. You will recall that the blood pressure in the glomerular capillaries establishes a hydrostatic pressure gradient between the glomerulus and lumen of the Bowman's capsule and that this pressure gradient forces an ultrafiltrate of the blood plasma into the capsule, overcoming the opposing osmotic gradient resulting from the colloid osmotic pressure of the plasma. The system is adapted to assuring a relatively rapid rate of flow into the nephrons. An important factor is that resistance in the blood vessels afferent to the glomeruli is relatively low, so that blood arrives in the glomerular capillaries at a comparatively high pressure of 60–70 mm Hg. Precapillary resistance is higher in other parts of the systemic circulation, and blood accordingly arrives in other capillary beds at a substantially lower pressure. The pressure, nonetheless, is still sufficient to cause some ultrafiltration of fluid from the capillaries into the tissue spaces, and considerable attention has been devoted to this phenomenon and its implications.

In the general systemic capillaries, as in the glomerulus, two forces affect the movement of fluid across the capillary membrane: osmotic forces and forces of hydrostatic pressure. As a generalization, the hydrostatic effect all along the capillaries is to cause ultrafiltrational loss of fluid from the capillaries to the tissue spaces, whereas the osmotic effect all along the capillaries is to cause osmotic entry of water from the tissue spaces into the capillaries. The net movement of water depends on the relative strengths of these opposing forces. We shall examine net movements in the systemic capillaries after briefly discussing the hydrostatic and osmotic forces individually.

The critical factor in ultrafiltration is the difference in hydrostatic pressure between the capillaries and tissue spaces, that is, the capillary blood pressure minus the tissue pressure. As an idealization, we may state that the excess of capillary over tissue pressure drops from about 30–35 mm Hg at the arterial end of capillary beds to about 15 mm Hg at the venous end in resting man.

The critical factor in osmotic movement of water is the difference in osmotic pressure between the blood plasma and the tissue fluids. The plasma osmotic pressure is greater than that in the tissue fluids because the plasma contains high concentrations of large protein solutes that cannot readily diffuse across the capillary membrane. The osmotic effect of these solutes cannot be equalized on the two sides of the membrane by diffusion. Ions and other small solutes diffuse across the membrane freely, and were it not for the unequal distribution of protein solutes, these smaller chemical species would all distribute themselves at equal concentrations on either side of the membrane, with the result that they would not contribute at all to a differential of osmotic pressure. Because the plasma proteins are anionic, however, they affect the equilibrium distribution of small ionic solutes according to the Donnan principle (see Chapter 4). As a result sodium and chloride ions are distributed unequally across the capillary membrane, sodium being more concentrated in the plasma than in the tissue fluids and chloride being more concentrated in the tissue fluids than in the plasma. In the balance, the total osmotic pressure due to sodium and chloride is greater in the plasma than in the tissue fluids. In summary, then, the plasma osmotic pressure exceeds the osmotic pressure of the tissue fluids because of the greater concentration of nondiffusible protein solutes in the plasma and because of the unequal distribution of small ionic solutes that results from the unequal distribution of the protein solutes. The difference in osmotic pressure, that is, the colloid osmotic pressure, is about 25 mm Hg at the arterial end of human capillaries.

Our present concept of the interplay of forces in the capillaries still follows the reasoning outlined by Starling in 1896. According to the Starling hypothesis, there is a net loss of fluid from the capillaries into the tissue spaces at the arterial end of the capillaries because the hydrostatic force favoring loss of water from the capillaries (estimated earlier as about 30–35 mm Hg) exceeds the osmotic force favoring entry of water into the capillaries (about 25 mm Hg). As the blood flows through the capillaries and the blood pressure declines, the difference between the hydrostatic force and osmotic force falls. Where the forces become equal, there is no net movement of water. As the blood moves beyond this point toward the venous end of the capillaries, the continuing fall in capillary blood pressure swings the balance of forces to favor reabsorption of water into the capillaries. Note that the hydrostatic force of about 15 mm Hg at the venous end is less than the osmotic force, meaning that net movement of water is from the tissue spaces into the capillaries. In review, then, the Starling hypothesis envisions a loss of water from the capillaries at the arterial end and a return of water at the venous end. Whether the hypothesis provides a complete understanding of fluid movements across the capillary walls remains uncertain, for not only is it difficult to measure all the applicable parameters at the minute level of the capillary bed, but these parameters (such as capillary blood pressure, osmotic pressure of the tissue fluids, and so on) vary temporally and spatially.

The end result of the outward and inward movement of fluid across the capillary walls is generally a net loss of fluid from the plasma as it traverses the capillaries. This fluid, which in total may amount to 2–4 liters per day in man, cannot be allowed to accumulate in the tissue spaces. The responsibility for returning it to the blood vascular system falls on the lymphatic circulation.

The pulmonary circulation

Discussion of the pulmonary circulation has been postponed to this point because an understanding of the forces affecting fluid movement across capillary walls is important to appreciating an interesting problem associated with perfusion of the lung. The systemic circulation is a high-pressure, high-resistance system. We have seen that even in the presence of large precapillary resistances, pressures in the systemic capillaries are still sufficient to cause appreciable net losses of fluid into the tissue spaces. Efficient gas exchange in the lung requires that the pulmonary capillaries be in close apposition to the membranes of the alveoli or air capillaries. Loss of fluid across the pulmonary capillaries at the rate seen in the systemic capillaries would flood the terminal air spaces of the lung with fluid, thus increasing the aqueous diffusion path between air and blood and impairing adequate gas exchange. The pulmonary circulation of mammals and birds is adapted to preventing this problem. The importance of preventing it is seen in the dire consequences of pulmonary edema in man: shortness of breath, limited capacity for exercise, and possibly death.

The systolic pressure developed in the pulmonary arteries by the right ventricle in man averages about 22 mm Hg, and the diastolic pressure averages about 9 mm Hg. Mean blood pressure in the pulmonary arteries is about 14 mm Hg. These pressures are vastly lower than those in the systemic aorta, and as every student of comparative anatomy knows, the right ventricle is less muscular than the left. The mean blood pressure in the pulmonary capillaries is about 6 mm Hg, as compared to a mean pressure of about 25 mm Hg in the systemic capillaries. The pulmonary capillary pressure is so much less than the colloid osmotic pressure of the blood (about 25 mm Hg) that there is no net loss of fluid from the plasma across the capillary walls. In short, then, the pulmonary circulation, by contrast to the systemic, is a low-pressure circulation; and the development of the low-pressure pulmonary system in mammals and birds can be viewed at least in part as an adaptation to preventing ultrafiltrational influx of water into their aerial respiratory organs.

It is important to note that the pulmonary circulation, because it is in series with the systemic circulation, must carry the same volume of blood per unit time as the systemic circulation. Because it must do so with a far lower arterial pressure head, the pulmonary circuit must present a substantially reduced resistance to flow. We may note certain anatomical correlates of this lowered resistance in mammals. The vascular path of the pulmonary circuit is short by comparison to the systemic circuit. The arterial supply branches rapidly and profusely, and the terminal arteries lead directly to capillaries without intervening arterioles. The pulmonary capillaries, on the average, are somewhat larger than those of the systemic circulation. The pulmonary circuit is relatively distensible. Passive distention of the arteries and increasing patency of the capillary beds allow the circulation to accommodate large increases in blood flow with only minor changes in pressure. This is important during exercise. In man, for example, mean pulmonary arterial pressure increases by only 2–3 mm Hg when cardiac output is increased to three times the resting level. The difference at rest between the capillary blood pressure and the colloid osmotic pressure of the plasma is large enough to allow for substantial increases in blood pressure before filtration of fluid becomes a problem.

The circulation during exercise

In closing the discussion of avian and mammalian circulation, it is appropriate to examine responses to exercise. It is during exercise that demands on the circulation reach their zenith. The full potentialities of the circulatory system are called forth, and the limits on those potentialities may impose limits on the performance of the animal.

We saw in Chapter 9 that elevation of cardiac output is one of the major ways in which oxygen delivery to the tissues is increased during exercise. In man, for example, increased deoxygenation of the blood in the tissues can increase oxygen delivery to about 2.5 to 3 times the resting delivery, but delivery can be increased to a far greater extent by a combination of increased cardiac output and increased deoxygenation. Average young people can increase cardiac output by at least a factor of 4, and outputs of six or seven times the resting level have been measured in trained athletes during heavy exertion. The increase in cardiac output is generally realized by an increase in both heart rate and stroke volume.

Exercise is accompanied by a substantial drop in the total resistance presented by the systemic vasculature. Were it not for this, aortic blood pressure would have to increase exorbitantly to drive blood through the circulation at the rate implied by the increase in cardiac output. Mean pressure in the human systemic aorta may rise by only 10–20 mm Hg in heavy exertion. Systolic pressure increases much more, to as much as 50% higher than the resting level, but diastolic pressure characteristically remains almost unchanged. Clearly, there is an increased difference between the systolic and diastolic pressures during exercise. This reflects the decreased resistance of the peripheral vasculature, which allows a greater runoff of blood from the great arteries between heart beats and thus permits pressure to fall to usual diastolic levels despite greatly increased systolic levels.

Much of the decrease in peripheral resistance during exercise results from vasodilation in the vascular beds of the active muscles, including respiratory muscles. Blood flow through exercising muscles may increase to 15 or 20 times the resting level. As noted earlier, there is a great increase in the percentage of patent capillaries. The opening of the capillary bed effectively reduces the average diffusion distance between capillaries and muscle fibers and permits the capillary bed to carry a greatly increased flow of blood without major changes in the rate of flow through any given capillary. The latter effect is important in allowing a sufficient residence time for exchange between the blood and tissue. The response of the peripheral vasculature is a highly coordinated and adaptive one assuring a preferential distribution of the increased cardiac output to the tissues that require increased blood perfusion. Vasoconstriction may decrease blood flow to below resting rates in such organs as the intestines, kidneys, and liver and in muscles not involved in the exercise. The brain receives about the same amount of blood regardless of exercise state. Flow to the skin is often re-

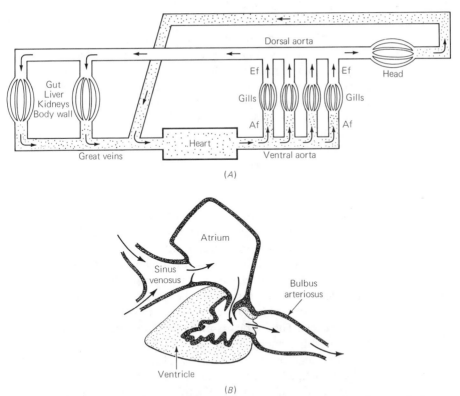

Figure 10–3. (*A*) Schematic representation of the circulatory plan in teleost and elasmo-branch fish. Stippled parts carry relatively deoxygenated blood. The heart, consisting of four chambers in series, pumps blood anteriorly into the ventral aorta, which gives off afferent branchial vessels (Af) to the gill arches. After perfusing the gills, blood is collected into efferent branchial vessels (Ef) that connect to the dorsal aorta. Blood is distributed to the major systemic circuits along the length of the body by the dorsal aorta. (*B*) The arrangement of the four heart chambers in a trout. (*B* from Randall, D. J. 1968. Amer. Zool. 8: 179–189.)

duced at the start of exercise. This, of course, permits a greater percentage of the cardiac output to be diverted to the active muscles. With continued exercise, however, the increased metabolic heat production often demands augmentation of cutaneous flow to aid in heat dissipation.

In total, the vascular response to exercise provides an excellent example of the fine spatial and temporal control of blood distribution possible in the avian and mammalian circulatory system.

GENERAL CIRCULATORY FEATURES OF FISH

The circulatory systems of elasmobranch and teleost fish, amphibians, and reptiles are closed systems like those of birds and mammals in which exchange between the blood and tissues occurs across capillary beds interposed between the arterial and venous vasculature.

The circulatory plan of fish is illustrated in Figure 10–3*A*. Blood is pumped anteriorly by the heart into the ventral aorta, which distributes it to the afferent gill vasculature. The blood passes through the gill capillaries and is then brought by the efferent gill vasculature to the dorsal aorta, a large dorsal artery that distributes the blood to the systemic tissues. After perfusing the systemic capillaries, the blood returns in the veins to the heart. As in mammals and birds, the circulatory plan places the respiratory organs in series with the systemic tissues, thus assuring efficient oxygen transport. Unlike mammals and birds, there is no heart between the respiratory circulation and systemic circulation to impart fresh energy to the blood before it leaves for the systemic tissues. The energy imparted by the heart to the blood entering the ventral aorta must thus be sufficient to produce an adequate rate of circulation through the resistances of both the branchial and systemic circuits.

The fish heart (Figure 10–3B) consists of four chambers arranged in series: a sinus venosus into which the great veins empty, an atrium, a ventricle, and a conus arteriosus (elasmobranchs) or bulbus arteriosus (teleosts) which empties into the ventral aorta. The main propulsive force is developed by the ventricle. The conus arteriosus of elasmobranchs is a muscular chamber that contracts sequentially with the rest of the heart and may function to continue the pressure phase of the heart beyond ventricular systole. The bulbus arteriosus of teleosts is a muscular, elastic chamber that does not contract along with the rest of the heart. It functions as an important pressure reservoir, smoothing out oscillations in ventral aortic pressure over the ventricular cycle. The bulbus is inflated under the force of ventricular systole, thus absorbing energy as elastic potential energy that in turn helps to maintain the head of pressure during diastole. This function of the bulbus is significant in maintaining an appreciable flow of blood through the branchial vasculature well into the diastolic phase of the heart.

Measurements of circulatory function in fish have been hindered by the difficulties of introducing suitable measurement devices into the animals while still maintaining them under reasonably natural conditions. Many of the available measures of blood pressure have been taken under distinctly unnatural circumstances, but recently data on unrestrained fish have accumulated. Mean ventral aortic pressures determined in teleosts and elasmobranchs have ranged from about 20 mm Hg to about 70 mm Hg, with most being toward the lower end of this range. The difference between systolic and diastolic pressure is generally around 10–20 mm Hg. Blood pressure drops as the blood perfuses the branchial circulation, and considerable interest has centered on the magnitude of this decline inasmuch as the pressure remaining in the dorsal aorta is that available for perfusion of the systemic circulation. In general, the mean dorsal aortic pressure is 20–45% lower than the mean ventral aortic pressure.

Cardiac outputs measured in fish at temperatures of 10°–20°C have generally proved to be far lower than those of mammals of comparable size. This is to be expected from the fact of the lower metabolic demands of fish.

A synchrony between pulsation of the heart and ventilation of the gills has been noted in a number of fish and has attracted considerable interest. In many fish the heart rate is lower than the rate of ventilatory movements, and in the elasmobranchs *Squalus acanthias* and *Mustelus antarcticus*, for example, the heart tends to beat in synchrony with every second, third, or fourth ventilatory cycle. Experiments on these species have shown that the cardiorespiratory synchrony is mediated by nerves that influence the initiation of the heart beat according to sensory input on some, as yet unidentified, parameter in the buccopharyngeal region (perhaps respiratory movements, perhaps oxygen tension in the water, and so on). The advantage, if any, of cardiorespiratory synchrony has not been demonstrated. An interesting suggestion is that the phase of maximal rate of blood flow through the gills may be synchronized with the phase of maximal water flow across the gill lamellae. This would favor effective oxygenation of the blood by assuring maximal availability of oxygen at the time when blood is passing the respiratory surfaces most rapidly.

Recent experiments by Stevens, Randall, and others on exercising rainbow trout (*Salmo gairdneri*) not only offer information on the circulatory response to exercise, but also provide some of the best available insight into possibilities of peripheral vascular response in fish. In the following summary the quantitative data refer to trout swimming at a rate (1.7 ft/sec) that increases their oxygen consumption about fivefold above the resting level. Under these conditions cardiac output also increases nearly fivefold. This suggests that oxygen delivery to the tissues during exercise is increased more by an increase in blood flow than by an increase in oxygen extraction from each unit volume of blood; confirmation comes from data indicating that the percent utilization of arterial oxygen increases only from about 62% during rest to about 71% during exercise. The increase in cardiac output in exercising trout is due almost entirely to an increase in stroke volume. The heart rate increases only about 15% (from a mean of 47 beats/min at rest to a mean of 54 beats/min during exercise), whereas the stroke volume increases about fivefold. Other fish have also been found to show only minor changes in heart rate during exercise, though there are exceptions. In mammals the heart receives both parasympathetic cardioinhibitory nerve fibers and sympathetic cardioaccelerator fibers, and the latter are strongly involved in quickening the heart rate during exercise. The sympathetic innervation of fish hearts, to the extent that it is present at all, appears usually to be dominated by the parasympathetic innervation, and neurally mediated increases in heart rate are achieved largely by diminution of parasympathetic inhibition rather than by direct stimulation. In rainbow trout, there is little or no parasympathetic inhibition at rest, leaving little room for neurally mediated cardioacceleration during exercise.

The increase in cardiac output during exercise in trout is accompanied by decreases in branchial and systemic vascular resistance and by increases in blood pressure in the major arteries. At rest, blood pressure in the ventral aorta is 40/32 mm Hg, and that in the dorsal aorta is 29/25 mm Hg. During exercise, the pressures in the ventral and dorsal aortae are, respectively, about 56/43 mm Hg and about 33/29 mm Hg. The resistance posed by the branchial vasculature decreases by about 70% with exercise, as does that posed by the systemic vasculature. There is some evidence that flow through the viscera is reduced during exercise, suggesting that the major decreases in systemic peripheral resistance occur in the muscles.

These results on the rainbow trout and less-extensive results on other species indicate clearly that there is considerable adaptive flexibility in the function of the cardiovascular system. Unfortunately the mechanisms by which changes are mediated in such critical parameters as stroke volume and peripheral resistance are essentially unknown, though there is a substantial amount of suggestive data in the literature.

AIR-BREATHING FISH

As discussed in Chapter 8, many fish are adapted to air breathing. In most cases their aerial respiratory organs are derived from structures such as the pharyngeal membranes, gut, or swimbladder that primitively are serviced by the systemic circulation. It is thus not surprising to find that these structures, even when adapted to respiration, commonly drain into the systemic venous vasculature, rather than the systemic arterial vasculature. The electric eel, *Electrophorus*, to illustrate, is an obligate air breather. Its gills are much reduced, and gas exchange is carried out largely across the wall of the mouth cavity, which is thrown into an elaborate system of well-vascularized papillae. The circulatory plan of *Electrophorus* is illustrated in Figure 10–4A. The afferent vessels to the buccal respiratory surfaces arise from the gill vasculature. The efferent vessels drain into the systemic venous vasculature. In contrast, then, to the usual teleost pattern and to the avian and mammalian pattern, the respiratory circulation is placed in parallel with the systemic circulation. The implications of this arrangement are profound. Anatomically, it appears that oxygenated blood from the buccal respiratory surfaces must mix in the systemic veins with deoxygenated blood from other tissues, meaning that the heart will deliver a mixture of deoxygenated and oxygenated blood to both the respiratory surfaces and the systemic arteries (note Figure 10–4A). This consequence of the parallel circulatory arrangement has been confirmed experimentally in *Electrophorus*. When blood draining the buccal respiratory surfaces is at least 90% saturated with oxygen, for example, blood pumped to the systemic arteries is only about 65% saturated because of the admixture of systemic venous blood with the freshly oxygenated blood before the blood is delivered to the systemic circulation. The mixing of oxygenated and deoxygenated blood reduces the efficiency of oxygen transport. The systemic tissues do not receive blood at the full oxygen tension established in the respiratory organs, and the respiratory organs do not receive blood at the full state of deoxygenation realized in the tissues. Instead, blood deoxygenated in the tissues is in part recycled directly back to the tissues, and blood oxygenated in the respiratory organ is in part recycled directly back to the respiratory surfaces. The energy used by the heart to pump the blood is not optimally directed to meeting the oxygen demands of the tissues, for the heart is in part simply recycling deoxygenated blood. A similar parallelism of the aerial respiratory circuit and systemic circuit is observed in fish in which the gastrointestinal tract or swimbladder is adapted for air breathing (Figure 10–4B, C).

Although caution is necessary in drawing evolutionary implications from the condition of modern species, it appears that the early evolution of aerial respiration in piscine vertebrates, because it involved the respiratory adaptation of structures with a primitively systemic venous drainage, entailed the abandonment of a series circulation for a less-efficient parallel circulation. In the evolutionary progression of the vertebrates it is not until the level of birds and mammals that we again find the respiratory and systemic circuits connected inextricably in series by a circulatory anatomy that cannot permit mixing of oxygenated and deoxygenated blood. In the lungfish, amphibians, and reptiles, however, we find significant developments in

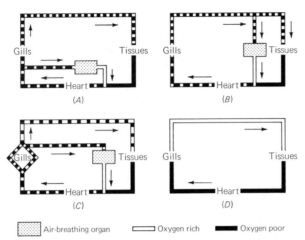

Figure 10-4. Schematic representations of the circulatory plans of certain air-breathing fish. The general piscine arrangement is shown in *D* for comparison. The amount of white and black in the circulatory tracts represents only the approximate amount of oxygenated and deoxygenated blood carried in the vessels. (*A*) The circulatory plan in *Electrophorus* and certain other fish with air-breathing organs derived from the pharyngeal and/or opercular mucosa. Afferent vessels to the air-breathing organ are derived from the afferent branchial vasculature. Efferent vessels from the air-breathing organ connect to systemic veins. (*B*) The circulatory plan in armored catfish (*Plecostomus*) and certain other fish in which the air-breathing organ is associated with the stomach or intestinal tract. Afferent vessels to the air-breathing organ are derived from the dorsal aorta; efferent vessels enter systemic veins. (*C*) The circulatory plan in the bowfin (*Amia*) and certain other fish that utilize the swimbladder as an air-breathing organ. Afferent vessels to the swimbladder are specialized, arising from the efferent branchial vasculature of the sixth aortic arch (similar to the condition in lungfish). Efferent vessels from the swimbladder enter systemic veins (unlike lungfish). The gill vasculature is represented in two tracts to emphasize that afferent vessels to the swimbladder arise from efferent vessels of only the posterior pair of gill arches. [From Johansen, K. 1970. Air breathing in fishes. *In:* W. S. Hoar and D. J. Randall (eds.), *Fish Physiology.* Vol. IV. Academic, New York.]

circulatory anatomy and physiology that, at least in some species, assure a considerable separation of oxygenated blood returning from the lungs and deoxygenated blood returning from the systemic tissues.

Anatomical features of the circulation in lungfish

The heart of lungfish is very different from that of other fish. The atrial and ventricular chambers are partly divided into right and left halves by septa. This anatomical division of the heart is more developed in the obligate air breathers *Protopterus* and *Lepidosiren* than in *Neoceratodus*, which is strongly dependent on aquatic respiration, though it does use its lung under hypoxic conditions. The conus arteriosus, here called the bulbus cordis, possesses two longitudinal ridges that project toward each other from opposite sides of the lumen, thus partially dividing the lumen into two tracts. These ridges are fused at the anterior end, and there the lumen is completely divided.

The lungs of lungfish are supplied by arteries that arise from the efferent vasculature of the most posterior pair of gills (sixth aortic arches). A similar condition is seen in some fish that utilize the swimbladder as an aerial respiratory organ (see Figure 10-4C). This arterial supply is specialized, for primitively the swimbladder is supplied from the dorsal aorta, and the arteries supplying the swimbladder or lungs from the branchial efferent vasculature are homologous to the pulmonary arteries of higher vertebrates. The venous vasculature of the lungs in lungfish is quite unlike the venous vasculature of any other piscine air-breathing organs, for the veins lead directly into the left side of the atrium of the heart rather than connecting with the systemic venous vasculature. This is a very significant development. On simple anatomical grounds, blood from the lungs is kept separate from systemic venous blood at least into the heart. The sinus venosus, which now receives only systemic venous blood, connects to the right side of the atrium of the heart.

The heart in lungfish is positioned quite anteriorly by comparison to other fish, and the ventral aorta, which primitively serves as a common conduit to all the afferent gill vessels (Figure 10–3), is virtually or entirely eliminated. The afferent branchial vessels thus arise immediately from the anterior end of the bulbus cordis, a condition presaging that of the homologous vessels in amphibians. As in most teleosts, there are four pairs of afferent branchial arteries, which supply the four pairs of gill arches and which correspond, from anterior to posterior, to aortic arches 3 to 6. In *Neoceratodus* all the gill arches bear well-developed gills, and blood flowing to the gill arches in the afferent branchial vessels perfuses the gills on its way to the efferent branchial vessels and dorsal aorta. In *Lepidosiren* and *Protopterus* the two anterior pairs of gill arches lack gills, and the posterior pairs of arches bear only rudimentary gills. The branchial vessels to the two anterior pairs of gill arches (aortic arches 3 and 4) do not perfuse gills and, unusually among fish, do not break up into capillaries but, rather, form direct through connections to the dorsal aorta. The branchial vessels to the posterior gill arches (aortic arches 5 and 6) perfuse the rudimentary gills on their way to the dorsal aorta. As noted earlier, the arteries to the lungs arise from the branchial efferent vessels of the sixth aortic arch.

Physiological studies of *Protopterus*

The numerous modifications of circulatory anatomy in lungfish stimulate contemplation of the idea that some separation of deoxygenated and oxygenated blood may be realized. Certainly we would not expect such pervasive alterations of circulatory morphology without some adaptive advantage, and one immediately apparent potential advantage would be improved efficiency of oxygen transport through preferential delivery of oxygenated blood from the lungs to the systemic tissues and preferential delivery of deoxygenated blood from the tissues to the lungs. Whether this occurs cannot be ascertained from anatomical considerations alone. The incompleteness of the division of the heart chambers into right and left halves, for example, could potentially permit thorough mixing of pulmonary and systemic venous blood in the heart and thereby result in delivery of mixed blood to both the systemic and pulmonary arteries. The issue must be resolved by studies of actual blood flow in living animals.

Recently such studies have been performed using two basic techniques. In one technique, termed radiography, a material that is opaque to x-radiation is injected into the circulation and its path followed by x-ray cinematography. Injection of a radiopaque medium into a pulmonary vein, for example, makes it possible to follow the course of the pulmonary venous blood as it traverses the heart. This technique has the advantage of permitting visual observation of the path of blood flow but does not allow rigorous quantification of the degree to which pulmonary and systemic venous bloods are kept separate in their passage through the central circulation.* Such quantification can be realized by another technique, that of monitoring oxygen tension simultaneously in critical blood vessels on the arterial and venous sides of the heart. If, for example, we know the oxygen tensions in pulmonary and systemic venous bloods prior to entry into the heart and we also know the oxygen tension of blood in the pulmonary arteries, we can compute, with knowledge of the oxygen dissociation curve, what percentage of the pulmonary arterial blood flow is contributed from systemic venous blood and what percentage is contributed from pulmonary venous blood.

Radiographic studies of *Protopterus* reveal a clear tendency for pulmonary and systemic venous bloods to follow different paths through the central circulation. Pulmonary venous blood tends to follow a course through the left atrium and left ventricle and is delivered preferentially to the ventral channel of the bulbus cordis and thence to aortic arches 3 and 4, which provide direct through channels to the dorsal aorta. Thus oxygenated blood from the lungs is delivered preferentially to the systemic arterial circulation. Systemic venous blood tends to pass through the right atrium and right ventricle. Then it appears, in radiographic plates, to be distributed about evenly to the two channels of the bulbus cordis, to the four pairs of aortic arches, and to the dorsal aorta and pulmonary arteries. Thus systemic venous blood is directed to both the systemic and pulmonary circulations.

Oxygen analyses of blood in major veins and arteries provide further information. In experiments on *Protopterus* in aerated water, the oxygen tension of blood coming from the lungs in the pulmonary veins averaged 46 mm Hg, whereas that of blood in the systemic veins averaged only 2 mm Hg. The tension of blood in the anterior pairs of branchial arteries (aortic arches 3 and 4) was 38 mm Hg, and calculations indicated that about 90% of this blood came from the pulmonary veins. This result indicates a strong bias toward perfusion of the branchial through channels with pulmonary, rather than systemic, venous blood. The oxygen tension in a major systemic artery arising from the dorsal aorta was 30 mm Hg. This tension, which was lower than that in the branchial through channels, reflected the admixture of relatively deoxygenated blood traveling to the dorsal aorta through the posterior pairs of branchial vessels. The oxygen tension in the pulmonary arteries averaged 25 mm Hg. This was lower than that in the systemic arteries but still reflected a significant degree of recirculation of pulmonary venous blood into the pulmonary arteries. The difference in oxygen tension between systemic and pulmonary arterial bloods (30 mm Hg versus 25 mm Hg) is worthy of some special note, for it provides a holistic indicator of the extent to which selective flow is realized in the central circulation. It shows clearly that the systemic arteries receive blood that is richer in oxygenated pulmonary venous blood than the blood delivered to the pulmonary arteries.

Protopterus exhibits interesting and significant changes in circulatory function over the breathing cycle. Cardiac output tends to increase after

---

*The term *central circulation* is used throughout this chapter to refer to the heart and the veins and arteries leading to and from the heart.

each filling of the lungs with air, and the rate of flow in the pulmonary arteries is about four times greater after a breath than just before a breath. Because the latter effect occurs even when there is no increase in cardiac output, it clearly reflects a redistribution of cardiac outflow between shunts to the dorsal aorta and the pulmonary arteries, presumably mediated by vasomotor controls in the branchial arteries. Calculations indicate that the percentage of cardiac outflow directed to the pulmonary arteries may be as great as 70% just after a breath and as low as 20% just before a breath. These adjustments in circulation result in maximal perfusion of the lungs when the oxygen tension in the lungs is highest.

We are not in a position as yet to provide a complete description of circulatory dynamics in *Protopterus*. Different experimental approaches provide different insights, and each approach has its own potential for altering the condition of the fish from that in the natural habitat. There is no doubt, however, that *Protopterus* can realize a significant degree of selective distribution of deoxygenated and oxygenated blood and that circulatory dynamics can change in adaptive ways according to demands for circulatory transport. These conclusions are of great interest in the study of the evolution of vertebrate aerial respiration.

Physiological studies of Lepidosiren and Neoceratodus

Lepidosiren, which, like *Protopterus*, is an obligate air breather, also exhibits selective passage of deoxygenated and oxygenated venous bloods to the pulmonary and systemic circuits, respectively. In one set of experiments, average pulmonary arterial oxygen tension was 10 mm Hg less than average systemic arterial tension, and in another set the pulmonary value was fully 16 mm Hg lower.

Neoceratodus is interesting because it is predominantly an aquatically respiring animal and, unlike the other lungfish, has fully functional gills. In aerated water it relies entirely on gill breathing and does not ventilate its lung; blood going to the lung from the efferent branchial vessels is already virtually fully saturated from exchange in the gills and does not increase in oxygen tension during perfusion of the lung. In hypoxic water *Neoceratodus* commences air breathing, and blood is oxygenated to a marked extent in the lung. It is in this situation that preferential distribution of pulmonary venous blood to the systemic arterial circulation becomes meaningful. Interestingly, in aerated water blood delivered to the anterior branchial arteries (which bypass the pulmonary circulation) is about evenly composed of pulmonary and systemic venous blood, but in hypoxic water there is a strong bias toward perfusion of the anterior branchial arteries with pulmonary venous blood rather than systemic venous blood. Moreover, blood flow to the lung accounts for 25% more of the cardiac output in hypoxic water than in aerated water.

Pulmonary vascular resistance in lungfish

A final noteworthy observation from the experiments on lungfish is that the pulmonary vascular resistance seems clearly to be lower than the systemic vascular resistance. We have earlier discussed the importance of relatively low vascular resistance in the pulmonary circuit of mammals and birds, and the results on lungfish indicate that this development is phylogenetically old. Pulmonary resistance has been found to be lower than systemic resistance in a variety of amphibians and reptiles as well.

AMPHIBIANS

In the heart of typical lung-breathing amphibians, the atrium is completely divided into right and left halves by a septum. Pulmonary venous blood enters the left atrium, and systemic venous blood enters the right atrium via the sinus venosus. Separation of oxygenated and deoxygenated blood is assured on simple anatomical grounds until the blood enters the ventricle. The ventricle lacks a septum altogether. The ventricular lumen is intersected by many muscular cords, termed trabeculae, which impart to it something of the appearance of a sponge. It is commonly hypothesized that the trabeculae form an anatomical basis for the direction of blood from the left and right atria along different paths through the ventricle, though it is by no means obvious how they would do this inasmuch as they do not divide the ventricle into two distinct channels. The ventricle discharges into a bulbus cordis, which leads to the paired systemic and pulmonary arteries. Running along the inside of the bulbus is a complexly twisted membrane, termed the spiral fold or spiral "valve," which incompletely divides the bulbar lumen.

Studies of selective blood distribution

Studies of central blood flow in amphibians have indicated everything from highly significant selective distribution of deoxygenated and oxygenated blood to complete mixing. Recent experiments on several species have shown unequivocally that selective distribution can occur. Radiographic analysis of the salamander *Amphiuma tridactylum*, for example, shows that systemic venous blood tends to be restricted to the right side of the ventricle and to be delivered to the pulmonary aortic arches. Pulmonary venous blood tends to follow a course through the left portion of the ventricle into the systemic aortic arches. The selective distribution in the ventricle is striking in view of the lack of a ventricular septum and lends strong support to the hypothesis that the trabeculae in some way act to guide the flow of left and right atrial blood. The spiral fold in the bulbus is thought to play an important role in directing left ventricular blood into the systemic aortae and right ventricular blood into the pulmonary aortae. There are several hypotheses concerning how the fold could act in this way, but as yet no firm conclusion can be reached.

Oxygen analyses in *Amphiuma* confirm that selective distribution is realized in the central circulation. In one animal the oxygen content of pulmonary venous blood was 8.5 vol %; that of systemic venous blood was 5.0 vol %; that of systemic arterial blood was 8.5 vol %; and that of pulmonary arterial blood was 6.9 vol %. These results and similar results on other animals show clearly that pulmonary venous blood is preferentially distributed to the systemic arteries, whereas systemic venous blood is preferentially distributed to the pulmonary arteries; some admixture of oxygenated and deoxygenated blood does occur. To cite another example, oxygen analyses in the toad *Bufo paracnemis* yielded the following average percent saturation values: pulmonary vein, 96; systemic artery, 94; systemic vein, 14; pulmonary artery, 76. The results on both the salamander and toad indicate that the systemic arteries receive most of their blood from the pulmonary veins, whereas the pulmonary arteries receive substantial amounts of pulmonary venous blood as well as systemic venous blood. These results are indicative of left-to-right mixing in the heart. That is, the predominant departure from complete selective distribution appears as a contribution of left venous blood to right (pulmonary) cardiac outflow. (The pulmonary outflow is termed the right outflow through analogy with the mammalian heart.)

As noted earlier, some workers have reported complete mixing of pulmonary and systemic venous blood in the heart or far less complete separation than indicated thus far. In some cases the results may be indicative of interspecific differences, but workers on the same or closely related species have sometimes arrived at different conclusions. In the latter context it is noteworthy that the workers on *Amphiuma* and some other species in which strong selective distribution has been observed report that the pattern of circulation is physiologically labile and is easily disturbed during experimentation. It is important to recall that the various species of amphibians exhibit great diversity in respiratory physiology and that gas exchange in any particular species can commonly be realized by two or more routes, such as pulmonary and cutaneous. It would hardly be surprising, then, to find different circulatory patterns in different species or to find modulations of circulatory pattern in individuals of a given species according to changes in their physiology of external respiration.

Studies of distribution of cardiac output

Recently studies have been performed on the distribution of cardiac output over the pulmonary ventilatory cycle in anesthetized clawed toads, *Xenopus laevis*, and frogs, *Rana pipiens* and *R. temporaria*. Although flow to the systemic aortae was found to remain fairly constant over the breathing cycle, flow to the pulmonary aortae tended to increase considerably when the lungs were filled and then decline over the apneic period between breaths. In *Xenopus*, for example, pulmonary aortic flow was about equal to systemic flow shortly after filling of the lungs but decreased to one fifth or less of systemic flow over long respiratory pauses. These results indicate not only that the distribution of cardiac output varies over the breathing cycle, but also that the total cardiac output varies, being greatest after filling of the lungs. An important governing factor in the distribution of cardiac output is variation in the resistance posed by the pulmonary circuit. In *Xenopus* pulmonary resistance was observed to increase by a factor of almost 6 over a long respiratory pause following a breath and, by the end of the pause, became far greater than systemic resistance. This process acted to direct cardiac output preferentially to the systemic circuit. Decreased pulmonary resistance following pulmonary ventilation acted to promote a more even distribution of cardiac output to the two circuits.

The variation in flow to the pulmonary circuit over the breathing cycle is probably important to circulatory efficiency, assuring high pulmonary blood flow when the oxygen supply in the lungs is rich and limiting flow when the supply has become depleted over apneic periods. It is hypothesized that pulmonary blood flow is gradually diminished during periods of underwater diving and that this circulatory response acts to meter the transfer of oxygen from the pulmonary store to the systemic tissues over the period of submergence.

Circulatory features of cutaneous gas exchange

Cutaneous gas exchange is important in many amphibians. The major cutaneous arteries arise from the pulmonary arteries (which for this reason are often called pulmocutaneous arteries). Cutaneous venous blood, however, is not returned to the left atrium but, rather, is delivered to the sinus venosus and right atrium along with blood from other systemic tissues. This places the cutaneous circuit in parallel with the general systemic circuit, and the mixing of oxygenated cutaneous venous blood with deoxygenated blood prior to entry into the heart denies the possibility of selective distribution. In amphibians that rely strongly on cutaneous respiration, this circumstance could lead to decreased evolutionary selective pressures for development or maintenance of structures contributing to selective blood distribution in the heart. Interestingly, the lungless salamanders are reported to have a poorly developed or fenestrated atrial septum and to have a simplified spiral fold in the bulbus cordis. In amphibians, such as frogs, that can rely predominantly on either pulmonary or cutaneous respiration, it would not be surprising to find changes in the cardiac flow pattern when cutaneous respiration is dominant. This demands study. When frogs (*Rana esculenta*) are submerged in water and must rely on cutaneous respiration, there is a marked increase in perfusion of the skin capillary beds, indicating vasomotor control of the cutaneous circulation according to respiratory demands.

REPTILES

In reptiles there is a complete atrial septum, with pulmonary venous blood returning, as usual, to the left atrium and systemic venous blood returning to the right atrium. The sinus venosus, which in mammals and birds is reduced into the right atrial wall, remains. The bulbus cordis is gone, and the systemic and pulmonary arteries arise directly and separately from the ventricle. The ventricle of turtles, lizards, and snakes is complexly divided into three intercommunicating chambers, whereas that of crocodilians is divided into two chambers by a complete septum.

Circulation in noncrocodilian reptiles

Many studies of turtles, snakes, and lizards have shown that a very effective selective distribution of pulmonary and systemic venous blood can occur in the heart. This phenomenon directs attention to the dynamics of blood flow in the ventricle, for the ventricle is the one place where mixing can occur.

Unfortunately, there is no simple way to depict the structure of the ventricle. Figure 10–5 shows a cross section of the ventricle near the atrioventricular apertures, looking toward the atria. The openings of the atria and of the arteries emanating from the ventricle are behind the section shown and are depicted as dashed lines. The three ventricular chambers are

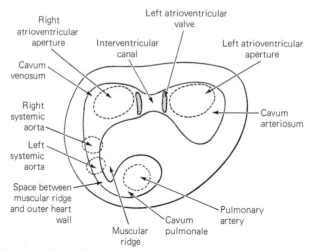

Figure 10–5. Cross section of the ventricle of a noncrocodilian reptile near the atrioventricular apertures looking toward the atria. The atrioventricular apertures and openings of the systemic and pulmonary arteries are behind the section, and their positions are indicated by dashed circles. Cavum arteriosum and cavum venosum are largely separated by a septum; the interventricular canal, shown in the section, is a localized passage between the two chambers, positioned near the atrioventricular apertures. The atrioventricular valves are single flaps attached at the medial margins of the atrioventricular apertures. It is believed that they are alternately forced against the interventricular canal during atrial systole and against the atrioventricular apertures during ventricular systole. They are shown here as they might be positioned during atrial systole. See text for further explanation. (After White, F. N. 1968. Amer. Zool. 8: 211–219.)

termed the cavum arteriosum, cavum venosum, and cavum pulmonale. The cavum arteriosum and cavum venosum are separated entirely by a septum except for a single, localized fenestration that is shown in the section and termed the interventricular canal. This canal is positioned immediately between the two atrioventricular apertures. The cavum pulmonale is a small chamber separated incompletely from the cavum venosum by a muscular ridge. Note that the right atrium opens into the cavum venosum, and the left atrium opens into the cavum arteriosum. The right and left systemic arteries originate from the cavum venosum, whereas the pulmonary artery originates from the cavum pulmonale. There are no arteries originating from the cavum arteriosum.

The best current hypothesis explaining the selective distribution of blood in the ventricle may be outlined as follows. The two atrioventricular valves are single flaps attached at the medial margins of the atrioventricular apertures and on either side of the interventricular canal. When the atria contract and fill the ventricular chambers, the valvular flaps are forced toward the ventricle and medially and are believed to obstruct the interventricular canal. As a consequence of the obstruction of the canal, oxygenated blood entering the heart from the left atrium is sequestered in the cavum arteriosum, whereas deoxygenated blood entering from the right atrium is limited to the cavum venosum and the cavum pulmonale. During ventricular systole, the elevated pressure in the ventricle forces the atrioventricular valves against the atrioventricular apertures, preventing backflow of blood into the atria and reversing the obstruction of the interventricular canal. There is evidence that outflow into the pulmonary artery commences before outflow into the systemic arteries, this being attributed to lower vascular resistance in the pulmonary circuit. In this manner the deoxygenated blood occupying the cavum venosum and the cavum pulmonale is directed primarily to the pulmonary artery. As systole proceeds, oxygenated blood from the cavum arteriosum enters the cavum venosum through the open interventricular canal, and outflow into the systemic arteries commences. The sequence of events contributes to selective distribution of oxygenated blood to the systemic circulation. Further, the muscular ridge separating the cavum venosum from the cavum pulmonale is believed to be appressed against the adjacent outer wall of the ventricle during the latter part of systole, thus limiting flow of oxygenated blood into the cavum pulmonale and the pulmonary artery.

This conception of ventricular events, although hypothetical, represents the integration of a variety of experimental observations on ventricular function and provides a rational explanation for selective distribution in the noncrocodilian heart. Most modern investigations of lizards, snakes, and turtles have revealed that a significant degree of selective distribution occurs. In broad outline three types of flow pattern may be recognized. (1) There may be nearly complete separation of flows, with the pulmonary artery receiving almost exclusively systemic venous blood and the systemic aortae receiving almost exclusively pulmonary venous blood. This has been observed in several species of snakes and lizards. (2) The pulmonary artery may receive mostly systemic venous blood, but a significant amount of systemic venous blood may pass to the systemic aortae along with the pulmonary venous blood, resulting in mixed flow to the systemic circulation. This condition is characterized as a right-to-left shunt in reference to the fact that right atrial blood is being partially diverted to the systemic arteries, which (in analogy with mammals) are considered to represent the left ventricular outflow. Several studies on lizards have indicated this pattern of flow. (3) The systemic aortae may receive mostly pulmonary venous blood, but a significant amount of pulmonary venous blood may be recirculated to the pulmonary artery along with the systemic venous blood. This left-to-right shunt appears to be the usual pattern in turtles during normal aerial respiration.

In painted turtles, *Pseudemys scripta*, the direction of shunting is reversed during diving. There is a left-to-right shunt in air, with some 60% of the cardiac output being delivered to the pulmonary artery and about 40% to the systemic arteries. During diving the heart rate falls (bradycardia), and the cardiac output declines to about 5% of the predive level. Simultaneously a right-to-left shunt develops, and the proportion of cardiac output delivered to the systemic circulation increases to around 60%. The shift in ventricular flow pattern is attributed to a change in relative vascular resistance in the pulmonary and systemic circuits. Pulmonary resistance increases during the dive, and this is believed to impair the preferential delivery of blood from the cavum venosum and the cavum pulmonale to the pulmonary artery during early ventricular systole, resulting in diversion of some of this blood to the systemic aortae along with blood from the cavum arteriosum. In view of the fact that the lungs of turtles may become depleted of oxygen during a long dive, the increase in pulmonary resistance and right-to-left shunt may act to improve circulatory efficiency by limiting perfusion of the lungs and diverting cardiac work more to the perfusion of the systemic circulation.

Circulation in crocodilian reptiles

In crocodilians the ventricle is completely divided into two chambers. The right systemic aorta originates from the left ventricle, but the left systemic aorta arises from the right ventricle, as does the pulmonary artery (see Figure 10–6A). The two systemic aortae are in communication near their origins via an extraventricular aperture known as the foramen Panizzae. In view of the origin of the left systemic aorta from the right ventricle and the presence of the foramen between the two systemic aortae, it is impossible to deduce the pattern of flow in the crocodilian heart from anatomical considerations alone.

Physiological investigations have revealed that virtually complete selective distribution can occur. When this is the case (Figure 10–6B), systemic venous blood from the right ventricle enters the pulmonary artery but not the left systemic aorta. Pulmonary venous blood from the left ventricle is delivered not only to the right systemic aorta but also to the left via the foramen Panizzae. The reason that right ventricular blood does not enter the left systemic aorta is that right ventricular pressure does not reach the level of pressure prevailing in the aorta. Under this circumstance the valve into the aorta will not be forced open, and flow will not occur. Right ventricular pressure is sufficient to produce flow into the lower-pressure, lower-resistance pulmonary circuit.

The pattern of flow is significantly altered during diving in alligators (Figure 10–6C). Pulmonary resistance increases during the dive, thus hindering runoff of blood into the pulmonary artery and allowing right ventricular pressure to rise to higher levels than when the animal is in air. The right ventricular pressure in this case does become high enough to produce flow into the left systemic aorta, and a significant right-to-left shunt of systemic venous blood into the systemic arteries develops. This type of shunt may hold advantages for circulatory efficiency, as noted earlier in the discussion of diving turtles.

CONCLUDING COMMENTS ON VERTEBRATE CIRCULATION

In concluding this examination of vertebrate circulation, we see that among lungfish, amphibians, and reptiles, significant advances have been made toward the establishment of a series connection between the pulmonary and systemic circuits despite the fact that this pattern of flow is not guaranteed on simple anatomical grounds. The morphological and physiological properties of the central circulation that provide the capability of selective blood distribution have presumably been selected in evolution according to the advantage that series circulation has for efficiency of respiratory gas transport. Many important questions remain. The functional properties of the peripheral circulation are poorly understood in all but a few animals. Studies of the central circulation in lungfish, amphibians, and

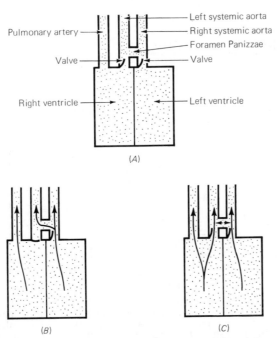

Figure 10–6. A schematic representation of the pattern of blood flow in the ventricles and systemic and pulmonary arteries during ventricular systole in crocodilians. (*B*) Pattern during normal pulmonary ventilation. (*C*) Pattern in alligators during diving. Parts are labeled in (*A*).

reptiles have gone far in elucidating possibilities for selective blood distribution but do not provide a synthetic understanding of these groups. One cautionary note needs to be sounded. Many texts in the past have said or implied that there is an almost orthogenetic progression in the development of the central circulation from lungfish to mammals and have said or implied that the central circulation in "lower" air-breathing vertebrates is "defective" in its inability to guarantee selective distribution of the blood anatomically. These arguments place an unwarranted and philosophically undesirable emphasis on the mammalian or avian condition as a standard by which to judge other animals. True, a series connection between the pulmonary and systemic circuits is believed to have distinct advantages during pulmonary respiration, but this does not mean that an anatomically rigid series arrangement is necessarily ideal. Different animals place different demands on the circulation, and it may well be that the shunts that are possible in an incompletely divided central circulation provide an important and adaptive flexibility to some animals. *Protopterus* can redistribute its cardiac work between perfusion of the systemic and pulmonary circuits according to the oxygen content of its lungs, and a similar phenomenon has been observed in several amphibians. Alligators and turtles are known to bypass the lungs to some extent during diving, and, again, there is suggestive evidence of a similar response in amphibians. These things are not possible when an anatomically rigid series arrangement requires that pulmonary blood flow always equal systemic flow.

Figure 10–7 provides a recent attempt at summarizing certain features

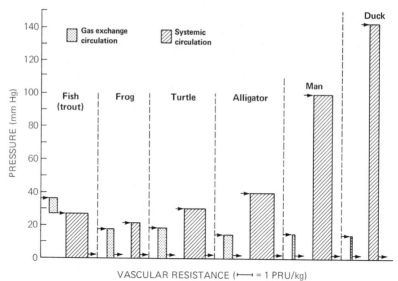

Figure 10–7. A summary of perfusion pressures and vascular resistances observed in representatives of vertebrate classes. Arrows indicate pressures in major vessels going to and coming from the systemic circulation and gas exchange circulation (pulmonary circulation except in the fish). The height of each bar provides an indication of the pressure drop across the circulation. The width of each bar expresses vascular resistance measured in peripheral resistance units (PRU) per kilogram of body weight. One PRU is a resistance requiring a pressure gradient of 1 mm Hg to achieve a rate of flow of 1 cc/min; thus resistance in PRU is obtained by dividing the pressure drop across a circulatory bed by the rate of flow of blood in cc/min. A summarizing diagram of this sort cannot possibly represent the range of variation observed even within a single species. This diagram is meant only to provide an overview of salient features as elucidated in selected species that have received study. (From Johansen, K. 1972. Resp. Physiol. 14: 193–210.)

of the circulation in the vertebrate classes. Note that systemic vascular resistance shows little, if any, systematic change in phylogeny. Birds and mammals have considerably higher cardiac outputs and rates of systemic blood flow than the other animals depicted and must develop higher systemic pressures accordingly; certain lizards (not represented in the figure) exhibit weight-specific rates of blood flow similar to those in birds and mammals when at high body temperatures and then also show similar systemic pressures. The figure shows that the resistance of the respiratory circuit, be it a gill or lung circuit, is typically lower than that of the systemic circuit. The pressure drop required to produce a given rate of flow is accordingly lower. In amphibians and reptiles the cardiac output to the lungs may be lower or higher than that to the systemic circulation owing to the possibilities of shunting, but in birds, mammals, and most fish flow to the two circuits must be the same. One of the dramatic features of birds and mammals is their very low pulmonary resistance. The fact that pulmonary capillary pressure must remain below the colloid osmotic pressure of the plasma imposes limits on the pressure that can be advantageously developed in the pulmonary circuit. Birds and mammals have thus increased the rate of flow through the lungs not by increasing pulmonary pressure relative to other terrestrial vertebrates, but by lowering pulmonary resistance.

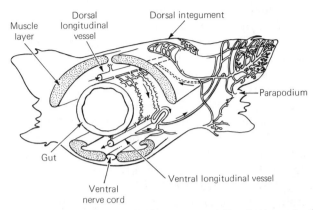

Figure 10–8. Major blood vessels in a body segment of *Nereis virens*, looking anteriorly. Arrows show directions of blood flow. [From Dales, R. P. 1963. *Annelids*. Hutchinson, London; in turn, based on Nicoll, P. A. 1954. Biol. Bull. (Woods Hole) 106: 69–82, and Lindroth, A. 1938. Zool. Bidr. Upps. 17: 367–497.]

INVERTEBRATE CIRCULATION

Some invertebrate groups lack a blood-vascular system, and in some others the system is poorly developed. Internal transport in many of these forms has been discussed in Chapter 8, and the present treatment will be restricted to three major phyla, the annelids, arthropods, and molluscs.

Annelids

Most oligochaetes and polychaetes have a well-developed blood-vascular system modeled on a common plan, though exhibiting diverse specialized features in various species. This basic plan will receive emphasis here. Characteristically, there are two major longitudinal vessels which run the length of the body, a dorsal vessel above the gut and a ventral vessel below the gut. The connections between these longitudinal vessels reflect the strongly segmented body plan of these animals. In general, there are two lateral circulatory arcs in each body segment, which run between the dorsal and ventral longitudinal vessels on either side of the segment. These circulatory arcs vary greatly in complexity not only in different species but often in different body segments of a single species. Figure 10–8 depicts the lateral vascular system on one side of a body segment in a fairly generalized polychaete, *Nereis virens*. Note first the large longitudinal vessels above and below the gut. These are connected, quite typically for polychaetes and oligochaetes, via vessels in the gut wall. There is also a far-flung system of vessels, with connections to both longitudinal vessels, that supplies the parapodium, integument, nephridium, musculature of the body wall, and other structures on the side of the body segment shown. A similar vascular arrangement is repeated in most body segments. *Nereis* undergoes external respiration chiefly across the parapodia and dorsal integument, and the major vasculature of these structures is emphasized in Figure 10–8. Some other polychaetes have specialized gills, which then are also supplied by the segmental circulation. Most oligochaetes lack gills and respire across the general integument; in these the segmental circulation supplies a dense network of integumentary vessels.

The blood-vascular system of polychaetes and oligochaetes is a closed system. By and large, the major vessels are connected across beds of true capillaries in which exchange of gases, nutrients, and other commodities occurs. The vasculature of the gut wall may be of a capillary nature or may consist of large, anastomosing blood spaces or sinuses.

Blood characteristically flows anteriorly in the dorsal longitudinal vessel and returns posteriorly in the ventral longitudinal vessel. Superimposed on this longitudinal flow pattern is a lateral flow pattern in each body segment. The details of the lateral flow differ in different species, and the case in *Nereis* is presented here as a somewhat generalized example. As shown in Figure 10–8, blood flows ventrally from the dorsal longitudinal vessel into the gut vasculature. This flow is combined with outflow from the ventral longitudinal vessel to supply the segmental circulation to the parapodium, integument, and other lateral structures. Blood draining the lateral segmental vasculature is returned to the dorsal longitudinal vessel. Ventral-to-dorsal flow through the lateral segmental vasculature is of considerable generality in polychaetes and oligochaetes. When, as in *Nereis*, this is combined with dorsal-to-ventral flow in the gut vessels, a complete segmental circulatory loop is established, allowing recycling of blood within a segment. In the common earthworm *Lumbricus* flow in the gut vessels as well as that in the lateral segmental vessels is ventral to dorsal.

Annelids have no central heart. Rather, the propulsive force for circulation of the blood is developed more or less diffusely by pulsatile contractions in major vessels. Characteristically, the dorsal longitudinal vessel undergoes peristaltic contractions that drive the blood forward. Commonly, it is observed that all or most of the other major vessels are also contractile, though contractions in some appear to be more significant to the circulation than those in others. In *Nereis*, for example, there are important peristaltic contractions in the ventral, afferent vessels in each segment that drive blood to the vasculature of the parapodium and integument (see Figure 10–8). In oligochaetes certain of the lateral vascular arcs in the esophageal region have become specialized into strongly contractile, direct through connections between the dorsal and ventral longitudinal vessels. These specialized lateral vessels, of which there are five pairs in *Lumbricus*, a common dissection preparation, are often called "hearts" but do not warrant this special designation on either anatomical or physiological grounds. They contribute an important component to the longitudinal circulation, driving blood from the anterior dorsal longitudinal vessel into the ventral longitudinal vessel, but the peristaltic contractions in the dorsal vessel are at least as important to the longitudinal circulation. Characteristically there is little evidence of coordination among the various pulsatile vessels that propel the blood in polychaetes and oligochaetes. In *Lumbricus*, for example, the two lateral through vessels ("hearts") in a given esophageal segment tend to contract together, but the five pairs of these vessels do not contract in a common rhythm, nor are they coordinated with the contractions of the dorsal longitudinal vessel.

The overall picture that emerges in annelid circulation is one of considerable lack of centralization. Blood does not flow out from a central location in the body and then return. Rather, there is a longitudinal path that carries blood along the length of the body, and there are virtually as many lateral paths as there are body segments, which carry blood to the gut, nephridia, musculature of the body wall, and integument. Pulsatile propulsion of the blood may be developed almost anywhere. In vertebrates regional blood flow is controlled by varying resistance to flow against a central head of high pressure. To the extent that regional blood flow is controlled in annelids, it is probably done largely by varying the activity of local pulsatile vessels.

The course of freshly oxygenated blood draining the integument, parapodia, or gills has not been analyzed in detail but would appear to be under little control. In *Nereis*, for example, blood oxygenated in a parapodium may be recycled within the same body segment, perhaps traveling to muscle or other oxygen-demanding tissue or perhaps being returned to the parapodial exchange surface. Alternatively, the oxygenated blood may travel some length along the dorsal longitudinal vessel to enter another body segment. It would appear that oxygen transport rests mainly on the principle that the constant motion of the blood will mix enough oxygenated blood into the bloodstream to assure maintenance of an adequate oxygen tension throughout.

Basic features of open circulatory systems

The arthropods and most groups of molluscs have open circulatory systems. Before looking at these systems in detail, it is appropriate to describe some of their basic features.

Animals with open circulatory systems commonly have a well-developed central heart. This may discharge into an extensive arterial network, and arteries may service capillary beds of discrete, minute vessels lined only with a single layer of endothelial cells. Ultimately, however, the blood leaves discrete vessels to enter lacunae and sinuses. *Lacunae* are small spaces in among tissue cells. The tissue may be thoroughly permeated by an anastomosing network of lacunar spaces that brings blood close to all the cells. In this respect lacunar networks are not dissimilar to capillary networks, but the lacunar channels are characteristically irregular in shape and are not in the form of discrete vessels lined with endothelium. *Sinuses* are more capacious spaces, commonly representing thoroughfare channels for the blood. Lacunae and sinuses are sometimes bounded by a membrane of some type but are sometimes simply bounded by ordinary tissue cells. This latter point is of considerable importance, for it means that the blood bathes the tissue cells directly, and there is, in fact, no clear distinction between blood and intercellular tissue fluid. In recognition of this, some authors prefer to refer to the circulating fluid by the compound term *hemolymph*, rather than calling it blood.

Now it must be frankly admitted and recognized that the boundaries between blood spaces that are lined with membrane and those that are lined with ordinary tissue cells are generally only dimly known. It is not always easy, for example, to distinguish between capillary beds and minute lacunar networks, and an argument has been made that there is in truth some type of barrier between blood and tissue fluid throughout the circulatory system in certain groups that have traditionally been claimed to have an open circulatory system. Regardless of the ultimate decision on these points, it is clear that the design of the circulatory system in arthropods and in most molluscs is significantly different from that in vertebrates or earthworms. In the latter groups blood arriving at the tissues in major vessels undergoes a short passage through discrete exchange vessels (capillaries) and then is collected from each tissue into a system of efferent vessels (veins in vertebrates). In arthropods and most molluscs there is no organized system of efferent vessels that collects blood directly from each tissue. Once blood has left the arteries (or sometimes capillaries) it is left to follow a relatively long and ill-defined path through lacunae and sinus spaces before again being drawn into discrete vessels. This fact has important implications for the control of blood flow and the dynamics of the circulation.

Crustaceans

Some small or sessile crustaceans lack a heart and blood vessels altogether. The circulation is exclusively through sinuses and lacunae, and the propulsive force is provided by ordinary body movements. In others there is a heart, but the arteries end abruptly after only a short distance and thus do little to control the flow of blood to discrete parts of the body. The present treatment will emphasize the decapods. In these the crustacean circulatory system reaches one of its highest levels of organization. Many of the basic features seen in decapods have their counterparts in other members of the class.

The heart of decapods is a single-chambered saccular structure positioned in the dorsal thorax. Typically for crustaceans, all the vessels connected to the heart are arteries, and these leave in a number of directions. The arrangement of arteries in a crayfish heart is illustrated in Figure 10–9A as an example. Blood enters the heart not through vessels, but through slits in the heart wall, termed *ostia*. Decapods usually have three pairs of ostia; other crustaceans may have 13 or more pairs or as few as a single pair. As depicted schematically in Figure 10–9B, the heart is suspended by elastic ligaments within a bounded sinus, the pericardial sinus. In decapods and other malacostracans the only vessels entering the pericardial sinus are veins draining the gills or other respiratory surfaces (discussed later). In entomostracans, however, the pericardial sinus may communicate with other sinuses from which systemic blood may enter. When the heart contracts during systole, the ostia are closed by muscular tension, flap valves, or both; blood within the heart is driven into the arteries and cannot regurgitate into the pericardial sinus. Contraction of the heart stretches the suspensory ligaments attached to the heart wall, and elastic rebound of these ligaments during diastole expands the heart back to its presystolic volume. It is this elastic rebound that is the primary force for refilling of the heart, for as the heart is stretched open, the pressure within is reduced below that in the pericardial sinus, and blood is sucked inward from the sinus through the ostia.

The arteries of crustaceans are typically nonmuscular and thus are neither pulsatile nor involved in vasomotor control. The arterial network in decapods is extensive. Branching arteries lead the blood from the heart to most regions of the body: the brain and other parts of the head, the various parts of the gut, the end sacs of the green glands, the gonads, the ventral nerve cords, regions of the body wall, and so forth. Sometimes the blood is discharged from the arteries directly to lacunar networks.

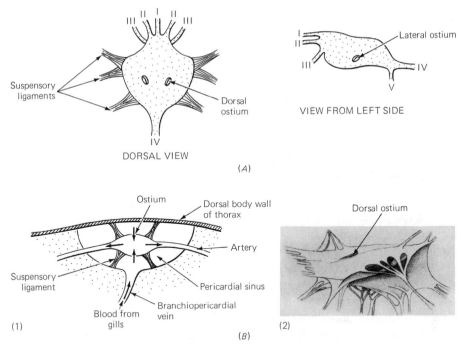

Figure 10–9. (*A*) Diagrams of the heart of a crayfish (*Astacus fluviatilis*). The seven arteries emanating from the heart are indicated by Roman numerals: I, one ophthalmic (median anterior) artery; II, two antennary arteries; III, two hepatic arteries; IV, one abdominal artery; V, one sternal artery. There are six ostia: two dorsal, two lateral, and two ventral. The heart is suspended in a cavity, the pericardial sinus, by fibrous bands, the suspensory ligaments. (*B*) The suspension of the decapod heart. *B1:* Highly schematic lateral view of heart suspended by suspensory ligaments in the pericardial sinus. The sinus is located in the dorsal thorax. Arrows show the direction of blood flow. Blood enters the pericardial sinus from the gills through branchiopericardial veins, is aspirated into the heart through the ostia during diastole, and is driven into the arteries during systole. *B2:* Three-dimensional lateral view of the major suspensory ligaments of the heart of a lobster, showing the actual complexity of the system. (*A* from Huxley, T. H. 1884. *The Crayfish.* 4th ed. Kegan Paul, Trench & Company, London; *B2* from Plateau, F. 1880. Arch. Biol. 1: 595–695.)

Sometimes capillary beds intervene. Prominent capillary beds are observed, for example, in association with the brain and ventral ganglia. It is interesting that in many decapods and other Malacostraca there is an accessory heart, known as the cor frontale, positioned along the median anterior artery just before this artery gives rise to the extensive vascular networks of the brain. This heart typifies the nature of accessory hearts in crustaceans. It consists of an enlargement in the artery, the volume of which is increased and decreased not by the contraction of vascular musculature, but by the action of extrinsic, somatic muscles that have become specialized for this circulatory role. The cor frontale contributes added energy to the blood and is believed to be an adaptation to assuring adequate perfusion of the capillary beds of the brain.

Having been delivered to lacunar networks throughout the body, the blood ultimately drains into a system of sinus thoroughfare channels located ventrally along the length of the animal. In these sinuses blood from the posterior regions of the body flows anteriorly and that from the anterior regions of the body flows posteriorly, both flows converging on a thoracic sinus termed the median ventral sinus. From this sinus the blood enters the gills and is then returned to the pericardial sinus and heart. Running between the pericardial sinus and the gills is a series of discrete vessels termed branchiopericardial veins. The central axis of each gill has a hollow core divided into two longitudinal channels by a septum. One of these channels connects to a branchiopericardial vein, whereas the other communicates with the median ventral sinus. Blood is drawn from the median ventral sinus along the afferent channel, and after traversing a system of small sinuses or lacunae in the filaments or lamellae of the gill, it is discharged into the efferent channel and branchiopericardial vein. Thus we see that after a period of seemingly poorly controlled flow through the nonvascular part of the circulation, the blood is again channeled in a most important way prior to its return to the heart. The gills are placed in series with the rest of the circulation, and oxygenation of the blood pumped by the heart is assured. This arrangement is not limited to decapods but, in its essentials, is quite general in the Crustacea.

Functional attributes of open circulatory systems

Relatively little is known about the functional implications of circulation through sinuses and lacunae rather than discrete vessels. Certain speculations can be made from hemodynamic principles and from our understanding of the anatomy of such systems, though it should be obvious from the course of debate about flow patterns in vertebrate hearts that deductions from anatomy do not always hold up well when exposed to the glare of direct physiological observations. Some data are available, and because information on decapods is perhaps as extensive as can be found, a discussion at this point is appropriate.

Open circulatory systems, in comparison with closed systems, are commonly characterized by relatively large blood volumes. Blood volume in vertebrates typically represents 2–10% of body weight (that is, 2–10 cc of blood per 100 g body weight). Ranges of values observed in several groups of animals with open circulatory systems are: decapod crustaceans, 10–50% of body weight; insects, 1–45%; and noncephalopod molluscs, 35–80%. The higher blood volumes of animals with open circulatory systems are in part explained by the fact that their blood includes fluid that in animals with closed systems is distinguished as intercellular tissue fluid. In man, for example, blood volume is near 7% of body weight, but intercellular tissue fluid amounts to about 12% of body weight, so that the total of blood and intercellular fluid is near 19% of body weight, a value not unlike the blood volume of some decapods.

The pressure gradients developed across the circulatory system by the heart are typically far lower in animals with open circulatory systems than in vertebrates. Some investigators have measured mean arterial pressures, whereas others have reported systolic and diastolic pressures. The mean or systolic difference in pressure between the major arteries and the major venous sinuses has been found to range from less than 1 mm Hg to about 15 mm Hg in a variety of crustaceans, insects, and noncephalopod molluscs that have received study. Usually the pressure difference has proved to be in the lower half of this range.

It is commonly observed in animals with open circulatory systems that body movements affect the pressures in the blood spaces. When such movements create pressure differences from one part of the body to another, blood flow will be promoted. This is probably often an important element in the circulation of these animals. Measures on crayfish have revealed that the pressure in a moving appendage may exceed that in inactive appendages, and the pressure in a leg may briefly exceed that at the heart. A pressure gradient such as the latter could aid return of blood to the heart. In American lobsters flexion of the abdomen, which is an important locomotory movement, can cause the rate of blood flow toward the thorax in the ventral abdominal sinuses to increase by a factor of 10 or more. In addition to causing pressure gradients in the blood space, body movements and changes in body conformity can affect the overall level of pressure throughout the blood space. A change in posture, for example, that tends to reduce the volume of the blood space will pressurize the blood throughout the body much in the way that pushing on a water-filled rubber ball will raise the pressure of the water. Insofar as a pressure change is transmitted throughout the blood space, it will not in any immediate sense affect blood flow, for only *differences* in pressure from one place to another will cause flow. Thus it is important to distinguish the effects of body movements on overall pressure from the effects on pressure gradients. In various decapods, pressure in the ventral thoracic sinus has been found to range from 2 mm Hg to 19 mm Hg. This pressure provides a measure of the "base-level" or static pressure in the body fluids, and pressure gradients, such as the gradient established by the heart, are superimposed on this static pressure. To illustrate schematically, the mean arterial pressure in a crab may at one time be 6 mm Hg and the mean thoracic sinus pressure, 2 mm Hg. At another time, perhaps in a different posture, these pressures may be, respectively, 16 mm Hg and 12 mm Hg. Note that in the latter case the absolute pressures are higher, but the critical circulatory factor, the pressure gradient, is still only 4 mm Hg. When we speak of "high-pressure" and "low-pressure" circulatory systems, we in fact refer to the pressure gradient required to produce adequate blood flow. Thus the crab has a "low-pressure" circulatory system whether its absolute pressures are high or low. Static pressures in insects are usually low, but there are reports of astounding pressures at certain times. In certain blowflies, for example, static pressure can rise transiently to 90 mm Hg or more during certain phases of molting and emergence of the adult from the pupa. In freshwater mussels the static pressure is normally near 1 mm Hg but can double when the body contracts. To conclude, it is often found that pressures are quite variable in open circulatory systems. Static pressure rises and falls and pressure gradients come and go with movements of the body. These effects can make it very difficult to analyze circulatory dynamics. It has been particularly difficult to quantify the effects of body movement on circulation of the blood, and our knowledge of this potentially important dimension of circulation is meager.

Some of the most thorough studies on cardiodynamics in an animal with an open circulatory system have been performed on the American lobster, *Homarus americanus*. Blood pressure in the ventral thoracic sinus is usually 2–6 mm Hg. The heart, beating around 100 times per minute, develops systolic pressures in the major arteries that are 7–15 mm Hg higher. During diastole, arterial pressure normally falls only 3–5 mm Hg below systolic pressure. The pressure gradient from the major arteries to the ventral sinus is thus seen to be relatively high for an animal with an open circulation. The reduction of heart volume during systole essentially enlarges the volume of the pericardial sinus space surrounding the heart and results in a drop in pericardial sinus pressure to 2–3 mm Hg below the pressure in the ventral thoracic sinus. Blood is therefore aspirated into the pericardial sinus during systole via the gills and branchiopericardial veins. During diastole the elastic expansion of the heart reduces intra-cardiac pressure to a few mm Hg below pericardial sinus pressure and blood is aspirated into the heart through the ostia. The dual action of the heart during systole, forcing blood into the arteries and causing suction of blood into the pericardial sinus, is probably quite general among crustaceans.

It is important to understand the implications of circulatory pressures for circulatory function. Not uncommonly, it is implied that low pressure gradients mean low rates of blood flow and that open circulatory systems, because they are characterized by low pressure gradients, must be "sluggish." This reasoning ignores the fact that the rate of blood flow is a function of circulatory resistance as well as the pressure gradient. From anatomical considerations we would expect resistance in an open type of circulation to be less than that in a closed type, and available evidence indicates that this is the case. Thus pressures that would support only a sluggish rate of flow in a vertebrate may support a considerably greater rate of flow in an animal with an open circulation. The mean pressure gradient established by the heart in the American lobster is near 10 mm Hg. This is well below values observed in teleost fish, yet the cardiac output of the lobster turns out to be toward the upper end of the range for fish. Thus we see that although some animals with an open circulation probably do have a rather sluggish rate of blood flow, this is emphatically not always the case.

Evaluating the relative merits of open and closed circulatory systems is no simple matter. There is no obvious single criterion to use, and critical functional data are available for only a few species with open systems. The late D. M. Maynard calculated a series of important parameters for the American lobster and a typical bony fish. He frankly acknowledged that, given uncertainties in available data, some of his figures might be in error by 50% or more, but his analysis is enlightening and is duplicated in part in Table 10–1. Arterial pressure is lower in the lobster, but cardiac output is considerably higher than in this "typical" fish. Because the lobster's blood volume is much greater than the fish's, the lobster requires a greater time to pump a volume of blood equivalent to its total blood volume (turn-over time) despite its greater cardiac output. The heart of the lobster, working against a lower circulatory resistance, requires less energy to pump a given volume of blood than the heart of the fish. In this sense the efficiency of the lobster's circulation is greater than that of the fish's circulation. Because, however, the lobster's blood delivers far less oxygen on a circulatory cycle, the heart of the lobster must do more work than that of the fish to provide a given rate of oxygen transport to the tissues. In this sense the lobster's circulation is less efficient. If nothing else, this

Table 10–1. Comparative circulatory parameters in the American lobster and a bony fish. Data are for 500-g animals at 15°–20°C.

	Lobster	Fish
Blood volume	85 cc	15 cc
Cardiac output	40 cc/min	12 cc/min
Blood turnover time	128 sec	75 sec
Mean arterial pressure	11 mm Hg	37 mm Hg
Cardiac work per 100 cc blood moved	0.035 cal	0.12 cal
Cardiac work per cc O_2 consumed by animal	0.07 cal	0.023 cal
Oxygen extraction during circulation of blood	0.5 cc O_2/100 cc blood	5 cc O_2/100 cc blood

SOURCE: Maynard, D. M. 1960. Circulation and heart function. *In:* T. H. Waterman (ed.), *The Physiology of Crustacea.* Vol. I. Academic, New York.

analysis demonstrates the difficulty of coming to any "clean" decision on the relative merits of the two systems. In fact, based on this type of information, we may be more impressed with similarity than with dissimilarity in the overall performance of the two.

If the open type of circulation has an inherent weakness relative to the closed type, it is perhaps most likely to be found in the peripheral realm. At first sight, flow through a system of lacunae and sinuses rather than discrete vessels would appear to be rather haphazard. It is easy to imagine that blood flow through any particular region of the body could be erratic and that stagnant pools could develop, denying a steady renewal of oxygen supplies to the tissues. The extent to which these deleterious possibilities are realized cannot be appraised at present, though there is evidence of more regularity to the peripheral flow than might at first be supposed. Large sinuses are often subdivided by septa or body organs such that more or less defined channels for flow are established. Students of insects have closely observed the circulation through minute sinuses in the head, for example, and have been able to recognize well-defined patterns of flow. A critical factor of which we know essentially nothing is the modulation of blood flow and distribution in the periphery. In a closed system of continuous vessels we have seen that flow through particular regions of the body can be controlled by the action of vascular valves. This is important to circulatory efficiency, for the cardiac output is readily redistributed according to tissue demands. During muscular exercise in mammals, for example, the increased cardiac output is directed preferentially to the muscles rather than being "wasted" in increased perfusion of tissues that have no requirement for enhanced blood flow. Once the blood has left the arterial vessels in an open type of circulation, it is difficult to see how the distribution of flow could be controlled in any general sense, though there can be little doubt that body movements, contraction of somatic muscles, and the activity of peripheral accessory hearts influence peripheral flow. Also, one is led to wonder from the design of the open circulation whether an increase in cardiac output, as during exercise, would result in a commensurate increase of flow through lacunar exchange networks or whether some of the increased cardiac output, following paths of least resistance, might skirt around the lacunar networks and enter sinus channels rather directly. Deficiencies in the control of peripheral flow would limit the flexibility and efficiency of the circulation by comparison to closed systems and may well represent a serious inherent weakness in the open type of circulation.

Whatever weaknesses the open circulation may have, they have not been sufficient to prevent some crustaceans, even large ones, from assuming a relatively active and metabolically demanding way of life. This is obvious to anyone who has ever watched a blue crab swim through the sea, a ghost crab dash about on the beach, or a crayfish avoid the net of a biologist. It is noteworthy, on the other hand, that animals with open circulatory systems have never attained the highest levels of metabolic intensity—with the singular exception of the tracheate arthropods, in which the circulatory system is relieved of the function of oxygen transport.

Insects

The circulatory system of insects follows the basic arthropodal plan. A dorsal vessel extends along most of the body (Figure 10–10) and in many insects is the only vessel. It is divisible, often indistinctly, into a posterior portion, the heart, and an anterior portion, the dorsal aorta. The heart is usually restricted to the abdomen but sometimes extends into the posterior thorax. It is perforated by 1 to 13 pairs of ostia, which typically are slit-shaped and valved. The dorsal aorta leads forward from the heart into the thorax and head and lacks ostia. The heart may be directly bound to the dorsal body wall or may be attached to it by threads of connective tissue. A septum of muscular and connective tissue, termed the dorsal or pericardial septum, divides the abdomen, to a greater or lesser extent, into a dorsal pericardial sinus and ventral perivisceral sinus. This septum may be attached to the heart, extending laterally to the body wall on each side, or it may be separate, running across the abdomen underneath the heart. In the former case the ventral wall of the heart may face into the perivisceral sinus. In the latter the heart rests entirely in the pericardial sinus and is attached to the septum by threads of connective tissue. The ostia usually open into the pericardial sinus in either situation.

Refilling of the heart is accomplished much as it is in crustaceans. Cardiac contraction stretches the elastic elements to which the heart is attached, and rebound of these elements pulls the heart wall out again during diastole, resulting in aspiration of blood into the heart from the pericardial sinus through the ostia. Often blood can enter the pericardial sinus from the perivisceral sinus through fenestrations in the dorsal septum. When the septum is imperforate, blood enters from the perivisceral sinus by a more discrete course around the posterior margin of the septum.

The posterior end of the heart is usually closed. Not uncommonly, there are segmentally arranged lateral vessels emanating from the dorsal vessel. In the common cockroach *Periplaneta*, for example, five pairs of lateral vessels leave the heart in the abdomen and two pairs in the thorax. The heart typically contracts in a peristaltic wave from posterior to anterior, forcing blood into the dorsal aorta and into lateral vessels if they are present. The dorsal aorta is itself contractile and commonly continues the peristaltic wave of the heart. Lateral vessels may also be vigorously con-

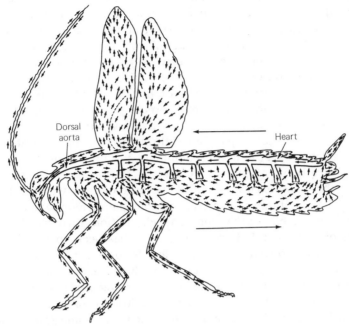

Figure 10–10. Diagrammatic representation of blood flow in an insect. (After Jones, J. C. 1964. The circulatory system of insects. *In:* M. Rockstein (ed.), *The Physiology of Insecta.* Vol. III. Academic, New York.)

tractile. The major vessels end abruptly with little branching, discharging blood directly to the lacunar circulation. The dorsal aorta, for example, runs to the head, where it sends off, at most, only a few branches before terminating. Once blood has exited the vessels, it is left to pass through lacunae and sinuses, ultimately to arrive in the perivisceral sinus, from which it returns to the pericardial sinus and heart. Figure 10–10 portrays the general course of the circulation.

Accessory pulsatile structures are widespread in insects, commonly serving to enhance the flow of blood through appendages. In a number of groups, for example, there are pulsatile organs at the bases of the antennae that pump blood into antennal vessels. The antennal vessel of each antenna typically extends the length of the antenna and is fenestrated; blood flows outward in the vessel, emerges through the fenestrations at various points, and flows back to the head in the space surrounding the vessel. The antennal pulsatile organs often receive blood from branches of the dorsal aorta. In contrast, most accessory pulsatile structures in insects are not supplied by the vascular part of the circulation but instead draw blood from sinuses. The legs are usually divided longitudinally into two sinus channels, a quite typical arthropodal condition. The two channels communicate with thoracic sinuses, one channel serving to carry blood out into the leg and the other serving for return. In a number of groups, pulsatile organs are found within the legs, often in the tibia or near the femoro-tibial joint.

Blood is circulated through the wings in many insects, commonly in an exceedingly elaborate system of fine channels. The major blood channels are within the wing veins (these are not veins in the vascular sense but rather are structural tubes formed as thickenings of the cuticle). Veins communicate with thoracic sinuses at the base of the wing. The gross pattern of flow through the wing is variable, but in a common pattern blood flows in through anterior veins and follows an outward course through veins of the anterior portion of the wing, then returns through veins of the posterior portion of the wing (see Figure 10–10). Between the major veins there are extensive networks of fine blood channels in which the pattern of flow is highly variable. Thoracic pulsatile organs that aid the wing circulation are common. Often they are attached to the efferent, posterior veins and act to pull blood through the wings. Minute accessory pulsatile structures also occur commonly within the wings themselves, characteristically in major vein channels. In muscid flies, for example, there is a thoracic pulsatile organ at the base of each wing, and there are four pulsatile membranes situated at various points along the large efferent vein channels within each wing. The circulation through the wings in insects is of considerable importance, for it has been shown that parts of the wing that are deprived of blood flow become dry and brittle and are easily broken away. The circulation of the blood is also believed to be important in expanding the wing membranes during development.

A final accessory pulsatile structure worthy of note is the ventral septum, a fibromuscular structure that occurs in some groups of insects but not others. This septum, which is often perforated, traverses the body cavity just above the ventral nerve cord, delimiting a perineural sinus below from the perivisceral sinus above. It is observed to undergo undulatory contractions that are believed to be important in circulating blood over the nerve cord and, sometimes, in aiding the posterior movement of blood from thorax to abdomen.

The circulation in insects deserves serious consideration by the student of comparative physiology. Many insects are active animals with relatively high oxygen demands. During sustained flight, strong flyers attain weight-specific rates of oxygen consumption that not only are many times higher than any reported for crustaceans, but are among the highest in the entire animal kingdom. In other phyla (notably chordates and molluscs) we are accustomed to finding that groups with a relatively intense metabolic capacity show clear circulatory refinements in comparison to groups with a more modest metabolic capacity. This is understandable in view of the fact that circulatory transport is an indispensable element in meeting the oxygen demands of the tissues. In comparing insects and crustaceans, however, we find that the circulatory system in insects is not more advanced either anatomically or, so far as we know, physiologically. In certain anatomical respects the insect system appears less advanced: witness the far less extensive development of the arterial vasculature and the complete lack of capillary beds. Of course, it is highly relevant that the tracheal respiratory system of insects itself supplies oxygen to all tissues. The insects indicate, then, that the circulatory system can remain relatively simple in even highly active animals provided it is relieved of the burden of oxygen transport. This in turn supports the contention that selection for respiratory gas transport has been the primary factor in the evolution of refined circulatory systems in groups of other phyla that are characterized by an intense metabolic capacity. The insect circulatory system has been subject to selective pressures to meet the less immediately demanding requirements for transport of nutrients, metabolic wastes, hormones, and the like. It is an interesting observation that some insects survive for considerable periods with their heart removed. In cockroaches (*Periplaneta*) from which most or all of the dorsal vessel has been removed, circulation of blood cells virtually ceases; yet the animals not only survive, but remain quite active for a time. If only the abdominal portion of the dorsal vessel is removed, the cockroaches can still successfully molt. Already in 1628, William Harvey observed that the heart occasionally stops for a period in some insects, and it has often been noted that circulation in peripheral regions can vary widely and sometimes slow to the point of stopping. Such observations indicate that insects are less reliant on the steady operation of the circulation than nontracheate animals with similar metabolic demands.

Arachnids

The arachnid heart resembles the insect heart in its basic form, being tubular, perforated by ostia, and suspended within a dorsal pericardial sinus. Typically both posterior and anterior aortae emanate from the heart. The arachnids have received but little attention from circulatory physiologists, which is regrettable because some members of the group respire by book lungs whereas others are entirely tracheate, and there is thus opportunity for more or less direct comparative study of circulatory systems that are responsible for respiratory gas transport and ones that are not. Spiders that possess book lungs and scorpions (all of which rely entirely on book lungs) have substantially more-extensive arterial systems than insects. In scorpions, for example, the anterior aorta branches repeatedly, giving rise to vessels to the head and appendages and to a ventral vessel that extends posteriorly along much of the body length. Lateral arteries emanating from the heart supply the viscera below the heart, and the posterior aorta and ventral branch of the anterior aorta supply the viscera and other structures within the "tail" (metasoma). It is characteristic of arachnids with book lungs that the lungs are drained by discrete vessels, the branchiopericardial or pulmonary veins, which lead to the pericardial sinus. The pattern of circulation is thus reminiscent of that in crustaceans: after being discharged to the lacunar circulation, blood makes its way to ventral sinuses and then follows a channelized path back to the heart via the book lungs and pulmonary veins. Blood from the book lungs may be the only blood to enter the pericardial sinus, in which case the heart pumps only freshly oxygenated blood, or there may also be opportunity for direct entry of blood from systemic sinuses. Tracheate arachnids typically exhibit considerably less arterial branching than those with book lungs. It is not altogether clear how much this reduction of vascular development in tracheate forms should be attributed to their mode of respiration because many of them are relatively small, a condition that could also lead to circulatory reduction. Even in solpugids (sun spiders) and other larger tracheate forms, however, the arteries terminate rather quickly.

It is interesting that extensor muscles are poorly developed or lacking at certain major joints of the walking legs in spiders. The legs are flexed by flexor muscles but are extended under hydraulic pressure. The legs are attached to the prosoma (cephalothorax) and receive arteries (pedal arteries) from the heart, which is located in the abdomen. Recent studies on the tarantula *Dugesiella hentzi* have indicated that both the prosomal and abdominal cavities are compressed by muscular activity during walking, thus raising the base-line or static pressure throughout the blood spaces; base-line pressure reaches 40–60 mm Hg in the prosoma, and a relatively modest additional pressure may be developed in the pedal arteries by the beating of the heart. It appears that the prosomal head of pressure serves to extend the legs whenever their flexor muscles are relaxed. When tarantulas are at rest, prosomal pressure can fall to around 10 mm Hg. Experiments showed that in resting animals, systolic-diastolic pressures in the heart could be as low as 12/8 mm Hg and indicated that pressure gradients of only a few mm Hg are required to drive blood from the heart through the arteries and lacunar circulation to the major sinuses and then return the blood via the book lungs to the heart.

Bivalve and gastropod molluscs

The circulatory system in bivalve and gastropod molluscs exhibits many differences in detail in the various groups and species but a remarkable degree of basic similarity. The present treatment will emphasize only the fundamental and common properties.

The heart is chambered and not unlike that of vertebrates in its basic form. There is a single, relatively muscular ventricle that receives blood from one or two less-muscular atria. The atria receive blood from the gills, and usually there is a one-to-one relationship between gills and atria, with the number of atria following the number of gills. In the bivalves there are two gills and, correspondingly, two atria. Gastropods that have two gills have two atria; but in the majority the right gill is lacking and the right atrium is reduced or, more usually, absent. The ventricle gives rise to an anterior aorta and not uncommonly to a posterior aorta as well. The heart is enclosed in a pericardial cavity by a thin membrane, the pericardium. As in vertebrates there are no direct connections between the heart lumen and pericardial cavity, and the pericardial cavity does not receive blood vessels. Usually in bivalves and occasionally in gastropods the gut passes through the ventricle of the heart; the significance of this arrangement is unknown. The heart is positioned dorsally in the body with the atria in more or less close juxtaposition to the bases of the gills. The exact arrangement of the heart chambers and efferent and afferent vessels is variable, being affected, for example, by such important developments as torsion in the gastropods. Figure 10–11 depicts the central vascular system in a typical bivalve.

The arterial system is often extensive, branching repeatedly before discharging the blood into lacunar networks or sometimes into capillary or capillary-like beds that in turn lead to lacunae. Typically, blood is carried to most major regions of the body in discrete arterial vessels. In the common marine snail *Littorina littorea*, for example, the anterior aorta runs forward, giving off branches to the stomach, foot, radular area, and buccal area; whereas the posterior aorta branches to supply the kidney, intestine, stomach, digestive gland, and gonad. In the freshwater mussel *Anodonta* the anterior aorta sends branches to the viscera, foot, anterior adductor muscle, and mantle wall; the posterior aorta supplies the posterior adductor muscle and region of the siphons.

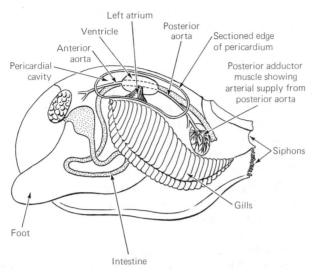

Figure 10–11. A freshwater mussel (*Anodonta*) with the left shell removed, semidiagrammatic. The pericardium has been sectioned to reveal the heart and aortae. The two atria (of which only the left is shown) are positioned laterally on either side of the ventricle. The gut passes through the ventricle. [After Parker, T. J. and W. A. Haswell. 1940. *A Text-book of Zoology.* Vol. I. 6th ed. (as revised by O. Lowenstein) Macmillan and Company, Ltd., London. Used by permission of Macmillan London and Basingstoke.]

From the lacunar circulation, blood enters a system of sinus thoroughfare channels that sooner or later lead to discrete, lined vessels that are properly called veins. One of the striking commonalities in the circulatory plan of these animals is a tendency for the blood to return to the heart via the kidneys and gills. In its simplest form this route may be outlined as follows. Sinus channels or veins from the major regions of the body converge on vessels that lead to extensive lacunar networks in the walls of the kidneys. After passing through these networks, blood is collected into afferent branchial vessels, passes through lacunae or capillaries in the gills, and then travels to the atria of the heart via efferent branchial vessels. The kidney circulation is a renal portal system and provides opportunity for removal of wastes and other renal functions prior to repumping of the blood by the heart. The circulation through the gills just upstream from the heart is reminiscent of the condition in crustaceans and is of the greatest importance, for, again, it places the respiratory organs in a defined series arrangement relative to the systemic circulation.

Departures from the simple path through the kidneys and gills to the heart are common but often minor. Some blood may enter the gills and travel to the heart without having first passed the renal portal system, or some blood from the kidneys may flow directly to the heart without traversing the gills. The latter circuit, which sometimes assumes major significance, introduces blood into the heart that has not been exposed to the branchial respiratory surfaces immediately beforehand.

The mantle wall presents possibilities for gas exchange and sometimes is provided with a more or less specialized circulation. Bivalves have a large mantle surface, extending over the entire inner faces of the shells. Commonly, it is supplied elaborately by arterial branches. In a number of forms, such as *Anodonta* and the scallop *Pecten*, some blood from the mantle is drained directly into the atria, bypassing the kidneys and gills. Thus there is a separate mantle circuit in parallel with the usual systemic-branchial circuit. In the intertidal mussel *Mytilus* all blood from the mantle is drained into the kidneys along with general systemic blood, but there is an appreciable direct flow from the kidneys to the heart, bypassing the branchial circulation. In this case there is also some degree of parallelism in the mantle circuit. The importance of the mantle circuit to respiratory gas exchange requires quantitative evaluation.

In terrestrial pulmonate gastropods, gills have been lost, and the principal site of external respiration is a lung derived from the mantle (see Chapter 8). Blood from the systemic sinuses is now collected in vessels that supply the heavily vascularized mantle wall and, after passing the respiratory surfaces, is returned directly to the single atrium of the heart. Some blood circulates through the kidney on its way back to the heart, but this is a subsidiary route in the pulmonates.

Blood volume in bivalves and gastropods is large, typically representing 40–50% or more of body weight. Displacements of blood from one region of the body to another often play an important role in movement and in altering body form. Anyone who has ever observed a clam, for example, is aware that the foot can be expanded to a sometimes remarkable volume when it is extended from the shell. This expansion is achieved by displacing blood into the pedal sinus from other sinuses, and the entrance to the pedal sinus is guarded by a sphincter muscle (Keber's valve) that, when constricted, acts to retain blood in the expanded foot. When clams dig, their foot is extended into the substrate under the force of blood pressure; during this extension, circular or transverse muscles in the foot hold it in a tapered or cylindrical shape, permitting ready penetration of the sand or mud. Once the foot is extended, distal circular or transverse muscles relax, allowing the distal end of the foot to swell under the force of blood pressure. This dilation anchors the tip of the foot so that when longitudinal muscles contract and shorten the foot, the main body of the animal is pulled toward the tip. By repetitions of this process many clams can move through the substrate at an amazingly rapid rate, as any weekend clam digger is well aware. The blood, acting as a hydraulic skeleton, is also instrumental in the extension of the siphons in clams and of the foot and penis in gastropods, to cite some examples.

Cephalopod molluscs

The cephalopods include not only the most active molluscs but some of the more active aquatic animals in general. The giant squids are the largest of all invertebrates. The circulatory system of cephalopods differs from that of other molluscs in striking and important ways that presumably are adaptive to meeting their heightened demands for internal transport, especially respiratory gas transport.

The arterial and venous systems are very extensive and are in large part joined across true capillary beds, giving the circulatory system a closed character. This is a most remarkable development in a phylum whose other members have typical open systems. Some large blood sinuses are present, but they are lined, and the blood is confined to the circulatory system, distinct from intercellular tissue fluid. The basic circulatory plan remains typically molluscan: Blood enters the systemic heart from the gills, is

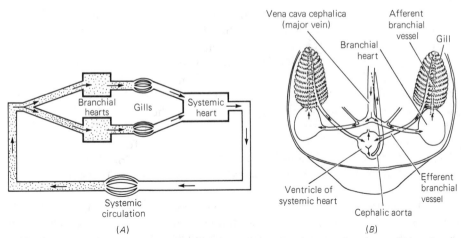

Figure 10–12. (*A*) Schematic representation of the circulatory plan in squids and octopuses. Stippled parts carry relatively deoxygenated blood. (*B*) Diagram of the major features of the central cardiovascular system in an octopus. (*B* from Johansen, K. and C. Lenfant. 1966. Amer. J. Physiol. 210: 910–918.)

pumped out to the body, and on its return to the heart passes first through the kidneys and then through the gills. The systemic heart consists of a muscular ventricle that receives blood from one or two pairs of atria; each atrium in turn receives blood from a gill via a distinct efferent branchial vessel. In *Nautilus* there are two pairs of gills and a corresponding number of atria. All other living cephalopods, that is, the squids and octopuses, have one pair of gills and one pair of atria. In squids and octopuses, blood returning from the systemic tissues in the major veins is split into two symmetrical paths to be directed to the gills. Much of the returning blood follows the two lateral vena cavae through the kidneys before arriving at the bases of the gills, as described in Chapter 7 (see Figure 7–14). Near the base of each gill is a bulbous accessory pulsatile organ, the branchial heart, which receives the systemic venous blood. This heart pumps the blood into an afferent branchial vessel, from which it passes through capillaries in the gill to arrive in the efferent branchial vessel and return to the systemic heart. The circulatory plan is diagramed in Figure 10–12. The arrangement of hearts, you will note, is much "closer to home" than appears at first, being similar to that in mammals except for the fact that the respiratory pumps are anatomically separate from the systemic pump.

Physiologically, the circulation of cephalopods resembles that of vertebrates more than it does that of other molluscs. Several excellent studies have recently been performed on unrestrained *Octopus dofleini*, and the data discussed here are for that species. The blood volume represents about 6% of body weight, a value within the usual range for vertebrates and far lower than values observed in noncephalopod molluscs. Systolic pressure in the systemic heart ranges approximately from 33 mm Hg to 52 mm Hg; diastolic pressure is typically about 15 mm Hg lower than systolic. These pressures are not dissimilar to those developed in the hearts of teleost fish. As a base of comparison with other molluscs, heart pressure in *Anodonta* is near 5 mm Hg and that in the snail *Helix* is near 1 mm Hg. In resting or mildly active octopuses, cardiac output is typically around 10–20 ml/kg/min. These values are in the lower end of the range for fish.

The pressures developed in the systemic circuit considered together with the cardiac output indicate that peripheral resistance is appreciable, as would be expected from the closed character of the circulation. As in vertebrates, then, we find that octopuses achieve relatively rapid flow through a high-resistance systemic circuit by virtue of a heart that maintains high pressures. Blood pressure has fallen to 0–12 mm Hg by the time the blood arrives in the great veins but is elevated considerably by the branchial hearts. Systolic pressures developed by these hearts are 18–37 mm Hg; diastolic pressure is about 11 mm Hg. The head of pressure produced by the branchial hearts is important not only to the perfusion of the gills, but also to the process of ultrafiltration into the kidneys, for ultrafiltration occurs across a protuberance of the heart wall, the branchial heart appendage (Chapter 7).

The arteries and veins of cephalopods are muscular. With high arterial pressures and the closed character of the circulation, there is an obvious potential for fine vasomotor control of blood distribution such as that observed in vertebrates. We know little of the extent to which this potential is realized. It is interesting that in *Octopus* blood loss from a wound is stopped not by clotting of the blood, but by vascoconstriction in vessels supplying the wounded area. The muscularity of the blood vessels also establishes the potential for vascular pulsatile activity. Peristaltic contractions have been observed in various veins of a number of species and are believed to be important in aiding the return of the blood toward the branchial hearts.

In comparing the circulation of these active invertebrates with that in vertebrates, a striking degree of evolutionary convergence is evident. There is perhaps no more compelling illustration of the interplay between the evolution of the circulation and that of a metabolically demanding way of life.

SELECTED READINGS

Burton, A. C. 1972. *Physiology and Biophysics of the Circulation*. 2nd ed. Year Book Medical, Chicago.

Fretter, V. and A. Graham. 1962. *British Prosobranch Molluscs*. The Ray Society, London.

Functional Morphology of the Heart of Vertebrates. (A Symposium.) 1968. Amer. Zool. 8: 177–229.

Horvath, S. M. and C. D. Howell. 1964. Organ systems in adaptation: the cardiovascular system. *In*: D. B. Dill (ed.), *Handbook of Physiology. Section 4: Adaptation to the Environment*. American Physiological Society, Washington, D.C.

Johansen, K. 1972. Heart and circulation in gill, skin and lung breathing. Resp. Physiol. 14: 193–210.

Johansen, K. and A. W. Martin. 1965. Comparative aspects of cardiovascular function in vertebrates. *In*: W. F. Hamilton (ed.), *Handbook of Physiology. Section 2: Circulation*. Vol. III. American Physiological Society, Washington, D.C.

Martin, A. W. and K. Johansen. 1965. Adaptations of the circulation in invertebrate animals. *In*: W. F. Hamilton (ed.), *Handbook of Physiology. Section 2: Circulation*. Vol. III. American Physiological Society, Washington, D.C.

Maynard, D. M. 1960. Circulation and heart function. *In*: T. H. Waterman (ed.), *The Physiology of Crustacea*. Vol. I. Academic, New York.

Satchell, G. H. 1971. *Circulation in Fishes*. Cambridge University Press, London.

See also references in Appendix.

11 METABOLIC RESPONSES TO OXYGEN DEFICIENCY AND LOWERED AVAILABILITY OF OXYGEN

This chapter takes up two aspects of metabolic response to lowered oxygen. The first section discusses anaerobic metabolism, the mechanisms by which the energy requirements of tissues can be met when there is insufficient oxygen to support energy metabolism entirely through aerobic catabolic pathways. The second section discusses the effects of lowered ambient oxygen levels on the rate of aerobic metabolism.

ANAEROBIC METABOLISM: MECHANISMS AND INTERRELATIONS WITH OVERALL ENERGY METABOLISM

The energy-demanding processes in cells draw their energy from the high-energy phosphate bonds of adenosine triphosphate (ATP). Cleavage of ATP to form adenosine diphosphate (ADP) and inorganic phosphate releases chemical energy that can then be directed to activating a muscle fiber, constructing a protein, transporting an ion across a membrane, or any of many other energy-demanding events. Other nucleoside phosphates, such as guanosine triphosphate (GTP) and cytidine triphosphate (CTP), may intervene in the utilization of ATP energy; thus, for example, GTP as well as ATP is required for protein synthesis, and the GTP is formed by transfer of high-energy phosphate from ATP to GDP: ATP + GDP → ADP + GTP. ATP, although not always the immediate source of energy for energy-demanding processes, is the fundamental source and is indispensable to life.

Animals ultimately receive the energy they require from the chemical bonds of ingested foodstuffs. Energy-demanding processes, however, cannot draw the energy they need directly from the bonds of these foodstuffs; instead they must have it in the form of ATP. Accordingly, the cell possesses intricate biochemical machinery to transfer energy from the bonds of ingested foods to the bonds of ATP. If these catabolic pathways fail to transfer energy to ATP at a sufficient rate to meet demands, the cell can literally starve in the midst of plenty: no matter how much energy is available in the bonds of glycogen, lipids, or other such compounds, the vital energy-demanding processes will halt if the transfer of energy to ATP cannot proceed.

The vast majority of animals possess mechanisms for the complete oxidation of foodstuffs to carbon dioxide and water. These pathways have great advantages in that a large proportion of the bond energy of the foodstuffs is liberated in the process, and much of this energy is harnessed in the bonds of ATP for subsequent utilization. The complete oxidation of foodstuffs is always dependent on oxygen, and the fundamental question raised in this section is: How can organisms continue to tap the bond energy of foodstuffs when oxygen is unavailable or else supplied at an insufficient rate for aerobic pathways to meet ATP demands? To appreciate fully the problems of oxygen deficiency we shall first examine the pathways of aerobic catabolism and the reasons for their dependency on oxygen.

Aerobic catabolism:
glycolysis and the Krebs cycle

In the presence of oxygen carbohydrates are catabolized along two basic pathways in most animals: the glycolytic (or Embden-Meyerhof) pathway and the Krebs citric acid cycle (or tricarboxylic acid cycle). Because these pathways are discussed in detail in general biology and chemistry texts, only certain aspects will be presented here.

Figure 11–1 depicts the basic reactions of glycolysis and the Krebs cycle. (Some intermediate steps have been omitted for simplicity.) The glycolytic pathway begins with glucose and terminates with pyruvic acid. Glucose is first phosphorylated at the *cost* of an ATP molecule to form glucose-6-phosphate. Glucose-6-phosphate is then converted to fructose-6-phosphate, and the latter is phosphorylated to form fructose-1,6-diphosphate. This phosphorylation also is accomplished at the *cost* of an ATP molecule, so that at this point two ATP molecules have been invested in the reactions. Fructose-1,6-diphosphate is cleaved to form two three-carbon molecules, dihydroxyacetone phosphate and glyceraldehyde-3-phosphate. These compounds are interconvertible, and when glucose is being catabolized for release of energy, dihydroxyacetone phosphate is converted to glyceraldehyde-3-phosphate, so that the net effect is conversion of each molecule of fructose-1,6-diphosphate to two molecules of glyceraldehyde-3-phosphate. The reactions subsequent to glyceraldehyde-3-phosphate in Figure 11–1*A* are all multiplied by 2 to emphasize that two molecules follow these pathways for each glucose molecule that enters the system. Glyceraldehyde-3-phosphate is next oxidized, with the addition of inorganic phosphate, to the three-carbon diphosphate 1,3-diphosphoglyceric acid. This reaction is the *only* oxidative reaction in the entire glycolytic pathway and

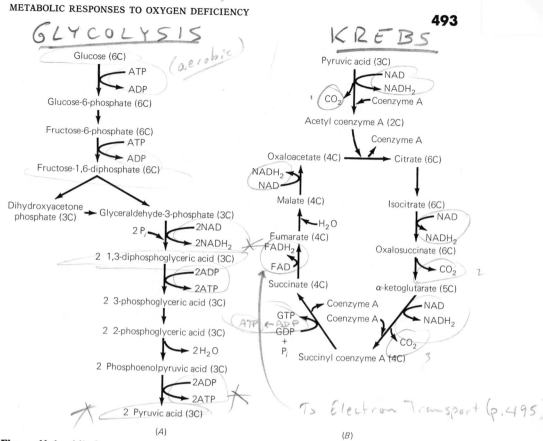

Figure 11–1. (*A*) Summary of the major reactions of glycolysis. (*B*) Summary of the major reactions of the Krebs citric acid cycle. The number of carbon atoms in each compound is indicated in parentheses. P_i = inorganic phosphate. The expression for reduction of NAD, NAD → NADH₂, is a shorthand expression; the actual reaction is NAD+ + 2H → NADH + H+.

will prove most significant in our subsequent discussion of anaerobic metabolism. The reaction does not in itself require oxygen; rather, it involves the concomitant reduction of one molecule of the enzyme cofactor, nicotinamide adenine dinucleotide (NAD), per molecule of glyceraldehyde-3-phosphate. The reduction of NAD is symbolized as NAD → NADH₂, and the fate of the NADH₂ will be discussed subsequently. The 1,3-diphosphoglyceric acid formed in the oxidation reaction is next converted to the monophosphate 3-phosphoglyceric acid, with the *formation* of one ATP per molecule. The 3-phosphoglyceric acid is converted in two steps to phosphoenolpyruvic acid, and the latter reacts to form pyruvic acid, again with the *formation* of one ATP per molecule. At this point the Embden-Meyerhof or glycolytic pathway is complete. In all, we have seen, one molecule of glucose is converted to two molecules of pyruvic acid with the following important results. First, two molecules of NAD have been reduced to NADH₂ per molecule of glucose. Second, two molecules of ATP have been used and four have been formed, yielding a net increase of two molecules of ATP per molecule of glucose.

During aerobic metabolism the pyruvic acid formed in glycolysis is oxidized by the Krebs citric acid cycle, which is depicted in Figure 11–1*B*. For simplicity the reactions involved in the oxidation of just one molecule of pyruvic acid are shown, but remember that two molecules are processed for each molecule of glucose. Pyruvic acid enters the Krebs cycle through a complex set of reactions in which it is oxidatively decarboxylated, forming carbon dioxide and a two-carbon acetyl group that emerges in combination with coenzyme A as acetyl coenzyme A. In the process, a molecule of NAD is reduced. Acetyl coenzyme A then reacts with oxaloacetate, the end result being that coenzyme A is released and the acetyl group is condensed with oxaloacetate (four-carbon) to form citrate (six-carbon). In the ensuing series of reactions oxaloacetate is ultimately regenerated and then again can combine with acetyl coenzyme A. The reactions are well known and need not be reviewed stepwise here, though the following points deserve emphasis. Decarboxylations occur at two points, in the conversion of oxalosuccinate to α-ketoglutarate and in the conversion of α-ketoglutarate to succinyl coenzyme A. These two decarboxylations plus the one in the reaction of pyruvic acid to form acetyl coenzyme A account for the formation of three molecules of carbon dioxide for every molecule of pyruvic acid processed; in effect the three carbons of pyruvic acid thus emerge as carbon dioxide. Oxidations occur at four points. At three of these NAD is reduced, whereas at one (the oxidation of succinate to form fumarate) another coenzyme, flavin adenine dinucleotide (FAD), is reduced. Counting the formation of one $NADH_2$ in the reaction of pyruvic acid to form acetyl coenzyme A, the processing of each pyruvic acid molecule thus results in formation of four $NADH_2$ and one $FADH_2$. Finally, note that the reaction of succinyl coenzyme A to form succinate is accompanied by the formation of guanosine triphosphate (GTP) from guanosine diphosphate (GDP). The phosphate bond formed is a high-energy bond, and GTP subsequently donates its terminal phosphate group to ADP, resulting in GDP and ATP. In essence, a molecule of ATP is generated.

Aerobic catabolism: the electron transport chain and the role of oxygen

Thus far it may appear paradoxical that in a discussion of aerobic catabolism we have not mentioned the involvement of molecular oxygen. Molecular oxygen is not a participant in any of the reactions of glycolysis or the Krebs cycle, and in a very narrow sense all the reactions can proceed without it. Oxygen nonetheless is essential; the reason for this lies in the disposition of the reduced cofactors, $NADH_2$ and $FADH_2$

Oxidation reactions involve the removal of electrons, and when electrons are removed from one compound, they must be transferred to another. In the several oxidation reactions of glycolysis and the Krebs cycle, the immediate electron acceptors are the enzyme cofactors, NAD and FAD. These cofactors, however, can never serve as the ultimate electron acceptors, for they are present in only limited quantities. If $NADH_2$ and $FADH_2$ were simply allowed to accumulate, the cell would soon run out of NAD and FAD, and the reactions of glycolysis and the Krebs cycle would come to a halt for lack of these vital, immediate electron acceptors. $NADH_2$ and $FADH_2$ are oxidized back to NAD and FAD in reactions with the electron transport chain in the mitochondria, as illustrated in Figure 11–2. The electrons taken from $NADH_2$ and $FADH_2$ are passed sequentially from one member of the chain to another in a series of reductions and oxidations,

ELECTRON TRANSPORT CHAIN

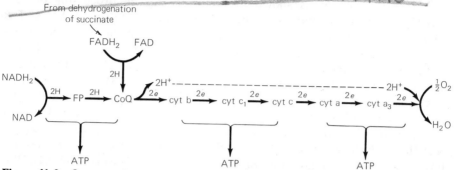

Figure 11–2. Conceptual scheme of major features of the electron transport chain. Shown are the probable locations of respiratory-chain phosphorylations resulting in formation of ATP. There is a large decline in free energy from $NADH_2$ to H_2O; the chain is believed to be a device for breaking up the total decline into a series of smaller energy drops, some of which are of sufficient size to support phosphorylation of ADP. The biochemical mechanisms of harnessing energy released in the chain to produce ATP (coupling mechanisms) are incompletely understood. cyt = cytochrome; e = electron; CoQ = coenzyme Q. FP stands for the flavoprotein NADH dehydrogenase. Succinate dehydrogenase is also a flavoprotein, containing FAD as its flavin cofactor.

ultimately to be donated to the final electron acceptor, oxygen. Protons (H^+ ions) that "accompany" the electrons also combine with oxygen, the final product being water. The essential role of oxygen is clear. The cytochromes and other members of the electron transport chain, like NAD and FAD, are present in more or less fixed quantities in the cell and therefore cannot act as terminal electron acceptors. Oxygen, on the other hand, is continuously supplied to the cell, and the product of its reduction, water, can be dissipated to the environment. If oxygen is denied, the electron transport chain is transformed from a route for the dissipation of electrons to a dead end. With the free flow of electrons stopped, the members of the chain enter a statically reduced state; the chain cannot accept electrons from $NADH_2$ and $FADH_2$, and the vital resupply of oxidized NAD and FAD is blocked.

The electron transport chain, in addition to serving simply as a route for the oxidation of $NADH_2$ and $FADH_2$, is much involved in the transfer of energy from the bonds of foodstuffs to ATP. Molecular oxygen has a much higher affinity for electrons than the enzyme cofactors, NAD and FAD, and there is a large decline in free energy as the electrons originally taken from nutrient molecules are passed through the electron transport chain. Some of this energy is harnessed in the bonds of ATP at more or less defined points in the chain, as indicated in Figure 11–2. At each of these points, one ADP is phosphorylated to form ATP for each pair of electrons transported. The process is known as *respiratory-chain phosphorylation* or *oxidative phosphorylation*.

Aerobic catabolism: the yield of ATP

It is now appropriate to compute the total yield of ATP for each molecule of glucose that is aerobically catabolized. First, the yield from respiratory-chain phosphorylation can be calculated from knowledge of the reactions of $NADH_2$ and $FADH_2$. Each molecule of glucose results in two molecules of pyruvic acid, and the subsequent catabolism of these pyruvic acid molecules results in eight molecules of $NADH_2$, which enter the electron transport chain as shown in Figure 11–2. The processing of each $NADH_2$ results in formation of three ATP's, for a total of 24 ATP's. Two $FADH_2$ are produced in the citric acid cycle for each molecule of glucose. $FADH_2$ enters the electron transport chain at a different point than $NADH_2$, and only two ATP's are produced for each $FADH_2$, the yield thus being an additional four molecules of ATP. Two molecules of $NADH_2$ are generated in glycolysis for each molecule of glucose. These $NADH_2$ are formed in the cell solution and follow a different course from the other $NADH_2$, which are formed in the mitochondria. The electrons of glycolytically produced $NADH_2$ enter the electron transport chain at a point analogous to those of $FADH_2$, and only two molecules of ATP are generated for each $NADH_2$. With two such $NADH_2$ per molecule of glucose, four ATP's result, and the total number of ATP's formed by respiratory-chain phosphorylation per molecule of glucose is $24 + 4 + 4 = 32$.

In addition to these ATP's, others are generated in the reactions of 1,3-diphosphoglyceric acid to form 3-phosphoglyceric acid (Figure 11–1A), phosphoenolpyruvic acid to form pyruvic acid (Figure 11–1A), and succinyl coenzyme A to form succinate (Figure 11–1B). These phosphorylations, in contrast to the respiratory-chain phosphorylations, are known as *substrate-level phosphorylations* because they occur immediately in the reactions of substrates of glycolysis and the Krebs cycle rather than depending on the subsequent oxidation of reduced cofactors. In all, six ATP's are formed by substrate-level phosphorylation per molecule of glucose.

In total, 38 molecules of ATP are generated for each glucose molecule catabolized. Because two ATP's are consumed in the phosphorylation of glucose and fructose-6-phosphate (see Figure 11–1A), the *net* yield is 36 molecules of ATP.

The efficiency of the catabolic pathways in transferring energy from bonds of glucose to ATP can be calculated by comparing the amount of energy incorporated into ATP with the total amount released in the oxidation of glucose to carbon dioxide and water. This calculation can be approximated using standard free energies for the appropriate reactions at 37°C and pH 7.0. The complete oxidation of glucose is accompanied by a decline in free energy of about 680 kcal/mole. The formation of a mole of ATP from a mole of ADP is accompanied by an increase in free energy of about 7.3 kcal. Because 36 moles of ATP are formed for each mole of glucose oxidized, some 263 kcal of the energy released from glucose is harnessed in bonds of ATP. The efficiency of energy transfer is thus 263/680, or about 40%. The remaining 60% of the energy released from glucose is lost as heat and is unavailable for the performance of biological work.

The problem of oxygen deficiency and anaerobic pathways in vertebrate skeletal muscle

The preceding review of aerobic catabolism sets the stage for an understanding of the problems presented by oxygen deficiency and some of the adaptive solutions that are possible. It is appropriate first to consider the generation of ATP by anaerobic routes in vertebrate skeletal muscle inasmuch as this is one of the most thoroughly understood systems. For simplicity we shall begin with discussion of a muscle that is entirely de-

ANAEROBIC GLYCOLYSIS

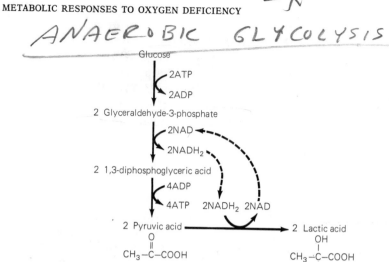

Figure 11–3. Principal features of the anaerobic glycolysis of glucose in vertebrate skeletal muscle. See Figure 11–1A for intermediate reactions of the glycolytic chain.

prived of oxygen, though this is in fact a rather unrealistic condition for most vertebrates.

As noted earlier, oxygen is necessary for the functioning of the electron transport chain. Oxygen deprivation thus eliminates the production of ATP by respiratory-chain phosphorylation, a consideration that in itself denies 32 of the 36 ATP's available in the aerobic catabolism of glucose. A second important implication of oxygen deprivation is that the electron transport chain can no longer act as a route for the oxidation of the reduced cofactors $NADH_2$ and $FADH_2$. As stressed earlier, NAD and FAD are essential electron acceptors in a number of reactions of glycolysis and the Krebs cycle, but because of the limited quantities available, they must be able to pass their electrons on to other compounds. Though the cell without oxygen is deprived of respiratory-chain phosphorylations to produce ATP, it can potentially make use of the substrate-level phosphorylations—but only if an alternative mechanism is available to reoxidize the reduced cofactors.

In vertebrate skeletal muscle the substrate-level phosphorylations of glycolysis are made possible in the absence of oxygen through the use of pyruvic acid as an electron acceptor for reoxidation of the $NADH_2$ produced in the oxidation of glyceraldehyde-3-phosphate. Pyruvic acid is reduced to lactic acid in this reaction, as illustrated in Figure 11–3, and the entire sequence from glucose to lactic acid is known as *anaerobic glycolysis*. Because one molecule of pyruvic acid is produced for each $NADH_2$ generated in the oxidation of glyceraldehyde-3-phosphate, the supply of pyruvic acid keeps pace with the need for it as an electron acceptor. Accordingly, the pathway from glucose to lactic acid is said to be in *oxidation-reduction (redox) balance*; all NAD that is reduced can be reoxidized, the supply of NAD is not depleted, and in principle the conversion of glucose to lactic acid could proceed indefinitely. Through anaerobic glycolysis, the muscle cell is able to utilize the substrate-level phosphorylations of the glycolytic pathway to transfer energy from the bonds of glucose to ATP.

As seen in Figure 11–1B, another substrate-level phosphorylation occurs in the Krebs cycle in the reaction of succinyl coenzyme A to form succinate. For pyruvic acid to be channeled to this reaction in the absence of oxygen, however, additional mechanisms for the anaerobic reoxidation of $NADH_2$ would have to be available inasmuch as NAD is utilized as an electron acceptor in a number of the reactions leading to succinyl coenzyme A. As will be discussed later, some animals have apparently developed means of maintaining redox balance while utilizing the succinyl coenzyme A reaction. This is not generally possible in vertebrate skeletal muscle, however, and the anaerobic catabolism of glucose accordingly terminates with the conversion of pyruvic acid to lactic acid.

The energetics of anaerobic glycolysis

What is the yield of ATP from anaerobic glycolysis? The conversion of a molecule of glucose to lactic acid involves the utilization of two ATP's and the formation of four ATP's—the net yield being two ATP's. This is vastly less than the yield from aerobic catabolism of glucose, but as we shall see, it can be a vitally important source of ATP under certain circumstances in vertebrate muscle. You will recall that the release of free energy in the aerobic catabolism of glucose is 680 kcal/mole (standard free energy at 37°C and pH 7.0). Because the free energy harnessed in 2 moles of ATP is 14.6 kcal, only about 2% of the energy obtainable from glucose is made available for biological work through anaerobic glycolysis. The true efficiency of the process is considerably higher, for all the energy available from glucose is not released in the conversion to lactic acid. Instead, much of the energy remains in the bonds of the lactic acid formed. The standard decline in free energy in the conversion of glucose to lactate is 47 kcal/mole, and of this 14.6/47, or 31%, is harnessed in ATP—an efficiency similar to that of aerobic catabolism. When 1 mole of glucose is converted to 2 moles of lactate, it can be seen that about 630 kcal of the original 680 kcal available remains in the lactate. This free energy is released if lactate is oxidized to carbon dioxide and water and then can be partially harnessed in ATP, but the oxidation of lactate must proceed aerobically and is therefore dependent on restoration of adequate oxygen supplies.

Glycogen as well as glucose can act as an initial substrate for glycolysis, and it is important here to recognize a distinction between the energetics of glycogen catabolism and that of glucose catabolism. Glycogen is, of course, a polymerized form of glucose, and during glycogen catabolism individual glucose residues are removed from the glycogen molecule as glucose-1-phosphate by a reaction that does not require ATP. Glucose-1-phosphate is then transformed to glucose-6-phosphate, which in turn enters the glycolytic pathway (see Figure 11–1A). When free glucose is the initial substrate for glycolysis, an ATP molecule is required to produce each molecule of glucose-6-phosphate, but no ATP is required to generate glucose-6-phosphate when glycogen is the substrate. Thus with glycogen the net yield of ATP in anaerobic glycolysis is three molecules per unit of glucose rather than two, a 50% increase. Although it may appear that the cell is getting something for nothing in utilizing glycogen rather than glucose, this is hardly the case. In structuring the glycogen molecule, not one but two ATP's are required for each glucose molecule added. Thus, in fact, it is energetically more costly to convert glucose to glycogen and then use the glycogen as a substrate for glycolysis rather than simply using glucose directly. However, glycogen can be synthesized from glucose at times when oxygen is available and demands for ATP do not tax the ability of the cell to produce it. Then, when oxygen supplies become inadequate and there is a premium on ATP production, the cell can utilize the glycogen and realize a significant increase in the ATP yield permitted by anaerobic glycolysis.

The roles played by anaerobic catabolism in vertebrates

Sometimes the muscles and other tissues of vertebrates are literally deprived of oxygen by interruption of their circulatory oxygen supply. This can occur when ambient oxygen tensions are low (see later parts of this chapter) and is observed during diving in a variety of air-breathing forms (see Chapter 13). Anaerobic catabolism then becomes the principal or only route available for generation of ATP. There are other situations in which anaerobic catabolism is of great importance, and two of these deserve note.

When animals commence exercise, the circulatory and respiratory systems require appreciable amounts of time (one to several minutes in man) to attain peak oxygen delivery; the rate of oxygen supply to the muscles increases gradually over this period. The rate of aerobic production of ATP, being limited by oxygen supply, increases more or less in parallel. Because the intensity of activity of the myofibrils is dependent on the rate at which ATP is made available, it follows that the intensity of exercise would also be more or less limited to a gradual increase if aerobic catabolism were the only source of ATP. Anaerobic catabolism, however, being independent of oxygen supplies, is not subject to the limitations of circulatory and respiratory oxygen transport. Significantly, it can be brought into play immediately as activity begins and can nearly instantaneously attain a high rate of ATP production. This is often a vital factor in allowing vertebrates to reach and sustain intense levels of exercise within a matter of seconds. When man engages in abrupt, vigorous exercise, such as running a sprint, anaerobic catabolism assumes a pivotal role in supplying ATP.

Another significant attribute of having an anaerobic capability is that even when the circulatory and respiratory systems have attained high rates of oxygen delivery and the rate of aerobic ATP production is accordingly high, anaerobic catabolism, occurring in tandem with aerobic catabolism, can provide for a still greater rate of ATP production. Combined production of ATP by aerobic and anaerobic routes allows the animal to attain a higher level of performance than would be permitted by aerobic catabolism alone.

Anaerobic catabolism also has important limitations, and the interplay between aerobic and anaerobic production of ATP varies in significant ways with the type of exercise. These factors will be discussed in subsequent sections.

Some features of intense exercise in man

It will now be useful to consider an example of exercise in which anaerobic catabolism is brought into play. Taking the case of a man running a half-mile race, we may first note a number of more or less obvious physiological responses. During the race the man's circulatory and respiratory systems greatly increase the rate of oxygen delivery to the muscles, and his rate of oxygen consumption rises well above the resting level. If we take a blood sample at the end of the race, we will find that the lactate concentration is substantially elevated. It is a matter of common observation that the man will breathe hard for a period after the race, and if we measure his oxygen consumption, we will find that it does not fall to the resting level for at least 30 minutes after the race. All these observations reflect important aspects of the energetics of this form of exercise.

It is perhaps first to be stressed that oxygen delivery to the muscles not only continues throughout the exercise but reaches a level far above the resting level. Aerobic catabolism can proceed apace with the oxygen supply and generates much of the ATP required for the contractile process.

The accumulation of lactate indicates that anaerobic catabolism also occurs. This is an example, then, of exercise that is supported both aerobically and anaerobically. The anaerobic production of ATP is important in allowing the racer to attain high speeds abruptly and continues to contribute to the total rate of ATP production even after the circulatory and respiratory systems have increased oxygen delivery sufficiently to support a high rate of aerobic ATP production. In fact, approximately half of the ATP utilized by the muscles during a half-mile race is produced anaerobically. Without the anaerobic production of ATP, the muscles would be limited to a substantially lower average work intensity, and the racer would take far longer to reach the finish line.

The lactic acid formed during exercise is retained in the body and metabolized once the race is completed. In this process lies much of the explanation for the prolonged elevation of oxygen consumption after the race.

Pathways of lactic acid
dissipation and the concept of oxygen debt

Once exercise entailing anaerobic catabolism is completed, there are advantages, as we shall see below, to clearing the body of lactic acid fairly quickly. There are two basic biochemical routes along which lactic acid can be channeled after exercise. First, it can be returned to the stores of carbohydrate (glucose and glycogen) that were drained during exercise. Second, it can be directed into the Krebs cycle and oxidized to carbon dioxide and water. We shall briefly examine these two pathways before considering their quantitative interrelationships.

In converting lactic acid back to glucose or glycogen, glucose-6-phosphate is first resynthesized. The principal intermediates are the same as those of glycolysis (Figure 11–1A), being formed, of course, in reverse order. Most of the reactions of glycolysis are reversible and are catalyzed in reverse direction by the same enzymes that catalyze them in the forward direction. At least two, and perhaps three, of the reactions between glucose-6-phosphate and pyruvic acid are not reversible, however. In glycolysis, for instance, fructose-6-phosphate is phosphorylated to fructose-1,6-diphosphate in a reaction that consumes one ATP and is catalyzed by phosphofructokinase. The reverse reaction, which would involve formation of an ATP, cannot occur because the free energy released in the dephosphorylation of fructose-1,6-diphosphate is less than that required for the formation of a high-energy phosphate bond. Accordingly, in the synthesis of glucose-6-phosphate from lactic acid, the transformation of fructose-1,6-diphosphate to fructose-6-phosphate takes place by simple hydrolysis under catalysis of fructose diphosphatase, a reaction that liberates inorganic phosphate rather than phosphorylating ADP. Because some of the reactions of glycolysis are irreversible, the formation of glucose-6-phosphate from lactic acid requires some enzymes that are not involved in glycolysis and is not the stoichiometric opposite of glycolysis. The latter point is evident in the net changes in ATP in the two processes. Going from glucose-6-phosphate to lactic acid, there is a net yield of three ATP's. It is intuitively clear that the reverse reactions must consume ATP; the consumption, however, is not three ATP's per molecule of glucose-6-phosphate formed, but six ATP's. As will be discussed in more detail later, glucose-6-phosphate formed from lactic acid may be hydrolyzed to glucose or may be converted to glucose-1-phosphate and subsequently incorporated into glycogen.

Some lactic acid, rather than being returned to carbohydrate stores, is oxidized in the Krebs cycle after exercise. This results in release of energy, which can be harnessed in ATP. Lactic acid is first oxidized to pyruvic acid, with NAD serving as the electron acceptor; this reaction is simply the reverse of the reaction involved in formation of lactic acid from pyruvic acid (Figure 11–3) and also occurs in the synthesis of carbohydrate from lactic acid. The pyruvic acid thus formed is then channeled into the Krebs cycle in the usual manner. With oxygen freely available, the $NADH_2$ formed in the initial oxidation of lactic acid and the reduced cofactors produced in the Krebs cycle are oxidized by the electron transport chain. The potential yield of ATP is 34 molecules per pair of lactic acid molecules catabolized.

We now see that one of the paths followed by lactic acid after exercise (conversion to carbohydrate) requires ATP, whereas the other (oxidation) yields ATP. Pioneering experiments on amphibian muscle revealed that these two processes are quantitatively interrelated such that enough lactic acid is oxidized to produce approximately the amount of ATP required to return the remainder of the lactic acid to carbohydrate stores. To elaborate, most of the lactic acid produced during exercise is returned to stores at a cost in ATP. A lesser amount is oxidized to yield ATP. The fact that oxidation of a molecule of lactic acid yields much more ATP than it takes to convert a molecule of lactic acid to carbohydrate means that oxidation of a relatively small portion of the lactic acid present after exercise can produce the amount of ATP required to convert the greater proportion to carbohydrate.

This concept of the interrelationship between the two paths of lactic acid dissipation after exercise, developed initially on amphibian muscle, has been applied generally to other vertebrate muscle as well. Recently, however, experiments on mammals have indicated that as much as 80% of the lactic acid present after exercise may be oxidized; if this oxidation results in the usual ATP production per molecule of lactic acid, far more ATP would be produced than is required to convert the remaining 20% to carbohydrate. At present, a full comparative understanding of quantitative interrelationships in the dissipation of lactic acid after exercise must await further investigation.

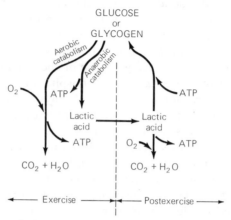

Figure 11–4. Summary of the energetics of aerobic and anaerobic support of exercise in vertebrates.

A significant attribute of the dissipation of lactic acid after exercise is that it requires oxygen. Oxidation of lactic acid requires oxygen directly; and the conversion of lactic acid to carbohydrate requires it indirectly, for the ATP required must be generated aerobically (as by oxidation of other lactic acid molecules). The consequence is that the processing of lactic acid after exercise causes the animal's rate of oxygen consumption to remain above the resting level for a period even though the animal is behaviorally at rest. Essentially, the oxygen demand for the metabolic disposal of lactic acid is superimposed upon the ordinary requirement for physiological maintenance during rest, and oxygen consumption remains above the usual resting level until the lactic acid has been dissipated. We see, then, that once our half-mile racer has produced lactic acid in supporting his activity anaerobically, he is committed to an excess oxygen consumption after exercise. Thus over the period of anaerobic metabolism he is said to *incur an oxygen debt*, and after exercise he is said to *repay the oxygen debt*.

Figure 11–4 is an attempt to summarize the energetics of the aerobic and anaerobic support of exercise. You will note that the *net* effect of the production and dissipation of lactic acid is very simply the aerobic catabolism of a certain amount of glycogen or glucose. This emphasizes that, in the ultimate sense, the entire cost of exercise is supported aerobically, with part of the oxygen being supplied during exercise and part afterward.

The fuels of exercise

The major fuels of short-term exercise are glucose and glycogen, both of which can be catabolized either aerobically or anaerobically. The source of glycogen is that stored in the muscle cells themselves, and experiments on humans and other mammals have demonstrated that there is a steady decline in muscle glycogen over a period of exercise. Glucose is drawn from the blood. The liver is much involved in maintaining blood glucose levels through the process of breaking down liver glycogen to form free glucose. Through this indirect route the exercising muscle gains access to the store of liver glycogen, and a decline in liver glycogen has been measured during exercise.

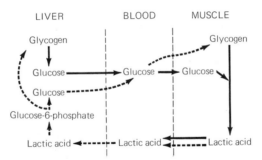

Figure 11–5. Major pathways of interconversion of carbohydrates and lactic acid during and after exercise in mammals. Solid arrows depict major events during exercise. Dashed arrows depict major events after exercise.

The carbon atoms of glucose and glycogen catabolized aerobically are, of course, dissipated as carbon dioxide. As discussed above, some of the carbon atoms appearing anaerobically in lactic acid are returned to carbohydrate stores after exercise. For the most part the conversion of lactic acid to glucose-6-phosphate occurs in the liver. Lactic acid diffuses into the blood from the muscle cells during and after exercise and is taken from the blood by the liver for this important process. Glucose-6-phosphate formed in the liver can be used to synthesize liver glycogen, or, alternatively, it can be hydrolyzed to glucose and enter the bloodstream, thus being made available for reconstitution of muscle glycogen. These events are summarized in Figure 11–5.

Fatty acids are important fuels, particularly in prolonged exercise. They are first broken down to acetyl coenzyme A, and acetyl coenzyme A enters the Krebs cycle (see Figure 11–1*B*). The fatty acids do not pass through the glycolytic chain, and their catabolism is strictly aerobic.

Two additional mechanisms of oxygen debt

Besides anaerobic glycolysis, there are two principal additional mechanisms that provide for the production of ATP during exercise without contemporaneous input of oxygen from the environment: (1) the use of high-energy phosphate bond reservoirs and (2) the use of internal stores of oxygen. These mechanisms, like anaerobic glycolysis, permit a greater intensity of muscular activity than would be possible if ATP production were strictly dependent on transport of oxygen from the environment to the muscle cells by the respiratory and circulatory systems. Similarly, they impose an oxygen debt.

The skeletal muscles of vertebrates and muscles of many invertebrates contain compounds termed *phosphagens* that serve as reservoirs of high-energy phosphate bonds. Creatine phosphate (Figure 11–6) is the phosphagen of vertebrate muscle and also occurs in some groups of invertebrates. The most widespread phosphagen of invertebrates is arginine phosphate (Figure 11–6), and other phosphagens are known, especially among annelids. These compounds contain high-energy phosphate bonds. They are synthesized in reactions with ATP and subsequently can donate their phosphate bonds to ADP to form ATP, as illustrated here for creatine phosphate:

$$\text{creatine phosphate} + \text{ADP} \rightleftharpoons \text{creatine} + \text{ATP}$$

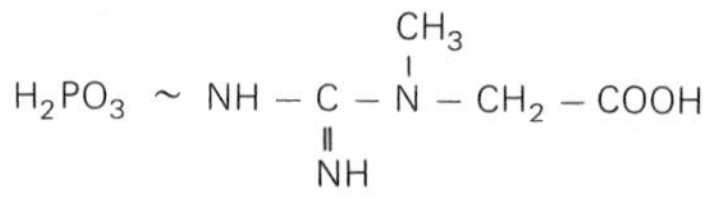

$$H_2PO_3 \sim NH - C - N - CH_2 - COOH$$

with CH_3 above the second carbon (on N) and $\| NH$ below the first carbon.

Creatine
phosphate

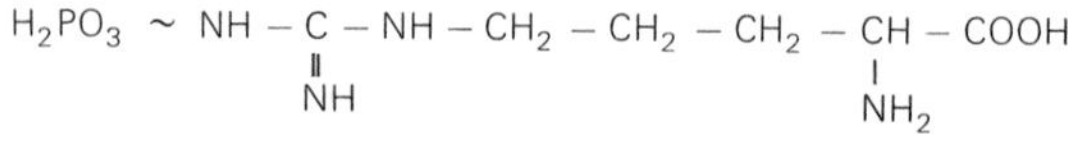

$$H_2PO_3 \sim NH - C - NH - CH_2 - CH_2 - CH_2 - CH - COOH$$

with $\| NH$ below the carbon and NH_2 below the rightmost CH.

Arginine
phosphate

Figure 11–6. Two important phosphagens. High-energy phosphate bonds are indicated by sigmoid bond symbols. Creatine phosphate is also known as phosphocreatine or phosphorylcreatine. Arginine phosphate is also termed phosphoarginine or phosphorylarginine.

When ATP concentrations in the muscle cell are high, formation of creatine phosphate is favored. However, when ATP concentrations fall, there is a strong tendency for the reaction to proceed to the right, with formation of ATP. The formation of creatine phosphate thus provides a mechanism for the temporary storage of phosphate bond energy.

The significance of creatine phosphate to the energetic support of exercise may be outlined by returning to the example of a half-mile racer. Over the period of rest prior to the race, when ATP supplies are relatively untaxed, most of the creatine in the muscle cells comes to be phosphorylated to creatine phosphate. Thus there is an appreciable supply of phosphagen at the start of the race, perhaps four to five times the molar concentration of ATP itself, or, in other words, enough to rephosphorylate each ATP several times over as the ATP is utilized to support work. During exercise, when there is a premium on ATP production, creatine phosphate donates its high-energy phosphate bonds to ADP, effectively generating ATP without the need for contemporaneous oxygen consumption. After the race, creatine phosphate must be resynthesized from creatine, a process that is supported by aerobic production of ATP and that imposes an oxygen debt.

Another mechanism that allows for production of ATP during exercise without the need for contemporaneous input of oxygen *from the environment* is the use of internal oxygen "stores." At any given time the body contains a considerable quantity of molecular oxygen bound to hemoglobin and myoglobin. There is evidence that over a bout of vigorous exercise, myoglobin becomes relatively deoxygenated and remains so until after the activity is completed. Also, as shown by the decrease in venous oxygen tension during exercise, the average oxygen content of the blood falls during activity. Through these mechanisms, oxygen that is present in the body at the start of exercise in the form of oxyhemoglobin and oxymyoglobin is donated to the support of catabolism, permitting aerobic production of ATP to proceed more rapidly than accounted for by the contemporaneous respiratory input of oxygen. The use of internal stores of oxygen contributes to the respiratory oxygen debt, for after exercise the oxygen molecules taken from stores must effectively be replaced from the environ-

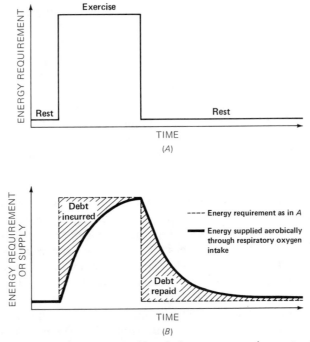

Figure 11-7. (*A*) Schematic representation of the energy requirement, strictly as governed by level of exertion, in a man performing a two- or three-minute bout of steady, vigorous exercise. (*B*) The curve of part *A* is repeated. Also shown, in somewhat stylized form, is the energy supply provided by aerobic catabolism utilizing respired oxygen, a curve that mirrors the rate of oxygen consumption.

ment, bringing the average oxygen content of the blood and the level of oxymyoglobin back to ordinary resting values.

Concluding comments on oxygen debt

It is now appropriate to summarize the concept of oxygen debt and introduce some additional concepts and experimental observations.

In Figure 11-7 we consider a man performing a two- or three-minute bout of steady, vigorous exercise. Part *A* depicts his energy requirement strictly as it is governed by his level of exertion; energy demand rises rapidly as the exercise is started and falls rapidly to the resting level when the exercise is stopped. If the systems responsible for oxygen transport were instantaneously responsive, the man's rate of respiratory oxygen consumption would similarly rise and fall in stepwise fashion. This, of course, is not the case. Part *B* presents a somewhat stylized picture of the energy supplied aerobically through respiratory oxygen intake, mirroring changes in the rate of oxygen consumption. As exercise begins, the systems responsible for oxygen transport respond to increase delivery, but respiratory oxygen supply does not become adequate to support the muscular activity entirely aerobically for a matter of minutes. Once it has become adequate the man could continue the exercise aerobically on a "pay-as-you-go" basis for some time. During the first minutes of exercise, actual metabolic rate exceeds that predicated by oxygen consumption, with the difference being made up through utilization of phosphagen and internal oxygen stores and through anaerobic glycolysis. The magnitude of the *oxygen debt incurred* is defined to be the amount of extra respiratory oxygen that would have had to be consumed over this period to support the exercise entirely aerobically; it is the amount of respiratory oxygen required to produce as much ATP as was actually generated through anaerobic glycolysis and utilization of phosphagen and stored oxygen. Once the exercise is completed, the rate of oxygen consumption remains above that ordinarily associated with rest until lactic acid has been metabolized, phosphagen regenerated, and oxygen stores reconstituted. The magnitude of the *oxygen debt repaid* is the total excess oxygen consumption after exercise, integrated from the cessation of activity to the time that oxygen intake returns to the ordinary resting level.

The debt repaid need not equal the debt incurred. Sometimes they are approximately equal, but often the debt repaid has been found to exceed the debt incurred, even by as much as a factor of 2. The biochemical and physiological parameters that influence the relative magnitudes of the incurred and repaid debts are incompletely understood. Repayment of a debt in excess of that incurred has the effect of increasing the total energetic cost of the exercise.

A given individual in a certain state of training has a maximal rate of oxygen consumption. This, in turn, establishes the maximal intensity of exercise that he can support aerobically on a "pay-as-you-go" basis. Work that does not require utilization of the full aerobic capacity is termed *submaximal*, whereas work that requires a greater rate of energy expenditure than can possibly be supported aerobically is often called by the convenient but somewhat incongruous term *supramaximal*. To illustrate, suppose a given man has a maximal oxygen consumption of 4 liters/min. Work requiring 3 liters/min (such as running at about 7.5 mph) would be strenuous but submaximal and could be supported aerobically for a substantial time. Work requiring 6 liters/min (running at about 12.5 mph) would be supramaximal and could not be performed at all if it had to be supported entirely through respiratory oxygen uptake.

An oxygen debt is incurred over the first minutes of even light submaximal exercise because of the initial lag between oxygen delivery and oxygen demand. Once the delivery systems have responded adequately, the remainder of this type of exercise is supported entirely aerobically; but, interestingly, even though the full capacity for oxygen delivery is not utilized in such exercise and delivery could therefore be increased sufficiently to repay the debt during the period of work itself, much or all of the debt is in fact repaid after exercise has stopped. In light exercise the debt incurred is small, and there is little or no accumulation of lactic acid. The debt has been called an *alactacid debt* and is believed to be attributable to utilization of phosphagen and internal oxygen stores. The alactacid debt is repaid quickly after exercise, being half dissipated in about 30 seconds and fully dissipated within a few minutes in man.

As the intensity of work is increased toward the maximum that can be sustained aerobically, the magnitude of the oxygen debt incurred over the initial lag period increases, and accumulation of lactic acid enters the picture. Thus anaerobic glycolysis is utilized to support the initial stages of vigorous submaximal work, and a *lactacid debt* is superimposed on the alactacid debt. By far the greatest accumulations of lactic acid occur in supramaximal work, work that could not be performed were it not for anaerobic mechanisms of generating ATP. Repayment of a lactacid debt requires far more time than repayment of an alactacid debt. An appreciable accumulation of lactic acid requires about 15 minutes for half dissipation and may require an hour or two for full dissipation in man.

Studies of all vertebrate classes have shown that there are limits to the magnitude of the oxygen debt that can be incurred during exercise. As a corollary, there are also limits to the accumulation of lactic acid (reflecting the lactacid debt). Characteristically, exhaustion sets in when the maximal oxygen debt has been attained. In man there is a clear tendency for the magnitude of the maximum debt to increase with training. Trained athletes can incur debts of 10–15 liters of oxygen, or even more in exceptional individuals; young nonathletes in good physical condition can incur debts of 5–10 liters. The capacity to incur an oxygen debt is not usually a limiting factor for submaximal exercise, for only a relatively modest debt need be incurred over the first minutes of work and thereafter the effort is supported aerobically. Supramaximal exercise, on the other hand, requires, by definition, an ongoing energetic contribution from anaerobic mechanisms and cannot be continued once the limits of oxygen debt have been reached. As will be discussed more in the next section, short- and middle-distance races are supramaximal efforts. The ideal for the racer is to take virtually full advantage of his capacity for anaerobic energy production but not to approach his limits of oxygen debt until the finish line, for an undeniable sense of exhaustion sets in once the limits are reached. The final "kick" in a mile race draws on the racer's remaining capacity for oxygen debt, and the racer hopes that, through good training, he will know how to pace himself so that the limits of debt are approached at the finish line and not in the middle of the home stretch.

The reasons for the limitation on oxygen debt and the exhaustion encountered at maximal debt are not fully clear, though the effects of accumulated lactic acid are frequently implicated. Lactic acid tends to acidify the body fluids; arterial pH's as low as 7.0 have been reported in humans following heavy, exhausting work. Changes in pH in the cells could adversely affect many enzymatic reactions. The buffering of lactic acid by bicarbonate can potentially affect the carbon dioxide equilibrium curve sufficiently to create problems in carbon dioxide transport (Chapter 9). Also, a decline in the pH of the blood will alter the oxygen equilibrium curve. In man, the Bohr shift to lower affinity is believed to aid oxygen transport because, within limits, unloading at the tissues is promoted while loading at the lungs is only slightly reduced. In some other animals, however, oxygen transport may be impaired; reduced affinity of hemoglobin may seriously lower the degree of loading at the lungs or gills, and a Root effect, if present, will reduce carrying capacity (see Chapter 9). Although these and other potential effects of lactic acid are readily enumerated, it is more difficult to determine if, in fact, accumulation of lactic acid is a critical limiting factor for performance and development of oxygen debt. At one time lactic acid was widely viewed as a most critical fatigue factor in humans; today some investigators hold to this hypothesis, but others dispute it. It is postulated by some authorities that training improves tolerance of acidosis and that this is the underlying basis for the improvement in maximal oxygen debt.

Because there are absolute limits to the accumulation of lactic acid and the magnitude of the lactacid debt, an animal that has sustained a debt in one bout of exercise displays a reduced capacity to incur a debt in subsequent exercise until the first debt has been repaid. This is made clear in studies of racers who enter several events in one track meet. Measures on one champion racer revealed that he required 40 to 60 minutes to dissipate his lactacid debt after a race. He ran the mile, 880, and 440 in one meet; the events were spaced adequately for him to repay his debt after each race, and he won all three races. Another racer required about the same time to repay his debt, and when required to run the low hurdles only 20 minutes after the 220, he started the hurdles with a high blood lactate concentration and turned in a subpar performance. It is a matter of common observation that if we perform an exhausting bout of exercise, we are not ready to do the same exercise again for a substantial time. This does not hold true for light or moderate submaximal exercise; such exercise does not demand a large debt, and the debt, being substantially alactic, is largely repaid within a few minutes.

Finally, we may note here that although terrestrial vertebrates retain virtually all the lactic acid produced in exercise, some animals excrete part or all of the lactic acid or other products of anaerobic metabolism. When these compounds are excreted, they need not be metabolized aerobically after exercise, and the need for excess oxygen consumption is reduced accordingly. Such animals are said not to repay part of the oxygen debt incurred. In excreting their anaerobic end products, they lose metabolites of high energy value. More will be said of this later.

Aerobic and anaerobic components in vertebrate exercise

Anaerobic catabolism and other mechanisms of oxygen debt assume different importance in the support of exercise in various species of vertebrates and in various types of exercise within a species. The latter point is nicely illustrated by information on man. In submaximal exercise (e.g., walking) that lasts an appreciable length of time, aerobic energy supply is overwhelmingly dominant, with the anaerobic mechanisms contributing just to the first minutes. At the other extreme are sprints such as the 100-yd dash. Running at 10 yd/sec would require an oxygen consumption of over 20 liters/min to be supported aerobically. The highest oxygen consumption ever reported for a man is only 5.9 liters/min, and, furthermore, several minutes are required to reach maximal oxygen delivery. At least 90% of the cost of the 100-yd dash is met anaerobically and paid for after the race.

In middle-distance races, such as the mile, the aerobic and anaerobic contributions are about equal. Suppose, to illustrate, that a miler can average an oxygen consumption of 4.0 liters/min, taking into account the initial lag in oxygen delivery. This would permit him to run at about 14.5 ft/sec and complete the race in about six minutes on a strictly aerobic basis. If, though, he can also contract an oxygen debt of 17 liters at a rate of 4.3 liters/ min, then his total energy supply—aerobic plus anaerobic—will amount to

8.3 liters/min. This will permit a pace of about 22 ft/sec and will allow him to complete the race in four minutes—sustaining half the cost anaerobically during the race and paying for it aerobically afterward.

In long-distance races the capacity to develop oxygen debt may be fully utilized, but the proportionate contribution of anaerobic metabolism to the support of exercise falls off with the length of the race simply because the total cost increases while the energy available anaerobically remains fixed. Running the marathon might require energy equivalent to 650 liters of oxygen; only about 2% of this can be met anaerobically even if the racer fully exploits his capacity for oxygen debt. The marathon must be run on a "pay-as-you-go" basis, and the pace, being limited by capacity for oxygen consumption, is considerably slower than in the mile.

Other species of vertebrates also show characteristic types of inter-play between the aerobic and anaerobic support of exercise. Here we may briefly note some recent and interesting results gathered by Bennett, Licht, Dawson, and Moberly on amphibians and reptiles stimulated to maximal activity by manual or electrical irritation. When toads, *Bufo boreas*, were stimulated, they responded by walking at a moderate pace and continued the activity without overt fatigue for 10 minutes. By contrast, tree frogs, *Hyla regilla*, initially reacted much more vigorously, jumping about rapidly; but they showed fatigue within a few minutes and, by the end of 10 min-utes, were capable of only slow and uncoordinated jumping. Measures re-vealed that the toads attained a far higher rate of oxygen delivery than the tree frogs, but, on the other hand, the tree frogs accumulated a far greater lactacid debt than the toads. Over the first two minutes of activity, when most of the oxygen debt was incurred, over 70% of ATP production in the toads was *aerobic*, whereas over 80% in the tree frogs was *anaerobic*. It can be said, then, that the toads, with a modest anaerobic capacity but relatively large aerobic capacity, depended mostly on aerobic catabolism for their exercise. Initially they performed less vigorously than the tree frogs, but they did not fatigue. The tree frogs, with only a modest aerobic capacity, utilized their relatively great anaerobic capability to support very vigorous exercise at first, but once the limits of oxygen debt were ap-proached, fatigue set in, and their endurance was markedly lower than that of the toads. Experiments on tree frogs exercised for just two minutes revealed that they did not reach peak oxygen consumption until 5 to 10 minutes after the exercise (that is, during the period of oxygen debt repay-ment); the response of their oxygen delivery systems is thus quite slow. An even more extreme example of dependence on anaerobiosis was found in the lungless salamander *Batrachoseps attenuatus*. This amphibian first responded with very rapid jumping but showed fatigue in just 1 to 2 min-utes and was capable of only a slow walk after 10 minutes. Experiments involving just two minutes of stimulation showed that over 95% of ATP pro-duction during this initial period of vigorous activity was anaerobic. Again, oxygen consumption did not reach a peak until 5 to 10 minutes after exercise had stopped, and the maximal oxygen consumption was only half of that in *Hyla*. The salamander exhibited the highest capacity to accumulate a lactacid debt of all three species. Taken together, these results indicate an inverse correlation between aerobic and anaerobic competence in these am-phibians. The toad had the lowest anaerobic competence but the highest aerobic; the salamander had the highest anaerobic and lowest aerobic; and the tree frog was intermediate in both respects. The results also suggest a property of amphibians that has been confirmed in studies of several other species, namely, that rapid activity is dependent on a strong anaerobic contribution. Species, such as the toad, which do not incur large lactacid debts respond relatively sluggishly to stimulation. The common leopard frog, *Rana pipiens*, responds with great vigor, incurs a large debt, and becomes totally immobile within minutes.

Most lizards that have received study also rely heavily on anaerobic catabolism to support vigorous activity. All of our North American lizards investigated so far have exhibited a relatively high capacity to incur lactacid debt—comparable to that observed in the more anaerobically competent of amphibians. The interspecific similarity in anaerobic capacity among lizards contrasts sharply with the considerable range of capacities seen in amphibians. As in amphibians that rely strongly on anaerobic catabolism for vigorous exertion, the lizards fatigue rapidly, being capable of perhaps two minutes of peak running and generally being exhausted within about five minutes.

In lizards and many amphibians—as, for example, in man and trout—the ability to produce ATP rapidly by anaerobic mechanisms allows the animal to engage abruptly in rapid locomotion, without having to wait for oxygen delivery to respond. This can be of considerable survival value during pursuit of prey or escape from predators. The fatigue imposed by the development of a large oxygen debt, however, has its implications for survival also. An animal, for instance, that calls heavily on anaerobic catabolism during escape from a predator must find suitable cover before exhaustion sets in and, once exhausted, is committed to a period of debt repayment before it can again perform at the same level.

A recent comparative study of two large lizards, the spiny chuckwalla (*Sauromalus hispidus*) and a varanid (*Varanus gouldii*), has provided some significant insights. The chuckwalla is a North American iguanid. It is herbivorous and ordinarily slow-moving, though it can engage briefly in rapid running. Its lungs, like those of most lizards, are relatively simple. The varanid is an Australian monitor lizard. The monitors are exceptional in their life history and general level of activity. They are carnivores, preying principally on rodents and other lizards. They forage over wide areas, are active and wary, and flee rapidly from danger. Their lungs are the most complex found among reptiles.

The chuckwallas and varanids were stimulated to vigorous activity for seven minutes. The chuckwallas, like most lizards, showed a tendency to exhaust over this period of time; often they ceased to respond to stimulation before the end of the experiment. The varanids, on the other hand, did not exhaust, and more-extended tests revealed that they could continue intense struggling for over an hour. The following physiological differences were observed during seven-minute periods of exercise at 35°–40°C, the range of body temperatures preferred by both species. Although the standard rate of oxygen consumption was similar in both lizards, the peak oxygen consumption during exertion was about twice as high in the varanids as in the chuckwallas; the active rate of oxygen consumption in varanids was about 10 times the standard rate, whereas that in chuckwallas was only about 5 times the standard rate. The chuckwallas incurred a considerably (22–50%) higher total oxygen debt and a higher lactacid debt than the varanids and, at 40°C, required more than three times as long to repay their debt after exercise. The varanids exhibited little or no change in blood pH during exercise, but the chuckwallas suffered a large decline in pH (from 7.37 to 7.06). This acidosis in the chuckwallas caused great decreases in both hemoglobin oxygen affinity and oxygen-carrying capacity; at 40°C, P_{50} was increased from 62 mm Hg to 87 mm Hg, and carrying capacity was decreased by 17%. These effects of acidosis in the chuckwallas undoubtedly interfere greatly with oxygen loading in the lungs during intense activity. The fact that the varanids suffered no acidosis whereas the chuckwallas displayed a great decrease in pH is explicable on three grounds. First, the varanids have a far superior blood buffer capacity; second, they produce less lactic acid; and, third, because they are able to take in oxygen at a far greater rate, they probably also are more effective in dissipating carbon dioxide and thus limiting the accumulation of carbonic acid.

From an energetic point of view, the chuckwallas resemble most lizards that have been studied. Intense activity is strongly dependent on anaerobic mechanisms and, accordingly, cannot be sustained very long. This dependency is at least partially attributable to a relatively limited ability to increase the rate of oxygen delivery and aerobic catabolism; in turn, the acidosis caused by accumulation of lactic acid further impairs oxygen delivery. On the other hand, the varanids have an exceptional ability to support exertion aerobically and do not experience rapid fatigue. Their especially well-developed lungs probably play a pivotal role in providing a relatively high aerobic capacity and in permitting comparatively rapid repayment of oxygen debt. Also, their high blood buffer capacity helps to prevent acidosis and its adverse consequences. The physiological differences between chuckwallas and varanids correlate well with attributes of their life history. The varanid can lead its active life without having to go through a continuous cycle of fatigue and recovery. The chuckwalla, on the other hand, being herbivorous and not greatly besieged with enemies, can ordinarily move about at the modest rate permitted by its more limited aerobic capacity and can call on anaerobic mechanisms for brief periods of vigorous exertion when necessary.

Some additional information on the support of exercise by anaerobic mechanisms will be discussed in Chapter 14. Also, air-breathing vertebrates that utilize anaerobic catabolism during diving will be discussed in Chapter 13.

Comparative tolerance to anoxia in animals

Many animals are potentially exposed to anoxic conditions in nature, and there is a great deal of information from the laboratory on length of survival without oxygen. Prolonged survival is generally indicative of substantial ability to support metabolism by anaerobic pathways, and for many animals our only knowledge of anaerobic capabilities comes from study of anoxic survival.

When placed in an anaerobic environment, some animals die within minutes; others survive for many hours, days, or weeks; and still others survive indefinitely. The diversity of responses is so great as to defy simple classification, but the following broad types of response can be recognized. First, many animals clearly have an essential, steady demand for oxygen. Under anoxia they die within minutes or, at most, a few hours. These are classed as *obligate aerobes*. Second, many animals survive anoxia for a substantial period—from a day to many weeks, depending on species— yet do not survive indefinitely. These are among the animals classed as *facultative anaerobes*. Third, a few multicellular animals and a good number of Protozoa can survive anaerobically indefinitely. Some live equally well with or without oxygen and are also classed as *facultative anaerobes*. Others are debilitated or killed in the presence of oxygen and are *obligate anaerobes*.

Most of the vertebrates die quickly without oxygen, for reasons that are reasonably well known. First, certain of the vital tissues of most vertebrates—notably the heart and central nervous system—have an absolute dependence on oxygen; they are incapable of functioning anaerobically. In an anaerobic environment, these tissues fail as soon as their metabolic needs can no longer be met using internal stores of oxygen. A man denied oxygen dies within minutes as the oxygen available in his blood is depleted. Some diving mammals survive submergence for over an hour by virtue of having relatively large internal stores of oxygen and by having mechanisms that direct some of these stores preferentially to the oxygen-dependent tissues (see Chapter 13). Another factor of significance in most vertebrates is that the anaerobic capacity is limited in those tissues, such as skeletal muscle, that can function anaerobically. A man, lizard, or trout can accumulate only so much oxygen debt before anaerobic metabolism comes to a halt. Again, this places distinct limits on anaerobic survival. Many invertebrates, including representatives of probably every phylum, are also obligate aerobes. Some of the groups whose members are predominantly or entirely in this category are the cephalopod molluscs, adult insects, and decapod crustaceans. Some of these animals have been shown to develop an oxygen debt during exercise or during brief exposure to anoxic conditions, indicating that they, like the vertebrates, have a short-term anaerobic capability.

At the other extreme are those species that can survive anoxia indefinitely. In these, anaerobic routes of catabolism must be capable of meeting the needs of all tissues. The best-known examples among multicellular animals are some of the helminth parasites of the mammalian intestine. Worms such as the nematode *Ascaris lumbricoides*, which occupies the small intestine of man and pigs, and the rat tapeworm *Hymenolepis diminuta* are known to prosper under anaerobic conditions and are thought to spend their entire adult lives under virtual anoxia inasmuch as oxygen tensions in their intestinal habitat are very low. Both of these species should probably be classed as obligate anaerobes inasmuch as they are adversely affected by culture under aerobic conditions, but some other species appear to be facultative anaerobes. There seem to be no demonstrated examples of free-living multicellular animals that survive anoxia indefinitely, though, as will be discussed later, there are some cases in which this may prove to be possible.

There are many species of invertebrates that survive anoxia for anywhere from a day to many weeks. They are collectively termed facultative anaerobes, but they may or may not form a natural grouping physiologically. Although species capable of relatively long anoxic survival almost certainly can support all their vital functions anaerobically, it is possible that some species that survive for only a day or so have oxygen-dependent tissues that are supported using internal oxygen stores, much in the way that diving mammals support heart and central nervous functions using stores. Unfortunately we know much more about the length of survival in most invertebrates than about the internal events during survival. It is also important to note that most or all of these animals do not survive indefinitely under anoxia; even if all vital functions can be supported anaerobically for a time, long-term survival is dependent on restoration of oxygen supplies. Why this is so is largely unknown.

Species that can survive for at least a day without oxygen uptake are known from many invertebrate groups, including the coelenterates, polychaete and oligochaete annelids, gastropod and bivalve molluscs, crustaceans, echinoderms, and urochordates. Their responses are so diverse as to defy easy summarization, and the emphasis here will be placed on illustrative examples. If sea anemones, *Diadumene*, are allowed to deplete the oxygen in a closed container, they cease oxygen uptake when the ambient oxygen tension has fallen to a certain level and then survive for up to a day with no measurable oxygen consumption. A similar shutdown of oxygen uptake, followed by survival for one to several days, has been observed in invertebrates of at least seven other phyla. It is seen, for example, in the sea cucumber *Thyone briareus*, which is found buried in mud flats at low tide and may well depend on anaerobic metabolism until the tide returns. Many bivalve molluscs can survive anaerobic conditions for several days. The ribbed mussel, *Modiolus demissus*, for example, lives in the mud of salt marshes and often becomes buried; it survives in an atmosphere of nitrogen for a median of five days. Numerous intertidal bivalves close their shells tightly when exposed to the air at low tide; some are known to survive without oxygen for substantial periods, and some have been shown to consume oxygen at an elevated rate after low tide, indicating repayment of an oxygen debt. Some of the most remarkable reports of survival without oxygen come from studies of animals that live in or on the bottom sediments of lakes, ponds, and other relatively still bodies of fresh water; as noted in Chapter 8, severe and prolonged oxygen depletion can develop in such regions. The freshwater clam *Pisidium idahoense*, some tubificid worms (*Tubifex*), and several species of midge larvae (*Chironomus*), for example, have been found to survive anaerobically for over 10 weeks in the laboratory, and field observations have shown that their population numbers are not adversely affected over substantial periods of anoxia in nature. It is conceivable that some such animals are capable of indefinite anaerobic survival, especially at low temperatures. (Survival of facultative anaerobes is often lengthened at low temperatures, probably because of the depression of metabolic rate.)

Some cases of prolonged survival without oxygen are known among vertebrates. European carp can survive under anoxic conditions for two or three months at low temperatures during the winter; they inhabit small ponds, and when ice blocks exchange with the atmosphere, the waters of the ponds often become depleted of oxygen and remain so until the spring thaw. Ecological evidence suggests that benthic fish of Lake Tanganyika are capable of protracted anaerobic existence inasmuch as the lower waters of the lake are virtually anoxic; similar reasoning can be applied to the fish of certain other deep lakes. Recent studies of turtles, to be discussed in detail in Chapter 13, have shown that some species can survive anaerobically for several days at low temperatures and for 20 to 30 hours even at 22°C.

Products of anaerobic metabolism in facultative anaerobes

The products of anaerobiosis in invertebrate facultative anaerobes are diverse. Some produce lactic acid in quantity, and the evidence is that the reactions are similar or identical to those discussed earlier for vertebrate skeletal muscle. Often, however, other compounds are produced either along with or to the exclusion of lactic acid. Succinic acid is a particularly common end product, and other acids, such as acetic, propionic, isobutyric, and isovaleric, are widely reported. The amino acids alanine and proline are also produced in quantity in some forms. The accumulation of compounds other than lactic acid indicates that the routes of anaerobic metabolism in many invertebrate facultative anaerobes are different from those typical of vertebrate skeletal muscle, and there is much current interest in the biochemical pathways. We shall return to this topic shortly.

Diving turtles seem to rely predominantly on anaerobic glycolysis, producing large quantities of lactic acid during prolonged submergence. Although many fish, such as trout and other salmonids, are apparently limited to anaerobic glycolysis (and utilize it on a strictly short-term basis), there is good evidence that other pathways are employed by some of the fish that are capable of prolonged anaerobiosis. Earlier we noted the ability of European carp to survive for months without oxygen at low temperatures. The closely related goldfish has been shown to rely partially on anaerobic metabolism for long periods at low oxygen tensions. In both carp and goldfish it is known that carbon dioxide is produced in anaerobic metabolism. Although the biochemical pathways are unknown, the production of carbon dioxide indicates that mechanisms unlike those of most vertebrates are utilized, for carbon dioxide is not generated in glycolysis (see Figure 11–3).

Very commonly, the invertebrate facultative anaerobes excrete the products of anaerobic metabolism either partly or wholly. After protracted anaerobiosis, some species show no elevation of oxygen consumption on return to aerobic conditions; others repay less of an oxygen debt than would be expected were the anaerobic end products retained entirely within the body. European carp exhibit little or no repayment of oxygen debt after anaerobiosis.

The excretion of anaerobic end products seems profligate energetically; the compounds have a high energy value if retained and catabolized aerobically upon restoration of oxygen supplies. On the other hand, accumulation of all the material produced over a long period of anaerobiosis would probably not be possible because of the physiological effects exerted

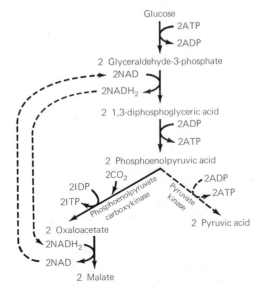

Figure 11-8. The metabolism of glucose to malate during anaerobiosis in bivalve molluscs. Phosphoenolpyruvic acid reacts to form pyruvic acid under aerobic conditions but is carboxylated to form oxaloacetate under anaerobic conditions. IDP = inosine diphosphate.

by the compounds (such as the effects resulting from their acid nature). Probably prolonged anaerobiosis is dependent on being able to rid the body of at least some of the anaerobic end products.

In light of the fact that so many bivalve molluscs are facultative anaerobes, it is worth noting the interesting observation that, at least in some species, the carbonates of the shell are available to buffer acid end products of anaerobiosis (forming HCO_3^-), thus augmenting buffer capacity above that provided by buffers of the tissues and body fluids. The process can result in a detectable erosion of the shell if anaerobiosis is prolonged.

Anaerobic pathways in some of the invertebrate facultative anaerobes

We may now turn to a brief examination of the biochemical pathways in invertebrate facultative anaerobes. This is an area of most active current interest, with an especially large amount of information being accumulated on the bivalve molluscs and parasitic helminths. The pathways are incompletely known, but the data available have recently been integrated by Hochachka and his coworkers to provide plausible and most interesting hypotheses.

During anaerobiosis various bivalves accumulate succinate and alanine, either along with or to the virtual exclusion of lactate. Major questions are: What pathways lead to these products? and Do these pathways have advantages over simple anaerobic glycolysis?

Glucose (or glycogen) is catabolized to phosphoenolpyruvic acid by the usual glycolytic reactions (Figure 11-1A). Under aerobic conditions phosphoenolpyruvic acid is converted to pyruvic acid under catalysis of the enzyme pyruvate kinase and then enters the Krebs cycle. As depicted in Figure 11–8, however, phosphoenolpyruvic acid predominantly reacts to form oxaloacetate under anaerobic conditions. This is a carboxylation reaction, catalyzed by phosphoenolpyruvate carboxykinase, and proceeds with a sufficient decline in free energy that a molecule of inosine diphosphate (IDP) is phosphorylated to inosine triphosphate (ITP); alternatively, GDP may be phosphorylated to GTP. Of course, either ITP or GTP can donate its terminal high-energy phosphate bond to ADP, forming ATP. The oxaloacetate formed from phosphoenolpyruvic acid is then reduced to malate in a reaction that utilizes $NADH_2$ as the electron donor. Up to this point the process in bivalves has important similarities to anaerobic glycolysis in vertebrate skeletal muscle. The conversion of glucose to malate has the same net yield of ATP as the conversion of glucose to lactic acid. Furthermore, the formation of malate from oxaloacetate serves to regenerate NAD, as does the formation of lactic acid from pyruvic acid (compare Figure 11–3); thus, as indicated in Figure 11–8, the essential condition of redox balance is maintained. A very important difference between the process in bivalves and anaerobic glycolysis in vertebrates is that the product in bivalves, malate, is in the mainstream of carbon flow and can undergo further conversions. By contrast, the formation of lactic acid is essentially a metabolic dead end. In vertebrates and apparently in animals generally, lactic acid cannot be utilized to any appreciable extent in anabolic or catabolic reactions except by first being reconverted to pyruvic acid, from which it was originally formed. Thus lactic acid must be allowed to accumulate (or be excreted) until such time as restoration of oxygen supplies establishes favorable conditions for the resynthesis and utilization of pyruvic acid.

Before exploring the fate of malate, it is necessary to examine another route by which malate can be formed. Much evidence indicates that at least some bivalves utilize amino acids as substrates for anaerobic catabolism; that is, they simultaneously mobilize both amino acids and carbohydrates (glucose or glycogen). The routes of mobilization of amino acids are not entirely clear, but, on the basis of available evidence, Hochachka and his coworkers have suggested the pathway shown on the right side of Figure 11–9A. The reactions are transaminations, in which an amino acid donates its amino group to a keto acid, thus generating an amino acid from the original keto acid and a keto acid from the original amino acid:

$$R\!-\!\underset{\underset{\text{amino acid}}{|}}{\overset{}{\underset{NH_2}{CH}}}\!-\!COOH + R'\!-\!\underset{\underset{\text{keto acid}}{||}}{\overset{}{\underset{O}{C}}}\!-\!COOH \rightarrow R\!-\!\underset{\underset{\text{keto acid}}{||}}{\overset{}{\underset{O}{C}}}\!-\!COOH + R'\!-\!\underset{\underset{\text{amino acid}}{|}}{\overset{}{\underset{NH_2}{CH}}}\!-\!COOH$$

The amino acid glutamate is hypothesized to react with various keto acids to form its corresponding keto acid, α-ketoglutarate. The amino acid aspartate is then hypothesized to undergo transamination with α-ketoglutarate, generating the corresponding keto acid of aspartate, oxaloacetate, and also generating glutamate. The fate of glutamate is of much importance, but first we must consider that of oxaloacetate.

As we have seen earlier (Figure 11–8), oxaloacetate is formed from

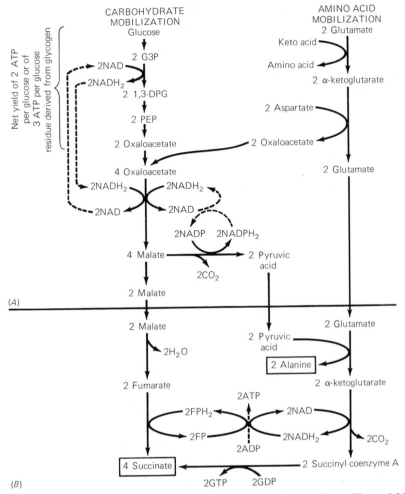

Figure 11-9. Proposed routes of anaerobic catabolism in bivalve molluscs. The stoichiometry is that dictated by the requirement of redox balance. G3P: glyceraldehyde-3-phosphate; 1,3-DPG: 1,3-diphosphoglyceric acid; PEP: phosphoenolpyruvic acid; NADP: the enzyme cofactor, nicotinamide adenine dinucleotide phosphate; FP and FPH$_2$: oxidized and reduced forms of flavoprotein, respectively. The transfer of electrons from NADPH$_2$ to NAD, forming NADP and NADH$_2$, is assumed in part A. The redox couple between the oxidation of α-ketoglutarate and the reduction of fumarate in part B occurs through a modified and truncated version of the electron transport chain, and the formation of ATP is a respiratory-chain phosphorylation; the actual sequence of events in this couple is probably more complex than shown. See text for discussion. (Based on Hochachka, P. W., J. Fields, and T. Mustafa. 1973. Amer. Zool. 13: 543–555.)

glucose as well as from aspartate. Because all the reactions occur within the cell solution, oxaloacetate from both sources enters a common pool, as is emphasized in Figure 11–9A. If we envision the simultaneous mobilization of one glucose, two α-ketoglutarates, and two aspartates, then four oxaloacetates result. These are all reduced to malate; thus malate derives from both carbohydrate and amino acids. In terms of redox balance, two of the NAD's produced in the reduction of oxaloacetate are required to provide reducing potential for the oxidation of glyceraldehyde-3-phosphate in glycolysis, as emphasized earlier. The other two NAD's are believed to provide for a most important reaction of malate, its oxidative decarboxylation to pyruvic acid. According to considerations of redox balance, two of the four malates can follow this path. The subsequent fate of the two pyruvic acid molecules produced and the fate of the remaining two malates are shown in Figure 11–9B.

Pyruvic acid undergoes transamination with glutamate, resulting in α-ketoglutarate and the amino acid alanine. As noted earlier, alanine is an important end product of anaerobic metabolism and accumulates in quantity. The production of α-ketoglutarate sets the stage for the formation of succinate by the usual reactions of the Krebs cycle, reactions that involve substrate-level phosphorylation of GDP. First, α-ketoglutarate is oxidatively decarboxylated to succinyl coenzyme A; NAD is required as an electron acceptor, and we shall return shortly to the problem of regenerating NAD for this reaction. Succinyl coenzyme A then reacts to form succinate, an end product, with formation of GTP.

Returning to the fate of malate, we see that some malate is dehydrated to form fumarate. Fumarate, in turn, is believed to act as the ultimate electron acceptor for electrons removed from α-ketoglutarate. The transfer of electrons to fumarate is believed to occur by a modified and truncated version of the electron transport chain, allowing the animal to capitalize on its capacity for respiratory-chain phosphorylation of ADP. In essence it is hypothesized that $NADH_2$ resulting from the oxidation of α-ketoglutarate passes its electrons to a flavoprotein. This reaction is analogous to or identical to the first step in ordinary electron transport (see Figure 11–2) and proceeds with sufficient decline in free energy to provide for phosphorylation of ADP. Fumarate has a greater affinity for electrons than flavoprotein, and, accordingly, reduced flavoprotein can pass its electrons to fumarate. Fumarate is reduced to succinate, which accumulates as an end product, and we see that redox balance is maintained.

As indicated frequently, the scheme outlined for anaerobic metabolism in bivalves is partly hypothetical. It accounts, however, for the formation of the known end products, succinate and alanine. It is compatible with available information on bivalve enzyme kinetics. And it is in perfect redox balance, meaning that, just as in vertebrate skeletal muscle, the animal could utilize these anaerobic pathways indefinitely without passing into a state of redox imbalance. The energetic advantages of this scheme over anaerobic glycolysis are clear and of great significance. If for each molecule of glucose metabolized, two molecules of aspartate and two of α-ketoglutarate are also mobilized, it can be seen from Figure 11–9 that six ATP's will be produced. This should be compared with the yield of only two ATP's per glucose molecule in anaerobic glycolysis. If glycogen is the initial carbohydrate substrate, anaerobic glycolysis generates three ATP's per glucose residue, whereas the scheme in bivalves generates seven ATP's. The increased production of ATP in bivalves, as provided by their simultaneous mobilization of carbohydrate and amino acids, is probably of much importance to their capacity for facultative anaerobiosis. Over periods of days or even weeks they must draw all the ATP needed for their vital processes from anaerobic catabolism.

Many of the parasitic helminths and at least one free-living oligochaete annelid (the African swamp worm, *Alma*) produce propionate in

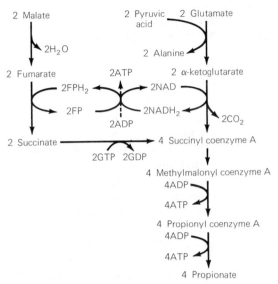

Figure 11-10. Proposed anaerobic pathways in an animal that utilizes the sequences of part *A* in Figure 11-9 but then produces propionate to the exclusion of succinate. When succinate is the end product, as in Figure 11-9, 6 ATP's are formed in the catabolism of one molecule of glucose + 2 aspartates + 2 α-ketoglutarates. When propionate is the end product, as illustrated here, 10 ATP's are formed. Animals producing a mixture of succinate and propionate realize intermediate yields of ATP. ATP is generated by substrate-level phosphorylation in the reaction of methylmalonyl coenzyme A to form propionyl coenzyme A and also in the reaction of propionyl coenzyme A to form propionate. (Based on Hochachka, P. W., J. Fields, and T. Mustafa. 1973. Amer. Zool. 13: 543–555.)

quantity as an end product of anaerobiosis. The reactions leading to propionate are hypothesized to be extensions of reaction sequences like those outlined for bivalves, and significant energetic advantages result. As shown in Figure 11–10, succinyl coenzyme A formed from α-ketoglutarate can react to form methylmalonyl coenzyme A, then propionyl coenzyme A, and finally propionate. This sequence yields two molecules of ATP per molecule of succinyl coenzyme A, in contrast to the yield of just one ATP when succinyl coenzyme A reacts to form succinate. It is also possible for succinate originating from the reduction of fumarate to be converted to propionate. Succinate first reacts to form succinyl coenzyme A, with utilization of one ATP. The subsequent conversion of succinyl coenzyme A to propionate yields two molecules of ATP so that there is a net yield of one ATP per succinate. It is easy to see that an animal producing propionate to the complete exclusion of succinate would gain 10 ATP's upon catabolism of one molecule of glucose plus two aspartates plus two α-ketoglutarates. This appears to occur in some helminth parasites, and the high energy yield is undoubtedly relevant to their ability to live anaerobically for indefinite periods.

Much remains to be learned about metabolism in facultative anaerobes. The pathways presented here may be common, but it is clear that many species utilize other pathways either along with or instead of those discussed. Interest in the subject has been increasing in recent years, and new insights should be expected in the near future.

THE RELATIONSHIP
OF OXYGEN CONSUMPTION
TO AMBIENT OXYGEN TENSION

Basic concepts

In the effort to understand the relationships of animals to environmental oxygen, physiologists have gathered a great deal of information on the interplay between oxygen consumption and ambient oxygen tension. The results are expressed graphically as in Figures 11–11 and 11–12. Commonly, it is found that the response of oxygen consumption to declining oxygen tension can be divided into two phases, as illustrated for the goldfish data in Figure 11–11B. Over a certain range of tensions, oxygen consumption is relatively unaffected by tension; this pertains in the goldfish at tensions of about 34 mm Hg and higher. The animal is said to be an oxygen regulator (or nonconformer) in this range, and the span of tensions involved is termed the *range of oxygen regulation* or *oxygen independence*. Below this range oxygen consumption falls sharply with decreasing tension. Here the animal is said to be an oxygen conformer, and the span of tensions over which this occurs is termed the *range of oxygen conformity* or *oxygen dependence*. The tension at which regulation ceases and conformity begins (34 mm Hg in the case of the goldfish) is termed the *critical tension* or *critical pressure*, abbreviated T_c or P_c. The critical tension is not always so clearly defined as it is in the goldfish data, for sometimes there is a gradual transition between true regulation and strong conformity (see, for example, *Salvelinus* in Figure 11–11). Also, although oxygen consumption sometimes shows little or no systematic change with tension in the range of regulation (see goldfish data), this is not always the case. Sometimes consumption increases with decreasing tension over part of the range; and sometimes consumption falls slightly with tension, though the rate of decline is far less than in the range of conformity.

In nature, the peak oxygen tension encountered in air or water is commonly about 160 mm Hg (see Chapter 8). Thus although responses to higher tensions may be studied in the laboratory, particular interest centers on the range from 0 mm Hg to 160 mm Hg. Animals are often classified holistically as *oxygen conformers* or *oxygen regulators* on the basis of their response within this range of tensions. Oxygen conformers, by this definition, are animals whose oxygen consumption varies directly with tension over the entire range from 0 mm Hg to 160 mm Hg. Examples include *Salvelinus* in Figure 11–11 and *Baetis* and *Ephemera* in Figure 11–12. Conformers generally do display a zone of regulation, but their critical pressure is above 160 mm Hg. Oxygen regulators are animals that have a critical pressure below 160 mm Hg and that therefore display regulation over part of the range from 0 mm Hg to 160 mm Hg. Examples include the goldfish in Figure 11–11 and *Leptophlebia* and *Cloeon* in Figure 11–12.

In nature, the oxygen consumption of conformers is routinely affected by oxygen tension inasmuch as the zone of conformity encompasses the entire usual range of ambient tensions. Regulators, on the other hand, are able to maintain a stable oxygen consumption despite variations in oxygen tension over part of the usual range of ambient tensions.

The holistic classification of animals as oxygen regulators or con-

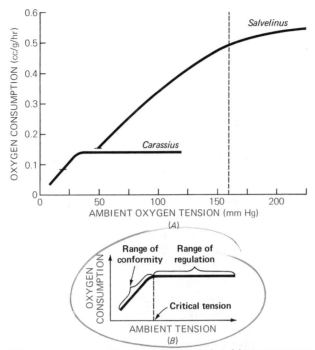

Figure 11–11. (*A*) Oxygen consumption as a function of ambient oxygen tension in young, exercising goldfish (*Carassius auratus*) and speckled trout (*Salvelinus fontinalis*) at 20°C. Fish were compelled to swim near their maximal rate. Horizontal bars toward the left on each curve indicate the standard oxygen consumption of each species in well-aerated water at 20°C. The vertical dashed line marks the oxygen tension of water in equilibrium with dry air at standard pressure. The goldfish averaged 3.8 g in weight; the trout data are for 5-g individuals. The animals were maintained in aerated water except during experimentation. (*B*) Data for goldfish indicating the range of regulation, range of conformity, and critical tension. See text for definitions. [Data from Fry, F. E. J. and J. S. Hart. 1948. Biol. Bull. (Woods Hole) 94: 66–77; Job, S. V. 1955. Univ. Toronto Biol. Ser. No. 61, Publ. Ontario Fish. Res. Lab. No. 73, pp. 1–39.]

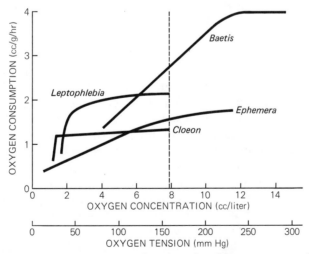

Figure 11–12. Oxygen consumption as a function of ambient oxygen tension at 10°C in the aquatic nymphs of four species of British mayflies: *Baetis* sp., *Cloeon dipterum*, *Ephemera vulgata*, and *Leptophlebia marginata*. The animals were studied in groups and probably were stimulated to at least moderate activity. The vertical dashed line depicts the oxygen content of water in equilibrium with the atmosphere. (From Fox, H. M., C. A. Wingfield, and B. G. Simmonds. 1937. J. Exp. Biol. 14: 210–218.)

formers has only limited utility. In various species, critical pressures ranging from less than 5 mm Hg to over 250 mm Hg have been found, and the responses of animals fall into a rather thorough continuum between these widely separated extremes. One "regulator" may have a critical pressure of 130 mm Hg, whereas another may have a P_c of 10 mm Hg; obviously these responses are similar in only a very qualitative sense.

Some determinants of response

The response of a given species to declining oxygen tension depends on the conditions of measurement. Thus the critical pressure and other parameters of response are not singular species characters; rather, a given species can exhibit a variety of different critical pressures and other response parameters, depending on experimental circumstances.

Some of the important experimental variables can be illustrated with regard to the goldfish. Figure 11–13 presents standard (resting) and active rates of oxygen consumption in 100-g goldfish under several conditions. Consider first the solid "standard" curve at 20°C; this represents the response of fish that were acclimated to high oxygen tensions, being maintained in aerated water except during experimentation. As oxygen tension was lowered from 160 mm Hg, the oxygen consumption of these fish showed no systematic change until the tension reached about 90 mm Hg. Oxygen consumption then rose as the tension was dropped from 90 mm Hg to 65 mm Hg; as will be discussed subsequently, this is believed to reflect the cost of increased ventilation needed to obtain sufficient oxygen at reduced tensions. Oxygen consumption fell as the tension was lowered from 65 mm Hg. With a curve of this shape it is not a simple matter to decide what tension should be considered the critical tension. One procedure is to recognize the stable rate of oxygen consumption between 90 mm Hg and 160 mm Hg as the unstressed resting rate and define the critical tension to be that tension at which oxygen consumption falls below this level. With this approach the critical tension is about 27 mm Hg. (There is nothing magic about this number. More than anything the example shows that animals do not always adhere to our simple conceptual schemes.) The dashed standard curve in Figure 11–13A refers to fish that were acclimated at each oxygen tension prior to being tested; these fish, unlike those held in aerated water, thus had opportunity to adapt to lowered tensions before their response was determined. The fish acclimated to the test tensions showed a similar response to those held in aerated water at high tensions; the curves for the two groups are simply superimposed between 90 mm Hg and 160 mm Hg. However, the fish acclimated to test tensions showed substantially lower rates of oxygen consumption between 25 mm Hg and 90 mm Hg. This interesting result will be examined further below. Here it is sufficient to note that the tension of acclimation can influence the response.

In studies of standard oxygen consumption at 10°C (Figure 11–13B), fish acclimated to aerated water and fish acclimated to test tensions gave similar responses over the entire range of tensions. The increase in oxygen consumption at low tensions was barely discernible, in contrast to the results at 20°C. The critical tension was a few mm Hg lower than at 20°C. In all these respects we see that temperature can influence the response.

Active fish at both 10°C and 20°C displayed oxygen conformity over

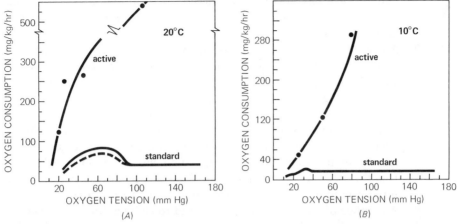

Figure 11–13. Standard and active rates of oxygen consumption of 100-g goldfish (*Carassius auratus*) as functions of oxygen tension at two temperatures. Standard oxygen consumption was determined for two groups of fish. One group was kept in well-aerated water except during tests (solid curve in part *A*), and the other was acclimated to the tensions at which tests were performed (dashed curve in part *A*). The responses of the two groups were the same above a tension of 90 mm Hg at 20°C (part *A*) and were identical over the entire range of tensions at 10°C (part *B*). Active oxygen consumption was measured only on fish acclimated to well-aerated water. Standard rates were obtained by monitoring both oxygen consumption and spontaneous activity in quiet water and extrapolating from these data to obtain oxygen consumption at zero activity. Active rates were measured on fish that were compelled to swim in a circular chamber. The experimental animals weighed from 17 g to 290 g, but all results were adjusted to refer to 100-g fish. (From Beamish, F. W. H. 1964. Can. J. Zool. 42: 355–366. Reproduced by permission of the National Research Council of Canada from the Canadian Journal of Zoology.)

the entire range of tensions studied. The critical pressures for the active fish were not determined but clearly were well above those for resting fish. These results show that state of activity can strongly influence the response.

It is of interest to compare the results for active fish in Figure 11–13*A* with those shown for goldfish in Figure 11–11. In both experiments the fish were at 20°C and active; the critical tension, however, was about 34 mm Hg for the fish in Figure 11–11 and was greater than 100 mm Hg for the fish in Figure 11–13*A*. There were several differences in experimental design in the two studies, and it is not possible to say which differences were most important in yielding the different results. One factor that could be significant is the size of the fish. Those studied in Figure 11–11 averaged 3.8 g in weight, whereas those studied in Figure 11–13*A* ranged from 85 g to 193 g, and results were adjusted to apply to 100-g individuals. It has sometimes been found that small individuals exhibit lower critical tensions than large individuals.

Some of the effects illustrated in the goldfish deserve some further elaboration.

1. A decrease in P_c with decrease in test temperature has been reported in a number of studies on poikilotherms. In the investigation of active goldfish from which Figure 11–11 was derived, P_c was 40 mm Hg in fish acclimated and tested at 35°C, 34 mm Hg at 20°C, and 15 mm Hg at 5°C. An appreciable reduction in P_c with

temperature has also been observed, for example, in trout (*Salvelinus fontinalis*), bluegill sunfish (*Lepomis*), and crayfish (*Orconectes immunis*).

2. A decrease in P_c with decrease in the size of the animal has sometimes been observed, sometimes not. In crayfish (*Orconectes*), P_c was 65 mm Hg in 17-g individuals but 32 mm Hg in 4-g individuals. In exercising trout (*S. fontinalis*), however, no systematic difference was observed in animals ranging from 5 g to 1000 g in weight.

3. P_c is typically elevated by activity; that is, active oxygen consumption becomes dependent on oxygen tension at higher tensions than resting metabolism. This effect, which has been observed many times, can be large. In trout, for example, P_c during active swimming was found to be 70 mm Hg higher than that during rest.

4. A number of animals have been found to show a different response when acclimated to lowered oxygen tensions than when acclimated to high tensions. Among fish this has been shown, for example, in trout, carp, and eel, in addition to goldfish. Chironomid larvae and a leech (*Erpobdella*) exhibit lower P_c's (increased ranges of regulation) after acclimation to reduced tensions.

The first and third of the effects just enumerated may, at least in part, have a common mechanistic basis. Activity raises oxygen demand, and among poikilotherms, increased temperature typically has a similar effect. The animal that has a heightened oxygen requirement presumably faces an increased circulatory and respiratory challenge to sustain its oxygen consumption when oxygen tensions are lowered. It is logical that such an animal should display a smaller range of regulatory compensation than one with relatively low oxygen requirements. An additional consideration at high temperatures is that water of a given tension has a lower oxygen concentration than at low temperatures (Chapter 8). This would be expected to increase the difficulty of maintaining oxygen supplies at a given tension and, again, would tend to diminish the range of compensation as tension is lowered.

Comparative considerations

Because the response of a particular species to changes in oxygen tension is dependent on a number of factors, caution must be exercised in making interspecific comparisons. Specifically, it is important that species be studied under similar conditions if their responses are to be compared in a rigorous manner. A great many species have been investigated, but relatively few synthetic principles have emerged. In part the fact that different species have often been studied under different conditions may have tended to obscure orderly relationships. This cannot be the whole story, however, for there are many cases in which experimental conditions have been kept fairly uniform, yet the results defy systematic comprehension. With these points in mind we may briefly review the responses of animals in a comparative frame of reference.

When at rest or moderately active, most or all of the terrestrial insects and vertebrates display regulation over part of the usual range of ambient tensions; that is, they are classed as regulators. A majority of fish are regulators when near rest, but many become conformers when exercising vigorously. Some fish are conformers even when relatively inactive; this is more likely to be the case at high than at low temperatures. Active species of fish tend to have higher critical pressures than sluggish species; active fish have relatively higher rates of oxygen consumption, and, as discussed earlier, this would be expected to limit the range of regulation. Note in Figure 11–11 that swimming trout consume more oxygen at high tensions than swimming goldfish; the trout also display a higher P_c. Most of the aquatic invertebrate groups exhibit a diversity of responses to oxygen tension. Some coelenterates, annelids, molluscs, crustaceans, aquatic insects, and echinoderms are regulators; others are conformers. Conformity appears to be particularly common in large, sluggish species.

Some studies of related species have indicated a tendency for those from oxygen-poor habitats to show a greater range of regulation than those from oxygen-rich habitats; that is, those from the oxygen-rich habitats are the more likely to show oxygen dependence at relatively high tensions. In the mayfly nymphs represented in Figure 11–12, *Baetis* was collected from swift streams, *Leptophlebia* from a lake, *Ephemera* from pond mud, and *Cloeon* from a pond that was known to undergo oxygen depletion at night. *Baetis*, from an oxygen-rich environment, shows a high critical pressure and is a conformer. *Cloeon*, from an oxygen-poor environment, is a strong regulator, and the lake-dwelling *Leptophlebia* is intermediate in its response. (*Ephemera* will be discussed later.) In a study of midge larvae, *Calopsectra* from a stream were found to be conformers, whereas *Chironomus* from a ditch were strong regulators. Among crustaceans, lobsters (*Homarus*) and blue crabs (*Callinectes*), which come from waters that are unlikely to become depleted of oxygen, are conformers; on the other hand, spider crabs (*Pugettia*), which are often found in tide pools, and crayfish (*Orconectes*), which inhabit slowly flowing rivers and streams, are regulators down to moderate oxygen tensions. This list of examples could be extended. Often the species from oxygen-rich environments exhibit higher rates of oxygen consumption at high tensions than their relatives from low-oxygen environments (compare, for example, *Baetis* and *Cloeon*). In these cases, the greater degree of conformity in the species from high-oxygen environments has at least superficial similarity to the increased conformity observed during activity within a single species. Indeed, the higher oxygen consumption in species from oxygen-rich environments is sometimes attributable to their displaying a higher level of spontaneous activity or higher capacity for activity than their relatives from oxygen-poor environments. Evolutionarily it is logical that a preference for oxygen-rich habitats and the development of an oxygen-demanding way of life should coevolve.

Although a number of studies have indicated the preceding tendency toward a direct correlation between critical pressure and oxygen availability in the habitat, the correlation is not always a good one, sometimes is not evident at all, and sometimes is reversed. Earlier we noted that *Ephemera* in Figure 11–12 was collected from pond mud, an environment that would ostensibly be likely to develop oxygen deficiency. Rather than being a regulator, however, *Ephemera* is a conformer, being substantially more sensitive to oxygen tension than the lake-dwelling *Leptophlebia*. Another group of *Baetis* was collected from swift streams; these were about 2.5 times larger than the *Baetis* shown in Figure 11–12 and may or may not have been more mature individuals of the same species. However that may be, this second group of *Baetis* showed regulation down to at least 100 mm Hg, a response more like that of *Leptophlebia* than the other *Baetis*. In a recent study of 31 marine invertebrates, representing a number of phyla, there was no compelling correlation between habitat and oxygen response. Again, the list of examples could be continued.

One obvious need in this area of investigation is for more complete information on microenvironmental oxygen tensions. Whatever correlations may or may not be present, rigorous analysis requires that the habitat be understood quantitatively and not simply be assessed on the basis of more or less subjective judgments of the differences between streams, ponds, ditches, deep waters, shallow waters, and so on.

Implications for the species

Whether or not correlations with habitat or other characters exist, the response of a particular species does provide insight into the relationship of that species to environmental oxygen, and, by extension, differences in response between species may aid in understanding differences in the effects of lowered oxygen tensions. Because a certain level of activity is ordinarily necessary for survival in the natural habitat, the response curve for the animal during at least moderate activity is probably more ecologically meaningful than the curve for resting oxygen consumption. Fish physiologists have offered the hypothesis that oxygen tensions that impair oxygen consumption during peak sustained activity (termed, for short, "active oxygen consumption") are likely in the long run to be detrimental to survival. The tension at which active oxygen consumption begins to decline (P_c for the active curve) has accordingly been termed the *incipient limiting tension*. For small trout and goldfish at 20°C, as depicted in Figure 11–11, this tension is, respectively, about 150 mm Hg and about 34 mm Hg. These results correlate with the fact that goldfish prosper in waters of lower oxygen content than trout. Experiments on several species of fish have shown that capacity for activity falls off below the incipient limiting tension. At 25°C, for example, young yellow perch can cruise at about 60 ft/min when the oxygen tension is near 105 mm Hg, 50 ft/min when the tension is 50 mm Hg, and only about 20 ft/min when the tension is 27 mm Hg. Such results indicate that activities that are of importance to survival are indeed limited by low oxygen tensions, but we need further work to find out how far tension can fall before the probability of survival is demonstrably reduced.

Another bench mark in oxygen relationships is the tension at which oxygen consumption falls below the unstressed standard rate (that is, the standard rate at high oxygen tensions). This tension may be different, depending on whether experiments are performed on active or resting animals (see Figure 11–13). When determined on active animals, the tension has been termed the *tension of no excess activity* in reference to the fact that the animal cannot support activity beyond the level ordinarily associated with a nearly resting condition. The standard rates of oxygen consumption of trout and goldfish are indicated on Figure 11–11, and it can be seen that the tension of no excess activity is about 18 mm Hg in the goldfish and about 48 mm Hg in the trout. Note that the goldfish, which is better adapted to waters of low oxygen content, has suffered no limitation on its oxygen consumption at the tension where trout are constrained to only their standard rate. The tension of no excess activity is sometimes useful as an index of the incipient asphyxial level, because below it the animal cannot maintain even its standard oxygen consumption. Much depends, however, on the capability of the species for anaerobiosis. Remember that only the aerobic component of metabolism is reflected by oxygen consumption. Species with only a limited or short-term anaerobic capacity are undoubtedly highly threatened at the tension of no excess activity, but species that are capable of prolonged anaerobiosis may survive indefinitely. We shall return to this issue shortly. It is worth emphasizing here that curves of *oxygen consumption* as a function of tension are not necessarily curves of *metabolism* as a function of tension.

The functional basis of oxygen response

Curves of oxygen consumption in relation to tension are relatively easy to obtain experimentally and have accumulated in the literature in great numbers. Unfortunately, knowledge of the underlying physiology has not increased at a comparable pace, and for many species almost nothing is known.

The fundamental site of oxygen limitation is at the cytochromes in the mitochondria, for it is the supply of oxygen at this level that determines the rate at which aerobic metabolism can proceed. As we have seen in earlier chapters, diverse processes are involved in the movement of oxygen from the environment to the mitochondria. In a teleost fish, for example, oxygen must be transported to the gill membranes through ventilation, must diffuse across the gill membranes into the blood, must dissolve in the plasma and combine with hemoglobin, must be transported to the tissues by the circulation, and finally must diffuse from the capillaries through the tissues and across the cellular and mitochondrial membranes. Clearly, then, the ambient oxygen tension is only one of many factors that influence mitochondrial oxygen supply. In other words, true hypoxia—hypoxia at the mitochondrial level—is a function of both ambient oxygen tension and various physiological responses of the animal.

Modulation of the function of both the respiratory and circulatory systems can provide for active compensation for reduced ambient tensions. Increased ventilation, for example, can assure continued saturation of the blood in the respiratory organs despite a decline in ambient tension, and once ventilatory compensation has been maximized, an increase in the rate of circulation can compensate for reductions in arterial saturation that are sustained. These and other compensations have their limits, but they can provide for a wide range of oxygen regulation.

Often the type of ventilatory response to lowered oxygen provides a partial explanation for interspecific differences in oxygen regulation. Species that display a vigorous increase in ventilation tend to be better regulators than those that show little or no increase. Of course, this is only a proximal explanation; it is another question why some species have evolved a strong compensatory ventilatory response, whereas others have not. Most elasmobranch and teleost fish are regulators when resting or moderately active at reasonably low temperatures, and correspondingly, they show a pronounced increase in gill ventilation as oxygen tension is lowered. The cost of this ventilation is often reflected in the oxygen consumption–oxygen tension curve (see Figure 11–13). Tench (*Tinca*) regulate their oxygen consumption between 150 mm Hg and 80 mm Hg and nearly double their rate of ventilation as the tension is lowered. Rainbow trout (*Salmo*) can increase ventilation by a factor of 13 and can regulate oxygen consumption down to 40 mm Hg. By contrast, sharks, *Scyliorhinus stellaris*, exhibit only a weak ventilatory response, increasing the rate of water flow by only 25% between 140 mm Hg and 50 mm Hg; they are conformers below 130 mm Hg. Among crustaceans, the green crab (*Carcinus*) shows appreciable powers of oxygen regulation; ventilatory frequency and the oxygen utilization coefficient increase as ambient tension is lowered. On the other hand, the American lobster (*Homarus*) does not increase ventilation when the tension is lowered and is a conformer.

The circulatory response to lowered oxygen has received much less attention than the ventilatory response. Very often oxygen transport pigments saturate at relatively low tensions; so long as ventilatory compensations can maintain those tensions in the respiratory exchange vessels, there will be no great need for an increase in circulatory rate. In fish the ventilatory response appears typically to be the first line of defense, and this may perhaps be anticipated in other groups with substantial ventilatory flexibility. In rainbow trout showing oxygen regulation down to 40 mm Hg, there was little or no change in cardiac output, but there was a strong ventilatory response. Below 80–100 mm Hg the trout exhibited a marked decline in heart rate, termed bradycardia; but the stroke volume increased commensurately. Development of bradycardia is a very common response to hypoxia in fish, but its significance is not entirely clear; in the trout the ventilatory and cardiac rhythms become more closely synchronized as the oxygen tension is lowered, and this may enhance oxygen exchange across the gills by promoting favorable flow relationships between the water and blood. In the shark *Scyliorhinus stellaris*, bradycardia is accompanied by a more than fourfold reduction in cardiac output as the oxygen tension is lowered from 140 mm Hg to 60 mm Hg; this species, as noted earlier, is an oxygen conformer below 130 mm Hg.

As discussed in Chapter 9, oxygen transport pigments of relatively high affinity are better suited to function at low ambient tensions than pigments of lower affinity. Possession of a high-affinity pigment favors oxygen regulation down to relatively low tensions by permitting more extensive compensation at the level of respiratory exchange.

An important parameter in the oxygen relationships of animals is the capacity for anaerobiosis. Species with only a limited anaerobic capability will perish if unable to support their energy demands aerobically for any length of time; these animals would be expected to make every possible effort to maintain at least the minimal oxygen uptake required for life. Species that can sustain themselves anaerobically for prolonged periods are not threatened in this immediate sense by low oxygen levels. They can substitute anaerobic for aerobic metabolism as oxygen tension falls rather than having to struggle to maintain oxygen uptake. A gradual substitution may, in fact, be what occurs in many invertebrate oxygen conformers as the oxygen tension is lowered. If so, these animals could have a much more stable metabolic rate than the oxygen response would suggest, and they might, on these simple grounds, exhibit little or no compensatory response in the systems responsible for oxygen uptake. Also, once oxygen tensions fall low enough that the benefits of aerobic catabolism do not adequately counterbalance the cost of obtaining oxygen, such animals might cease oxygen uptake altogether. Unfortunately, little is known of the relative contributions of anaerobic and aerobic metabolism during the phase of oxygen conformity in invertebrates; it would be most interesting to see if anaerobic metabolism increases as oxygen consumption falls with oxygen tension. It *is* known that a number of invertebrates of several phyla cease oxygen consumption at tensions well above zero and that they can survive anaerobically for substantial periods thereafter. To cite some examples, anemones (*Diadumene*) were found to stop consuming oxygen at 15–55 mm Hg; clams (*Rangia*) stopped at 4–13 mm Hg; and tunicates (*Ciona*) stopped at 14–17 mm Hg. All of these survived for at least a day before resuming oxygen uptake.

Among fish, accumulation of lactic acid has been reported in a number of species under hypoxic conditions. In species such as trout, in which there is no evidence of a prolonged anaerobic capability, this probably represents only a short-term and limited solution to the problems of oxygen deficiency. On the other hand, as noted in the previous section of this chapter, goldfish are known to be capable of prolonged partial anaerobiosis at low oxygen tensions, and European carp can survive anaerobically for months at low temperatures. Such anaerobic capacities are of obvious relevance to oxygen relationships.

As noted earlier, some animals alter their response to lowered oxygen when they are acclimated to low oxygen tensions. In some cases the mechanisms involved have received study, or potential mechanisms have been suggested. As shown in Figure 11–13, goldfish at 20°C exhibit a reduced standard rate of oxygen consumption at low tensions when they have been acclimated to low tensions; a similar response in the standard rate of oxygen consumption has also been observed in carp (*Cyprinus*) at 20°C and in brook trout (*Salvelinus*) at 10°C and 15°C. In the experiments on these three species a device was utilized to measure spontaneous activity, and it was also observed in tests at low tensions that fish acclimated to low tensions consumed less oxygen at a given level of activity than fish acclimated to high tensions. It has been suggested (but not demonstrated) that the fish acclimated to low tensions come to rely more heavily on anaerobic metabolism and that it is this that permits their lower levels of oxygen utilization.

An increase in the oxygen-carrying capacity of the blood resulting from increased hemoglobin levels has been reported in a number of fish when acclimated to low oxygen levels. This would be expected to improve capabilities for oxygen regulation by facilitating circulatory transport. A recent study of eels (*Anguilla anguilla*) is of great interest and deserves special note. As shown in Figure 11–14, acclimation to low oxygen tensions for seven days resulted not only in a great increase in carrying capacity but also in a substantial increase in the affinity of hemoglobin for oxygen. The increase in affinity was apparently due to a reduction in red cell ATP, for

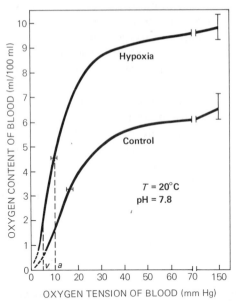

Figure 11-14. Oxygen dissociation curves for whole blood of eels (*Anguilla anguilla*). Controls were kept in water of high oxygen tension (140 mm Hg), whereas individuals acclimated to hypoxia were kept in water of low tension (15–40 mm Hg) for a week. Vertical bars at the right indicate oxygen-carrying capacity in the two groups as the mean ± twice the standard error; the mean in controls was 6.6 vol %; that in hypoxic animals was 9.8 vol %. Hemoglobin content of the blood showed a comparable increase in the hypoxic animals. Horizontal bars toward the left indicate P_{50}, again as the mean ± 2SE; mean P_{50} of controls was 16.6 mm Hg; that of hypoxic individuals was 10.6 mm Hg. Vertical dashed lines indicate mean arterial (*a*) and venous (*v*) oxygen tensions in eels living in water of low oxygen tension (30 mm Hg). (From Wood, S. C. and K. Johansen. 1973. Amer. J. Physiol. 225: 849–851.)

electrophoresis indicated no change in the type of hemoglobin, but the molar ratio of ATP to hemoglobin fell by about 45%. As discussed in Chapter 9, red cell organic phosphates are receiving increasing attention as modulators of oxygen affinity in vertebrates. Tests of oxygen transport in eels at low ambient oxygen tensions (about 30 mm Hg) revealed that the changes in carrying capacity and affinity with acclimation made a substantial contribution to oxygen transport. Arterial and venous oxygen tensions are indicated in Figure 11-14, and it can be seen that the blood of control animals delivered only about 1.0 cc of oxygen per 100 cc on each circulatory cycle, whereas that of animals acclimated to hypoxic conditions delivered about 2.5 cc/100 cc. As noted earlier, the acclimation of blood parameters required about a week. The responses of the eels over this period are significant. On the first day after transfer to hypoxic waters, they exhibited a 50% increase in ventilatory frequency and gave no evidence of accumulation of lactic acid. On the second and third days, however, ventilatory frequency fell nearly fourfold, and oxygen uptake was insufficient to meet metabolic demands as indicated by a great accumulation of lactic acid, accompanied by pronounced acidosis. By the seventh day lactic acid levels and blood pH had returned to control values in most of the eels, indicating that adequate oxygen delivery had been restored. It seems likely that the facilitation of circulatory oxygen transport resulting from the acclimation of blood parameters was a central factor in this adjustment to life at low oxygen tensions.

SELECTED READINGS

Bennett, A. F. and P. Licht. 1972. Anaerobic metabolism during activity in lizards. J. Comp. Physiol. 81: 277–288.

Bennett, A. F. and P. Licht. 1974. Anaerobic metabolism during activity in amphibians. Comp. Biochem. Physiol. 48A: 319–327.

Bullard, R. W. 1966. Physiology of exercise. *In:* E. E. Selkurt (ed.), *Physiology.* 2nd ed. Little, Brown, Boston, Mass.

Fry, F. E. J. 1957. The aquatic respiration of fish. *In:* M. E. Brown (ed.), *The Physiology of Fishes.* Vol. I. Academic, New York.

Hochachka, P. W., J. Fields, and T. Mustafa. 1973. Animal life without oxygen: basic biochemical mechanisms. Amer. Zool. 13: 543–555.

Hochachka, P. W. and G. N. Somero. 1973. *Strategies of Biochemical Adaptation.* Saunders, Philadelphia.

Hughes, G. M. 1973. Respiratory responses to hypoxia in fish. Amer. Zool. 13: 475–489.

Mangum, C. and W. Van Winkle. 1973. Responses of aquatic invertebrates to declining oxygen conditions. Amer. Zool. 13: 529–541.

Margaria, R. 1967. Aerobic and anaerobic energy sources in muscular exercise. *In:* R. Margaria (ed.), *Exercise at Altitude.* Excerpta Medica Foundation, Amsterdam.

Robinson, S. 1974. Physiology of muscular exercise. *In:* V. B. Mountcastle (ed.), *Medical Physiology.* 13th ed. Vol. II. Mosby, St. Louis, Mo.

Satchell, G. H. 1971. *Circulation in Fishes.* Cambridge University Press, London.

Simonson, E. (ed.). 1971. *Physiology of Work Capacity and Fatigue.* C. C Thomas, Springfield, Ill.

Von Brand, T. 1946. *Anaerobiosis in Invertebrates.* Biodynamica, Normandy, Mo.

See also references in Appendix.

12 THE PHYSIOLOGY OF LIFE AT HIGH ALTITUDE

Many groups of animals are known to occur at high elevations. Unfortunately, in the case of most taxa we have little information except records of occurrence. Many groups of insects and arachnids occur in the alpine zone; butterflies, bees, and spiders, for example, have been observed above 6000 m (19,700 ft) in the Himalayas. Snails can be found above 4000 m (13,100 ft) on Mount Kilimanjaro. A substantial number of amphibians and reptiles are found above 3000 m; certain frogs occur as high as 4500–5200 m, and the highest report of a reptile is 5500 m. Mountaineers have reported avian activity at 8000 m (26,200 ft), and the bar-headed goose migrates over the Himalayas at an astounding altitude of 10,000 m. Eggs of the snow partridge have been found at 5800 m (19,000 ft); if young are successfully raised, this is the highest known elevation for vertebrate reproduction. A good number of mammals, representing several orders, occur above 4000 m. In South America, chinchilla and vicuna, for example, range to at least 5000 m, and in Asia, yaks, wild sheep, and wolves are reported from over 5500 m. As a group the mammals do not penetrate to as great altitudes as birds. Some of the species of animals found at high elevations are endemic to such regions, whereas others are also found in the lowlands. Many mobile animals roam to higher elevations than those at which they can successfully reproduce.

The number of species of all animal groups decreases at high altitude. The environment changes in a number of dimensions with increasing elevation, and it is not always clear which factors are central in limiting distribution.

As noted in Chapter 8, atmospheric pressure decreases by about half for every 5500 m (18,000 ft) of elevation, and the partial pressure of oxygen decreases in proportion. At the top of Mount Everest (8848 m), the mean atmospheric pressure is about 250 mm Hg, and the partial pressure of oxygen in dry air is only about 50 mm Hg—just one third of the value at sea level. The lowered oxygen pressure at altitude affects not only terrestrial animals but aquatic ones as well, for streams, ponds, and lakes tend to equilibrate with the atmospheric tension.

Another factor at high altitudes is decreased temperature: air temperature falls about 1°C for every 150–200 m of elevation. The cold tends to depress the metabolism of poikilotherms and to increase the energetic demands of homeothermy. In terrestrial habitats the decreased atmospheric absorption of solar radiation at altitude leads to increased daily variation of temperature. The ground and animals basking in the sun are warmed considerably during the day—a factor of particular importance to the activity of poikilotherms. Bodies of water at high elevations tend to be perpetually cold; they are fed from melting ice, and the thermal inertia of water hinders solar warming by comparison to the air. The coldness of the water may help to ameliorate the effects of low oxygen tensions, for the concentration of oxygen at a given tension increases with decreasing temperature. Summer is short in the alpine region; in the Alps at 3100 m, for example, average daily air temperature is above 0°C for only two months out of the year. Lakes and ponds at high elevation may be free from ice for only a few months, or even a few weeks. The short summer and the cold, perhaps in combination with other factors such as hypoxia, can interfere with the reproduction and productivity of animals and plants. A number of species are known to show slowed development or a reduced number of broods in the highlands as compared to the lowlands. Many butterflies of the Alps, for instance, produce only one brood per year at altitude but two in the lowlands, and alpine cladocerans are also usually limited to one brood. Some frogs pass more years as tadpoles at altitude, and some butterflies require an added year to reach adulthood. There seems to be a tendency among both amphibians and insects for species with a relatively brief development time to extend to higher altitudes than those with a relatively prolonged development.

Winds can be strong and persistent at altitude. High winds hinder the flight of insects and birds and tend to cool and desiccate exposed terrestrial animals.

As indicated earlier, decreased atmospheric absorption at high altitude allows greater penetration of solar radiation. A particular implication that often receives note is the increase in ultraviolet radiation; probably every mountain hiker has received an unexpected sunburn at one time or another. Whether the increase in ultraviolet radiation is sufficient to be potentially damaging to high-altitude residents remains unclear. A tendency toward melanism is apparent at altitude in many groups, among them the insects, amphibians, reptiles, and rodents. Melanism has at least two potential adaptive roles at high elevation: it acts to increase the absorption of visible solar radiation as heat and could be protective against ultraviolet radiation.

Another factor to be considered in the study of animals at high altitude is that all species must eat and find suitable cover. There is little doubt that many animals are limited in their altitudinal distribution not by intrinsic physiological factors, but by the inability of their prey, plant or animal, to survive or reproduce satisfactorily. The zonation of vegetation is well known to any hiker; as elevation increases, hardwood forests give way to dense conifer forests, and the latter are replaced by scrub, then tundra, and finally barren, sparsely vegetated mountain tops. Many animals are more or less restricted to certain vegetational zones because of the food base provided or the physical structure or both. Insectivorous birds are uncommon at high altitude, in correlation with the reduced populations of insects. Glacial lakes and ponds of the alpine region are sometimes very low in dissolved nutrients and support little plant growth; such bodies of water correspondingly have a limited fauna.

With the multitude of interacting factors that can influence altitudinal distribution, the problem of analyzing distributional limits is a complex one involving both physiological and ecological considerations. Although we have a qualitative appreciation of the important factors, it is to be hoped that more biologists will devote their talents to this topic and provide the data for a more rigorous understanding. It is an enlightening and disappointing commentary on the field that a review of reptiles and amphibians at altitude published by the late Raymond Hock in 1964 was two pages long, and in 1971 Robert Bullard had to conclude that very little new information had accumulated (see, in Selected Readings, volumes edited by Dill and by Yousef, Horvath, and Bullard).

Man at high altitude: introduction

Most work on man at high altitude has centered on the implications of hypoxia. Serious scientific investigation dates from the late nineteenth century, and the literature, in contrast to that on most animals, is huge. Research has been sparked not only by a pure interest in the problem, but also by applied considerations. Studies of man at altitude have contributed to understanding of hypoxic disease states, have helped to elucidate the particular afflictions of resident mountain populations, and have supplied knowledge of importance to mountaineering and aviation. In recent decades considerable information has accumulated on the responses of other mammals at altitude; this will be considered toward the end of the chapter.

Permanent human settlements occur as high as 4500–4850 m (14,800–15,900 ft) in both the Andes and Himalayas. Herdsmen in Tibet take their flocks as high as 5500 m (18,000 ft), and settlements occur at a similar elevation in the Andes, though the women are reported to descend to lower altitudes to bear their children. Natives of the Andes ascend to 6000 m (20,000 ft) to work in sulfur mines. The peoples of the Himalayas and Andes have lived at high altitudes for thousands of years and show particularly effective adaptations to hypoxia. The greatest elevation reached by man without supplemental oxygen is about 8600 m (28,200 ft)—just short of the peak of Mount Everest. This feat has been achieved by several parties of acclimatized lowlanders. When lowlanders first ascend to high altitude, they experience weakness, especially during exertion, and may suffer difficulty in sleeping, nightmares, headaches, nausea, or other discomforts. At least at altitudes up to 7000 m, however, well-being improves over a period of acclimatization, and, as we shall see, important physiological adjustments occur. Lowlanders do not, nonetheless, appear to reach the same levels of performance as natives even over weeks or months of acclimatization, and it remains an open question whether a lifetime of acclimatization could produce adjustments in lowlanders comparable to those in natives; the natives may have adaptive hereditary characteristics that have evolved over their generations of life at altitude. From this brief introduction it is apparent that we must distinguish the physiological responses of (1) lowlanders soon after their ascent to altitude, (2) lowlanders after acclimatization, and (3) native highlanders.

The oxygen consumption of man at altitude is not greatly different from that of man at sea level, either at rest or at any particular intensity of exercise. As we shall discuss later, newcomers to altitude cannot exercise as vigorously as natives or acclimatized lowlanders, but this is a somewhat different matter; within the range of exercise that can be attained, the oxygen consumption is about the same. This information tells us that there is no fundamental reduction in tissue oxygen demand at altitude. Of course, a man can reduce his demand behaviorally by limiting activity, and this is a common response of newcomers; but still the tissues consume oxygen at their same basic rate for the prevailing level of exertion.

It is apparent that, to survive or to support a given level of activity, man at altitude must be able to maintain mitochondrial oxygen tensions compatible with an unimpaired rate of oxidative metabolism despite the fact that the ambient oxygen tension is reduced. The central question in altitude physiology is how this is accomplished.

Man at high altitude: respiratory and circulatory transport

Respiratory and circulatory function are major determinants of tissue oxygen tension and are intimately involved in the adjustment to high altitude.

Figure 12–1 depicts the decline in oxygen tension from inspired air to mixed venous blood in a group of native lowlanders studied at sea level (Lima) and in two groups of native highlanders studied at 3700 m and 4500 m (La Oroya and Morococha). In the men from Lima we see a fairly typical picture for lowland man (see Chapters 8, 9, and 10). The oxygen tension in alveolar air is about 106 mm Hg—43 mm Hg lower than in inspired air. Blood leaves the lungs for the tissues at a tension of 94 mm Hg and, after perfusing the systemic capillaries, returns to the heart at 39.4 mm Hg. The mean capillary tension is between the arterial and mixed venous tensions.

The decline in tension from the inspired air to the venous blood has been likened to a cascade, with tension falling off in a number of steps. First, there is the decline from inspired to alveolar air; this depends on the rate of ventilation and the rate at which oxygen is removed from the alveoli by the blood. Second, there is a tension drop between the alveoli and arterial blood; this depends on the many factors that affect the degree of equilibration in the alveolar capillaries. Finally there is a decline in tension across the systemic capillary beds, dependent on the circulatory rate, arterial oxygen content, rate of oxygen extraction, and oxygen dissociation curve. Recognizing that oxygen must diffuse from the systemic capillaries

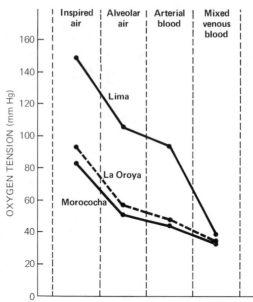

Figure 12–1. Mean oxygen tension in inspired air, alveolar air, arterial blood, and mixed venous blood during rest in three groups of native male Peruvians studied at their altitude of residence: (1) students and scientists of Lima (altitude: 10 m), (2) hospital workers of La Oroya (3735 m), and (3) miners of Morococha (4509 m). (Data from Torrance, J. D., C. Lenfant, J. Cruz, and E. Marticorena. 1970–1971. Resp. Physiol. 11: 1–15.)

to the mitochondria and that the capillary oxygen tension constitutes the head of diffusion pressure for this process, it is apparent that circulatory and respiratory function must be adequate to limit the tension drops in the oxygen cascade so that a suitable capillary oxygen tension is maintained.

Turning to the natives of La Oroya and Morococha, we see that they start off with an inspired oxygen tension 56–66 mm Hg lower than that at sea level, but their mixed venous tensions are only 5.3–6.6 mm Hg lower than those of Lima residents. The conservation of venous tension results from significant reductions in two of the tension drops of the oxygen cascade. The inspired-alveolar drop is 32–36 mm Hg, as compared with 43 mm Hg in Lima residents; and the arterial-venous drop is 11–14 mm Hg, as compared to 55 mm Hg at Lima. In part these advantages are realized through adjustments of circulatory and respiratory function that act to maintain tissue oxygen tensions despite the reduced ambient tensions at altitude. It is apparent from Figure 12–1 that mean capillary tension is reduced in the high-altitude natives. Because capillary tension cannot be measured, we do not know exactly how much it is lowered, but by using standard formulas, the drop can be estimated as around 20 mm Hg. This is quite significant, for, other things being equal, a reduced capillary tension implies a reduced rate of diffusion from the capillaries to the mitochondria. The circulatory and respiratory systems assure a high degree of compensation for the reduced ambient tension, but it is not complete. We shall return to this problem later in the chapter. First, we shall look in detail at the changes in respiratory and circulatory function, dealing initially with native highlanders and then comparing the responses of lowlanders at altitude.

Natives of high altitude show an elevated resting rate of pulmonary ventilation by comparison with lowlanders at sea level, the extent of hyperventilation increasing with altitude. At 3700–4500 m the ratio of ventilation to oxygen consumption at rest averages approximately 30–40% higher than that at sea level, but there is considerable individual variation. The increase in ventilation accounts for the diminished fall in oxygen tension from the inspired air to the alveolar air in highlanders. On the average, at 3700–4500 m alveolar oxygen tension would be about 7 mm Hg lower if hyperventilation did not occur.

The increase in ventilation, while augmenting alveolar oxygen tension, acts to reduce alveolar carbon dioxide tension, implying also a decline in the carbon dioxide tension of the blood and other body fluids. At the average degree of hyperventilation displayed at 3700–4500 m, alveolar carbon dioxide pressure is reduced to around 33–34 mm Hg, as compared to 40 mm Hg at sea level. As discussed in Chapter 9, a decrease in blood carbon dioxide tension, if uncompensated, would lead to a rise in pH. Native highlanders do not, however, suffer from such alkalosis, for renal compensations intervene. The kidneys tend to conserve metabolically produced noncarbonic H^+, thus increasing the body's noncarbonic acid content in compensation for the reduced carbonic acid content. Plasma bicarbonate concentration is reduced, and pH is maintained within normal limits (see Chapter 9).

Turning to the arterial-venous drop in oxygen tension, it must first be recognized that the diminished decline at high altitude is in good part a simple function of the oxygen-combining kinetics of human hemoglobin, as reflected in the oxygen equilibrium curve of Figure 9–2. At La Oroya and Morococha (Figure 12–1), arterial blood leaves for the tissues at a tension of 44–48 mm Hg, as compared with 94 mm Hg at Lima. As seen in Figure 9–2, the blood of the highlanders starts off on the steep portion of the oxygen equilibrium curve, whereas that of lowlanders starts off high on the plateau of the curve. Consequently, highlanders will realize a given oxygen yield with a far smaller drop in tension than lowlanders even without any special hematological or circulatory adjustments. Put another way, a lowlander could ascend to altitude and immediately realize a substantially diminished arterial-venous tension drop by virtue of the intrinsic characteristics of human hemoglobin.

Native highlanders exhibit characteristics that further diminish the arterial-venous tension difference and thus help to maintain adequate capillary tensions. First we must note that the resting cardiac output of highlanders is approximately the same as that of lowlanders. Because oxygen consumption is also the same, the blood yields about the same amount of oxygen over a circulatory cycle in both groups; that is, the arterial-venous difference in oxygen *content* is unchanged. The most obvious hematological difference between highlanders and lowlanders—and, historically, one of the first altitude-related adaptations to be discovered—is a marked increase in the oxygen-carrying capacity of highlanders. Although carrying capacity in man at sea level is about 20 vol %, natives of Morococha, for example, exhibit carrying capacities near 28 vol %. The extent of the increase in carrying capacity in high-altitude natives is positively correlated with their altitude of residence, tending to be greater the higher the altitude. The augmentation of carrying capacity is due to an increase in the number of red blood cells per unit volume of blood (polycythemia), the amount of hemoglobin per red blood cell being little changed. Increased carrying capacity significantly reduces the drop in oxygen tension across the systemic capillaries, for as illustrated in Figure 12–2A, the blood will yield a given amount of oxygen with less reduction in its tension.

The polycythemia of high altitude is believed to be mediated by a hormone, erythropoietin, which stimulates acceleration of red cell production in the bone marrow. Present evidence implicates the kidneys as the major sources of erythropoietin, and it is possible that the immediate stimulus to increased erythropoietin secretion is reduction of renal oxygen tension.

As a result of the progressive increase in carrying capacity with increase in altitude, the oxygen content of arterial blood shows remarkably little change with altitude despite the reduction of arterial tension—and percent saturation—at high elevation. In the experiments of Figure 12–1, arterial percent saturation decreased with altitude, averaging 97%, 80%, and 73% in the residents of Lima, La Oroya, and Morococha, respectively. Conversely, carrying capacity increased, averaging, in the same order, 21 vol %, 24 vol %, and 28 vol %. The oxygen content of arterial blood ranged only from a low of 19.2 vol % at La Oroya to a high of 20.7 vol % at Morococha, showing that the increase in carrying capacity effectively cancels the effect of reduced saturation. A significant implication is that because residents of all three locations have about the same arteriovenous drop in oxygen content at rest, their resting percent utilization of arterial oxygen is also virtually the same, ranging only from 24% to 27%. Thus high-altitude residents have as great an arterial reserve to call upon during exercise as do low-altitude residents, though, of course, they must tolerate somewhat lower capillary tensions than lowlanders in order to make use of any given fraction of their reserve.

A second hematological adjustment seen in highlanders is a reduction in the oxygen affinity of their hemoglobin, reflected in a rightward shift of the oxygen equilibrium curve and an increase in P_{50} (the oxygen tension required for half saturation). In the experiments of Figure 12–1, the average P_{50} of Lima residents (at 37°C and a pH of 7.4) was 26.7 mm Hg—quite ordinary for lowland man. Residents of Morococha showed a sufficient reduction in oxygen affinity to shift their mean P_{50} to 29.8 mm Hg, and residents of La Oroya were intermediate with a P_{50} of 28.8 mm Hg. These results are typical of those obtained in other studies of highland man. The reduced affinity at altitude is mediated by an increase in red cell 2,3-diphosphoglycerate, as discussed in Chapter 9. Morococha residents showed a 16% increase in 2,3-DPG as compared to Lima residents.

The significance that reduced affinity has for oxygen transport at altitude is as yet a matter of some uncertainty and debate. One important principle is illustrated in Figure 12–2B. The P_{50} of curve L is typical for lowland man, whereas that of curve H is 5 mm Hg higher, as might be observed in highland man. Both curves have been drawn according to a carrying capacity of 30 vol % so that the effect of reduced affinity alone can be perceived. Tension changes are shown for release of 5 vol % oxygen over the circulatory cycle. It can be seen that when the arterial tension is 90 mm Hg, the mixed venous tension commensurate with release of 5 vol % oxygen is significantly higher for curve H than for curve L;

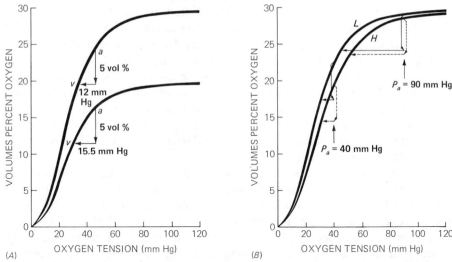

Figure 12–2. (*A*) Oxygen equilibrium curves for human blood having a carrying capacity of 30 vol % (upper curve) and 20 vol % (lower curve). P_{50} for both curves is the same (26.5 mm Hg). If arterial tension (*a*) is 46 mm Hg and the arteriovenous difference in oxygen content is to be 5 vol %, then tension must fall by 15.5 mm Hg when the carrying capacity is 20 vol % but only by 12 mm Hg when the carrying capacity is 30 vol %. Accordingly the venous tension (*v*) is higher when carrying capacity is 30 vol %. This example illustrates that increased carrying capacity reduces the arteriovenous decrement in tension required for a given oxygen yield. (*B*) Two human oxygen equilibrium curves, the P_{50} of curve *H* being 5 mm Hg higher than that of curve *L*. In both cases, the carrying capacity is 30 vol %. The format of the vertical and horizontal arrows is the same as in part *A* of this figure; vertical arrows depict an arteriovenous decline in oxygen content of 5 vol %, and horizontal arrows depict the drop in tension required for this oxygen yield. Changes in oxygen content and tension are shown for both curves assuming arterial tensions (P_a) of 90 mm Hg and 40 mm Hg. See text for discussion.

thus reduced affinity contributes to maintenance of high tissue oxygen tensions. On the other hand, when the arterial tension is 40 mm Hg, the mixed venous tension is virtually identical for both curves, and reduced affinity does not appear to hold distinct advantages. Basically, as discussed in Chapter 9, a reduction of affinity tends to hinder oxygen loading at the lungs and promote unloading in the tissues. When the arterial oxygen tension is high enough to be on the plateau of the oxygen equilibrium curve (e.g., 90 mm Hg), the effect of reduced affinity on percent loading is small; thus when venous tension falls to be on the steep portion of the curve, where unloading is greatly facilitated by reduced affinity, the advantage of reduced affinity for unloading outweighs its disadvantage for loading, and the net effect is a higher mixed venous tension. On the other hand, when the arterial tension itself is on the steep portion of the curve (e.g., 40 mm Hg), the effect on loading is large and can completely offset the advantage in unloading. In short, then, a reduction in oxygen affinity helps significantly to maintain high tissue oxygen tensions when arterial tension is high enough to be on the plateau of the equilibrium curve, but the benefit decreases as arterial tension falls and becomes very small or negligible once the arterial tension is low enough to be on the steep portion of the curve. Because arterial tension decreases with altitude, we see that, by this analysis, reduced affinity is mainly advantageous at relatively low altitudes; above 3500 m the advantage is small or absent. In the experiments of Figure 12–1, the residents of La Oroya (3735 m) showed a mean P_{50} of 28.8 mm Hg, and it was calculated that this shift in affinity from the lowland value would raise mixed venous oxygen tension by about 1.2 mm Hg at the altitude of La Oroya. The residents of Morococha (4509 m) had an average P_{50} of 29.8 mm Hg; this increase over the lowland value resulted in an improvement of 0.8 mm Hg in mixed venous tension at the altitude of Morococha. The analysis of the effects of affinity changes presented here is somewhat limited in that it deals only with mixed venous tension. A more refined and realistic analysis will be possible when we know the tensions prevailing in the capillaries of various tissues, where unloading actually occurs.

Table 12–1 represents an attempt to quantify the effects of hyperventilation, increase in carrying capacity, and reduction in oxygen affinity in natives of La Oroya and Morococha. The table shows the amount by which mixed venous oxygen tension would have been lowered if each of these responses had not occurred. Hyperventilation emerges as being most significant in maintaining high tensions, and, together, the three responses have a considerable adaptive effect.

Having reviewed the major circulatory and respiratory adaptations of native highlanders, we may now turn briefly to the responses of lowlanders at high altitude. As noted earlier, lowlanders undergo progressive acclimatization to altitude, a process that involves temporal changes in their individual response parameters.

When lowland man first ascends to altitude, there is little or no increase in resting ventilation up to about 3000 m. At this elevation his alveolar oxygen tension without hyperventilation is about 60 mm Hg, still sufficient to assure a high degree of arterial saturation (see Figure 9–2). Above about 3000 m there is an immediate increase in ventilation that helps to maintain alveolar tension as the ambient tension is further reduced. The intensity of this immediate hyperventilation increases with altitude such that the ratio of ventilation to oxygen consumption is about 60% higher than normal at 6000 m.

At altitudes in excess of a few hundred meters, ventilation shows a gradual increase over the first few days of residence. Thus the lowlander who exhibits no immediate increase in ventilation upon ascent to altitudes of 3000 m or less will show a very clear increase after three to four days, and the immediate hyperventilation at altitudes in excess of 3000 m is supplanted by an even more vigorous response over this period. After four days at 6000 m, the ratio of ventilation to oxygen consumption may approach twice the normal value. At all altitudes the increase in ventilation over the first few days of residence is accompanied by an elevation of alveolar oxygen tension and is one of the most important of the early responses serving to protect the tissues against hypoxia. Once ventilation has reached its peak, it remains at a high level for a long period. It shows some tendency to decline but the reduction is small even after a number of years.

It is instructive to compare the resting ventilatory rate of the acclimatized high-altitude sojourner with that of native highlanders. The rate of the sojourner increases with altitude and at all altitudes is higher than that of the native. At the elevation of Morococha, for example, the

Table 12–1. Mean mixed venous oxygen tension in resting natives of La Oroya and Morococha and amount by which mixed venous tension would have been reduced by the lack of each of three responses to altitude. In calculating the effect of lack of a particular response, it was assumed that all other responses remained intact.

	La Oroya	Morococha
Actual mixed venous tension (mm Hg)	34.1	32.8
Amount (mm Hg) by which mixed venous tension would have been reduced by		
Lack of hyperventilation	4.6	5.1
Lack of increase in carrying capacity	2.1	2.3
Lack of shift in oxygen affinity	1.2	0.8

SOURCE: Data from Torrance, J. D., C. Lenfant, J. Cruz, and E. Marticorena. 1970–1971. Resp. Physiol. 11: 1–15.

ratio of ventilation to oxygen consumption in natives averages about 40% higher than that of man at sea level, whereas the same ratio in sojourners averages about 50% higher. As a consequence, the sojourner has a higher alveolar oxygen tension than the native and a lower alveolar carbon dioxide tension. As noted earlier, there is a tendency for the ventilation of the sojourner to decline slowly over years of residence at altitude. Possibly it would fall to the level of natives after a number of decades.

The mechanisms of control of ventilation at altitude have received much attention. The main stimulus to hyperventilation in the newly arrived lowlander is believed to be the reduced blood oxygen tension, which is sensed by the aortic and carotid bodies (see Chapter 8). The increase in ventilation reduces carbon dioxide tension in the body fluids and initially induces a state of alkalosis, for the renal compensatory mechanisms require many days to correct the pH imbalance. You will recall that the medulla oblongata senses H^+ concentration and modulates ventilation according to the concentration, stimulating ventilation in response to increased H^+ and depressing it when H^+ concentration falls. The cerebrospinal fluids bathing the medulla become alkalotic during the initial period of exposure to altitude, and it is believed that this sets up the interesting situation of an alkalotic depression of ventilation that opposes the hypoxic stimulation, thus limiting the actual ventilatory response. Cerebrospinal pH is returned to normal over the first few days at altitude even though the blood remains alkalotic for a considerably longer time. The correction of cerebrospinal pH, it is thought, is the underlying basis of the gradual increase in ventilation observed during this period: as the depressing effect of elevated pH is relieved, the hypoxic stimulation is expressed more and more fully. Why do native highlanders ventilate at a lower rate than acclimatized lowlanders? It has been shown in both Andean and Himalayan natives that highlanders exhibit a lower ventilatory sensitivity to hypoxia than lowlanders; thus at any given level of arterial oxygen tension, the highlanders experience less stimulation of ventilation. This blunted sensitivity to hypoxia appears to be acquired during the first few years after birth, for natives who spend their infant years at high altitude continue to show reduced sensitivity even after prolonged periods at sea level, yet people of highland descent who are born and raised at sea level show the same sensitivity as other lowlanders. There is some evidence that lowlanders acquire a reduced sensitivity to hypoxia if they spend many years at altitude. Some native highlanders and acclimatized lowlanders develop a disease termed chronic mountain sickness, which results from loss of adaptation to the high-altitude environment. The fundamental cause of the disease is a loss of adequate respiratory drive, believed to result from a pathological drop in hypoxic sensitivity. Ventilation becomes inadequate to support tissue oxygen tensions, and the person may experience somnolence, decreased tolerance for exercise, dizziness, cyanosis, and other symptoms. Generally those afflicted must descend to low altitude.

We may now examine the circulatory and hematological responses of lowlanders at altitude. Immediately after ascent the lowlander exhibits a marked increase in resting cardiac output; but this response subsides within a few days, and thereafter the resting cardiac output of the lowlander at altitude is the same as that of men at sea level. In this respect the acclimatized lowlander resembles the native highlander.

The lowlander at altitude shows a progressive increase in red cell count and oxygen-carrying capacity and may ultimately attain about the same levels as native highlanders. There is considerable individual variability in response over the first days or weeks—some showing a marked increase in carrying capacity, others showing little or none. The increase in carrying capacity may not reach its peak for months or even a year or more. Several mechanisms underlie the changes in carrying capacity. Erythropoietin levels rise soon after ascent, and acceleration of red cell production in the bone marrow is central to the long-term increase in carrying capacity. Plasma volume often decreases considerably over the first weeks at altitude; this increases the carrying capacity by acting to concentrate the red cells. Later, plasma volume increases, but it may not reach its original level even after months. Soon after ascent red cell count may be increased slightly by release of cells from stores, notably the spleen.

Lowlanders also show a decrease in the oxygen affinity of hemoglobin. This response is due to an increase in red cell 2,3-DPG and occurs quickly, within the first several days at altitude.

In summary, we see that the sojourner at altitude displays much the same adaptive responses as the native highlander. Because each individual response shows its own time course of development, the physiological state of the sojourner depends on his length of residence. Figure 12–3 summarizes some recent data on natives and three-day sojourners at Morococha; the data on the natives are the same as those in Figure 12–1. After three days the sojourners have had sufficient time to develop a rather full ventilatory response. Their alveolar ventilation, expressed relative to oxygen consumption, is greater than that of the natives, and they consequently realize higher alveolar and arterial oxygen tensions. The decrease in oxygen affinity, which occurs quickly after ascent, is fully expressed in the sojourners; but the increase in carrying capacity, which requires a longer time for full development, is only partially expressed. The sojourners, because of their higher arterial oxygen tension, had a slightly higher percent arterial saturation than the natives; but because their carrying

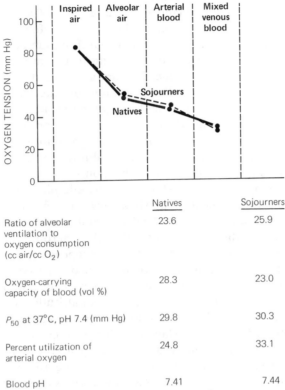

Figure 12–3. Mean resting circulatory and respiratory data on two groups of men at Morococha (4509 m): (1) native miners and (2) lowland sojourners after a stay of three days. (Data from Torrance, J. D., C. Lenfant, J. Cruz, and E. Marticorena. 1970–1971. Resp. Physiol. 11: 1–15.)

	Natives	Sojourners
Ratio of alveolar ventilation to oxygen consumption (cc air/cc O_2)	23.6	25.9
Oxygen-carrying capacity of blood (vol %)	28.3	23.0
P_{50} at 37°C, pH 7.4 (mm Hg)	29.8	30.3
Percent utilization of arterial oxygen	24.8	33.1
Blood pH	7.41	7.44

capacity was lower, they had a lower arterial oxygen content. Because cardiac output, oxygen consumption, and the arteriovenous decrement in oxygen content showed no significant differences between the two groups, the sojourners, with their lower arterial oxygen content, suffered a larger arteriovenous drop in tension than the natives and had slightly lower mixed venous tensions. Further, the resting percent utilization of arterial oxygen was considerably higher in the sojourners. In total, the sojourners maintained almost the same mixed venous tension as the natives, but they realized this result in a somewhat different manner. Their inspired-alveolar tension drop was lower owing to greater hyperventilation, but their arterial-venous drop was greater owing to their lower carrying capacity. Their greater resting percent utilization of arterial oxygen signifies that they had a lesser arterial reserve to exploit during exercise. As a final note, the sojourners had a higher blood pH, indicating that renal compensation for their respiratory alkalosis was not complete after three days.

Man at high altitude: exercise

Thus far we have emphasized the resting state. It is now appropriate to examine some features of exercising man at altitude.

The newcomer to high altitude is typically aware of a reduction in his tolerance for exercise. This situation improves over days or weeks of acclimatization; subjectively the individual feels better and is not tired so easily by exertion. Native highlanders living at 4500–5000 m do not appear to suffer any significant reduction in their capacity for exercise by comparison to man at sea level. It is not unusual, for example, for a resident of Morococha to spend many hours working in the mines and then play a game of soccer in his free time. Mountaineers in the Himalayas have often commented on the remarkable stamina of their native guides and porters. It is an open question whether acclimatized lowlanders gain the full competence of native highlanders over months or years of residence.

An individual's maximal rate of oxygen consumption (aerobic capacity) provides a holistic indicator of the capabilities of all systems responsible for the uptake, transport, and utilization of oxygen. This parameter has received much attention in studies of altitude physiology. Figure 12–4 shows the relationship between maximal oxygen consumption and atmospheric pressure in a group of lowland mountaineers acclimatized at 4000 m for about two months. You will note that their aerobic capacity declined with altitude, especially above 4000 m. The shape of the curve for these mountaineers is qualitatively fairly typical; maximal oxygen consumption is usually depressed more and more for a given decrease in atmospheric pressure, the lower the pressure. Newcomers to altitude, after several days of residence, exhibit a substantially lower aerobic capacity at given altitude than well-acclimatized lowlanders. There is evidence that the performance of newcomers actually deteriorates over the first few days, but subsequently there is gradual improvement. Some reports indicate that lowlanders, after long residence, achieve about the same aerobic capacities as native highlanders, whereas others indicate that lowlanders remain inferior. The performance of highlanders is itself a matter of some controversy. Studies, for example, of Peruvian Indians at 4000 m and of Himalayan Sherpas at 4900 m have indicated that aerobic capacity remains very close to sea-level values up to their altitude of residence and declines only at higher altitudes. This performance is superior to that of the acclimatized mountaineers in Figure 12–4. On the other hand, several studies of residents of Morococha at 4500 m have found aerobic capacities of 36–40 cc O_2/kg/min—values not unlike those observed in the mountaineers at equivalent altitude (Figure 12–4, barometric pressure: 450 mm Hg).

Because man at altitude requires about the same rate of oxygen consumption as man at sea level to sustain a given intensity of exercise, the decline in maximal oxygen consumption at altitude implies a similar decline in maximal work intensity. At any altitude work requiring maximal oxygen consumption elicits peak responses from the respiratory and circulatory systems and is accompanied by a subjective impression of extreme exertion. The demand of a given work load at altitude should thus be gauged by comparing its oxygen requirement to the maximal possible oxygen consumption at that altitude. Exercise that requires 20 cc O_2/kg/min (e.g., riding a bicycle) is accomplished without stress at sea level, for its oxygen demand falls far short of the maximal possible oxygen uptake and entails only a modest acceleration of circulatory and respiratory function. However, at 7500 m (see Figure 12–4, barometric pressure: 300 mm Hg) exercise requiring the same oxygen consumption is extremely taxing, for it represents a maximal intensity of work and elicits peak circulatory and

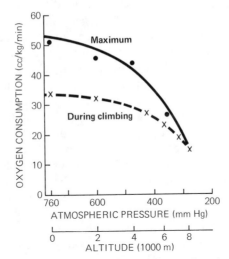

Figure 12–4. Oxygen consumption of mountaineers as a function of atmospheric pressure after about two months of acclimatization to high altitude. Solid curve depicts maximum oxygen consumption; symbols (●) are mean experimental values. Dashed curve depicts oxygen consumption during normal mountain climbing; symbols (X) are mean experimental values—either measured directly, or calculated from body weight and rate of climbing. Note that oxygen consumption for normal climbing approaches or equals the maximum possible consumption at high altitudes. (From Pugh, L. G. C. E. 1964. Animals in high altitudes: man above 5,000 meters—mountain exploration. *In:* D. B. Dill (ed.), *Handbook of Physiology. Section 4: Adaptation to the Environment.* American Physiological Society, Washington, D.C.)

respiratory responses. E. F. Norton, climbing at 8500 m on Mt. Everest, reported his attempts to take 20 consecutive uphill steps; he never could do so. Mountaineers at such altitudes make very slow progress, yet they are working near their maximal capacity and find the effort to be greatly taxing (see Figure 12–4).

As noted in the preceding section, resting ventilatory rate is higher at elevation than at sea level in both highlanders and acclimatized lowlanders, and it increases with altitude. Similarly, the ventilatory rate associated with a given rate of oxygen consumption during exercise is considerably higher than at sea level, the degree of hyperventilation again increasing with the altitude. Peak ventilatory rate is attained when an individual is exercising near his aerobic capacity and is approximately the same regardless of the altitude. In a group of mountaineers, for example, aerobic capacity was about 60% lower at 7400 m than at sea level, but the ventilatory rate during exercise at the aerobic capacity was about the same at both altitudes. It is significant to note that the amount of oxygen transported to the lungs at a given ventilatory rate decreases with altitude because of the decreasing oxygen concentration of the air. Thus an individual ventilating at his peak rate at 7400 m inspires only about 40% as much oxygen as at sea level, for the ambient partial pressure of oxygen is only about 60 mm Hg.

A recent analysis of acclimatized lowlanders has suggested a most interesting relationship between oxygen consumption, aerobic capacity, and ventilation—namely, when oxygen consumption equals a certain percentage of aerobic capacity, the ventilatory rate is about the same regardless of altitude. To illustrate, a group of lowlanders exercising at 60% of aerobic capacity displayed about the same ventilatory rate at 5800 m and at sea level even though their rate of oxygen consumption at 5800 m was only half of that at sea level in accordance with the reduction in aerobic capacity at altitude. The maximal possible oxygen consumption at altitude, even though it may be much reduced by comparison to sea level, is thus the bench mark by which to gauge the ventilatory demand of exercise entailing a certain level of oxygen uptake.

Available information on native highlanders indicates that, at any particular altitude, their ventilatory rate at a given rate of oxygen consumption is about 20% lower than that of short-term sojourners. Thus the natives support their metabolism with a lower alveolar oxygen tension than the sojourners—a similar condition to that seen at rest.

As discussed earlier, cardiac output at rest is about the same at altitude as at sea level in both native highlanders and acclimatized lowlanders. During exercise, cardiac output, in contrast to ventilation, increases in about the same relation to oxygen consumption at altitude as it does at sea level. That is, a given oxygen consumption elicits about the same cardiac response whether the altitude is 0 m, 3000 m, or 6000 m. As a consequence, the arteriovenous decrement in oxygen content at a given level of exertion is also about the same regardless of altitude. Peak cardiac output is elicited when oxygen consumption is at the aerobic capacity, and numerous investigations have shown that the peak cardiac output at altitude is less than at sea level. In acclimatized mountaineers at 5800 m, for example, maximum output was 16–17 liters/min, in contrast to 22–25 liters/min at sea level. This is a most interesting phenomenon and has yet to be explained satisfactorily. Teleologically, the limitation on cardiac output would appear to be a deficit at high altitude, for, other things being equal, a higher output would aid in the maintenance of adequate tissue oxygen tensions. It would also demand increased oxygen delivery to the heart.

Because the aerobic capacity of man at altitude is reduced, one might expect an increased anaerobic capacity. Work on acclimatized lowlanders and native highlanders, however, has revealed that intense exercise is accompanied by a lesser accumulation of lactic acid and smaller oxygen debt at altitude than at sea level. The reasons for the reduction in anaerobic metabolism are not clear. As part of the acid-base compensation for reduced carbon dioxide tension, high-altitude man has a reduced blood bicarbonate concentration. As noted in Chapter 9, bicarbonate plays a very important role in buffering the acid end products of anaerobic metabolism, and it is often pointed out that in the presence of a reduced bicarbonate reserve at altitude, a normal production of lactic acid could lower pH to a detrimental extent. This may be the limiting factor for anaerobic metabolism at altitude.

Tissue-level adaptations

Up to this point we have been concerned with those systems that supply oxygen to the tissue capillaries. As stressed in the preceding chapter, the fundamental site at which the adequacy of oxygen supply and utilization is determined is still further along the path followed by oxygen in the body, namely, at the level of the mitochondria within the cells. Thus we must consider not only respiratory and circulatory transport, but also the factors that affect the flow of oxygen from the capillaries to the mitochondria and the factors that affect its utilization once it has reached the mitochondria. The integrated activity of all these systems determines, in large measure, the range of altitudes over which man can live and the intensity of exercise of which he is capable at any particular altitude.

Most studies of tissue-level changes at high altitude have been performed on mammals other than man, and it remains to be seen if many of the changes observed in these animals also occur in humans. The topic is introduced at this point, nonetheless, as a logical continuation of the preceding sections.

We have seen earlier in Figures 12–1 and 12–3 that mean capillary oxygen tension and mixed venous tension are reduced in resting man at high altitude despite significant adjustments in respiratory and circulatory function. Although there is little information on these tensions during exercise at altitude, there is small doubt that they fall even further during moderate to heavy exertion. The decrease in capillary tension tends to reduce the capillary-mitochondrial tension gradient that provides the driving force for diffusion of oxygen to the mitochondria. This fact has stimulated the search for tissue-level changes that might tend to improve the rate of diffusion or permit the mitochondria to function at lower oxygen tensions than at sea level.

There is evidence for two types of changes that would tend to improve the rate of diffusion: increase in capillary density and increase in muscle myoglobin. Experiments on rats, rabbits, and puppies acclimated to high altitudes have revealed significant increases in the density and diameter of capillaries in the brain. In the rats, for example, the mean distance between capillaries was 12% less at altitude than at sea level, and the mean capillary diameter was twice as great at altitude. The increase in capillary density acts to reduce the average diffusion distance between capillaries and tissue mitochondria, and the increase in capillary diameter acts to increase the area for diffusion across the capillary membrane. Both factors should improve the rate of diffusion. That these changes are observed in the brain is significant in that the central nervous system is particularly sensitive to oxygen deficiency. Increases in the density of patent capillaries and in capillary diameter have also been observed in the heart and skeletal muscles of rats, guinea pigs, and puppies at altitude. In the rats, for instance, the density of patent capillaries was increased by 39% in the myocardium and 30% in skeletal muscles. The changes in the myocardium are perhaps particularly worthy of note not only because the heart is relatively sensitive to oxygen deficiency, but also because its continued ability to generate high cardiac outputs at altitude is a basic prerequisite to exercise. For the most part it is not known whether increases in the density of patent capillaries result from the opening of pre-existing capillaries or the development of new capillaries. There is some evidence for both processes.

The hypothesis that myoglobin facilitates diffusion of oxygen through muscle tissue was reviewed in Chapter 9, and recent evidence suggests that the degree of facilitation increases with the concentration of myoglobin. It is proposed that increases in myoglobin at altitude could thus aid the flow of oxygen from the capillaries to the mitochondria. Studies of dogs in the Andes were the first to demonstrate increases in myoglobin at altitude. Elevated concentrations were found in the heart, diaphragm, and skeletal muscles, the latter, for example, being 60–70% richer in myoglobin than muscles of dogs at sea level. The most marked increase was found in the diaphragm, which is interesting because this structure is much involved in the ventilatory response that is such a central feature in high-altitude life. Since the first studies of dogs, increased myoglobin concentrations have been reported in native and acclimatized man, hamsters, laboratory rats, guinea pigs, llamas, and some rodents native to the Peruvian mountains. In various of these studies the heart, diaphragm, and skeletal muscles have all, again, been implicated.

As noted in Chapter 9, the mitochondria require a certain oxygen tension, 0.2–2 mm Hg according to present estimates, if their rate of oxidative catabolism is not to be oxygen-limited. Up to this point we have been discussing features of the oxygen delivery process that might help to maintain adequate mitochondrial tensions. A considerable amount of work has also been devoted to the possibility that changes in the mitochondria at altitude might allow a normal rate of oxidative catabolism at tensions that would ordinarily be limiting. This could be achieved, for example, by an increase in mitochondrial cytochrome oxidase (cytochrome a_3; see Figure 11–2). With a higher concentration of this terminal cytochrome, the transfer of electrons to oxygen could occur unabated despite some reduction in oxygen tension. Because the ordinary limiting oxygen tension is quite low to begin with, such a change can only lower the limiting mitochondrial oxygen tension by a modest amount, probably 1 mm Hg or less. But even this small improvement could be of important advantage in extending the limits of exercise at altitude.

A recent investigation of cattle has revealed that Herefords born and raised at 4250 m in the Andes show an increase in cytochrome oxidase of about 25% per mitochondrion in the myocardial mitochondria by comparison to Herefords raised near sea level in New England. Similarly, an increased activity of cytochrome oxidase has been reported in skeletal muscles of cats and in various tissues of mice and rats that were acclimated to low oxygen pressures in a hypobaric chamber. However, other studies of rats and guinea pigs have failed to show any change in cytochrome oxidase at altitude. Many other enzymes of oxidative catabolism have also been studied for potential changes during acclimation to altitude. The activity of cytochrome c, for example, has variously been reported to increase, remain unchanged, or decrease slightly. Succinic dehydrogenase activity was recently shown to increase in the hearts of dogs, guinea pigs, and rabbits born and raised at 4400 m. In the Hereford cattle discussed earlier the rate of succinate oxidation by heart mitochondria proved to be 25% higher in animals born and raised at 4250 m, and the rate of NADH oxidation was 50% higher. Mitochondrial suspensions prepared from the hearts of the high-altitude cattle also consumed oxygen at a greater rate than suspensions from low-altitude animals. These and other results suggest that significant changes in mitochondrial enzyme activities occur in at least some tissues of some species.

In conclusion, many lines of evidence point to important, potentially adaptive developments at the tissue level in animals at high altitude. Because our abilities to monitor function at the tissue level in living animals are presently very limited, it has not been possible to quantify the contribution of these developments to the support of aerobic metabolism. Many investigators have expressed the opinion that tissue-level adaptations may be every bit as important to the success of man and other animals at altitude as the more experimentally accessible adjustments in circulatory and respiratory function. Indeed, it is possible that we will find some of the most significant differences between species and between different groups within a species at this level. Many observers, for example, have gained at least a subjective impression that men native to high altitude are distinctly superior in performance to even well-acclimatized lowlanders. It is difficult to explain this on the basis of known circulatory and respiratory differences, and differences at the tissue level may prove critical.

Respiratory and circulatory function in mammals other than man

Three groups of mammals at high altitude may be recognized. (1) First, there are many wild species endemic to mountainous regions, including, for example, pikas and marmots in North America and vicuna and guanaco in South America. Such species are especially likely to have evolved special adaptations to high-altitude life. We may include in this group certain domesticated forms, such as the llama and alpaca, that are descended from highland species and that to this day have primarily a highland distribution. (2) Second, there are many wild species that occur at both high and low elevations. Depending on the extent of genetic admixture, lowland and highland populations of these species may represent distinct physiological races, or they may be genetically similar. (3) Third, man has taken a number of domesticated animals to high altitudes as livestock, pets, or experimental subjects—for example, cattle, mules, dogs, and laboratory rats. These forms have a pre-existing history of lowland life, but some have reproduced at altitude for generations, and depending on genetic admixture with lowland populations, they may have begun a process of genetic adaptation.

Information on these animals at high altitude is limited in a number of respects. There are very few species for which we have a complete set of data on respiratory, circulatory, and hematological characters. Usually data are available on only one or two characters, and because the adjustment to altitude is such a highly integrated affair, it is difficult to interpret the information on these characters without knowledge of the others. Another limiting factor is a taxonomic bias in the animals selected for investigation. Probably at least half of the available information comes from study of domestic animals of lowland descent; and though mammals of many groups occur natively at altitude, the only highland forms to receive concerted attention represent just one order, the rodents, and one family of another order, the camelids.

We shall first review some of the comparative information available on individual features of respiratory and circulatory function and then close with an integrated treatment of a native highland form that has received particular attention, the llama.

A number of studies have approached the basic question of whether highland species have a superior resistance to hypoxic conditions. Twenty-three South American rodents, for example, were placed at 5°–10°C, temperatures that elicit a marked increase in metabolism; and the ambient oxygen tension was then lowered until oxygen consumption was reduced to twice the basal level. The tension that caused this reduction in oxygen consumption ranged from 54 mm Hg to 74 mm Hg in fourteen highland forms and from 68 mm Hg to 122 mm Hg in nine lowland forms, indicating little overlap between the groups and a distinct superiority of the forms native to high altitude. Because the lowland forms had not been acclimatized to altitude, these results do not say whether the lowland forms could achieve the same performance as highland forms with some experience at altitude. This question was approached in a comparison of golden-mantled ground squirrels trapped at 3800 m and laboratory rats that had been born and reared at the same elevation. The rats did well at 3800 m (oxygen pressure: 100 mm Hg), but when the animals were exposed to an ambient oxygen tension of 49 mm Hg for an hour, simulating an altitude of 10,000 m, the rats proved to be distinctly inferior to the ground squirrels. Both species showed reduced oxygen consumption and a steady decline in body temperature, but the loss of temperature was substantially greater in the rats. Although the ground squirrels maintained cardiac output, the rats exhibited a precipitous decline. In one set of tests most of the rats died before the hour of exposure was over.

Knowledge of the dramatic increase in red cell count and oxygen-carrying capacity in man at altitude has stimulated much interest in whether the same phenomenon occurs in other mammals. There is probably more information on this than on any other single character, and the results are noteworthy. Studies of laboratory rats, feral and laboratory house mice, domestic rabbits, guinea pigs, deer mice, Russian field mice (*Apodemus*), and dogs, for example, have all indicated considerable increases in carrying capacity at altitude. These species are either native lowland forms or forms that have both a lowland and highland distribution. In stark contrast, many forms that are more or less restricted to highlands show no increase in carrying capacity. Studies of a dozen species and subspecies of rodents in Peru and Chile indicated that the highland forms had no greater carrying capacities than the lowland forms. In the United States both the golden-mantled ground squirrel and yellow-bellied marmot at 3800 m were found to have unremarkable oxygen capacities, 19–21 vol %. Similarly, the llama, vicuna, and alpaca have modest carrying capacities, 14–22 vol %. In a recent study three llamas were taken from sea level to 3400 m and showed no significant increase in carrying capacity after 10 weeks. It is clear, then, that many mammals native to high altitude do not exploit this hematological adjustment that is so pronounced in many lowland species. Polycythemia is believed to have certain disadvantages, for it increases the effective viscosity of the blood and presumably increases the circulatory effort of the heart. Although this disadvantage is tolerated in many lowland forms at altitude for the sake of increased carrying capacity, it appears that many highland forms have found other ways to circumvent the problems of hypoxia and thus need not cope with the disadvantageous aspects of increased red cell count.

Additional contrasts are found in the realm of oxygen affinity. In man at altitude oxygen affinity decreases, a change believed to increase tissue oxygen tensions at low to moderate elevations. In contrast to what this might lead us to expect, many high-altitude species have relatively high oxygen affinities. As noted in Chapter 9, there is an inverse correlation between P_{50} and body size in mammals, large species tending to have lower P_{50}'s and higher affinities than small species. For their size, the llama, vicuna, alpaca, guanaco, American marmot, yellow-bellied marmot, and golden-mantled ground squirrel all exhibit a distinctly high affinity. The llama and alpaca, for example, would be expected to have P_{50}'s near 26–27 mm Hg according to the standard relationship between P_{50} and size, yet they actually have P_{50}'s of 18–24 mm Hg. The marmots would be expected to have P_{50}'s near 33 mm Hg, but their actual P_{50}'s are 20–22 mm Hg. Interestingly, certain birds resident at high altitude, such as the Andean goose, have also been found to have relatively high affinities by comparison to sea-level birds.

There is comparatively little information on changes in P_{50} with changes in altitude in native highland species. So far as is now known, the Camelidae (camels, llamas, vicuna, and so on) possess little or no red cell 2,3-DPG, and their hemoglobin affinity is not modulated by 2,3-DPG. Although other modulators could be present, this suggests that affinity may not be modified as a function of altitude, and, indeed, recent studies of llamas indicated no significant change in P_{50} upon transfer from sea level to 3400 m. Sheep and goats, both of which are successful at high altitudes, have normal P_{50}'s for their size but, again, do not exhibit 2,3-DPG modulation. Some species are known to behave very much like man. Laboratory rats and guinea pigs, for example, have rather normal P_{50}'s for their size at sea level and exhibit a decrease in affinity and increase in 2,3-DPG at altitude.

The high hemoglobin affinity of many native highland species helps to assure a high degree of loading in the lungs, but it also implies that relatively low tissue oxygen tensions must prevail for unloading. We shall return shortly to the question of how this situation, so different from the response in man and some other lowland species, might fit into the overall adaptation of these animals to their mountain habitat.

Compared to information on hematological characters, our knowledge of respiratory and cardiac function in mammals other than man is limited. Studies of Chilean and Peruvian rodents indicated that highland forms had higher breathing rates than lowland forms, that the rates of the lowland forms tended to increase at altitude, but that the rates of the highland forms did not change significantly on transfer to sea level. Because these were measures of breathing rate only, they may or may not reflect changes in ventilation rate with accuracy. There is good evidence that goats, sheep, cattle, dogs, cats, yaks, and llamas show increased ventilation at altitude. Further, individuals of all these species that have been born and reared at altitude fail to show the blunting of hypoxic sensitivity that occurs in man born and reared at altitude. The absence of blunted sensitivity in yaks and llamas is perhaps particularly significant inasmuch as both species have a long history of life at altitude. At present, then, the blunting of response in native man must be viewed as unique. Studies of llamas, sheep, cattle, and cats have revealed that, as in man, the resting cardiac output is little changed at altitudes of 3400–4300 m.

As we have seen earlier, many species native to altitude exhibit high oxygen affinities and modest oxygen-carrying capacities, whereas man and many other lowland species exhibit normal or reduced oxygen affinities and elevated capacities. At first sight these discrepancies seem to strike at one of the fundamental rationales of comparative physiology. One of the ways in which we try to elucidate adaptive characteristics is by correlational analysis. If we find that a particular characteristic occurs in diverse animals in a given environment, then we are strengthened in our belief that the characteristic is an adaptive trait and not just a result of happenstance. Here, however, we find some animals at high altitude with a lowered oxygen affinity and others with just the opposite. Some have high carrying capacities, and others have normal carrying capacities. Indeed, some authors, noting the contrasting characters of some native highland species, have concluded that the responses of man and some other lowland species are not adaptations to altitude at all. Some, for example, have viewed the polycythemia of man as a pathological state. Also, because man at sea level shows a reduced oxygen affinity in response to anemia and other hypoxic disease states, it could be argued that the lowering of affinity at altitude is not an adaptive response to the reduced ambient oxygen tension, but simply a misplaced manifestation of a protective response against disease.

If we make one fundamental assumption, however, it is not difficult to appreciate how the different responses of various species could all be adaptive in the animals in which they occur. Let us presume that native highland species have evolved tissue-level adaptations permitting normal function at lower capillary oxygen tensions than those required in lowland species. Then we can trace two scenarios: In *lowland species* the circulatory, respiratory, and hematological responses must, first and foremost, maintain a relatively high capillary tension. This is accomplished in part by an increase in carrying capacity even though the polycythemia might impose added demands on the heart. Further, a normal or even lowered oxygen affinity would be favored because, despite its negative implications for loading in the lungs, it assures unloading at relatively high tensions. The reduced percent loading resulting from lowered ambient tensions at altitude is in good part offset by the increased carrying capacity. In *highland species*, with the tissues able to function at lower capillary tensions, the animal can take advantage of the increased loading that results from a relatively high oxygen affinity. (High affinity helps to preserve a high percent saturation in the lungs at lowered ambient tensions, but it does imply lower unloading tensions.) If, through high affinity, the percent loading can be maintained near normal, then there really is no need for increased carrying capacity because arterial oxygen content will remain nearly normal without a change in capacity. By not increasing capacity, the potentially negative implications of polycythemia are avoided.

These scenarios are entirely hypothetical at the moment and place the onus on tissue-level differences of which we know almost nothing. In

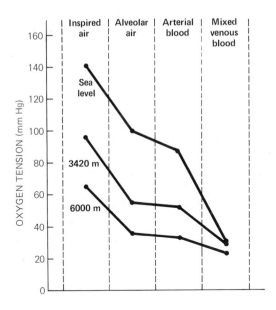

	Sea level	3420 m	6000 m
Carrying capacity, vol %	14.2	15.8	
P_{50} (37°C, pH 7.4), mm Hg	22.7	23.7	
Arterial saturation, %	97.2	92.9	76.8
Venous saturation, %	67.2	63.9	47.0
Cardiac output, cc/kg/min	118	128	130

Figure 12–5. Summary of resting circulatory and respiratory transport in three male llamas (*Lama glama*) born and raised at sea level and studied under three conditions: (1) at sea level, (2) at 3420 m after 10 weeks of acclimatization, and (3) during acute exposure to a simulated altitude of 6000 m following acclimatization to 3420 m. All values are means. The differences in mean carrying capacity and oxygen affinity at sea level and 3420 m are not statistically significant. Alveolar oxygen tensions were estimated from respiratory data. (From Banchero, N., R. F. Grover, and J. A. Will. 1971. Resp. Physiol. 13: 102–115.)

any case they should serve to illustrate and emphasize the principle that all individual responses to altitude must ultimately be interpreted in the context of all other responses, such is the tightly integrated nature of the process of oxygen delivery and utilization.

We shall close with a brief review of one of the few thorough studies of oxygen transport in a native highland species. Figure 12–5 depicts parameters of respiratory and circulatory function in three llamas that were born and reared near sea level and then taken to Climax, Colorado, for study at 3420 m. They were allowed 10 weeks of acclimatization at Climax and while there were exposed acutely to a simulated altitude of 6000 m.

A notable feature of their function at sea level is the relatively low resting mixed venous oxygen tension, 30 mm Hg. From Chapter 9 you will recall that resting man at sea level has a mixed venous tension near 40 mm Hg. Both man and the llama realize about the same percent arterial saturation at sea level, and both show about the same arteriovenous decrement in oxygen content. The lower venous tension of the llama results from two factors. First, the llama has a lower carrying capacity, signifying that venous percent saturation must fall more to maintain a given yield of oxygen per circulatory cycle; venous saturation in resting man is near 75%, but in the llama it is 67%. Second, the llama displays a higher oxygen affinity, meaning that a lower venous tension is required to realize a given fall in percent saturation.

After 10 weeks of acclimatization to 3420 m the llamas showed no significant changes in carrying capacity or oxygen affinity. In further contrast to the human response to this altitude, resting ventilation did not increase at all, and the tension drop between inspired and alveolar air was the same as at sea level. Nonetheless, because of the high affinity of the llama's blood, arterial saturation fell only slightly, to 92.9%, and arterial oxygen content remained high. We see, then, that the high affinity of the blood protects arterial loading up to a fairly high altitude. At the arterial tension of the llamas, 52.6 mm Hg, human blood of normal affinity is about 87% saturated; man ameliorates his condition by an increase in ventilation and increase in carrying capacity. The cardiac output of the llamas (as in man) remained virtually unchanged at 3420 m. With arterial saturation only slightly reduced, venous saturation and venous tension had to fall only slightly to maintain the same oxygen delivery as at sea level. Again there is a significant contrast between mixed venous tension in man and the llama, for man at 3450–3700 m maintains venous tensions of 34–35 mm Hg.

At 6000 m the llamas exhibited an increase in ventilatory rate, and the inspired-alveolar tension drop was reduced. Arterial saturation fell, and, with cardiac output unchanged, venous saturation declined to a comparable extent. The venous tension was quite low, 23 mm Hg. Man at this altitude exhibits a greater ventilatory response than the llama and maintains a higher alveolar oxygen tension, about 42 mm Hg in comparison to about 36 mm Hg in the llama. Yet because of the reduced affinity of man's blood, his percent arterial saturation is somewhat lower than the llama's.

All of these results indicate that the llama routinely functions at lower capillary tensions than man. This important feature, coupled with a high blood oxygen affinity, allows the llama to fare well at altitude with a very modest carrying capacity and with a lesser ventilatory response than man.

SELECTED READINGS

Dill, D. B. (ed.). 1964. *Handbook of Physiology. Section 4: Adaptation to the Environment.* American Physiological Society, Washington, D.C.

Dill, D. B. 1968. Physiological adjustments to altitude changes. J. Amer. Med. Assoc. 205: 747–753.

Lenfant, C. 1973. High altitude adaptation in mammals. Amer. Zool. 13: 447–456.

Lenfant, C. and K. Sullivan. 1971. Adaptation to high altitude. N. Engl. J. Med. 284: 1298–1309.

Margaria, R. (ed.). 1967. *Exercise at Altitude.* Excerpta Medica Foundation, Amsterdam.

Porter, R. and J. Knight (eds.). 1971. *High Altitude Physiology: Cardiac and Respiratory Aspects.* A Ciba Foundation Symposium. Churchill Livingstone, Edinburgh.

Yousef, M. K., S. M. Horvath, and R. W. Bullard (eds.). 1972. *Physiological Adaptations. Desert and Mountain.* Academic, New York.

See also references in Appendix.

13 THE PHYSIOLOGY OF DIVING IN VERTEBRATES

Accomplished divers are found among all the classes of terrestrial vertebrates. The ability of some to remain submerged for long periods has fascinated man since times unknown and has attracted the interest of physiologists for over a century. We shall first center on the physiology of diving in mammals and birds, citing comparative information on the other classes as appropriate. The amphibians and reptiles will be treated more specifically toward the end of the chapter.

Among birds and mammals, the longest and deepest dives are performed by certain of the marine cetaceans and pinnipeds. Many whales and seals have evolved diving abilities that enable them to forage for food for considerable periods and often at great depths. They are thus able to tap resources that would be unavailable to strictly surface-dwelling forms. Because of the difficulty of observing these animals, there is a paucity of data on their natural diving behavior; nonetheless we have sufficient information to show that their capabilities are remarkable. Perhaps the most complete information on any one species comes from a recent study of over 900 dives performed by 31 Weddell seals (*Leptonychotes weddelli*) under the antarctic ice. The animals carried instruments to record the depth of their dives and had to surface at holes cut in the ice, but otherwise they were entirely free to display their spontaneous behavior. Most dives lasted five minutes or less and extended to depths of 100 m or less. However, there were many dives lasting 5 to 20 minutes and extending to 100–400 m. The longest dive observed lasted over 40 minutes, and the deepest dive was to around 600 m (about a third of a mile). In subsequent work Weddell seals have been found to dive for as long as 70 minutes. Data on other pinnipeds are less complete. A ringed seal (*Pusa*) was observed in the field to dive for 21 minutes, and in the laboratory harbor seals (*Phoca*) survived forced submergence for 23 minutes without apparent ill effects. Stellar sea lion (*Eumetopias*), harp seal (*Pagophilus*), and northern elephant seal (*Mirounga*) have been hooked on lines in the ocean at depths of 180 m or more. The capacities of some whales appear to exceed those of the seals. Sperm whales (*Physeter*) and bottlenose whales (*Hyperoodon*) are reported to dive regularly for periods of about an hour, and there are observations of dives lasting up to two hours in the latter species. A harpooned fin whale (*Balaenoptera*) dove to 355 m without apparent ill effect, and sperm whales have been found entangled in deep-sea cables at astounding depths of 900–1100 m, having apparently drowned after becoming entrapped. Compared to many seals and whales, the porpoises and dolphins dive for relatively brief periods, but some can reach considerable depths. Bottlenosed porpoises (*Tursiops*) trained to hold their breath have been found to remain submerged for a maximum of 5 to 7 minutes. A porpoise named Tuffy was trained to dive in the open ocean and was able to reach at least 300 m. In one set of experiments Tuffy made 10 dives to 200 m in a span of just 34 minutes, spending over a third of the period at depths in excess of 100 m.

The pressures experienced by marine mammals at depth are enormous, and their ability to withstand these pressures is of just as much interest as their ability to remain submerged for long periods. Pressure increases by about 1 atm for every 10 m of depth.

Information on diving mammals of fresh waters and the inshore coastal waters is relatively sparse. Because such waters are, in general, comparatively shallow, these animals need not be capable of exceptionally deep dives even to be able to forage on the bottom. The sea otter (*Enhydra*) of our West Coast is one of the most aquatic of the mustelids. It collects sea urchins, shellfish, crabs, and even fish in relatively brief, but often frequent, dives in shallow waters (5–40 m usually). When forceably submerged, muskrat (*Ondatra*) were found to survive for 12 minutes and beaver (*Castor*) for 15 minutes. In contrast, nondiving mammals such as laboratory rats, cats, and dogs survived for only 2–4 minutes.

A good number of birds dive for food. Typically, they do not descend to particularly great depths, but some can remain submerged for long periods. Domestic ducks, for example, can stay under water for 10 to 20 minutes, and penguins have been observed to dive for 5 to 7 minutes.

Studies of various groups of trained human divers have indicated that man is limited to about 3 minutes of submergence when at rest and about 90 seconds of submergence when swimming or diving; some exceptional individuals can do slightly better. In certain parts of the world, diving for pearls or food is adopted as a career, and these divers have attracted considerable interest. In Korea and Japan people (mostly women) known as ama dive in the offshore waters for shellfish and edible seaweeds. Many begin diving in their adolescence and continue for the rest of their working lives. Their abilities are typical of those exhibited by other diving peoples. Some ama dive in the shallow waters. Their dives are not particularly impressive in themselves, ordinarily lasting only about 30 seconds and being to depths of 4–6 m. However, it is remarkable that they dive repeatedly with only about 30 seconds of rest (and breathing) between dives, averaging around 60 dives per hour. Other ama dive in deeper waters, typically with assistance during descent and ascent so that their time at the bottom can be maximized. They carry weights to aid descent and are pulled back to the water's surface on a rope by an assistant in a boat. These ama routinely dive for periods of 60 to 80 seconds and reach depths of 15–25 m; with about a minute of rest between dives, they average about 30 dives per hour. The extremes recorded for human performance have been attained by competitive divers. The record holders have reached depths in excess of 70 m (using assistance in descent and ascent) in dives lasting about 2 minutes. Dives of this length and depth tax man to his limits but are routine and unremarkable in many marine mammals.

Some basic physiological responses during prolonged dives

Most or all of the accomplished mammalian and avian divers exhibit a fundamentally similar set of responses during the dive. A review of this common response pattern will set the stage for a more thorough examination of its individual elements.

By 1870 it was clear that diving species survive submergence for a longer period than related, nondiving species, and by the turn of the century it was recognized that the internal oxygen stores of divers are inadequate to meet their full metabolic demands aerobically over the duration of the dive. In 1934 Laurence Irving presented a hypothesis to explain how metabolism could be supported over a prolonged dive, and he and P. F. Scholander then performed pioneering experiments that gave support to the hypothesis. It is of great credit to these men that they drove to the core of the matter with remarkable insight. Irving recognized that certain tissues—notably the heart and central nervous system—are quickly damaged if their oxygen supply is inadequate, but other tissues—notably the skeletal muscles—are much more tolerant of oxygen deprivation and have well-developed anaerobic capabilities. Upon submergence, the animal carries a certain amount of oxygen in internal stores. If all tissues were simply to draw on these stores according to their demand for oxygen, oxygen tensions throughout the body would fall quickly to levels that would impair the oxygen-sensitive tissues, and the animal would have to surface despite the fact that the skeletal muscles and other less-sensitive tissues could continue to function for some time. As Irving pointed out, the animal can lengthen its period of submergence by using some of its stores preferentially to meet the demands of the oxygen-sensitive tissues, thus maintaining relatively high oxygen tensions in those tissues while allowing tensions in the less-sensitive tissues to fall to low levels. The preferential distribution of some oxygen to the oxygen-sensitive tissues is achieved by adjustments of circulatory function.

During the dive, circulation to the appendages, trunk muscles, gut, kidneys, and certain other tissues is substantially curtailed through vasoconstriction in the vessels supplying these parts. The skeletal muscles, deprived of active blood flow, can make use of the hemoglobin-bound oxygen sequestered in their capillaries and of myoglobin-bound oxygen, but as these stores are depleted, they resort to anaerobic catabolism, and lactic acid accumulates in quantity. With the circulation to many parts of the body curtailed, the heart primarily pumps blood between itself and the lungs and head. The oxygen stores of this blood are thereby reserved primarily for the oxygen-sensitive tissues, and whatever oxygen is extracted from the air in the lungs is likewise delivered preferentially to these tissues. Consequently, adequate oxygen tensions can be maintained for a long period. Because circulation to the skeletal muscles is strongly limited, lactic acid tends to remain sequestered in the muscles during the dive. This is important in limiting the buildup of acid in the actively circulated parts of the body. Once the animal surfaces for air, circulation to the muscles is restored, and there is a sudden rise of lactic acid in the blood. The observation that circulating lactic acid increases *after* the dive was one of the early indications of the circulatory adjustments during diving.

The oxygen stores of divers

Although amphibians and some reptiles can draw appreciable amounts of oxygen from the water when diving, birds and mammals are limited to internal stores carried at the time of submergence. The three major stores are oxyhemoglobin, oxymyoglobin, and oxygen within the lungs. Much interest has centered on whether these stores are larger in diving species than in nondiving species. In this discussion and throughout the chapter it is essential to remember that different species of diving mammals and birds have different life histories and have been subject to different evolutionary pressures. Thus we should not be surprised if easy generalities do not emerge.

The amount of oxygen available as oxyhemoglobin depends on the carrying capacity of the blood, the volume of blood, and the average percent saturation at the time of submergence. The former two parameters can be measured relatively easily, but the latter is difficult to determine inasmuch as the animal must be studied under natural conditions and one must know the percent saturation of both arterial and venous blood and the proportion of each in the bloodstream. Data on carrying capacity and blood volume indicate the potential magnitude of the oxygen store and provide a basis for interspecific comparison if it is assumed that different species exhibit about the same average percentage of loading at submergence. Most estimates of the actual oxyhemoglobin store are based on assumptions regarding the level of saturation in various parts of the bloodstream.

Some species of diving mammals have exceptionally high carrying capacities, but others have capacities that are well within the ordinary range for terrestrial mammals. A high carrying capacity cannot be called a typical feature of diving species. Bottlenosed dolphins, north Pacific pilot whales, gray whales, northern fur seals, stellar sea lions, and sea otters, to cite some examples, have carrying capacities between 16.5 vol % and 22 vol %—quite ordinary for mammals. On the other hand, species with especially high carrying capacities include the pygmy sperm whale (32 vol %), harbor seal (26–29 vol %), Weddell seal (29–36 vol %), and ribbon seal (34 vol %). Among pinnipeds there is a tendency for species that undergo long dives to have higher carrying capacities than those that perform shorter dives. Thus, as noted above, the Weddell seal, harbor seal, and ribbon seal have especially high carrying capacities, whereas the northern fur seal and stellar sea lion have only modest capacities. A study of carrying capacities in porpoises yielded the following average values for three species: the Dall porpoise, 28 vol %; Pacific white-sided porpoise, 23 vol %; bottlenosed porpoise, 20 vol %. These differences were correlated with other characteristics of the species. Maximum swimming speed, for example, decreases in the same order as carrying capacity, and there is some evidence that Dall porpoise habitually dive to greater depths than, at least, the Pacific white-sided porpoise. Although information on oxygen-carrying capacities of diving birds is relatively meager, it is clear that there is again a considerable range of variation. The domestic duck, for example, has a modest carrying capacity of about 17 vol %, whereas the guillemot *Uria troile* has a capacity of 26 vol %.

There is a tendency for the hemoglobin of diving mammals to have a somewhat lower affinity for oxygen than that of comparably sized terrestrial mammals. This may be significant during diving inasmuch as the blood will release its oxygen at relatively higher oxygen tensions. The Bohr effect in diving mammals presents a very mixed picture. Some species exhibit a Bohr effect of ordinary magnitude for mammals; others display an especially large effect. In the latter the buildup of carbon dioxide and lactic acid during the dive will cause an unusually great reduction in oxygen affinity and thus promote unloading to an especially large degree.

Accurate determinations of blood volume have been performed on only a relatively few species of diving birds and mammals, but the evidence indicates a tendency toward high values. Man, dogs, horses, and rabbits have average blood volumes of about 60–110 cc/kg of body weight. By contrast, such accomplished divers as the harbor seal, ribbon seal, and Weddell seal have volumes of 130–160 cc/kg. The northern fur seal, stellar sea lion, and sea otter have more modest values of 90–110 cc/kg, but these are still relatively high. Blood volumes of 40–90 cc/kg have been reported in pigeons, pheasants, and chickens. In the domestic duck, a diver, the value is near 100 cc/kg, and values of 90–140 cc/kg are reported for a guillemot, puffin, and penguin.

By multiplying the blood volume by the carrying capacity of the blood, the maximum possible oxyhemoglobin store is obtained. This figure is only tangentially meaningful to the normal physiology of the animal because the entire volume of blood is never fully oxygenated. However, as noted earlier, it can be used in interspecific comparisons if we assume that various species dive with their blood at about the same average level of saturation; in this case a species with twice the possible oxyhemoglobin store of another, for example, would carry twice as much hemoglobin-bound oxygen upon submergence. With the understanding that this assumption is purely an assumption and likely, at best, to be only approximately true, we can look briefly at the data on maximum oxyhemoglobin capacity. It tends to be relatively large in diving mammals and birds. In man and horses the value is around 14–15 cc O_2/kg. Very much higher capacities are found in those seals that combine the advantages of both high blood volume and high carrying capacity. Average reported values for the Weddell seal, harbor seal, and ribbon seal range between 39 cc O_2/kg and 47 cc O_2/kg—as much as three times higher than in man and the horse. In the northern fur seal and stellar sea lion, which have only modest carrying capacities but have blood volumes toward the high end of the range for terrestrial mammals, maximum oxyhemoglobin capacity is 21 cc O_2/kg and 16 cc O_2/kg, respectively. Among porpoises, Dall porpoise have a very high blood volume and high carrying capacity and possess maximum blood oxygen stores of about 39 cc O_2/kg. In contrast, the bottlenosed porpoise can store only about 14 cc O_2/kg. The Pacific white-sided porpoise is intermediate at 25 cc O_2/kg. Earlier, some of the ecological differences among these species were noted. In chickens and pigeons oxyhemoglobin stores can amount to 5–17 cc O_2/kg. Among diving species, the domestic duck and a penguin gave values of 17–18 cc O_2/kg, but much higher values were obtained for a guillemot and puffin, 28–34 cc O_2/kg.

All in all, it is clear that diving birds and mammals tend to have higher capacities for storage of oxygen as oxyhemoglobin than terrestrial species, and in some cases their capacities are extraordinarily greater. The blood store of oxygen is made available in some measure to all tissues during the dive. As noted earlier, some of it is preferentially restricted for use by the oxygen-sensitive tissues through circulatory adjustments.

The store of oxygen in oxymyoglobin is almost certainly available only to the muscles. On superficial grounds this is so because the blood does not circulate freely between the major skeletal muscles and other parts of the body during submergence, but there is a more basic reason. Even if circulation were unaltered during the dive, the myoglobins, because of their very high affinity for oxygen, would not yield appreciable amounts of oxygen to the blood for transport elsewhere until blood tensions had fallen to low and threatening levels. In oxymyoglobin, then, the muscles have their own private store of oxygen.

Myoglobin concentrations tend to be high in the skeletal muscles of diving mammals and birds—so high that the muscles of some proficient divers are reported to appear almost black. In man skeletal muscles contain about 4–5 mg of myoglobin per gram of wet weight, and in horses values of 4–9 mg/g are reported. By contrast, harbor seals have 55 mg/g and ribbon seals, about 80 mg/g. Even such less effective divers as the northern fur seal, stellar sea lion, and sea otter have 24–35 mg/g. Interestingly, however, the sea cow (*Trichechus*), which is sluggish and dives only to shallow depths, has about the same myoglobin levels as man.

Discussion of the usable oxygen store in the lungs is complicated by a number of factors. First, there is the question of how much air is contained in the lungs at the time of submergence, and, second, there is uncertainty about how much oxygen can be extracted from the lungs during the dive. Some of the important features of external respiration will be discussed in more detail in a subsequent section. Here we shall simply introduce some of the important considerations regarding the lung as an oxygen store.

At first it might appear that a large store of oxygen in the lungs would typically be advantageous to diving mammals. However, there are a number of factors that can militate against this conclusion. First, the density of air is so much less than that of water that the volume of air contained in the lungs exerts a strong effect on buoyancy. The buoyant effect of a large store of air will increase the effort of diving and swimming below the surface. This effect will be particularly important during shallow dives and will diminish as the air volume is compressed by the increasing pressures at depth. The fact that the lungs are compressed at depth leads to another consideration. It is believed that the alveoli are typically the first elements of the lung to collapse as the air volume is reduced; in other words, at depth the air comes to be contained largely in the anatomical dead space (bronchioles, bronchi, and trachea). Once this occurs, whatever oxygen is contained in the lungs is largely unavailable, for it is in the alveoli that oxygen is transferred from the lungs to the blood. Thus although diving mammals can readily make use of their pulmonary oxygen store during shallow diving, this is not true at depth; and in many species this again calls into question the utility of carrying a large amount of air in the lungs. A final consideration, which will be examined in detail later, is that a large pulmonary air store implies not only a large oxygen store, but a fourfold greater nitrogen store. As the air is compressed at depth, the partial pressure of the nitrogen (and oxygen) is increased, and there is the possibility that a large amount of nitrogen will become dissolved in the blood and tissue fluids. This sets up the potential for developing the bends when the animal resurfaces, and the greater the initial store of air (and nitrogen), the greater is the potential. With all these factors to be considered, it is not surprising to find quite a bit of variation in the use of the lungs as an oxygen store in diving mammals.

Lung capacity is defined to be the amount of air contained in the lungs at full inflation. Capacity per unit of body weight does not show consistent differences in diving mammals as compared to terrestrial mammals. This comparison may not, however, be the most meaningful one. It may be more appropriate to consider lung capacity relative to the weight of the tissues that are comparatively active metabolically, and, because fat is a relatively inactive tissue, this would imply consideration of lung capacity relative to lean body weight. This makes a difference, for the marine mammals tend to have particularly large amounts of fat in the form of blubber. Lung capacity per unit of lean weight does tend to be somewhat higher in diving mammals than in terrestrial mammals. One species in particular has an indisputably large lung capacity—more than twice as great as that of terrestrial mammals of equivalent size. This is the sea otter. Most marine mammals spend considerable time floating at the water's surface, and most are aided in this by the fact that they have large amounts of blubber, a tissue that is less dense than water and thus contributes significantly to buoyancy. The sea otter, however, has no blubber and not only floats but is dependent on floating for its mode of feeding. It rests on its back at the surface, places a rock on its abdomen, and cracks open sea urchins, shellfish, and other prey by striking them against the rock. The extraordinary lung capacity of this animal is undoubtedly important in providing the buoyancy needed for this habit.

The size of the oxygen store carried upon submergence depends not only on lung capacity, but also on the extent to which the lungs are inflated. Some diving mammals dive on inspiration and thus may carry something approaching their lung capacity. This appears to be true, for example, in sea lions and porpoises. Deep-diving pinnipeds dive on expiration. Measures on a number of seals have shown that their lungs are filled to only 20–60% of capacity upon submergence. This is very interesting, for even on this simple ground these animals are not making full use of their potential pulmonary oxygen store. It should also be noted that quite a few of the deep divers have relatively small lungs. Per unit of body weight, the harbor seal, ribbon seal, Weddell seal, and some of the whales have significantly lower lung capacities than more shallow-diving species such as the northern fur seal, stellar sea lion, and walrus. It thus appears that there is little premium placed on a large pulmonary store in many deep divers. Most of these animals have particularly large amounts of blubber, and it may be that the added buoyancy of large, fully inflated lungs would pose too great an impediment to diving. But it may also be that they habitually dive to such depths that alveolar collapse under the hydrostatic pressure largely voids the utility of a large pulmonary store, which leads to the next important consideration, namely, how much do diving mammals actually draw on whatever pulmonary oxygen they carry. Unfortunately this is a matter of which we know relatively little.

One of the most elegant experiments was performed on the bottle-nosed porpoise Tuffy. Tuffy would submerge after inhalation and, when breath holding just under the surface or swimming at a modest depth of 20 m, would make thorough use of his pulmonary oxygen, reducing the oxygen concentration in the lung air from about 13% to 3% in a matter of a few minutes. During deep dives, however, utilization of pulmonary oxygen virtually ceased below 100 m. If Tuffy dove to 100 m and returned, pulmonary oxygen fell from about 13% to 7%, but in dives to 300 m, which lasted almost three times as long, pulmonary oxygen fell only slightly more, to 5%. The limited availability of pulmonary oxygen below 100 m probably resulted from alveolar collapse under the great pressures. Qualitatively similar results were obtained on Weddell seals. During dives to depth the average rate of oxygen extraction from the lungs of the seals was about three times lower than during breath holding at the surface; further, lower alveolar tensions were reached during simple breath holding than during dives. Both observations indicate impairment of oxygen uptake during parts of the dives. These experiments on Tuffy and the seals show clearly that the use of the pulmonary oxygen store is a complex matter requiring much further investigation and suggest that the store is more likely to be utilized extensively in shallow divers than in deep divers.

Despite all the uncertainties involved in appraising the total usable oxygen stores of diving animals, at least two points seem clear. First, some diving mammals have very much greater stores than terrestrial mammals. This is illustrated in Figure 13-1, in which total oxygen stores have been estimated for man and five species of marine mammals by applying certain assumptions to available data. Note that in such proficient divers as the harbor seal and ribbon seal (and Weddell seal as well), the blood and myoglobin stores alone far exceed the total stores of man. It is also interesting to compare these seals with the fur seal and sea lion, which are believed to be less-accomplished divers. The immense lung capacity of the sea otter is vividly reflected in the data; unfortunately, we do not know how much of this capacity is actually used during diving.

The second point made clear by available information is that the oxygen stores of diving mammals and birds, despite their magnitude, are entirely inadequate to sustain a rate of oxygen consumption during submer-

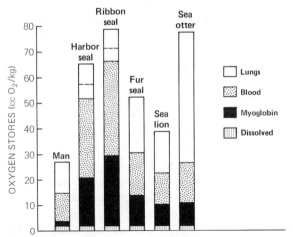

Figure 13–1. Total estimated oxygen stores of six species: man, harbor seal (*Phoca vitulina*), ribbon seal (*Histriophoca fasciata*), northern fur seal (*Callorhinus ursinus*), stellar sea lion (*Eumetopias jubata*), and sea otter (*Enhydra lutris*). Oxygen dissolved in tissues and body fluids other than blood was estimated as 2 cc/kg for all species. Oxygen present as oxymyoglobin was calculated from known myoglobin concentrations of skeletal muscle under the assumption that muscle constitutes three tenths of body weight. Oxygen present in blood was calculated by assuming that one third of the blood is arterial and 95% saturated and that two thirds of the blood is venous and at 5 vol % lower oxygen content than arterial blood. Lung oxygen stores were calculated by assuming that the lungs are fully inflated and contain 15% oxygen by volume. The assumption of full inflation is probably reasonably realistic for man and the sea lion, which dive on inspiration. Ribbon seal and harbor seal dive on expiration, and dashed lines show lung stores assuming the lungs to be 40% inflated. The degree of inflation upon submergence is unknown in the sea otter and fur seal. (Data for marine mammals from Lenfant, C., K. Johansen, and J. D. Torrance. 1970. Resp. Physiol. 9: 277–286.)

gence equivalent to that observed in these animals while they are at rest and breathing air. In his pioneering work P. F. Scholander calculated the total oxygen stores of a 29-kg hooded seal (*Cystophora*) to be 1520 cc O_2. The resting oxygen consumption of this seal on land was 250 cc/min. Thus the seal could sustain its resting, aerial rate of oxygen consumption for about six minutes during submergence if it could completely utilize all stores. Actually, the seal could remain submerged for at least 15 to 18 minutes. Similar calculations have been performed for many other species of diving birds and mammals, and even with generous assumptions concerning the magnitude and utilization of stores, dives turn out to last from two to several times longer than would be predicted if the animal were to function aerobically at a rate comparable to that during rest in air. Among non-diving species the length of survival during submergence is more closely predictable from the magnitude of oxygen stores. The fact that the increased oxygen stores of diving mammals do not entirely account for their diving abilities is illustrated by comparison of man and the harbor seal. Although the weight-specific oxygen stores of the seal are 2 to 2.5 times those of man (Figure 13–1), the seal can dive for over 12 times as long.

Circulatory adjustments during diving

The earliest observation of cardiovascular involvement in the diving response was the demonstration by Paul Bert in 1870 that the heart rate of ducks decreased from about 100 to 14 beats/min during diving. Such bradycardia (slowing of the heart rate) has since been observed in numerous other species and is one of the most characteristic elements of diving physiology. To cite some further examples, harbor seals trained to dive in a pool showed an almost immediate drop from 120–140 beats/min to 20–40 beats/min upon submergence. Weddell seals averaged 64 beats/min when resting in the sea and breathing air; during short, shallow dives, their heart rate fell quickly to a mean of about 27 beats/min. Heart rate was also monitored over the first 30 to 45 seconds of deep dives in Weddell seals, and there was a clear trend for the rate during this initial period to be lower, the longer the subsequent duration of the dive. Thus in dives lasting 4 to 5 minutes, the initial decline in heart rate was to about 30 beats/min, but in dives lasting over 40 minutes the decline was to about 15 beats/min. This is one of the few studies on animals diving spontaneously in their natural habitat and suggests that the cardiac response anticipates the length of the dive.

Diving bradycardia has been observed in probably every tetrapod vertebrate studied, including species that are not habitual divers, such as dogs, armadillos, pigs, and man. Typically, the onset of bradycardia is not so abrupt and the depression of heart rate not so profound in primarily terrestrial species as in diving species. A 30–50% reduction in heart rate is fairly commonly observed in diving humans; some individuals show an even more pronounced effect. Among the ama divers, heart rate was found to fall from about 100 beats/min to about 70 beats/min over the first 20 seconds of submergence and to about 60 beats/min after 30 seconds. Interestingly, it has been shown that a number of fish exhibit bradycardia when out of water, a situation that impairs external respiration in the fish just as diving does in a bird or mammal. Grunion, which crawl out of the water for brief periods to breed in the spring, exhibit an abrupt and profound bradycardia when in air, and flying fish show a similar response during simulated flights. As noted in Chapter 11, it is also becoming increasingly clear that bradycardia is a common response among fish when the oxygen tension in their water is low. All this evidence indicates that bradycardia is a widespread and probably primitive response to asphyxia in the vertebrates.

The existence of profound bradycardia in diving animals was known for a long period before its significance became clear. In the 1930s and 1940s evidence accumulated to show that bradycardia is simply one element of a much more pervasive cardiovascular response, and as we shall examine in more depth shortly, it is believed to be an adjustment to the decreased dimensions of the active circulatory path during the dive. First we shall discuss the peripheral vascular response in diving mammals and birds.

As noted earlier, circulation to many parts of the body is curtailed during submergence. This was first studied directly by Irving using hot-wire flowmeters; in this method heated wires are placed in various parts of the body, and the degree of cooling of the wires is taken as an indication of the rate at which circulation of the blood carries heat away from the site of the wires. In studies of several species, including beaver, muskrat, and rabbits, he observed that during asphyxia blood flow to the muscles is curtailed whereas that to the brain is maintained. Irving also exposed the intestinal mesentery of seals and observed intense vasoconstriction during immersion of the animals in water. Scholander made incisions in the pectoral muscles of penguins and the toes of seals and penguins and observed that bleeding, although profuse in air, occurred slowly or stopped altogether during submersion. These and other early studies indicated a pervasive and selective redistribution of blood flow during diving. More recent studies have expanded our knowledge of this phenomenon.

Rubidium-86 injected into the bloodstream is taken up by the various tissues of the body at a rate that depends partly on their blood supply. In 1964 Johansen evaluated the distribution of blood during submergence in ducks by analyzing the uptake of rubidium-86. He found evidence of curtailed circulation in pectoral muscles, leg muscles, neck muscles, the intestine, the gizzard, much of the skin, the kidneys, and the pancreas. On the other hand, blood flow was maintained or increased in the brain, myocardium, some feeding muscles, esophagus, eye, liver, thyroid, and adrenal glands. The duck dives to forage for food, and it is of interest that blood flow was maintained in some of the tissues directly involved in feeding.

More recently, modern ultrasonic flow detectors have been chronically implanted around certain major arteries of harbor seals. Such detectors are inserted surgically and, after the animal has recovered, provide a direct and continuous measure of blood flow. During both voluntary and forced submergence, flow in the renal artery and abdominal aorta fell almost to zero. Flow in the common carotid artery to the head was reduced during diving but remained steady at a substantially higher rate than in the renal and abdominal arteries. One of the insights provided by this study was observation of the extreme rapidity of these vascular responses. Flow in the renal and abdominal arteries fell virtually instantaneously upon submergence and recovered equally rapidly on return to air breathing. Similar vascular responses, though not so profound, were observed in dogs. Flow in the superior mesenteric, renal, and abdominal arteries fell to low values when the dog's nose and mouth were forced under water, but flow in a coronary artery was maintained. Reduced blood flow in the legs of man has also been shown during simulated diving.

One of the most striking illustrations of the vascular response in diving animals comes from studies using radiopaque (contrast) materials injected into the bloodstream. Once administered these materials flow with the blood, and because they can be visualized by x-ray, their distribution in the blood vessels can be monitored and used to determine blood distribution. Figure 13–2 shows the distribution of contrast material in arteries of the posterior trunk of a harbor seal before and after diving. Note that many of the arteries that have received contrast material in the nondiving situation receive no perceivable amounts during diving, reflecting the profound vasoconstriction of the vascular beds. In these studies it was found that cerebral circulation was maintained without apparent alteration during diving, but circulation to the skeletal muscles, skin, flippers, kidneys, spleen, and liver was drastically curtailed or stopped.

In Chapter 10 we emphasized the fact that the arterioles are the typical sites of vasomotor control in birds and mammals. Quite a different picture emerges in the diving birds and mammals during submergence, for a number of studies have indicated that pronounced vasoconstriction occurs in sizable arteries, well upstream of the arterioles. This is evident in

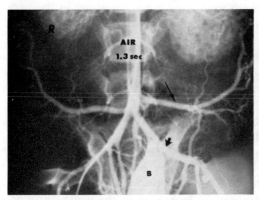
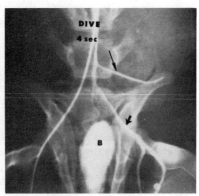

Figure 13–2. Angiograms of the posterior trunk of a harbor seal (*Phoca vitulina*) during air breathing and during submergence. Contrast material was injected into the aorta and arrived at the posterior trunk of the animal in the major artery positioned at the top-center of the photographs. The thin arrows point to an artery supplying the flanks, and the thick arrows point to an artery supplying the hind flippers. *B* marks the bladder. Note the severe restriction of peripheral blood flow during diving. Times indicate the interval between injection of contrast medium and exposure of the angiogram. The diving angiogram depicts the maximal amount of arterial filling observed during submergence. (From Bron, K. M., H. V. Murdaugh, Jr., J. E. Millen, R. Lenthall, P. Raskin, and E. D. Robin. 1966. Science 152: 540–543. Copyright 1966 by the American Association for the Advancement of Science.)

Figure 13–2; the vessels visualized during air breathing are arteries of macroscopic size, and it can be seen that flow to many of them is cut off virtually at their connections with the major trunk vessels during diving. Similar events are observed in other parts of the vascular tree in harbor seals, such as in the renal arteries.

The vasoconstrictive response is believed, at least in part, to be mediated by the sympathetic nervous system, and a recent study has revealed important differences in the innervation of the arteries in the harbor seal and a nondiving species, the cat. In the cat arterial sympathetic fibers do not penetrate into the muscular layers of the arteries. In the seal the same is true of the carotid arteries, pulmonary arteries, coronary arteries, and systemic aorta; these are vessels that remain open during diving. However, sympathetic fibers do penetrate into the muscular layers of many arteries that are constricted when the seal dives; included, for example, are the renal arteries and arteries that supply the gut and hind flippers. The direct sympathetic innervation of the arterial muscle layers is postulated to be an important element in the vigorous vasoconstrictive response of these vessels.

What are the possible advantages to diving animals of cutting off blood flow at the level of major arteries rather than at the arterioles? There may be benefits in centralizing vascular control; the circulation to a major muscle mass, for example, can be curtailed by constricting only a relatively few major arteries, but to achieve the same result at the arteriolar level would require constriction of tens of thousands of arterioles. Another factor is that metabolically produced substances that exert a vasodilatory effect are known to accumulate in vertebrate tissues during exercise or periods of oxygen deprivation. These substances are believed to act at the arterioles and would compete with neural vasoconstrictive signals at that level. It may be significant that the diving mammals, which must maintain a vasoconstricted state for long periods, have removed the site of vasoconstriction to vessels that are less likely to be affected by the metabolic dilators.

In the accomplished divers, characterized by profound vasoconstriction in the arteries supplying many parts of the body, the active circulation is more or less limited to the heart, lungs, and head during diving. We may now examine briefly the physiology of this circulation. Cardiac output is a function of heart rate and stroke volume. Studies of several birds and mammals have indicated that stroke volume changes little, if at all, during diving. Thus cardiac output falls profoundly during the dive, approximately in proportion to the decline in heart rate (bradycardia). Another important line of investigation deals with blood pressure in the central circulation. Work on a number of mammals and birds has shown that blood pressure in the great systemic arteries either remains virtually unaltered during diving or falls by only a modest amount. The maintenance of central blood pressure despite profound decreases in cardiac output is in itself strong evidence for vasoconstriction of much of the peripheral vasculature, for if the overall resistance of the vascular bed did not increase, the reduction in cardiac output would lead to a large drop in central pressure.

We can now see the highly integrated nature of the cardiovascular response to diving. The dimensions of the active circulatory path are reduced. An approximately normal rate of flow can be maintained to the head and other circulated parts with a much lower cardiac output than is required to maintain flow throughout the entire body, and cardiac output is reduced through a reduction of heart rate with little or no change in stroke volume. Cardiac activity is related to the resistance of the circulated vascular beds in such a way that approximately normal arterial pressures are maintained, which, of course, is an essential condition to the continuance of adequate perfusion.

Metabolism during the dive and some related topics

As we have noted earlier, oxygen supplies to the heart, and, particularly, the central nervous system must be maintained for survival. It is generally believed that the key to the prolonged underwater excursions of diving mammals and birds lies in their ability to preserve certain oxygen stores for the exclusive use of these tissues. Although the circulatory adjustments that achieve this result are seen in their basic form in terrestrial species, they occur in particularly pronounced and refined form in the diving animals. In fact, the responses of diving species are probably the most profound cardiovascular adjustments observed anywhere in the animal kingdom during ordinary events of the life history. The importance of these responses to the prolongation of submergence is indicated by experiments in which the cardiovascular adjustments have been blocked by drug therapy. Interference with the vasoconstrictor response was found to limit harbor seals to submergence of about four minutes; these seals have been known to survive for over 20 minutes with their vascular response intact.

The metabolism of the heart, brain, and other circulated tissues is believed to be largely aerobic during the dive, and it has been shown in several species that oxygen tensions in the active circulation fall steadily over the dive and reach low levels by the time the animal returns to the surface to breathe. Three factors determine the limits of submergence insofar as the oxygen-sensitive tissues are concerned: the magnitude of the oxygen store available, the rate of utilization of oxygen, and the degree to which oxygen tension can fall before impairment of function sets in. We shall discuss these briefly in order. (1) The magnitude of the oxygen store made available to the actively circulated parts of the body is not well known. Basically the oxygen comes from two sources, the blood store and the pulmonary store; thus the question becomes: How much of the total blood store is made available to these particular parts of the body and how much is the pulmonary store is actually utilized? The latter issue has been discussed earlier. Insofar as the blood store goes, we have seen that the total store tends to be particularly large in diving species, and presumably this is reflected in an enhancement of the blood-bound oxygen made available to the oxygen-sensitive tissues. To a large extent a portion of the total blood volume becomes sequestered in the active, central circulation during the dive, and its oxygen becomes reserved for these tissues. However, despite peripheral vasoconstriction, there is probably some exchange between the central circulation and the periphery. Some blood that is initially in the central circulation "leaks" away into peripheral vessels, and other blood correspondingly enters the central circulation from the great veins of the lower body. It is interesting in this context that the venous system of the abdomen in whales and seals has a particularly large capacity in comparison to terrestrial mammals. The inferior vena cava, for example, sometimes attains truly immense proportions. At the start of a dive the blood in the abdominal veins will, of course, not be as saturated as arterial blood, but it still can contain a great deal of oxygen. Recent evidence on the northern elephant seal indicates that blood from the venous reservoir is fed into the central circulation during the dive and that the venous system of the lower body therefore acts as a store of relatively oxygenated blood for us by the tissues that are actively circulated. (2) Concerning the rate of oxygen utilization of the circulated tissues, little can be said. It has often been pointed out that the slowed beating of the heart implies a reduced rate of oxygen uptake by the cardiac muscle, a factor that should help to postpone the time when oxygen tensions in the circulating blood become limiting. The central nervous system is presumed to continue oxygen uptake at its usual nondiving rate. (3) There is some recent evidence that the oxygen-dependent tissues can continue to function down to lower oxygen tensions in diving species as compared to nondiving species. This would help to lengthen the dive. In the Weddell seal, for example, electroencephalographic patterns indicative of cerebral impairment appear at lower oxygen tensions than in terrestrial mammals.

We may now turn to a consideration of metabolism in those tissues that are denied active blood flow during diving, with emphasis on the skeletal muscles inasmuch as these have received the most attention experimentally. As noted earlier, the muscles possess an essentially private store of oxygen in the form of oxymyoglobin, and they can draw on whatever oxyhemoglobin is sequestered in their capillaries. As these oxygen stores are exhausted, metabolism becomes anaerobic. Scholander, Irving, and their colleagues examined these events in harbor seals that were strapped

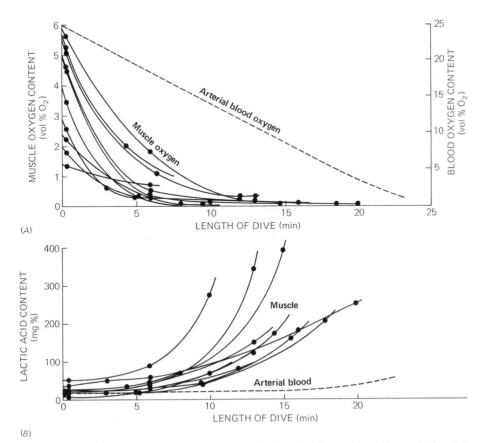

Figure 13–3. Oxygen and lactic acid content in the dorsal trunk muscles and circulating arterial blood of harbor seals (*Phoca vitulina*) during submergence in a bathtub, expressed as a function of time of submergence. (*A*) Changes in muscle oxygen content (see left ordinate) in each of ten seals and the average course of oxygen depletion in circulating arterial blood (see right ordinate). (*B*) Changes in muscle lactic acid content in ten seals and the average course of accumulation of lactic acid in circulating arterial blood during the dive. (From Scholander, P. F., L. Irving, and S. W. Grinnell. 1942. J. Biol. Chem. 142: 431–440.)

to a board and lowered into water, and their results are illustrated in Figure 13–3. In part *A* it is seen that muscle oxygen fell close to zero within the first 5 or 10 minutes of submergence, and in part *B* it is seen that the lactic acid content of the muscle began to increase markedly at about the same time that oxygen stores became exhausted. The three animals that reached the highest levels of lactic acid struggled vigorously during the experiment. Average changes in the oxygen and lactic acid content of arterial blood in the central circulation are also shown in Figure 13–3 and provide evidence of the isolation of the peripheral tissues from the actively circulated vascular bed. Arterial oxygen content fell progressively throughout the period of submergence and approached very low values only toward the end of the dive. Lactic acid rose to only a small extent in the arterial blood during the dive, showing that the acid produced in the muscles remains more or less isolated there. Now it must be emphasized that the seals in these experiments were restrained, and most remained relatively quiet. There is little question that the time course of the events in the muscles would be hastened during a normal dive involving active swimming. Unfortunately data of this type are not available from animals performing free dives. Presumably the muscles are limited to some maximum accumulation of oxygen debt, as in terrestrial vertebrates, but we do not know whether this or the decline in central oxygen tension is the factor that sets the limit on the length of a dive. Probably the answer depends on the type of dive. In a very vigorous dive the muscles might exhaust their anaerobic capacity before the central tissues had drained their oxygen supply, but the converse might hold in a more leisurely dive.

When the animal returns to the water's surface, bradycardia is rapidly abolished, circulation is restored throughout the body, and the lactic acid content of the blood rises steeply as acid accumulated in the muscles is liberated into the general circulation. An acute rise in blood lactate after diving is in itself presumptive evidence of curtailed muscle circulation during the dive and has been observed not only in diving mammals and birds, but also in alligators and water snakes—and, incidentally, in grunion upon return to water after an excursion into air. To cite one quantitative example, blood lactate increased from about 20 mg % to 40 mg % over a 15-minute dive in a gray seal but then rose steeply to about 140 mg % within 4 minutes after return to air; the excess lactate accumulated over the dive was then dissipated at a quasi-exponential rate over a period of about 50 minutes.

Both the aerobic and anaerobic metabolism of diving animals present a challenge to acid-base homeostasis. Aerobic catabolism produces carbon dioxide that cannot be voided during the dive, and anaerobic catabolism generates lactic acid. These acid loads must be buffered sufficiently during the dive and the immediate postdive period to prevent a deleterious decline in the pH of the body fluids. In this light, it is noteworthy that diving animals tend to possess particularly high blood buffer capacities. Among mammals, in fact, buffer capacity, of all the basic blood parameters, is perhaps the one that most consistently differs between diving and nondiving species. Arterial pH typically falls several tenths of a unit in diving birds and mammals during the dive and the period of lactic acid release after the dive, but this pH disturbance is limited by their high buffer capacity.

The final question to be considered here is the magnitude of the metabolic rate during the dive. There is no doubt that the overall rate of oxygen utilization (that is, the aerobic component of metabolism) is reduced, for, as we have noted earlier, the oxygen stores of divers are inadequate to meet their ordinary resting rate of oxygen consumption over the length of the dive. A full evaluation of the total metabolic rate—aerobic plus anaerobic—is difficult at this time because critical experiments have been performed only in laboratory situations. Much interest has centered on whether the metabolic rate during *resting* submersion is lower than that during rest in air, and there is now considerable evidence that it is. Scholander and Irving produced three types of evidence for this conclusion. First, in seals and ducks the size of the oxygen debt repaid after a period of quiet submergence was only a quarter or a half as great as that which would have been expected had metabolism remained at its resting, predive level. Second, by measuring the amount of oxygen consumed from stores during the dive and the amount of lactic acid produced, heat production could be calculated from knowledge of the energetics of the aerobic and anaerobic pathways and proved to be reduced by comparison to the predive state. Third, temperatures of various parts of the body were followed during submergence. In harbor seals the temperatures of the brain, abdomen, back musculature, and other deep tissues fell by 1°–2°C over the diving and immediately postdiving period (13- to 15-minute dives). It was pointed out that because peripheral vasoconstriction should improve body insulation during the dive, a decline in temperature would have to be interpreted as a consequence of decreased heat production. Since these pioneering experiments, other work has further substantiated a decrease in metabolism during quiet submergence in various species. Thus, for example, alligators have been shown to repay a "less-than-expected" oxygen debt; ducks exhibit a marked decline in body temperature during simulated dives in which just the head is immersed; and toads and turtles show decreased metabolism as indicated by direct measures of heat output during submergence.

A fall in metabolic rate could be mediated in at least two different ways. First, there could be a primary reduction in the metabolic activity of some tissues, triggered by the act of submergence and presumably controlled nervously or hormonally. Alternatively, metabolic activity could be depressed as a secondary consequence of the altered tissue environment during the dive, perhaps particularly because of the depletion of oxygen, switchover to anaerobic metabolism, and accumulation of lactic acid in some tissues. There has as yet been no clear resolution of this issue in mammals and birds.

The only information on the metabolism of actively diving birds and mammals comes from observations of individuals undergoing vigorous struggling during laboratory experiments. Scholander observed that active harbor seals often repaid an oxygen debt several times larger than would have been predicted from maintenance of their predive, resting metabolic rate. It seems likely that animals in nature commonly support an elevated rate of metabolism during dives despite their limited aerobic resources.

Pulmonary function and external respiration

As noted earlier, maximum lung capacity per unit body weight is not consistently different in diving mammals as compared to terrestrial mammals; there is a tendency for capacity per unit of lean weight to be greater in the diving forms. As a group, the whales, porpoises, and seals do tend to exhibit certain ventilatory peculiarities. (1) Resting tidal volume tends to be high relative to lung capacity. An average man might have a maximal lung capacity of 6 liters and a resting tidal volume of 500 cc, meaning that tidal volume represents near 10% of capacity. By comparison, measures on certain porpoises and whales indicate that resting tidal volume occupies 80–90% of capacity, and in seals values near 40% have been reported. (2) Resting breathing rate in the diving mammals tends to be relatively low; rates of 1–4 breaths/min are common. (3) Oxygen utilization from the respired air tends to be high—twice as great in whales and porpoises as in man. Such high oxygen utilization is facilitated by the slow breathing rate, which provides for a long residence time of the air in the lungs, and by the large tidal volume, which enhances the fraction of inspired air that reaches the respiratory exchange spaces rather than the dead spaces. (4) Resting respiratory minute volume tends to be low. This feature is complementary to the high oxygen utilization, for less volume needs to be ventilated to gain a certain amount of oxygen, the higher the utilization. (5) The ability to enhance minute volume above the resting level tends to be low. At peak ventilatory effort, man can exchange 15 to 20 times as much air per minute as at rest. By contrast, gray, bladdernose, and Weddell seals seem capable of only a four- to sixfold increase, and the capabilities of porpoises may be even lower. The relatively poor abilities of marine mammals to increase ventilation are, in part, related to their high resting tidal volume. Man can greatly augment both tidal volume and ventilatory frequency, and with this combination he can achieve a large increase in minute volume. Whales and porpoises, by contrast, can hardly increase tidal volume at all and are therefore almost totally dependent on increases in breathing rate to raise minute volume. Seals have some ability to increase tidal volume, but recent studies of Weddell seals show an increase of less than 50% even during the heavy breathing following a long dive.

The significance of these manifold and integrated ventilatory differences in the marine mammals remains largely obscure and poses an interesting challenge for future research. Many whales tend to surface only briefly between dives, taking just one or a few breaths. In this context it is noteworthy that they can achieve almost complete turnover of their pulmonary air in a single breath owing to their exceedingly large tidal volume, and they apparently can do this very rapidly. Most or all marine mammals maintain their lungs in the inspiratory position between breaths when floating at the water's surface. This habit provides buoyancy, and it may be significant that the buoyant effect of the lungs is interrupted only infrequently owing to their low breathing rate. Living as they do in water, marine mammals face special problems of maintaining a high body temperature. Their high oxygen utilization coefficient allows them a relatively low minute volume, and because respiratory evaporative heat loss is in part a function of minute volume, their pattern of ventilation has advantages in thermoregulation. Likewise, the reduced respiratory water loss contributes to maintenance of water balance in the marine habitat. Whether these or other considerations have been central factors in the evolution of the ventilatory pattern in these animals remains uncertain.

A topic that has attracted interest since the early days of research on diving mammals and birds is the sensitivity of their respiratory center to carbon dioxide. As discussed in Chapter 8, increased carbon dioxide tensions and/or decreased pH potently stimulate increased ventilation in man and other terrestrial species. Increase in blood carbon dioxide is believed to be the central factor eliciting the irresistible urge to breathe in diving humans, and it seemed logical to postulate that species that dive for long periods would have a reduced sensitivity to carbon dioxide. This has proved to be the case. During normal breathing in man, an increase of 4 mm Hg in arterial carbon dioxide tension will cause a doubling of ventilation rate, but in harbor seals, for example, a similar ventilatory response requires a much larger increase in arterial carbon dioxide tension, about 20 mm Hg. Similarly, a much-reduced sensitivity to carbon dioxide has been demonstrated in ducks.

Another topic that has long attracted attention is the problem of how mammals that dive to great depths avoid the illnesses that sometimes afflict human divers. The list of "divers' diseases" is long, but students of marine mammals have devoted most attention to the particular affliction known as decompression sickness, caisson's disease, or the bends. It is first important to review something of what we know of the etiology of decompression sickness in man.

Undisputed cases of decompression sickness occur in humans diving with a compressed-air source. With a compressed-air source, the air pressure in the lungs is maintained equal to ambient pressure at depth. This prevents the lungs from collapsing under the force of the ambient pressure and allows continued breathing. The elevation of total pressure in the lungs implies an increase in the partial pressure of each individual pulmonary gas. Of particular interest is the increase in alveolar nitrogen tension, which may rise from about 570 mm Hg at sea level to many atmospheres at depth. When a man commences a dive, his tissues and body fluids are at equilibrium with the normal alveolar nitrogen tension; each contains dissolved nitrogen in accordance with its absorption coefficient for a tension of 570 mm Hg (see Chapter 8). At depth the tissues and body fluids take up nitrogen because their nitrogen tension is below the new, elevated alveolar tension. As nitrogen is extracted from the lungs, the alveolar tension does not fall because the man is breathing from a constantly renewed air source, and if the dive is continued for long enough, the tissues will dissolve sufficient nitrogen to come to a nitrogen tension as high as that maintained in the alveoli. Now if the man is suddenly brought back to the surface ("decompressed"), alveolar tension will fall to its ordinary value, and the nitrogen-charged tissues will lose nitrogen across the lungs in a reversal of the processes that occurred at depth. Problems can arise because the tissue nitrogen tension, being as high as many atmospheres, exceeds the hydrostatic pressures prevailing in the body at sea level, meaning that bubbles of nitrogen can be formed within the body. This is not difficult to understand in principle. If the hydrostatic pressure in a particular tissue is 760 mm Hg and a minute gas space develops there, the pressure in the gas space will also be 760 mm Hg. If the body fluids surrounding the gas space contain dissolved nitrogen at, say, 5000 mm Hg, then they will lose nitrogen into the gas space even if the space is filled purely with nitrogen, for gases always move from regions of higher tension to regions of lower tension. Accordingly the minute gas space will grow into a bubble. Such bubbles are believed to be primarily responsible for the symptoms of decompression sickness. Most commonly, throbbing pains develop in the joints and muscles of the arms and legs (the bends). Additionally the person may suffer neurological symptoms, such as paralysis, and severe breathing problems (the chokes). Exactly how the bubbles cause these effects is not entirely clear. Bubbles forming in the blood are believed to clog vascular beds, thus denying adequate circulation, and bubbles forming in connective tissue, muscle, and other extravascular spaces are believed to press on nerve endings and, if large, could cause cell rupture or other tissue damage.

The factors that determine whether decompression sickness will occur are not completely known. The first and essential condition is that tissue gas tensions must be elevated. In addition, minute (microscopic) gas spaces must "get started" in the tissues; this process is subject to many influences. Finally, excess gas is steadily eliminated across the lungs, and if the gas overload is not too great, this process may lower tensions sufficiently rapidly that even if bubbles do start to form, their proliferation will be halted before external symptoms are manifest. In general, humans can surface immediately without fear of the bends if their tissue nitrogen tension does not exceed 2 atm. At higher tensions there is considerable variability of response, but precautions against the bends are indicated.

When we consider humans undergoing breath-hold diving, a crucial difference from diving with compressed air is immediately apparent. The breath-hold diver carries only the limited amount of nitrogen contained within his lungs upon submergence; the nitrogen supply is not steadily renewed. During descent to depth the lungs of the breath-hold diver are compressed under the force of increasing ambient pressure, and the nitrogen tension increases initially to high levels, just as in diving with compressed air. This establishes a tension gradient favorable to the transfer of nitrogen from the lungs to the tissues and body fluids, but because the quantity of nitrogen in the lungs is limited, pulmonary tension falls as the transfer occurs and ultimately the tension gradient is abolished. The increment in tissue nitrogen tension depends on (1) the total quantity of nitrogen transferred from the lungs, (2) the mass of the tissues to which it is distributed, and (3) the absorption coefficient of the tissues (which relates tension to concentration). In man it is clear that the nitrogen transfer during a single breath-hold dive is insufficient to produce decompression sickness. If dives are repeated, however, with insufficient time between dives for the tissues to release accumulated nitrogen, it is conceivable that tissue nitrogen tension could be elevated in increments to a threatening level. Some people undergoing serial breath-hold dives have complained of symptoms resembling decompression sickness.

The marine mammals are, of course, breath-hold divers, and the first question to be asked is whether they face a potential problem of decompression sickness. Various calculations indicate that they do. If all the pulmonary nitrogen carried by a Weddell seal on a single dive were transferred only to its actively circulated blood and tissues, tensions of 7–8 atm could be developed in those tissues. Now it is unrealistic to assume that all pulmonary nitrogen would be transferred during the dive, and the opening of the whole circulatory bed after the dive would act to distribute the nitrogen more evenly throughout the body and lower tensions considerably. But against these ameliorating factors lies the fact that these seals undergo serial dives, with the potential for incremental buildup of nitrogen. Also, seals probably carry less nitrogen per unit of body weight on submergence than other groups of marine mammals; the calculations for seals therefore underestimate the potential problem in the other groups. It seems clear, then, that the marine mammals do require some type of defense against decompression sickness. Although various hypotheses have been made, the one for which the evidence is most compelling is that special features of their lungs act to limit the transfer of pulmonary nitrogen to the blood. This was originally proposed by Scholander in 1940 and has since gained increasing support.

Increasing depth and pressure

Figure 13–4. Diagram of the hypothesis of preferential collapse of the alveolar sacs at depth. The heavy lines represent the trachea, bronchi, and bronchioles. The circles at the left represent the alveolar sacs. As the pulmonary air is compressed to a smaller volume at depth, it is postulated that the alveolar sacs collapse preferentially and that all the air thus comes to be contained in the anatomical dead spaces, where exchange with the blood cannot occur.

Scholander suggested that as the pulmonary air is compressed to a smaller and smaller volume at depth, the respiratory air spaces (alveolar sacs) collapse preferentially and that the air thus ultimately comes to be held entirely in the anatomical dead space. This is illustrated diagrammatically in Figure 13–4. Below a certain depth, nitrogen invasion from the lungs into the tissues therefore cannot occur because the nitrogen is safely sequestered from the respiratory exchange membranes. Oxygen is also sequestered, but this may be a price that the animal must pay to avoid the bends. The hypothesis of alveolar collapse at depth is supported by various studies of gas exchange. Earlier, evidence that oxygen extraction from the lungs ceases below a certain depth in porpoises and seals was reviewed. Recent studies of seals in a compression chamber revealed that vascular nitrogen tensions remained well below pulmonary tensions in simulated dives to depth, indicating, again, that the blood is not in free communication with the pulmonary air.

The depth at which the alveolar sacs will become completely collapsed is that where the pressure is sufficient to reduce the initial volume of pulmonary air to the volume of the pulmonary dead space. For example, if a species has a dead-space volume of 2 liters and carries 30 liters of pulmonary air on submergence, then alveolar collapse should be complete at about 150 m, for at that depth the pressure is near 15 atm—sufficient to reduce the initial volume by a factor of 15 to 2 liters. Of course, alveolar collapse will occur gradually, and exchange across the alveoli will be impaired before the animal reaches the depth of complete closure. One reason that the deep-diving seals dive on expiration may be to hasten alveolar collapse by reducing the initial lung volume relative to the dead-space volume. Recent studies of Weddell and northern elephant seals indicate that the alveolar exchange area is much reduced at depths as shallow as 30 m. According to measures of oxygen exchange discussed earlier, alveolar collapse is essentially complete at about 100 m in bottlenosed porpoises, which dive on inspiration.

Alveolar collapse is promoted in the marine mammals by special features of their thoracic and pulmonary anatomy. The thorax is flexible and highly compressible; thus the thoracic walls are freely pushed inward as pressure increases at depth, and the lungs within are readily compressed. In contrast, man has a relatively inflexible thoracic cage. As he dives, his lungs are compressed in proportion to ambient pressure only down to a certain depth. Upon further descent, thoracic resistance impairs further pulmonary compression, with potentially dire consequences. At these depths pulmonary pressure remains below ambient pressure because of the lack of compensatory air compression; this increases the difference in pressure between the alveolar air and the alveolar capillaries, and pulmonary edema or even capillary rupture can occur. Another important feature of the marine mammals is that their pulmonary airways are reinforced with cartilage or muscle right down to the openings to the alveolar sacs. In terrestrial mammals such reinforcement stops in the terminal bronchioles, and the alveolar sacs are separated from the reinforced portions of the airways by several millimeters of delicate respiratory and terminal bronchioles. The significance of this difference is made clear by recent experiments on dogs and sea lions. When the dog lungs were exposed to pressure, the delicate terminal and respiratory bronchioles collapsed shut before the alveolar sacs had emptied. Thus a significant portion of the pulmonary air became trapped in the alveolar sacs, and during a dive this air would remain freely available for exchange with the blood. In the sea lions, however, the reinforced bronchioles remained patent until the alveolar sacs had emptied. The more-extensive reinforcement of the airways in marine mammals thus assures that the alveolar sacs will be free to discharge their contents completely into the pulmonary dead space as pressure is increased. The adaptive value of the extended reinforcement is indicated by the fact that it cuts across taxonomic lines. It is present in both pinnipeds and cetaceans and, interestingly, occurs in the sea otter but not the closely related river otter.

Some features of diving
in amphibians and reptiles

Any attempt at summarization of the diving physiology of these groups is hampered, first, by the extreme diversity of their habits and, second, by the fact that data are available on only a few species. We have a patchwork of information that indicates certain potentialities but does not permit synthetic understanding.

An important factor to be recognized from the outset is that many amphibians and some reptiles have appreciable abilities to extract dissolved oxygen from water. Unlike birds and mammals, these species can renew their supply of oxygen while diving. Some features of underwater cutaneous respiration in frogs were discussed in Chapter 8. At the low temperatures prevailing during hibernation in temperate regions, many frogs appear to support their metabolism aerobically through cutaneous exchange over long periods while submerged, and some species are capable of meeting a substantial fraction of their metabolic demands through cutaneous oxygen uptake during diving in aerated waters even at 15°–20°C. Some turtles also have well-developed abilities to extract oxygen from the water, but others do not. Soft-shelled turtles (*Trionyx*), for example, take up oxygen at a high rate, perhaps sufficient to meet their resting demands at moderate temperatures in aerated water. Their principal site of aquatic respiration is the buccopharyngeal cavity; the cavity is ventilated rapidly, and the pharynx bears highly vascularized villi. These turtles also exchange oxygen across the cloaca, which is ventilated, and the skin and cartilaginous plastron, both of which are well vascularized. At the other extreme are species such as the yellow-bellied or red-eared slider, *Pseudemys scripta*, which has only a meager capacity for aquatic respiration. Submerged in well-aerated water at 22°C, *Pseudemys* was found to take up oxygen at less than 5% of its resting rate in air, and it survived little longer in aerated water than in water devoid of oxygen.

Diving amphibians and reptiles exhibit the bradycardia typical of other vertebrates. This has been shown in several species of frogs and toads, alligators, water snakes, iguanas, and a number of turtles. Other cardiovascular changes are incompletely understood, and it is therefore difficult to appraise the role of bradycardia in terms of the integrated response to diving. In Chapter 10 changes in central blood flow were discussed. You will recall that in several amphibians, *Pseudemys scripta*, and alligators there is evidence of an increase in pulmonary vascular resistance during diving that results in a decrease in the portion of the cardiac output directed to the lungs and an increase in the portion directed to the systemic circuit. The decreased pulmonary perfusion may initially help to meter out gradually the available pulmonary reserve of oxygen; once the pulmonary reserve is exhausted, preferential perfusion of the systemic circuit would seem clearly to be the most efficient utilization of the energy invested in cardiac contraction. Peripheral vascular responses are virtually unknown. Work on water snakes and alligators has shown that lactic acid floods the arterial circulation after diving—strong evidence for peripheral vasoconstriction in the muscles. The integrated cardiovascular response in alligators may well closely resemble that in birds and mammals, with the added feature of the redistribution of cardiac output that is permitted by their incompletely divided central circulation. Studies of frogs and turtles have revealed that bradycardia is associated (as expected) with a decline in systemic cardiac output. Associated alterations of arterial pressure are indicative of an increase in systemic peripheral resistance, but which vascular beds are constricted remains unknown. In diving mammals the skin is typically one site of vasoconstriction, but frogs and certain turtles almost certainly differ in this respect inasmuch as they utilize cutaneous respiration. Observations of *Rana esculenta* have indicated cutaneous vasodilation during diving.

The remainder of this discussion will be devoted to recent investigations of turtles, some of which have added new dimensions to our understanding of vertebrate capabilities. A study of 70 species of reptiles breathing pure nitrogen at 22°C established that, as a group, freshwater and terrestrial turtles have remarkable abilities to survive under anoxic conditions. The 25 species of these turtles investigated survived for 6 to 33 hours, whereas 45 species of lizards, snakes, crocodilians, and sea turtles endured for only 20 minutes to 2 hours. Other studies have shown that turtles have very low critical oxygen pressures when breathing air mixtures; in a number of species, normal resting oxygen consumption is maintained down to ambient oxygen tensions as low as 8–38 mm Hg. These observations have two implications for diving turtles. First, if turtles can maintain normal oxygen consumption down to such low ambient tensions, they must be able to do so at even lower internal tensions and therefore should be capable of a very thorough utilization of oxygen stores during diving. Second, their tolerance to anoxia indicates a strong ability to support vital functions anaerobically, a feature that should prolong dives considerably when oxygen stores have been exhausted and any oxygen up-take from the water is inadequate to support function entirely aerobically.

With these points in mind, we can examine several studies of diving in turtles. River turtles, *Pseudemys concinna*, undergoing voluntary dives in an aquarium were investigated by Daniel Belkin. Dives averaged about an hour in length, but some lasted over two hours. In between dives the turtles would surface and breathe for 30 seconds to 4 minutes. On the average, 98% of their time was spent under water. The parameters of oxygen storage in this species are unremarkable for turtles; oxygen stores, if fully charged at submergence, are estimated to be about 33 cc/kg. The resting oxygen consumption in air is 10–15 cc/kg/hr. Because turtles can maintain their resting oxygen consumption down to low tensions, it would appear that the stores of this species are adequate to meet the resting demands of all tissues for two or three hours, and over an average one-hour dive, there should be ample oxygen for some degree of activity. These data lead to the interesting conclusion that the ordinary voluntary dives of this species can be supported entirely aerobically from oxygen stores, with no need for cardiovascular adjustments to conserve oxygen for par-ticular tissues. The dives are long by mammalian or avian standards, but the oxygen demands of turtles are relatively low.

P. concinna was found to undergo a "typical" diving bradycardia. However, because most of the turtle's life is spent under water and be-cause special cardiovascular adjustments to diving seem unnecessary, Belkin suggests that the low heart rate during submergence is, in fact, the *normal* heart rate and that the heightened heart rate during surfacing represents the departure from usual conditions. According to this reason-ing, this turtle would exhibit an *emergence tachycardia* (increase in heart rate) rather than a diving bradycardia. The distinction is far from simply semantic and, at the least, should challenge our common inclination to approach "lower" animals with points of view established in work on mammals. From studies on the closely related *Pseudemys scripta*, it seems likely that emergence is accompanied not only by an increase in heart rate, but also by (1) increased cardiac output, (2) decreased pulmonary and systemic vascular resistance, and (3) an increase in the proportion of cardiac output directed to the lungs (see Chapter 10). The heightened blood flow, the opening up of the circulation, and the redistribution of cardiac output would all be compatible with the notion that the circulatory adjustments that occur on emergence act to hasten the recharging of bodily oxygen stores.

Another instructive study of diving physiology was performed by Donald Jackson on the slider *Pseudemys scripta*. This species has been shown to survive for about 20 hours while constrained to water devoid of oxygen at 22°C. Under these circumstances oxygen uptake from the water is impossible, but, as noted earlier, sliders have only meager abilities to extract oxygen from the water even when it is well aerated. Jackson fol-lowed pulmonary oxygen, blood oxygen, and metabolic heat production in sliders undergoing forced dives in oxygen-depleted water and divided their responses into three phases (Figure 13–5). In phase I, lasting about 20 to 25 minutes, heat production (metabolic rate) remained at the resting predive level. At the same time, blood and pulmonary oxygen stores were

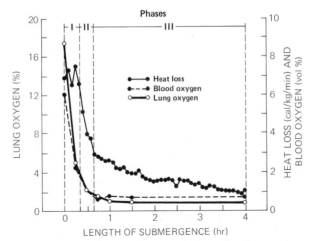

Figure 13-5. Rate of heat loss, blood oxygen content, and pulmonary oxygen content as functions of time in turtles, *Pseudemys scripta elegans*, during forced submergence for four hours in oxygen-free water at 24°C. Values are averages for 5 to 10 turtles. Three phases of response are recognized and indicated, as discussed in the text. The rate of heat loss to the surrounding water by the turtles was determined by direct calorimetry and provides a measure of the rate of heat production and, thus, metabolic rate. Blood oxygen content was determined on samples drawn from the heart. (From Jackson, D. C. 1968. J. Appl. Physiol. 24: 503-509.)

rapidly depleted. Because *P. scripta* is known to have a low critical oxygen pressure, oxygen utilization would probably not be impaired even at the lowest oxygen tensions reached in phase I, and it may well be that all tissues were respiring aerobically at their normal rate through this phase. During phase II, oxygen was still being extracted from stores, though at a lower rate than that in phase I, and metabolic rate underwent a precipitous decline. Almost certainly, at least some tissues were turning to anaerobic metabolism at this stage. Phase III is the most remarkable. No further oxygen was extracted from either the pulmonary or the vascular store. Heat production had fallen to 40% of the initial rate at the start of this phase and continued to fall to 15% at the end of four hours and to 4% at the end of seven hours (not shown). Although the muscles may have had oxymyoglobin stores at the start of phase III, there would seem to be little doubt that metabolism was entirely anaerobic by the end. This result and the earlier results on the prolonged survival of turtles in a nitrogen atmosphere indicate that the heart and central nervous system of turtles do not have the absolute requirement for oxygen that characterizes mammals, birds, and, probably, other reptiles. This property presents a fascinating topic of study to cellular and neural biologists and permits the turtles to undergo dives of remarkable duration.

So far as we now know the anaerobic pathways of turtles are similar to those of most vertebrates. Lactic acid has been shown to accumulate in high concentrations in the blood during anoxia. The importance of anaerobic glycolysis to prolonged survival has been demonstrated by injecting turtles with iodoacetate, a metabolic poison that interrupts the oxidation of glyceraldehyde-3-phosphate to 1,3-diphosphoglyceric acid (see Figure 11-1A). *P. scripta* survive for only an hour in oxygen-depleted water when so treated.

Despite their great tolerance to anoxia, turtles are not capable of indefinite anoxic survival, at least at moderately high temperatures. It is noteworthy in Jackson's data (Figure 13–5) that metabolism declined steadily over phase III, indicating a progressive impairment of function. The factors that ultimately limit endurance are unclear. The accumulation of lactic acid may gradually diminish the rate of anaerobic energy production through end-product inhibition. Also, though turtles have exceptional buffer capacities, the decline in pH may become deleterious.

Prior to Jackson's work it had been demonstrated by direct calorimetry that toads and turtles show a greatly reduced metabolic rate during prolonged submergence, and one of Jackson's chief concerns was the question raised earlier in the discussion of mammals. Is this reduction in metabolism a reflex response to diving or is it a secondary result of the switchover to anaerobic metabolism? The close parallel between the fall in metabolism and the decline of oxygen reserves in *Pseudemys* (Figure 13–5) strongly indicates that, at least in turtles, the conversion to anaerobiosis is the factor involved.

Finally a few comments on diving in nature are appropriate. Muscular activity was recorded in *P. scripta* during Jackson's experiments, and it was found that bursts of activity, although frequent and intense in phase I, had become infrequent and less intense by phase III. In nature, these turtles can probably be reasonably active only while they retain some oxygen reserves and thus would likely limit dives for feeding and other active pursuits to phases I and II. While at rest under water, however, they could exploit their great capacity for anoxic survival; even late in phase III they can develop enough muscular force to propel themselves back to the water's surface.

SELECTED READINGS

Andersen, H. T. 1966. Physiological adaptations in diving vertebrates. Physiol. Rev. 46: 212–243.

Andersen, H. T. (ed.). 1969. *The Biology of Marine Mammals.* Academic, New York.

Elsner, R., D. L. Franklin, R. L. Van Citters, and D. W. Kenney. 1966. Cardiovascular defense against asphyxia. Science 153: 941–949.

Hong, S. K. and H. Rahn. 1967. The diving women of Korea and Japan. Sci. Amer. 216: 34–43.

Kooyman, G. L. 1973. Respiratory adaptations in marine mammals. Amer. Zool. 13: 457–468.

Lenfant, C., K. Johansen, and J. D. Torrance. 1970. Gas transport and oxygen storage capacity in some pinnipeds and the sea otter. Resp. Physiol. 9: 277–286.

Ridgway, S. H. (ed.). 1972. *Mammals of the Sea. Biology and Medicine.* C. C Thomas, Springfield, Ill.

Schaefer, K. E. 1965. Circulatory adaptation to the requirements of life under more than one atmosphere of pressure. *In*: W. F. Hamilton (ed.), *Handbook of Physiology. Section 2: Circulation.* Vol. III. American Physiological Society, Washington, D.C.

Scholander, P. F. 1963. The master switch of life. Sci. Amer. 209: 92–106.

Scholander, P. F. 1964. Animals in aquatic environments: diving mammals and birds. *In*: D. B. Dill (ed.), *Handbook of Physiology. Section 4: Adaptation to the Environment.* American Physiological Society, Washington, D.C.

See also references in Appendix.

14 THE ACTIVE ANIMAL

Activity is an essential dimension of the physiology of the species and of the interplay between physiology and ecology. Examples of the importance of activity to an understanding of physioecology are numerous. An animal's activities, because they impose a metabolic cost, are an important factor in determining how much food it must find and consume, and, in turn, they influence the energetic demand of the animal on the ecological community it occupies. A fish that can survive in waters of low oxygen content so long as it can remain nearly at rest may not be able to obtain sufficient food in the same waters if the activity of gathering food is energetically demanding (see Chapter 11); such a fish would require waters of higher oxygen content for survival in its natural habitat than in a protective laboratory environment. We are all well aware of the importance of behavior to the ecological relationships of species: some species move at great speed to capture prey or avoid predators, whereas others lead a comparatively sluggish way of life; some migrate over large distances, others spend their entire lives within a restricted area; some swim, some run, and some fly. It is important to recognize that each of these behavioral attributes has physiological concomitants. Species that have assumed an ecological role demanding sustained, vigorous exercise have had to evolve circulatory and respiratory systems capable of supplying high oxygen demands at the tissue level. Species that have assumed a sluggish or sedentary way of life have not experienced selective pressures for such high-capacity circulatory and respiratory systems; or, conversely, groups with a fundamentally limited capacity for oxygen transport have been restricted to a relatively inactive way of life by these physiological constraints.

Though it is obvious that we cannot genuinely understand the functional biology of a species without understanding the physiology of activity, difficulties in studying actively moving animals have limited the amount of information available. Until fairly recently, measurement of parameters such as heart rate and blood pressure, for example, required that wires be attached to the animal. This, in turn, precluded measurements on free-roaming animals in the wild and placed constraints on the type of activity that could be studied in the laboratory. The standard methods of measuring metabolism that have been in widespread use for decades require that the animal be confined, either so that gas exchange can be monitored or so that consumption of food and production of excrement can be measured. Although much insight concerning the metabolic demands of activity has been obtained from carefully designed laboratory experiments, the methods are, again, not applicable to free-roaming animals. Another methodological problem has been posed by the need for quantifying the amount of activity. One recourse in the case of terrestrial animals is to train them to run on a treadmill so that running speed ·and the angle of ascent can be controlled and measured while still keeping the animal sufficiently confined to measure physiological parameters by standard methods. Treadmills and other such devices, however, have sometimes serious limitations: training an individual to exercise on the treadmill can be a time-consuming and tedious process, and the type of exercise performed on a treadmill may only vaguely resemble the exercise actually performed by the species in nature. In view of the numerous methodological problems of studying the active animal, it is not surprising to find that the literature on active animals is far more limited than that on resting, restrained, or anesthetized animals. Although studies under the latter conditions have provided great insight into many dimensions of physiological ecology, they give only a limited perspective on the animal in nature. Recognition of this fact and important improvements in the technology of physiological measurement have prompted an increasingly rapid accumulation of good work on exercising animals over the last two decades.

The performance of exercise entails adjustments in many physiological systems. Many facets of the circulatory and respiratory responses have been discussed in earlier chapters and will not be dealt with in detail here. It is worth reiterating the important point made earlier that the circulatory and respiratory systems of many animals operate at a relatively leisurely pace when the animal is at rest, and the evolution of the full potentialities seen in these systems can often only be understood in the context of the demands imposed during exercise. The emphasis in this chapter will be on energy metabolism during exercise.

METHODOLOGY IN THE STUDY OF EXERCISE

Before turning to a discussion of our knowledge of the energetics of exercise, it is appropriate to examine in more detail some of the methodological problems encountered in the study of exercise and some of the techniques that are available and have been put to use.

Controlling and
measuring the intensity of exercise

A requisite in laboratory studies is to elicit exercise in the animal under circumstances that permit measurement of the physiological parameters under consideration. Usually this has meant that the animal must exercise within a more or less limited area of movement. In the simplest approach the experimenter either waits for spontaneous episodes of activity or provokes the animal to a rather uncontrolled level of activity by simple visual, aural, tactile, or electrical stimulation. Animals, for example, have been prodded to move vigorously about in their cage or aquarium, and physiologists have been known to amuse their colleagues and students as they chased frogs or lizards about the laboratory or down hallways. A number of methods that allow for more exacting control of the intensity and type of exercise have been devised. The treadmill, as noted earlier, has frequently been put to use in studies of running animals. Flying animals have sometimes been trained to fly in circles on the end of a tether. Here the speed of flight is not easily controlled, but it can be measured accurately. Control of the speed and angle of flight has been achieved by training animals to fly against the air current in a wind tunnel. The speed of flight can be controlled by varying the velocity of the air current, and the tunnel can be tilted to simulate flight at an upward or downward angle. A device analogous to a wind tunnel has been used to study swimming in fish. The fish is placed in a chamber and trained or forced to swim against a water current that is driven through the chamber. Another technique used in studies of fish involves a doughnut-shaped aquarium rotated at known speed on a turntable. The fish is stimulated to keep its position by swimming counter to the direction of rotation of the aquarium, and the rate of swimming is controlled by varying the rate of rotation. Other methods have been devised to induce controlled levels of exercise in animals, and the student who searches the literature will be impressed with the ingenuity displayed by investigators—and sometimes also by the ingenuity of the animal in flaunting attempts to make it perform. Vance Tucker, who has pioneered the use of wind tunnels in the study of avian flight, placed an electrical grid in the bottom of the test chamber to prevent birds from landing during experiments. He reports on one parakeet that successfully avoided being shocked by rolling over on its back and keeping both feet in the air.

The laboratory devices described have the advantage of permitting systematic collection of data under well-defined conditions. This has been an indispensable element in the development of exercise physiology as a quantitative science. A disadvantage of the use of these laboratory devices is that they produce a steady and generally unidirectional form of exercise that may only rarely be displayed by the species in nature. The flight of a migratory bird in a wind tunnel may closely resemble flight during actual migration. But if we induce a deer mouse to run at 100 ft/min on a treadmill, we must wonder how often mice actually run steadily at 100 ft/min when living in their natural habitat. They probably do so rarely, and confronted with the highly variable pattern of activity displayed by mice in nature, we face a real challenge in estimating the energetic cost of a night's activity from data on exercise under regimented conditions in the laboratory.

Recent methodological developments have made it possible to monitor a number of physiological parameters in free-ranging animals, thus permitting collection of data from animals in the wild. Data collected on free-ranging animals have the advantage of being immediately relevant to the natural situation. We can, for example, measure the total metabolic expenditure of a deer mouse in its natural habitat over a number of days, a parameter of considerable interest to the physiologist and ecologist alike. The deficiency of data collected in nature is that the conditions of measurement frequently lack quantitative definition. The heart rate of an animal, for instance, may be known at a particular time, but often the animal will not be within view, meaning that its behavior at the time is unknown; or even if the animal can be seen, it will generally be possible to provide only a subjective, qualitative description of its activity level. The lack of quantitative definition of the conditions of measurement makes it difficult to develop a systematic understanding of the species or compare the exercise physiology of one species with that of another. This is where laboratory studies under carefully controlled conditions have much value.

In summary, both laboratory and field studies have their advantages and disadvantages, and ultimately a thorough understanding of the active animal will usually depend on integration of the two.

Electrical transducers and radio telemetry

Many physiological parameters of interest can be measured by electrical methods, including heart rate, body temperature, blood flow rate, blood pressure, acceleration of the body, breathing rate, and blood oxygen tension. Heart rate is monitored electrically by amplifying differences of electrical potential that are set up in the body by the depolarization and repolarization of the cardiac muscle fibers, yielding the well-known electrocardiogram (EKG). In this case signals of an electrical nature are present in the animal, and all that is needed is to detect and record them. With many other parameters of interest, there is no pre-existing signal of an electrical nature in the animal, and measurement by electrical methods requires a transducer that will produce electrical signals corresponding quantitatively to the parameter under consideration. Body temperature, for example, does not produce an electrical signal that we can simply amplify and record. It is nonetheless possible to record body temperature electrically by use of a suitable transducer. One type of transducer, the thermistor, consists of a small sintered bead of nickel and manganese oxides, the electrical resistance of which changes in an orderly way with temperature. The bead is connected between two wires and inserted in the animal; then the temperature of the body can be monitored by measuring the resistance across the wires.

Electrical techniques for monitoring physiological parameters have many advantages, in recognition of which there has been much effort devoted to their development. A particular advantage in studies of exercise is that once the animal has been outfitted with suitable transducers, it can be allowed some freedom of movement while measurements are being made. The only external attachments required are the electrical

Figure 14–1. A self-contained miniature transmitter for highly accurate transmission of the electrocardiogram of a man or animal. In actual use the transmitter would be encased in metal or plastic. It could then be attached externally or implanted surgically. (From Fryer, T. B. 1970. *Implantable Biotelemetry Systems.* National Aeronautics and Space Administration, Washington, D.C.)

leads to the amplifying, decoding, and recording equipment, and it is often possible to arrange these leads so that the animal can exercise reasonably normally within the restricted area of a treadmill, wind tunnel, or aquarium.

Although the use of electrical transducers in this manner has contributed significantly to our understanding of exercise physiology in a number of animal groups, the methods have limitations. Maintenance of hard-wire connections with the animal is impractical for studies of free-roaming individuals in nature and implies that the range of movement in the laboratory must be reasonably restricted. Further, some animals become so irritated by the presence of the wires that they constantly attempt to remove them; this behavior interferes with the study of exercise, and often one vital connection or another is ultimately broken. It is thus that the development of methods for radio telemetry over the past two decades has been accompanied by considerable excitement. In telemetry the transducer is connected to a small radio transmitter that is either implanted internally in the animal or attached externally on the body surface. Data are transmitted from the animal by radio waves, thus obviating the need for hard-wire connections. This enhances freedom of movement in laboratory studies and permits monitoring of free-roaming animals in nature.

Figure 14–1 shows a telemeter capable of highly accurate transmission of the electrocardiogram and illustrates the degree of miniaturization that can be achieved using small solid-state components. The completed unit is the size of three pennies stacked on top of each other and weighs only 2 g. Powered by a hearing-aid battery, it can transmit a suitable signal over a distance of 100 ft for about two days, or over a distance of 10 ft for seven weeks. For certain applications telemeters considerably smaller than the one in Figure 14–1 have been designed.

Figure 14–2 shows a giraffe that has been equipped with a blood pressure telemeter and then released to roam freely. The pressure transducer was placed in the right carotid artery, and the wires from the trans-

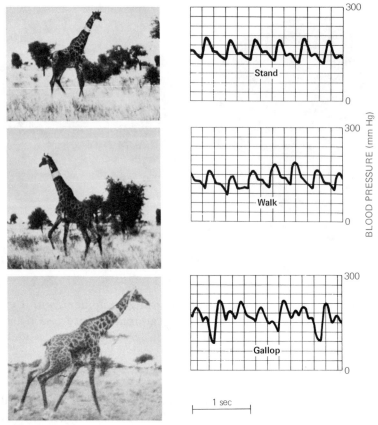

Figure 14–2. A giraffe equipped with a blood pressure telemeter. The transmitter was taped externally to the neck and the pressure transducer inserted surgically in the right carotid artery. Records were then obtained while the giraffe was free in the wild. The sharp drops in pressure during galloping coincided with the animal's front-hoof beats. (From Van Citters, R. L., W. S. Kemper, and D. L. Franklin. 1966. Science 152: 384–386. Copyright 1966 by the American Association for the Advancement of Science.)

ducer were led through the surgical incision to an external, relatively long-range transmitter taped to the animal's neck. As is evident from the records, this arrangement permitted measurement of both blood pressure and heart rate in a variety of exercise states. The relatively high blood pressures of giraffes, which were discussed in Chapter 10, were confirmed in these studies.

With the advantages that radio telemetry offers, many types of studies are now possible that could not be performed in the past owing to methodological limitations. Despite great progress in miniaturization, conventional telemeters unfortunately remain too large for use in many smaller animals. Continuing efforts at size reduction promise some amelioration of this problem.

Methods of measuring energy metabolism

In concluding this discussion of methodology, it is appropriate to give some special emphasis to techniques of measuring energy utilization during exercise because much of the remainder of the chapter will deal with energetics.

For determining the energetic demand of a particular form of exercise, analysis of respiratory gas exchange is typically the method of choice. The advantage of this type of analysis is that it is not only accurate, but can be carried out within a short time frame. If an animal, for example, can be induced to run on a treadmill or fly in a wind tunnel for just 10 or 20 minutes, measures of steady-state oxygen consumption and carbon dioxide production can be obtained. If the exercise is supported aerobically on a "pay-as-you-go" basis, then the rates of gas exchange will indicate the total rate of energy utilization required by the exercise. If, on the other hand, the exercise is supramaximal, requiring a continuous anaerobic contribution (see Chapter 11), the rate of anaerobic catabolism as well as that of aerobic catabolism must be determined to obtain the total cost. Various approaches have been used to determine the anaerobic contribution, including measurement of the postexercise oxygen debt and, in many vertebrates and some invertebrates, measurement of lactate accumulation.

Often the total daily energy expenditure is of interest. In the laboratory this is generally measured by one of two methods. First, oxygen consumption can simply be monitored over the entire day and integrated to yield the total daily consumption. Alternatively, as described in Chapter 2, daily energy expenditure can be measured by material balance studies. However measured, total daily energy expenditure includes components due to activity; a mouse, for example, may sleep, walk, eat, groom, and run over the experimental period, and the energetic demands of all these behavioral states will be included in the metabolic rate obtained.

Measurement of the energy metabolism of animals in the field has proved to be a difficult challenge. There is no simple transducer that can be implanted in the animal to radio out the metabolic rate, and the methods of determining oxygen consumption that are so elegantly useful in the laboratory cannot generally be put to use in the field.

Most of what we know about the energetics of particular forms of exercise in nature has been derived by extrapolation from laboratory studies. Natural activities are simulated in the laboratory, and their energetic demands are determined by measures of gas exchange. The confidence that can be placed in laboratory measures as indicative of the natural situation depends on the accuracy with which the natural activities can be simulated for a long enough period to allow a measure of oxygen consumption. The steady flight of a migrating bird can be simulated well, but simulation of the activity of a bird flitting about in the trees is much more challenging.

Probably the most common way in which the total daily energy expenditure of animals in nature has been estimated is through integration of laboratory data on the energetic cost of various activities and field data on the time that the animal spends in each type of activity. To illustrate, the amount of time that a bird spends resting, singing, foraging, defending its territory, and flying might be estimated by observing the bird in the field (this would be termed a time budget). The energetic cost of each of these activities could then be estimated by simulating the activity in the laboratory and the total daily energy demand in the field computed by multiplying the cost of each activity by the time spent in the activity and summing the products. The problems encountered in this method are several: (1) The entire 24 hours of the animal's day in nature must be assigned to one category of activity or another. Even if the animal can be observed steadily, it is often impractical to set up a separate category for each and every activity observed. Accordingly, major activities are generally recognized, and time spent in minor activities (e.g., preening, nest maintenance) is lumped with whichever major activity seems to involve about the same amount of effort. (2) The development of an accurate time budget depends on being able to observe the species for a substantial fraction of the day in nature. Thus animals such as diurnal birds that are relatively easily observed during their active period in the field are more readily analyzed than secretive or nocturnal species. The more difficult the animal is to observe, the more coarse and uncertain becomes the time budget. (3) The method depends on being able to obtain an accurate measure of the cost of each type of activity in the laboratory. This is no easy matter and only rarely has actually been done. Many authors have drawn cost estimates from the literature rather than determining them on the actual species under study, and it is not uncommon to find rather straightforward guesses employed for activities difficult to examine in the laboratory.

Estimates of total daily energy expenditure based on time budget analysis are common in the literature. Because there are many potential problems involved, however, it is essential to examine the methods and assumptions that lie behind any given estimate so that the accuracy attributed to the estimate will be that which is warranted.

Techniques involving extrapolation from the laboratory to estimate average daily metabolic rate are always subject to the uncertainties inherent in extrapolation itself. Accordingly, considerable thought has gone into the development of techniques that would permit more direct measurement of metabolism in free-ranging animals. Several of these deserve mention.

One approach has been to seek parameters that can readily be monitored by radio telemetry and that, in turn, might provide an accurate index of metabolism. Heart rate is one candidate that has received substantial attention. In many animals delivery of oxygen to the tissues is dependent on the circulatory system, and the rate of oxygen transport by the circulation provides a measure of the rate of oxygen utilization by the organism. To determine oxygen consumption by measures of circulatory parameters, the most straightforward approach, in theory, is to measure circulatory oxygen transport itself. This is, in fact, sometimes feasible using telemetry. In a mammal, for example, the rate of blood flow could be monitored by placing a flow detector around the systemic aorta, and the arteriovenous change in oxygen content could be monitored using oxygen electrodes implanted in the great arteries and great veins. Data from these transducers could be transmitted from a free-roaming animal by telemetry and, when integrated, would indicate the rate of oxygen transport to the systemic tissues. The problems with this approach are practical. Insertion of the transducers would require delicate surgery, and the number of transducers and size of the transmitter required would limit application of the method to large animals. An alternative approach is to monitor only a single circulatory parameter, such as heart rate, in the hope that it might be sufficiently well correlated with oxygen transport to provide a suitably accurate indication of oxygen utilization. Measurement of heart rate has practical appeal, for heart rate telemeters can be quite small and the connections to the animal are simple. But because heart rate is only one of several parameters that determine the rate of circulatory oxygen transport, it is essential first to evaluate whether heart rate in fact provides a suitable index of oxygen transport. If it should, a "calibration" curve relating oxygen consumption to heart rate could be developed in the laboratory, and the oxygen consumption of a free-ranging animal could then be estimated from telemetric data on heart rate.

The method of estimating metabolism from measures of heart rate is still in the testing stage. Available data suggest that the method may have promise, but indicate that if it is to be successful at all, it must be used with much care. The relationship between oxygen consumption and heart rate in birds and mammals not only varies among species, but also varies among individuals within a species. Accordingly, a calibration curve must be developed for each individual studied. More troublesome is the fact that heart rate is sometimes found to be a fundamentally poor indicator of oxygen consumption inasmuch as oxygen consumption can assume widely different values at one and the same heart rate. When the potential range in rate of oxygen consumption at a given heart rate is large, heart rate certainly cannot be used as a moment-to-moment indicator of metabolism. However, there is perhaps ground for hope that the average heart rate over a long period, such as many hours or a day, will prove to be a suitable indicator of the average oxygen consumption over that period— the idea being that some of the errors involved on a moment-by-moment basis will be averaged out over a long period. Unfortunately, very few studies have actually tested the utility of heart rate as an indicator of average metabolism over long periods of time. Of those, studies on man and penned blue-winged teal have found an acceptable correlation between average metabolism and average heart rate, whereas studies on sheep found heart rate to be a wholly unreliable indicator of metabolism in at least half the experimental animals.

Of all the potential indicators of metabolism that can be monitored by telemetry, heart rate has received the most attention. Breathing rate has attracted some interest, but the data available indicate that it is unlikely to prove suitable.

Another approach to monitoring the average metabolic rate of free-ranging animals has involved measuring the rate of elimination of radioactively tagged elements such as ^{131}I, ^{65}Zn, ^{32}P, or ^{134}Cs. The hypothesis behind this method is that certain elements that are processed biochemically by the animal might be eliminated from the body at a rate varying systematically with the rate of biochemical activity involved in energy metabolism. If this should be so, the rate of elimination of such an element over a period of time would be correlated with and provide an indication of the rate of metabolism over the period. The rate of elimination can be measured by using radioactively tagged isotopes. The animal is "loaded" with a measured amount of isotope by feeding or injection; then, at a subsequent time, the amount of isotope remaining is measured and subtracted from the initial amount to determine how much isotope was eliminated over the time period. The method has potential application in measuring the average metabolism of free-ranging animals because all that is needed to determine the rate of elimination is a measure of the amount of isotope present in the body at the beginning and end of the experimental period. An animal loaded with isotope can thus be released in its natural habitat for a period and then recaptured for the final isotope determination.

Whether this method can actually be useful obviously depends on whether elements can be found that are eliminated in a predictable relationship to the rate of energy metabolism. Knowledge of the metabolic processing of elements such as iodine, zinc, and cesium is insufficient to point the way to elements that will meet this requirement, and it has been necessary to evaluate the suitability of various elements on a strictly empirical basis, by studying their rate of elimination in relationship to energy metabolism in the laboratory. So far, at least seven elements have been tested in various species, and tests on additional elements are either under way or planned. Results to date are conflicting. The first tests involved ^{65}Zn in several species of invertebrates and indicated that the method had very real promise. Several elements have been tested in mammals. Some have proved to be useless, but others seem to be potentially useful. Even the more promising of the elements tested in mammals cannot yet be used to provide a reliable index of metabolism, but there is hope that refinements of technique and analysis may improve reliability.

The final method of measuring the average metabolism of free-ranging animals to be discussed is the so-called $D_2{}^{18}O$ method. This method has proved successful in laboratory tests and has been put to use in the field in several studies of mammals and birds. The name of the method derives from the fact that the animal is simultaneously administered measured amounts of heavy water, D_2O, and water composed of oxygen-18, $H_2{}^{18}O$. The rates of elimination of the deuterium and oxygen-18 are then measured, and from these data the rate of carbon dioxide production can be calculated. The rate of carbon dioxide production in turn provides a measure of metabolic rate. Once the D_2O and $H_2{}^{18}O$ have been administered, the animal can be released in its natural habitat, subsequently to be recaptured for determination of the amounts of deuterium and oxygen-18 eliminated during the experimental period. The metabolic rate computed is the average rate over the period.

The basic rationale behind the $D_2{}^{18}O$ method is not difficult to understand, though there are a number of complexities that must be taken into account in actual application. The initial observation that led to the development of the method was that the oxygen of expired carbon dioxide is in isotopic equilibrium with the oxygen of body water. Thus if the body water consists of given proportions of $H_2{}^{16}O$ (ordinary water) and $H_2{}^{18}O$, the carbon dioxide expired by the animal will contain both oxygen-16 and oxygen-18 in approximately the same proportions. Accordingly, when the concentration of $H_2{}^{18}O$ is experimentally elevated, the excess oxygen-18 will gradually be voided to the environment in expired carbon dioxide, and the rate at which this occurs will depend on the rate of carbon dioxide production and elimination. Because the rate of carbon dioxide production depends on metabolic rate, the rate of dissipation of oxygen-18 also depends on metabolic rate.

There is a second major route by which excess oxygen-18 is voided from the body water. Water lost through evaporation, urination, and other mechanisms consists of both $H_2^{16}O$ and $H_2^{18}O$, meaning that some of the excess $H_2^{18}O$ will be carried away in general water losses. Accordingly, the total rate of dissipation of excess oxygen-18 is in fact a function of *both* the rate of carbon dioxide production *and* the rate of water loss. If an investigator were to measure only the rate of oxygen-18 elimination, he would not know how much this reflected dissipation of oxygen-18 in either water or carbon dioxide alone. However, if an independent measure of the rate of water loss is obtained, the fraction of oxygen-18 lost in water can be calculated and subtracted from the total loss of oxygen-18 to obtain the amount lost in carbon dioxide. The reason that D_2O is administered along with $H_2^{18}O$ in the $D_2^{18}O$ method is that the rate of elimination of deuterium from the body water provides the required independent measure of rate of water loss. Excess deuterium in the body water cannot be lost in expired carbon dioxide and is dissipated primarily in the general processes of water loss through evaporation, urination, and so forth.

To summarize, the $D_2^{18}O$ method involves the addition of labeled oxygen and labeled hydrogen to the body water. The labeled oxygen is lost primarily in the two oxygen-containing compounds, CO_2 and H_2O, and the rate of loss therefore depends on the rate of carbon dioxide production and the rate of water loss. The labeled hydrogen is lost primarily in the hydrogen-containing compound, H_2O, and its rate of loss therefore depends on the rate of water loss alone. Knowing the rate of water loss from the deuterium data, the rate of carbon dioxide loss can be calculated from the data on oxygen-18.

The $D_2^{18}O$ method has been tested in the laboratory on about 10 species of birds and mammals and has proved to yield most satisfactory measures of carbon dioxide production. In these tests carbon dioxide production has been measured simultaneously by the $D_2^{18}O$ method and by traditional methods of determining the carbon dioxide in exhaled air, and the results obtained by the $D_2^{18}O$ method have been in error by a mean of only 3–8% in the various studies. In Chapter 2 we noted certain limitations of using measures of carbon dioxide production to estimate metabolic rate. The $D_2^{18}O$ method is subject to those limitations but still appears to be the most accurate of the methods now available for the estimation of metabolic rate in free-ranging animals. It has been used in a number of enlightening field studies, some of which will be discussed subsequently.

THE ENERGETICS OF EXERCISE

Metabolism as a function of the type and intensity of exercise

The most extensive investigations of exercise have been performed on vertebrates and insects, and these groups will be the subjects of this and the following sections. It must be recognized that our understanding of animal activity may be limited in significant ways by the taxonomic bias in available data.

In the study of the energetics of exercise we are concerned with two basic processes: (1) the utilization of nutrient energy by the organism and

(2) the external result of the energy utilization. These functions are coupled in a complex manner. A man utilizing energy at a rate of 13 kcal/min might be able to walk at 8 ft/sec, run at 10 ft/sec, or swim the crawl stroke at 2 ft/sec. Each of these forms of exercise bears its own relationship to energy utilization. Only part of the total energy utilization of the organism goes to support of activity in the exercising muscles themselves; some is directed to maintaining the ordinary basal functions of the body, and some goes to support of the increased cardiopulmonary activity demanded by exercise. Of the energy actually used by the exercising muscles, some is lost as heat in the inefficiencies of catabolism and energy coupling to the myofibrils. Some is lost internally as heat in overcoming the viscous and frictional resistances of the muscles, tendons, and joints in the moving body. After these losses, some energy appears as mechanical energy of motion of the limbs. It is this fraction that can actually be put to use to set the body in motion and keep it moving. In running, the kinetic energy of the legs is translated into forward motion of the body with each stride, and in swimming both the arms and legs of man participate. The resistances of the environment to movement must be overcome with this energy; the high resistance of water, for example, greatly impedes progress during swimming in such a poorly streamlined beast as man. Also, the rate of progress depends on how effectively the person can translate the energy of his limb motions into movement of his body. At one extreme, a runner can run in place, going through all the motions but making no progress because of failure to apply the force developed by his legs against the ground in such a way as to accelerate his whole body. At the other extreme, he can adopt the techniques of champion racers to maximize the speed permitted by his level of muscular activity. The significance of effective technique is nowhere more vividly illustrated than in the swimming pool. The novice may thrash about, metabolizing and moving his limbs just as much as a seasoned swimmer, and yet make only a small fraction of the latter's progress. All these considerations illustrate the complexity of the relationship between energy metabolism and performance. The nature of the coupling between the two is a fascinating study in its own right, but, for the most part, physiological ecologists have been content to describe the empirical relationships between metabolism and the rate of swimming, flying, running, or other activities. This type of information permits an understanding of the energetic implications of activity in nature, and in this section we shall review several illustrative examples.

Figure 14–3 shows the relationship between oxygen consumption and swimming speed in yearling sockeye salmon (*Oncorhynchus nerka*) at 15°C. It will serve not only as an example of energy metabolism in fish, but as a basis for reviewing certain general concepts introduced in Chapter 11. Note that metabolic demand increases approximately exponentially with speed. The fish displays a maximal sustained oxygen consumption (*aerobic capacity*) of nearly 900 mg O_2/kg/hr (630 cc/kg/hr). This is often referred to as the "active rate of oxygen consumption"—a shorthand expression that can be misleading; the "active rate" is the maximal oxygen consumption that can be elicited by activity, but, obviously, lower oxygen consumptions are associated with lower levels of activity. Swimming speeds that demand the aerobic capacity or less (0–4 lengths/sec) can be sustained for long periods,

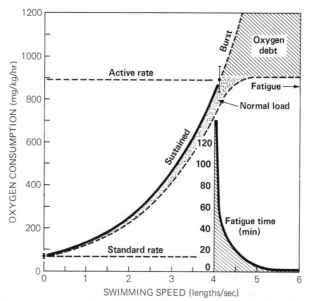

Figure 14-3. Oxygen consumption (solid "sustained" line) as a function of swimming speed in yearling sockeye salmon (*Oncorhynchus nerka*) acclimated and tested at 15°C. Fish weighed about 50 g and measured about 18 cm in length. They were tested in a "water tunnel" in aerated fresh water; water was driven through the experimental chamber at a known velocity, and fish were compelled to swim against the current. The "active rate" is the maximal sustained oxygen consumption. Speeds requiring less than the active rate could be sustained for long periods. There was some accumulation of oxygen debt at such speeds, but the debt was insufficient to interfere with performance. The debt was accordingly termed a "normal load," and its magnitude at each speed is indicated by the height of the shaded area below the curve for oxygen consumption. Burst speeds, being supramaximal, required a continuing anaerobic contribution. The theoretical oxygen demand of such speeds was determined by extrapolation of the curve for sustained speeds and is indicated by the dashed "burst" line. Actual oxygen consumption at burst speeds was at the active rate, and the remainder of the energy requirement was met by "oxygen debt." Time to fatigue at burst speeds and at nearly maximal sustained speeds is indicated in the lower right. (From Brett, J. R. 1964. J. Fish. Res. Board Can. 21: 1183–1226.)

for though there is some accumulation of oxygen debt, the activity can be supported aerobically on a "pay-as-you-go" basis. Such exercise is maximal or submaximal according to the terminology introduced in Chapter 11. Swimming at speeds greater than 4 lengths/sec requires more energy than can possibly be supplied aerobically. Being supramaximal, it can only be supported by a continuing anaerobic contribution and is limited in duration by the magnitude of the oxygen debt that can be incurred. Such swimming speeds are termed *burst speeds* in reference to their limited duration, and if continued they lead to profound fatigue as peak oxygen debt is approached. As shown in Figure 14-3, the time to fatigue decreases as the speed increases. Speeds of 5–6 lengths/sec (25–50% greater than the maximal sustained speed of 4 lengths/sec) can only be maintained for a matter of a few minutes. Studies of recovery after exercise in the salmon revealed that oxygen debts incurred during submaximal swimming were repaid within an hour; additionally, there was evidence that the fish could continue exercise unimpeded while retaining (not repaying) debts of the magnitude incurred in submaximal exercise. On the other hand, fish exercised supramaximally to exhaustion would rest quietly, with little spontaneous activity except hyperventilation, for as long as two to four hours after their burst swimming. Tests indicated that full repayment of their oxygen debt required about five hours; but the debt was largely repaid within three hours, and at that time the fish were again capable of burst swimming. It is clear that the salmon are greatly impaired for some time after exhausting bursts—a character of obvious significance in nature. Swimming at 6 lengths/sec for a minute or two may permit the fish to avoid one predator, but the outcome may not be so favorable if another predator happens along within the ensuing half hour.*

Figure 14–4 shows the relationship between oxygen consumption and swimming speed in five species of fish and that between oxygen consumption and running speed in six species of mammals. In both cases the data are limited to speeds that can be sustained aerobically on a pay-as-you-go basis (submaximal or maximal speeds). First it must be noted that the data for fish are plotted on semilogarithmic coordinates, whereas those for mammals are plotted on rectangular coordinates. The fish display an exponential increase in oxygen consumption with increase in speed, and the data are therefore linearized by a semilogarithmic plot. This is illustrated by comparing the data for sockeye salmon at 15°C in Figures 14–3 and 14–4A. The data in both figures are the same; the exponential relationship is evident in the rectangular plot of Figure 14–3, and the linearizing effect of the semilogarithmic plot is evident in Figure 14–4A. In contrast to the fish, the mammals display a linear relationship between oxygen consumption and speed (compare Figures 14–3 and 14–4B). Linear relationships have been reported for quite a few mammals, including man, but it must be noted that there are exceptions to this general type of behavior.

The fish in Figure 14–4A were exercised to their maximal sustained speed. It can be seen that aerobic capacity as well as the relationship between metabolism and speed varied with species and with temperature within a species. Bass at 20°C, for example, had an aerobic capacity only 37% as great as sockeye salmon at the same temperature and could attain speeds of only 2.4 lengths/sec, as compared to 3.7 lengths/sec in the salmon. The bass were able to swim faster than the salmon at a given oxygen consumption; otherwise the difference in peak swimming speed would have been much larger. The effects of temperature are noteworthy. From the data on bass and salmon it can be seen that not only did the resting metabolic rate (zero speed) increase with temperature, but also the metabolic rate required for a given speed increased. We shall return to this point later.

The mammals in Figure 14–4B were not exercised to peak speed, and the maximal rates of oxygen consumption recorded are therefore not indicative of aerobic capacities. It can be seen again, though, that the relationship of metabolism to speed depends on species. In particular, the slope of the relationship tended to increase with decreasing body size—indicating that small mammals must increase their weight-specific metabolism to a considerably greater extent than large mammals to attain a certain absolute velocity (km/hr). A similar relationship is evident in fish. A 150-g salmon

---

*The interplay between anaerobic and aerobic metabolism in supporting vigorous exercise is discussed in depth in Chapter 11, and the reader is referred there in particular for comparative information on amphibians, reptiles, and man.

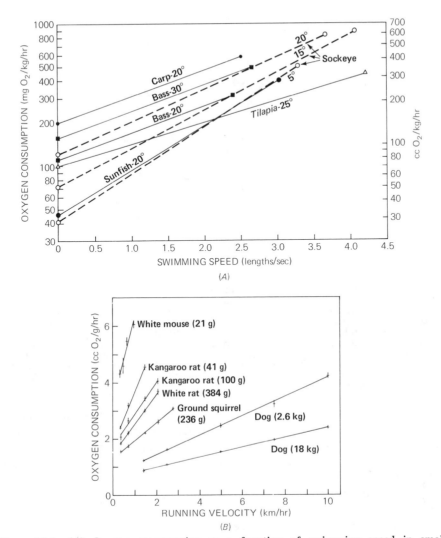

Figure 14–4. (*A*) Oxygen consumption as a function of swimming speed in small (50 ± 15 g) fish of five species: carp (*Cyprinus carpio*) at 20°C; largemouth black bass (*Micropterus salmoides*) at 20°C and 30°C; sockeye salmon (*Oncorhynchus nerka*) at 5°C, 15°C, and 20°C; pumpkinseed sunfish (*Lepomis gibbosus*) at 20°C; and *Tilapia nilotica* at 25°C. Fish were exercised up to maximum sustained speeds. Note that oxygen consumption is expressed on a logarithmic scale. Data for sockeyes are the same as in Figures 14–3 and 14–7. (From Brett, J. R. 1972. Resp. Physiol. 14: 151–170.) (*B*) Oxygen consumption as a function of running speed at 22°–27°C in six species of mammals: laboratory mice, laboratory rats, Merriam's kangaroo rats (*Dipodomys merriami*, average weight: 41 g), bannertailed kangaroo rats (*D. spectabilis*, average weight: 100 g), roundtail ground squirrels (*Citellus tereticaudus*), and domestic dogs (mongrels weighing 2.6 kg and Walker foxhounds weighing 18 kg). Animals were studied during horizontal running on treadmills. They were not necessarily exercised to full aerobic capacity. Horizontal bars indicate mean oxygen consumption, and vertical bars delimit twice the standard error on either side of the mean. By extrapolation, the oxygen consumption at zero speed is obtained as the *Y*-intercept. *Y*-intercepts exceed basal metabolic rates for the various species by a factor of 1.7 on the average. The greater-than-basal oxygen consumption at zero speed probably at least partly reflects the energetic cost of maintaining a running posture. (From Taylor, C. R., K. Schmidt-Nielsen, and J. L. Raab. 1970. Amer. J. Physiol. 219: 1104–1107.)

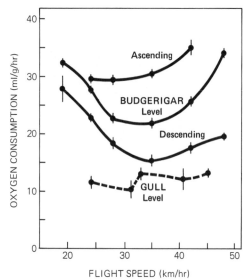

Figure 14–5. Oxygen consumption as a function of flight speed in budgerigars, *Melopsittacus undulatus*, (solid curves) and laughing gulls, *Larus atricilla*, (dashed curve). The birds were trained to fly in a wind tunnel and wore transparent masks, through which air was circulated, so that their respiratory gas exchange could be monitored. The gulls were studied during horizontal (level) flight only. The budgerigars flew horizontally and also at ascending and descending angles of 5°. Vertical bars delimit twice the standard error on either side of the mean response. The cost of flight was elevated to some extent by the extra drag contributed by the mask and tubing attached to the mask. (From Tucker, V. A. 1968. J. Exp. Biol. 48: 67–87; data for gulls from Tucker, V. A. 1969. Sci. Amer. 220: 70–78.)

that increases its weight-specific oxygen consumption by the same amount as a 50-g salmon will be able to swim at about the same number of body lengths per second—meaning that because it has a longer body, it can swim at a greater absolute speed than the 50-g fish for the same weight-specific metabolic expenditure.

As a final illustration of the relationship between metabolism and performance we can turn to data on budgerigars (parakeets) and laughing gulls flying in a wind tunnel. As seen in Figure 14–5, the relations between oxygen consumption and speed in these birds are of still another nature than those seen in fish or mammals. Although oxygen consumption during horizontal swimming in fish increases steadily and exponentially with speed and although that during horizontal running in mammals typically increases linearly with speed, oxygen consumption during horizontal flight in budgerigars decreased as speed was increased from 20 km/hr to 35 km/hr and then increased as speed was elevated above 35 km/hr. In the laughing gulls the relation between metabolic rate and speed during horizontal flight was more complex. Again we see evidence that larger animals can make more rapid progress for a given weight-specific metabolic rate than smaller animals. The data on budgerigars show that ascending flight is more costly than horizontal flight, and descending flight is less costly.

Scope for activity

The concept of scope for activity may be introduced by using the results of experiments on goldfish shown in Figure 14–6A. The fish were acclimated to temperatures of 5°–35°C and tested at their acclimation temper-

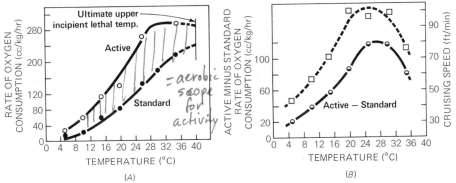

Figure 14–6. (*A*) Oxygen consumption as a function of temperature in goldfish (*Carassius auratus*) acclimated to test temperatures. The lower curve shows the standard rate of oxygen consumption, and the upper curve shows the rate during maximal sustained swimming (active rate). The fish averaged 3.8 g in weight. (*B*) The solid line depicts the difference between active and standard oxygen consumption at each temperature, termed the aerobic scope for activity; compare with part *A*. The dashed line depicts maximal sustained swimming speed as a function of temperature. The left ordinate gives units for aerobic scope, whereas the right ordinate gives units for swimming speed. [From Fry, F. E. J. and J. S. Hart. 1948. Biol. Bull. (Woods Hole) 94: 66–77.]

atures. First, oxygen consumption was measured at rest, and then the fish were exercised at maximal sustained speed to determine the aerobic capacity or "active" oxygen consumption. The difference between maximal oxygen consumption and standard oxygen consumption at each temperature is shown by the solid curve in Figure 14–6*B*. This difference is termed the *aerobic scope for activity.* It indicates the extent to which oxygen consumption can be increased above the resting level to support activity at each temperature.

The scope for activity has received a great deal of attention from physiologists, and before proceeding to a comparative discussion it is important to recognize some basic principles. The scope tells us, in a very immediate sense, the degree to which oxygen consumption can be increased by activity. As such, it is a holistic indicator of the flexibility and capability of the systems responsible for the uptake and utilization of oxygen. We see from Figure 14–6*B* that the goldfish can augment its oxygen consumption six times more at 25°C than at 5°C—quite a remarkable difference. How does this translate into capability for exercise performance? The answer to this question depends on the relationship between oxygen consumption and performance; that is, the scope alone does not immediately reflect performance. This is an important point: scope does indicate the capacity to increase aerobic catabolism, but it does not in itself necessarily indicate the capacity to perform. Within a single species or a group of related species, oxygen consumption and performance may be related in such a way that changes in scope do at least qualitatively parallel changes in performance. But it would clearly be absurd to assume that an insect with a greater weight-specific scope than a horse would be able, on those grounds alone, to fly faster than the horse could run. Having made this simple but important point, we can note in Figure 14–6*B* that maximal sustained swimming speed does parallel aerobic scope in the goldfish, not a surprising result within a single species. Note, however, that although the scope at 25°C is six times that at 5°C, the maximal speed at 25°C is just over twice as great. The

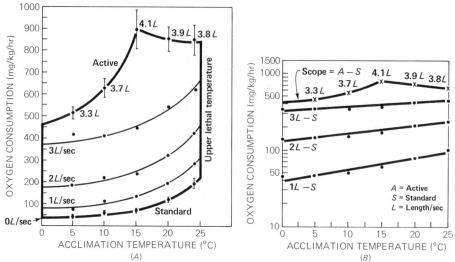

Figure 14–7. (*A*) The relationship between oxygen consumption, swimming speed, and temperature in yearling sockeye salmon (50-g, 18-cm). Fish were acclimated to test temperatures prior to experimentation. Lower lines show the standard rate of oxygen consumption and consumption at speeds of 1, 2, and 3 body lengths/sec. Upper line shows the maximal oxygen consumption elicited by activity (active consumption) with an indication of the maximal speed (lengths/sec) that could be maintained for one hour. Vertical bars on the active and standard curves delimit twice the standard error on either side of the mean. (*B*) The same data as in part *A*, replotted. Upper line depicts aerobic scope for activity (active oxygen consumption minus standard oxygen consumption). Lower lines show the net metabolic cost at 1, 2, and 3 lengths/sec. Net cost is the actual oxygen consumption minus the standard oxygen consumption. (From Brett, J. R. 1964. J. Fish. Res. Board Can. 21: 1183–1226.)

changes in scope and performance are not proportional, a result that follows in part from the basically exponential relationship between oxygen consumption and speed in fish (see Figure 14–3).

Figure 14–7*A* depicts standard and active oxygen consumption as well as oxygen consumption at various submaximal speeds in yearling sockeye salmon; the data come from the same study as portrayed in Figures 14–3 and 14–4*A*. The upper line in Figure 14–7*B* shows the scope for activity as a function of temperature, and again there is evidence of a parallel between scope and peak sustained swimming speed. The other lines in Figure 14–7*B* show the net energetic cost for three submaximal swimming speeds; net cost is the difference between actual oxygen consumption and standard oxygen consumption—the amount by which metabolism must be increased above the standard level to support activity. Note that the net cost of swimming at 1 length/sec is greater at 25°C than at 0°C. The same holds true at 2 and 3 lengths/sec, but the proportionate effect of temperature becomes less as the speed is increased. Extrapolating to still higher speeds, it appears that the net cost of burst speeds would be nearly constant regardless of temperature, an interesting conclusion that has also been indicated in a number of other species of fish.

It is noteworthy that both the goldfish and salmon, as well as some other fish that have received study, exhibit a maximal aerobic scope at some intermediate temperature, above and below which scope is lower. Thus the goldfish show peak scope at 25°–30°C and also attain their maximal sustained swimming speeds at about the same temperatures. The salmon

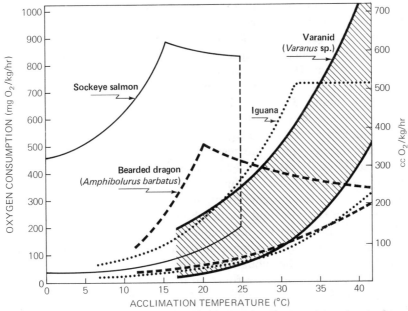

Figure 14–8. Oxygen consumption as a function of temperature in sockeye salmon and representatives of three genera of lizards, the bearded dragon (*Amphibolurus barbatus*), the iguana (*Iguana iguana*), and varanids (*Varanus* sp.). For each species the lower line shows standard oxygen consumption and the upper line, active oxygen consumption. The data for sockeyes are the same as in Figure 14–7*A*. (From Brett, J. R. 1972. Resp. Physiol. 14: 151–170.)

reach peak scope and maximal sustained swimming speed at about 15°C. The increase in scope with temperature at low temperatures is believed to reflect the facilitating effect of warmth on the metabolic activity of the muscles and circulatory and respiratory systems. The decrease in scope at high temperatures is not well understood. One potential limiting factor is the decrease in the oxygen concentration of water at high temperature. It has also been postulated that the metabolic demands of the circulatory and respiratory systems increase with temperature in such a way that it becomes energetically impractical to increase oxygen delivery by as much at high temperatures as at more moderate temperatures.

Figure 14–8 compares the active and standard metabolic rates of salmon and three groups of lizards. The most notable feature illustrated by these data is that the lizards, despite their being air-breathing animals, do not display remarkably different aerobic capacities or scopes for activity than fish. The salmon, which are among the more aerobically capable fish, have a higher capacity and scope than either the bearded dragon or iguana. The varanid lizards, which are particularly active reptiles, have a high aerobic capacity at high temperatures; recent studies indicate that in some species aerobic capacity is even higher than that shown here (over 1000 cc/kg/hr at 42°C). Even this is not remarkably higher than what some fish can attain, however.

As noted in Chapter 3, the basal metabolic rates of birds and mammals are 5 to 10 times higher than the standard rates of similarly sized lizards at body temperatures of 37°–40°C. It is clear, then, from Figure 14–8 that the *aerobic capacities* of fish and reptiles (and amphibians as well) are typically lower than the *basal* metabolic rates of birds and mammals of similar size. Thus, in considering the vertebrate homeotherms, we come to an entirely different order of aerobic competence. At 15°C, the temperature of maximal scope, a 350-g sockeye salmon has a standard metabolic rate of about 0.05 cc O_2/g/hr and an aerobic scope of about 0.5 cc O_2/g/hr. By contrast, a 350-g laboratory rat tested at 30°C will have a basal metabolic rate of about 0.8 cc O_2/g/hr and an aerobic scope of about 3.8 cc O_2/g/hr. At 40°C a 700-g varanid lizard (*Varanus gouldii*) will have a standard metabolic rate near 0.1 cc O_2/g/hr and a scope of about 0.9 cc O_2/g/hr. A comparably sized guinea pig tested at 30°C will have a basal metabolic rate of about 0.7 cc O_2/g/hr (lower than the aerobic capacity of the varanid) and a scope of about 3 cc O_2/g/hr. Trained human athletes have scopes of about 4 cc O_2/g/hr. Budgerigars flying in wind tunnels have exhibited scopes as high as 40 cc O_2/g/hr, and Costa's hummingbirds use oxygen about 40 cc/g/hr faster when hovering than when resting at thermoneutrality. In all, we see that aerobic scopes in mammals and birds are several- to manyfold greater than in poikilothermous vertebrates.

The analysis of scope as a function of ambient temperature in homeotherms is complicated by a number of factors and has not received extensive attention. At temperatures below thermoneutrality, resting metabolism is elevated above the basal level because of thermoregulatory heat production (shivering and nonshivering thermogenesis). Exercise also produces heat, and the effects of exercise on oxygen consumption depend on whether the heat produced in exercise can be substituted partly or wholly for shivering or nonshivering thermogenesis (see Chapter 3). In large organisms such as man, a considerable degree of substitution is typically possible. Thus a man at cold temperatures can exercise to a certain extent with no increase in oxygen consumption at all; the heat production of the exercise simply relieves the need for specific thermoregulatory heat production. In small mammals there may be little or no substitution. Exercise commonly increases the overall thermal conductance of the body in these animals to such an extent that, essentially, the increased rate of heat production resulting from exercise is simply counterbalanced by the increased rate of heat loss, and shivering or nonshivering thermogenesis must continue unabated to provide a sufficient total heat production to maintain body temperature. When antelope ground squirrels, for example, run at a given submaximal rate, their oxygen consumption is increased by the same amount at 10°C as at 25°C. Because the resting consumption at 10°C is about 2 cc/g/hr higher than at 25°C, the running consumption is also about 2 cc/g/hr higher. Similarly, white laboratory mice showed the same increment in oxygen consumption for a given intensity of submaximal work at temperatures ranging from −10°C to 30°C, indicating no substitution of exercise heat production for specific thermoregulatory heat production. Oxygen consumption at −10°C was over twice as great as at 30°C when the mice were running at 6 m/min.

In some, but not all, mammals, the maximal oxygen consumption during exercise is about the same at all temperatures. When this is the case, performance at low temperatures can be affected in different ways depending on the degree of substitution. When full substitution relieves the need for shivering or nonshivering thermogenesis, then the full aerobic capacity is available to support exercise even at cold temperatures. However, if there is no substitution or only a partial substitution, then some of the maximal oxygen consumption at low temperatures must go to support shivering or nonshivering thermogenesis. Correspondingly, the fraction of aerobic capacity that can be directed to support of exercise will be less at low temperatures than at high temperatures, implying diminished performance at low temperatures. In recent studies of laboratory rats and guinea pigs it was found that the maximal oxygen consumption elicited by running on a treadmill was the same at low and high temperatures, but the intensity of work that could be maintained was reduced at low temperatures. This is what is to be expected if part of the oxygen intake must be diverted at low temperatures to shivering and nonshivering thermogenesis.

The highest known weight-specific aerobic scopes for activity are reported from certain insects. Some bees and butterflies, for example, which have resting metabolic rates of less than 1 cc O_2/g/hr, have been shown to reach metabolic rates of 90–100 cc O_2/g/hr during flight. Migratory locusts reach 10–30 cc O_2/g/hr, and flying fruit flies, about 20 cc O_2/g/hr. These latter values are within the range known for birds and mammals.

Thus far we have considered the absolute magnitude of aerobic scope. It also is of interest to ask by how many times the maximal oxygen consumption can exceed resting consumption. This parameter provides some comparative indication of aerobic flexibility. A fish, for example, might have a resting metabolic rate of 0.05 cc O_2/g/hr and a maximal rate of 0.30 cc O_2/g/hr; and a mammal might have a resting rate of 2 cc O_2/g/hr and a maximal rate of 12 cc O_2/g/hr. The aerobic scope of the fish is far lower than that of the mammal, but it is noteworthy that both animals can increase aerobic metabolism by a factor of 6. That is, the systems of oxygen delivery and utilization in both species have about the same *relative* ability to increase their rate of activity. This hypothetical example illustrates the significance of expressing maximal oxygen consumption as a ratio of resting consumption. The ratio is often termed *aerobic expansibility*.

Untrained, healthy young people typically can increase oxygen consumption by a factor of from 8 to 12 above the basal level during exercise; training improves aerobic capacity, and athletes may increase oxygen consumption by a factor of 20. These performances are fairly typical of other large mammals that have received study; dogs and horses, for example, have been shown to increase consumption by 11 to 24 times. In small mammals a combination of cold and exercise may be required to elicit peak oxygen consumption. A recent study employing both stimuli has affirmed that small *running* mammals have a more-limited relative ability to augment oxygen consumption than large mammals. Laboratory mice, rats, hamsters, and guinea pigs could attain only six- to sevenfold increases of oxygen consumption above the basal level. In another study, however, it was found that bats weighing 70–110 g increased their oxygen consumption above the basal level by a factor of about 20 when flying. Costa's hummingbirds can maintain an oxygen consumption at least 14 times the basal level while hovering, and budgerigars can reach at least 25 times their basal rate during flight in a wind tunnel. In sockeye salmon (Figure 14–7) the largest factorial increase in oxygen consumption at any one temperature is 12.6 (at 15°C). Iguanas (*Iguana iguana*) can increase oxygen consumption by a fac-

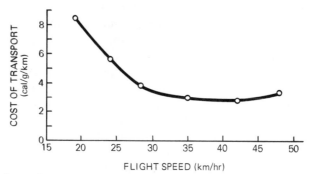

Figure 14-9. Cost of transport as a function of speed during horizontal flight in budgerigars. Data are the same as those for horizontal flight in Figure 14-5. (From Tucker, V. A. 1968. J. Exp. Biol. 48: 67–87.)

tor of 5 at 32°C, and varanids (*Varanus gouldii*), by a factor of 10 at 40°C. The highest factorial increases in aerobic metabolism are reported from insects, a number of which have been shown to consume oxygen 50 to 200 times faster when flying than when at rest. These values for insects, however, are not entirely comparable to those given for other animal groups inasmuch as many insects increase their body temperature during flight (see Chapter 3), and a thermal effect as well as an effect of activity itself is present.

Cost of transport

The cost of transport is defined to be the metabolic expenditure required to cover a unit of distance. If a 50-g salmon, for example, with an oxygen consumption of 30 cc/hr can cover 2400 meters per hour, then its cost of transport is $30/2400 = 0.0125$ cc O_2/meter—the ratio of the rate of oxygen consumption over the rate of movement. Cost of transport is a measure of the effectiveness of the animal in using its metabolic resources to cover distance, those with a high cost being less effective than those with a low cost. The measure is familiar to all of us in its inverse form for expressing the performance of automobiles. We often are concerned about our car's miles per gallon. If we spoke of gallons per mile instead, we would be using an expression of cost of transport, and clearly we would want to minimize the cost. Often cost of transport is expressed in weight-specific terms. Returning to the salmon mentioned earlier, we see that its weight-specific metabolic rate while covering 2400 m/hr is 0.6 cc O_2/g/hr. Thus its weight-specific cost of transport is $0.6/2400 = 0.00025$ cc O_2/g/m. This is the cost of moving one gram of body weight over a distance of one meter.

In Figure 14-5 we have seen the relationship between oxygen consumption and horizontal flight speed in budgerigars. The *cost per hour* (cc O_2/g/hr) is minimal at a speed of 35 km/hr. Thus a budgerigar with given nutrient reserves could stay aloft for the *longest time* by flying at this speed. The *cost of transport* (cc O_2/g/km) is shown in Figure 14-9. Here we see that the cost of covering a given distance is minimal at a speed of 42 km/hr. Thus a budgerigar with given nutrient reserves could cover the *longest distance* by flying at this latter speed. At 42 km/hr the bird uses more than the minimal amount of oxygen *per unit time*, but by flying at 42 km/hr rather than 35 km/hr, it also covers more distance per unit time.

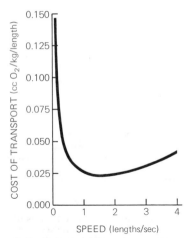

Figure 14-10. Cost of transport as a function of speed in yearling sockeye salmon at 15°C. Both parameters are expressed relative to the body length of the fish. Data are the same as those in Figure 14-3.

It is the interaction of these two functions that determines the speed at which cost per unit of distance is minimal. Note from Figure 14-5 that the cost per unit time is about the same at flight speeds of 19 km/hr and 47 km/hr. Obviously, in terms of distance covered, a speed of 19 km/hr is less economical than a speed of 47 km/hr, and this is reflected in Figure 14-9.

Figure 14-10 depicts the cost of transport in yearling sockeye salmon at 15°C. At low speeds the cost is very high. Though the increment in oxygen consumption caused by activity at such speeds is quite small (see Figure 14-3), the animal must also support its standard maintenance oxygen consumption, and the latter does not contribute to locomotion; all told, then, the rate of metabolism per unit distance covered is large. There is a broad range of speeds, from about 1 length/sec to 2.5 lengths/sec, where the cost of transport is minimal. At high speeds the cost increases again owing to the exponential increase in oxygen consumption required for each increment in speed.

Before looking briefly at mammals, it will be helpful to introduce the distinction between *net* and *gross* cost of transport. The gross cost is the total cost per unit of distance, including the resting oxygen consumption as well as the increment caused by activity. All the costs discussed thus far are gross costs. The net cost is obtained by subtracting the resting oxygen consumption from the gross cost. It expresses the *increment* in oxygen consumption required to traverse a unit of distance. To illustrate, if a mammal has a resting oxygen consumption of 2 cc/g/hr and a total consumption of 4 cc/g/hr when running at 4 km/hr, the gross cost is 4/4 = 1 cc/g/km, but the net cost is (4-2)/4 = 0.5 cc/g/km.

As seen in Figure 14-4B, mammals often display a linear relationship between rate of oxygen consumption and running speed. For any given species, the Y-intercept of the line relating oxygen consumption to speed represents the oxygen consumption of an animal in the running posture at zero speed. If we treat the Y-intercept as the resting rate of oxygen consumption, then it turns out that in mammals, the net cost of transport is a constant regardless of speed. To see why, consider that the relationship

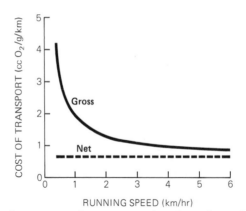

Figure 14–11. Net and gross cost of transport as functions of speed in roundtail ground squirrels (*Citellus tereticaudus*). Data are the same as those in Figure 14–4B. Net cost was calculated by subtracting the oxygen consumption at zero speed (extrapolated *Y*-intercept, Figure 14–4B) from the total oxygen consumption while running, then dividing by speed.

between total oxygen consumption, *M*, and speed, *S*, as shown in Figure 14–4B, can be described by a linear equation: $M = aS + b$, where a is the slope and b is the *Y*-intercept (resting oxygen consumption). If we subtract the resting oxygen consumption (b) from both sides of this equation, we get $(M - b) = aS$, where $(M - b)$ represents the increment in oxygen consumption above the resting level required at each speed. Now if we divide both sides by speed, *S*, we get $(M - b)/S = a$. The expression $(M - b)/S$ is the net cost of transport and is seen to be a constant and equal to the slope a (a has dimensions of cc O_2/g/hr per km/hr = cc O_2/g/km). In Figure 14–11 the net cost of transport in the roundtail ground squirrel is shown as a dashed line, and the gross cost of transport is also plotted. At low speeds the gross cost is very high, for the same reason it is high in salmon: the resting oxygen consumption, which in an immediate sense contributes nothing to locomotion, is included in the cost and, in fact, dominates because the increment caused by running is small at low speeds. However, as running speed is increased, the gross cost approaches the net cost asymptotically, for as the increment in oxygen consumption caused by running becomes greater and greater, the resting oxygen consumption becomes less and less significant in the total gross cost. In mammals we see that the gross cost of transport becomes very nearly independent of speed at moderate to high speeds.

Figure 14–12 shows the results of Schmidt-Nielsen's recent synthetic analysis of cost of transport in running, flying, and swimming animals. Cost of transport is expressed as a function of body size, and we see confirmation of the point suggested earlier, that larger animals can cover a given distance at less weight-specific cost than smaller animals. The line for swimming animals was constructed from data for fish only and that for running animals from data for mammals only. The line for flying animals, however, includes both birds and insects, and it is striking that these disparate groups seem to follow the same relationship between cost of transport and weight.

Figure 14–12 brings to light a number of interesting considerations.

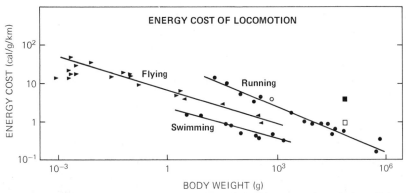

Figure 14-12. Cost of transport as a function of body weight in running, flying, and swimming animals. The line for *running* animals is constructed from data on the net cost in mammals; the gross cost approaches the net cost at relatively high speeds. Solid circles are individual data points, some of them derived from Figure 14-4*B*. The line for *flying* animals is constructed from data on birds and insects, using the minimal gross cost (gross cost at the most economical speed). Arrowheads are individual data points, including results for budgerigars and laughing gulls (Figure 14-5). Points below 1 g are all for insects; points above 10 g are all for birds; data for hummingbirds and large insects fall close to each other at 2-5 g. The line for *swimming* animals is constructed from data for fish, using the gross cost at three fourths of maximal sustained speed. Solid circles are individual data points and include information on sockeye salmon (e.g., Figure 14-3) as well as on five other species of fish. The open square depicts the average cost of transport in running man. The solid square gives the cost for swimming man. The open circle is for swimming mallard ducks. (From Schmidt-Nielsen, K. 1972. Science 177: 222-228. Original figure copyright 1972 by the American Association for the Advancement of Science.)

First, for an animal of a given body size, running is the most expensive form of locomotion and swimming, the least expensive. It is important to note that cost of transport is expressed on a logarithmic scale; thus the actual differences in cost are larger than might be gathered from a quick visual inspection of the graph. For a 100-g animal the cost of running a unit of distance is about 3.5 times as great as that of flying the same distance, and the cost of flying is about 2.5 times greater than that of swimming. The cost of running is about nine times greater than that of swimming. The fact that the graph is on logarithmic coordinates should also be kept in mind in interpreting the effect of body size on cost of transport. For a 0.01-g insect the weight-specific cost of transport is about 15 times greater than that for a 100-g bird.

One factor that is probably important to the relatively low cost of swimming is the buoyant effect of water. Unlike the running or flying animal, the fish needs to spend little, if any, energy to support its body in the medium. When airborne, insects and birds need to expend energy not only to make horizontal progress, but also to stay in the air, and it is not altogether clear why flying is so much less costly than running.

The energetics of migration will be discussed mainly in the next section, but here it is worth noting that although certain small and medium-sized fish and birds undergo long migrations, the same is not true of small and medium-sized running mammals. This makes sense energetically when we see that flying and swimming are far less costly than running. Some large mammals, such as caribou, migrate over long distances, but as we see, their cost of transport is comparable to that of fish of modest size.

We may now consider a number of additional interesting observations that have arisen from studies of cost of transport. The open square in

Figure 14–12 shows the mean cost of transport for running man, and it can be seen that the cost is about twice that expected for a 70-kg animal. Recognizing that the "standard" relationship between cost of running and weight is derived from data on mammals that run on all four legs, it has been suggested that bipedal running might be basically less economical than quadrupedal running. Recent experiments on chimpanzees and monkeys, however, argue against this conclusion. The chimpanzees and capuchin monkeys were trained to run on a treadmill on either two or four legs. In each species the relationship between oxygen consumption and speed turned out to be the same whether running was bipedal or quadrupedal. The chimpanzees, like man, exhibited a higher cost of transport than expected for their size, but the capuchin monkeys did not deviate from the expected cost. Also, spider monkeys running on two legs showed a cost similar to that predicted for their size. These data do not explain the relatively high cost of transport in man but indicate that it is unlikely to be due simply to his adopting a bipedal rather than quadrupedal habit.

The closed square in Figure 14–12 shows the cost of transport in swimming man. It is about four times greater than the cost of running in man and, based on extrapolation of the curve for swimming fish, is about 30 times greater than the cost that would be incurred by a fish of similar weight. The open circle in Figure 14–12 gives the cost of swimming in a mallard duck. The duck paddles about on the surface of the water and must expend nearly 20 times more energy to travel a given distance than a fish of comparable size.

Migration

Many animals undertake lengthy migrations that are energetically demanding. Those that do not feed along the way have attracted particular interest, for they must accomplish the entire feat using stored nutrient reserves. It may seem a bit fatuous to ask if, given their known reserves and their measured cost of locomotion, they can accomplish their migrations; obviously at least a portion of the individuals get where they are going. By asking (and attempting to answer) the question, though, we can gain insight into the energetic challenge that migration presents and delineate some of the parameters that affect success or failure.

Studies of migrating insects and vertebrates have shown that fat is the pre-eminent fuel, and it is instructive first to examine the advantages that fat holds. In Chapter 2 it was pointed out that, on the average, fat yields over 9 kcal of energy per gram on complete oxidation, whereas carbohydrate yields just 4.1 kcal/g and protein, 4.2 kcal/g. On these simple grounds it is evident that fat is energetically the most advantageous form of stored nutrient for the migrant. Each gram of stored food must be transported by the migrating animal—at an energetic cost—until it is used. Thus it is most economical to carry the type of nutrient that will yield the most energy when it is catabolized. The advantage of fat over the other common form of stored nutrient, carbohydrate, is in fact considerably greater than this analysis indicates. The energy yields given above refer to the oxidation of a gram of *pure* foodstuff. Fat is stored by animals in pure form, but glycogen is typically stored with an appreciable amount of water. In carrying a certain amount of glycogen, the animal must also carry this water, and water, of course, has no energy value whatsoever. From measures on vertebrates and insects it appears that each gram of glycogen is characteristically accompanied by 2.7 g of water. Thus 1 g of glycogen encumbers 3.7 g of body weight, giving stored glycogen an effective energy value of just 4.1 kcal/3.7 g—or 1.1 kcal/g. Glycogen is not without its advantages as a fuel; it can be mobilized rapidly and, unlike fat, can be used to support anaerobic glycolysis. In many animals it, or glucose, is the pre-eminent fuel of short-term exertion. But storage of glycogen is an uneconomical use of body weight for the migrant.

It has long been recognized that migratory birds typically lay down extensive deposits of fat prior to migration and that much of this fat is utilized in their travels. Indeed, the energetic cost of migration has often been estimated by comparing the fat content of pre- and postmigrants. Further evidence that fat is the dominant fuel of flight has come from studies of gas exchange. In both budgerigars and laughing gulls flying in wind tunnels the respiratory quotient has proved to be between 0.7 and 0.8—direct evidence that fat is the main substrate of catabolism. In many birds fat accounts for near 25% of body weight at the start of migration, and in some species values as high as 50% have been observed. Commonly, fat content is reduced to near 5% of body weight after migration. Tucker has recently estimated the nonstop distance that birds can travel by using these values for fat depletion in tandem with measures of minimal cost of transport gathered in laboratory studies. The basic rationale of his computations may be illustrated by example. If a 40-g bird uses fat equivalent to 25% of its initial body weight during migration, then the total energy yield is 93 kcal (10 g of fat catabolized, 9.3 kcal per gram of fat). From studies of budgerigars in wind tunnels, the minimal cost of transport for a 40-g bird is about 0.116 kcal/km (Figure 14–9). Thus a bird flying through still air at the speed that minimizes cost of transport could travel 93/0.116 = 800 km. Using a relationship between cost of transport and body weight like that in Figure 14–12, Tucker calculated flight range as a function of weight as depicted in Figure 14–13. The lower line shows the range when the bird uses fat equivalent to 25% of its initial weight, and the upper line shows the range using fat equivalent to 50% of initial weight. For given fat depletion, range increases with weight because the weight-specific cost of transport is lower in large birds than in small birds (Figure 14–12).

How do the actual flight ranges of birds compare with those estimated in Figure 14–13? Tucker cites three remarkable examples of birds that migrate over water and cannot stop to feed along the way. Ruby-throated hummingbirds fly 800 km (500 miles) across the Gulf of Mexico. Starting at a body weight of about 4.7 g, their flight range estimated from Figure 14–13 for 50% fat is about 1100 km. Because hummingbirds are known to commence migration with 40% or more of fat, we see that the estimate of Figure 14–13 is compatible with the actual performance of the birds, though it appears that an 800-km flight is near the limits of their capability. Blackpoll warblers weighing about 20 g fly 1300 km from New England to Bermuda and have an estimated range (Figure 14–13) of about 1500 km for 50% fat. Golden plovers, which weigh near 200 g, migrate 3900 km from the Aleutian to the Hawaiian Islands. This far exceeds their estimated range of 2600 km for 50% fat. In all three of these species, rang-

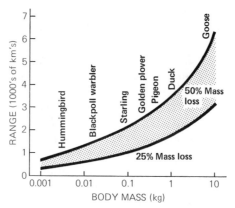

Figure 14–13. Nonstop flight range in still air (or range relative to the air) as a function of body weight at takeoff in birds. Fat is presumed to be the fuel of flight, and range is calculated assuming that fat equivalent to 25% (lower line) or 50% (upper line) of initial body weight is utilized. It is assumed that birds fly at speeds that minimize their cost of transport. See text for further discussion. (From Tucker, V. A. 1971. Amer. Zool. 11: 115–124.)

ing in weight from 4.7 g to 200 g, we see that actual flights are close to or exceed the estimated range. In interpreting this result a number of factors must be considered. First, cost of transport data are available from only a few species; thus the estimated cost of transport, which is an important factor in computing range, may be in error for the migrants considered. Second, ranges have been calculated under the assumption that the bird is flying in still air. In nature, flight into a head wind would increase the cost of transport, whereas flight with a tail wind would decrease the cost. Further, vertical updrafts can act to reduce the cost of transport by decreasing the metabolic effort required to stay aloft (providing lift). There is insufficient knowledge of ambient winds during migration and of the ways in which birds adjust their flight path in response to winds for these factors to be entered into the estimation of range. Despite the inadequacies in our present understanding of natural flight, it does appear from the comparisons made earlier that some species undertake migrations that are remarkably close to their limits. (Remember that flight at the *minimal* cost of transport is assumed in Figure 14–13, and 50% fat is near the upper limit reported for birds. The assumptions in Figure 14–13 are thus generous.) As Tucker points out, a land bird migrating over water has no second chance if it fails, and if the margin of safety is as small as it seems for some species, a high evolutionary premium would be placed on developing abilities for accurate navigation and for utilizing ambient winds to maximum advantage.

Birds migrating over land, of course, can stop to replenish their fuel reserves. The white-crowned sparrow, for example, migrates as far as 4000 km along the west coast of the Northern Hemisphere. The migration is accomplished in a series of nocturnal flights covering 100–600 km. Between flights the birds feed for one to several days, restoring their fat supplies. Weighing about 30 g and carrying about 20% fat on takeoff, their maximum range for a single night's flight is estimated at about 700 km by Tucker's method of calculation—a figure that agrees favorably with the peak distances the birds are known to cover.

Among fish, the salmon undertake some of the most spectacular migrations. Spawned in freshwater lakes and streams, they travel to the ocean and later return to fresh water to breed as adults. Their migrations against the current in rivers and streams as adults are sometimes of remarkable length, and they must overcome natural waterfalls and rapids as well as man-made obstacles. Chinook salmon, after entering the mouths of rivers, are known to travel as far as 3500 km (2200 miles) to spawning grounds in the Yukon. Sockeye salmon frequently travel upriver for distances of 800–1600 km (500–1000 miles). During their freshwater journey, salmon do not eat and thus must accomplish the entire task using stored nutrients.

Studies of sockeyes indicate that they, like some birds, very nearly reach their energetic limits during migration. The body composition of sockeyes was determined before and after a 1000-km migration up the Fraser River from Vancouver to Stuart Lake in British Columbia. It was found that their initial fat reserves were depleted by over 90%, and in the females protein was depleted by 50–60%. Taking into account that the females invest considerable energy in the development of their ovaries, it was calculated that the daily energy demand of swimming in males and females was near 75–80% of aerobic capacity. It seems likely, then, that the demands of migration have been a principal factor in the evolution of the relatively high aerobic competence of salmon; or, conversely, their high aerobic capacities have permitted their migratory habit. Laboratory studies indicate that adult salmon realize a minimal cost of transport at swimming speeds of about 1.8 km/hr. In the Fraser River they appear from energetic considerations to average about 4.2 km/hr—a speed at which the cost of transport is nearly twice the minimal value. Because they require about 20 days to complete their 1000-km journey, their average speed relative to the land is 50 km/day or 2.1 km/hr. The average speed of the opposing river current is about 4.8 km/hr, and a simple calculation would therefore indicate that the salmon must maintain a swimming speed of 4.8 + 2.1 = 6.9 km/hr. Their maximal speed in laboratory studies is just 4.7 km/hr, and, as noted earlier, energetic considerations indicate that they actually migrate at 4.2 km/hr. By some means, then, the salmon must maneuver in the river in such a way that they do not "buck" the full average speed of the river current.

As a final example of the energetics of migration, we may briefly consider the migratory locusts of Africa, *Schistocerca gregaria*. At takeoff the 2-g locust carries an average of about 0.2 g of fat, and fat serves as the predominant fuel of flight. Utilization of all this fat would yield about 1900 cal of energy, and the cost of flying is about 130 cal/hr. Thus flights would appear to be limited energetically to durations of about 14 hours. Actual flights last six to eight hours and are interspersed with periods of feeding that are a scourge to man (a swarm can weigh up to 10,000–20,000 tons and consume as much food in a day as over a million people). If locusts carried 0.2 g of "hydrated" glycogen instead of fat, the available energy yield would be just 230 cal, and flights would be limited to just two hours in duration. It is noteworthy that other groups of insects that include species that undergo long migrations, such as the butterflies, also use fat almost exclusively as a fuel. Two groups, the Diptera (flies) and

Hymenoptera (ants, bees, wasps), seem generally to be restricted to use of carbohydrate during flight. Some of their members gain an advantage over glycogen storage by carrying significant quantities of honey or nectar in their crops. These concentrated sugar solutions can provide a considerably higher energy yield than hydrated glycogen per unit of weight, but still no more than a third as high as fat.

Daily energy expenditure in nature

To the ecologist the energetic parameter of most interest is often the average daily metabolic rate. This determines the food requirements of the individual and is central to the trophic relationships of populations within an ecosystem. Average daily metabolic rate is affected by many factors, among which temperature and activity are often dominant. It is not an easy parameter to measure; the methods used and their advantages and shortcomings were discussed earlier in this chapter. Because there is immense variability not only between species, but also within a species from one time of year to another, the information available cannot be readily generalized or summarized. Thus emphasis will be given to illustrative examples that, more than anything, will indicate the types of analysis presently possible.

As noted earlier, the most common approach to estimating daily energy expenditure in nature has been the integration of a time budget gathered in the field with data on energy demand gathered in the laboratory. This method is illustrated by Pearson's early study of male Anna hummingbirds, summarized in Table 14–1. He determined the metabolic rate associated with three types of behavior—hovering, perching, and roosting—using measures of oxygen consumption in the laboratory. Care was taken to adjust these measures to temperatures comparable to those in the field. He estimated the time that a single wild animal spent in each type of behavior by observing a male hummingbird on two days in early September. With these data, the daily energy demand of each type of behavior was estimated by multiplying the time devoted to the behavior by the energetic cost of the behavior. Pearson did not know whether the hummingbird spent the night simply sleeping or whether it entered torpor. Accordingly, the energy demand over the period of roosting was estimated under both assumptions. The total daily energy demand was calculated by summing the daily costs of all three behaviors and turned out to be about 10.3 kcal under the assumption of simple sleep overnight, or about 7.6 kcal under the assumption that the animal entered torpor for most of the night. It is interesting that flight, though it occupied just 10% of the day, accounted for nearly as much energy expenditure as either perching or sleep. Subsequent studies of hummingbird hovering have suggested that the cost per hour is perhaps only 60% as great as Pearson measured; this would lower the estimates of daily cost of hovering and of total daily energy expenditure, but hovering would still account for a relatively large portion of the total daily expenditure. The average daily rate of metabolism estimated by Pearson for a hummingbird that sleeps at night is about 5.6 times the basal metabolic rate.

Summaries of the type seen in Table 14–1 are often termed *time-energy budgets*, or simply *energy budgets*.

Table 14-1. Energy budget for a male Anna hummingbird (*Calypte anna*). The hourly energetic cost of each behavior was determined in studies of several animals in the laboratory. Time devoted to each behavior was determined for a single wild male observed on the campus of the University of California over two days in September, 1953. See text for further discussion.

Behavior	*A* Hours per Day Devoted to Behavior	*B* Hourly Cost of Behavior (cc O₂/hr)	*C* Daily Cost of Behavior: *A* × *B* (cc O₂)
Perching	10.53	75.4	794
Flying (hovering)	2.35	272	639
Roosting	11.13	64.5 - - - - simple sleep - - - - - - - - 718	
		12.6 - - - - - torpor - - - - - - - - - 140	

$$\text{TOTALS (total daily energy expenditure):}$$
$$\text{simple sleep: } 2151 \text{ cc O}_2 = 10.3 \text{ kcal}$$
$$\text{torpor: } 1573 \text{ cc O}_2 = 7.6 \text{ kcal}$$

SOURCE: Data from Pearson, O. P. 1954. Condor 56: 317–322.

Figure 14–14*A* depicts certain aspects of the behavior and temperature relationships of young sockeye salmon in Babine Lake, British Columbia. In the summer this lake is thermally stratified, as may be seen by comparing the left (temperature) and right (depth) ordinates. The young fish spend most of the daylight hours in the cold, deeper waters (hypolimnion) but ascend to the surface at dusk to feed on plankton. Later in the night they descend to 10–12 m, and then they return to the surface waters at dawn for another bout of feeding. Afterward they descend into the hypolimnion. This pattern of behavior affects energy demand in at least two important dimensions. First, there are large changes in ambient temperature, the mean diurnal temperature being about 5°C and the nocturnal temperature being 14°–17°C; as we have seen earlier, ambient temperature affects both the standard metabolic rate and the cost of activity. Second, there are changes in the level of activity, the fish presumably being more active during their feeding periods than at other times. From laboratory studies on the energetic effects of temperature (see Figure 14–7*A*) and feeding activity, the cycle of metabolic demand in Figure 14–14*B* was constructed. Demand is lowest during the period of relative rest at cold temperatures during the day. Peaks are reached during the bouts of feeding in the warm surface waters at dawn and dusk. Over the middle of the night, when the fish are not feeding but remain in warm waters, the metabolic demand is intermediate. Integrating mean demand over the entire 24 hours, the estimated daily energetic cost comes out to be about 2800 cc O₂/kg = 13.5 kcal/kg (or 27 cal for a 2-g fish). This is the rate at which energy is utilized (degraded to heat); the fish must actually consume more energy equivalents because they are growing and thus directing some ingested food to the production of new biomass. J. R. Brett, who worked out this daily energy cycle for young salmon, pointed out the most interesting feature that these animals, by descending to cold waters for a large portion of the day, realize energetic advantages analogous to those gained by homeotherms that undergo daily torpor. Whereas the homeotherm lowers

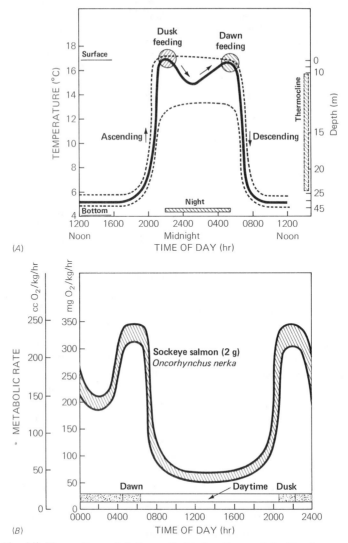

Figure 14–14. (*A*) The pattern of daily vertical migration and feeding in young sockeye salmon in Babine Lake, British Columbia, during midsummer. Dashed lines indicate the general range of temperatures experienced by the fish at various times of day, and the heavy line indicates the approximate mean response. The lake is thermally stratified, as reflected by comparison of the right and left ordinates. (*B*) The estimated daily cycle of energy demand in young sockeyes. Note that the time axis is shifted relative to that in part *A*. (From Brett, J. R. 1971. Amer. Zool. 11: 99–113.)

its body temperature and metabolic rate through internal adjustments in thermoregulatory control, the young salmon does so by the behavioral expedient of entering the hypolimnion. Because the metabolic rate is profoundly reduced for about 10 hours, the total daily cost, including that of active feeding in warmer waters, turns out not to be greatly different from the standard (resting) demand at 15°C. Experiments have shown that when young salmon consume submaximal amounts of food per day, their growth rate is maximized at relatively low temperatures; the reduced catabolic demand at such temperatures allows more of the ingested food to be directed to anabolism. Brett has suggested that the descent to cold

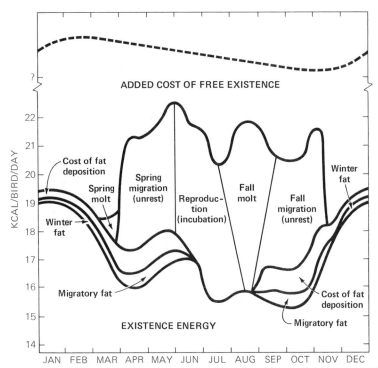

Figure 14–15. An estimated partial annual energy budget for the tree sparrow (*Spizella arborea*). Lowermost curve depicts existence energy utilization for birds overwintering in Illinois and spending the summer in northern Manitoba. Other energetic costs have been added to the existence energy curve, as discussed in the text. Existence energy utilization was determined in laboratory cages and is probably greater in the wild, especially in winter. The upper dashed line, although purely hypothetical, reflects the original investigator's belief that, with the additional costs of free existence added in, total daily energy demand might well be fairly constant over the year. (From West, G. C. 1960. Auk 77: 306–329. Used by permission of the American Ornithologists' Union.)

waters during the day might permit the fish to maximize the growth potential of their morning meal. As to why they do not descend after the evening meal, he has pointed out that digestion is slowed at low temperatures, and because the interval between dusk and dawn feeding is relatively short, residence in the warmer upper waters may be necessary for the fish to be able to take good advantage of the feeding opportunity at dawn.

Figure 14–15 depicts a partial estimated annual energy budget constructed by West for tree sparrows (*Spizella arborea*). As in the previous two examples, energetic costs were estimated from laboratory data. The first step was to determine the so-called existence rate of energy utilization, indicated by the lowest line in the figure. This was measured at appropriate temperatures by subtracting the energy content of urine and feces from the energy content of ingested food for birds housed in cages. Existence energy utilization includes the basal energy requirement plus components due to thermoregulation, specific dynamic action, and the activity of the caged birds. With the curve for existence energy demand in hand, several other measurable costs were added. (1) Tree sparrows lay on fat over the winter (November to February) and particularly during the spring and fall premigratory periods. The rate of fat deposition was measured in caged birds, and both the caloric value of the fat and the metabolic cost of forming fat were superimposed upon the existence energy requirement. (2) Caged birds exhibited increased metabolic rates when molting, and the cost of molting was estimated by subtracting existence costs and other known costs from the total metabolic rate during these periods. The costs thus determined for the spring and fall molts were then added into the energy budget. (3) Caged birds become nocturnally restive at the time of migration. The cost of this increased activity was added to the energy budget, but the cost of migration itself was not known and could not be included. (4) The cost of incubation for females during the reproductive season was estimated by assuming that the heat production needed to keep the eggs warm was the same per unit of weight as that needed to maintain an equivalent body temperature in the parent bird.

As West noted, many of the cost estimates in Figure 14–15 are preliminary or even hypothetical. Further, there are a number of known omissions or postulated misestimations. The cost of migration, as noted, is not included; nor are the costs of producing eggs or feeding the young. Although the existence energy demand includes a component resulting from feeding activity in the experimental cages, there is little doubt that the cost of feeding is greater in the wild, where the birds must fly about to find and collect their food. West postulated that the cost of feeding is probably particularly high in winter owing to difficulties of finding food. These considerations, if nothing else, illustrate the multitude of challenges faced by the investigator who seeks to construct a complete energy budget for a species. Although the budget for tree sparrows is incomplete and subject to a number of uncertainties, it is sobering to realize that West attempted actual measurements of more parameters than have been measured in many other studies of this kind.

Figure 14–15 expresses some general principles that also pertain in many other species of birds. The energetic costs of migration, reproduction, and molting are timed to occur during the warmer months of the year, when existence metabolism is lowest. Also, these costs occur more or less sequentially. First there is the spring migration, then reproduction, then the major molt, then fall migration. The life cycle of the bird is thus timed so that two or more major energy-demanding events do not directly coincide.

The recent development of the $D_2{}^{18}O$ method for measuring total metabolism in free-ranging animals has made it possible, for the first time, to evaluate the accuracy of energy budgets constructed by extrapolation from laboratory data. Recently such an evaluation has been performed by Mullen and Chew for longtail pocket mice, *Perognathus formosus*, living in the Nevada desert. Three types of results are available. (1) Animals were placed in metabolism chambers for an entire day at 25°C and their oxygen consumption integrated to give the total daily metabolic rate within the confines of this laboratory environment. (2) Daily metabolic rate in nature was estimated by integration of field and laboratory data, taking into account the effects of temperature and activity. This method followed the lines of the studies discussed earlier. Data were available from the field on burrow temperature, desert surface temperature, and the time that the animals spent in their burrows and on the desert surface. Laboratory data were available on the resting metabolic rate as a function of temperature and the increment in metabolic rate caused by ordinary nocturnal activity. Assuming that the animals were at rest in their nest when in their burrows, the metabolic expenditure over the period in the burrows was estimated by first calculating nest temperature from the measured burrow temperature, then determining resting metabolic rate at the nest temperature, and finally multiplying this metabolic rate by the number of hours spent in the burrow. The metabolic expenditure for the period on the desert surface was calculated by first determining the resting metabolic rate at the temperature of the desert surface, then adding the laboratory-measured activity increment to get total metabolic rate, and finally multiplying this metabolic rate by the number of hours spent on the desert surface. The expenditures for the periods in the burrow and on the desert surface were then added to get total daily expenditure. (3) Mice living in the desert were loaded with D_2O and $H_2^{18}O$, released for a period of one to several days, and then recaptured for determination of their total energy expenditure over the experimental period in the wild.

The total metabolic rate of the mice studied in metabolism chambers at 25°C turned out to be about 59 cc O_2/g/day. In September, the month when temperatures in the field most closely approximated 25°C, the $D_2^{18}O$ method indicated a total metabolic rate for wild animals of about 100 cc O_2/g/day—about 70% higher than the laboratory figure. The underestimate in the laboratory study is not at all surprising; the animal in a metabolism chamber does not have to search for food and is immensely restricted in its range of movement. The lesson of this comparison is clear: daily metabolic expenditure within the confines of a laboratory cage may hardly resemble expenditure in the wild. The daily metabolic rate estimated for wild animals by the time-energy budget technique (method 2 above) was about 78 cc O_2/g/day in the month of September. This is about 20% lower than the rate obtained by the $D_2^{18}O$ method in the same month but much closer to the mark than the estimate gained from the simple laboratory study. Results from $D_2^{18}O$ analysis and time-energy budget analysis were obtained for eight different months, and September was the month of the greatest discrepancy between the two. The average discrepancy over all eight months was just 12%. This indicates that, in *P. formosus*, a carefully applied time-energy analysis provides reasonably accurate estimates of energy expenditure in the wild. The success of the method probably rests in large part on the fact that ambient temperature and activity are the major modulators of metabolic rate in these animals. Because the method of time-energy analysis takes both of these factors into account, it is realistic in its results. Both the time-energy and $D_2^{18}O$ analyses showed that the energy demand of *P. formosus* is far greater in winter than summer, the daily demand in February and March being about twice as great as in July and August.

A recent study of purple martins has also indicated good agreement between daily energy costs estimated by the $D_2^{18}O$ method and time-energy analysis. On the other hand, a study of Merriam's kangaroo rats (*Dipodomys merriami*) in the Nevada desert using $D_2^{18}O$ has yielded quite unexpected results. Over the period of spring to fall (March to October), daily energy expenditure was systematically and inversely related to mean ambient temperature. In March, for example, when the mean temperature was about 10°C, daily energy expenditure was over three times greater than in August, when the mean temperature was near 30°C. During winter (November to February), unexpectedly low values for daily energy expenditure were found. Thus, for example, the energy expenditure of four animals in November at a mean temperature of 12°C was about the same as that of five animals in September, when the mean temperature was near 25°C. It was hypothesized that the animals in winter were undergoing daily torpor, a phenomenon that would reduce energy demand. This species has been shown to enter torpor in the laboratory, but only when starved. Given that the laboratory data indicate that starvation is necessary to induce torpor, an investigator seeking to estimate daily energy demand in nature by time-energy budget analysis would probably not have taken torpor into account as a routine phenomenon, and, as a result, he would have greatly over-estimated the energy demand during winter months. This illustrates one of the potential weaknesses in the time-energy type of analysis. If all energetically significant processes are not known and built into the analysis, the results can obviously be seriously in error. The $D_2{}^{18}O$ method gives a direct measure of *total* energy demand but does not tell us the nature of the individual processes that, taken together, determine the total demand. Thus although we now know that Merriam's kangaroo rat can exhibit a remarkably low energy demand in winter, further experimental work will be needed to determine if torpor is, as hypothesized, responsible.

Returning to the example of *Perognathus formosus*, we can say that the success of the time-energy analysis probably indicates that our laboratory understanding of the species is accurate. That is, the species in nature does not appear to be doing unexpected things of great energetic consequence.

SELECTED READINGS

Brett, J. R. 1965. The swimming energetics of salmon. Sci. Amer. 213: 80–85.

Brett, J. R. 1972. The metabolic demand for oxygen in fish, particularly salmonids, and a comparison with other vertebrates. Resp. Physiol. 14: 151–170.

Gessaman, J. A. (ed.). 1973. *Ecological Energetics of Homeotherms*. Utah State University Press, Logan.

Karpovich, P. V. and W. E. Sinning. 1971. *Physiology of Muscular Activity*. 7th ed. Saunders, Philadelphia.

Mackay, R. S. 1970. *Bio-medical Telemetry*. 2nd ed. Wiley, New York.

Robinson, S. 1974. Physiology of muscular exercise. *In*: V. B. Mountcastle (ed.), *Medical Physiology*. 13th ed. Vol. II. Mosby, St. Louis, Mo.

Schmidt-Nielsen, K. 1972. Locomotion: energy cost of swimming, flying, and running. Science 177: 222–228.

Tucker, V. A. 1969. The energetics of bird flight. Sci. Amer. 220: 70–78.

Tucker, V. A. 1970. Energetic cost of locomotion in animals. Comp. Biochem. Physiol. 34: 841–846.

Weis-Fogh, T. 1968. Metabolism and weight economy in migrating animals, particularly birds and insects. *In*: J. W. L. Beament and J. E. Treherne (eds.), *Insects and Physiology*. Elsevier, New York.

See also references in Appendix.

15 RHYTHMICITY IN ANIMAL FUNCTIONS

As is well known from common experience, many functions of animals vary with the time of day or time of year. Activity, for example, is more or less restricted to the day or night in many species, and reproduction is often confined to certain seasons. Such regular variations in functions are termed *biological rhythms*. It is the purpose of this chapter to discuss the diversity of rhythms and some of the mechanisms involved in their control. Rhythms are not only significant attributes of animals, but also can be important sources of variation in physiological studies. To exemplify the latter point, suppose that a given species of mouse displays a higher resting metabolic rate at night than during the day, and suppose that an investigator is studying the effects of temperature on metabolic rate in this species. Clearly, the time of day is an important factor to control in the studies of temperature. If the investigator were, by chance, to perform most studies of high temperatures at night and most studies of low temperatures during the day, his results would reflect not only the variation of metabolic rate with temperature, but also the superimposed variation of metabolic rate with time of day. To see the effect of temperature only, experiments at different temperatures should be performed at the same time of day. It is also possible that the effect of temperature will depend on the time of day, meaning that a full understanding of the effect of temperature will require that thermal comparisons be made at several different times of day. Biological rhythms can be most important considerations in both the design and interpretation of many types of physiological experiments.

The diversity of biological rhythms

Since its beginning life has evolved in the context of a rhythmically varying physical environment. Day and night come and go; the seasons change; and on the seashore the tides rise and fall. From the outset living organisms were subject to the effects of these physical rhythms. Photosynthesis, for example, could be carried out only during the day, implying a daily rhythm, and the amount of daylight varied over the year, implying a seasonal rhythm. Temperature in the surface waters of the seas rose and fell with the seasons, affecting both the animals and plants living there.

Although organisms may at first have responded rather passively to the cyclic changes in their physical environment, there seems to be little doubt that they soon evolved active responses that paralleled environmental rhythms. Looking at modern species we find, for example, that some planktonic algae exhibit 24-hour rhythms in their capacity for photosynthesis; their rate of photosynthesis increases during the day not only because light intensity increases, but also because their photosynthetic apparatus is more competent during the day than during the night. In the same vein, some diatoms seasonally form resting spores; some possess distinct summer and winter phenotypes; and some of the green and brown algae exhibit monthly or semimonthly rhythms in the release of gametes or propagative spores. With the evolution of life, rhythmicity—whether as a passive response to environmental fluctuations or as an active physiological response—became a property of the biotic as well as the physical environment. Thus each species evolved within the context of other species whose functions exhibited rhythmic fluctuations. Competition could be reduced by feeding at a different time of day or by reproducing at a different time of year than competitors, and predation could be minimized by avoiding exposure during the predator's period of feeding. These and other factors have made powerful contributions to the nature of the world's biota as we see it today. In the delineation of ecological niches, temporal features are every bit as important as spatial ones.

With these basic considerations in mind, we may proceed to examine some of the basic types of rhythms that are known. There are so many rhythms reported in the literature that a thorough review is impossible, and the purpose here is only to give some impression of the great variety of functions that can exhibit rhythmic fluctuations. In some cases it is possible to posit a selective value for rhythmic variation, but in others the selective value remains obscure or, at best, strongly hypothetical.

Among the most obvious rhythms are daily variations in activity. (Here *daily* will be used to refer to 24-hour cycles, and *day* will be used to refer to the 24-hour day. *Diurnal* will be used to refer to activities that occur specifically during daylight.) Many animals restrict their activities to certain times of day and are inactive, or relatively so, during the remainder of the day. Thus some mammals are diurnal; others are nocturnal; and still others are best termed crepuscular, being most active around the times of dawn and dusk. Similarly, diurnal, nocturnal, and crepuscular species are known among birds, lizards, fish, insects, and crustaceans, to cite just some groups in which such phenomena have been demonstrated. An area of physiology in which activity rhythms have obvious importance is the study of energetics. Activity contributes a component to metabolism, and students of many groups of animals have recognized that studies of resting metabolism are best carried out during the time of day when the animals tend to be inactive. There are other types of daily rhythms in behavioral activity. For example, some schooling species of fish tend to form schools most strongly at certain times of day and to disperse at other times. Invertebrate planktonic animals of several phyla and some fish are known to move up and down in the water in regular daily patterns. Such vertical migrations have been particularly well studied in planktonic crustaceans, which characteristically occupy waters closer to the surface at night than during the day. An example of daily vertical migration in a fish, the sockeye salmon, is provided in Figure 14–14.

Many functions besides activity can vary in a daily pattern. Examples are known even among unicellular plants and animals. The photosynthetic dinoflagellate *Gonyaulax polyedra*, for instance, shows daily rhythms in photosynthetic capacity, intensity of bioluminescence, and cell division. The latter rhythm deserves some comment because it introduces a basic concept that applies to a good many other rhythms. In a population of *Gonyaulax*, cell divisions tend to occur toward the end of the night phase of each day; thus the population grows in staircase fashion, with a marked increase in numbers occurring near dawn and then little increase occurring until the next dawn. What is important to recognize is that each individual in the population does not divide each day. Rather, the rhythm operates like a "gateway" such that if an individual is going to divide on a certain day, it will do so near dawn when the "gate" is open. The rhythmic opening of the gate, although it does not always culminate in division in any one individual, is evident when individuals are studied together as a population. In some species of *Paramecium*, there is a daily rhythm in the mating reaction and conjugation that again operates in the manner of a gateway; those individuals that are going to mate on a certain day tend to do so during the daylight phase.

Diverse insects exhibit daily rhythms of oviposition and mating behavior, tending to copulate and lay eggs within restricted time intervals. Recent experiments have revealed that male moths of several species display daily rhythms in their responsiveness to female sex pheromones (sex attractants). Another process that commonly occurs rhythmically in insects is emergence of the adult stage from the pupa or preadult nymph. These so-called eclosion rhythms are of the gateway type; individuals that are going to emerge on a certain day tend to do so within a restricted time interval. In some species mating occurs soon after emergence, and one function of the eclosion rhythm appears to be to assure the simultaneous presence of many males and females. Species of *Drosophila* emerge near dawn in a rhythm thought to assure favorable climatic conditions for the emergent adults; the emergent animals are particularly susceptible to desiccation, and at dawn temperatures are typically low and the relative humidity is typically high. A number of species of fish are known to show daily rhythms in spawning, and rhythms of color change mediated by chromatophore activity are known among fish, amphibians, and reptiles.

Rhythms in the rate of energy metabolism are very widespread. These variations in metabolism are often, to a considerable extent, secondary to other known rhythms. Thus, as noted earlier, rhythms in activity tend to be reflected in rhythms of metabolic rate as well. Also, many animals display rhythms of feeding, and the specific dynamic action of ingested food (see Chapter 2) then tends to elevate metabolism over the period following feeding. There is uncertainty whether metabolism also exhibits primary rhythmic fluctuations that occur independently of such other rhythms as activity and feeding; some evidence on mammals and insects suggests that an underlying, primary metabolic rhythm is present. Daily rhythms of body temperature have been reported in a considerable number of birds and mammals, including man. A large variety of other physiological functions are also known to show daily rhythms; only a few examples drawn from the literature on mammals and insects will be mentioned. As is evident from common experience, the rate of urine production in man varies rhythmically over the day, being relatively low at night and higher during daylight. The rates of urinary excretion of potassium, sodium, and chloride ordinarily vary more or less in parallel with the rate of volume production, but interestingly the rhythm of potassium excretion can be dissociated from that of volume production, suggesting independent control mechanisms. A considerable number of blood parameters in man show daily rhythms, including eosinophil count, lymphocyte count, hematocrit, serum iron concentration, serum sodium concentration, and carbon dioxide tension. Heart rate and blood pressure exhibit daily rhythms, as do various waveforms in the electroencephalogram. In mice, mitosis occurs in a daily rhythm in both epidermal and adrenal cells, the periods of peak activity being different in the two tissues. In certain insects, glycogen stores and circulating concentrations of the sugar trehalose have been found to vary rhythmically over the day. Studies of both insects and mammals have revealed daily rhythms in susceptibility to toxins or other damaging agents. Thus certain insects vary rhythmically in their sensitivity to insecticides or cyanide; laboratory mice vary in their susceptibility to bacterial endotoxins; and a number of rodents are more adversely affected by large doses of ionizing radiation at some times of day than others.

With this brief review it is clear why some authorities have concluded that virtually all physiological processes may be subject to daily rhythms.

Animals of the intertidal zone are exposed to water movements that follow two basic lunar periodicities. The lunar day of 24.8 hours is the time elapsing between two successive ascents of the moon to its zenith in the sky. The tides in most parts of the world rise and fall twice in each lunar day, meaning that each tidal cycle takes about 12.4 hours. The lunar month of about 29.5 days is the period between one full moon and the next. This period can also be of ecological importance to animals because the height of the tides varies over the month. At new and full moon, the sun and moon are lined up in such a way that their gravitational attractions reinforce each other, and the tides, termed "spring" tides, tend to show a large displacement; that is, the high tides are especially high and the low tides are especially low. Contrariwise, at the first and third quarters of the moon, the solar and lunar forces work against each other, and the tides, termed neap tides, show a relatively small displacement. Spring tides occur about every 15 days, and the upper reaches of the intertidal zone are regularly inundated only during the spring tides.

Among intertidal organisms there are many biological rhythms that parallel the tidal rhythm of 12.4 hours (half of a lunar day). In general, animals that feed in the water become active or show other feeding behavior during the periods of high tide. Thus sea anemones expand their tentacles at high tide but contract at low tide; starfish actively wander at high tide but remain quiet in secluded microhabitats at low tide; intertidal mussels exhibit maximum water-pumping activity at times of high tide; oysters and quahog clams open their shells at high tide; and snails (*Nassarius*) move about most at high tide. Animals such as fiddler crabs, which scavenge for food on the exposed mud or sand, become most active at low tide. Rhythms of oxygen consumption that parallel the tidal cycle are common among intertidal species. In fiddler crabs (*Uca*) there is a rhythm of chromatophore activity that, like some other rhythms in intertidal animals, includes both solar-day and lunar-day periodicities. These crabs darken during daylight hours and become light in color at night in a typical solar-day rhythm. In addition, however, they become particularly dark during certain hours of daylight, and these periods of particular darkness coincide with the times of low tide. If low tide occurs in the morning on a certain day, the crabs become darkest in the morning; but if low tide occurs in the late afternoon, the crabs become darkest in the late afternoon.

A number of organisms are also known to show rhythms tuned to the lunar month. An interesting example is provided by the midge *Clunio marinus* of western Europe. This species lays its eggs in clumps of seaweed that are exposed only during the ebb of spring tides; further, the adult females have a reproductive life of only a few hours after their emergence from the pupa. *Clunio* emerge and swarm in great numbers every 15 days at the ebb of spring tides, and within a short time they mate and deposit their eggs in the exposed seaweeds. A dramatic case of reproductive timing according to the lunar monthly rhythm is found in grunion, small fish of our western coast. These fish come onto the beaches at times of high spring tides and deposit their eggs and sperm in the sands at the upper reaches of the intertidal zone. Barring storms or other such events, these sands are next inundated about 15 days later at the next spring tides, and at that time the eggs hatch and the young are carried out to sea. Here, as also in the case of the midges mentioned earlier, we have an instance of lunar timing being superimposed on an annual cycle. The grunion spawn only for a few months out of the year and then do so at times of spring tides. A considerable number of invertebrates that do not inhabit the intertidal zone time their spawning according to the lunar cycle. For example, the polychaete Palolo worm of the Pacific swarms and breeds near dawn at the third quarter of the moon in October and November, and the Atlantic fireworm swarms shortly after sunset for three or four days near full moon during summer months. Because these worms do not occupy the shore, there is no obvious physical advantage to spawning at a particular time of the lunar cycle; rather they seem to have seized on the lunar cycle as a timing mechanism to assure synchrony of their spawning with that of other individuals in the population. The same can be said of many other species.

Finally we may briefly note some well-known annual rhythms. Many species of birds reproduce, lay down migratory fat, migrate, and molt at well-defined times of year (see Figure 14–15). Similarly, mammals molt, mate, bear their young, lay down winter fat, and hibernate in annual rhythms. Annual reproductive rhythms are virtually ubiquitous in the animal kingdom, being known in all major groups. Almost as widespread are physiological adjustments to difficult seasons. Some desert mammals and birds enter estivation in the heat and aridity of the summer; surface-dwelling fish of the Arctic develop protective antifreezes in winter (see Chapter 3); many terrestrial animals hibernate in winter; many insects enter diapause during winter or periods of summer heat. Life cycles are so attuned to seasonal variations that, in general, it is unlikely that individuals collected in one season will be physiologically identical to individuals collected at another.

The control of rhythms

As early as 1729 it was discovered that certain plants that raise and lower their leaves in a daily rhythm would continue to do so even when placed under constant darkness at relatively constant temperature. That is, the rhythm continued in a more or less daily pattern even without obvious environmental cues about the time of day. Especially in the last 40 years, experiments of a similar nature have been performed on many different plant and animal rhythms, and often these rhythms are found to persist in a constant laboratory environment. Rhythms that continue in the absence of *obvious* environmental timing cues are generally termed *endogenous*, whereas those that fail to persist are termed *exogenous*, being dependent on such environmental cues for maintenance of their periodicity.

In setting up an experiment to determine if a rhythm is endogenous it is common to place the organism under constant illumination and constant temperature, and such other parameters as humidity and barometric pressure may be controlled as well; these are the obvious environmental cues. Many workers have concluded that rhythms that persist under such conditions are endogenous in the truest sense of the word; it is postulated that the organism has an entirely self-contained "biological clock" that provides information on time and thus permits the rhythm to persist in a more or less regular pattern. As will be discussed in more detail later, however, some workers hypothesize that organisms under the standard controlled conditions may still be receptive to subtle environmental cues such as geomagnetic forces that provide external timing information. If this were to be the case, "endogenous" rhythms would be endogenous only in a relative sense; although not requiring the obvious environmental timing cues for persistence, they would nonetheless be dependent on information from the environment and not reflect the operation of entirely self-contained biological clocks. Putting this debate aside for the moment, we may look at some of the properties of endogenous animal rhythms under standard controlled conditions.

Figure 15–1 provides a record of activity in a flying squirrel under conditions of steady darkness at 20°C. It is apparent that the squirrel displayed a clear rhythm of activity, but the episode of activity shifted to an earlier time with each succeeding day. The time elapsing between the manifestation of a particular part of the rhythm on one day and the manifestation of that same part on the next day is defined to be the *period* of the rhythm. One part of the squirrel's rhythm that is particularly easy to define is the very start of activity. Because, on the average, one episode of activity

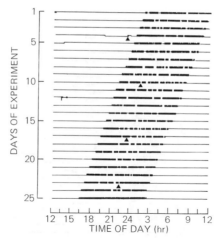

Figure 15–1. Activity of a flying squirrel (*Glaucomys volans*) as recorded on a running wheel over a period of 25 days in continuous darkness at 20°C. Each horizontal line corresponds to one day. Turning of the running wheel activated a pen to record a short vertical line for each rotation of the wheel; these vertical blips are often blended to give the appearance of a heavy, continuous line during periods of steady running. Note that activity started earlier with each passing day. Triangles indicate times at which food was added to the activity cage. (From DeCoursey, P. J. 1960. Cold Spring Harbor Symp. Quant. Biol. 25: 49–55. Copyright 1960 by the Cold Spring Harbor Laboratory.)

started 23.6 hours after the preceding episode, the period of this rhythm was 23.6 hours. When, as in this case, the period is less than 24 hours, episodes of activity will shift to an earlier and earlier time with each succeeding day. If the period were greater than 24 hours, activity would shift to a later time on each successive day.

Figure 15–2 shows a record for another flying squirrel over a series of 53 days. For the first seven days the squirrel was in continuous darkness (except for one entire day when continuous light was provided). On day 8 the squirrel was placed on a photoperiodic cycle of 12 hours of light and 12 hours of dark per day (12L:12D). At first the squirrel commenced its activity in the middle of the light phase, but gradually activity began at a later time each day until, on day 21, activity began at the start of the dark phase, the "natural" time for a nocturnal animal. Thereafter activity continued to start at the beginning of the dark phase until, on day 49, the squirrel was returned to continuous darkness. Over days 49–53 the period of the activity rhythm reverted to being less than 24 hours, and activity started at an earlier time with each passing day. This experiment illustrates a number of important and general principles. When the squirrel was first placed on the 12L:12D light cycle, its activity rhythm was *out of phase* with the environmental cycle, but gradually the activity rhythm shifted to be in phase and then it remained in phase so long as the environmental cycle was provided. The process of bringing a biological rhythm into phase with an environmental rhythm is termed *entrainment,* and the biological rhythm is said to be *entrained* by the environmental cues. An environmental cue that is capable of setting the phase of a biological rhythm is termed a *phasing factor* or *Zeitgeber* (German for "time-giver"). When obvious phasing factors or Zeitgeber are removed, as in Figure 15–1 or on days 49–53 of Figure 15–2, the biological rhythm that persists is said to *free-run* or to be a *free-*

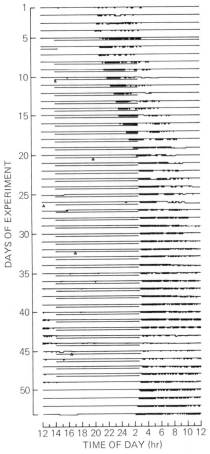

Figure 15–2. Activity of a flying squirrel (*Glaucomys volans*) as recorded on a running wheel over a period of 53 days at 20°C. See legend of Figure 15–1 for description of format. For each day the thin horizontal line running across the entire figure is the tracing of the unactivated recording pen attached to the running wheel. Horizontal lines just under the pen tracings indicate times that lights were on. Note that the animal was in continuous darkness over days 1–4; lights were on continuously from hour 14 on day 5 to hour 14 on day 6; the animal was returned to continuous darkness for most of day 6 and all of day 7; from day 8 to day 48, the animal was placed on a light cycle of 12 hours of light and 12 hours of dark each day; and the animal was returned to continuous darkness on days 49–53. (From DeCoursey, P. J. 1960. Cold Spring Harbor Symp. Quant. Biol. 25: 49–55. Copyright 1960 by the Cold Spring Harbor Laboratory.)

running rhythm. Because free-running rhythms usually have periods close to but not exactly equal to 24 hours, they are termed *circadian rhythms*, meaning that they are *about* a day in length (*circa* = "about," *dies* = "a day"). When an animal displaying a free-running rhythm is suddenly provided with an external phasing stimulus, as on day 8 of Figure 15–2, entrainment commonly requires a number of days if the phase shift required is large. The total phase shift is accomplished in a series of smaller daily shifts (days 8–21 in Figure 15–2) termed *transients.*

Experiments such as those on the flying squirrels provide a great deal of information on the control of biological rhythms. We see that these squirrels have an endogenous rhythm of activity that persists even in the absence of obvious environmental information about the time of day. However, this endogenous circadian rhythm is not precisely timed to the 24-hour day, and episodes of activity drift to occur at earlier and earlier times of day. When the squirrels are provided with a Zeitgeber in the form of a daily light-dark cycle, their activity is timed according to the external cues and does occur at 24-hour intervals. The onset of darkness in nature serves to cue the beginning of activity, but we see that, teleologically speaking, the squirrels do not have to wait in a state of total ignorance each day to see when darkness will arrive. Rather, they have an endogenous sense of the time of day, and the onset of darkness simply serves as a cue which maintains a *precise* daily rhythm in a system which, in and of itself, will maintain an *approximate* daily rhythm. Over the seasons the time of sunset varies considerably in temperate latitudes, and flying squirrels always commence their activity near sunset; in July, for example, activity might start at 7:30 in the evening, whereas in November it might start at 4:30 in the afternoon. By using nightfall as a Zeitgeber, the squirrels adjust the phasing of their activity rhythm to correspond to the seasonally varying photoperiodic cycle. Nonetheless, if light cues are suddenly denied at any given time of year, the activity rhythm can free-run in approximate phase with the external light cycle, for the effect of nightfall is to set the phase of an endogenous circadian rhythm.

Not all daily rhythms prove to be endogenous when tested in the absence of obvious environmental cues; many processes lose their rhythmic nature under such conditions, and are said to be arrhythmic. Rhythms that continue in the absence of obvious cues are so common, however, and are distributed so thoroughly among the animal and plant phyla that circadian rhythmicity is deemed to be an almost universal and probably fairly primitive feature of life. In the photosynthetic dinoflagellate *Gonyaulax polyedra*, the rhythms of bioluminescence, photosynthetic capacity, and cell division all persist under conditions of constant temperature and illumination. Similarly, to cite another example in a unicellular form, mating in some species of *Paramecium* continues to occur rhythmically under constant conditions. Known endogenous circadian rhythms in multicellular animals are so numerous that a list of them all would occupy several pages. Among them are activity rhythms in a variety of mammals, birds, lizards, fish, insects, and members of some other groups; metabolic rhythms in an equally diverse set of species; chromatophore rhythms in some fish and crabs; rhythms of oviposition, mating, and eclosion in a number of insects; rhythms of female pheromone release and male pheromone sensitivity in some insects; rhythms of body temperature in a number of birds and mammals; and rhythms of urinary excretion in man.

The period of a free-running rhythm exhibits individual variation and is often a function of test conditions. In laboratory mice held in constant light at constant temperature, for example, the average period of the activity rhythm was found to be 23.5 hours at low light intensity and 25.8 hours at high light intensity. There is a generalization, known as Aschoff's rule, that diurnally active species tend to decrease their free-running period as intensity of illumination is increased under constant conditions; whereas nocturnally active species, such as the laboratory mouse, tend to increase their free-running period as the intensity of illumination is increased. The generalization, although useful, is known to have a number of exceptions.

Sometimes the intensity of illumination is found to be a critical factor in whether an endogenous rhythm will be displayed at all; thus steady intense light is known to suppress activity virtually completely in some nocturnal insects.

One of the most remarkable properties of free-running periodicities is their low sensitivity to temperature changes. Defining the *frequency* of a rhythm to be the inverse of its period, it is possible to study the rhythm at two different constant body temperatures and, using frequency values, compute a Q_{10} for the rhythm. Frequency, of course, is a measure of the rate at which the rhythm moves through an entire cycle. If an increase in temperature causes the period to shorten, meaning that the rhythm progresses more rapidly, then a Q_{10} value greater than 1.0 will be obtained; on the other hand, if the period lengthens with increase in temperature, a Q_{10} of less than 1.0 will be obtained. You will remember from Chapter 3 that processes such as heart rate, metabolic rate, and breathing rate commonly exhibit Q_{10} values near 2.0 in poikilothermic animals; that is, such processes often approximately double their rate when body temperature is raised by 10°C. In striking contrast, free-running circadian rhythms seldom exhibit Q_{10} values of more than 1.2 and often have Q_{10}'s near 1.0, indicating near independence of body temperature. Some rhythms slow down with increased temperature, but, again, thermal sensitivity is not large, and Q_{10} values are seldom below 0.8. As has often been pointed out, an internal biological clock would be of relatively little use if it were highly sensitive to temperature; imagine the chaos if our wristwatches were to double their rate for every increase in temperature of 10°C. In some manner the clocks of plants and animals are able to keep time despite changes in temperature. If, as many believe, these clocks are entirely innate and not dependent on external cues of any kind, then the clocks must be part of the metabolism of cells. It is a major challenge to find out how this particular metabolic apparatus has been rendered rather immune to the thermal effects that so strongly influence most metabolic processes.

We may now examine what types of environmental cues can act as phasing factors or Zeitgeber for endogenous circadian rhythms. Many of these same factors are also known to act as timing cues for exogenous daily rhythms (rhythms that *require* environmental cues to maintain their periodicity). Of all the known Zeitgeber, the ones most commonly successful in entraining endogenous circadian rhythms are daily light cycles or changes in light intensity. There seems to be no known circadian rhythm that cannot be entrained by a daily light cycle. It is hardly surprising to find that animals commonly use light signals to synchronize their rhythms to the daily environmental cycle, for in nature changes in illumination are generally the most reliable indicators of the time of day. Another parameter that commonly varies systematically with time of day is temperature. It is, of course, not as reliable a cue as light; the intensity of light, for example, always decreases with nightfall, but temperature, although usually decreasing at night, sometimes increases owing to local weather conditions. When animals in the laboratory that have been free-running are exposed to a daily temperature cycle with illumination held constant, it is sometimes found that their circadian rhythms come into phase with the temperature cycle. Temperature is typically a weaker Zeitgeber than light, and some rhythms that entrain to light fail to entrain to temperature. On the other hand, rhythms in diverse processes in a variety of phyletic groups are known to entrain to temperature, and in some cases temperature cycles prove to be more effective phasing factors than light cycles. Another type of cue to which rhythms sometimes become entrained is the presence of food. An interesting example is provided by honey bees, which, if they are fed at a particular time and particular place each day, will continue to return at the proper time for a number of days after the food has been removed. In nature, bees learn the times of day at which various flowers secrete nectar and visit flowers at the appropriate times in a daily rhythm.

A number of lunar and annual rhythms, as well as daily rhythms, are known to persist endogenously in the absence of obvious environmental cues. In nature fiddler crabs display a tidal rhythm of activity, being active during low tide when their burrows are exposed and they can scavenge on the exposed sand or mud. This tidal rhythm of activity persists in the laboratory in the absence of tidal water movements. Other tidal rhythms that continue in the absence of obvious tidal cues include the expansion and retraction of sea anemones (*Actinia*); the activity rhythm of snails (*Nassarius*); the rhythm of shell opening in oysters (*Crassostrea*); the pumping rhythm of mussels (*Mytilus*); tidal metabolic rhythms in a variety of species; and the tidal rhythm of darkening in fiddler crabs. Sometimes the free-running period of tidal rhythms is very close to 12.4 hours or 24.8 hours; but in other cases the period is significantly longer or shorter, and the rhythms are best termed *circatidal*.

Although free-running tidal rhythms may be affected by changes in the daily light cycle, they cannot be brought into phase with a daily light cycle. The failure of the daily light cycle to act as a true phasing factor is hardly surprising because tidal rhythms do not have 24-hour periods. Studies of a number of species have shown that their tidal rhythms are in some manner entrained to the ebb and flow of the tides in their home habitat. When fiddler crabs, for example, are collected from beaches with different tidal cycles, they continue for some time in the laboratory to stay in phase with the tides in their respective habitats. When animals are transferred to a shoreline with a strange tidal cycle, they gradually come into phase with the new cycle. The environmental cues that act as Zeitgeber are not fully understood. The most obvious cue is the water movement itself; in addition there are tidal cycles of temperature in the microhabitats of intertidal animals, and there are changes in nocturnal illumination and geometeorological forces resulting from the relative motions of the earth and moon. Fiddler crabs were collected from two places around Cape Cod that, because of different topography, experience low tide at different times of day despite the fact that the motions of the moon in the sky are virtually the same at both places. The rhythms of the crabs were found to be in phase with the local tides at both places, indicating that the rhythms were cued to the tides themselves and not directly to the motions of the moon. Because topographic and other earthbound factors commonly and strongly affect the timing of tidal cycles, the motion of the moon in the sky cannot be used as a general and predictable indicator of the timing of the tides at any particular location. At the least, therefore, it is most parsimonious to expect that intertidal animals would use water movements or other direct tidal cues to set the phase of their tidal rhythms. Interestingly, however, when fiddler crabs and oysters were transported to Illinois from the East Coast and placed under constant temperature and illumination, with no obvious tidal cues, they at first displayed rhythms in phase with the tides in their home habitat but gradually shifted to be in phase with what the tidal cycle would be on an open beach in Illinois if Illinois were on the sea coast. Although direct tidal cues are probably the dominant Zeitgeber in nature, it was concluded that these animals are also able to use subtle geometeorological cues arising from the motions of the moon. The precise nature of these cues is unknown. There are lunar tides in the atmosphere, but there is reason to believe that these are not the cues being used.

Recently it has become clear that some animals possess endogenous rhythms of approximately a year in length. Such *circannual rhythms*, by their very nature, cannot be elucidated without lengthy experimentation, and the number studied is far lower than the number of circadian rhythms that have received attention. Golden-mantled ground squirrels (*Citellus lateralis*) have been found to continue their annual cycle of autumnal fat accumulation and subsequent hibernation over a period of four years when exposed to a constant temperature and constant daily light cycle. The free-running period of their hibernation cycle is approximately, but not exactly, a year. Wood warblers, *Phylloscopus trochilus*, are a migratory species that spend the summer in Europe and the winter in Africa. They accumulate fat at the time of both migrations and undergo two molts each year, one in the summer and one in the winter. In the laboratory they become restless at night at the times of migration. When these warblers are kept at constant temperature on a constant 12L:12D light cycle, they continue approximately annual rhythms of nocturnal restlessness, molting, and fat accumulation. Cave crayfish, *Orconectes pellucidus*, show a persistent circannual rhythm of reproductive readiness when housed at constant temperature in constant darkness, and other examples of annual rhythms that continue under steady conditions could be cited. It should be noted that some annual rhythms either do not persist or lose much of their regularity when tested. Further, the test conditions are sometimes an important factor in determining whether a rhythm will persist. The wood warblers discussed earlier, for example, exhibit clear circannual rhythmicity on a 12L:12D cycle but not on an 18L:6D cycle. The reasons for this latter phenomenon are unknown.

The environmental cues to which circannual rhythms are entrained in nature are not, in a specific sense, well known because few entrainment studies have yet been performed. However, a great deal of attention has been directed to the control of annual rhythms in general, and such factors as changes in photoperiod and temperature are widely recognized as important. Endogenous rhythms, when present, are presumably entrained to such factors.

Changes in the length of day and night over the year are widely used as cues for the timing of events in the life cycle. In mammals, for example, some species, such as sheep, goats, and mink, come into estrus and mate in response to the shortening days of autumn, whereas others, such as a variety of mice, squirrels, and ferrets, enter their mating period in response to the lengthening days of spring. Among the other processes known to be responsive to photoperiod are hibernation and molting in some mammals; reproduction, molting, and migration in many birds; reproduction in some lizards and a variety of fish; developmental rate and the development of body form and coloration in some insects; and diapause in many insects.

Generally, photoperiodic cues are not the sole determinants of such processes. Often there is an indication of some type of endogenous timing mechanism that is entrained or reinforced by appropriate photoperiodic stimuli; the evidence for this extends well beyond the relatively few species in which true circannual rhythms have been rigorously established. Some birds, for example, exhibit at least a modest degree of gonadal development in the spring even when artifically denied vernal photoperiods, and some hibernating species of mammals appear to be programed to enter hibernation well in advance of the fall and winter, for they hibernate even if denied the usual autumnal conditions of photoperiod and temperature. There is also much evidence that environmental factors besides photoperiod often act in concert with photoperiod. Many species of birds, for example, require long days for full gonadal development, but, provided long days are present, both temperature and social interactions are also found to be influential. The gonads may develop more rapidly when temperatures are mild rather than cold, and social interactions with conspecifics may hasten development or even be required for full development. In this example, as in some others, it seems appropriate to conclude that photoperiod is used as a cue to time reproduction to the proper time of year, but other cues are used to "fine-tune" the time of reproduction according to the actual ecological conditions in which the bird finds itself.

A final point worthy of note is that when a species possesses an endogenous timing mechanism or when it otherwise goes through a highly ordered life cycle, photoperiodic cues received at one time may determine the timing of events that occur substantially later. The commercial silkworm (*Bombyx mori*), for example, diapauses as an egg over winter. Whether eggs will diapause is determined by day length during the late embryonic and early larval stages of the animals that lay the eggs. Eggs that develop under the short days of spring give rise to adults that lay nondiapausing eggs, but eggs that develop during the long days of summer develop into adults that lay diapausing eggs. The onset of diapause in the fall is thus determined during the summer, when the fall-laying adults are themselves developing. Similarly, some workers have hypothesized that hibernation in some mammals is in fact programed by the experience of long days in the spring and summer.

Near the equator, annual changes in photoperiod are small, and there is considerable debate whether equatorial animals utilize photoperiodic cues to time their life cycles. Also, although photoperiod is a highly reliable indicator of time of year, there are parts of the world where time of year is itself not a particularly reliable indicator of ecological conditions. In this context it is interesting to note that birds of the arid Australian interior do not exhibit the clear annual cycles of reproduction to which we in temperate climates are so accustomed. Rather, they breed irregularly in response to local conditions of moisture and plant growth. Because rains occur irregularly, these immediate indicators of ecological conditions are far more reliable cues for the timing of successful reproduction than is day length.

It is always important to recognize that responses to photoperiod, temperature, moisture, and other cues have developed through evolution, and a particular type of responsiveness to a particular cue will have been subject to positive selective pressures only if it has helped to promote survival and reproduction.

Before closing this chapter it is appropriate to look briefly at the adaptive value and physiological basis of endogenous rhythmicity. There can be little question that daily, tidal, and seasonal variations in such processes as feeding and reproduction are of advantage to many animals; this is indicated by the ubiquity of such variations in the animal kingdom. There are, however, at least three ways in which such variations could be kept in phase with environmental changes. Taking as an example an animal that retires to its burrow at dawn and later comes out to feed in the evening, one possibility would be for the animal to keep its activity rhythm in phase with the daily light cycle strictly through the use of exogenous cues. That is, it could possess no internal timing mechanism whatsoever and use dawn as a cue for entering its burrow and dusk as a cue for exiting its burrow. In this case, if night were artificially prolonged, the animal would not return to its burrow, for it would be strictly dependent on the external cue of dawn. Alternatively, the animal could possess what has been called an "hour-glass" internal clock. This clock would be set in motion by an external cue each day and then would run for the remainder of that day, endogenously timing the remainder of the day's activities; but the clock would "run out" after a single day and would have to be restarted by an external cue on the next day. Thus dawn could set the hour-glass timer in motion, and the animal, using its internal clock, would know to come out at the time of dusk without requiring the external cue of dusk. On the next day dawn would again set the timer in motion; but if dawn were denied, the clock would not be started, and the animal would not endogenously know when to come out at dusk. The third possibility is that the animal could have a constantly running clock that would stay in motion for days on end without external cues to start it but that would be entrained to external cues so that it would be kept in phase with the external world. This is the type of clock we have discussed earlier. It is the type of clock reflected by free-running circadian rhythmicity and is illustrated in Figure 15–1. There are well-known examples of all three types of timing mechanisms. Some animal rhythms are controlled strictly by exogenous cues; some act as if controlled by hour-glass timers; and some exhibit true free-running circadian rhythmicity. The question arises, why do animals have internal clocks at all? Why do they not all simply rely on exogenous cues?

There are no simple answers to these questions, and our present answers are undoubtedly incomplete. One general advantage hypothesized for clocks is that they permit anticipation of forthcoming events. The animal that is strictly dependent on external cues must wait until appropriate cues appear before it knows to act, but the animal with an internal clock can anticipate approximately when action will be necessary and thus can undertake whatever preparations may be appropriate. Bats in caves and rodents in burrows, for example, cannot see that dusk has arrived except by coming to the opening of their cave or burrow. With no internal sense of time, they would presumably have to come to the opening regularly to ascertain if dusk had come. With internal clocks, they know approximately when to arouse for their nightly activities, and it in fact has been shown that bats begin to fly about deep in their caves just shortly in advance of nightfall.

Another potential advantage of clocks is that they permit rhythms to persist in proper phase with the external world even when the external cues are abnormal. If, for example, an animal were to become active in response to lowered temperatures at night, it would possibly fail to become active on an unusual night when temperatures did not fall if its activity were strictly dependent on external cues. With a circadian clock, however, activity would occur as usual over several nights of unusual temperatures; whatever phase drift occurred during this free-running period would be corrected when cool nights returned and again entrained the rhythm.

Clocks also permit well-ordered rhythms during times when external cues are either unreliable or difficult to use. During night, for instance, changes in illumination may be small, and changes in temperature and humidity may be erratic from day to day. An animal might be hard-pressed to maintain a regular nightly rhythm using external cues, but a clock entrained to some more reliable and obvious cue, such as dusk, would permit accurate timing of activities throughout the remainder of the night. This might, for example, assure that the animal would always be present for a feeding opportunity that occurred at a particular time of night, whereas an animal using external cues would undoubtedly sometimes be misled and arrive too late.

Although these and other potential advantages of clocks are clear, it should be noted that our knowledge of the existence of clocks is far wider than our concrete knowledge of their adaptive value. There is a need for careful study of the natural biology of animals with clocks to give us a more rigorous understanding of the contribution of clocks to ecological success.

The physiological nature of internal clocks is a subject on which there is a vast literature, and it would be beyond the scope of this text to do more than discuss some of the more salient results and debates. Many of the volumes listed in the Selected Readings go into this subject in detail. One important question is whether overt processes such as activity are themselves part of the clock mechanism. Possibly, for example, a bout of activity might fatigue the animal, leading then to a period of recovery prior to the next bout of activity, and this recurring cycle could potentially produce a circadian rhythmicity in activity. Much evidence indicates, however, that overt processes are, in general, not part of the clock itself. A useful analogy may be drawn with the clock mechanism and the hands of an ordinary watch. The operation of the clock mechanism is independent of the hands and continues even if the hands are removed. In organisms the clock mechanism likewise operates independently of the overt processes it controls, and the overt processes, like the hands on a watch, simply reflect the underlying operation of the clock.

Experiments on unicellular organisms, such as those discussed earlier, demonstrate unequivocally that circadian clocks can exist within the confines of a single cell. There is also considerable evidence for the existence of circadian clocks in isolated cells and tissues of multicellular animals. It is widely believed that a large proportion, or even all, of the cells in a multicellular animal possess circadian clocks. Efforts at finding glands or nervous centers that control rhythmicity have borne some fruit. In cockroaches, for example, there is accumulating evidence that the optic lobe region of the brain is responsible for controlling the circadian rhythm of activity, and in mammals rhythmic activity of the adrenal gland controls the circadian rhythms of epidermal mitosis and eosinophil level in the blood. Such nervous and glandular controllers do not contain the only biological clocks of the organism. There is often reason to believe that they act to synchronize, or hold in phase, the activities of cells and tissues that themselves are endogenously rhythmic, somewhat in the way that the sinoatrial node of the mammalian heart dominates and coordinates the contractions of many cardiac muscle fibers that in turn prove to be rhythmically contractile if isolated. Central control centers in the brain are responsible in insects and vertebrates for mediating the entrainment of circadian rhythms throughout the body to daily light cycles.

The biochemical or biophysical nature of the circadian clock has yet to be elucidated in any organism, though some important features of the clock are well established. As discussed earlier, the period of the clock is remarkably resistant to thermal effects. It also is impressively resistant to many metabolic poisons ranging from cyanide to some of the inhibitors of protein synthesis. Some chemical treatments, such as exposure to heavy water, do significantly affect the period of the clock, and these effects presumably provide clues to the fundamental nature of the clock; but as yet there is no biochemical-biophysical model of the clock mechanism that can account for the known chemical effects. Several types of experiments have indicated clearly that the clock is inherited and innate; the capacity for endogenous rhythmicity is not learned from parents or imprinted from exposure to obvious daily environmental cycles. Fruit flies (*Drosophila*), for example, can be reared through many generations in darkness and will still display their daily rhythm of pupal eclosion. Although eclosion occurs arrhythmically so long as steady darkness is maintained, a brief, single flash of light will cause eclosion to occur in a daily rhythm thereafter. Apparently each individual has an endogenous clock which times eclosion, and the flash of light simply serves to synchronize (entrain) the rhythms of all the individuals in the population so that the population will then display the daily eclosion rhythm.

One of the most vigorous and persistent debates in the study of biological clocks is whether they are entirely endogenous or are dependent on subtle environmental cues to maintain their remarkably stable and consistent periods. According to the endogenous-clock hypothesis, organisms possess biochemical or biophysical mechanisms that keep time without the need for any external cues, though they can, of course, be entrained to external cues. According to the exogenous-clock hypothesis, the biological clock mechanism cannot, in and of itself, keep time but rather is cued to solar-day or lunar-day fluctuations in environmental factors. The fluctuating factors are postulated to be such subtle parameters as gravity, geomagnetism, geoelectrostatic fields, or background radiation. Proponents of the exogenous-clock hypothesis argue that even though organisms may be placed under constant conditions of illumination, temperature, humidity, and other obvious factors, they are still receiving and using information about the time of day from fluctuations in subtle geophysical factors. Thus the rhythms that persist in the absence of obvious cues—and that can be entrained to obvious cues—in fact reflect receptivity to the subtle cues; and if the subtle cues could be denied, the rhythms would lose their precise timing.

The exogenous-clock hypothesis perhaps offers a simpler explanation than the endogenous-clock hypothesis for the ability of organisms to keep time. However, the exogenous-clock hypothesis also involves greater logical complexities. To appreciate this, recall, for example, that the period of a free-running daily rhythm is seldom exactly 24 hours; further, the period commonly varies among individuals within a species and may be altered in any one individual by changing the intensity of illumination or body temperature. The endogenous-clock hypothesis can explain these observations by postulating that the biochemical or biophysical clock is somewhat imprecise and affected by individual differences and environmental conditions. However, proponents of the exogenous-clock hypothesis argue that the clock is tuned to 24-hour environmental rhythms and accordingly keeps precise time. It is then necessary to postulate that, under free-running conditions, the manifestation of overt processes such as activity or color change is not precisely tied to the operation of the clock. It is argued that the clock ticks away precisely in tune with the 24-hour day, but the overt processes controlled by the clock progressively lag behind or get ahead of the clock so that they do not cycle precisely in a 24-hour rhythm. An analogy can be drawn with a watch in which the hands are not firmly attached to the watch drive. The watch mechanism would turn in precise time, but the hands might turn a little more slowly and then would not accurately reflect the operation of the clock mechanism itself.

Neither the endogenous nor the exogenous hypothesis can be discounted on the basis of available data, and the interested reader will find extensive arguments for and against the two in several of the Selected Readings.

SELECTED READINGS

Beck, S. D. 1968. *Insect Photoperiodism*. Academic, New York.

Biological Clocks. 1960. Cold Spring Harbor Symposia on Quantitative Biology. Volume XXV. The Biological Laboratory, Cold Spring Harbor, New York.

Brown, F. A., Jr., J. W. Hastings, and J. D. Palmer. 1970. *The Biological Clock: Two Views*. Academic, New York.

Bünning, E. 1967. *The Physiological Clock*. 2nd ed., revised. Springer-Verlag, New York.

Circadian Rhythmicity. 1972. Proceedings of the International Symposium on Circadian Rhythmicity, 1971. Centre for Agricultural Publishing and Documentation, Wageningen, Netherlands.

Cloudsley-Thompson, J. L. 1961. *Rhythmic Activity in Animal Physiology and Behaviour*. Academic, New York.

Farner, D. S. and B. K. Follett. 1966. Light and other environmental factors affecting avian reproduction. J. Animal Sci. 25: 90–118.

Harker, J. E. 1964. *The Physiology of Diurnal Rhythms*. Cambridge University Press, London.

Menaker, M. (ed.). 1971. *Biochronometry.* Proceedings of a Symposium. National
    Academy of Sciences, Washington, D.C.
Pengelley, E. T. and S. J. Asmundson. 1971. Annual biological clocks. Sci. Amer.
    224: 72–79.
Pittendrigh, C. 1961. On temporal organization in living systems. Harvey Lecture
    Series 56: 93–125.

See also references in Appendix.

APPENDIX

Each of the works listed below covers a variety of aspects of physiology. These are good reference sources. They have generally not been listed with the individual chapters of this text inasmuch as many of them would have to be listed under virtually all chapters such is the breadth of their coverage.

GENERAL WORKS

Barrington, E. J. W. 1967. *Invertebrate Structure and Function*. Houghton Mifflin, Boston.

Bolis, L., K. Schmidt-Nielsen, and S. H. P. Maddrell (eds.). 1973. *Comparative Physiology*. North-Holland, Amsterdam.

Dill, D. B. (ed.). 1964. *Handbook of Physiology. Section 4: Adaptation to the Environment*. American Physiological Society, Washington, D.C.

Florkin, M. and B. T. Scheer (eds.). 1967–1974. *Chemical Zoology*. 9 vol. Academic, New York.

Hochachka, P. W. and G. N. Somero. 1973. *Strategies of Biochemical Adaptation*. Saunders, Philadelphia.

Newell, R. C. 1970. *Biology of Intertidal Animals*. Elsevier, New York.

Prosser, C. L. (ed.). 1973. *Comparative Animal Physiology*. 3rd ed. Saunders, Philadelphia.

Ruch, T. C. and H. D. Patton (eds.). 1973. *Physiology and Biophysics*. 3 vol. Saunders, Philadelphia.

Scheer, B. T. (ed.). 1957. *Recent Advances in Invertebrate Physiology*. University of Oregon Publications, Eugene.

Schmidt-Nielsen, K. 1972. *How Animals Work*. Cambridge University Press, London.

WORKS DEALING WITH SPECIFIC GROUPS OF INVERTEBRATES

Binyon, J. 1972. *Physiology of Echinoderms*. Pergamon, New York.

Boolootian, R. A. (ed.). 1966. *Physiology of Echinodermata*. Wiley, New York.

Bursell, E. 1970. *An Introduction to Insect Physiology*. Academic, New York.

Dales, R. P. 1963. *Annelids*. Hutchinson, London.

Laverack, M. S. 1963. *The Physiology of Earthworms.* Pergamon, Oxford.

Lee, D. L. 1965. *The Physiology of Nematodes.* Freeman, San Francisco.

Lockwood, A. P. M. 1967. *Aspects of the Physiology of Crustacea.* Freeman, San Francisco.

Rockstein, M. (ed.). 1973–1974. *The Physiology of Insecta.* 2nd ed. 6 vol. Academic, New York.

Waterman, T. H. (ed.). 1960. *The Physiology of Crustacea.* 2 vol. Academic, New York.

Wigglesworth, V. B. 1972. *The Principles of Insect Physiology.* 7th ed. Chapman & Hall, London.

Wilbur, K. M. and C. M. Yonge (eds.). 1964, 1966. *Physiology of Mollusca.* 2 vol. Academic, New York.

WORKS DEALING WITH SPECIFIC GROUPS OF VERTEBRATES

Brown, M. E. (ed.). 1957. *The Physiology of Fishes.* Vol. I. Academic, New York.

Farner, D. S. 1970. Some glimpses of comparative avian physiology. Fed. Proc. 29: 1649–1663.

Farner, D. S. and J. R. King (eds.). 1971–1974. *Avian Biology.* 4 vol. Academic, New York.

Hoar, W. S. and D. J. Randall (eds.). 1969–1971. *Fish Physiology.* 6 vol. Academic, New York.

Marshall, A. J. (ed.). 1960–1961. *Biology and Comparative Physiology of Birds.* 2 vol. Academic, New York.

Moore, J. A. (ed.). 1964. *Physiology of the Amphibia.* Academic, New York.

Sturkie, P. D. 1965. *Avian Physiology.* 2nd ed. Comstock, Ithaca, N.Y.

INDEX

77 78 79   9 8 7 6 5 4 3 2